Manual of British Botany

Containing the Flowering Plants and Ferns Arranged According to the Natural Orders

Charles Cardale Babington
Edited by Henry Groves
and James Groves

Cambridge
University Press

CAMBRIDGE UNIVERSITY PRESS

Cambridge, New York, Melbourne, Madrid, Cape Town,
Singapore, São Paolo, Delhi, Mexico City

Published in the United States of America by Cambridge University Press, New York

www.cambridge.org
Information on this title: www.cambridge.org/9781108055666

This edition first published 1904
This digitally printed version 2013

ISBN 978-1-108-05566-6 Paperback

CAMBRIDGE LIBRARY COLLECTION

Books of enduring scholarly value

Life Sciences

Until the nineteenth century, the various subjects now known as the life sciences were regarded either as arcane studies which had little impact on ordinary daily life, or as a genteel hobby for the leisured classes. The increasing academic rigour and systematisation brought to the study of botany, zoology and other disciplines, and their adoption in university curricula, are reflected in the books reissued in this series.

Manual of British Botany

First published in 1843, this book ran to eleven editions, with two published posthumously. Compiled by Cambridge botanist Charles Cardale Babington (1808–95) over the course of nine years, this was the first comprehensive catalogue of British plants for nearly a century and was conveniently pocket-sized for fieldwork. Babington was by this time the leader in the taxonomical research of higher plants. Providing both the Latin nomenclature assigned at the time and the common English or anglicised name, he divides plants according to the Linnaean natural orders and describes them in great technical detail. A useful glossary is also included to help the reader navigate the descriptions. As demonstrated in *Memorials, Journal and Botanical Correspondence* (also reissued in this series), Babington was a highly esteemed and influential scientist. This is the expanded 1904 ninth edition of his invaluable and enduring compendium.

Cambridge University Press has long been a pioneer in the reissuing of out-of-print titles from its own backlist, producing digital reprints of books that are still sought after by scholars and students but could not be reprinted economically using traditional technology. The Cambridge Library Collection extends this activity to a wider range of books which are still of importance to researchers and professionals, either for the source material they contain, or as landmarks in the history of their academic discipline.

Drawing from the world-renowned collections in the Cambridge University Library and other partner libraries, and guided by the advice of experts in each subject area, Cambridge University Press is using state-of-the-art scanning machines in its own Printing House to capture the content of each book selected for inclusion. The files are processed to give a consistently clear, crisp image, and the books finished to the high quality standard for which the Press is recognised around the world. The latest print-on-demand technology ensures that the books will remain available indefinitely, and that orders for single or multiple copies can quickly be supplied.

The Cambridge Library Collection brings back to life books of enduring scholarly value (including out-of-copyright works originally issued by other publishers) across a wide range of disciplines in the humanities and social sciences and in science and technology.

MANUAL OF
BRITISH BOTANY

CONTAINING THE FLOWERING PLANTS
AND FERNS ARRANGED ACCORDING TO
THE NATURAL ORDERS

BY THE LATE
CHARLES CARDALE BABINGTON
M.A., F.R.S., F.L.S.
Professor of Botany in the University of Cambridge

NINTH EDITION
ENLARGED FROM THE AUTHOR'S MANUSCRIPTS
AND OTHER SOURCES

EDITED BY
HENRY AND JAMES GROVES

LONDON
GURNEY & JACKSON, PATERNOSTER ROW
(SUCCESSORS TO MR VAN VOORST)
MDCCCCIV

"Quod ad me attinet, ingenue fatear, me in rebus dubiis de specifica differentia numquam consulere Auctores, qui in herbariis plantis multis, sed eos modo, qui in natura plantis multum student. At iis, qui ad præceptas opiniones experientiam suam concinnant et in singulo externæ faciei lusu, neglectis notis essentialibus, formas transitorias vident, parum fido."—FRIES.

PRINTED BY TAYLOR AND FRANCIS, RED LION COURT, FLEET STREET.

PREFACE

[*to the Eighth Edition*].

IN this work it has been the Author's wish to adopt in all cases those names which have the claim of priority, unless good cause should be shown for a contrary proceeding; and with this object he has carefully examined nearly all the best European Floras, comparing our plants with the descriptions contained in them, and in most cases with foreign specimens of undoubted authenticity. In the adoption of genera and species an endeavour has been made, by the examination of the plants themselves, to determine which are to be regarded as truly distinct,—thus, it is hoped, taking Nature as a guide. Still, let it not be supposed that any claim is made to peculiar accuracy, or that the Author considers himself qualified to dictate to any student of botany; for he is well aware that there are many points upon which persons who have carefully studied the subject form different conclusions from those to which he has been led.

The progress of our knowledge has caused changes in the nomenclature in successive editions of this book and in the Author's views of the value of forms—as species or varieties. The inconvenience of these alterations to all, especially to statistical botanists, is fully admitted; but the Author does not know of any mode by which it can be avoided if each edition is to be brought up as completely as is in his power to the contemporary knowledge of our plants. No alterations have been admitted until careful study has convinced the Author that they are required. He may have fallen into error, but has earnestly endeavoured to discover the truth.

Attempts have been made greatly to reduce the number of recognized species found in Britain; but the results obtained seem to be so totally opposed to the teaching of the plants themselves, and the evidence adduced in their favour is so seldom more than a statement of opinion, that they cannot safely be adopted; nor does the plan of the present work admit of a discussion of the many questions raised by them. Also it has been laid down as a rule by some botanists, that no plant can be a species whose distinctive characters are not as manifest in an herbarium as when it is alive. We are told that our business as descriptive botanists is not "to determine what is a species," but simply to describe plants so that they may be easily recognized from the dry specimen. The Author cannot agree to this rule. Although he, in common with other naturalists, is unable to define what is a species, he believes that species exist, and that they may often be easily distinguished amongst living plants, although sometimes separated with difficulty when dried specimens alone are examined. He thinks that it is our duty as botanists to study the living plants whenever it is possible to do so, and to describe from them; to write for the use and instruction of field- rather than cabinet-naturalists; for the advancement of a knowledge of the plants rather than for the convenience of possessors of herbaria: also that the differences which we are able to describe as distinguishing plants being taken from their more minute organs, does not invalidate their claim to distinction. It seems to be our business to decide upon the probable distinctness of plants before we attempt to define them—to make the species afford the character, not the character define the species.

This volume being intended as a field-book or travelling companion for botanists, it is advisable to restrict the space allotted to each species as much as possible; and accordingly the characters and observations are only such as appear to be necessary for an accurate discrimination of the plants. Facts relating to their geographical distribution are therefore usually omitted. Synonyms have been almost wholly omitted; but the plates of the original *English Botany* or some other British plates are quoted. *Syme's English Botany* may well be used by those who desire full descriptions of the plants; and the plates in that work have often had valuable additions made to those of the old *English Botany*, from which most of them are taken. Localities are only given for new or rare plants; Mr. Watson's works and the numerous local floras render it unnecessary inconveniently to swell the present volume by their introduction. But in order to convey some idea of the distribution of plants throughout the United Kingdom, the letters E., S., or I. have been appended to the descriptions of such species as have, it is believed, been found in England, Scotland, or Ireland. The descriptions of a considerable number of plants which only occur in the Channel Islands, and are, therefore, not properly parts of the British Flora; or which, although included in our lists, there is reason to suppose have never been really detected in Britain; or, although naturalized, have very slender claims to be considered aboriginal natives; or which are now supposed to be lost by the alterations made in the places where they were found by our predecessors; are included within []: and notices of a few plants concerning which more accurate information is requisite, are distinguished in a similar manner. It is hoped that by this arrangement the truly indigenous species will be clearly distinguished from those which have little or no claim to be considered aboriginal or even thoroughly naturalized. The attempt to do this is necessary for two seemingly contradictory reasons, namely :—the great tendency of many collectors to consider native any plant found growing upon a spot where it is not cultivated; and the peculiar scepticism of some of our botanists concerning the claims of many local or thinly scattered species to be admitted as indigenous, even when their distribution upon the European continent is not unfavourable to the belief that they may inhabit Britain. It has been recommended that the descriptions of these excluded species should be placed in an Appendix or even omitted; but as some of them are not unlikely to be observed by collectors, it is more convenient that they should be arranged with their allies. Those who desire to obtain a complete knowledge of the distribution of our plants should consult Watson's *Cybele Britannica*, and Moore & More's *Cybele Hibernica*.

Full characters of the Natural Orders are to be found in most of the best 'Introductions to Botany'; and it has therefore not been considered advisable to give them in detail in the present volume. In his definitions, the Author has endeavoured to point out the characteristic marks, more especially as far as British plants are concerned.

In using this book the student will find it convenient to pay attention to the *italicized* parts of the generic and specific characters, and, if they are found to agree with the plant under examination, then to compare it with the other parts of those characters, and also with those of allied genera and species.

It is most desirable that the students of our native flora should not confine their attention to books published in this country. Owing to such an unavoidable restriction we fell far behind our continental brethren during the earlier part of the present century. A few modern works may be named which will assist them in their studies. Koch's *Synopsis Floræ Germanicæ*, ed. 2, and Grenier and Godron's *Flore de France* are strongly recommended,—also, although in a rather less degree, Lloyd's *Flore de l'Ouest de la France*, Brébisson's *Flore de la Normandie*, and Cosson and Germain's *Flore des environs de Paris*, ed. 2. Boreau's *Flore du Centre de la France*, ed. 3, and the scattered papers of M. Alexis Jordan of Lyons are valuable for the study of varieties: for many of their species can claim no higher rank. But, above all, the works of Fries deserve careful study—especially his *Novitiæ Floræ Suecicæ*, with its three *Mantissæ*, and *Summa Vegetabilium Scandinaviæ*. It is necessary to warn students against the very common error of supposing that they have found one of the plants described in a foreign Flora when in reality they have only gathered a variety of some well-known British plant. The risk of falling into such errors renders it necessary to consult such works as those of Messrs. Boreau and Jordan with great caution, lest we should be misled by descriptions, most accurate, indeed, but often rather those of individuals than species. Amongst plants so closely allied as are many of those called species in some continental works, it is scarcely possible to arrive at a certain conclusion without the inspection of authentic specimens.

The Author takes this opportunity of returning most sincere thanks to his botanical friends and correspondents (far too numerous to record by name) for the great assistance they have again rendered to him by the communication of valuable suggestions, observations, and specimens.

The book has been again carefully revised throughout, so as, if possible, to keep pace with the rapidly advancing knowledge of British plants.

As many as possible of the real English names are given. All the genera and species could not be thus named, owing to the absence of any recognized English terms which have been applied to them. It does not seem desirable to invent or adopt new English names, known only to botanists, for the few genera which have them not, the Latin name being sufficient in those cases, and better in the original than in an Anglicized form. . . .

It is hoped that those who use this book will favour the author with information of any (even the slightest) addition, correction or alteration that may appear to be necessary, in order that it may be employed in the preparation of a future edition, as it is only through such assistance that the flora of an extensive country can attain to even a moderate degree of perfection.

Cambridge, July 15, 1881.

PREFACE TO THE NINTH EDITION.

The primary object of this edition is to publish the notes in the late Author's interleaved copy of the former edition, but some years having elapsed since the last of these entries, it has been necessary to make further additions, so as to include some of the results of more recent work.

At the urgent request of Mrs. Babington we agreed to act as Editors, feeling that it would be a great misfortune if so useful a work—for more than half a century practially the only critical handbook to our Flora—were not republished. Owing to the claims of business and interruptions from other causes there has been a considerable delay in the completion of our task.

Of the late Professor Babington's notes, some were alterations and additions ready for insertion in the book, while others were merely references and memoranda for further investigation. The latter class we have tried to deal with as nearly as possible in the spirit of the rest of the work. It was Mrs. Babington's particular wish that the text as amended by the Author should not be interfered with. Owing to this limitation, we have been unable to make alterations in the treatment of some of the critical genera which might perhaps have been desirable.

Those species and varieties and additional characters and remarks which we have inserted are printed in smaller type, and, where interjected in the text, are included in square brackets. Many varieties recently described or indicated as British we have thought too trivial to

a 5

particularise. We have inserted a few well-authenticated additional localities, but have not attempted any general revision of this part of the work.

In the case of *Hieracium*, the number of species recognized in this country has so greatly increased during the last twenty years, that it has been thought desirable to substitute an entirely fresh account of the genus, and this has been drawn up by Miss Rachel F. Thompson, under the guidance of Mr. F. J. Hanbury, and forms a very valuable addition to the book.

In the somewhat similar case of *Rubus*, Professor Babington having for so many years been the recognized authority on the genus in this country, it was felt that it would not be fitting for his account to be removed from the book, although it was clear that he had intended to rewrite it. It was therefore arranged by the kind permission of the Rev. W. Moyle Rogers to reprint, as an appendix, a conspectus of the groups and species, from that gentleman's admirable '*Handbook of British Rubi*.'

During the past twenty years great changes have taken place in the matter of nomenclature. The publication of the '*Index Kewensis*' and other lesser works bearing on the subject, besides greatly facilitating the study of botanical literature, has demonstrated how imperfectly the laws of priority as regards names have been observed in the past. It is only by a strict observance of these laws that any approach to finality is likely to be reached, and we have therefore adopted the earliest names as far as we have been able to ascertain them, taking 1753 as the starting-point for both genera and species. In doing this we have had to make a large number of alterations in names and authorities.

It has only been possible for us to carry out this work by the kindness of the many friends and correspondents, too numerous to mention individually, who have been ever ready to help us with information and by the loan of specimens and books, and this assistance we gratefully acknowledge. We are especially indebted to Mr. F. J. Hanbury for placing his own and the Boswell herbarium entirely at our service, to Mr. Arthur Bennett for his frequent help with specimens and notes, to Mr. J. Britten and Mr. W. P. Hiern for valuable advice and assistance especially in matters of nomenclature, to Mr. E. G. Baker and Dr. Rendle for kind help in referring to books and specimens at the Natural History Museum, to Messrs. A. Fryer and F. Townsend for assistance with *Potamogeton* and *Euphrasia* respectively, and to Messrs. Colgan and Scully for kindly furnishing us with a list of additions to the Irish Flora.

H. & J. G.

London, May 1904.

ERRATA.

P. 26, line 31, for "RADIC'ULA" read "RADI'CULA."
P. 64, line 29, after "538" add "*Cerastium*, L."
P. 67, line 18, after "viii" read "*Spergularia*, Pers."
P. 224, line 7, after "*Carduus*" add "*marianus*."
P. 413, line 23, for "Buchan" read "Buchen."

A GLOSSARY

OF

THE TERMS USED IN THE MANUAL.

Accumbent; used to express the application of the edges of the cotyledons to the radicle in the seeds of Crucifers.

Achene; a hard dry one-seeded superior pericarp.

Acicular; needle-shaped; very slender from a slightly broader base.

Acotyledonous; without distinct cotyledons.

Acuminate; drawn out into a long point, but with the sides slightly hollowed.

Acute; sharp; forming an angle less than a right angle at the tip.

Adhering; uniting together of two different parts, as a calyx to an ovary.

Adnate; attached throughout their whole length. Adnate anthers have their lobes so attached to the filament. Stipules are often adnate to the petiole by one of their edges.

Adpressed; pressed close to any thing.

Adpressed-serrate; serrate with the teeth lying closely on each other or to the edge of the leaf.

Æstival; produced in summer.

Æstivation; the arrangement of the floral organs in the bud.

Albumen; nutritious matter contained in the seed to feed the young plant; more correctly called *perisperm*.

Alternate; placed successively on the opposite sides of an axis, as in the case of leaves; or opposite to the spaces between the parts of the next whorl in flowers.

Amplexicaul; clasping the stem with their base.

Anastomosing; veins combining with each other at their ends.

Annual plants rise from the seed, flower, and die in the same year.

Annular; forming a ring.

Anterior; the part of a flower next the bract or in front.

Anther; the part of the stamen which contains the pollen.

Apex; the end furthest from the point of attachment.

Apical; at or relating to the apex.

Apiculate; having a very small hard point at the end, usually formed by the tip of the midvein.

Apocarpous; fruit formed of carpels which are quite separate.
Approximate; close together.
Arching; curved into the form of an arch.
Arcuate; curved so as to form a considerable part of a circle.
Aril; an aftergrowth from the placenta or seedstalk surrounding the seed.
Arillode; an aftergrowth from the lips of the foramen (or terminal opening of seed).
Ascending; curving upwards into a vertical position.
Asperous; rough with short raised points.
Attenuate; narrowing gradually to a point.
Auricled; having *auricles*, or appendages at the base of the leaves.
Awn; a long-pointed bristle-like appendage, as the beard of Barley.
Awned; having awns.
Axil; the upper angle formed by the union of the stem and leaf.
Axillary; placed in an axil.
Axis; the line passing through the centre of any thing; the common stalk of the flowers in a spikelet of Grasses.

Baccate; pulpy like a berry.
Base; the end nearest the point of attachment.
Beak; a long pointed projection.
Bearded; having long hair like a beard.
Berry; a pulpy fruit containing several seeds; a true *bacca* when inferior, a *uva* when superior.
Biennial plants spring from the seed in one year, flower in the following year, and then die.
Bifariously; arranged in two rows, one on each side of any thing.
Bifid; divided halfway down into two parts.
Bipartite; divided nearly to its base into two parts.
Bipinnate; when the divisions of a pinnate leaf are themselves pinnate.
Bipinnatifid; when the divisions of a pinnatifid leaf are themselves pinnatifid.
Biternate; when the divisions of a ternate leaf are themselves ternate.
Boatshaped; resembling a small boat.
Bracteoles; minute bracts.
Bracts; small leaves somewhat different from the others, seated on the inflorescence.
Bulb; a leaf-bud with fleshy scales, usually placed underground.
Bulbiferous; bearing bulbs on its stem.
Bulblike; resembling a bulb in appearance, but solid.

Bulbous; having radical bulbs.
Bulbous hairs have a round swelling at their base.

Cæspitose; growing in tufts from the root.
Calyx; the outer whorl of leaflike organs forming the flower, usually green, called sepals.
Capillary; like very slender threads.
Capitate; growing in heads or close clusters; having a knob like the head of a pin.
Capsular; like a capsule.
Capsule; a dry usually many-seeded seed-vessel.
Carpel; the divisions of the ovary or capsule: sometimes one carpel forms an ovary, being rolled up so that its edges meet.
Carpophore; the stalk of the ovary or capsule within the outer whorls of the flower.
Catkin; a deciduous unisexual spike of crowded flowers in which the perianths are replaced by bracts.
Cauline; growing from the stem, not radical.
Cellular tissue; a collection of minute vesicles filled with fluid;
Chaffy; covered with minute membranous scales.
Channelled; hollowed somewhat like a gutter.
Cilia; hairs placed like eyelashes on the edge of any thing.
Ciliate; with cilia.
Circinate; rolled up from the top towards the base like a crosier.
Clavate; clubshaped.
Claw; the narrow base of a petal.
Clawed; having a claw.
Cleft; deeply cut, but not to the midrib.
Clubshaped; a long solid body which is slender at the base and gradually thickens upwards.
Cluster; a kind of dense cyme; also the patches of capsules in Ferns.
Cœsious; with a fine pale-blue bloom.
Cohering; the attachment to each other of similar parts; as the petals forming a gamopetalous corolla.
Collapsing; shrinking together. The submersed and much-divided leaves of aquatic plants often collapse into a form like a painter's pencil, when removed from the water.
Columella; a cylindrical central placenta.
Commissure; the inner faces of the carpels (mericarps) of Umbelliferæ, by which they join.
Compound; formed of many similar parts which ultimately and naturally separate from each other. *A compound umbel* has small umbels on its branches.
Compressed; when flattened laterally.
Conduplicate; folded upon each other lengthwise.

Cone; fruit of a fir-tree.
Conical; a solid figure narrowing to a point from a circular base.
Connate; when two similar parts, as leaves, are slightly connected round the stem.
Connective; the continuation of the filament between the cells of an anther.
Connivent; converging.
Constricted; narrowed at some point as if by the pressure of a string.
Contiguous petals touch or overlap by their edges.
Converging; their points gradually approaching.
Convolute; rolled together lengthwise.
Cordate; ovate, acute, with two rounded lobes at the base; like the figure of the heart on cards: a *cordate-based* leaf is of any shape, but has the two lobes at its base.
Coriaceous; leathery; firm, dry, tough.
Corm; a fleshy bulblike, but solid, not scaly underground stem.
Corneous; like horn.
Corolla; the whorl of floral leaves between the calyx and stamens, usually coloured, called petals.
Corymb; a raceme with the peduncles becoming gradually shorter as they approach the top, so that all the flowers are about on a level.
Corymbose; in the form of a corymb.
Cotyledons; the seed-lobes, often forming the first leaves of the plant.
Crenate; with rounded marginal teeth. When these are again crenate, the whole is *doubly crenate*; not *bicrenate*, which means having two such teeth.
Crenatures; the blunt rounded teeth of a crenate leaf.
Crenulate; minutely crenate.
Crested; having an appendage like a crest.
Crowned; having an appendage on the upper side at the base of the limb, as some petals.
Cruciform; four parts, as petals, arranged so as to form a cross.
Crustaceous; hard, thin and brittle.
Cuneate; like a wedge, but attached by its point.
Cuspidate; abrupt, but with a point starting suddenly from the middle of its end.
Cuticle; the external skin.
Cylindrical; nearly in the form of a cylinder.
Cyme; inflorescence formed of a terminal flower, beneath which are lateral branches each having a terminal flower and lateral branches again similarly dividing, and so on. A

globose cyme has flowers so placed as to form a globose mass. A *scorpioid* cyme produces only the external branch of each pair, except the first.

Cymose; arranged in a cyme.

Deciduous; falling off.

Declining; straight, but pointing downwards.

Decumbent; lying on the ground, but tending to rise at the end.

Decurrent; when the limb of a leaf is prolonged down the stem below the point of attachment of the midrib.

Decussate; opposite leaves, but the successive pairs placed at right angles to each other.

Deflexed; curved downwards or towards the back.

Dehiscence; the mode in which an organ opens.

Deltoid; fleshy with a triangular transverse section.

Dentate; with short equilateral triangular teeth. When these are again dentate, the whole is *doubly dentate*, not *bidentate*, which means having two teeth.

Denticulate; finely dentate.

Depressed; when flattened vertically or at the top.

Determinate inflorescence ends in a flower.

Dicotyledonous; with two opposite cotyledons.

Didymous; formed of two similar parts attached to each other by a small portion of their margin.

Diffuse; widely spreading.

Digitate; fingered; of several leaves all starting from the top of the petioles.

Diœcious; with the sexes on different plants.

Disk; a fleshy space from which the stamens and pistils spring, or between them; the central part of a head (capitulum).

Dissepiments; vertical plates dividing an ovary into parts; septa.

Distichous; arranged above each other in two rows on opposite sides of an axis.

Distinct; separate from its neighbours.

Divaricate; spreading at an obtuse angle.

Diverging; gradually separating.

Dorsal; attached to, or on the back.

Drupe; a one-celled superior fruit, not bursting, fleshy externally, stony within, containing one or two seeds.

Echinate; armed with straight slender prickles like a hedgehog.

Elliptic; oval but acute at each end.

Elongate; much lengthened.

Emarginate; slightly notched at the end.

Embryo; the young plant as first seen in the seed.

Entire; not toothed nor lobed at the edge.
Epidermis; the skin.
Epigynous; apparently seated upon the ovary.
Epipetalous; borne on the petals.
Epiphytes; plants growing upon others, but not deriving nourishment from their juices.
Equalling; when the ends of organs rise to the same height even though their relative lengths are different.
Equitant; when a conduplicate organ covers the edges of another similarly folded, and that covers a third, and so on.
Erect; standing nearly perpendicular to that from which it grows, as a seed rising from the base of an ovary; at right angles to its support.
Exceeding; when an organ extends beyond an adjoining organ, but is not necessarily itself longer than it.
Excurrent; extending beyond the edge or point.
Exserted; projecting beyond that which surrounds its base.
Extrorse anthers have the slit by which the pollen escapes directed from the ovary.

Falcate; like a sickle.
Falling short of; the reverse of exceeding.
Fasciculate; when several similar parts are collected into a bundle and spring from the same spot; often the developed leaves of an undeveloped axillary branch form a fascicle.
Fastigiate; when all the branches are parallel and point upwards.
Feathery; like a feather in structure.
Felted; tomentose.
Fibre; a hair-like kind of elementary structure.
Fibrous; having many threadlike parts.
Filament; the stalk usually found supporting an anther.
Filiform; like a thread.
Flaccid; weak.
Flexuose; zigzag, usually changing its direction at each joining.
Floccose; with little tufts like wool.
Follicle; an inflated 1-celled carpel, opening by only one suture to which several seeds are attached.
Forked; like a fork of two prongs.
Frond; the leaflike part of Ferns.
Fruit; the seed-vessel with its ripe contents and any external appendages.
Fruit-bearing; the state of the inflorescence when the fruit is ripe or nearly so, contradistinguished from *flower-bearing*.
Fruticose; shrubby.

Fugacious; soon falling off.
Funnel-shaped; tubular below, but gradually enlarging upwards.
Furcate; forked.
Fusiform; spindle-shaped; thick, tapering to each end.

Gamosepalous; *gamopetalous*; when the sepals or petals are joined by their edges so as apparently to form one.
Germen; the ovary.
Gibbous; swollen on one side.
Glabrous; without hairs or other clothing.
Gland; a wartlike cellular secreting organ usually raised above the surface.
Glandular; having glands.
Glandular-hairy; having hairs tipped with glands.
Glandular-serrate; having short teeth tipped with glands.
Glaucous; green with a whitish-blue lustre.
Globose; round like a globe.
Glumes; the scales enclosing the spikelet of flowers in Grasses; the imbricate bracts enclosing the flowers of Sedges.
Glumiferous; having flowers covered by glumes.
Granular; covered with minute projecting points.

Habit; the general appearance of a plant.
Haft; a winged leaf-stalk; the linear part of a spathulate leaf or petal.
Hastate; enlarged at the base into two lobes directed nearly horizontally.
Head; a close terminal collection of flowers surrounded by an involucre.
Helmet; the hooded upper part of a flower.
Helmet-shaped; arched and concave like a helmet.
Herbaceous; the parts of plants which are not woody; also organs, or parts of them, of a green colour.
Hermaphrodite; having both sexes in one flower.
Hilum or *hile*; the mark on a seed which indicates its place of attachment.
Hispid; covered with stiff hairs.
Hoary; with greyish-white down.
Hooded; formed into a hood at the end.
Horizontal; spreading at right angles to their support, as leaves on a stem.
Hybrid; a mule.
Hypogynous; springing from below the base of the ovary and not attached to the calyx.

Imbricate; arranged over each other like the tiles of a roof.
Imparipinnate; pinnate with a single terminal leaflet.

Incise; deeply cut.
Included; not extending beyond the organs surrounding it.
Incumbent; when the radicle is applied to the back, not edges, of the cotyledons.
Incurved; curved inwards.
Indefinite; many but uncertain in number.
Indehiscent; not bursting.
Indeterminate; inflorescence having always a terminal leaf-bud.
Induplicate; when the edges of organs arranged in a valvate manner are folded inwards.
Indusium; a thin membrane often covering the clusters of capsules of Ferns.
Inferior; an inferior calyx or corolla is wholly free from the ovary; the reverse of superior.
Inflexed; curved inwards.
Inflorescence; arrangement of the flowers.
Innate; attached by their base to the apex of a stalk as are some anthers.
Inserted; growing upon.
Internode; the space between two nodes; a joint.
Interruptedly pinnate; when pairs of small pinnæ (leaflets) alternate with large pinnæ.
Introrse; anthers having the slit by which the pollen escapes directed towards the ovary.
Inverse; *inverted.* An embryo is so called when its radicle is directed towards a point at the opposite end of the seed from the hile.
Involucels; the involucres of secondary umbels.
Involucre; the whorled bracts at the base of an umbel or head; or sometimes below a single flower.
Involute; rolled from the back of any thing, as towards the upper side of a leaf.

Joinings; the places where the parts of the stems are attached to each other; the nodes.
Joints; the spaces between the knots, nodes, or joinings; the parts joined.

Keel; a prominent ridge. The two lower petals of a Pea-flower, within the others and united more or less by their anterior edge, form the keel.
Kneed; bent like the knee.
Knots; the joinings or nodes of the stem in Grasses.

Label; the terminal segment of the lip in Orchids.
Laciniate; divided into narrow irregular lobes.

Lanceolate; narrowly elliptic and tapering to each end.
Lancet-shaped; shortly and bluntly lanceolate.
Lax; loosely arranged.
Leaflets; the subdivisions of compound leaves.
Legume; a one-celled and two-valved seed-vessel with the seeds arranged along the inner angle, as the pod of a Pea.
Lenticular; like a doubly convex lens.
Ligulate; strap-shaped; not very narrow nor long, and with nearly parallel sides.
Ligule; a membrane at the base of the limb of the leaf of Grasses.
Limb; the flattened expanded part of a leaf or petal.
Linear; very narrow and long, with parallel sides until near the end.
Lingulate; tongueshaped; long, fleshy, convex, blunt.
Lipped; applied to a corolla or calyx appearing to consist of two lips.
Lobate; lobed; with large divisions.
Loculicidal; opening down the back (or midrib) of the carpel.
Lower part of a floral whorl; that furthest from the main axis; anterior.
Lunate; shaped like the new moon.
Lyrate; a pinnatifid leaf with the lobes successively and gradually enlarging from the petiole, and ending in one still larger lobe.

Marcescent; fading but remaining in its place.
Medullary; relating to the pith. Medullary rays are plates of cells which connect the pith with the growing part next to the bark.
Membranous; of the texture of membrane; thin and flexible.
Mericarps; the carpels of Umbelliferæ.
Midrib; the large vein extending along the middle of a leaf from its petiole nearly or quite to the other end.
Moniliform; cylindrical but constricted at regular intervals.
Monocotyledonous; having one sheathing cotyledon.
Monœcious; with the sexes in separate flowers on the same plant.
Monosepalous; *monopetalous*; when the sepals or petals are joined by their edges so as apparently to form one.
Mucronate; abruptly tipped with a short point of the same texture.
Multifid; divided into many parts.
Muricate; covered with short sharp points.
Mono-, *di-*, &c. *androus*; with 1, 2, &c. stamens.
Mono-, *di-*, &c. *gynous*; with 1, 2, &c. free styles or stigmas.

Naturalized; introduced but propagating itself freely by seed.
Nectary; an organ which secretes honey.
Netted; covered with lines connected together like network.
Node; a point in a stem where a leaf is produced; a joining.
Nut; a hard dry 1-seeded superior pericarp; also used for a *glans*, a hard dry 1- or few-seeded inferior pericarp not bursting and seated in a cup-like involucre, *e. g.* acorn.

Ob; in conjunction with terms means inverted; as *obovate* is ovate with the attachment at the narrow end.
Oblong; long oval, equally broad at each end.
Ocrea; a tubular membranous stipule surrounding the stem.
Opaque; not shining.
Opposite; when two similar organs grow one on each side of some body; or different organs are opposed to each other with a stem between them.
Orbicular; nearly round and flat.
Oval; an ellipse; a figure rounded at each end, not broader at one end than at the other; and about twice as long as broad.
Ovary; the young seed-vessel.
Ovate; egg-shaped; a short flat figure (thin like a leaf) rather broader below the middle of its length.
Ovoid; a solid eggshaped figure.
Ovule; the young seed.

Palate; the prominent part of the base of the lower lip which closes the mouth of a ringent corolla.
Pales; the leaflike parts of the flower of Grasses, enclosing the stamens, pistils, and hypogynous scales.
Palmate; with lobes spreading like the fingers of a hand from the same point.
Panicle; a raceme with branching pedicels; hence *paniculate*.
Papilionaceous; like the flower of a Pea.
Papillæ; small elongated protuberances.
Papillose; with small long protuberances.
Pappus; the crest of the fruit in Composites, formed of the altered limb of the calyx.
Parabolic; starting from a broad base and gradually narrowing with curved sides to a blunt point, as the divisions of a calyx.
Parallel veins start several together from the base of a leaf, diverge slightly, then proceed parallel and simple, and converge at the apex.
Parietal; on the inner surface of an ovary.
Patent; spreading widely.
Pedate; palmate of three lobes with the lateral lobes having similar large lobes on their upper edge.

Pedicel; the branch of a peduncle.
Peduncle; flowerstalk.
Pellucid; nearly transparent.
Peltate; when its point of attachment is on the face, not at the edge of a leaf or other organ.
Pendulous; seeds hanging from the top of an ovary.
Pentagonal; with five angles having convex spaces between them.
Pentangular; with five angles and five flat or concave faces.
Perennial plants live several years and flower more than once, usually many times.
Perfect flowers have both stamens and pistils in an efficient state.
Perfoliate; when the leaf completely surrounds the stem so that the latter seems to pass through it.
Perianth; the floral whorls when the calyx and corolla are not distinguishable.
Pericarp; seed-vessel, including adhering calyx if present.
Perigynous; when the corolla and stamens are borne on the calyx but free from the ovary.
Perisperm; the so-called albumen.
Persistent; not soon falling off.
Personate; a gamopetalous two-lipped corolla of which the lower lip is pressed upwards so as to close the opening.
Petal-like; resembling petals in texture and colour.
Petals; the divisions of the corolla.
Petiolate; having a petiole.
Petiole; the stalk of a leaf: *petiolule*; of a leaflet.
Phænogamous; visibly furnished with stamens and pistils
Phanerogamous; phænogamous.
Phyllaries; the scales or bracts of the involucre of Composites.
Pilose; with scattered rather stiff hairs.
Pinnæ; the segments of a pinnate leaf.
Pinnate; when leaflets are arranged on opposite sides of a common stalk. A leaf is 2- or 3-pinnate when its primary or secondary divisions are pinnate.
Pinnatifid; a leaf deeply cut into segments nearly to the midrib. A 2- or 3-pinnatifid leaf corresponds to a 2- or 3-pinnate leaf.
Pinnules; the segments of a bipinnate leaf.
Pistil; the ovary, style, and stigma taken together.
Pith; a column of cellular tissue in the centre of the stem and branches of Dicotyledons.
Pitted; covered with small depressed spots.
Placenta; the part of the carpel from which the ovules spring.
Plane; flat; also an imaginary flat surface in which things are placed.

Plicate; plaited.
Plumule; the ascending leafy part of the embryo.
Pod; a 1-celled and 2-valved seed-vessel with the seeds arranged along the inner angle.
Pollen; the dust in the anther.
Polygonal; with many angles.
Polypetalous; with many separate petals.
Polysepalous; with many separate sepals.
Pome; a compound fleshy many-seeded fruit, an apple or fruit resembling it.
Pores; small, often roundish, holes.
Porrect; extending forwards.
Posterior; the part of the flower nearest to the axis.
Prickles; hardened epidermal appendages resembling thorns, but not woody.
Primordial; the first flower of inflorescence.
Procumbent, prostrate; lying on the ground.
Prolonged; drawn out into a long point, like acuminate, but with no hollowing at the sides.
Pubescence; closely adpressed down.
Pubescent; with pubescence.
Pulverulent; covered with fine powdery matter.
Punctate; having minute spots like pin-holes, real or apparent.
Pyramidal; nearly in the shape of a pyramid.
Pyriform; pear-shaped.

Quadrate; squarish.

Raceme; a spike with stalked flowers: hence
Racemose; flowering in a raceme.
Rachis; the central stem of some kinds of inflorescence; as the stalk common to several spikelets of Grasses; the stalk of the frond of Ferns above the lowest pinnæ.
Radiate flowers; those at the margin of a head or other inflorescence which are long and spreading like rays.
Radical; springing from just above the root.
Radicle; the end of the embryo from which the root grows; also small roots.
Raphides; minute needle-shaped crystals found in the cells of some plants.
Rays (see *Radiant*); parts diverging in a circle from a central point.
Receptacle; the dilated top of the stalk bearing the flowers in Composites; the common support of the parts of a flower.
Reclinate and *reclining*; curved downwards.
Recurved; bent moderately backwards.

Reflexed; bent considerably backwards.
Reniform; transversely oval, but broadly cordate at the base.
Repand; with a rather wavy margin.
Reticulate; forming a network.
Retrorse; directed from the point of an organ.
Retuse; abruptly blunt with a notch in the middle.
Revolute; rolled back, as towards the underside of a leaf.
Rhizomatous; having rhizomes.
Rhizome; a prostrate more or less subterranean stem producing roots and leafy shoots.
Rhomboidal; approaching a quadrangular, not square, figure attached by one of its more acute angles.
Ringent; a 2-lipped widely open corolla.
Rootstock; a thick short rhizome or tuber.
Rosette; a collection of leaves growing close together, like the petals of a double rose.
Rosulate; arranged in a rosette.
Rotate; a monopetalous corolla with a short tube and very spreading limb.
Rudimentary; imperfectly developed.
Rugose; covered with a network of lines enclosing convex spaces.
Rugulose; finely rugose.
Runcinate; where the lobes of leaves are directed towards the base.
Runner; a prostrate shoot rooting at its end; a stole.

Sagittate; like the barbed head of an arrow, the auricles or lobes pointing backwards.
Salvershaped; a corolla with a long slender tube and flat limb.
Sarmentose; having a prostrate stem, starting with a very small arch from its root.
Scabrous; rough like a blacksmith's hand.
Scales; minute rudimentary leaves; very small flat semidetached parts of the cuticle.
Scape; a leafless radical peduncle.
Scarious; very thin, dry, and semitransparent.
Scorpioid; said of the branches of a cyme curved in a circinate manner, and the flowers produced only on the upper side.
Secund; all turned towards one side.
Seed; the ovule arrived at maturity.
Seedstalk; the stalk connecting the hilum of a seed with the placenta.
Sepals; the divisions of the calyx.
Septicidal; when a fruit splits through the middle of the septa or partitions.

Septifragal; when a fruit splits by the separation of the backs of the carpels from the septa.

Septum; the division of an ovary formed by the inflexed edges of the carpels.

Serrate; toothed like a saw.

Serratures; teeth like those of a saw.

Serrulate; with very small sawlike teeth.

Sessile; without a stalk.

Seta; a bristle; a bristle tipped with a gland; a slender straight prickle.

Setaceous; like a bristle.

Setose; bearing bristles or setæ usually ending in glands.

Sheath; the lower part of a leaf or its petiole, which forms a vertical sheath surrounding the stem. It is sometimes found alone.

Silicle; a silique not four times as long as broad.

Silique; a long podlike fruit of Crucifers having its edges connected by an internal membrane.

Simple; not compound; not branched.

Sinuate; having many large blunt lobes and notches.

Slashed; with deep tapering incisions.

Smooth; free from all kinds of roughness.

Sobole; a creeping underground stem producing roots and leaf-buds at intervals; an underground stole.

Soboliferous; having soboles, or long underground shoots ending in suckers.

Solitary; growing singly.

Spadix; a succulent spike bearing many sessile closely placed flowers.

Spath; a large bract often enclosing a spadix.

Spathulate; oblong, with a long linear claw or haft.

Spike; a long simple axis with many sessile flowers; hence *spicate* flowers.

Spikelet; the small group of flowers in Grasses enclosed within one or more glumes.

Spine; a stiff sharp woody persistent thorn.

Spinous; furnished with spines.

Spinulose; with small, often very minute spines or prickles.

Spiral vessels: fine tubes composed of membrane with spirally twisted fibres internally.

Sporules; the seedlike reproductive bodies of flowerless plants.

Spur; a tubular extension of the lower part of a petal or gamopetalous corolla; a loose prolongation of the base of a leaf beyond its point of attachment.

Spurred; furnished with a spur.

Squarrose; covered with appendages spreading at right angles or more.

Stamen; the male organ of a flower, usually formed of a filament and anther.
Staminode; a scale on the inside of the upper lip of some Scrophulariaceæ.
Standard; the upper or posterior petal of a Pea-flower, which is outside the others in the bud.
Starlike; applied to flowers of which the petals are narrow and distant and radiant like a star.
Stellate; radiating from a centre like a star.
Stellulate; like minute stars.
Stigma; the cellular part at the top of a carpel or style to which the pollen adheres.
Stigmatic disk; a broad surface at the top of the style, or forming the whole of it, upon which the stigmas are placed.
Stipe; the stalk of Ferns up to the lowest pinna.
Stipules; leaflike appendages at the base of the petiole.
Stipulodes; spines beneath the whorls of branches in *Chara*.
Stole; a lax trailing shoot from the crown of the root, rooting at intervals.
Stoloniferous; having stoles.
Stomates; minute organic openings in the skin of plants.
Strapshaped; not very narrow nor long, and with nearly parallel sides.
Streak; a straight line of peculiar colour or structure, or a furrow.
Striæ; very slight furrows or ridges.
Striate; with slender streaks or furrows.
Striped, having coloured streaks.
Stripes; the vittæ of Umbellifers.
Style; the space between the ovary and stigma.
Stylopode; a fleshy disk crowning the ovary and supporting the style of Umbellifers.
Sub; in composition means a near approach to; as subrotund is nearly round.
Subulate; awlshaped, tapering from the base to a fine point, a long narrow triangle.
Sucker; a stem produced at the end of an underground shoot.
Superior; above any thing; a calyx is superior when its tube is wholly attached to the ovary, half-superior when attached only to the lower half of it; an ovary is superior when wholly free from the calyx; a part of a flower placed next to the axis.
Suspended ovules hang down from near the top of the ovary.
Suture; the line of junction of similar organs cohering.
Swordshaped; very long, narrow, nearly parallel-sided, sharp-edged, acute.

Sympode; a stem formed of a series of superposed branches so as to resemble a continuous axis.
Syncarpous; fruit formed of cohering carpels.
Syngenesious flowers form a head and have 5 stamens with united anthers.

Tailed; having a long slender point.
Tassel-like; resembling a silken tassel or painter's camel's-hair pencil.
Tendril; a twisting slender organ for laying hold of objects.
Terete; having a nearly round transverse section.
Ternate; growing in threes about the same point of a stem.
Testa; the outer coat of a seed.
Testaceous; brownish yellow.
Tetragonous; with four angles and four convex faces.
Thorn; an abortive branch with a sharp point; distinguished from a prickle by being woody.
Three-veined; having three veins, usually of nearly equal size, proceeding from the base. Sometimes leaves are *falsely three-veined* when the ends of a series of lateral veins combine to form a submarginal vein near each edge of the leaf.
Throat; the orifice of the tube of a gamopetalous corolla or gamosepalous calyx.
Thyrsoid; having a close-branched raceme of which the middle is broader than the ends.
Tomentose; covered with cottony entangled hairs, forming a matted shagginess called *tomentum*; felted.
Torulose; uneven, alternately elevated and depressed like a knotted cord.
Torus; the part within the calyx to which the floral organs are attached.
Transversely; applied to forms like oval when attached by one of their longer sides.
Triangular; with three angles and three flat faces.
Trichotomous; in forks of three prongs.
Trifarious; arranged in three rows.
Trifid; dividing about halfway down into three parts.
Trifurcate; forked with three nearly equal prongs.
Trigonous; with three angles and three convex faces.
Tripartite; divided into three parts nearly to its base.
Triquetrous; having three angles and three concave faces.
Truncate; blunt as if cut off at the end.
Tube; the pipe formed by the cohesion of the parts of a floral whorl.
Tuber; a thickened underground fleshy part of the stem.
Tubercles; little round knobs.
Tubercular; *tubercled*; covered with little knobs.

Tuberous; like a tuber, but not part of the stem.
Tubular; hollow and nearly cylindrical.
Tumid; swollen.
Turbinate; topshaped, conical, and attached by its long point.
Two-edged; compressed so as to have two sharp edges.

Umbel; when many stalked flowers spring from one point and reach about the same level. *Partial* umbels are umbels seated upon the branches of an umbel, when the umbel forms a *compound* umbel.
Umbilicate; peltate, but having the attached organ hollowed to receive the top of the stalk.
Unilateral; turned to one side.
Upper part of a floral whorl; that next the main axis of the stem; posterior.
Urceolate; like a pitcher contracted at the mouth.
Utricle; a bladder-like covering as in *Chenopodiaceæ*, i. e. an achene with a membranous pericarp. The envelope of the nut of *Carex*.

Valvate; having *valves* or parts of an organ opening like little doors; or organs touching only along their edges.
Veins; bundles of vessels in leaves and their modifications.
Ventricose; swelling unequally on one side.
Vernation; the arrangement of leaves in a bud.
Versatile; swinging freely on its support, as an anther attached by one point of its back.
Villose; shaggy with loose long soft hair.
Viscous; clammy.
Vittæ; linear receptacles of oil in the fruits of Umbelliferæ; stripes.
Viviparous; bearing young plants in the place of flowers.

Wedgeshaped; like a wedge, but attached by its point.
Whorl; formed of similar organs arranged in a circle round an axis.
Whorled; arranged in whorls.
Winged; having leaflike or membranous expansions.
Wings; the lateral petals of a Pea-flower; the flat membranous appendages of some seeds.

When two terms are combined, as *ovate-lanceolate*, it means that the form or structure is compounded of the two, or lies between then.

SYNOPSIS

OF THE

NATURAL ORDERS OF BRITISH PLANTS.

This Synopsis has been prepared for the purpose of facilitating the discovery of the Order to which an unknown British Plant belongs. *It must be used with caution, as a very slight error will totally mislead,* and often the character used is not quite determinable, or is inconstant in some degree: I therefore am very far from advising the use of this Synopsis. The student must always commence with the pair of characters numbered 1 in the left-hand margin; and having determined with which of these his plant agrees, proceed similarly with the group of characters referred to by the number on the right-hand side of the page, and so on.

For instance, having gathered a *Hawthorn*, he finds it to agree with the *second* character of number 1, the *first* of number 3, the *second* of number 4, the *third* of number 29, and the *first* of number 30. It therefore belongs to Order xxvi. Rosaceæ and Suborder Pomeæ. Then turning to the body of the *Manual* (p. 104), he will examine the specimen by the characters given for that Order and its Suborder. Finding it to agree with them, a perusal of the generic definitions placed under Pomeæ will show that it is a *Cratægus*. After a little experience in the examination of plants, the eye becomes so familiar with the principal Orders as to render this process unnecessary, except in doubtful cases.

1 Leaves straight- or parallel-veined (belonging generally to plants having 3-parted floral whorls or a 6-parted perianth) * 2
Leaves net-veined (belonging generally to plants having 5- or 4-parted floral whorls) † 3
2 Seeds in a seed-vessel 62
Seeds apparently naked on an axillary scale or in a fleshy cup and solitary. Male fl. in catkins. L. linear or subulate ....lxxix. *Coniferæ.*
3 Fl. with a calyx and corolla 4
Fl. with a perianth (calyx undistinguishable from corolla) or none 50
4 Corolla polypetalous, inferior 5
Corolla polypetalous, superior 29
Corolla gamopetalous (petals cohering), superior.... 34
Corolla gamopetalous, inferior 39
5 Ovaries many, distinct or united, each bearing a style; or solitary with one lateral placenta 6
Ovary solitary; placentas 2 or more, parietal or on the dissepiments, not forming a central axis........ 13
Ovary solitary; placentas central 17
6 Corolla regular 7
Corolla irregular 10
7 Sepals distinct. Stamens hypogynous 8
Sepals more or less combined below 9
8 Stamens few. Anth. adnate; connective extending beyond them ....lxxx. *Trilliaceæ.*
Stamens indefinite, usually many (when few, alternate with the petals). Anth. at top of filament, opening by two longitudinal clefts.... i. *Ranunculaceæ.*
Stamens as many as and opposite to the petals. Anth. at top of filament............ ii. *Berberidaceæ.*

* The leaves of these plants (*Monocotyledones*) are nearly always parallel-veined. A few exceptions are *Dioscoreaceæ* (*Tamus*), *Trilliaceæ* (*Paris*), and *Araceæ* (*Arum*), which will be found by following either series of characters. Care must be taken not to confound leaves having *parallel veins which are connected by simple transverse veins* with *net-veined* leaves. The broad leaves of *Alisma, Potamogeton*, and *Hydrocharis* are examples of the former. Pinnatifid leaves with linear lobes as in *Anemone Pulsatilla*, may be mistaken for parallel-veined leaves; also the submersed leaves of *Ranunculi.*

† The narrow leaves of some of these plants (*Dicotyledones*) are apparently parallel-veined. They are chiefly aquatics:

Leaves divided in a pinnatifid way. xxx. *Haloragaceæ* (Myriophyllum).
Leaves repeatedly forked ....lxxiii. *Ceratophyllaceæ.*
Leaves simple ....xxx. *Haloragaceæ* (Hippuris).

9 Stamens as many as, or twice the number of the petals, inserted at the base of the calyx. No stipules. xxxiv. *Crassulaceæ.*
Stamens 20 or more, inserted on the calyx. Leaves with stipules.......... xxvi. *Rosaceæ* (in part).
10 Leaves with stipules 11
Leaves without stipules 12
11 Stamens 10, mon- or diadelphous..xxv. *Leguminosæ.*
12 Stamens many free. Fruit of 1 or more follicles. i. *Ranunculaceæ* (in part).
Stamens 6, in two bundles........ v. *Fumariaceæ.*
13 Corolla regular. Petals 4 14
Corolla regular. Petals 5 15
Corolla regular. Sepals and petals many, gradually passing into each other...... iii. *Nymphæaceæ.*
Corolla irregular 16
14 Sepals 2. Stamens many........ iv. *Papaveraceæ.*
Sepals 4. Stamens tetradynamous.. vi. *Cruciferæ.*
15 Sepals equal, distinct, imbricate. Stamens 5. x. *Droseraceæ.*
Sepals distinct; 3 inner twisted in the bud; 2 outer smaller or wanting. Stamens indefinite, many. viii. *Cistaceæ.*
Sepals equal, more or less united below. Stamens as many or twice as many as the petals, from a shield-like disk................xxviii. *Tamariscaceæ.*
16 Leaves with stipules. Sepals 5. Stamens 5, free. ix. *Violaceæ.*
No stipules. Sepals 2 or wanting. Stamens 6, in two bundles. Ovary closed........ v. *Fumariaceæ.*
17 Calyx imbricate in the bud.......................... 20
Calyx valvate in the bud, or with distant lobes. Calyx and corolla regular.......................... 18
18 Stamens united into a column ......xv. *Malvaceæ.*
Stamens free.......................... 19
19 Stamens hypogynous............. xvi. *Tiliaceæ.*
Stamens perigynous, opposite the petals and equalling them in number............ xxiv. *Rhamnaceæ.*
Stamens inserted in the tube of the calyx, alternating with, or twice as many as, but below the petals. xxvii. *Lythraceæ.*
20 Corolla regular.......................... 21
Corolla irregular ....xx. *Balsaminaceæ.*
21 Calyx tubular...... xiv. *Caryophyllaceæ* (Sileneæ).
Sepals distinct or slightly connected below 22
22 Ovary 1-celled.......................... 23
Ovary many-celled 25

23 Stamens opposite the petals. Sepals 2. xxxii. *Portulaceæ*.

Stamens opposite the petals. Sepals 3—5 24

24 Stamens 10 or fewer. Stipules none. xiv. *Caryophyllaceæ* (Alsineæ).

Stamens 5. Stipules present. Petals distinct. xiv. *Caryophyllaceæ* (Polycarpeæ).

Stamens 5. Stipules present. Petals subulate. xxxiii. *Paronychiaceæ*.

25 Stamens free 27

Stamens polyadelphous ....xvii. *Hypericaceæ*.

Stamens monadelphous or inserted in an hypogynous ring 26

26 Stamens 4—5. Stipules none ....xxii. *Linaceæ*.

Stamens 10. Stipules none or united to the base of the petioles. Cells of ovary 5; each many-seeded. xxi. *Oxalidaceæ*.

Stamens 10. Stipules present. Cells of ovary 5; each 1-seeded ....xix. *Geraniaceæ*.

27 Style 1 28

Styles 3—5. Anthers terminal ....xiii. *Elatinaceæ*.

Styles 3—5. Filaments extending beyond the anthers. lxxx. *Trilliaceæ*.

28 Stamens 3. Petals 3. Fruit fleshy. lxxi. *Empetraceæ*.

Stamens and petals 4 or 5, inserted in an hypogynous disk ....xxiii. *Celastraceæ*.

Stamens 5 or more. Petals 5. Caps. 1-celled, 3—4-valved ....xi. *Frankeniaceæ*.

Stamens usually 8. Petals 5. Fruit winged, separating into two capsules ....xviii. *Aceraceæ*.

Stamens 8—10. Capsules 5-celled, 5-valved. xlvii. *Ericaceæ* (Pyroleæ and Monotropeæ).

29 Ovary 1-celled with one pendulous ovule. Stamens as many as and opposite to the petals. xl. *Loranthaceæ*.

Ovary 1-celled; placentas 2 parietal; ovules many. Stamens and petals 4—5, alternating. xxxv. *Ribesiaceæ*.

Ovary 2—many-celled; placentas central. Or a pome. 30

30 Stamens many, indefinite. Fruit a 1—5-seeded pome. xxvi. *Rosaceæ* (Pomeæ).

Stamens as many as and alternating with the petals, or twice as many 31

31 Petals imbricate in the bud 32

Petals valvate in the bud 33

Petals twisted in the bud. Sepals valvate. Style 1. xxix. *Onagraceæ*.

32 Petals 5. Stamens 5. Styles 2. Inflorescence umbellate. Seeds solitary ....xxxvii. *Umbelliferæ*.

b 5

Petals 4—5. Stamens 5—10. Styles 2. (Calyx sometimes inferior.) Capsule 2-valved. Seeds many. (Fruit with two horns.) ....xxxvi. *Saxifragaceæ.*

Petals 4. Stamens 4—8. Styles 4. Fruit of 4 hard nuts ....xxx. *Haloragaceæ.*

33 Fruit a berry. Styles more than 2. Leaves alternate. xxxviii. *Hederaceæ.*

Fruit a drupe. Style 1. Leaves opposite. xxxix. *Cornaceæ.*

Fruit dry. Styles 2. Leaves alternate. xxxvii. *Umbelliferæ* (in part).

34 Stamens inserted beneath an epigynous disk. xlvii. *Ericaceæ* (Vaccinieæ).

Stamens inserted with the corolla and free from it .. 35

Stamens inserted on the tube of the corolla or between its lobes 36

35 Filaments free ....xlvi. *Campanulaceæ.*

Filaments united into 3 bundles ..xxxi. *Cucurbitaceæ.*

36 Anthers united. Flowers in a head. xlv. *Compositæ.*

Anthers free 37

37 Flowers in a head. Calyx double ..xliv. *Dipsacaceæ.*

Flowers corymbose or cymose 38

38 Fruit a double indehiscent pericarp, 2-celled, 2-seeded. xlii. *Rubiaceæ.*

Fruit dry, with 1 perfect cell, 1-seeded; and often 2 empty cells. Stamens 1—3 ..xliii. *Valerianaceæ.*

Fruit fleshy, with 1 or several seeds. Stamens 4—5. xli. *Caprifoliaceæ.*

39 Ovary and fruit 4-lobed, separating into 4 small 1-seeded nuts. Style from base of ovary 40

Fruit of several follicles. xxxiv. *Crassulaceæ* (Cotyledon).

Ovary and fruit simple. Style terminal 41

40 Stamens 5. Corolla regular. Leaves alternate. liv. *Boraginaceæ.*

Stamens 4, didynamous, or 2. Corolla 2-lipped. Leaves opposite ....lviii. *Labiatæ.*

41 Ovary 1-celled, 1-seeded. Calyx tubular. Stamens 5. Styles 5 ....lxii. *Plumbaginaceæ.*

Ovary 1-celled, many-seeded 42

Ovary with 2 or more cells (but fruit sometimes 1-celled) 44

42 Corolla scarious, regular, 4-parted. Stamens 4. lxiii. *Plantaginaceæ.*

Corolla coloured 43

43 Corolla irregular. Stamens 2 ....lx. *Lentibulariaceæ.*

Corolla regular. Stamens 4—5, opposite to the segments of the corolla ....lxi. *Primulaceæ.*

Corolla irregular. Stamens 4, didynamous. lvi. *Orobanchaceæ.*

44 Stamens hypogynous, scarcely attached to the corolla, distinct....xlvii. *Ericaceæ.* (Ericeæ and Arbuteæ).
Stamens upon the corolla, filaments connected. xii. *Polygalaceæ.*
Stamens upon the corolla, distinct.................. 45
45 Stamens 2. Corolla regular ....xlix. *Oleaceæ.*
Stamens 2 or 4, and didynamous. Corolla irregular .. 46
Stamens 4 or 5, not didynamous 47
46 Ovary 2-celled, not lobed; placentas central. lvii. *Scrophulariaceæ.*
Ovary 2—4-celled, lobed............lix. *Verbenaceæ.*
47 Cells of ovary each with 1 or 2 ovules 48
Cells of ovary each with many ovules 49
48 Fruit fleshy, not bursting. Stigmas sessile. xlviii. *Aquifoliaceæ.*
Fruit a capsule, bursting. Styles manifest. liii. *Convolvulaceæ.*
49 Fruit a double follicle ....l. *Apocynaceæ.*
Fruit 2- or imperfectly 4-celled. Leaves alternate. lv. *Solanaceæ.*
Fruit 1- or imperfectly 2-celled, 2-valved. Leaves opposite ....li. *Gentianaceæ.*
Fruit 3-celled, 3-valved............lii. *Polemoniaceæ.*
50 Flowers not in catkins........................... 51
Male flowers in catkins ....lxxviii. *Amentiferæ.*
Flowers inconspicuous (rarely found). Plant formed of minute leaflike fronds, floating freely. xciv. *Lemnaceæ.*
51 Ovary one, superior. Perianth sometimes wanting .. 54
Ovary inferior 52
Ovaries many, distinct 53
52 Stamens 1 or 8. Fruit not bursting, 1- or 4-celled, 1- or 4-seeded. Limb of perianth of female flower very minute ....xxx. *Haloragaceæ.*
Stamens 4 or 5. Fruit not bursting, 1-celled, 1-seeded. Limb of perianth manifest ......lxix. *Santalaceæ.*
Stamens 4. Style filiform. Ovary 4-celled, many-seeded ....xxix. *Onagraceæ.*
Stamens 5, syngenesious ....xlv. *Compositæ.*
Stamens 6, on base of perianth. Style 1, trifid. Fl. diœcious ....lxxxi. *Dioscoreaceæ.*
Stamens 6—12, epigynous. Style short. Stigma radiant. Ovary 3—6-celled, many-seeded. lxx. *Aristolochiaceæ.*
Stamens 8—10. Styles 2. Ovary 1-celled, 2-beaked, many-seeded, opening like a cup. xxxvi. *Saxifragaceæ.*

53 Sepals petal-like, deciduous. Anthers extrorse.
i. *Ranunculaceæ* (Caltha).
Sepals petal-like, persistent. Anthers introrse.
i. *Ranunculaceæ* (Pæonia).
54 Fruit separating into several carpels. No perianth.... 55
Fruit not separating into carpels 56
55 Carpels 4, not bursting. Stamen 1.
lxxiv. *Callitrichaceæ.*
Carpels 3 or 2, opening, separating with elasticity.
lxxii. *Euphorbiaceæ.*
56 Leaves with stipules 57
Leaves without stipules 58
57 Stipules sheathing the stem ......lxvi. *Polygonaceæ.*
Stipules attached to the petiole.
xxvi. *Rosaceæ* (Sanguisorbeæ).
Stipules free, deciduous. Ovary 1-celled. Perianth 4—5-parted ....lxxv. *Urticaceæ.*
Stipules free, deciduous. Ovary 2-celled.
lxxvii. *Ulmaceæ.*
Stipules free, deciduous. Ovary 1-celled. Female perianth scale-like, open ....lxxvi. *Cannabinaceæ.*
58 Flowers monœcious or diœcious.................. 59
Flowers perfect or polygamous.................... 60
59 Fruit fleshy. Male fl. each of one 2-celled naked stamen crowded together on a spadix.
xciii. *Araceæ* (Areæ).
Fruit fleshy. Stamens 2 or 3. Perianth of scales imbricated in several rows ....lxxi. *Empetraceæ.*
Fruit dry. Stamens 3 or more. Perianth tubular.
lxvii. *Elæagnaceæ.*
Fruit dry. Stamens 12—20. Perianth 10—12-cleft.
lxxiii. *Ceratophyllaceæ.*
60 Perianth hardened over the fruit.
xiv. *Caryophyllaceæ* (Sclerantheæ).
Perianth not hardened over the fruit 61
Perianth none. Fruit compressed linear and leaflike at the end................xlix. *Oleaceæ* (Fraxinus).
61 Perianth 3—5-cleft, herbaceous ..lxv. *Chenopodiaceæ.*
Perianth 3-cleft, scarious........lxiv. *Amaranthaceæ.*
Perianth tubular. Stamens perigynous.
lxviii. *Thymelaceæ.*
Perianth 6—8-parted, herbaceous. Stamens hypogynous.......................lxxx. *Trilliaceæ.*
62 Leaves net-veined. Floral envelopes whorled 63
Leaves net-veined. Perianth none. Fl. monœcious, on a spadix ....xciii. *Araceæ* (Areæ).

Leaves parallel-veined, persistent. Floral envelopes whorled, or none.............................. 64
Leaves parallel-veined, persistent. Floral envelopes imbricate, bractlike 75
63 Perianth inferior. Flowers perfect ..lxxx. *Trilliaceæ.*
Perianth superior. Flowers diœcious.
lxxxi. *Dioscoreaceæ.*
64 Ovaries several, each bearing a style or stigma 65
Ovary 1, inferior 66
Ovary 1, superior 69
65 Perianth 4-parted, inconspicuous or none.
xcv *Potamogetonaceæ.*
Perianth 6-parted, conspicuous. Carpels not opening.
lxxxvi. *Alismaceæ.*
Perianth 6-parted, conspicuous. Carpels opening at their inner edge............lxxxix. *Melanthaceæ.*
66 Stamens and style united in a central column.
lxxxiii. *Orchidaceæ.*
Stamens free from the style 67
67 Perianth wholly petal-like 68
Three outer segments of perianth resembling an herbaceous calyx ....lxxxii. *Hydrocharidaceæ.*
All the 4 segments of perianth resembling an herbaceous calyx ....xl. *Loranthaceæ.*
68 Stamens 3. Anthers bursting outwards.
lxxxiv. *Iridaceæ.*
Stamens 6. Anthers bursting inwards.
lxxxv. *Amaryllidaceæ.*
69 Perianth like a corolla of 6 petals. Flowers perfect .. 70
Perianth more or less glumaceous, or partly coloured or wanting 71
70 Perianth petal-like, conspicuous. Fruit dry, bursting with 3 valves .... lxxxviii. *Liliaceæ.*
Perianth petal-like, conspicuous. Fruit succulent, not bursting ....lxxxvii. *Asparagaceæ,*
Perianth herbaceous, scale-like. Fruit baccate, not bursting ....xciii. *Araceæ* (Oronteæ).
71 Flowers perfect. Perianth 6-parted. Stamens 6.
xc. *Juncaceæ.*
Flowers monœcious 72
72 Flowers on a spadix............................ 73
Spadix wanting 74
Flowers solitary ....xcvi. *Naiadaceæ.*
73 Flower on a thick spadix. Spath wanting. Anthers wedgeshaped, erect; filaments long.
xcii. *Typhaceæ.*

Flowers on a thick spadix in a spath. Perianth none. Anthers ovate; filaments very short. xciii. *Araceæ* (Areæ).

74 Flowers capitate. Stamens 2—5 . . xci. *Eriocaulaceæ*.

Flowers 2 in a spath, one male, one female. (Floating). xciv. *Lemnaceæ*.

Flowers in two rows on one side of a spath, enclosed by a fold of the leaf . . . . xcvi. *Naiadaceæ*.

75 Leaves with entire sheaths. Anthers entire at the ends . . . . xcvii. *Cyperaceæ*.

Leaves with split sheaths. Anthers notched at both ends . . . . xcviii. *Gramineæ*.

CLASSES, DIVISIONS, AND ORDERS

OF

BRITISH PLANTS.

Class I. DICOTYLEDONES [1].

Stems when perennial composed of bark, wood, and pith. The wood furnished with medullary rays and increasing by the addition of concentric layers externally. Leaves usually net-veined. Cotyledons 2 or more, opposite or whorled.—Each floral whorl composed of 5 or 4 parts.

Division 1. THALAMIFLORÆ.

* *Apocarpous.*

[1] The characters are drawn to suit our plants.

** *Appearing syncarpous from the fleshy disk enveloping the ovaries.*

*** *Syncarpous with parietal placentas.*

**** *Syncarpous with axile placentas.*

† Fl. irregular.

†† Fl. regular (except in Ord. 20).

Division 2. CALYCIFLORÆ.

* *Stam. and petals mostly perigynous.*

Division 3. COROLLIFLORÆ.

*** *Corolla hypogynous. Stam. epipetalous.*

† Cor. nearly or quite regular.

†† Corolla more or less irregular.

Division 5. GYMNOSPERMEÆ.

Class II. MONOCOTYLEDONES.

Stem not separable into bark, wood, and pith; of cellular tissue with vascular bundles embedded irregularly in it. Leaves mostly alternate and sheathing, with parallel simple veins connected by smaller transverse veins, rarely net-veined. Cotyledon one, or, if more, alternate. Floral whorls composed of 3 or 6 parts.

Division 1. DICTYOGENÆ.

Division 2. FLORIDÆ.

* *Perianth superior, petal-like. Syncarpous.*

** *Perianth inferior, apocarpous, or carpels separable when ripe.*

*** *Perianth inferior, syncarpous.*

Class III. CRYPTOGAMEÆ.

NATURAL ORDERS OF BRITISH PLANTS.

Class I. *Dicotyledones.*

Division I. THALAMIFLORÆ.

Order I. RANUNCULACEÆ.
II. BERBERIDACEÆ.
III. NYMPHÆACEÆ.
IV. PAPAVERACEÆ.
V. FUMARIACEÆ.
VI. CRUCIFERÆ.
VII. RESEDACEÆ.
VIII. CISTACEÆ.
IX. VIOLACEÆ.
X. DROSERACEÆ.
XI. FRANKENIACEÆ.
XII. POLYGALACEÆ.
XIII. ELATINACEÆ.
XIV. CARYOPHYLLACEÆ.
XV. MALVACEÆ.
XVI. TILIACEÆ.
XVII. HYPERICACEÆ.
XVIII. ACERACEÆ.
XIX. GERANIACEÆ.
XX. BALSAMINACEÆ.
XXI. OXALIDACEÆ.
XXII. LINACEÆ.

Div. II. CALYCIFLORÆ.

XXIII. CELASTRACEÆ.
XXIV. RHAMNACEÆ.
XXV. LEGUMINOSÆ.
XXVI. ROSACEÆ.
XXVII. LYTHRACEÆ.
XXVIII. TAMARISCACEÆ.
Order XXIX. ONAGRACEÆ.
XXX. HALORAGACEÆ.
XXXI. CUCURBITACEÆ.
XXXII. PORTULACEÆ.
XXXIII. PARONYCHIACEÆ.
XXXIV. CRAS ULACEÆ.
XXXV. RIBESIACEÆ.
XXXVI. SAXIFRAGACEÆ.
XXXVII. UMBELLIFERÆ.
XXXVIII. HEDERACEÆ.
XXXIX. CORNACEÆ.

Div. III. COROLLIFLORÆ.

XL. LORANTHACEÆ.
XLI. CAPRIFOLIACEÆ.
XLII. RUBIACEÆ.
XLIII. VALERIANACEÆ.
XLIV. DIPSACACEÆ.
XLV. COMPOSITÆ.
XLVI. CAMPANULACEÆ.
XLVII. ERICACEÆ.
XLVIII. AQUIFOLIACEÆ.
XLIX. OLEACEÆ.
L. APOCYNACEÆ.
LI. GENTIANACEÆ.
LII. POLEMONIACEÆ.
LIII. CONVOLVULACEÆ.
LIV. BORAGINACEÆ.
LV. SOLANACEÆ.
LVI. OROBANCHACEÆ.
LVII. SCROPHULARIACEÆ.
LVIII. LABIATÆ.
LIX. VERBENACEÆ.

Order LX. LENTIBULARIACEÆ.
LXI. PRIMULACEÆ.
LXII. PLUMBAGINACEÆ.
LXIII. PLANTAGINACEÆ.

Div. IV. MONOCHLAMYDEÆ.

LXIV. AMARANTHACEÆ.
LXV. CHENOPODIACEÆ.
LXVI. POLYGONACEÆ.
LXVII. ELÆAGNACEÆ.
LXVIII. THYMELACEÆ.
LXIX. SANTALACEÆ.
LXX. ARISTOLOCHIACEÆ.
LXXI. EMPETRACEÆ.
LXXII. EUPHORBIACEÆ.
LXXIII. CERATOPHYLLACEÆ.
LXXIV. CALLITRICHACEÆ.
LXXV. URTICACEÆ.
LXXVI. CANNABINACEÆ.
LXXVII. ULMACEÆ.
LXXVIII. AMENTIFERÆ.

Div. V. GYMNOSPERMÆ.

LXXIX. CONIFERÆ.

Class. II *Monocotyledones.*

Division I. DICTYOGENÆ.

LXXX. TRILLIACEÆ.
LXXXI. DIOSCOREACEÆ.

Division II. FLORIDÆ.

LXXXII. HYDROCHARIDACEÆ.
LXXXIII. ORCHIDACEÆ.
LXXXIV. IRIDACEÆ.
LXXXV. AMARYLLIDACEÆ.
LXXXVI. ALISMACEÆ.
LXXXVII. ASPARAGACEÆ.
LXXXVIII. LILIACEÆ.
LXXXIX. MELANTHACEÆ.
XC. JUNCACEÆ.
XCI. ERIOCAULACEÆ.
XCII. TYPHACEÆ.
XCIII. ARACEÆ.
XCIV. LEMNACEÆ.
XCV. POTAMOGETONACEÆ.
XCVI. NAIADACEÆ.

Division III. GLUMIFERÆ.

XCVII. CYPERACEÆ.
XCVIII. GRAMINEÆ.

Class III. *Cryptogameæ.*

XCIX. EQUISETACEÆ.
C. FILICES.
CI. MARSILEACEÆ.
CII. LYCOPODIACEÆ.
CIII. CHARACEÆ.

ABBREVIATIONS.

In the descriptions.

anth.	.... anther.	*ped.*	.... peduncle.
cal.	.... calyx.	*per.*	.... perianth.
caps.	.... capsule.	*pet.*	.... petal.
carp.	.... carpel.	*phyll.*	.... phyllary.
cor.	.... corolla.	*pr.*	.... prickle.
fl.	.... flower.	*segm.*	.... segment.
fr.	.... fruit.	*sep.*	.... sepal.
ft.	.... feet.	*st.*	.... stem.
gl.	.... glume.	*stam.*	.... stamen.
in.	.... inch.	*stig.*	.... stigma.
interm.	.... intermediate.	*stip.*	.... stipule.
inv.	.... involucre.	*t.*	.... plate.
l., l., ls.	.... leaves.	*term.*	.... terminal.
lt., lts.	.... leaflets.	*Tr.*	.... tribe.
nect.	.... nectary.	*var.*	.... variety.
pan.	.... panicle.		

Books.

A.N.H. ..Annals of Natural History.
Curt. ..Curtis's Flora Londinensis.
E.B. ..English Botany.
E.B.S. ..Supplement to E. B.
Fl. Dan. .Flora Danica.
J. of B. ..Journal of Botany.
Sy. E.B. .Engl. Botany, Syme's Edition.
P. ..Parnell's Grasses of Britain.
Phytol. ..The Phytologist.
R. ..Reichenbach's Icones Floræ Germanicæ.
R.I. ..Reichenbach's Iconographia Botanica.
St. ..Sturm's Deutschlands Flora.

Duration and Native Country.

A. ..Annual.
B. ..Biennial.
P. ..Perennial.
Sh. ..Shrub.
T. ..Tree.

E. ..England.
S. ..Scotland.
I. ..Ireland.

† ..Possibly introduced, but now appearing like a true native.
‡ ..Probably introduced, but admitting of some slight doubt on the subject.
* ..Certainly naturalized.
! ..After the name of a plant shows that an authentic specimen has been seen.

I. II. III. &c. represent the months of flowering, viz. Jan., Feb., March, &c.; but they differ so much in different parts of the kingdom that only an approximation to the true time can be given.

When the Initial letter of the generic name is prefixed to that of a variety, it is intended to show that the author quoted considered it to be a species.

MANUAL

OF

BRITISH BOTANY.

FLOWERING PLANTS.

Substance composed of cellular tissue, woody fibre, and spiral vessels. Epidermis with stomates. Flowers with stamens and pistils. Embryo with cotyledons.

Class I. DICOTYLEDONES.

Stems formed of bark, wood, and pith. The wood furnished with medullary rays and increasing by the addition of concentric layers externally. Leaves mostly with netted veins. Cotyledons 2 or more, opposite or whorled.—Each floral whorl composed of 5 or 4 parts.

Division I. THALAMIFLORÆ.

(Orders I.—XXI.)

Petals distinct (rarely 0), and as well as the stamens growing separately from the sepals on the top of the peduncle below the ovary (hypogynous).

Order I. RANUNCULACEÆ.

Sep. 3—6, often petal-like. Pet. 5 or more, rarely 0. Stam. usually many; anth. adnate, opening lengthwise. Carp. many, distinct, or forming a single pistil. Seeds erect or pendulous, albuminous.—Stip. 0, or adnate to petiole.

A. *Anthers extrorse.*

† Fruits (achenes) many, 1-seeded, short.

‡ *Sepals valvate in the bud.*

Tribe I. *CLEMATIDEÆ.* Achenes with feathery persistent styles. Seed pendulous. Leaves opposite.

1. CLEMATIS. Cal. of 4 or 5 sepals. Pet. 0. Carp. not bursting, awned. Stam. and styles many. (St. woody.)

‡‡ *Sepals imbricate in the bud.*

Tr. II. *ANEMONEÆ.* Seed pendulous. Leaves radical or alternate.

2. THALICTRUM. Cal. of 4 or 5 sepals. Pet. 0. Carp. not bursting, without awns. Stam. and styles many.

3. ANEMONE. Cal. petal-like, sep. 5—9. Pet. 0. *Carp.* not bursting, *tipped with the persistent* sometimes feathery *styles*, placed upon a thickened hemispherical or conical receptacle. Stam. and styles many. (*Fl. involucrate.*)

4. ADONIS. Cal. of 5 sepals. *Pet.* 5—10, without a honey-bearing pore. *Carp.* not bursting, *without awns.* Stam. and styles many.

Tr. III. *RANUNCULEÆ.* *Seed erect* (except in *Myosurus*). Pet. with a honey-bearing pore at their base.

5. MYOSURUS. Cal. of 5 sepals, prolonged into a spur at the base. Pet. 5, with a filiform tubular claw. Stam. 5. Styles many. *Carp.* not bursting, *closely imbricate upon a long filiform receptacle.* Seed pendulous.

6. RANUNCULUS. Cal. of 5, rarely 3, sepals. Pet. 5, rarely many, with a honey-bearing pore naked or covered by a scale. *Carp.* not bursting, *collected into a globular or oblong mass.* Stam. and styles many.

†† Fruits (follicles) many-seeded, bursting, long.

Tr. IV. *HELLEBOREÆ.* Stam. many. (Pet. small, often abnormal or wanting.)

* Flowers regular.

7. CALTHA. Cal. of 5 petal-like deciduous sepals. *Pet.* 0. Follicles 5—10.

8. TROLLIUS. Cal. of 5 or many petal-like deciduous sepals. *Pet. small, linear, flat,* clawed. Follicles many, sessile.

[9. CAMMARUM. Cal. of 5—8 petal-like deciduous sepals. *Pet. small, tubular,* with a long claw, *2-lipped*; inner lip very short. Follicles many, *stalked.*]

10. HELLEBORUS. Cal. of 5 petal-like persistent sepals. Pet. small, tubular, 2-lipped, clawed. Follicles 3—10, *sessile.*

11. AQUILEGIA. Cal. of 5 petal-like deciduous sepals. *Pet.* 5, *funnel-shaped, with a long horn-like spur.* Follicles 5.

** Flowers irregular.

12. DELPHINIUM. Cal. of 5 petal-like deciduous sepals, *upper sep. with a long spur* at its base. Pet. 4 ; 2 *upper ones with spurs included in the spurred sepal, or all combined into one spurred petal.* Follicles 1, 3, or 5.

13. ACONITUM. Cal. of 5 petal-like sepals, upper sep. helmet-shaped. *Two upper pet. tubular, on long stalks, concealed in the helmet-shaped sepal.* Follicles 3—5.

B. *Anthers introrse.* (*Stam. arising from a glandular disk.*)

Tr. V. *PÆONIEÆ* or spurious Ranunculaceæ.

14. ACTÆA. Cal. of 4 petal-like deciduous sepals. Pet. 4, very small. *Carp.* 1, baccate, *not bursting,* many-seeded.

[15. PÆONIA. Cal. of 5 persistent sepals. Pet. 5 or more. *Follicles* 2—5, many-seeded, *bursting inwards.*]

Tribe I. *Clematideæ.*

1. CLEM'ATIS *Linn.* Traveller's Joy.

1. *C. Vital'ba* (L.) ; st. climbing, l. pinnate, leaflets ovate acuminate entire coarsely serrate or incise-lobate rounded or cordate below, petioles twining, sep. oblong downy on both sides, fr. with long feathery awns.—*E. B.* 612. *R.* iv. 64.—St. woody, angular, branched, very long. Petioles acting as tendrils.—Hedges and thickets on a calcareous soil. Sh. VI. E.

Tribe II. *Anemoneæ.*

2. THALIC'TRUM *Linn.* Meadow-Rue.

1. *T. alpi'num* (L.); st. perfectly simple and nearly leafless, raceme terminal simple, fruitstalks reflexed, carp. shortly stalked

tipped with the hooked style.—*E. B.* 262. *R.* iii. 26.—St. 3—6 in. high, quite smooth. L. mostly radical, upon long stalks, twice ternate.—Higher parts of mountains. P. VI. VII. E. S. I.

2. *T. mínus* (L.) ; st. striate branched *leafless* but sheathed *at the base*, stip. with spreading auricles, l. 2—3-pinnate, lts. ternate 3-cleft glaucous, petioles with angular ascending branches, fl. drooping, anth. apiculate, carp. fusiform 8-ribbed subcompressed ventricose below externally.—St. 1—1½ ft. high, usually solid; sheaths at its base rather lax. Fl. greenish yellow (as are those of Sp. 3 and 4).—*a. T. dunense* (Dum.) ; fr.-branches often horizontal or declining, pan. usually broad and short.—*β. T. montanum* (Wallr.) ; fr.-branches erect-patent, pan. much narrower than in var. *a.*—Sand-hills. *β.* Stony pastures. P. VI. VII. E. S. I.

3. *T. május* (Sm.) ; *st. leafy to the base* branched, l. 2—3-pinnate, lts. 3—5-cleft, *fl. drooping*, anth. apiculate, pan. with patent or reclinate branches.—*a. T. collinum* (Wallr.) ; petioles with divaricate branches, stip. with reflexed auricles, carp. narrowly elliptical. St. often 3—4 feet high, solid, striate. *T. flexuosum* R. not Bernh.—*β. T. Kochii* (Fr.) ; petioles with patent branches, stip. with horizontal auricles, carp. ovoid. *E. B.* 611. St. often 4 ft. high, hollow, striate only below the joinings.—*a.* Damp bushy and stony places. *β.* Lake District. P. VII. VIII. E. S. I.

4. *T. flávum* (L.) ; st. erect furrowed, l. bipinnate, lts. broadly obovate or wedgeshaped trifid, *panicle* compact *corymbose*, *fl. erect*, anth. not apiculate, carp. ovoid.—L. rather paler beneath. Rootstock creeping.—*a. T. sphærocarpum* (Lej.) ; pan. rather close, carp. ovoid. *R.* iii. 4639.—*β. T. riparium* (Jord.); pan. usually lax with ascending branches, carp. ovoid. —*γ. T. nigricans* (Jacq.) ; pan. interrupted with erect-patent branches, carp. elliptic. *R.* iii. 4640. *T. Morisonii* (Gmel.) ed. viii.—In wet fields. P. VII. VIII. *Common Meadow-Rue.* E. S. I.

3. ANEMO'NE *Linn.*

1. *A. Pulsatil'la* (L.) ; fl. solitary erect, involucre sessile in deep linear segments, l. doubly pinnate, leaflets pinnatifid with linear lobes, *carp. with feathery tails.*—*E. B.* 51.—Fl. bell-shaped, violet-purple, externally silky ; stalk 5—8 in. high. Inv. silky, close to the flower, but distant from the fruit.—Open calcareous pastures, rare. P. IV. V. *Pasque flower.* E.

2. *A. nemorósa* (L.) ; fl. solitary, *sep.* 6 oblong spreading, inv. of 3 ternate or quinate stalked leaves with lobed and cut leaflets, l. similar, *carp.* pubescent *keeled* not tailed.—*E. B.* 355. *R.* iv. 47.—Fl. white or purplish ; stalk 4—8 in high. Beak

about as long as the carp. Rootstock horizontal. Sep. glabrous on both sides.—Groves and thickets, common. P. III. —V. *Wind-flower.* E. S. I.

[*A. apennína* (L.); fl. solitary, *sep. many lanceolate*, involucre of 3 ternate stalked deeply cut leaves, l. similar, *carp. pointed* not tailed.—*E. B.* 1062. *R.* iv. 47.—Fl. bright blue. Rootstock tuberous and roundish.—Scarcely naturalized. P. IV.] E.

[*A. ranunculoïdes* (L.); fl. solitary or in pairs, *sep.* 5 *elliptic*, involucre of 3 nearly sessile ternate deeply-cut leaves, l. similar often quinate, *carp. pointed* downy not tailed.—*E. B.* 1484. *R.* iv. 47.—*Fl. bright yellow.* Sep. externally pubescent. Rootstock horizontal.—Not native. P. IV.] E.

4. Ado'nis *Linn.* Pheasant's Eye.

‡1. *A. autumnális* (L.); cal. glabrous patent, pet. connivent, carp. without teeth collected into an ovate head and tipped with a straight beak.—*E. B.* 308. *R.* iii. 24.—Pet. scarlet, black at the base, scarcely exceeding the sepals. L. triply and copiously pinnatifid, segments linear. St. about 8 in. high.—Corn-fields, rare. A. VII. E.

Tribe III. *Ranunculeæ.*

5. Myosu'rus *Linn.* Mousetail.

1. *M. min'imus* (L.).—*E. B.* 435. *R.* iii. 1.—Scapes many, single-flowered, 2—5 in. high. L. linear, fleshy. Receptacle becoming very long (1—3 in.) with many oblong carp. Seed attached to the upper part of the carp. and pendulous, the radicle pointing upwards.—In damp places in fields. A. V. VI. E.

6. Ranun'culus *Linn.* Crowfoot.

A. *Fr.-st. arching*; *carp. transversely wrinkled laterally attached, pet. white* (*with a yellow claw in all our plants*), *nectary naked.* Batrachium S. F. Gray. Water Crowfoot.[1]

* Submersed leaves twice or thrice 3-furcate with filiform segments spreading in the segment of a sphere, rarely wanting. Receptacle hispid.—(The submersed leaves become stiff when old. The younger ones should be examined).

---

[1] We have identified four hybrids in this section—*R. Baudotii* × *Drouetii*, *R. Baudotii*×*heterophyllus*, *R. peltatus*×*Lenormandi*, and *R. peltatus*×*trichophyllus*. No doubt further investigations will add many to the list.—H. & J. G.

1. *R. trichophyl'lus* (Chaix); *submersed l.* (blackish green) closely trifurcate *not collapsing* into a pencil (tassel-like) when taken from the water, segments short rigid, ped. not tapering equalling or slightly exceeding the l., fl. small, *pet.* obovate 5—7-*veined* not contiguous *evanescent,* stig. oblong, receptacle globular, *carp.* ½-obovate laterally apiculate *compressed.*—*E. B. S.* 2968. *St.* 67. 11. *R. divaricatus* Schrank.—St. not rising out of the water. Upper l. sessile, all dense. Floating l. very rare (when it is *R. Godronii* Gren. ! and also *R. radians* Rev. !). Stip. ½-adnate, large, round, auricled. Fr.-ped. short, thick, arching, most curved near their base. *Buds globose.* Fl. starlike. Stam. exceeding the pistils. Stig. short but lengthening. Receptacle as thick as peduncle. Carp. a little narrowed at the end.—Ponds and ditches. P. V. VI. *Water Fennel.* E. S. I.

2. *R. Drouet'ii* (F. Schultz); *submersed l.* (light green) rather loosely trifurcate, tassel-like segments flaccid, floating l. (rare) tripartite with subsessile or stalked wedgeshaped bifid segm., ped. not tapering about equalling the l., fl. small, pet. obovate 5—7-veined not contiguous evanescent, stig. oblong, receptacle oblong, *carp.* ½-obovate sublaterally apiculate *inflated at the end.*—*E. B. S.* 2967.—Bright green. St. not rising out of the water. Upper l. nearly or quite sessile. Floating l. rare, evanescent; lateral segm. stalked and in a different plane from the usually sessile deflexed middle segm. Stip. much adnate, large, auricled. Fr.-ped. short, slender, bent at the base, nearly straight above. *Buds oblong.* Fl. starlike. *Stam.* exceeding the pistils, *few.* Fr.-receptacle as thick as peduncle. Carp. with the edge flattish at the end, base of style small often subcentral.—The plant from Rescobie referred to *R. confervoides* appears to be a small wholly submersed form of this species. *J. of B.* xviii. 344.—Ponds and ditches. P. V. VI. E. S. I.

3. *R. heterophyl'lus* (Fries); *submersed l.* loosely trifurcate, segments long tassel-like, floating l. subpeltate tripartite with sessile or stalked wedgeshaped 3—5-lobed segm., ped. not tapering but narrowed scarcely exceeding the l., fl. large, *pet.* broadly cuneate-obovate 7—9-*veined* not contiguous *persistent,* stig. oblong, *receptacle conical,* carp. ½-obovate, laterally pointed.—*R. aquatilis* Sm. *E. B.* 101.—St. not rising out of the water. Fr.-ped. slender, curved near their base, nearly straight above. Floating l. nearly circular. Stip. much adnate. *Buds slightly depressed* and rather pentagonal. Fl. becoming starlike. *Stam. many,* exceeding pistils. Style hooked. Carp. blunt, inner edge straight.—β. *submersus,* floating l. 0, submersed l. more dense and often rather less flaccid, fl. large, stam. many.—Ponds and streams. P. V. VI. E. I.

4. *R. confúsus* (Godr.); *submersed l.* loosely trifurcate not tassel-like, segments long rather rigid, *floating l.* long-stalked subpeltate *subtripartite with sessile obovate* 3—5-lobed *segments,* ped. slender gradually tapering exceeding the l., fl. large, pet. cuneate-obovate 7—9-veined not contiguous persistent, stig. tongueshaped, receptacle ovoid-conic, *carp.* ½*-ovate compressed and narrowed upwards.*—*St.* 82. 2.—Floating l. semicircular, flat, nearly tripartite, outer base of lateral segments rounded. Stip. oblong, much adnate. Buds globular. Fl. starlike. Stam. many, exceeding pistils. Style recurved. Is *R. salsuginosus* (Hiern) when floating l. are wanting.—Ponds and ditches, especially near the sea; often in brackish water. P. VI.—IX.
E.

5. [*R. triphyl'los* (Wallr.); submersed l. loosely trifurcate not tassel-like, segm. long rather rigid, *floating l.* long-stalked *tripartite with* subsessile *wedgeshaped* 3—5-lobed *segments,* ped. slender tapering exceeding the l., fl. moderately large, pet. cuneate-obovate 5—7-veined contiguous persistent, stig. . . . ., receptacle spherical, carp. ½-ovate inflated.—Floating l. nearly or quite tripartite, acute, or blunt, or with deep linear lobes at the end.—Ponds and ditches. Guernsey. P. V.—VII.]

6. *R. Baudótii* (Godr.); submersed l. closely trifurcate, segments rather rigid not tassel-like, floating l. long-stalked tripartite with sessile or stalked *wedgeshaped* 3—4-lobed *segments,* ped. thick narrowed at the top exceeding l., pet. 7-veined not contiguous persistent, *stam. not exceeding pistils,* stigma tongueshaped, receptacle long-conic, *carp.* ½-obovate *inflated at the end.*—*E. B. S.* 2966.—Floating l. nearly or quite tripartite, the base of all the segments wedgeshaped; or often of many linear blunt segments. Stip. much adnate. Buds globular but a little flattened at the top. Fl. starlike. Stam. 15—20. Style straight, beaklike and persistent below, recurved above. Receptacle thicker than ped., very tall. Carp. very many.—*R. marinus* (Fr.) is a form which wants the floating leaves.—Slightly brackish water or near the sea. P. V.—VIII.
E. S. I.

7. *R. floribun'dus* (Bab.); submersed l. closely trifurcate, segments rather rigid divaricate not tassel-like, floating l. long-stalked subpeltate ½-trifid or tripartite with obovate 3—5-lobed segments, *ped. not tapering* scarcely exceeding the l., fl. large, *pet.* obovate-cuneate 9—*many-veined not contiguous* persistent, stam. many exceeding pistils, *stig. tongueshaped,* receptacle spherical, carp. ½-obovate very blunt.—*A. N. H.* ser. 2. xvi. 397. *E. B. S.* 2969.—Floating l. convex, divided more than

halfway down, more than semicircular; outer base of lateral segments much rounded; rarely with stalked segments. Stip. broad, with a free rounded end. Fl. starlike. Stam. 20—30. Style recurved. Receptacle as thick as peduncle. Inner edge of carp. nearly straight.—Ponds. P. V.—IX. E.

8. *R. penicillátus* (Hiern); submersed l. loosely trifurcate tassel-or whip-shaped subsessile, segments very long, floating l. long-stalked subpeltate ½-trifid or tripartite with obovate segments, each having two or three notches, *ped. very long exceeding the leaves,* fl. large, pet. broad becoming obovate-cuneate 9-veined contiguous persistent, stam. many exceeding the pistils, stigma . . . ., receptacle spherical, carp. ½-obovate very blunt.—*B. penicillatum* (Dum.), *R. pseudo-fluitans* (ed. vi.).—St. wholly submersed. Floating l. semicircular or broader than long, outer base of lateral segments much rounded. Often there are no floating leaves, when it seems to be the *B. aquatile, β. rivulare* (Schur), *R. pseudo-fluitans* (Hiern). Submersed l. often 3—4 in. long with the segm. lying close together almost as in *R. fluitans,* flaccid and whip-shaped.—In water, especially streams. P. V.—VIII. E. I.

9. *R. peltátus* (Fries); submersed l. loosely trifurcate, segments rather rigid divaricate not tassel-like, floating l. long-stalked subpeltate nearly half 3—5-fid with obovate segments having 2 or 3 notches, *ped. tapering exceeding the leaves,* fl. large, *pet.* round becoming obovate-cuneate 9-veined *contiguous* persistent, stam. many exceeding pistils, stigma clubshaped, receptacle ovoid, carp. ½-obovate very blunt.—*E. B. S.* 2965. *St.* 67. 7. *B. truncatum* Dum.—Floating l. ⅔-circular, convex, outer base of lateral segments much rounded. Stip. adnate nearly throughout. *Fl. sweet-scented,* very large. Stam. about 30. Style curved. Receptacle small. Inner edge of carp. nearly straight. —*R. elongatus* (Hiern!), *B. elongatum* (Schultz) has very much longer ped. but otherwise does not seem to differ, except that the l.-segm. are rather less rigid.—In water and wet places. P. V.—IX. E. S. I.

10. *R. tripartitus* (DC.); submersed l. loosely trifurcate, segments very slender somewhat collapsing, floating l. small deeply trifid with rounded 2—5-lobed segments, the central usually as long as the lateral, stip. roundish, upper free, ped. slender, about as long as the petioles, ultimately recurved, fl. very small, pet. scarcely exceeding the cal., stam. 5—8, stigma tapering, receptacle small roundish, carp. few obovate inflated with a very small beak glabrous.—*R.* iii. 2.—A small slender plant resembling Sp. 11 but at once distinguished by the well-developed capillary submersed l., the rare submersed l. of *R. intermedius* having the segments flattened.—Helston and Roche, Cornw. Baltimore, Cork, *Mr. R. A. Phillips.* P. IV—VI. E. I.

11. *R. lutárius* (Bouv.); *divided l. rare rather rigid not collapsing*, floating or aerial, floating l. subpeltate deeply trifid with cuneate-obovate 2—4-fid lobes, pet. scarcely exceeding the cal., style subulate terminal with a slender base, carp. unequally obovate much inflated with a nearly terminal point.—*R. tripartitus E. B. S.* 2946 [not DC.]. *R. intermedius*, ed. viii.—St. usually suberect, aerial. L. ¾-circular, lateral lobes with 3, middle with 2—4 crenatures. Upper stip. free. Pet. pinkish. Stam. 5—10. Style deciduous. Inner edge of carpels much rounded. Sometimes the petals are longer and 5-veined.—Damp ground and wet ditches, rare. A. V.—VIII. E.

[*R. ololeúcos* (Lloyd) having larger wholly white fl., a prominent nearly terminal beak to the obovate carp. and rare rigid not collapsing submersed l., should be looked for.]

** Submersed l. not as in Section *.—† Receptacle hispid.

12. *R. circinátus* (Sibth.); l. all submersed and sessile trifurcate with repeatedly and closely forked *rigid segments all placed in one roundish plane* not tassel-like, ped. tapering exceeding l., fl. large, pet. obovate many-veined nearly contiguous persistent, stam. exceeding pistils, stig. cylindrical, receptacle oblong, carp. ½-ovate compressed rather acute.—*E. B. S.* 2869.—St. submersed. L. sheathing, not auricled, forming a flat rigid disk. Buds flattened at the top. Stam. 15—20. Receptacle narrower than ped. Inner edge of carp. nearly straight.—Streams and ponds, but not common. P. VI.—VIII. E. S. I.

†† Receptacle not hispid.

13. *R. flúitans* (Lam.); l. all submersed about twice trifurcate with very *long linear twice or thrice forked nearly parallel segments*, ped. tapering, fl. large, pet. broadly obovate many-veined contiguous persistent, *stam. falling short of pistils*, stig. cylindrical, receptacle conical, carp. obovate inflated much rounded at the end laterally apiculate.—*E. B. S.* 2870.—St. submersed, usually very long. Petioles and stout segments of l. often very long, together a foot in length. Stip. broadly lanceolate. Buds shortly pyramidal, pentagonal. Pet. often more than five. Stam. very short, many. Inner edge of carpels slightly rounded.—*β. R. Bachii* (Wirtg.); slender, l. short almost sessile finely divided, pet. narrowly obovate. [*γ.* ? *cambricus* (*R. aquatilis* var. *cambricus* Ar. Benn.); small and slender, fl. small, l. with few short segm.]—Rivers. *β.* rare. [*γ.* Coron Lake, Anglesey.] P. VI. VII. E. S. I.

*** Usually no divided submersed leaves; receptacle not hispid.

14. *R. Lenorman'di* (F. Schultz); l. all roundish cordate with 3—5 rather *deeply divided lobes which widen from their base*, pet. exceeding cal., *style terminal* upon the ovate-conical ovary, carp. unequally obovate with a terminal point.—*E. B. S.* 2930. *R. cœnósus* Bab. (not Guss.).—St. floating or creeping upon mud. L. not spotted, often opposite before the plant flowers; lobes very broad at the top, mostly with 2—3 notches. Upper stip. very broad, ½-adnate. Pet. narrow, obovate, 5-veined. Stam. 8—10. Style nearly central on the ovary and usually so on the carpel. Inner edge of carp. much rounded towards the top.—Shallow ponds or mud. P. VI.—VIII. E. S. I.

15. *R. hederáceus* (L.); l. roundish cordate with 3—5 shallow rounded *lobes widening to their base*, pet. scarcely exceeding cal., *style prolonging the inner edge of the ovary*, carp. ½-oval or ½-obovate with a lateral point.—*E. B.* 2003.—St. floating or creeping upon mud. L. usually spotted; lobes usually entire or with a central notch, often rather triangular. Stip. narrow, much adnate. Pet. narrow, 3-veined. Stam. 6—8. Style lateral upon both ovary and carpel. Inner edge of carpel nearly straight.—When floating it is *R. omoiophyllus* (Ten.), *R. cœnosus* (Guss.).—Shallow ponds or mud. P. VI.—IX. E. S. I.

B. *Fr.-st. straight; carp. transversely wrinkled on the middle of each side basally attached, nectary naked.*

16. *R. sceleràtus* (L.); root fibrous, lower l. stalked tripartite, segments blunt crenate, upper l. trifid linear entire or incise-dentate, calyx reflexed, *head of fr. oblong*, carp. minute.—*E. B.* 681. *R.* iii. 11.—Fl. very small, pale yellow. Nect. round, open, bordered all round, rarely not bordered above. Lower l. broad, glabrous, shining. Stem 1—2 feet high, thick.—By and in ditches and ponds. A. VI.—IX. *Celery-leaved Crowfoot.* E. S. I.

C. *Fr.-st. straight; carp. not transversely wrinkled, basally attached.*

[*R. alpestris* and *R. gramineus* were probably recorded through mistakes.]

† L. undivided, fl. yellow, nectary nearly or quite naked bordered.—L. nearly parallel-veined.

17. *R. scot'icus* (Marshall); *st. solitary nearly erect* not rooting at the nodes, early root-l. numerous small deciduous *without laminæ*, later root- and lower stem-l. larger with or without *obtuse* oblong-lanceolate laminæ, upper stem-l. lanceolate rather obtuse sessile or subsessile, pet. obovate-truncate, carp. obovate pitted.—*R petiolaris* Marsh. (not H. B. K.), J. of B. 1892, p. 289, t. 328.—Glabrous, l. entire. Nearly related to Sp. 18, of which it is perhaps a variety.—Margins of Highland lakes. P. VI.—VIII. S. I.

18. *R. Flam'mula* (L.); *l.* ovate- or linear-lanceolate nearly entire *stalked, stem reclining* at the base *and rooting*, ped. furrowed, *carp.* obovate minutely *pitted pointed.*—*E. B.* 387. *R.* iii. 10.—Stem 6—18 in. high; sometimes procumbent and rooting. L. sometimes serrate, hairy or glabrous.—Small forms (var. *radicans* Nolte) are often mistaken for *R. reptans.* A large floating form from Cornwall may be var. *natans* Pers.—In wet places. P. VI.—VIII. *Lesser Spearwort.* E. S. I.

19. *R. rep'tans* (L.); *l. linear* entire, st. procumbent filiform creeping, *carp. ovate obtusely beaked.*—*Sy. E. B.* 30. *St.* 82. 14.—St. rooting at every joining. Fl. very small. A doubtful species.—Wet places. Sandy shore of Ulleswater, Westm. West end of Loch Leven, Kinross-shire. P. VI.—VIII. E. S.

20. *R. ophioglossifólius* (Vill.); *lower l. cordate-ovate stalked,* upper l. oblong sessile amplexicaul, *stem erect* hollow, *carp.* obliquely ovate margined *tubercled* with a short terminal point. —*E. B. S.* 2833. *R.* iii. 21.—Glabrous. Fl. small. St. about 1 foot high, branched, many-flowered, tapering below.—Hythe, Southampton, *Mr. Groves.* E. Gloster. St. Peter's Marsh, Jersey (said to be extinct). A. VI. E.

21. *R. Lin'gua* (L.); *l.* long-lanceolate acute somewhat serrate *sessile* amplexicaul, *stem erect,* ped. not furrowed, *carp.* margined minutely *pitted, with a* broad *swordshaped* beak.—*E. B.* 100. *R.* iii. 10.—Fl. large. St. 2—3 feet high. St. and l. glabrous or with adpressed hairs. Early submersed l. 8—9 in. long 3 in. broad ovate-oblong blunt cordate at base.—Marshes and ditches, rather rare. P. VI. VII. *Great Spearwort.* E. S. I.

†† L. undivided, fl. yellow, nectary with a scale.

22. *R. Ficária* (L.); *root* with *fasciculate* knobs, *l. cordate stalked* angular or crenate, st. with 1—3 l., single-flowered, sep. usually 3, *carp. smooth* blunt.—*E. B.* 584. *R.* iii. 1.—Fl. about 1 in. across; pet. usually 8; rarely apetalous. St. 3—8 in. long, weak, often producing axillary bulbs.—*a. divergens*; lobes of lowest l. separate at base, lowest sheaths narrow —*β. incumbens*;

lobes of lowest l. overlapping at the base, lowest sheaths broad clasping.—Damp rather shady places, common. P. IV. V. *Pilewort.* E. S. I.

††† L. divided, fl. yellow.

‡ Nectary without a scale, carp. smooth.

23. *R. auricomus* (L.); *root fibrous, radical l. reniform* 3—7-partite with crenate or cut lobes stalked, stem-l. sessile digitate, with linear or lanceolate more or less toothed segments, peduncles round, calyx pubescent, *carp. downy* ventricose, beak slender hooked.—*E. B.* 624. *R.* iii. 12.—Pet. often wanting. Sep. yellow. Receptacle covered with cylindrical tubercles upon which the carp. are seated. St. about 1 foot high. Radical l. sometimes very deeply divided (var. *incisifolius* R.).—Woods and thickets, common. P. IV. V. *Goldilocks.* E. S. I.

‡‡ Nectary with a scale, carp. smooth.

24. *R. ácris* (L.)[1]; *st. not bulblike,* root fibrous, radical l. palmately tripartite, segments trifid and deeply cut, uppermost stem-l. tripartite with linear segments, *ped.* terete, *calyx* pubescent *erect-patent, carp.* oval *glabrous* margined, beak short marginal recurved, *receptacle glabrous.*—*E. B.* 652. *R.* iii. 17.—Hairy. No stoles. St. 2—3 feet high. Beak about ¼ the length of the carpel.—Sometimes dwarf and 1-flowered on mountains. —The variations *R. Boræanus* (Jord.) having the base of st. glabrous, l.-segm. very narrow, *R. vulgatus* (Jord.) having usually an oblique or horizontal rhizome, and *R. tomophyllus* (Jord.) having a præmorse rootstock, have been found. *J. of B.* viii. 257, x. 238. A var. with glabrous shining l. much more bluntly cut than usual [var. *pumilus* Wahl.] grows on Cairngorm.—Meadows and pastures, common; mountains. P. VI. VII. *Upright Crowfoot.* E. S. I.

25. *R. répens* (L.); l. with three-lobed segments, lobes 3-fid and cut, *ped. furrowed, calyx* pubescent *erect-patent,* carp. oval glabrous margined minutely pitted, beak longish slightly curved, *receptacle hairy.*—*E. B.* 516.—Root fibrous. Stoles strong, leafy. Primary stem usually erect, 10—12 in. high.—Meadows and pastures, common. P. V.—VIII. *Creeping Crowfoot.* E. S. I.

26. *R. bulbósus* (L.); *stem bulblike* at the base, radical l. with 3 segments each tripartite trifid and cut, *ped. glabrous*

---

[1] For an account of the vars. and forms, see Townsend, *J. of B.* xxxviii. (1900) p. 379.—H. & J. G.

furrowed, calyx hairy *reflexed*, carp. round margined smooth, beak short, receptacle hairy.—*E. B.* 515. *R.* iii. 20.—St. about 1 foot high. Upper l. cut into narrow segments.—Meadows and pastures. P. V. *Bulbous Crowfoot.* E. S. I.

‡‡‡ Nectary with a scale, carp. rugose or tubercular.

[*R. flabellátus* (Desf.); *root of short ovoid knobs* and fibres, *stoles very slender* with minute scales and ending in a young plant, rt.-l. with 3 segments each 3—7-parted or 3-fid, st.-l. 1—2 with linear segments, ped. terete, cal. hairy spreading, carp. many roundish, beak acute, receptacle oblong glabrous.—*R. chærophyllus* (auct.) ed. viii. *J. of B.* x. t. 125.—St. 6—12 in. high, and, as well as the leaves, hairy.—Dry places, St. Aubin's, Jersey. P. V.]

27. *R. sardóus* (Crantz); root fibrous, radical l. with 3 stalked trifid and cut leaflets, peduncles furrowed, calyx reflexed, carp. round margined *with a series of tubercles near the margin*, beak short curved, receptacle hairy.—*R. hirsutus* (Curt.) ed. viii. *E. B.* 1504. *R.* iii. 23. *R. Philonotis* (Ehrh.) Koch.—St 4—18 in. high; the smaller specimens are *R. parvulus* L. Upper l. in narrow acute segments. Fl. pale yellow.—Waste land and corn-fields, rare. A. VI.—X. E. S.

28. *R. parviflórus* (L.); root fibrous, *stems spreading*, l. roundish-cordate 3—5-lobed cut, upper l. oblong undivided or 3-lobed, calyx at first erect afterwards reflexed, carp. orbicular muricate.—*E. B.* 120. *R.* iii. 22.—Peduncles opposite the leaves. Pet. narrow.—Corn-fields and dry banks, rare. A. V. VI. E. I.

†29. *R. arven'sis* (L.); root fibrous, radical l. 3-cleft dentate, stem-l. once or twice ternate with linear-lanceolate segments, calyx erect-patent, *carp.* margined *beaked and spinous.*—*E. B.* 135. *R.* iii. 21.—St. 6—18 in. high. Fl. pale yellow.—Corn-fields. A. VI. *Corn Crowfoot.* E. S. I.

Tribe IV. *Helleboreæ.*

7. Cal'tha *Linn.* Marsh Marigold.

1. *C. palus'tris* (L.); *st. ascending, l. cordate rounded* crenate. —*E. B.* 506. *R.* iv. 101.—About a foot high. Fl. large. Sep. bright yellow.—*a. vulgaris*; sep. roundish-ovate contiguous, carp. spreading their *beak very short.*—*β. C. cornúta* (Schott);

sep. oblong-ovate not contiguous, carp. spreading their *beak long*.—γ. *C. latifólia* (Schott); l. coarsely dentate throughout, st.-l. broadly reniform, beak short.—δ. *minor*; st. decumbent mostly 1-flowered, fl. small, sep. not contiguous, carp. erect their *beak very short*.—ε. *zetlandica* (Beeby)[1]; st. rooting, 5-8 in. long, fl. small.—Marshy places, common. δ. Mountains. ε. Shetland. P. III.—V. E. S. I.

2. ? *C. radicans* (Forst.); *st. creeping*, l. triangular, serrate-crenate small, sep. not contiguous narrow.—*E. B.* 2175.—Base of the l. almost at right angles with petiole.—Rescobie, Forfarsh. P. V. VI. S.

8. Trol lius *Linn.* Globe Flower.

1. *T. europæ'us* (L.); sep. 10-15 concave converging into a globe, pet. 10 about equalling the stam., l. palmately 5-parted, segments rhomboidal 3-partite incise-serrate.—*E. B.* 28. R. iv. 101.—Fl. bright yellow. Pet. ligulate. St. 1—2½ feet high.—Damp mountain pastures. P. VI. VII. E. S. I.

9. Cam marum *Hill.* (*Eranthis* Salisb. ed. viii.) Winter Aconite.

[* *C. hyemále* (Greene); sep. 6—8 oblong.—*Sy. E. B.* 43. *R.* iv. 101.—Radical l. on long stalks, 5—7-parted, deeply cut into linear-oblong segments. St. 4—6 in. high; invol. of 3 sessile leaflets just below the solitary yellow fl. Rhizome tuberous.—Naturalized in thickets. P. II. III.] E. S. ?

10. Helleb orus *Linn.* Hellebore.

1. *H. vir'idis* (L.); radical *l. digitate* stalked, stem-l. sessile at the ramifications, st. few-flowered, *calyx spreading*.—*E. B.* 200.—Veins of the l. prominent beneath. Stigma erect. St. 1 foot high, annual. Fl. greenish-yellow.—Thickets on a calcareous soil. P. III. IV. *Green Hellebore. Bear'sfoot.* E.

2. *H. fœt'idus* (L.); *l. pedate* stalked, st. leafy many-flowered, *calyx converging*.—*E. B.* 613.—L. successively contracting upwards into bracts. St. 2 feet high, perennial. Fl. globose, drooping, greenish tipped with purple.—Thickets in chalky districts. P. III. IV. *Stinking Hellebore. Setterwort.* E.

---

[1] Mr. Beeby now places this under *C. radicans* which he regards as a subsp. of *C. palustris*, distinguished from *palustris* proper by its rooting at the nodes.—H. & J. G.

11. Aquile'gia *Linn.* Columbine.

1. *A. vulgáris* (L.); spur of the pet. incurved, limb blunt falling short of the stamens, l. biternate, leaflets 3-lobed crenate. —*E. B.* 297. *R.* iv. 114.—St. 2—3 feet high, slightly leafy. Caps. cylindrical hairy. Inner stam. frequently imperfect.—Woods and thickets and heaths, not common. P. V. VI. E. I.

12. Delphin'ium *Linn.* Larkspur.

*1. *D. Ajácis* (Gay not R.); st. erect with spreading branches, racemes 4—16-fl., pet. combined, ovary abruptly narrowed into a subulate style, follicle downy obliquely acuminate, seeds with transverse contiguous wavy ridges.—*D. Consolida* Sm. *Sy. E. B.* 47.—St. loosely and sparingly branched, about a foot high. Fl. of a vivid and permanent blue, rarely red pink or white. L. deeply multifid. Style equal to about $\frac{1}{6}$ of carpel.—Sandy or chalky corn-fields. Nearly extirpated. A. VI. VII. E.

[*D. Consol'ida* (L.); st. erect with patent branches, racemes few-flowered, pet. combined, ovary narrowed into style, *follicle glabrous* truncate short, seeds with transverse interrupted ridges. —*R.* iv. 116.—Much like *D. Ajacis.* Style lateral.—Found once in Jersey and in Cornwall. A. VI. VII.]

13. Aconi'tum *Linn.* Monk's-hood.

†1. *A. Napel'lus* (L.); pet. horizontal upon curved stalks, spurs bent down, fl. racemose, young carpels diverging.—*E. B. S.* 2730. *R.* iv. 92.—Rootstock black, of 2 oblong knobs, very poisonous. St. 1—2 ft. high. Fl. purple. Filaments slightly hairy, with cuspidate wings. Pet. inflated above; lip broad. Helmet open, hemispherical. Pedicels erect, downy.—Banks of rivers and brooks, rare. P. VI. VII. E.

Tribe V. *Pæonieæ.*

14. Actæ'a *Linn.* Bane-berry.

1. *A. spicáta* (L.); raceme simple elongate, pet. as long as the stamens, berries oval.—*E. B.* 918. *R.* iv. 121.—L. stalked, biternate; lts. ovate, trifid, deeply cut. St. 1—2 feet high. Fl. white.—Mountainous limestone tracts in the North. P. V. E. S.

15. Pæo'nia *Linn.* Pæony.

[* *P. coral'lina* (Retz.); l. biternate, leaflets ovate entire glaucous beneath, caps. downy recurved from the base.—*E. B.*

1513. *R.* iv. 128.—Roots fleshy, knobbed. Herb 2 feet high. Fl. large, crimson with yellow anthers.—On the Steep Holmes Island in the Severn. P. V. VI.] E.

Order II. BERBERIDACEÆ.

Sep. 3 or 4 or 6 in a double row. Pet. the same or double that number. Stam. opposite to the petals. Anth. opening by valves attached at the top and turning upwards. Carpel 1, 1-celled; seeds attached to the bottom or on a lateral placenta, albuminous.—Stipules usually wanting.

1. Berberis. Sep. 6, deciduous. Pet. 6, each with 2 glands at the base within. Stam. 6. Berry 2—3-seeded.

[2. Epimedium. Sep. 4, deciduous. Pet. 4. Nectaries 4, cupshaped. Stam. 4. Caps. podlike, many-seeded.]

1. Berberis *Linn.* Barberry.

1. *B. vulgáris* (L.); spines 3-parted, l. obovate ciliate-serrate, racemes pendulous many-flowered, petals entire.—*E. B.* 49. *R.* iii. f. 4486.—Height 6—8 feet. Fl. yellow. Berries red, oblong, slightly curved. Filaments curiously elastic.—Hedges and thickets. Sh. V. VI. E. S. ? I.

2. Epimedium *Linn.* Barrenwort.

[*E. alpínum* (L.); rhizome producing leaves and stems, stem-l. twice ternate.—*E. B.* 438. *R.* iii. f. 4485.—Lts. ovate-heart-shaped, serrate. Pan. shorter than the l. and appearing to grow from the petiole. Fl. red with yellow nectaries.—Subalpine woods. Not a native. P. V.] E. S.

Order III. NYMPHÆACEÆ.

Sep. 4—6. Petals many, seated with the many stamens upon a fleshy disk more or less completely surrounding the ovary. Stigma peltate. Fruit many-celled; seeds many, in a gelatinous aril. Embryo in a bag on the outside of the base of the albumen.—St. prostrate, submersed. L. floating, falsely peltate.

1. Castalia. Cal. of 4 sepals. Pet. many, seated together with the stamens upon a fleshy disk enveloping the germen and passing gradually into them. Berry many-celled, many-seeded. Stigma sessile, of many rays.

2. NYMPHÆA. Cal. of 5 sepals. Pet. many, seated together with the stamens upon the receptacle. Berry many-celled, many-seeded. Stigma sessile, of many rays.

1. CASTALIA *Salisb.* White Water-Lily.

1. *C. speciósa* (Salisb.); l. roundish deeply cordate entire with approximate lobes, stigma of 12—20 rays.—*Nymphæa alba* (L.) ed. viii. *E. B.* 160. *R.* vii. 67.—Fl. large, white, floating. Stig. yellow. Ovary covered with the stam. almost to its top. Caps. dissolving into pulp. Notch in the leaves with nearly parallel sides.—There is a small-flowered form (*N. alba β. minor* DC.).—Slow rivers, lakes and clear ditches. P. VII. E. S. I.

2. NYMPHÆ'A *Linn.* (*Nuphar* Sm. ed. viii.) Yellow Water-Lily.

1. *N. lútea* (L.); l. oblong-cordate, sep. 5, *stigma with* 9—20 *rays not extending to the entire margin, anthers linear.*—*E. B.* 159. *R.* vii. 63.—Fl. 2½ in. across, yellow, smelling like brandy. Caps. bursting irregularly. Submersed l. thin, transparent, wavy, oblong-cordate.—*β. intermedia*; fl. 1½ in. across, margin of stigma wavy. *Sy. E. B.* 55. *Nuphar intermedium* Ledeb.—Lakes and ditches. P. VII. *Brandy-bottle.* E. S. I.

2. *N. púmila* (Hoffm.); l. oblong deeply cordate with distant lobes, sep. 5, *stigma with* 8—10 *rays* extending to the margin and *forming acute teeth, anth. subquadrate.*—*E. B.* 2292. *R.* vii. 65.—Fl. yellow, small. Caps. furrowed upwards. Anth. not twice as long as broad. Much smaller than *N. lutea.* Submersed l. thin, transparent, wavy, reniform.—In small Highland lakes, and at Ellesmere, Salop. P. VII. VIII. E. S.

Order IV. PAPAVERACEÆ.

Sep. 2, deciduous. Cor. regular, of 4 petals. Stam. generally many, free. Ovary free; placentas parietal, usually projecting; seeds many. Stigmas as many as the placentas, simple or lobed; the lobes of adjoining stigmas combining, thus appearing (falsely) to be opposite to the placentas. Seeds albuminous. —Stip. 0.

* *Fruit globular, oblong or clavate, opening by pores; stigmas radiant.*

1. PAPAVER. Pet. 4. Stam. many. *Style* 0. *Stigmas* 4—20, *radiant on a flattish disk.* Caps. opening by pores beneath the stigma. Placentas like dissepiments.

2. MECONOPSIS. Pet. 4. Stam. many. *Style short. Stigmas* 4—6, radiant, *free.* Caps. obovate, opening by pores beneath the style. Placentas filiform.

** *Fruit linear, opening by valves ; stigmas* 2—4.

3. ROEMERIA. Pet. 3. Stam. many. *Stigmas* 2—4, *sessile,* radiant. *Caps.* 2—4-*valved,* 1-celled ; *placentas distinct.* Seeds not crested.

4. GLAUCIUM. Pet. 4. Stam. many. Stigmas 2, sessile. *Caps.* 2-*valved* ; *placentas connected* by a spongy dis epiment. Seeds not crested.—A maritime plant with yellow flowers.

5. CHELIDONIUM. Pet. 4. Stam. many. Stigmas 2. Caps. 2-valved, 1-celled ; placentas distinct. Seeds crested.

1. PAPAVER *Linn.* Poppy.

†1. *P. Argemóne* (L.) ; filaments dilated upwards, *caps. clavate* hispid with erect bristles, stem leafy many-flowered, l. bipinnatifid.—*E. B.* 643. *R.* iii. f. 4475.—St. usually about 1 ft. high. Fl. small. Pet. distinct, fugacious, pale red, black at the base. Bristles sometimes spreading.—In corn-fields. A. VI. VII. E. S. I.

†2. *P. hyb'ridum* (L.); filaments dilated upwards, *caps. roundly ovoid hispid* with spreading bristles, stem leafy many-flowered, l. bipinnatifid.—*E. B.* 43. *R.* iii. f. 4476.—Sap milky. Fl. small. Pet. purplish, often with a dark spot at the base.—Sandy fields, rare. A. VI. VII. E. I.

†3. *P. Rhœ'as* (L.) : *filaments subulate, caps. roundly obovoid without bristles,* stem bristly many-flowered, l. pinnatifid cut.—*E. B.* 645. *R.* iii. f. 4479.—Fl. large. Pet. deep scarlet, often nearly black at the base. Peduncles with spreading hairs, or (β. *strigosum* Boenn.) with adpressed hairs.—In arable fields, common. A. VI. VII. *Corn Rose.* E. S. I.

†4. *P. dúbium* (L.) : filaments subulate, *caps. clavate narrowing gradually* from the top throughout without bristles, *stig-disk with patent edge,* l. pinnatifid with distant broad entire bluntish lobes, sap milky.—*Curt. Lond.* ii. 104. *P. Lamottei* Bor.—St. 1—2 ft. high. Sap not turning yellow. Fl. large. Pet. transversely oval, pale red. Ped. with adpressed hairs. *Stig.-disk of nearly ripe caps. like eaves,* obscurely lobed.—Sides of fields. A. VI. VII. E. S. I.

†5. *P. Lecoq'ii* (Lamot.); filaments subulate, *caps. clavate-oblong* suddenly narrowed near the base without bristles, *stig.-disk folded over the edge of the caps.*, l. bipinnatifid with distant narrow entire acute lobes. *Sap becoming dark yellow* in the air.—*Sy. E. B.* 60.—St. 1—2 ft. high. Fl. large. Pet. usually distinct, obovate-wedgeshaped, red. Ped. with adpressed hairs. Stig.-disk with bluntly triangular lobes.—Sides of fields chiefly on a calcareous soil. A. VI. VII. E. S. I.

‡6. *P. somnif'erum* (L.); *filaments dilated* upwards, caps. truly globular *without bristles*, l. oblong unequally toothed amplexicaul.—*E. B.* 2145. *R.* iii. f. 4481.—Fl. large. Pet. bluish white with a violet spot at the base. Whole herb smooth [var. *glabrum* Wats.] or sometimes a few rigid spreading bristles on the ped. and one tipping many of the teeth of the leaves [var. *hispidum* Wats.].—On sandy ground near the sea, and in the Fens, rare. A. VII. E. I.

2. Meconop'sis *Vig.* Welsh Poppy.

1. *M. cam'brica* (Vig.); caps. smooth, l. stalked.—*E. B.* 66.—Caps. elliptic-oblong, beaked. St. many-flowered, about 1 foot high. L. pinnate, cut, glaucous beneath. Fl. large, yellow, on long stalks.—Damp rocky and shady places. P. VI.—VIII. E. I.

3. Roeme'ria *Medic.*

‡1. *R. violácea* (Medic); pod 3-valved erect with a few rigid hairs at its top.—*E. B.* 201. *R. hybrida* (DC.) ed. viii.—Caps. linear, 2—3 in. long. L. 2—3-pinnatifid, with linear nearly smooth bristle-pointed lobes. St. about 1 foot high, usually slightly hairy. Fl. violet-blue. Pet. falling before noon.—Chalky corn-fields in Cambridgeshire and Norfolk, very rare. A. V. VI. E.

4. Glau'cium *Mill.* Horned Poppy.

1. *G. flávum* (Crantz); st. smooth, stem-l. ½-clasping sinuate, caps. minutely tubercular-asperous.—*G. luteum* (Scop.) ed. viii. *E. B.* 8.—Glaucous. St. 1—3 feet high, stout, much branched, glabrous or slightly hairy. Root-l. stalked, lyrate, lobed and cut, hairy. Pet. large, golden yellow. Caps. 6—12 in. long, curved.—Sandy sea-shores. B. VI.—VIII. E. S. I.

[*G. phœníceum* (Crantz); st. pilose, stem-l. pinnatifid cut, *caps. hispid.*—*E. B.* 1433—Pet. scarlet with a black spot at the base.—Rarely found, on cultivated ground. A. VI. VII.] E.

5. Chelido'nium *Linn.* Celandine.

†1. *C. május* (L.); ped. umbellate, l. deeply pinnatifid, segments rounded and bluntly lobed.—*E. B.* 1581.—Fl. yellow, small. Caps. linear. St. 1—2 feet high. All parts full of orange juice.—[β. *C. laciniatum* (DC.); segments of the leaves deeply pinnatifid, lobes incise-serrate.]—Waste places and old walls. β a very doubtful native. P. V.—VIII. E. S. I.

Order V. FUMARIACEÆ.

Sep. 2 or 0. Cor. irregular, of 4 parallel petals, one or both of the two outer pet. gibbous or spurred at the base. Stam. 6, in 2 bundles opposite to the outer petals; lateral stam. in each bundle 1-celled. Ovary free, 1-celled. Style filiform. Stigma with 2 or more points. Seeds albuminous.—Stipules 0.

1. Capnoides. Cal. of 2 sepals or wanting. Pet. 4, the upper one spurred at the base. Stam. diadelphous. *Pod 2-valved many-seeded*, compressed.
2. Fumaria. Cal. of 2 sepals. Pet. 4, the upper one spurred at the base. Stam. diadelphous. *Fruit a nut*, indehiscent, 1-seeded.

1. Capnoi'des *Mill.* (*Corydalis* DC. ed. viii. *Neckeria* Scop.)

[*C. sol'ida* (Moench); root tuberous solid, l. biternate cut, lowest petiole a leafless scale, bracts palmate.—*E. B.* 1471.—Lobes of the l. blunt. Fl. purplish. St. a span high.—Not a native, scarcely even naturalized. P. IV. V.] E.

*1. *C. lútea* (Gaert.); *root fibrous*, l. triternate, *bracts minute oblong* cuspidate, seeds shining granular-rugose with a patent denticulate crest.—*E. B.* 588.—Lts. obovate, trifid. Bracts shorter than pedicels. Fl. yellow. St. 1 ft. high, brittle, spreading.—Naturalized on old walls. P. V.—VIII. *Yellow Fumitory.* E. S.

2. *C. claviculáta* (Druce); root fibrous, *leaves pinnate*, pinnæ ternate, *footstalks ending in tendrils*, bracts oblong acuminate.—*E. B.* 103.—Leaflets entire, elliptic. Bracts rather longer than the pedicels. Fl. small, yellow or nearly white. St. slender, climbing, 1—4 feet long.—Bushy places in hilly districts. P. VI. VII. *White Climbing Fumitory.* E. S. I.

2. FUMA'RIA *Linn.* Fumitory.

* *Capreolatæ.* Lower pet. narrowing gradually and slightly from its middle upwards. Fr. not retuse.—St. erect, climbing or diffuse. Fr. subcompressed, smooth. Raceme lax, short, few-flowered. Sep. as broad as cor.-tube. L. flat.

1. *F. capreoláta* (L.); sep. ovate toothed [peltate] at least ½ as long as the cor., fr. blunt its *neck short and narrow*, fr.-st. patent or reflexed.—Sep. soon falling. Cor. large. Fr. not regularly rounded vertically but squarish; apical pits small and deep.—*a. speciosa* (Jord.); fr. longer than broad, its neck narrower than the tip of ped., fr.-st. recurved. Cor. cream-coloured, tipped with red or pink. *Sy. E. B.* 71.—*β. F. Boræi* (Jord.)[1]; fr. broader than long, its neck very narrow, fr.-st. patent, sep. large oval. Cor. purplish, tipped with dark purple. *Sy. E. B.* 72. *Curt.* ii. 145.—Borders of fields. A. VI.—IX. E. I.

2. *F. confúsa* (Jord.); sep. ovate toothed [attached at or very near the base] not ½ as long as cor., fr. rounded at the top its *neck very broad*, fr.-st. patent.—*E. B. S.* 2976. *Sy. E. B.* 73.—Sep. often persistent. Cor. large, dull white or pinkish, tipped with dark purple. Fr. regularly rounded vertically; apical pits broad and shallow; neck nearly as broad as the fruit.—Borders of fields, chiefly in the West. A. VI.—IX. E. S. I.

3. *F. murális* (Sond.); sep. ovate toothed [peltate] not ½ as long as the cor., *fr. obovate* rounded at the top its *neck obconic narrow*, fr.-st. erect-patent.—*Sy. E. B.* 74. *Fl. Dan.* 2473.—More lax than Sp. 1 and 2, cor. smaller and nearly black at the tip. Fr. uniformly rounded at the sides and top, nearly pyriform if taken with its neck; apical pits very faint.—Borders of fields. A. VI.—IX. E. S. I.

** *Officinales.* Lower petal spathulate.

4. *F. officinális* (L.); *sep.* ovate-lanceolate acute toothed *narrower than the cor.-tube, fr. obovate retuse*, bracts much shorter than the fruitstalks, lts. flat.—*Curt.* i. 14.—St. erect or diffuse. Raceme long, many-flowered. Sep. broader than the pedicel, about ⅓ the length of the corolla. Pet. rose-coloured. Fr. rugose, broader than long.—Common. A. V.—IX. E. S. I.

---

[1] Most authors agree in separating *F. Boræi* as a species. It seems to us more nearly allied to *F. muralis* and *F. confusa* than to *F. capreolata*.—H. & J. G.

†5. *F. densiflóra* (DC.); *sep.* large *roundish* dentate broader than the cor.-tube, fr. subglobose subapiculate, bracts longer than the fr.-stalks, lts. linear channelled.—*F. micrantha* Lag. *E. B. S.* 2876.—St. branched, diffuse. Pet. pale purple. Fr. rather longer than broad.—Fields. A. VI.—IX. E. S. I.

6. *F. parviflóra* (Lam.); sep. minute ovate acute cut, *fruit-stalk shorter than the* obovate pointed *fr. and equalling the bract*, lts. linear channelled.—*E. B.* 590. *R.* iii. 1.—Foliage yellowish green glaucous. Lts. ascending. Fl. whitish, afterwards pale purple.—Fields in Kent, Essex, Camb. [&c.]. A. VI.—IX. E.

7. *F. Vaillan'tii* (Lois.); sep. very minute triangular, *fr.-st. longer than* the obovate pointed *fr. and twice as long as the bract*, lts. narrow flat.—*E. B. S.* 2877. *R.* iii. 1.—Foliage greyish green glaucous. Lts. spreading. Fl. purplish, afterwards whitish.—Fields in the South and East, rare. A. VI.—IX. E.

Order VI. CRUCIFERÆ.

Sep. 4. Cor. cruciform, of 4 petals. Stam. 6, tetradynamous; 2 shorter opposite the lateral sepals. Ovary free, with marginal placentas connected by a false septum. Stigmas 2, opposite to the placentas (or, rather, alternate with them, lobed and combining). Fruit a 2-celled and 2-valved caps. (or pod), the valves opposite the shorter stamens, deciduous from the placentas.—Stipules 0.

Suborder I. SILIQUOSÆ.

Pod (silique) long, not dividing transversely, linear or linear-lanceolate, opening by 2 valves throughout; dissepiment narrow, but in the broadest diameter of the pod.

Tribe I. *ARABIDEÆ.* Cotyledons accumbent (radicle lateral), seed compressed.

* *Stigma 2-lobed.*

1. Matthiola. Pod round or compressed. *Lobes of stig. erect gibbous or horned at the back.*

2. Cheiranthus. *Pod* compressed or 2-edged, *with an elevated longitudinal rib upon each valve. Lobes of stig. patent.* Seeds in a single row in each cell.

** *Stigma a disk or head.*

3. RADICULA. Pod terete; *valves convex, veinless.* Seeds irregularly in 2 rows.

4. BARBAREA. Pod terete; *valves* convex, *with a prominent* longitudinal *rib.* Seeds in a single row.

5. ARABIS. Pod compressed; valves nearly flat, with a prominent longitudinal rib or with numerous longitudinal veins. *Seeds in a single or double row.*

6. CARDAMINE. Pod compressed; *valves flat, veinless.* Seeds in 1 row; seed-stalk simple, filiform.

7. DENTARIA. Pod lanceolate, compressed; valves flat, veinless. Seeds in a single row; seed-stalk *dilated, winged.*

Tr. II. *SISYMBRIEÆ.* Cotyledons incumbent (radicle dorsal), seed compressed. Seeds in one row.

* *Stigma of two closely converging erect ovate lobes.*

8. HESPERIS. Pod quadrangular or subcompressed; valves keeled, somewhat 3-veined. Seeds in a single row.

** *Stigma a disk with a thickened edge.*

9. SISYMBRIUM. Pod terete or rarely 4-edged; valves convex, with 3 longitudinal veins. Seeds in a single row, smooth; seed-stalk filiform.—*S. Thalianum* has a tetragonous pod with 1 strong conspicuous rib on the valves.

10. ALLIARIA. Pod terete; valves convex, with 3 longitudinal veins, the middle one prominent and strong, the 2 lateral slender and branching. *Seeds* in a single row, *striate*; *seed-stalk flattened, winged.*

11. ERYSIMUM. Pod 4-edged; *valves* prominently keeled, *with* 1 *longitudinal vein.* Seeds in a single row; seed-stalk filiform.

Tr. III. *BRASSICEÆ.* Cotyledons conduplicate (longitudinally folded in the middle); radicle dorsal, within the fold.

12. BRASSICA. Cal. erect. Pod terete or angular. Seeds globose, in a single row.

13. SINAPIS. Cal. spreading (in the flower). Pod terete or angular. Seeds in a single row.

14. DIPLOTAXIS. Cal. patent. Pod compressed. *Seeds* oval or oblong, *in 2 rows.*

Suborder II. LATISEPTÆ.

Pouch (silicle) short, not dividing transversely, opening with two valves; dissepiment in its broadest diameter.[1]

Tr. IV. *ALYSSINEÆ.* Cotyledons accumbent.

15. ALYSSUM. Pouch roundish or oval, compressed. Seeds 2—4 in each cell. Filaments simple or toothed or the shorter ones with a gland or subulate process on each side at the base.

16. DRABA. Pouch oval or oblong, slightly convex. *Seeds many* in each cell, in 2 rows, not margined. Filaments simple.

17. COCHLEARIA. Pouch globose; *valves very convex,* dorsal vein prominent. Seeds many. Filaments simple.

18. ARMORACIA. Pouch oblong or globose; *valves* very convex, *no vein.* Seeds many. Filaments simple.

[Tr. V. *CAMELINEÆ.* Cotyledons incumbent.

19. CAMELINA. Pouch subovate; valves ventricose, with a linear prolongation at the end which is confluent with the persistent style.]

Suborder III. ANGUSTISEPTÆ.

Pouch (silicle) short, laterally compressed, opening with 2 boatshaped valves keeled or winged on the back; dissepiment narrow, linear or lanceolate.

Tr. VI. *THLASPIDEÆ.* Cotyledons straight.

* *Cotyledons accumbent or rarely incumbent.*

20. THLASPI. Pouch roundish, notched; valves boatshaped winged at the back. *Seeds more than 2 in each cell.* Pet. equal, filaments simple.

---

[1] In *Cochlearia* the valves are sometimes so convex that the pouch is laterally compressed.

21. HUTCHINSIA. *Pouch* elliptic, *entire*; *valves* boat-shaped, keeled *not winged* at the back. *Seeds 2 in each cell.* Pet. equal. Filaments simple. Cotyledons rarely incumbent.

22. TEESDALIA. Pouch roundish, notched; *valves* boat-shaped, their back *keeled below narrowly winged above.* Seeds 2 in each cell. Pet. equal or two outer ones larger. *Filaments with a little scale at the base* of each within.

23. IBERIS. Pouch ovate or roundish, notched; valves boat-shaped, winged at the back. *Seeds 1 in each cell. Pet. unequal,* 2 outer ones much larger. Filaments simple.

** *Cotyledons usually incumbent.*

24. LEPIDIUM. Pouch roundish or oblong, entire or notched; valves compressed, keeled or winged at the back. *Seeds 1 in each cell.* Pet. equal. Cotyledons rarely accumbent.

25. BURSA. Pouch triangular-obcordate; *valves* compressed, *keeled but not winged.* Seeds many.

Tr. VII. *SUBULARIEÆ.* Pouch with a rather broad dissepiment. Cells many-seeded. Cotyledons incumbent, long, linear, curved back above their base, thus appearing like 4 in transverse section.

26. SUBULARIA. Pouch oval-oblong, laterally subcompressed; valves boatshaped.

Tr. VIII. *SENEBIEREÆ.* Valves not separating. Cells one-seeded. Cotyledons incumbent, long, linear, curved back above their base, as in Tr. VII.

27. SENEBIERA. Pouch broader than long, somewhat kidney-shaped, entire at the end, or notched above and below and almost 2-lobed.

Suborder IV. NUCUMENTACEÆ.

Pouch (silicle) scarcely dehiscent, often 1-celled owing to the absence of the septum.

Tr. IX. *ISATIDEÆ.*

28. ISATIS. Pouch laterally compressed, 1-celled, 1-seeded, valves keeled, eventually separating. Cotyledons incumbent.

Suborder V. LOMENTACEÆ.

Silicle or silique dividing transversely into 1-seeded cells; the true silique often barren, all the seeds being in the beak.

Tr. X. *CAKILINEÆ.* Silicle 2-jointed.

29. CAKILE. Silicle angular, of two 1-seeded indehiscent joints, upper joint deciduous with an erect seed, lower persistent seedless or with a pendent seed.

30. CRAMBE. Silicle 2-jointed, upper joint globose with 1 seed pendent from a long curved seed-stalk springing from the bottom of the cell, lower joint barren stalklike.

Tr. XI. *RAPHANEÆ.* Silique linear or oblong, terete or moniliform.

31. RAPHANUS. Silique linear or oblong, tapering upwards, smooth and indehiscent, or moniliform and dividing transversely into 1-seeded cells, lowermost cell barren imperfectly 2-valved stalklike.

Suborder I. *Siliquosæ.* Tribe I. *Arabideæ.*

1. MATTHI'OLA *R. Br.* Stock.

†1. *M. incána* (R. Br.); st. shrubby upright branched, *l.* lanceolate *entire* hoary, pods "cylindrical without glands."—*Sy. E. B.* 1935. *R.* ii. 45.—Fl. of a full purple.—Cliffs in the Isle of Wight. P. V. VI. *Hoary Stock.* E.

2. *M. sinuáta* (R. Br.); st. herbaceous diffuse, l. oblong downy, *lower l. sinuate, pods* compressed *muricated* with glands. —*E. B.* 462. *R.* ii. 45.—Fl. purple.—South and South-west coasts. B. VI.—VIII. *Sea Stock.* E. I.

2. CHEIRAN'THUS *Linn.* Wallflower.

*1. *C. Cheiri* (L.); st. shrubby, l. lanceolate acute entire with bipartite adpressed hairs, pods tetragonal.—*E. B.* 1934. *R.* ii. 45.—Fl. yellow or tinged with red.—Old walls. P. IV. V. E. S. I.

3. RADIC'ULA *Hill* (*Rorippa* Scop. *Nasturtium* R. Br., ed. viii.). Water-Cress.

1. *R. officinális* (Groves); l. pinnate, lts. ovate or oblong subcordate sinuate-dentate, pet. *white* twice as long as calyx, pods linear.—*E. B.* 855. *R.* ii. 50.—Pods patent. When

growing out of water it is slender with small leaves (*N. micro-phyllum* R.); when remarkably luxuriant, many feet in length, the stem often nearly an inch thick, and the leaves very large and resembling those of a *Sium*, it is *N. siifolium* (R.).—Running water. P. VI. VII. *Water-Cress.* E. S. I.

2. *R. pinnáta* (Mœnch); creeping, l. deeply pinnatifid, lts. oblong or lanceolate cut, uppermost l. often nearly entire, *pet. yellow twice as long as calyx*, pods linear.—*E. B.* 2324. *N sylvestre* (R. Br.) ed. viii.—Fr.-stalks patent; pods patent or ascending, variable in length, usually as long as their stalks, sometimes shorter.—River-banks and wet places. P. VI.—VIII. E. S. I.

3. *R. palus'tris* (Mœnch); *root fibrous*, lower l. lyrate, upper l. deeply pinnatifid, lts. oblong toothed, *pet. yellow not longer than the calyx*, pods oblong thick.—*N. terrestre* Sm. *E. B.* 1747.—Fl. small. Fr.-stalks patent or deflexed; pods ascending, short, about as long as their stalks.—Wet places. P. VI.—IX. E. S I.

4. Barbare'a *R. Br.* Rocket.

1. *B. vulgáris* (R. Br.); *lower l.* lyrate, *upper pair of lobes equalling the breadth of the* large roundish *subcordate terminal lobe*, uppermost l. undivided toothed, pods adpressed obliquely erect or patent with a subulate point.—*E. B.* 443. *R.* ii. 47.—Pet. twice as long as the calyx. Flowering raceme lax. Pods short.—I cannot separate *B. arcuata* (*Sy. E. B.* 121)[1].—In damp places. B. ? V.—VIII. *Yellow Rocket.* E. S. I.

2. *B. stric'ta* (Andrz.); *lower l.* lyrate, *upper pair of lobes* small much *shorter than the breadth of the* large oblong-ovate *terminal lobe*, uppermost l. undivided toothed, pods adpressed with a subulate point.—*Sy. E. B.* 122. *R.* ii. 47. *B. parviflora* Fries. —Pet. half as long again as the calyx. Flowering raceme close. Fl. much smaller than in *B. vulgaris.* Pods short. Lateral lobes of the lowermost l. very small, often obsolete.—By Thames near Kew. Yorkshire [&c.] B. ? V.—VIII. E.

†3. *B. intermédia* (Bor.); lower l. lyrate, upper pair of lobes equalling the breadth of the cordate-ovate term. lobe, *upper l. all pinnatifid, pods erect with a short conical point.*—*Sy. E. B.* 123.—St. with 2 or 3 angles, 1½—2 ft. high. Raceme close. Pet. exceeding calyx. Pods thick, short, angular.—Near Manchester. N. of Ireland. B. V.—VIII. E. I.

---

[1] Syme distinguished *B. arcuata* R. by the yellower-green colour, larger fl., more persistent pet., laxer raceme with longer spreading pods, longer styles and smaller darker and narrower seeds.—H. & J. G.

‡4. *B. præcox* (R. Br.); lower l. lyrate gradually larger upwards, upper pair of lobes equalling the breadth of the subcordate single terminal lobe, *uppermost l. pinnatifid with linear-oblong entire lobes, pods patent with a short thick point.*—*E. B.* 1129.—Fl. moderate. Raceme close. Pods long, torulose. Lower l. usually interruptedly pinnate.—South of England. (American.) B. V.—VII. E. I.

5. AR'ABIS *Linn.* Rock-Cress.

* *Seeds in one row.*

1. *A. hirsúta* (Scop.); l. hispid dentate, *stem-l.* truncate-*auricled or cordate at the base,* auricles patent, pods erect narrow straight. —*E. B.* 587.—St. 1 foot high, clothed with spreading mostly simple hairs and many erect leaves. Sometimes the hairs on the stem are adpressed and branched, or absent [var. *glabrata* Sy.]. Root-l. narrowed into a footstalk.—Walls and banks. B. VI—VIII. E. S. I.

2. *A. ciliáta* (R. Br.); l. glabrous and ciliate or hispid somewhat toothed nearly sessile, *stem-l. sessile with a rounded base,* pods erect narrow straight.—*E. B.* 1746.—St. erect, glabrous. L. glabrous, ciliate. Seeds without wings. St. and l. sometimes hispid [var. *hispida* Sy.].—Rocks by the sea. Connemara, Ireland. Lidstep, Pembrokeshire. B. VII. VIII. E. I.

3. *A. stric'ta* (Huds.); *l. hispid* and ciliate deeply *sinuate-dentate,* narrowed into a footstalk, *stem-l. sessile,* pods few distant erect patent straight.—*E. B.* 614.—St. erect, glabrous, hispid below, 6—8 in. high. Fl. rather large. Seeds winged at the apex.—Limestone cliffs near Bristol. P. IV. V. E.

4. *A. alpina* (L.); l. clothed or fringed with 3-4-forked hairs, rosette-l. sessile or narrowed downwards into a short winged petiole, stem-l. oblong acute *amplexicaul with distinct auricles,* irregularly dentate sometimes with small secondary teeth. *Pods spreading* or curved upwards, *broad* obtuse. Seeds roundish-oval, *surrounded by a narrow membranous wing.*—*E.B.* iii. *Suppl.* t. 117 A.—Stoloniferous. St. glabrous or hairy. Pet. white about twice the length of the sep. Outer sep. gibbous at the base.—Cuchullin Mountains, Skye. *Mr. H. C. Hart.* P. VII. VIII. S.

5. *A. petræ'a* (Lam.); *l. glabrous* or with forked hairs lyrate-pinnatifid or oblong-ovate nearly entire with long stalks, *stem-l. narrow nearly entire stalked,* pods spreading slender straight.—*E. B.* 469.—St. erect or decumbent, 3—8 in. long, *glabrous.* Fl. large, white tinged with purple. Seeds oblong with a straight

wing at the end. [Very variable as to the shape and degree of hairiness and serration of the l. The varietal names *hispida* (DC.) and *hirta* (Koch) have been given to the more hairy forms, and var. *grandifolia* (Druce) =*A. ambigua* (DC.)? to a form with large fl. and with l. broader and mostly entire.]—Alpine rocks in E. and S. Glenade Mountains, Co. Leitrim, Ireland. P. VII. VIII. E. S. I.

[**A. Turrita* (L.); l. clothed with short forked hairs dentate elliptic narrowed into a stalk, *stem-l. deeply cordate-amplexicaul* long, pods flat with a thickened margin recurved from an erect stalk, *seeds with a membranous margin*.—*E. B.* 178.—St. 1 foot high, erect hairy. Fl. yellowish. Pods 3—4 in. long, without any central vein but with many prominent longitudinal anastomosing veins.—On walls at Oxford, Cambridge, and Cleish Castle, Kinross. Become very rare if not extinct. B. V.] E. S.

** *Seeds in two rows.*

6. *A. perfoliáta* (Lam.); radical l. stalked toothed hairy, st.-l. glabrous entire clasping, pods slender erect.—*Turritis glabra* (L.) *E. B.* 777.—Very erect and straight, 1—3 ft. high. Fl. pale yellow.—Banks and cliffs, local. B. V.—VII. E. S.

6. Cardami'ne *Linn.* Bitter Cress.

1. *C. impátiens* (L.); *l. pinnate*, leaflets of the lower l. ovate 3-fid, of the upper l. oblong lanceolate toothed or entire, *petioles of the stem-l. with slender sagittate auricles*, pet. linear or wanting.—*E. B.* 80. *R.* ii. 26.—Pet. erect, white. Distinguished by having auricles at the base of its petioles.—Hilly districts, preferring limestone. A. VII. VIII. E. S.?

2. *C. flexuósa* (With.); l. pinnate, *leaflets of the lower l. roundish angled or toothed*, of the upper l. narrower, pet. twice as long as the calyx, pods erect upon patent pedicels, *stam.* 6, *style long*.—*E. B.* 492. *R.* ii. 26. *C. sylvatica* Link.—Fl. small. Pet. erect, white. *St. wavy*, more leafy than in *C. hirsuta*. Root oblique, covered with fibres. In shady places. A. or B. IV.—IX. E. S. I.

3. *C. hirsúta* (L.); l. pinnate, leaflets of the lower l. roundish angled or toothed, of the upper l. narrower oblong and entire or broadly oval and sometimes incised, pet. twice as long as the calyx, pods and pedicels erect, *stam.* 4, *style equal in length to about ½ the breadth of pod.—Sy. E. B.* 110. *R.* ii. 26.—Fl. small. Pet. erect, white. *St. nearly straight*, rather leafy.—In damp places. A. IV.—VIII. E. S. I.

4. *C. praten'sis* (L.); l. pinnate, leaflets of the lower l. roundish slightly angled, *of the upper l. linear-lanceolate entire, pet.* 3 *times as long as the calyx* spreading, stam. half the length of the petals, st. terete.—*E. B.* 776. *R.* ii. 28.—Fl. large, lilac. Anth. yellow. Style short. Lts. sometimes stalked and occasionally acutely angular (*C. dentata* Schultz). *C. Hayneana* R. having many small narrow lts. resembles Sp. 3.—Moist meadows, common. P. IV. *Lady's Smock.* E. S. I.

5. *C. amára* (L.); l. pinnate, *leaflets* of the lower l. roundish-ovate, of the upper l. oblong, *all angular*, pet. 3 times as long as the calyx erect, *stam. nearly as long as the petals*, st. angular.—*E. B.* 1000.—Fl. large, white. *Anthers purple.* Style long, slender. Stigma small. Stoloniferous.—Fl. sometimes purplish-lilac (var. *lilacina* F. B. White).—Moist meadows near streams, rare. P. V. VI. *Common Bitter Cress.* E. S. I.

7. Denta'ria *Linn.* Coralwort.

1. *D. bulbif'era* (L.); st. simple, l. alternate, lower l. pinnate, upper l. simple, axils of the l. producing bulbs.—*E. B.* 309. *Cardamine bulbifera* (Crantz).—Rhizome thick, with fleshy toothlike knobs. St. 1—1½ foot high. Lts. and l. lanceolate, serrate or entire. Fl. large, rose-coloured or purple.—Woods and shady places, rare. P. V. VI. E. S.

Tribe II. *Sisymbrieæ.*

8. Hes'peris *Linn.* Dames' Violet.

*1. *H. matronális* (L.); st. erect branched above, l. ovate-lanceolate acuminate toothed, pedicels about as long as the calyx, pet. obovate blunt apiculate, pods erect from a patent pedicel terete.—*E. B.* 731.—Fl. lilac, "fragrant," large and handsome.—Pastures. B. V. VI. E. S. I.

[*Malco'lmia marit'ima* (R. Br.) has been found in Kent, but is not a native.]

9. Sisym'brium *Linn.* Hedge-Mustard.

1. *S. officinále* (Scop.); *pods* subulate *adpressed to the stem*, l. runcinate-pinnatifid with 2 or 3 pairs of oblong dentate lobes and a large hastate terminal lobe.—*E. B.* 735. *R.* ii. 72.—St. 1—2 feet high, with divaricate branches, upper part leafless. Fl. small, pale yellow Pods on exceedingly short stalks, downy, sometimes glabrous (var. *leiocarpum* DC.).—Common. A. VI. VII. *Hedge-Mustard.* E. S. I.

[**S. polycerátium* (L.); *pods* subulate *spreading* sessile axillary *about* 3 *together*, l. lanceolate repand-dentate or subhastate.—*Sy. E. B.* 97. *R.* ii. 73.—St. leafy throughout, branched, "prostrate." Fl. small.—Bury St. Edmunds, naturalized. A. VII. VIII.] E.

2. *S. I'rio* (L.); *pods* terete 4 *times as long as their pedicels erect-patent*, when young exceeding the fl., seeds oblong, l. runcinate-pinnatifid, lobes dentate oblong the terminal lobe angular, lobes of the upper l. lanceolate with the terminal lobe hastate.—*E. B.* 1631. *R.* ii. 75.—Erect, branched, st. and l. glabrous. Fl. yellow. Pods narrow, linear. Pedicels slender.—Near old towns, rare. A. VII. VIII. *London Rocket.* E. S. I.

3. *S. Sophía* (L.); pods terete 3 times as long as the pedicels erect-patent, seeds oblong, *l. doubly or trebly pinnatifid.* segments linear or linear-lanceolate.—*E. B.* 963. *R.* ii. 84.—St. erect, branched, and as well as the l. slightly downy. Fl. yellow. Pet. short. Pods linear, narrow. Pedicels slender.—Waste places, not common. A. VI.—VIII. *Flixweed.* E. S. I.

4. *S. Thaliánum* (Gay); pods 4-angular linear ascending twice as long as their patent pedicels, seeds oblong not striate, *l. oblong-lanceolate undivided* toothed.—Arabis *Sm. E. B.* 901. Conringia *R.* ii. 60.—St. erect, slender, much branched, with few leaves which are nearly all radical. Fl. small, white. *Pods angular,* not convex on the back as in the other species nor with the lateral longitudinal veins so strongly marked.—On walls and banks. A. IV. V. and IX. X. E. S I.

[*S. pannon'icum* (Jacq.) (= *S. Sinapistrum* Crantz ?) is quite established near Crosley, Lanc. *J. of B.* x. 239.]

10. Allia'ria *Adans.* Sauce-alone.

1. *A. officinális* (Andrz.); l. heartshaped the lower ones reniform sinuate-dentate all stalked, pods erect-patent much longer than their stalks, seeds oblong subcylindrical striate.—*R.* ii. 60. *Sisymbrium Alliaria* Scop. *E. B.* 796.—St. erect, 1—3 feet high, slightly branched. L. large, thin, veined, smelling like garlic when bruised. Fl. white.—Hedge-banks. B. V. VI. *Jack-by-the-Hedge.* E. S. I.

11. Erys'imum *Linn.* Worm-seed.

1. *E. cheiranthoïdes* (L.); l. oblong-lanceolate slightly toothed with stellate-tripartite hairs, all narrowed into a slight footstalk, pedicels longer than the calyx 2- or 3-fold shorter than the pods, ped. patent, pods suberect, seeds small many.—*E. B.*

942. *R.* ii. 83.—Seeds very small, so many in the pod as to be nearly 2-rowed. Fl. small.—Cultivated ground, rare. Wild in the Fen country. B. VI.—VIII. E.

[*E. perfoliátum* (Crantz); *l. oval-heartshaped* blunt clasping the stem, radical l. obovate, all smooth glaucous undivided entire.—*E. B.* 1804. *R.* ii. 61. *E. orientale* ed. viii.—Fl. white or cream-coloured.—Fields. Introduced. A. V.—VII.] E.

Tribe III. *Brassiceæ.*

12. Bras'sica, *Linn.*

* *Valves of pod* 1-*ribbed.*

1. *B. oleràcea* (L.); l. glabrous glaucous waved and lobed, lower l. lyrate, *upper l. oblong sessile.*—*E. B.* 637. *R.* ii. 97.—Rootstock stout, branched. St. thick, persistent, usually decumbent. L. very large thick somewhat fleshy. Fl. large cream-coloured. Raceme elongated before the fl. expand. Cal. erect, adpressed. The wild state of the garden cabbage.—Sea-cliffs in South and West. B. VI.—VIII. *Wild Cabbage.* E.S.? I.

2. *B. campes'tris* (L.); radical l. glaucous hispid lyrate-dentate, st.-l. glabrous ovate-lanceolate auricled clasping, fl. subcorymbose.—*α*; fl. pale orange (*Rape*).—*β. B. Rapa* (L.); not glaucous, fl. bright yellow (*Turnip*).—*γ. B. Napus* (L.?); glaucous, fl. yellow (*Rape*), or (*B. rutabaga* L.) fl. buff (*Swede*). Root-l. of all hispid.—See Watson and Dyer in *J. of B.* vii. viii. and ix.—Fields and riverbanks. A. or B. VII. VIII. E. I.

** *Valves of pod* 3-*ribbed*; *beak* 1—3-*seeded.*

3. *B. monen'sis* (Huds.); l. stalked all deeply pinnatifid, lobes oblong unequally toothed those of the upper l. linear.—*E. B.* 962.—Fl. yellow. St. usually prostrate, glabrous. L. glabrous, mostly radical.—*β. B. Cheiranthos* (Vill.); st. 1—3 ft. high erect leafy hispid below, l. hispid. *E. B. S.* 2821.—On the western coasts, rare. B. or P. VI.—VIII. E. S.

13. Sina'pis *Linn.* Mustard.

1. *S. nigra* (L.); *pods quadrangular adpressed,* beak short sterile subulate, valves 1-veined, lower l. lyrate, terminal lobe large and lobed, upper l. lanceolate entire.—*E. B.* 969. *R.* ii. 88. *Brassica sinapioides* (Roth).—Fl. yellow. Lower l. large, rough.—Willowy riverbanks, not common. A. VI.—VIII. *Black Mustard.* E. S. I.

2. *S. arven'sis* (L.); *pods subcylindrical* knotty *longer than the conical compressed beak,* valves 3-veined, l. ovate the lowermost sublyrate stalked, upper l. sessile.—*E. B.* 1748. *R.* ii. 86. *B. sinapistrum* Boiss.—Scabrous. Fl. large, yellow. Pods glabrous or rough with deflexed bristles; beak with one seed; valves with faint intermediate veins. Stem 1—1½ foot high.—Corn-fields. A. VI.—VIII. *Charlock.* E. S. I.

†3. *S. al'ba* (L.); *pods cylindrical* knotty *shorter than the swordshaped beak,* valves 5-veined, l. lyrate pinnatifid irregularly lobed.—*E. B.* 1677. *R.* ii. 85. *B. alba* Boiss.—Fl. large, yellow. Pods hispid. St. 1—2 feet high.—Cultivated and waste calcareous land. A. VII. *White Mustard.* E. S. I.

[*S. incána* (L.); pods adpressed thick prominently veined with a *short* 1-*seeded beak,* l. lyrate hispid, stem-l. linear-lanceolate, st. much branched.—*E. B. S.* 2843. *B. adpressa* Boiss.—St. 1—3 feet high, branches divaricate with few very small leaves. Pods very short, glabrous or hairy, often scarcely longer than their glabrous beak.—Sandy places in Jersey and Alderney. B. VII. VIII.]

14. Diplotax'is *Cand.* Wall-Mustard.

1. *D. tenuifólia* (DC.); *st. woody below branched subglabrous* leafy, l. glaucous linear-lanceolate very acute sinuate-dentate or pinnatifid, segments linear remotely dentate, ped. very long, pet. roundish-obovate with a short claw.—*Sinapis* Sm. *E. B.* 525.—Fl. large; pet. blunt but slightly acuminate. St. 1—1½ foot high. Plant fœtid.—Old walls. P. VII.—IX. E. S. I.

2. *D. murális* (DC.); *st. herbaceous simple hispid* and leafy at the base, l. almost glabrous ovate-lanceolate sinuate-dentate or pinnatifid, ped. as long as expanded flower, pet. roundish-ovate with a short claw.—*Sinapis* Sm. *E. B.* 1090.—Pedicels as long as the flowers. Pet. abrupt or emarginate. L. often blunt never very acute.—We have two forms: (1) l. all radical in a rosette and st. simple; (2) *β. Babingtonii* (Sy.), base of st. leafy with axillary branches.—Waste ground. A. VIII. IX. E. S. ?I.

Suborder II. *Latiseptæ.* Tribe IV. *Alyssineæ.*

15. Alys'sum *Linn.*

*1. *A. alyssoides* (L.); herbaceous hoary with starry pubescence, l. obovate-lanceolate attenuate below, pods orbicular stellate-pubescent, calyx persistent, filaments all toothless, shorter ones from between 2 setaceous processes.—*E. B. S.* 2853. *R.* ii. 18. *A. calycinum* (L.) ed. viii.—Cells of the pod 2-seeded. Pet. yellow, becoming at length white.—Grassy commons, S.; ploughed land, E. A. V. VI. E. S.

[*A. marit'imum* (L.): procumbent, hairs bipartite, l. linear-lanceolate acute, cal. deciduous, pods oval pointed glabrous, cells 1-seeded.—*E. B.* 1729. *R.* ii 18.—St. rather woody below. Fl. white, sweet-scented.—Naturalized near the sea. P. VIII. IX.] E.

[*A. incánum* (L.); st. erect or ascending, hoary, l. lanceolate, cal. deciduous, pet. bifid, pods elliptic many-seeded.—*R.* ii. 22.—Escaping.]

16. Dra'ba *Linn.* Whitlow-grass.

[* Pet. almost entire.]

1. *D. aizoides* (L.); st. leafless glabrous, *l. linear rigid* acute *keeled glabrous* ciliate, stam. equalling the slightly notched petals, style as long as the breadth of the pouch.—*E. B.* 1271. *R.* ii. 15.—Fl. bright yellow. L. fringed with rigid hairs, densely collected into cushion-like tufts.—On rocks and walls at and near Pennard Castle near Swansea. P. III. IV. E.

2. *D. rupes'tris* (R. Br.); st. leafless or with 1 or 2 leaves pubescent, *l. lanceolate flat stellately pubescent,* stam. shorter than the slightly notched petals, style short.—*D. hirta* Sm. *E. B.* 1338.—Pouch long-oval, slightly hairy. St. very short, branched, each branch bearing a dense tuft of leaves and 1—3 short scapes. Fl. small. L. mostly entire—Tops of the Scottish mountains, rare. P. VII. S.

3. *D. incána* (L.); *stem-l. several,* l. lanceolate stellately pubescent toothed, pet. twice as long as the calyx, *pouch longer than its pedicel twisted,* style very short.—*E. B.* 388. *R.* ii. 14.—Pouch nearly or quite glabrous, erect, lanceolate-oblong. St. 4—12 in. high, simple or branched. Fl. white.—Extreme forms are often taken for species, as *D. contorta*(Ehrh.), st. simple or branched only at top densely leafy, pouch glabrous; and the more alpine state *D. confusa* (Ehrh.), st. branched only near its base, l. mostly rosulate, pods with stellate down.—Mountains and sand-hills by the sea. P. VI. VII. E. S. I.

4. *D. murális* (L.); *st. leafy* branched, l. ovate amplexicaul toothed hairy, pet. "entire," *pedicels spreading horizontally*

rather longer than the glabrous *pouch.—E. B.* 912. *R.* ii. 12.—Pouch elliptic. St. 5—12 in. high. Fl. white. Pubescence branched. Root-l. narrowed below.—Limestone hills. A. IV. V. E. S. I.

[** Pet. deeply 2-lobed. *Erophila* DC.]

5. *D. ver'na* (L.); *st. leafless* glabrous above, l. lanceolate acute narrowed below hairy, *pet. deeply cloven*, pouch oblong shorter than its pedicel.—*E. B.* 586. *R.* ii. 12.—Pouch compressed, [narrowed at both ends, or (*D. præcox* Stev.) shorter and rounded at the apex or (*D. inflata* Wats.) almost terete.]. Fl. white. L. entire or toothed. Scape sometimes rather pilose.—Very variable in minute points.—Very common on walls, banks, &c. A. III.—V. *Common Whitlow-grass.* E. S I.

17. Cochlea'ria *Linn.* Scurvy-Grass.

1. *C. officinālis* (L.); radical l. cordate-reniform stalked, *stem-l. sessile* oblong sinuate clasping, *pouch globose or obovoid, style short.—E. B.* 551.—Petioles long. Lower l. entire or sinuate. Seeds large. St. occasionally rooting and proliferous.—Sea-coast, mostly in muddy places. B.? VI.—VIII. *Common Scurvy-grass.* E. S. I.

2. *C. alpina* (Wats.); radical l. broadly cordate, obtuse entire, or sometimes obscurely lobed, upper stem-l. clasping triangular 3-lobed, or broadly ovate 4–6-toothed, pouch obovoid more or less *tapering at each end*, style short.—*Sy. E. B.* iii. 131. A more slender plant than sp. 1 with smaller l.—β. *C. micacea* (Marshall, *J. of B.* 1894, p. 289, tab. 345-6). Pouch narrower 1½–3 times as long as broad.—Mountains. B.? VI.–VIII. E. S. I.

3. *C. grœnland'ica* (L.); radical l. fleshy orbicular-reniform scarcely cordate at the base entire, upper stem-l. sessile or shortly stalked varying from elliptic-lanceolate to ovate-spathulate, pouch subglobose, style short.—*J. of B.* 1892, tab. 326 A.—Usually a small tufted fleshy plant with many root-l.—Sea-shores. Haddington and North of Scotland. Donegal. B.? S. I.

4. *C. dan'ica* (L.); *l. stalked*, radical l. cordate somewhat lobed, stem-l. 3—5-lobed rather triangular uppermost sub-sessile, *pouch ovoid*, style short.—*E. B.* 696. *R.* ii. 16.—Petioles of the root-l. very long, gradually shortening as they become more distant from the root. *Seeds small.*—Sea-coast. B. V.—VIII. E. S. I.

5. *C. ang'lica* (L.); *radical l.* stalked *ovate-oblong or obovoid* entire, stem-l. oblong entire or toothed mostly sessile the *upper ones clasping, pouch ovate-oblong*, style slender.—*Sy. E. B.* 133.—Pouch twice as large as that of *C. officinalis*, much compressed

laterally, usually deeply furrowed on each side, dissepiment very narrow. Seeds large. Lower l. rounded below or narrowed into a footstalk. Fl. large.—β. *Hortii* (Syme), radical l. rounded at the base, pouch smaller ellipsoid —Muddy sea-shores. A. V. *English Scurvy-grass.* E. S. I.

18. Armora'cia *Gaertn. M. & S.*

[*A. rustica'na* (G. M. & S.); rootstock long and thick, radical l. oblong crenate-serrate on long stalks, stem-l. long lanceolate incise-serrate or entire subsessile, pet. (white) twice as long as cal., pouch ovoid 4-seeded, stigma peltate.—*Coch. Armoracia* L. *E. B.* 2323. *R.* ii. 17.—St. 2—3 feet high. Rootstock cylindrical.—Waste ground, not native. P. V. *Horse-radish.*] E. S. I.

1. *A. amphib'ia* (Peterm.); roots fibrous, l. oblong narrowed at both ends serrate or pinnatifid, pet. (yellow) twice as long as the cal., pouch ovoid, "stigma capitate."—*Nasturtium* R. Br. *E. B.* 1840. *Radicula lancifolia* (Mœnch).—St. 2—3 feet high. Submersed l. deeply pinnatifid. Ped. usually deflexed.—[DC. divided *N. amphibium* into *a. indivisum* having all the l. nearly entire or serrate, and β. *variifolium* having some of the l. serrate, some pinnatifid, others multifid with capillary segments.] Watery places. P. VI.—VIII. E. S. ? I.

Tribe V. *Camelineæ.*

19. Cameli'na *Crantz.* Gold-of-pleasure.

[*C. sativa* (Crantz); pouches obovate inflated.—*E. B.* 1254.—α; pouches flaccid truncate.—β; pouches rounded at the end brittle.—In fields of flax and corn, introduced, but not naturalized.]

Suborder III. *Angustiseptæ.* Tribe VI. *Thlaspideæ.*

20. Thlas'pi *Linn.* Penny Cress.

1. *T. arven'se* (L.); fruitbearing raceme elongate, *pouch orbicular with a broad dorsal wing,* seeds concentrically rugose and striate 5—6 in each cell, stem-l. oblong sagittate toothed.—*E. B.* 1659. *R.* ii. 5.—Pouch very large ($\frac{1}{2}$–$\frac{3}{4}$ in.), stigma subsessile. St. often a foot high.—Fields and roadsides. A. V.—VII. *Penny Cress.* E. S. I.

2. *T. perfoliátum* (L.); fruitbearing raceme elongate, *pouch obcordate* broadly winged above, *style very short* included within the notch, seeds 3—4 in each cell smooth, *stem-l. deeply cordate-oblong*.—*E. B.* 2354. *R.* ii. 5.—About 6 in. high.—Limestone in Oxfordshire and Gloucestershire. A. V. E.

3. *T. alpes'tre* (L.); fruitbearing raceme elongate, pouch oblong-obovate narrowed below, *style equalling or exceeding the notch*, seeds 4—8 in each cell, stem-l. oblong cordate.—*R.* ii. 5. —Fl. white, often tinged with rose, small. Pet. about twice as long as the calyx.—*a. T. sylvestre* (Jord.), terminal lobes of pouch rounded. L. usually entire. *Sy. E. B.* 146.—*β. T. occitanum* (Jord.); fruitbearing raceme rather shorter, pouch triangular-obcordate with almost divaricate lobes, style much exserted. L. usually slightly toothed. *Sy. E. B.* 147.—Mountain pastures. Teesdale; Thornhaugh, Northumb.; Glen Isla, Forfar. *β*. Settle, York; Llanrwst, N. Wales. P. VI. —VIII. E. S.

4. *T. virens* (Jord.); fruitbearing *raceme oval or oblong*, pouch obovate scarcely notched, style much projecting, seeds 4—5 in each cell, stem-l. oblong cordate.—*Sy. E. B.* 148. *Jord. Obs. Pl. de Fr.* iii. t. 1. *T. alpestre* Sm.—Fl. white, small. Pet. about thrice as long as the calyx. Pouch with a very broad and shallow notch often almost truncate. L. usually entire.—Limestone rocks at Matlock. P. VI.—VIII. E.

21. Hutchins'ia *R. Br.*

1. *H. petræ'a* (R. Br.); l. pinnate, st. branched leafy, fr.-raceme lax long, pouch blunt at both ends.—*E. B.* 111. *R.* ii. 6. *St.* 65. 10.—St. 2—4 in. high. Fl. small; pet. scarcely exceeding the calyx. Cotyledons accumbent.—Limestone rocks, rare. A. III.—V. E.

22. Teesda'lia *R. Br.*

1. *T. nudicaúlis* (R. Br.); petals unequal.—*E. B.* 327. *R.* ii. 6.—L. many, spreading on the ground, lyrate-pinnatifid, rarely orbicular-spathulate and entire. St. 2—4 in. high, solitary or several from the crown of the root, sometimes bearing 1 or 2 small leaves. Stam. with remarkable scales. Pouch emarginate. —Sandy and gravelly places. A. V. VI. E. S. I.?

23. Ibe'ris *Linn.* Candytuft.

1. *I. amára* (L.); herbaceous, l. lanceolate somewhat toothed, pouches racemose orbicular notched with triangular porrect lobes.—*E. B.* 52. *R.* ii. 7.—L. usually with 1—3 blunt teeth on

each side. St. often 1 foot high, diffuse, branched. Fl. at first corymbose, afterwards in lengthened clusters. Outer pet. radiant. —Chalky fields in South and East. A. VII. *Bitter Candytuft.* E.

24. Lepid'ium *Linn.* Pepperwort.

* *Pouch cordate with turgid valves, style filiform.*

[*L. Drába* (L.); l. oblong entire or toothed lower ones narrowed into a footstalk, stem-l. sagittate and amplexicaul, style as long as the dissepiment.—*E. B. S.* 2683. *R.* ii. 9.—One foot or more in height, branched. Fl. many, small, white, upon long pedicels in a subumbellate corymb.—Established in many places, but not naturalized. P. V. VI.] E.

** *Pouch ovate or roundish winged notched.* † Style manifest.

1. *L. campes'tre* (R. Br.); l. downy toothed lower ones oblong narrowed into a footstalk, stem-l. lanceolate sagittate and clasping, *pouch* ovate *scaly* notched and rounded at the end, *style scarcely longer than the notch.*—*E. B.* 1385. *R.* ii. 9.—Anth. yellow. Scales on the pouch only minute globular blisters when fresh. St. one, upright, about a foot high, branched in the upper part, or (var. *longistylum* More) many with style about twice as long as notch.—Dry gravelly soil. B. VI.—VIII. E. S. I.

2. *L. heterophyl'lum* (Benth.); l. hairy toothed lower ones obcordate stalked, stem-l. lanceolate-sagittate clasping, *pouch* ovate *smooth* notched and rounded at the end, *style twice as long as the notch.*—*E. B.* 1803. *L. Smithii* (Hook.) ed. viii.—Anth. violet. Seeds ½ as long as in *L. campestre.* Pouch sometimes with a few scales, never hairy. St. several, 6—12 in. long, branched at the base; central st. erect, others diffuse. *β. alatostylum* (Towns., under *L. Smithii*); pouch not notched.—Hedge-banks. P. VI.—VIII. E. S. I.

†† Style minute.

[*L. sativum* (L.); lower l. lobed pinnate or bipinnate, upper l. sessile linear entire, pouch roundish oval blunt.—*R.* ii. 9.—Escaped from cultivation in many places.] E.

*** *Pouch oval or roundish notched, style minute.*

3. *L. ruderále* (L.); lower l. bipinnatifid, upper l. linear entire, pouch oval, diandrous, petals 0.—*E. B.* 1595. *R.* ii. 10.—End of pouch narrowed, winged. St. branched, often a foot high. Radicle dorsal.—Waste places near the sea, rare. A. V. VI. E. I.

**** *Pouch oval or roundish scarcely notched wingless, style minute.*

4. *L. latifólium* (L.); l. ovate-lanceolate serrate or entire undivided, pouch ovoid entire downy.—*E. B.* 182. *R.* ii. 10.—Fl. many, small, in compound leafy panicled clusters. St. 3 feet high, erect, branched. L. large, the lower ones upon long stalks, the upper nearly sessile and narrower.—In salt marshes. P. VII. VIII. *Dittander.* E. S. ? I.

25. Bur'sa *Weber* (*Capsella* Vent. ed. viii.). Shepherd's Purse.

1. *B. pastóris* (Weber); radical l. lanceolate pinnatifid or undivided toothed, upper l. clasping auricled, pouch triangular-obcordate.—*Thlaspi* Sm. *E. B.* 1485.—Varying greatly in size and the division of its leaves. Known by its peculiar pouches. —A common weed A. III.—X. E. S. I.

Tribe VII. *Subularieæ.*

26. Subula'ria *Linn.* Awlwort.

1. *S. aquat'ica* (L.); *E. B.* 732. *R.* ii. 12.—Plant small, submerged. Root of many long white fibres. St. 1—3 in. high. L. linear-subulate, radical. Fl. small, white, often perfected under water.—Margins of alpine lakes. P. VII. E. S. I.

Tribe VIII. *Senebiereæ.*

27. Senebie'ra *Cand.*

1. *S. Corónopus* (Poir.); *pouch undivided reniform* wrinkled and crested with little sharp points, style prominent, l. pinnatifid. —*E. B.* 1660. *R.* ii. 9. *Coronopus procumbens* (Gilib.), *C. Ruellii* (All.)—St. much branched, prostrate. Fl. small, white, in lateral clusters opposite to the leaves. Pouches large, in dense clusters.—Waste ground, common. A. VI.—IX. *Swine's Cress.* E. S. I.

2. *S. pinnatif'ida* (DC.); *pouch notched of two wrinkled lobes,* style very short, l. pinnatifid.—*E. B.* 248. *R.* ii. 9. *S. didyma* (Pers.) ed. viii.—St. spreading, prostrate, a foot or more in length. Fl. small, white, in long slender lax clusters. Pet. very short or none. Often only 2 stamens.—Waste ground near the sea chiefly in the South and South-west. A VI.—IX. E. I.

Suborder IV. *Nucumentaceæ.* Tribe IX. *Isatideæ.*

28. ISA'TIS *Linn.* Woad.

1. *I. tinctória* (L.); "radical leaves oblong crenate," pouch abrupt smooth thrice as long as broad.—*E. B.* 97. *R.* ii. 4.—St. 1—4 ft. high branched. Fl. yellow. Pouches ½ in. long, pendent.—Wild on cliffs by Severn, Tewkesbury (*Hooker*). B. VII. E.

Suborder V. *Lomentaceæ.* Tribe X. *Cakilineæ.*

The fruit consists of a very small 2-celled, stalklike, usually sterile pod, with a long moniliform beak bearing the seeds and dividing transversely into as many indehiscent cells as there are seeds.

29. CAKI'LE *Mill.* Sea-Rocket.

1. *C. marit'ima* (Scop.); joints of the pouch 2-edged, upper with 2 teeth at the base, l. fleshy pinnatifid or (*integrifolia* Horn.) somewhat toothed.—*E. B.* 231. *R.* ii. 1.—Fl. purplish. Pouches an inch long, erect, with 4 sharp angles, swordshaped in the upper part.—Sandy sea-shores. A. VI. VII. *Purple Sea-Rocket.* E. S. I.

30. CRAM'BE *Linn.* Sea-Kale.

1. *C. marit'ima* (L.); pouch without a style, l. roundish sinuate wavy toothed glaucous and as well as the st. glabrous. —*E. B.* 924. *R.* ii. 2.—Root thick, fleshy. St. 2 feet high. Fl. white. Longer filaments forked at the end. Pouches roundly ovoid, large.—Sandy sea-shores. P. VI. E. S. I.

Tribe XI. *Raphaneæ.*

31. RAPH'ANUS *Linn.* Radish.

†1. *R. Raphanis'trum* (L.); pods moniliform shorter than the very long beak with slender ribs, *l. simply lyrate.*—*E. B.* 856.—Root slender. Lobes of the leaves quite distinct. Pet. veined, yellow, white, or lilac.—Corn-fields. A. VI. VII. *Jointed Charlock.* E. S. I

2. *R. marit'imus* (Sm.); pods moniliform longer than the short beak with thick ribs, *radical l. interruptedly* pinnate.—*Sy. E. B.* 82.—Root thick. Lobes of l. usually so close as to overlap each other. Pet. yellow.—Sea-coast in South and West, rare. B.? VI.—VIII. *Sea-Radish.* E. S. I.

Order VII. RESEDACEÆ.

Sep. 4—8, persistent. Cor. irregular, pet. 4—8 entire or deeply cut. Stam. many, filaments variously united, inserted on a glandular irregular 1-sided hypogynous disk. Ovary 3- or 6-lobed, 1-celled, with 3 or 6 parietal many-seeded placentas; or of several 1-celled carpels. Fruit opening nearly at the end.—Stip. 0?

1. Reseda. Cal. many-parted. Pet. entire or variously cut, unequal. Stam. many. Fruit of one cell opening at the top. Styles 3—6.

1. Rese'da *Linn.* Mignonette.

1. *R. lútea* (L.); sep. 6 linear, *pet. very unequal,* ped. longer than the cal., *l. 3-cleft or pinnatifid.*—*E. B.* 321. *R.* ii. 100.—St. 2 feet high, branched, smooth. L. very variable. Two upper pet. with two wing-like lobes, lateral pet. with a single wing, lower ones nearly entire. Fl. yellow. Fr. oblong, wrinkled.—Waste chalky and limestone places. B. VI.—VIII. E. S I.

‡2. *R. suffruticulósa* (L.); sep. 5 linear-lanceolate, *pet.* 5 *nearly equal* 3-fid longer than the calyx, ped. shorter than the cal., *l. all pinnatifid,* segments linear acute sometimes wavy.—*E. B. S.* 2628. *R.* ii. 101.—St. 1½—2 feet high, rather shrubby below. Fl. white. Fr. oblong, wrinkled. Sometimes there are 6 sep. and pet., when it is *R. alba* (L.).—Waste sandy places near the sea, rare. B. or P. VII. VIII. E. S. I.

3. *R. Lutéola* (L.); sep. 4, pet. 4 or 5 very unequal longer than the calyx, *l.* long-lanceolate *undivided.*—*E. B.* 320. *R.* ii. 99.—St. 2 feet high. Pet. usually 4, upper one 3-, 4-, or 5-cleft, 2 lateral 3-cleft, segments linear, lower one (or two) linear entire. Fr. broad, depressed.—Waste places, particularly on chalk or limestone. B. VII. VIII. *Weld.* E. S. I.

Order VIII. CISTACEÆ.

Sep. 5, two outer smaller sometimes wanting, 3 inner twisted in the bud. Pet. 5, crumpled and twisted in the bud the contrary way to the sepals. Stam. many. Ovary 1, 1- or many-celled. Style simple. Stigmas 3. Capsule 3-, 5-, or 10-valved. Embryo spiral or curved, in the albumen.—Stip. small or 0.

1. HELIANTHEMUM. Cal. of 5 sepals, 2 outer smaller. Pet. 5, deciduous. Stam. many. Caps. 3-valved.

1. HELIAN'THEMUM *Mill.* Rock-rose.

1. *H. guttátum* (Mill.); erect herbaceous, l. oblong-lanceolate or linear, lower l. opposite without stipules, upper l. alternate stipulate, *stigma subsessile.*—St. mostly simple or branching from the base. Fruitstalks patent. Pubescence of long simple, and short stellate hairs. Fl. yellow, usually with a deep-red spot at the base of each petal.—*a*; racemes without bracts. *E. B.* 544.—*β. H. Breweri* (Planch.); racemes with or without bracts on the same plant. *Curt.* ii. 102. *J. of B.* iii. 21. *Sy: E. B.* 166.—Very rare. *a.* Three-Castle Head, Co. Cork. Jersey. *β.* Holyhead Mountain and Amlwch, Anglesea. Inish Boffin, Co. Galway. A. V.—VIII. E. I.

2. *H. marifólium* (Mill.); *shrubby, without stipules,* l. opposite ovate or oblong stalked flat hoary beneath, *racemes* terminal with bracts, "style twisted at the base reflexed but at the apex inflexed."—*H. canum* (Dun.) ed. viii. *Cistus marifolius* Sm. *E. B.* 396.—St. decumbent. L. hoary beneath, hairy above. Fl. yellow, small. "Anth. emarginate at both ends. Style longer than the stigma." The Teesdale plant, *H. vineale* (Pers.), has rather fewer hairs on the upperside of the leaves.—On limestone rocks, rare. P. V.—VII. E. I.

[*H. ledifólium* (Willd.); herbaceous, with stipules, ped. solitary, styles straight.—*E. B.* 2414.—Brean Down, Somerset. Probably an error. A. VI. VII.]

3. *H. Chamæcis'tus* (Mill.); procumbent, *shrubby, with stipules,* l. oval or linear-oblong opposite nearly flat green above hoary beneath, *racemes with bracts, style* longer than the germen *bent at the base, sep. subglabrous inner ones blunt apiculate.*—*H. vulgare* (Gaert.) ed. viii. *Cistus* L. *E. B.* 1321.—Fruitstalks contorted and deflexed. Varying much in the size and shape of its leaves and the amount of hoariness and pubescence. *Fl. yellow.* —*C. tomentosus* E. B. 2208 scarcely differs. *C. surrejanus* E. B. 2207 is a garden form. *H. vulg. pet. fl. perangustis* (Dill. H. Elth. 145) was again found near Croydon by the late Mr. Christy and is a monstrosity.—Common on dry hilly places. P. VII.—IX. *Common Rock-rose.* E. S.

4. *H. polifŏlium* (Mill.); shrubby, procumbent, hoary, with stipules, l. opposite ovate-oblong or oblong-linear more or less revolute at the edges, racemes with bracts, style bent at the base longer than the germen, *sep. tomentose inner ones blunt.*—*E. B.* 1322.—*Fl. white.*—Very rare. Brean Down, Som.; Torquay, Devon. P. VI. VII. E.

Order IX. VIOLACEÆ.

Sep. 5, imbricate. Pet. 5, regular or irregular. Stam. 5, filaments dilated, connective extended beyond the anthers as a flat membrane. Ovary 1-celled with 3 parietal placentas. Style with a hooded stigma. Caps. with 3 valves. Embryo straight, in fleshy albumen.—Stip. persistent.—Minute but fertile fl. which do not open are found on many species.

1. VIOLA. Sep. 5, extended at the base. Pet. 5, unequal, the lower one extended into a hollow spur behind. Stam. 5. Anth. connate, 2 lower ones spurred behind.

1. VI'OLA *Linn.*[1] Violet.

A. *Two interm. pet. patent laterally. Style from a slender base.*

* St. creeping. Stigma flat above. Fr.-st. erect. Caps. nodding.

1. *V. palus'tris* (L.); anth.-spurs short thick rounded, cor.-spur very short blunt, *l. reniform-cordate* glabrous.—*E. B.* 444. *R.* iii.—Anth.-cells nearly parallel. Fl. pale lilac with purple streaks. Scentless.—Wet and boggy places. P. IV.—VI. *Marsh-Violet.* E. S. I.

** Rhizome short. Stigma hooked; beak deflexed. Fruit-stalks prostrate. Caps. globular, downy.

a. *Stoles from axils of terminal rosettes.*

2. *V. odorāta* (L.); *anth.-spurs lancet-shaped* decurved blunt, cor.-spur blunt straight, lateral pet. entire lower one emarginate, l. cordate, *with stoles.*—*E. B.* 619. *R.* iii.—Anth.-cells diverging below. Spurs of the pet. inflated towards the end, slightly

---

[1] The following hybrids have been identified:—
V. odorata×*hirta* (*V. permixta* Jord.?); *V. canina*×*lactea*; *V. silvestris*×*Riviniana*; *V. canina*×*stagnina*; *V. Riviniana*×*canina*.—H. & J. G.

channelled above. Fl. purple, often white, sweet-scented. Fr.-sep. triangular, acute, twice as long as broad, not ciliate, glabrous. *Bracts above the middle of the flowerstalk. Petioles with deflexed hairs.*—Several species (?) allied to this are nearly or quite scentless: *V. permixta* (Jord.) has a large glabrous glandular-denticulate stip. ciliate towards the tip, short robust stoles usually not rooting, fl. pale blue: *V. sepincola* (Jord.) has large glandular-denticulate stip., longer somewhat rooting stoles, and much darker flowers. These plants and their allies seem quite different when growing, but are very difficult to define. Some authors join them to *V. hirta.*—Common. P. III. IV. *Sweet Violet.* E. S. ? I.

b. *Stoles wanting or very short.*

3. *V. hir'ta* (L.); *anth.-spurs nearly linear* blunt, spur of the cor. blunt hooked at the end, pet. entire or slightly emarginate, l. cordate-ovate, *stoles wanting.*—*E. B.* 894. *R.* iii.—Anth.-cells diverging below. Spur of the petals compressed, not channelled. Fl. pale blue sometimes white, scentless. Fr.-sep. roundly triangular, bluntish, as long as broad, more or less ciliate, mostly downy. *Bracts below the middle of the flowerstalk.* Stip. not hispid at the margin. *Petioles with spreading hairs.*—*β. calcarea* (Bab.); fl. smaller, ped. much longer than the leaves, sep. oblong-ovate. [*γ. glabrata* (Beeby); caps. glabrous.] Common on limestone. *β.* Gogmagog Hills. Portland. [etc.] [*γ.* Harston, near Cambridge.] P. IV. *Hairy Violet.* E. S. I.

*** Stigma hooked; beak horizontal. Having a stem. Fruitstalks erect.

† *Without a true sobole.*

4. *V. silves'tris* (Reich.); anth.-spurs narrowly lancet-shaped, pet. oblong narrow (lilac) lower with few parallel nearly simple veins not quite extending to the edge, spur compressed entire usually darker than the pet., cal.-appendages small becoming indistinct, caps. glabrous, l. cordate-prolonged, flowering branches axillary from a short flowerless central rosette of l.—*E. B. S.* 2986. *V. Reichenbachiana* (Bor.).—Fl. scentless.—Hedge-banks and thickets. P. IV. V. E. S. I.

5. *V. Rivin'iana* (Reich.); anth.-spurs narrowly lancet-shaped, pet. broadly ovate (blue) lower with many branched dark veins usually quite extending to the edge, spur thick usually yellowish-white, cal.-appendages broad and squarish persistent with fruit, caps. glabrous, l. broad cordate-acute, flowering branches

axillary from a short flowerless central rosette of l.—*E. B.* 620. *Curt.* i. 182.—Fl. scentless. *V. flavicornis* (Forst.) is a dwarf form having small l. and larger fl. *β. nemorosa* (Neum. W. & M.); cal.-appendages large much cut, veins of lower pet. not extending to the edge, spur thick not furrowed.—Hedge-banks, thickets, and heaths. Common. P. IV. V. *Wood Violet.* E. S. I.

6. *V. rupes'tris* (Schmidt); anth.-spur very narrowly lancet-shaped, cor.-spur blunt, l. roundly cordate, flowering branches axillary from a short flowerless central rosette of leaves, ped. young l. and acute *caps. downy*, pet. broadly obovate, lower pet. with many branched veins throughout, cal.-appendages broad squarish persistent.—*V. arenaria* (DC.) ed. viii. *Sy. E. B.* 174 b. *R.* iii. 9.—A small compact plant with large flowers.—Elevated pastures. Upper Teesdale and Westmoreland. B. V. VI. E.

7. *V. canina* (L.); anth.-spur lancet-shaped (3 times as long as broad), cor.-spur blunt, l. cordate-ovate roundly acute, *primary and lateral stems flowering and lengthening.*—*R.* iii. 10. *E. B. S.* 2984.—Rather cæspitose. L. always roundedly acute, longer than those of Sp. 4. Fl. bluish purple, scentless. Cor.-spur yellow, 1—3 times as long as cal.-appendages. Lower pet. spathulate.—*V. flavicornis* (Sm.) is a small form with cordate leaves. When the l. are cordate-oblong it is perhaps *V. montana* (Linn.).—*β. V. lancifolia* (Thore); *l. ovate-lanceolate rounded below*, stip. lanceolate incise-serrate. *V. lactea* Sm.[1] *E. B.* 445. *V. pumila* Fries (not Vill. which is *V. pratensis* Fr. and has a sobole). L. narrowing gradually from near their base, to a narrow but rounded point. [Var. *crassifolia* (Gronv.) with large fl. and thick fleshy l. has been found in Cambs.]—Sandy and peaty places. *β* in turf-bogs, rare. P. IV. V. *Dog-Violet.* E. S. I.

†† *Rhizome or rather sobole slender.*

8. *V. stagnina* (Kit.); *anth.-spur short* broadly lancet-shaped acute (not twice as long as broad), *cor.-spur very short* blunt, l. ovate-lanceolate subcordate below, petioles winged at the top, stip. linear-lanceolate incise-serrate shorter than the petioles, primary and lateral st. flowering and elongated.—*E. B. S.* 2985. *V. lactea* R. iii. 16. not *Sm.*—Sobole threadlike. St. erect. L. narrowing gradually from the base which in the lowest is some-

[1] Mr. Beeby separates *V. lactea* Sm. as a species, using the name *V. ericetorum* Schrad. for the type instead of *V. canina* L.—H. & J. G.

times rounded not cordate. Fl. pale blue, nearly white. Cor-spur scarcely longer than the cal.-appendages.—Rare. Turf-bogs. P. V. VI. E. I.

B. *Four upper pet. directed upwards and imbricate. Style clavate. Stigma inflated.*

9. *V. lútea* (Huds.); *anth.-cells nearly parallel, anth.-spurs long filiform,* spur of the cor. as long as or longer than the calycine appendages, sep. acute, l. crenate-serrate lower ones ovate-cordate, upper l. ovate or lanceolate, *stip. palmate-pinnatifid,* terminal lobe linear or linear-lanceolate, *st.* ascending *diffuse and filiform underground.*—*Sy. E. B.* 181.—Fl. wholly yellow, yellow with the 2 upper petals purple, or wholly purple [var. *amœna* Wats.], varying greatly in size. Caps. globose. All the lobes of the stip. of nearly equal size, lateral ones (usually 3 on one side and 1 on the other) all springing from near the base of the stip., the terminal lobe narrow and very nearly always quite entire but sometimes considerably larger than the others.—*β. V. Curtisii* (Forst.); stems angular rough, lower part of the stip. somewhat lengthened so as slightly to separate the lateral lobes. *E. B. S.* 2693. *V. sabulosa* Bor.—Mountainous pastures. *β.* Sands near the west coast. P. VI. VII. E. S. I.

10. *V. trícolor* (L.); *anth.-cells diverging below, anth.-spurs* long *subclavate-filiform,* spur of the corolla about equalling the calycine appendages, l. crenate-serrate, lower ones ovate-cordate, upper l. ovate or ovate-lanceolate, *stip. lyrate-pinnatifid,* terminal lobe spathulate crenate, st. ascending.—*E. B.* 1287. *R.* iii. 21. *V. Curtisii* (Mack.) from Portmarnock.—Root simple. Fl. with the upper pet. purple, lateral ones bluish, lower one yellow. Caps. ovoid. Terminal lobe of the stip. often having only one tooth on each side.—*β. V. arvensis* (Murr.); pet. shorter than the calyx whitish, caps. nearly globular. *E. B. S.* 2712.—A small form from Scilly is very like *V. parvula* Tin., and another small form from sandhills Jersey was referred by Mr. J. Lloyd to var. *nana* DC.—Common. A. V.—IX. *Heartsease. Pansy.* E. S. I.

Order X. DROSERACEÆ.

Sep. 5, imbricate. Pet. 5, regular. Stam. 5 or 10, free. Styles 3 or 5. Ovary free. Caps. 3—5-valved; valves bearing the seeds along their middle. Seeds without an aril.—L. with a circinate vernation.—Joined to *Saxifragaceæ* by some authors.

1. DROSERA. Cal. deeply 5-cleft. Pet. 5. *Stam.* 5, *hypogynous. Styles* 3—5, *deeply bifid.* Caps. 1-celled, with 3—5 valves, many-seeded.

1. DRO'SERA *Linn.* Sundew.

1. *D. rotundifólia* (L.); *l. orbicular* spreading flatly, petioles hairy, fl.-stalks erect from centre of rosette of leaves, seeds with a loose chaffy coat.—*E. B.* 868. *R.* iii. 24.—Flower-stalks 2—6 in. high. Stigmas white, clubbed, entire. Anth. white. L. covered, as in all other species, with hairs terminating in large glands secreting a viscid fluid which retains insects that settle upon them. Rachis recurved parallel to itself. Autumnal stoles with bulbous end.—Common in boggy places. P. VII. VIII. *Round-leaved Sundew.* E. S. I.

2. *D. longifólia* (L.); *l. spathulate* blunt erect, petioles glabrous, *fl.-stalks arcuate* or decumbent at the base *from base of rosette* of leaves, seeds with a close rough not chaffy coat.—*R.* iii. 24. *D. intermedia* (Hayne) ed. viii.—Stig. pink, bifid. Anth. yellow. Rachis not closely recurved. A variety with shorter leaves and the flower-stalks shorter than the leaves is sometimes found.—Common in boggy places. P. VII. VIII. E. S. I.

3. *D. ang'lica* (Huds.); l. obovate-lanceolate blunt erect, petioles glabrous, *fl.-stalks erect from centre of rosette* of leaves, seeds with a loose chaffy coat.—*E. B.* 869. D. longifolia *R.* iii. 24, *Koch, Fries.*—Much larger and taller than the last.—*β. D. obovata* (M. and K.)[1]; has broader leaves and the styles often, though not always, emarginate. *E. B.* 867.—In bogs, rather rare, common in Ireland. P. VII. VIII. E. S. I.

Order XI. FRANKENIACEÆ.

Sep. 4 or 5, in a furrowed tube below. Pet. 4 or 5, clawed, with appendages at the base of the limb. Stam. 4 or 5 or more, free, 2-celled, opening by 2 terminal pores or longitudinally. Caps. 1-celled, 2—4-valved; placentas 3, parietal. Style slender, simple or trifid. Seeds many, minute. Embryo in the albumen.—Stip. 0.

1. FRANKENIA. Style 3-fid; lobes oblong with the stigma on their inner side. Caps. 1-celled, 3—4-valved.

---

[1] Now considered a hybrid with Sp. 1.—H. & J. G.

1. Franke'nia *Linn.* Sea-heath.

1. *F. læ'vis* (L.); l. linear (or rather oblong with reflexed edges) glabrous ciliate at the base.—*E. B.* 205.—St. slightly downy, prostrate, wiry. Cal. slightly hispid between its prominent angles. Fl. terminal or from the forks of the stem, sessile, rose-coloured. L. sometimes pulverulent.—Salt marshes on the East and South coast. P. VIII. E.

[*F. pulverulen'ta* (L.); l. obovate retuse glabrous above pulverulent beneath, petiole ciliate.—*E. B.* 2222.—Formerly on the Sussex coast, now lost. A. VII.] E.

Order XII. POLYGALACEÆ.

Sep. 5, imbricate, irregular, 2 interior much larger petal-like. Pet. unequal, usually 3, 1 anterior and larger than the rest. Stam. subdiadelphous, in 2 equal opposite bundles. Anth. 1-celled, opening by a pore at their apex. Caps. 1—2-celled, with placentas in the axis. Seeds pendulous, usually with an aril at the base.—Stip. 0.

1. Polygala. Sep. 5, persistent; 2 inner (wings) broader and often petal-like. Cor. irregular. Pet. 3—5, connected together; the lower one keelshaped. Caps. compressed. Seeds solitary, with a 3-pointed basal aril.—Fl. crested.

1. Poly'gala *Linn.* Milkwort.

1. *P. vulgáris* (L.); *l. scattered, lower l. smaller* oblong, upper l. linear-lanceolate, cal.-wings elliptic mucronate their *veins branched the lateral joining a branch of the central vein*, caps. obcordate, bract equalling the pedicel.—*E. B.* 76.—St. prostrate, ascending. L. scattered. Racemes terminal. Fl. blue, pink, or white. Central vein of wings nearly simple ending in a mucro; lateral only branched externally; branches joining in loops and also with the upper ones of the central vein. Lobes of aril unequal, blunt, lateral ones ½ as long as the seed which is a little stalked within the aril. *P. oxyptera* (R.) is a state with smaller fl., and fr. broader than the wings. *E. B. S.* 2827.—β. *grandiflora* (Bab.); upper l. large lanceolate, cal.-wings oval apiculate, their *lateral veins rejoin the mostly simple central vein near its tip* and have many net-like veins externally. Fl. deep blue.—Dry pastures and peaty fens. β. Ben Bulben, I. P. VI.—IX. E. S. I.

2. *P. serpyllácea* (Weihe); *lower l. mostly opposite* and crowded, st. long prostrate wiry much branched, racemes ultimately lateral, cal.-wings as in Sp. 1, bract shorter than the pedicel.—*Sy. E. B.* 187. *P. depressa* (Wend.).—Upper part of st. and ped. sep. pet. and caps. sometimes pubescent (*P. ciliata* Lebel.).—Dry pastures. P. VI.–IX. E. S. I.

3. *P. calcárea* (F. Sch.); *l. chiefly in an irregular terminal rosette large* obovate blunt, those of fl.-st. short smaller lanceolate, cal.-wings oblong their veins branched the lateral looping with a branch from near the middle of the central vein, caps. oblong-obcordate, bract shorter than the pedicel.—*P. amara* Don, *E. B. S.* 2764. P. amarella *Coss. et Germ.* Atl. t. 7.—St. weak, procumbent or ascending, nearly naked below. *Fl.-st. several from the axils of the term. rosette,* simple, short; racemes terminal. Fl blue. Lobes of aril unequal, lateral ½ as long as seed.—Chalk hills, rare. P. V. E.

4. *P. amára* (L.); *lower l. larger* obovate blunt *in a basal rosette,* upper l. oblong-lanceolate, cal.-wings oblong or obovate blunt their *veins simple or slightly branched free,* caps. obcordate broader than the wings, lateral bracts shorter than the pedicels. —Rosette at crown of rootstock, its l. much the largest broad rounded at the end. Fl.-st. from the axils of the rosette, simple, short; upper l. acute; racemes terminal. Cal.-wings longer than caps. Lobes of aril nearly equal, blunt, ½ as long as seed.—*a. P. austriaca* (Cr.); caps. rounded at the base, fl. very small, pinkish? or pure white. *Sy. E. B.* 189.—*β. P. uliginosa* (R.); caps. rather wedgeshaped below, fl. larger, blue.—*a.* Wye Downs [and near Sevenoaks], Kent. Caterham, Surrey, *Mr. W. Whitwell.* *β.* Cronkley Fell, Yorkshire, *Mr. James Backhouse.* P. VI. VII. E.

Order XIII. ELATINACEÆ.

Sep. 3—5, distinct, or slightly connate. Pet. 3—5. Stam. as many or twice as many as the pet., free. Caps. 3—5-celled, 3—5-valved, loculicidal, dissepiments adhering to central axis. Styles 3—5; stigmas capitate. Seeds many, albumen 0, embryo curved with the seed.—L. opposite. Stip. minute or inconspicuous.

1. ELATINE. Cal. 3—4-parted. Pet. 3—4. Stam. 3—4 or 6—8. Styles 3—4. Caps. 3—4-celled, many-seeded. Seeds cylindrical, straight or bent.

1. Ela'tine *Linn.* Waterwort.

1. *E. hexan'dra* (DC.); l. opposite longer than their petioles, fl. slightly stalked with 6 stam. and 3 obovate pet., caps. turbinate concave at the summit 3-celled, seeds nearly straight ascending 8—12 in each cell.—*E. B.* 955. *R. I.* f. 599. *E. Hydropiper* and *E. tripetala* (Sm.).—Plant minute, creeping. Fl. alternate, axillary. Cal. 3-fid.—Forming small matted tufts under water, rare. A. VIII. E. S. I.

2. *E. Hydropiper* (L.); l. opposite shorter than their petioles, fl. sessile with 8 stam. and 4 ovate pet., caps. roundish depressed 4-celled, seeds bent almost double pendulous 4 in each cell.—*E. B. S.* 2670.—Plant minute, creeping. Cal. 4-fid.—Very rare, growing under water. A. VIII. E. I.

Order XIV. CARYOPHYLLACEÆ.

Sep. 5 or 4, distinct or connected into a tube. Pet. 5 or 4, clawed (rarely 0). Stam. usually twice as many, sometimes as many as the petals, free or connected at the base. Anth. opening longitudinally. Ovary one, often stalked. Stigmas 2—5, sessile, filiform. Caps. 1- or imperfectly 2—5-celled, opening by twice as many teeth as stigmas, sometimes valvular. Placenta central. Embryo generally curved round mealy albumen.—L. opposite, without or rarely with scarious stipules.

Suborder I. SILENEÆ.

Sep. connate, forming a tube. Stam. 10. Filaments connate into a tube below and adnate to the stalk of the ovary. Caps. usually stalked.—Pet. 5, clawed in all our plants. No stipules

* *Two or more imbricated opposite scales at the base of the calyx.*

1. Dianthus. Cal. 5-toothed. Styles 2. Caps. 1-celled, many-seeded, 4-valved at top. Seeds peltate, convex above, concave beneath and more or less keeled.

** *No scales at the base of the calyx.*

2. Saponaria. Cal. 5-toothed, terete. Styles 2. Caps. 1-celled, 4-valved at top. *Seeds globular or reniform.*

[3. Cucubalus. Cal. 5-toothed. *Styles* 3. *Caps. a globose 1-celled berry.* Seeds reniform.]

4. SILENE. Cal. 5-toothed. *Styles* 3. *Caps.* more or less completely 3-celled, 6-*valved at top.* Seeds reniform.—(Rarely 5 styles and 5-valved caps. in *S. maritima.*)

5. LYCHNIS. Cal. 5-toothed. *Styles* 5. *Caps.* 1- *or half* 5-*celled, opening at the top with* 5 *or* 10 *teeth.*

Suborder II. ALSINEÆ.

Sep. distinct. Stam. free, inserted into a more or less evident hypogynous ring. Caps. sessile. No stipules.

* *Valves of the caps. the same in number as the styles.*

6. SAGINA. Sep., entire pet. (or none), styles and valves of caps. each 4—5. Seeds reniform, wingless. Stam. 4—10.

7. HONKENEJA. Sep. 5. Pet. 5, large. Stam. 10, alternating with glands. Styles and valves 3. Seeds few, large.

8. MINUARTIA. Sep. and pet. 5 or 4. Styles and valves 3 (or 4). Seeds many, with a naked hile.—L. linear.

9. CHERLERIA. Sep. 5. Pet. 0 or 5, minute. Stam. 10, outer ones opposite to the sep. and springing from an oblong emarginate glandular base. Styles and valves 3. Seeds few, small.

** *Valves of the caps. bifid or twice as many as the styles.*

10. ARENARIA. Sep. 5. *Pet.* 5, *entire* or slightly emarginate. Stam. 10 or rarely 5. Styles 3. Caps. 6-valved. Seeds many.—L. broad.

11. HOLOSTEUM. Sep. 5. *Pet.* 5, *jagged at the end.* Stam. 5 or 3 or 4. Styles 3. Caps. subcylindrical, many-seeded, opening at the end with 6 teeth.

12. STELLARIA. Sep. 5. Pet. 5, bifid. *Stam.* 10 (or fewer). Styles 3. Caps. opening with 6 valves or teeth, many-seeded.

13. MALACHIUM. Sep. 5. Pet. 5, bifid or entire. Stam. 10 *Styles* 5. *Caps. opening with* 5 *bifid valves.*

14. CERASTIUM. Sep. 5. *Pet.* 5, *bifid.* Stam. 10 or 5 or 4. Styles 5 or 4. Caps. tubular, opening at the end with 10 teeth.—In *C. lapponicum* the styles are mostly 3.

15. MOENCHIA. Sep. 4, erect. *Pet.* 4, *entire.* Stam. Caps. many-seeded, opening at the end with 8 or 10 teetl

Suborder III. POLYCARPEÆ.

Sep. distinct. Stam. free. Caps. sessile. *Stipules scarious.*

16. POLYCARPON. Sep. keeled at the back, hooded at th end. *Pet.* 5, *emarginate.* Stam. 3—5. Styles 3, short Fr. 1-celled, many-seeded.

17. ALSINE. Sep. 5. Pet. 5, entire, usually as long a the calyx. Stam. 5—10. Styles 3 or 5. Fr. 3—5-valved, many-seeded; valves fewer than or alternate with the sepals.

18. SPERGULA. Sep. 5. Pet. 5, entire, as long as the calyx. Stam. 5—10. Styles 5. Fr. 5-valved, many-seeded; the valves opposite to the sepals.

Suborder IV. SCLERANTHEÆ.

Sep. connate, forming a hardened tube enclosing the 1-seeded capsule. No stipules.

19. SCLERANTHUS.[1] Tube of calyx vase-shaped, contracted at the mouth by a glandular ring; limb 5-fid. Pet. 0. Stam. 10 or rarely 5, inserted in the throat of the calyx. Styles 2. Fr. membranous.

Suborder I. *Sileneæ.*

1. DIAN'THUS *Linn.* Pink.

* *Fl. clustered.*

1. *D. prólifer* (L.); *fl.* in a dense cluster enveloped in membranous bracts, *cal.-scales membranous pellucid* the two outer ones shorter mucronate, inner ones blunt about equalling the calyx, st. glabrous, l. all linear, seeds boatshaped with a longitudinal membrane in the hollow rough pointed at one end.—*E. B.* 956. *R.* vi. 247.—St. 1—1½ ft. high, erect, usually simple. Fl. expanding one at a time, small. Pet. rose-coloured, obcordate. Cluster quite inclosed by brown dry scales.—Sandy and gravelly places, very rare. A. VII. E.

[1] Benth. & Hook. Gen. Plant. place *Scleranthus* in *Illecebraceæ*, which order they include in the Division *Monochlamydeæ*.—H. & J. G.

2. *D. Armēria* (L.); fl. close together, *cal.-scales and bracts lanceolate-subulate* herbaceous downy ribbed equalling the tube, st. and linear l. downy, seeds nearly flat on one side slightly hollowed and with a longitudinal keel in its middle rough pointed at one end.—*E. B.* 317. *R.* vi. 249.—St. 1—2 ft. high, erect, branched. Pet. rose-coloured, speckled with white dots, toothed. —Waste places, rare. A. VII. VIII. *Deptford Pink.* E.

** *Fl. solitary, one or more on the stem.*

*3. *D. plumārius* (L.); st. 2—5-flowered, *fl. solitary*, cal.-scales roundish-ovate shortly mucronate 4-fold shorter than the tube, *l. rough at the edge* linear-subulate, *pet. digitate multifid as far as the middle* with the central entire part obovate *downy*, barren st. procumbent rooting much branched, *seeds flat orbicular* with a point on one side.—*R.* vi. 257. *E. B. S.* 2979.—Flowering-stems 6—12 in. high. Calyx-teeth ciliate at the margin, slightly shorter than the capsule. Fl. pale pink, sometimes white, fragrant.—Old walls. P. VI. *Common Pink.* E.

*4. *D. Caryophyl'lus* (L.); fl. solitary, cal.-scales broadly obovate pointed three fourths shorter than the tube, *l. with smooth edges* linear, *pet. crenate-dentate* ovate *glabrous*, barren st. elongate procumbent branching, seeds pyriform nearly flat.—*E. B.* 214. *R.* vi. 268.—Fl.-stems 12—18 in. high. Calyx-teeth not ciliate, longer than the capsule. Fl. pink, fragrant.—Old walls, Kent. P. VII. VIII. *Clove Pink.* E. S.?

[*D. gal'licus* (Pers.), resembling Sp. 5 in habit and stature but with base of st. clothed with minute papillate hairs, cal. longer cylindrical, and pet. deeply and irregularly cut into sublinear obtuse lobes, occurs in Jersey, but its status is uncertain.]

5. *D. gratianopolitānus* (Vill.); st. mostly single-flowered, cal.-scales adpressed roundish shortly pointed three fourths shorter than the tube, *l. with rough edges* linear, *pet.* obovate *crenately-cut bearded*, barren stems long procumbent branching, seeds ovate pointed at one end.—*E. B.* 62. *R.* vi. 265.—*D. cæsius* (Sm.) ed. viii.—Flowering-stem 6—8 in. high. Fl. pale rose-colour, fragrant.—Limestone cliffs at Cheddar, Som. P. VI. VII. *Cheddar Pink.* E.

6. *D. deltoides* (L.); fl. solitary, *cal.-scales* usually 2 ovate with a subulate point ½ *as long as the tube*, l. linear-lanceolate the lower blunt rough at the edges and keel, *stem-l.* acute *and* as well as the st. *pubescent-asperous*, pet. obovate-dentate, barren st. short procumbent simple (?), seeds obovate flat netted-rugose.—*E. B.* 61. *R.* vi. 263.—Flowering-stems 6—12 in.

high, branched. Calyx-teeth lanceolate, minutely ciliate. Fl. rose-coloured, with a darker circle round the mouth, scentless. —Hilly pastures, rare. P. VI.--IX. *Maiden Pink.* E. S.

2. SAPONA'RIA *Linn.* Soapwort.

*1. *S. officinális* (L.); fl. in corymbose cymes, cal. cylindrical, pet. retuse crowned, l. elliptic-lanceolate ribbed glaucous, st. erect.—*E. B.* 1060. *R.* vi. 245. *St.* 6. 10.—St. 1—3 ft. high, stout, leafy. Fl. flesh-coloured or pale pink, large. Upper l. connate and sheathing. Upper part of st. sometimes pubescent (*puberula,* Wierzb.).—Hedges near villages, but probably introduced. Banks of streams on the borders of Wales, perhaps indigenous. P. VIII. E.

[*S. Vaccaria* (L.); st. much branched above, cal. with 5 angles, has been found as an escape.]

3. CUCU'BALUS *Linn.*

[*C. baccifer* (L.); st. branched spreading, l. ovate acute, cal. bellshaped, pet. distant.—*E. B.* 1577.—Fruit fleshy.—Isle of Dogs near London, not native.—P. VIII.]

4. SILE'NE *Linn.* Catchfly.

1. *S. ang'lica* (L.); *racemes terminal,* fl. alternate, cal. hairy with setaceous teeth ovate when in fruit, pet. slightly cloven or entire obovate, l. lanceolate lower ones spathulate.—*E. B.* 1178. —Hairy and viscid. St. 6—12 in. high, simple or branched, erect. Fl. solitary, secund, axillary, white or reddish. Fr.-st. often reflexed.—β. *S. quinquevulnera* (L.); fl. white with a large crimson spot on disk of each petal. *E. B.* 86.—Sandy and gravelly fields. About forms of this, see *J. of B.* xviii. 146. A. VI.—X. *English Catchfly.* E. S. I.

2. *S. nútans* (L.); pubescent, glandular-viscid above, *panicle* secund *with drooping* trichotomous *opposite* 3—7-*flowered branches, cal. ventricose* with acute teeth, pet. bifid crowned, segments linear, lower l. lanceolate-spathulate, stem-l. sessile lanceolate, carpophore scarcely half as long as the capsule, teeth of the capᵃ. reflexed.—*E. B.* 465.—St. $1\frac{1}{2}$ foot high. Fl. white, most expanded and sweetest in the evening.—β. *S. paradoxa* (Sm.): rt.-l. roundly spathulate mucronate with long hafts.— On limestone and chalky places. β. Dover Cliffs. P. VI. VII. *Nottingham Catchfly.* E. S.

[*S. ital'ica* (Pers.); pubescent, *panicle nearly erect* with opposite trichotomous viscid branches, *cal. clavate* with blunt teeth, pet. bifid not crowned, segments broad, lower l. lanceolate spathulate, stem-l. linear-lanceolate, *carpophore as long as capsule.*—*S. patens* E. B. S. 2748.—St. 2 ft. high.—An escape P. VI. VII.] E. S.

3. *S. Otites* (Wibel); *panicle long* with opposite tufted *whorled-racemose* branches, whorls many-flowered, ped. glabrous, cal. faintly veined smooth with blunt teeth, *pet. linear undivided* not crowned, l. lanceolate-spathulate, stem-l. small linear erect.—*E. B.* 85.—*Fl. subdiœcious,* small, yellowish. Caps. sessile. St. viscid at about the middle, 1 foot high. L. mostly radical.—Sandy and gravelly places in Suff., Norf. and Cambridgeshire. P. VI. E.

4. *S. Cucúbalus* (Wibel); panicle terminal, fl. many drooping, *cal. inflated* netted, pet. deeply cloven scarcely ever crowned, segments narrow, l. elliptic-lanceolate, stem erect.—*E. B.* 164. *S. inflata* (Sm.) ed. viii.—Glabrous, smooth. St. 2—3 ft. high. No barren procumbent stems. Inflorescence between corymbose and panicled. Pet. white. Cal. inflated especially below, its mouth narrower than its base. *Bracts scarious.* Branches of panicle unequal. Scented at night. Sometimes (*S. puberula,* Jord.) the st. and l. are rough with hairs and cal. downy.—Fields and roadsides. P. VI.—VIII. *Bladder Campion. White-bottle.* E. S. I.

5. *S. marit'ima* (With.); panicle terminal, *fl. few usually solitary erect,* cal. inflated netted, *pet. crowned,* segments broad, l. lanceolate or ovate-lanceolate, *barren st. spreading decumbent,* fl.-shoots ascending.—*E. B.* 957.—Barren procumbent shoots forming a cushion. Fl. larger than those of Sp. 4. Cal. elliptic, its mouth broader than its base, mostly inflated above the middle. Bracts herbaceous.—Near the sea, also by alpine rills. P. VI.—VIII. E. S. I.

6. *S. con'ica* (L.); *st. erect forked,* fl. from the forks or terminal, *cal. with* 30 *furrows* conical in fruit, teeth subulate acute, pet. obcordate crowned, l. linear subulate downy, *caps. oblong-ovate.*—*E. B.* 922.—St. 3—12 in. high, simple or branched. Cal. of the flowers conical-tubular, rounded below; of the fruit very broad at the base. Carpophore very short. Fl. reddish.—Sandy fields, rare. A. V. VI. E.

7. *S. noctiflóra* (L.); st. erect repeatedly forked, fl. from the forks or terminal, *cal.* veined and *with* 10 hairy glandular *ribs* in fruit elliptic-oblong, teeth long subulate, pet. deeply bifid

crowned, l. lanceolate lower ones obovate, *caps. ovate.*—*E. B.* 291. *St.* 3. 10.—Resembling *Lychnis vespertina.* St. about 1 foot high, downy and glandular. Caps. opening with 6 patent teeth. Carpophore very short. Fl. reddish white, rather large, sweet-scented in the evening; pet. rolled up by day; peduncles glandular.—Sandy and gravelly fields, rare. A. VII. VIII. E. S. I.

[*S. Arméria* (L.); pan. forked many-fl. level-topped, pet. notched each with a double awlshaped scale, cal. and l. smooth, caps. clavate, st. viscid.—*E. B.* 1398.—Yalding, Kent. Between Par and Fowey Point, Cornwall. A. VII.] E.

8. *S. acaúlis* (L.); *st. densely tufted* and much branched, *fl. solitary*, peduncles and cal. glabrous, cal. bellshaped with 10 striæ, teeth ovate blunt, pet. slightly notched crowned, l. linear ciliate below.—*E. B.* 1081. *R.* 5084. Forming broad dense tufts 2—3 in. high. Fl. purple or white, upon longish solitary stalks, sometimes nearly sessile. Caps. twice as long as the calyx. Plants somewhat diœcious.—On the higher mountains. P. VII. VIII. *Cushion Pink.* E. S. I.

5. Lych'nis *Linn.* Campion.

* *Cal. tubular, not inflated; teeth short.*

1. *L. Viscária* (L.); *pet. emarginate* crowned, *st.* glabrous *viscid* below the joinings, l. lanceolate glabrous the margins woolly at the base, fl. racemose-panicled somewhat whorled, *carpophore* ½ *the length of the capsule.*—*E. B.* 788.—St. simple, 1 foot high. Fl. large, rose-coloured. *Caps. 5-celled when young. Seeds* reniform, minute, *acutely tubercled.*—Dry rocks, very rare. P. VI. E. S.

2. *L. alpína* (L.); *pet. cloven* scarcely crowned, *st.* glabrous *not at all viscid*, l. linear-lanceolate glabrous sometimes minutely ciliate at the base, fl. corymbose, *carpophore* ½ *the length of the capsule.*—*E. B.* 2254.—St. simple, 5—6 in. high. Fl. small, rose-coloured, crown scarcely more than 2 small tubercles upon each petal. *Caps. 5-celled when young. Seeds* reniform, minute, *bluntly tubercled.*—Mountains. Glen Isla, Forfar (3200 ft.). Hobcarten Crag, Cumb. (2000 ft.). P. VI. VII. E. S.

3. *L. Flos-cucúli* (L.); *pet. deeply 4-cleft* crowned, segments linear palmately diverging, cal. with short teeth, st. with few deflexed hairs, l. lanceolate the lower ones narrowed below, fl. loosely panicled, caps. 1-celled, teeth 5, *carpophore very short.* —*E. B.* 573.—St. viscid and brownish olive, 1—2 ft. high.

Pet. rose-coloured, the crown bipartite; segm. subulate erect, usually with an acute tooth on the middle of the outer margin. Cal. 10-ribbed.—Moist places. P. V. VI. *Ragged Robin.* E. S. I.

** *Cal. inflated; teeth falling short of the petals.*

4. *L. al'ba* (Mill.); *pet.* half-*bifid* crowned, st. villose, l. peduncles and cal. hairy, l. ovate-lanceolate, fl. dichotomously panicled diœcious, *calyx-teeth of the fertile fl. linear-lanceolate, long, caps. conical, teeth* 10 *straight.—E. B.* 1580. *St.* 23. 9. *L. dioica* β L. *L. vespertina* (Sibth.) ed. viii.—St. 1—2 ft. high. Fl. white, very rarely reddish. Calyx of barren fl. obovate-oblong; of fertile fl. ovate, teeth twice as long as those of *L. dioica.* Carpophore broad, short.—Fields. B. (?) VI.—IX. *White Campion.* E. S. I.

5. *L. dioïca* (L.); pet. half-bifid crowned, st. l. and cal. villose, l. ovate acute, fl. dichotomously panicled subdiœcious, *calyx-teeth of the fertile fl. triangular, caps. nearly globular, teeth* 10 *recurved.—E. B.* 1579. *St.* 23. 8. *L. diurna* (Sibth.) ed. viii.—*Forms a tuft of decumbent leafy barren shoots.* St. 1—2 ft. high. Fl. red, very rarely nearly white. Carpophore narrow, short. Sp. 4 and 5 vary in colour from red to white and white to red.—Damp hedgebanks. P. V. VI. *Red Campion.* E. S. I.

*** *Cal. with long leaflike narrow coriaceous lobes exceeding the petals.*

6. *L. Githágo* (Scop.); *pet. entire or emarginate crownless, calyx-teeth longer than* the tube exceeding *the petals,* fl. solitary upon long stalks.—*Agrostemma* L. *E. B.* 741. *St.* 5. 6.—Fl. large, purple. St. dichotomous, 2—3 ft. high. Cal. ribbed, with 5 linear-lanceolate constantly erect-patent very long segments. Styles downy. Caps. 5-toothed.—Cornfields. A. VI.—VIII. *Corn-Cockle.* E. S. I.

Suborder II. *Alsineæ.*

6. Sagi'na *Linn.* Pearlwort.

* *Sep., stam., styles and valves of caps. usually* 4; *pet. very small or wanting.*

1. *S. procum'bens* (L.); branches long procumbent *from a central rosette, l. linear awned,* sep. blunt slightly shorter than the capsule, *apex of the ped. reflexed* after flowering *ultimately*

erect.—*E. B.* 880. *R.* v. 201. *St.* 30. 3.—Glabrous. Central st. never lengthening nor flowering; branches axillary, often rooting, usually with fasciculate leaves. Pet. small blunt, often wanting. A fifth part is occasionally added to the fl., in which case it is distinguished from *S. saxatilis* by its cal. spreading when in fruit and *styles reflexed* during flowering. A fleshy maritime form is *S. maritima* (Gren.).—*β. spinosa* (Gibs.); l. longer and narrower very minutely spinose-ciliate on the edges.—Waste ground. P. V.—IX. E. S. I.

2. *S. apet'ala* (Ard.); st. and branches ascending, *l. linear awned,* sep. blunt shorter than the capsule hooded ultimately spreading in the form of a cross, ped. always erect.—*E. B.* 881. *R.* v. 200.—Central st. lengthening flowering and together with the branches erect, never rooting; upper part of st., ped., and cal. often bearing glandular hairs. Pet. very minute, inversely wedge-shaped and truncate. Caps. conical-ovoid, subpeltate below, stalked.—*β. prostrata* (Gibs.); *prostrate,* branching from a rosette which lengthens into a stem and flowers. [*γ. S. Reuteri* (Boiss.); st. much branched, ped. short densely glandular, sep. usually appressed.]—Walls and dry places. *β.* Common on gravel walks. [*γ.* Worc. Heref. Pembr.] A. V.—IX. E. S. I.

3. *S. ciliáta* (Fr.); st. long, branches diffuse or spreading, l. linear awned, *outer sep. pointed scarcely shorter than and adpressed to the mature caps. their tips patent.*—*Sy. E. B.* 247. *R.* v. 200. *S. patula* Jord.—Glabrous. Central st. flowering and together with the branches erect or ascending, not rooting. Cal. and tips of ped. sometimes with gland-tipped hairs. Caps. ovate-attenuate, rounded below, stalked.—*S. ambigua* (Lloyd) is probably a maritime form.—Dry places and sandy heaths. A. V. VI. E. S. I.

4. *S. marit'ima* (Don); *central st. long forked,* branches ascending, *l. fleshy blunt* or apiculate rounded at the back glabrous, sep. blunt about equalling the capsule ultimately spreading slightly, *ped. erect.*—*E. B.* 2195. *S. stricta* Fries.—St. often purple, brittle. The central stem produces flowers and is erect, or in luxuriant plants more or less procumbent. Sep. concave with incurved tips. Caps. ovate, rounded below.—A much-branched prostrate form is (*β*) *S. debilis* (Jord.); its calyx usually exceeds the caps., and all its stems are often prostrate and spring from a false rosette.—Another very much branched state forming dense tufts, with short joints and shorter linear plane-convex l., is (*γ*) *S. densa* (Jord.).—On the seashore. Fries states that his plant sometimes occurs upon

mountains in Norway; and G. Don seems to have found it on Ben Nevis [var. *alpina* Sy.[1]]. A. V.—IX. *Sea Pearlwort.* E. S. I.

5. *S. Boyd'ii* (Buch.-White); cæspitose, st. erect, l. densely imbricate linear fleshy rigid strongly recurved shortly mucronate, fl. 4–5-merous, sep. broadly ovate blunt with narrow membranous margins, pet. 0, ovary globose flattened at the top, "caps. globose shorter [than] and covered by the sep.," ped. short stout slightly curved.—*J. of B.* 1892, t. 326 b.—Much branched below, forming dense tufts, glabrous dark green shiny. A remarkable plant which requires further study. We have not seen fruit.—Braemar, *Mr. W. B. Boyd.* P. VI. S.

** *Sep., pet., styles and valves usually* 5. *Stam.* 10. SPERGELLA.

6. *S. Linnæ'i* (Presl); *central st. short and barren*, l. linear mucronate glabrous, st. *ped. and cal. glabrous*, pet. shorter than caps. longer than the calyx.—*Spergula saginoïdes*, L. *E. B.* 2105. *S. saxatilis* (Wimm.) ed. viii.—St. prostrate, slightly rooting, many. Ped. long, their tips reflexed after flowering ultimately erect. Caps. rather longer than the calyx, sometimes twice the length. Closely resembling *S. procumbens*, but distinguished by the valves of its capsule being much more narrowed upwards, sep. adpressed and narrower, pet. longer, *styles not reflexed.*—Highland mountains. P. VI.—VIII. S.

7. *S. nivális* (Fries); central st. and branches ascending cæspitose, l. subulate mucronate glabrous, ped. short straight, sep. very blunt adpressed to the ripe capsules, pet. rather exceeding cal. but falling short of caps. entire.—*Sy. E. B.* 250 (bad).—St. and branches dividing repeatedly (no true rosette), not rooting, 1—1½ in. long, forming a dense tuft. Fl. divided in fours or fives. Ped. wholly straight. Sep. white with diaphanous edges.—Tops of Highland mountains, very rare. P. VIII. S.

8. *S. subuláta* (Presl); *l. awned* linear often ciliate, *ped. and calyx glandular-hairy*, pet. about as long as the caps. longer than the calyx.—*Spergula* Sw. *E. B.* 1082.—St. procumbent. Ped. very long, the tip slightly reflexed after flowering, ultimately erect. Caps. ovate-attenuate, rounded below, sessile.—Dry gravelly and sandy places. P. VI.—VIII. E. S. I.

9. *S. nodósa* (Fenzl); l. subulate glabrous *upper l.* shorter *fasciculate, pet. much longer than the calyx*, ped. always erect. —*Spergula* L. *E. B.* 694.—Primary stem short, not flowering;

[1] Mr. Druce's plant from Cairngorm (Ann. Scot. Nat. Hist. 1892, p. 273) which has been referred to this, has a central rosette, short ped. and pet. equalling the cal.—H. & J. G.

lateral stems procumbent at the base then ascending, 2—6 in. long. Fl. terminal, 1, 2 or 3 together, white, conspicuous. Whole plant often quite glabrous. Sometimes (*S. glandulosa* Bess.) the upper parts of the st., the connecting membrane of the l., and the base of the cal. are glandular-hairy.—Wet and sandy places. P. VII. VIII. *Knotted Spurrey.* E. S. I.

7. Honkene'ja *Ehrh.*

1. *H. peploïdes* (Ehrh.); l. sessile ovate acute fleshy glabrous 1-veined, pet. obovate, sep. ovate blunt 1-veined shorter than the petals.—*Arenaria* L. *Ammodenia* Gm. (name only). *Halianthus* Fr. *Minuartia* Hiern. *E. B.* 189.—St. forked, procumbent, rhizomatous. Fl. from the forks of the stem, frequently diœcious or polygamous. Caps. large, globose. Seeds few, large.—Sandy sea-coasts. P. VI.--IX. E. S. I.

8. Minuar'tia *Linn.* (*Alsine*, Wahl. ed. viii. not Linn.)

1. *M. stric'ta* (Hiern); *l.* filiform *veinless*, fl.-shoots erect naked above, pet. equalling the cal. oblong-oval attenuate below, *sep.* ovate-lanceolate *acute* 3-veined (when dry), *ped. terminal* 1—3 *very long.*—*Aren. uliginosa* Schlecht. *E. B. S.* 2890.—St. prostrate, cæspitose.—Widdy-bank Fell, Teesdale. P. VI. E.

2. *M. ver'na* (Hiern); *l.* linear-subulate *acute* 3-*veined, pet. exceeding the calyx* rounded-obovate attenuate below, sep. ovate-lanceolate acute 3-veined with a membranous margin, peduncles 1- or many-flowered.—*E. B.* 512.—St. 3—4 in. high. L. usually not adpressed and mostly with a minute point. Bracts acute.—*β. Aren. Gerardi* (Willd.); l. subulate bluntish not apiculate, pet. elliptic shortly clawed scarcely longer than the calyx. *R.* v. 208. L. usually pressed close to the stem.—Rocky places in mountainous districts. *β.* On the hills above Kynance Cove near the Lizard Point, Cornwall. P. V.—IX. E. S. I.

3. *M. rubel'la* (Hiern); *l.* linear-subulate *blunt* 3-veined, *pet.* obovate attenuate below *shorter than the calyx*, sep. ovate-lanceolate acute 3-veined with a membranous margin, peduncles 1-flowered.—*E. B. S.* 2638. *Aren. sulcata* Schlecht.—Like a *Sagina*. St. many, short. Bracts blunt. Flowering shoots terminal, downy, nearly always 1-fl., about 1 in. long, with 1—3 pairs of leaves. Ped. longer than calyx. Styles and valves of caps. 3—5.—Tops of Scottish mountains, very rare. [Serpentine hills, Shetland, *Mr. W. H. Beeby*.] P. VII. VIII. S.

4. *M. leptophyl'la* (Groves); l. subulate-acute 3-veined, *pet. ovate attenuate below shorter than the calyx,* sep. lanceolate-subulate 3-veined with a membranous margin.—*E. B.* 219. *Als. tenuifolia* (Wahl.) ed. viii.—St. slender, 4—6 in. high, much branched, forked, with flowers in the forks.—Glabrous. Sometimes the upper parts (*Als. laxa,* Jord.), or the cal. alone (*Als. hybrida* Vill.), bear patent gland-tipped hairs.—Sandy and chalky places, rare. A. V. VI. E. I.*

9. Cherle'ria *Linn.*

1. *C. sedoïdes* (L.).—*E. B.* 1212. *Minuartia* Hiern.—Pet. generally wanting. Fl. solitary, on short stalks. St. very many, forming a dense mass close to the ground. L. very many, linear-subulate, finely ciliate.—Summits of mountains. P. VI.—VIII. S.

10. Arena'ria *Linn.* Sandwort.

1. *A. triner'via* (L.); l. ovate acute ciliate stalked 3—5-veined the upper ones sessile, stam. 10, pet. shorter than the calyx, sep. long-lanceolate acute 3-ribbed the intermediate rib strongest and rough, *seeds* smooth *appendaged.*—*E. B.* 1483. *R.* v. 216.—St. about a foot high, weak, branched, downy. Fl. solitary from the forks of the stem and axils of the upper leaves. Ped. ultimately spreading and curved just below the fruit. Lateral veins of sep. often very faint. Distinguished by the appendage to the hile of its seeds.—Damp shady places. A. V. VI. E. S. I.

2. *A. serpyllifólia* (L.); *l. ovate* acute roughish sessile, *pet. shorter than the calyx, sep.* ovate-lanceolate acute 3—5-veined hairy on the veins, fr.-st. erect or patent straight longer than the *ampullaceous* caps. which exceed the sepals.—*E. B.* 923.—St. much branched, 3—6 in. long. Fl. from the forks of st. or axils of leaves. Pet. ovate, narrowed below. Ripe capsule brittle. Sometimes [var. *viscidula* Roth] with viscid hairs on the upper part, *A. viscida* Lois.—I cannot distinguish *A. Lloydii*[1].—Dry places and walls. A. VI.—VIII. E. S. I.

3. *A. leptoclados* (Guss.); l. small ovate acute sessile, pet. shorter than the cal., sep. lanceolate acute 3-veined hairy on the veins, *fr.-st.* patent *curved at the top* or ultimately straight longer than the *ovoid-oblong caps.* which exceed the sep.—*E.E.S.*

[1] *A. Lloydii* Jord. (*A. serpyllifolia* var. *macrocarpa* Lloyd) is a condensed seaside form with broader leaves and more strongly-veined eglandular sepals.—H. & J. G.

2972.—St. much branched. Fl. as in Sp. 2. Much more slender than in Sp. 2; caps. small and often nearly oblong flexible when ripe; seeds smaller.—Dry places and walls. A. VI.—VIII. E. I.

4. *A. ciliáta* (L.); *l. spathulate ciliate, pet. exceeding the calyx, sep.* ovate-lanceolate with 3 prominent ribs.—*E. B.* 1745.—St. long, much branched, prostrate, rough, with very short deflexed hairs, angular when dry. Fl. 1—5, terminal, somewhat panicled. Pet. ovate, slightly clawed.—Limestone cliffs of Ben Bulben range, especially King Mountain, co. Sligo. P. VI. VII. I.

5. *A. goth'ica* (Fr.); [l. ovate to ovate-lanceolate *acuminate* slightly ciliate towards the base], pet. exceeding the cal., sep. ovate-lanceolate keeled falling short of the caps. Caps. ovoid *constricted at the top* opening by revolute teeth.—[Fl. Dan. Suppl. i. t. 15.—St. short much branched slightly hairy. Fl. 1—3, terminal. Ped. covered with very short hairs. Sep. glabrous obscurely ribbed. Pet. oblong. Seeds dark brown tuberculated.—Ingleborough, Yorks. P. V.—IX. E.]

6. *A. norvégica* (Gunn.); *l. spathulate obovate fleshy not ciliate,* pet. exceeding the calyx, *sep.* ovate-acute *obscurely 3-ribbed* glabrous falling short of the oblong caps. which is *not constricted at the top* and opens by *erect* teeth.—*E. B. S.* 2852.—St. short, much branched, procumbent, nearly smooth, angular when dry. Fl. 1—3, terminal. Ped. with very short deflexed hairs. Pet. ovate, slightly clawed. Seed dark brown, tuberculated.—Inchnadamph, Sutherland. Unst, Shetland. P. VII. VIII. S.

11. HOLOS'TEUM *Linn.*

1. *H. umbellátum* (L.); fl. umbellate, peduncles pubescent viscid, pedicels reflexed after flowering, l. elliptic or long and acute.—*E. B.* 27. *R.* v. 221.—About 6 in. high.—On old walls and dry places in Norfolk and Suffolk. A. IV. E.

12. STELLA'RIA *Linn.* Stitchwort.

* *Seeds on a linear long columella.*

1. *S. nem'orum* (L.); st. ascending downy above, l. stalked heartshaped, upper l. ovate sessile, cyme lax panicled, pet. deeply bifid twice as long as the lanceolate sepals, caps. exceeding the calyx.—*E. B.* 92. *R.* v. 252.—St. 1—1½ foot high. L. large, rough on the upper surface, ciliate. Sep. with narrow scarious margins.—Damp woods, chiefly in the North. P. V. VI. *Wood Stitchwort.* E. S.

** *Caps. rounded below or scarcely if at all narrowed; columella very short.*

2. *S. média* (Vill.); *st.* procumbent and ascending *with a hairy line, l. ovate shortly pointed stalked, upper l. sessile,* fl. axillary and terminal, *fl.-pedicels usually not longer than the cal., fr.-ped. curved downwards and wavy,* pet. deeply bifid not exceeding the ovate-lanceolate single-ribbed glandular-pilose sepals, caps. oblong longer than the calyx.—*E. B.* 537. *R.* v. 222.—Very variable in length of st. and joints, size of l., number of stam. (3—10), and length of styles. Sep. with a narrow scarious margin, glabrous or with long hairs. Fr.-st. reflexed, often scarcely exceeding the leaves. Seeds with round tubercles. L. glabrous with broad ciliate petioles.—β. *Alsine pallida* (Dum.); pet. 0, styles 0, stig. short arcuate, stam. 3, seeds small bluntly and minutely tubercled. *S. Boræana* (Jord.).—γ. *S. neglecta* (Weihe)[1]; l. larger with longer stalks, upper l. sessile lower subcordate, stam. 10, seeds with prominent rounded tubercles. A. III.—IX. *Common Chickweed.* E. S. I.

3. *S. umbro'sa* (Opiz); st. procumbent and ascending with a hairy line, *l. narrowed gradually into long points, fl.-ped. twice the length of the calyx,* fr.-ped. deflexed at the base, otherwise *straight, ultimately erect,* cal. more narrowed below than in Sp. 2, sep. lanceolate acute tubercular, valves of the caps. narrower than in Sp. 2, seeds with prominent acute tubercles.—*S. Elizabethæ* Schultz.—St. much branched slender with autumnal barren shoots. Fr.-st. much exceeding l.—Moist and shady places. P. IV.—VII. E. S.

4. *S. Holos'tea* (L.); st. ascending angular with rough angles, *l. lanceolate-attenuate* acute with a rough margin and keel *all sessile,* cyme panicled, pet. half-bifid twice as long as the lanceolate very obscurely 3-veined sepals, caps. globose about as long as the calyx, bracts leaflike.—*E. B.* 511. *R.* v. 223.—St. 1—2 feet high, slender and procumbent at the base, thicker upwards. L. gradually narrowing from a little above the base to the very acute point. Fl. large, white, few, in a leafy cyme.—Woods and hedges. P. IV.—VI. *Greater Stitchwort.* E. S. I.

5. *S. palus'tris* (Retz.); *st.* erect weak angular *smooth, l. linear-lanceolate* acute quite *smooth sessile,* lower l. broader, fl. solitary or in a few-flowered lax panicled cyme, pet. bipartite exceeding the lanceolate 3-veined sepals, caps. oblong-ovate about as long

[1] Many authors combine this with Sp. 3. The name for the combined species would be *S. neglecta* (Weihe).—H. & J. G.

as the calyx, *bracts with* scarious and *glabrous margins*.—*E. B.* 825. *R.* v. 223. *S. glauca* (With.) ed. viii.—Usually glaucous. St. 6—12 in. high, leafy. Fl. rarely solitary. Pet. white, sometimes much exceeding the cal.; segments linear.—Marshy places, rather rare. P. V.—VII. E. S. I.

6. *S. gramin'ea* (L.); st. diffuse angular smooth, *l.* linear-lanceolate acute quite *smooth* ciliate below *sessile*, cyme lax panicled, pet. bipartite equalling or exceeding the 3-veined sepals, caps. oblong longer than the calyx, *bracts scarious ciliate*.—*E. B.* 803.—St. 1—2 feet high. Fl. smaller than those of Sp. 3 or 4, white. Shorter or longer pet. accompany an imperfection of the stam. or germen.—[*S. longifolia* (Fr.); *S. Friesiana* (Koch), has the upper part of its stem and the edges and keel of its leaves rough.]—Dry heathy and bushy places. P. V.—VIII. *Lesser Stitchwort.* E. S. I.

*** *Caps. narrowed below; hence the cal. has a funnelshaped base.*

7. *S. uliginósa* (Murr.); *st.* diffuse angular *glabrous*, *l.* oblong-lanceolate acute with a hard tip *glabrous* slightly *ciliate below*, fl. irregularly panicled lateral and terminal, pet. bipartite shorter than the lanceolate 3-veined sepals, caps. ovoid nearly equalling the calyx, *bracts scarious with glabrous* edges.—*E. B.* 1074.—Very variable in size, from about a foot long to 2 inches. Fl. in small cymes.—In wet places. P. V. VI. E. S. I.

13. MALA'CHIUM *Fries.* Great Chickweed.

1. *M. aquat'icum* (Fr.); st. decumbent and ascending angular covered with glandular hairs, l. cordate-ovate acuminate, fl. scattered solitary in the forks of the stem, pet. bipartite rather exceeding the calyx, caps. exceeding the calyx.—*R.* vi. 237. *E. B.* 538. *Stellaria* (Hook.).—Closely resembling *Stellaria nemorum.* Lowest l. and those of the barren branches stalked, others larger and sessile.—Usually in wet places. P. VII. VIII. E.

14. CERAS'TIUM *Linn.* Mouse-ear.

* *Root fibrous. Pet. not or but little exceeding the calyx.*

† Caps. curved.

1. *C. viscósum* (L.); l. ovate, *sep.* lanceolate very acute with a narrow membranous margin and *as well as the herbaceous bracts hairy throughout*, caps. cylindrical ascending twice as long as the calyx, fruitstalks about equalling the calyx.—*C. vulgatum*

Sm. *E. B.* 789. *R.* v. 220. *C. glomeratum* (Thuill.) ed. viii. —St. erect, glandular-hairy. Fl. in close cymes, longer than their stalks.—β. *C. apetalum* (Dum.); pet. 0, whole plant usually much more slender.—Fields and banks. A. IV.—IX. E. S. I.

2. *C. triviâle* (Link); l. oblong-lanceolate, *sep.* oblong-ovate bluntish and *as well as the bracts membranous at their margins and glabrous tips,* caps. cylindrical ascending twice as long as the calyx, fruitstalks at least as long as the calyx.—*R.* v. 229. *C. viscosum* Sm. *E. B.* 790. *C. vulgatum* Fries.—St. downy, mostly procumbent, some short and barren. Fl. larger than those of the last, in small terminal cymes the branches of which are often lengthened as the fr. ripens.—β. *C. holosteoides* (Fries); glabrous sides of st. alternately downy, l. dark smooth shining, fr. much larger. *St.* 63. 9.—In fields, and on mountains (var. *alpinum* M. & K.) with larger flowers; on seashores pentandrous and annual [var. *pentandrum* Sy.]. An extended form from Shetland (*C. longirostre* Wich.) has l. 1½ in. or more long and caps. nearly ¾ in. B. or P. IV.—IX. E. S. I.

†† Caps. nearly straight.

3. *C. semidecan'drum* (L.); l. broadly ovate, *sep.* lanceolate *broadly membranous at their margins and tips, bracts with their upper half membranous,* caps. cylindrical slightly inflated longer than the calyx, *fruitstalks* longer than the calyx *at first reflexed afterwards erect.*—*E. B.* 1630. *R.* v. 228.—St. erect or decumbent, downy, sometimes viscid. *Pet. with simple veins,* not distinctly notched. Lower l. with long linear hafts. Known by its half-membranous bracts.—Common in dry places. A. IV. V. E. S. I.

4. *C. pûmilum* (Curt. !); *l. spathulate, upper l. oblong,* sep. lanceolate acute with their tips and margins narrowly membranous, *uppermost bracts with an extremely narrow membranous margin,* caps. slightly curved upwards longer than the calyx, *fruitstalks short curved at the top declining from their base,* ultimately erect.—*Curt.* ii. 92. *C. glutinosum* Fr., *Fl. Dan.* 2537. —Viscid. St. branched at the root, afterwards nearly simple. *Pet. with branched veins,* notched. Lower l. with long linear hafts. Fl. in terminal forked cymes.—Dry banks in the South. A. IV. V. E.

5. *C. tetran'drum* (Curt.); l. oval or oblong, sep. lanceolate acute, their tips and margins narrowly membranous, bracts wholly herbaceous, caps. a little exceeding the cal. straight,

fruitstalks 2—4 *times as long as the caps. straight,* ultimately erect.—*E. B.* 166. *Curt.* ii. 93. *C. pumilum* Gren., Bor.—Viscid. St. cymose from the base. Bracts very broad, oval, rather acute or apiculate, or nearly round, leaflike. Pet. with branched veins, notched. Fl.-whorls of 4 or 5 parts.—Walls and sandy places near the sea. A. V.—VII. E. S. I.

** *Root truly perennial, with prostrate leafy shoots. Pet. much longer than the calyx.*

6. *C. arven'se* (L.); st. ascending prostrate below, *l. linear-lanceolate,* fl. many, sep. and *bracts* lanceolate slightly acute *with membranous margins and tips,* caps. at last longer than the calyx, seeds small acutely tubercled.—*E. B.* 93.—St. long. Fl. 3—14, in forked panicles. Fruitstalks erect, bent just under the calyx. St. and l. hairy.—β. *Andrewsii* (Syme); l. subglabrous rigid with a strong midrib; often 1-flowered.—In sandy, gravelly and chalky places, rare. β. Kerry, Aran I. P. IV.—VIII. E. S. I.

7. *C. arct'icum* (Lange)[1]; pubescence short, *st. prostrate cæspitose,* l. elliptic or lanceolate, fl. 1—3, sep. blunt with membranous margins, bracts herbaceous, caps. slightly narrowing straight, *seeds large rugose,* fruitstalks oblique patent.—*E. B.* 473. *C. latifolium* (L.) ed. viii.—Pubescence short, rigid, yellowish. Barren shoots usually long. L. variable in shape.—β. *Edmonstonii* (Beeby); l. roundish-ovate dark greenish purple, st. short densely leafy below.—Alpine parts of Wales and Scotland. β. Unst, Shetland. P. V. E. S.

8. *C. alpinum* (L.); hairy, st. ascending, l. ovate ovate-oblong or lanceolate, fl. few, sep. bluntish with membranous margins, bracts herbaceous their margins often narrowly membranous, caps. nearly cylindrical curved at the end, *seeds small acutely tubercled,* fruitstalks obliquely patent.—*E. B.* 472.—Pubescence long, simple. St. much branched below, then simple, elongated, prostrate or ascending. Fl. 1, 2 or 3 together, in a forked panicle, shorter than their stalks.—Alpine parts of Scotland and the North of England. P. VI.—VIII. E. S.

9. *C. lappon'icum* (Crantz); *st.* decumbent *with an alternate hairy line,* l. elliptic-oblong, ped. pubescent 1—3-flowered terminal, bracts herbaceous, styles mostly 3, caps. rather ex-

---

[1] It is not clear from the MS. whether it was the Author's final intention to combine this with *C. alpinum* as var. β. *compactum* or to keep it distinct. In view of more recent information we have thought it better to adopt the latter course, omitting the var. *compactum* of ed. viii. —H. & J. G.

ceeding the calyx.—*C. trigynum* (Vill.) ed. viii. *Stellaria cerastoides* L. *E. B.* 911.—St. 4—8 in. long, slender, leafless and much branched below. L. light green, glabrous (or hairy in *C. nivale* Don), subsecund and subfalcate. Fl. large, white. Teeth of caps. 6—10.—Highland mountains. P. VII. VIII. S.

15. Moen'chia *Ehrh.*

1. *M. quaternel'la* (Ehrh.); stam. 4.—*M. erec'ta* (Sm.) ed. viii. *Cerastium* Fenzl. *E. B.* 609. *R.* v. 227.—Glaucous. St. erect, glabrous, 1—4 in. high. L. opposite, linear-lanceolate, acute, rigid. Sep. with broad white membranous margins, acute.—Dry gravelly and sandy places. A. V. VI. E.

Suborder III. *Polycarpeæ.*

16. Polycar'pon *Loefl.*

1. *P. tetraphyl'lum* (L.); triandrous; pet. emarginate, stem-l. in fours, l. on the branches opposite.—*E. B.* 1031.—In young plants the l. are often all opposite.—Coasts of the South-west of England, rare. A. VI. VII. E.

17. Alsi'ne *Linn.*[1] [*Lepigonum* (Fries) ed. viii.]. Sand-Spurrey.

1. *A. rúbra* (Crantz); st. nearly terete, l. flat linear pointed, stip. triangular-ovate-prolonged mostly cut, *caps.* about equalling the cal. $\frac{1}{2}$—$\frac{2}{3}$ *shorter than the fr.-stalk, seeds* cuneate-obovate $\frac{2}{3}$ surrounded by a thickened border *none winged.*—*Sp. E. B.* 254.—St. procumbent (as in the other species). Pan. leafy. Pet. pink, about equalling the calyx. Seeds usually gibbous on one side. Smaller in all respects than the other species.—Sandy places. A. V.—IX. E. S. I.

2. *A. rupicola* (Hiern); st. terete, l. flattish fleshy pointed, stip. broadly ovate-prolonged mostly entire, *caps.* large equalling or slightly exceeding the cal. $\frac{1}{2}$—$\frac{2}{3}$ *shorter than the fr.-stalk, seeds* compressed pyriform nearly surrounded by a thickened border *none winged.*—*E. B. S.* 2977.—Root thick, woody. *L. fascicled*, short; pan.-l. very short, inconspicuous. Pet. pale pink exceeding cal. Stam. 10.—Near the sea. P. VI.—IX. E. I.

---

[1] See Mr. W. P. Hiern's paper in J. of B. xxxvii. (1899) p. 317, relative to the use of the name *Alsine* for this genus.

3. *A. salina* (Groves); st. compressed, l. flattish fleshy bluntish or slightly pointed, stip. broadly triangular-ovate entire, caps. exceeding the *cal. about as long as the fr.-stalk*, seeds compressed roundish nearly surrounded by a thickened border and some often with a broad scarious wing.—*E. B. S.* 2978. *L. salinum* (Kindb.) ed. viii.—L. long; pan.-l. sometimes long, sometimes inconspicuous. Ped. occasionally twice as long as the capsule. *Pet. pink with a white base* falling short of cal. Stam. less than 10.—*L. medium* (Fr.) has shortly ovate stip., caps. about equalling cal., seeds triquetrous smooth with a thickened border but rarely a few winged. *L. neglectum* (Kindb.) is glandular.—Near the sea. A. VI.—IX. E. S. I.

4. *A. marina* (Wahl.); st. compressed, l. fleshy bluntish, stip. broadly triangular usually entire, *caps. very large* often twice as long as the cal. but *scarcely ½ as long as the fr.-stalk, seeds* compressed roundish *nearly surrounded by a* thickened border within a *broad scarious wing.—E. B.* 958.—L. long, ½-terete; pan.-l. very short, inconspicuous. Caps. larger than in the other species. Seeds reddish. *Pet. pale pink,* with a white base. Stam. 10.—Muddy salt marshes. P. VI.—IX. E. S. I.

18. SPER'GULA *Linn.* Spurrey.

1. *S. arven'sis* (L.); l. linear convex above furrowed beneath *scarcely viscid,* fl. in forked panicles, fr.-stalks deflexed, seeds slightly compressed obscurely margined, *covered with club-shaped papillæ.—S. vulgaris* (Boenn.).—L. grass-green.—Cultivated land. A. VI.—VIII. E. S. I.

2. *S. sativa* (Boenn.); l. linear convex above furrowed beneath *very viscid,* fl. in forked panicles, fr.-stalks deflexed, seeds slightly compressed with a very narrow margin *covered with minute elevated points.*—L. grey-green. Seeds black.—Cultivated land. A. VI.—VIII. E. S. I.

[*S. pentan'dra* (L.); seeds broadly winged. A specimen in Dillenian Herb. supposed to have been collected by Sherard in Ireland. See Ann. Bot. iv. 378 and J. of B. 28. 302.]

Suborder IV. *Sclerantheæ.*

19. SCLERAN'THUS *Linn.* Knapwell.

1. *S. an'nuus* (L.); subdecandrous, *segm. of fr.-cal. patent acute* with a very narrow membranous margin, as long as their tube.—*E. B.* 351.—Styles exceeding the stamens. St. re-

peatedly dichotomous, green. Fl. green, often solitary in the forks of the stem, or densely corymbose.—A biennial state (*S. biennis* Reut.) is often taken for Sp. 2.—Sandy fields. A. VI.—VIII. E. S. I.

2. *S. peren'nis* (L.); decandrous, *segm. of fr.-cal. connivent blunt rounded* with a broad blunt membranous margin.—*E. B.* 352.—Styles usually falling short of the stamens. St. nearly simple or irregularly branched, procumbent, glaucous, at length reddish. Fl. variegated with green and white. L. erect, directed to one side.—Sandy fields in Norfolk and Suffolk. Stanner rocks, Radnorshire. P. VI.—VIII. E.

Order XV. MALVACEÆ.

Sep. 5 or 3 or 4, more or less connected below, often double, valvate in the bud. Pet. as many as the sepals, adnate to base of stam., twisted in the bud. Stam. many, connected at the base into a tube; anth. 1-celled, reniform, bursting transversely. Ovary formed by the union of several carpels round a common axis. Carp. 1- or many-seeded. Embryo curved with twisted or doubled cotyledons, albumen variable in quantity.—L. alternate, with deciduous stipules. Fl. axillary.

1. MALVA. Styles many. Cal. double, *outer 3-leaved,* inner 5-fid. Carpels in a ring round a thick axis, each 1-seeded.
2. ALTHÆA. Styles many. Cal. double, *outer 6—9-fid,* inner 5-fid. Carpels as in *Malva.*
3. LAVATERA. Styles many. Cal. double, *outer 3-lobed,* inner 5-fid. Carpels as in *Malva.*

1. MAL'VA *Linn.* Mallow.

1. *M. moschāta* (L.): st. erect, l. palmate *with* 5—7 *deep bipinnatifid lobes,* lower l. incise-crenate, stipules lanceolate acute, fruitstalks erect, outer sep. linear-lanceolate, *fruit hairy.*—*E. B.* 754. *R.* v. 169.—Fl. large, rose-coloured, on axillary single-flowered peduncles, crowded at the extremity of the stem and branches. Cal. hairy. St. 1—2 feet high. [Varies considerably in the cutting of the l.: the common British form, var. *laciniata* Lej., has all the l. deeply lobed with narrow segments; var. *heterophylla* Lej. has the lower roundish entire, the upper deeply lobed; var. *integrifolia* Lej. has all roundish and more or less entire.]—Gravelly places. P. VII. VIII. *Musk Mallow.* E. S. I.

[*M. Alcēa* (L.); outer sep. ovate acute, fr. glabrous; should be found in England.]

2. *M. sylves'tris* (L.); *st. erect, l.* palmate *with* 5--7 *deep crenate lobes,* stipules lanceolate, fruitstalks erect, outer sep. lanceolate, *fruit reticulate-rugose* its axis an acute cone with concave sides.—*E. B.* 671. *R.* v. 168.—Fl. large, in axillary clusters. Pet. much longer than the hairy calyx, purple. St. 2—4 ft. high. [Carp. sometimes hairy, var. *lasiocarpa* Druce.]—Roadsides and waste places. P. VI.—IX. *Common Mallow.* E. S. I.

[*M. nicæen'sis* (All.) resembling Sp. 2, but with pet. only twice as long as cal. and netted carp. with winged ridges, occurs as a casual.]

3. *M. rotundifólia* (L.); *st. decumbent,* l. roundish-heartshaped with 5—7 shallow acutely crenate lobes, stipules ovate acute, fruitstalks reflexed, *outer sepals* linear-lanceolate *shorter than the* ovate-acuminate *stellately hairy inner ones,* pet. 2 or 3 times as long as the calyx, fruit pubescent, carp. rounded on the edge smooth.—*E. B.* 1092. *M. vulgaris* Fries, *R.* v. 167.—Fl. small, purple. Carp. meeting each other with a straight line. Axis of the fr. rather large.—Waste places. P.? VI.—IX. *Dwarf Mallow.* E. S. I.

[*M. parviflóra* (L.), with short ped., broadly ovate-*mucronate* accrescent inner sep., pet. scarcely exceeding cal., carp. hairy transversely rugose with distinctly *winged edges,* and *M. verticilláta* (L.) E. B. S. 2953, with very short ped., broadly ovate-*acute* accrescent inner sep., pet. scarcely exceeding cal. and carp. glabrous nearly smooth with *squarish edges,* are occasionally found.]

*4. *M. pusil'la* (Sm.); outer as long as the glabrous but ciliate inner sepals, pet. scarcely exceeding cal., carp. margined netted rugose [with short hairs.]—*M. borealis* (Wallm.) ed. viii. *M. rotundifolia* Fr.—Carp. meeting each other with a toothed edge. Axis of fr. ½ as long as in Sp. 3. Much resembling Sp. 3.—Kent. Anthony, Cornw. Near London. A. VII.? E.

2. ALTHÆ'A *Linn.*

1. *A. officinális* (L.); l. soft on both sides crenate or crenate-serrate cordate or ovate 3—5-lobed, ped. axillary many-fl. falling short of the leaves, st. downy.—*E. B.* 147. *R.* v. 173.—St. 2—3 ft. high. Covered with soft velvety pubescence.—Marshes, particularly near the sea. P. VIII. IX. *Marsh-Mallow.* E. S.? I.

†2. *A. hirsúta* (L.); hispid cordate, lower l. reniform bluntly 5-lobed, upper l. palmate with 5 or 3 acute lobes, ped. axillary 1-flowered exceeding the leaves, st. hispid.—*E. B. S.* 2674. *R.* v. 172.—Stem and calyx very hispid.—Between Cobham and Cuxton, Kent. N. Somerset (*Hook.*). A. VI. VII. E.

3. LAVATE'RA *Linn.* Tree-Mallow.

1. *L. arbórea* (L.); *st. woody,* l. 7-angled plaited velvety, ped. axillary clustered 1-flowered shorter than the petioles.—*E. B.* 1841.—Fl. large, purplish rose-coloured with darker veins. St. 3—8 ft. high.—Maritime rocks, rare. B. VII.—IX. E. S. I.

*2. *L. sylves'tris* (Brot.); st. herbaceous erect or ascending hispid, lower l. orbicular-cordate, upper l. 5-lobed truncate below, l.-lobes triangular acute, ped. axillary 1-fl. shorter than petioles, fr. smooth its axis a cone with convex sides.—*J. of B.* xv. t. 191.—Fl. rose-purple. St. rarely prostrate. Much like *Malva sylvestris.*—Scilly Isles. Wareham, *Mansel Pleydell.* [Channel I.]. A. or B. VI. VII. E.

Order XVI. TILIACEÆ.

Sep. 4—5, valvate in the bud. Pet. 4—5. Stam. many, distinct, or slightly connected into bundles at the base; anth. 2-celled, bursting longitudinally. Glands 4—5 at the base of the petals. Fr. 4—10-celled, several seeds in each cell; or by abortion 1-celled 1-seeded. Embryo erect in the axis of fleshy albumen; cotyledons flat, leafy.—L. alternate, with stipules.

1. TILIA. Sep. 5, deciduous. Pet. 5, with or without a scale at the base. Stam. many, free or polyadelphous. Ovary globose, 5-celled, cells 2-seeded. Style 1. Fr. 1-celled, with 1 or 2 seeds.—No scale to the pet. in our plants.

1. TIL'IA *Linn.* Lime tree.

‡1. *T. europæ'a* (L.); *l.* obliquely cordate *glabrous* except woolly tufts at the branching of the veins beneath, ped. many-flowered, *ripe fr.* woody *not ribbed downy.*—*E. B.* 610. *T. intermedia* (DC.) ed. viii.—Fl. in a naked cyme springing from a lanceolate leaflike bract. L. thin membranous, light-transparent green, *twice the length of their petioles.*—In many old plantations. T. VII. E. S.

2. *T. cordáta* (Mill.); *l.* obliquely cordate *glabrous* except woolly tufts at the branching of the veins beneath, ped. many-flowered, *fr. oblique angular thin and brittle.*—*E. B.* 1705. *T. parvifolia* (Ehrh.) ed. viii.—L. thick, coriaceous, opaque above, *usually scarcely longer than their petioles,* with stellate hairs beneath. Lobes of the stigma ultimately spreading horizontally.—In old woods. T. VIII. E.

3. *T. platyphyl'los* (Scop.); *l.* obliquely cordate *downy beneath* with woolly tufts at the branching of the veins beneath, ped. mostly 3-flowered, *ripe fr. with 5 prominent angles* woody downy turbinate.—*E. B. S.* 2720. *T. grandifolia* (Ehrh.) ed. viii.—Young shoots hairy. L. thin, membranous, bright transparent green, longer than their petioles, with solitary hairs beneath. Lobes of the stigma erect.—*T. rubra* (Lindl.) is stated to have smooth fruit and to be *T. corallina* Sm. I have not seen it.—Old and rocky woods by the Wye, Teme, and Severn. T. VI. VII. E.

Order XVII. HYPERICACEÆ.

Sep. 4—5, distinct or cohering, persistent, with glandular dots, imbricate. Pet. 4—5, twisted in the bud. Stam. many, connected in 3 or 4 bundles at the base. Anthers versatile. Styles several, rarely connate. Fruit a dry or fleshy capsule of many cells and many valves, the valves curved inwards. Seeds small, many, on a central axis or the incurved margins of the valves, embryo straight with no albumen.—L. mostly opposite, with pellucid dots. Stip. 0. Fl. yellow.

1. HYPERICUM. Cal. 5-parted or of 5 sepals. Pet. 5. Styles 3 (in nearly all our plants) or 5. Caps. more or less perfectly 3-celled, many-seeded.—Fl. yellow.

1. HYPER'ICUM *Linn.* St. John's Wort.

* *Styles 5. Pet. unequal-sided.*

*1. *H. caly'cinum* (L.); *st. shrubby* square, l. oblong, fl. solitary, sep. unequal obovate blunt.—*E. B.* 2017.—Rootstock creeping. St. 1 ft. high. Fl. 3 or 4 in. across, yellow, as in all of this genus.—Established in bushy places. P. VII.—IX. E. S. I.

** *Styles 3, stam. in 5 sets, pet. deciduous.* ANDROSÆMUM.

2. *H. Androsæ'mum* (L.); st. shrubby compressed, l. broadly subcordate-ovate blunt, cymes trichotomous few-flowered, sep. broad unequal, pet. oval blunt, *styles much falling short of stam., caps. pulpy* imperfectly 3-celled blunt.—*Sy. E. B.* 264. *Curt.* i. 164.—St. very little branched, 2 ft. high. L. large, with a strong aromatic smell when rubbed. Pet. short. Styles much shorter than the black capsule, finally hooked.—Thickets and hedges. P. VII. VIII. *Tutsan.* E. S. I.

[*H. elátum* (Ait.); *st. shrubby*, 2-edged, *ped. 2-winged*, l. subcordate-ovate subacute, cymes few-flowered, sep. broad unequal ½ as long as petals persistent, *styles exceeding the stam.*, caps. oval.—*Wats. Dendr. Brit.* 85. *H. Androsæmum* Sm. *E. B.* 1225. *H. anglicum* Bert.—St. 3—4 ft. high, much branched. Fl. large, in terminal cymes. Sep. nearly as large as those of *H. Androsæmum*. Pet. about equalling stamens.—The remains of cultivation. P. VII.—IX.] E.

[*H. hircínum* (L.), *Wats. Dendr.* 86, *Sy. E. B.* 246, has lanceolate acute deciduous sepals, but otherwise much resembles *H. elatum*; it is established in some places.]

*** *Herbs. Styles* 3, *stam. shortly united in* 3 *sets, pet. persistent.*

3. *H. quadran'gulum* (L.); *st.* erect *with* 4 *wings*, l. ovate with pellucid dots and veins, *sep. erect lanceolate acuminate entire*, pet. lanceolate, styles half as long as the capsule.—*E. B.* 370. *R.* vi. 344.—*H. tetrapterum* (Fr.) ed. viii.—St. 1—2 ft. high. Fl. in terminal forked close many-flowered cymes, pale.—In wet places. P. VII. E. S. I.

4. *H. undulátum* (Schousb.); st. erect branched 4-edged l. oblong wavy at the edge with many pellucid dots netted with pellucid veins and (as well as the sep. and edges of the stem) with black marginal dots beneath, sep. erect ovate-lanceolate acute usually bluntly crenate, styles not ½ as long as the capsule. —*J. of B.* ii. 97. t. 16. H. bæticum *Boiss. Voy.* t. 34. *Sy. E. B.* 270 a.—St. 2—3 ft. high. L. wavy at the edge, much and uniformly dotted, declining. Fl. in very lax, much branched cymes. Pet. yellow, tinged externally on one longitudinal half with red. Anth. with a black spot. Styles divaricate.—Boggy places in Devon and Cornwall. P. VII. E.

5. *H. perforátum* (L.); st. erect 2-edged, l. oblong with pellucid dots, *sep. erect* lanceolate acute, pet. obliquely oblong, *styles as long as the capsule.*—*E. B.* 295. *R.* vi. 343.—St. 1—2 feet high. L. elliptic-oblong, varying much in form, and in the number and size of the pellucid dots; chief veins pellucid, but not forming a network. Sep. usually denticulate near the tip.—*β. angustifolium* DC.; l. linear-oblong, sep. lanceolate acute finely denticulate.—Woods, hedgebanks, &c. P. VII. VIII. E. S. I.

6. *H. maculátum* (Crantz); st. erect quadrangular, l. elliptic-ovate blunt with a few pellucid dots *netted with pellucid veins*, *sep. reflexed* with many black dots on the outside, pet. elliptic,

styles half as long as the capsule.—*E. B.* 296. *H. quadrangulum* Fries. *H. dubium* (Leers) ed. viii.—St. 1—2 ft. high. Fl. in forked terminal cymes. Sep. ovate, blunt, nearly entire. Caps. longitudinally striate.—β. *Babingtonii* (Groves) (var. *maculatum* ed. viii.); sep. oblong-lanceolate minutely denticulate, l. narrower.—Moist places by ditches, &c. P. VII. E. S. I.

7. *H. humifúsum* (L.); *st. prostrate* somewhat 2-edged, l. oval-oblong blunt minutely pellucid-punctate the margins with black dots beneath, fl. subcymose, *sep.* unequal, 3 *oblong blunt mucronate,* 2 lanceolate, all entire or [*H. decumbens* (Peterm.)] glandular-serrate and having a few black dots beneath, *stam.* 15—20, *styles very short.*—*E. B.* 1226. *R.* vi. 342.—St. slender, 3—6 in. long.—Gravelly and heathy places. P. VII. E. S. I.

8. *H. linarifólium* (Vahl); st. erect or ascending terete, l. linear blunt with revolute margins, fl. cymose, *sep.* rather unequal *lanceolate-acute* with glandular teeth and with many black dots beneath, *stam. about* 30, *styles half as long as the capsule.*—*E. B. S.* 2851.—St. 6—12 in. high. Fl. larger than in Sp. 7.—Channel Isles. Cape Cornwall. Banks of the Teign, Tamar and Tavy, Devon. Carnarvonsh. P. VII. E.

9. *H. hirsutúm* (L.); *st.* erect round *hairy, l.* ovate or oblong *slightly stalked* pellucid-punctate *pubescent, sep.* lanceolate acute *fringed with shortly stalked glands,* pet. linear oblong tipped with stalked glands, styles deciduous.—*E. B.* 1156. *R.* vi. 349.—St. about 2 ft. high, nearly simple. Fl. in axillary and terminal forked panicles.—Woods and thickets. P. VII. VIII. E. S. I.

10. *H. montánum* (L.); st. erect round *glabrous,* l. ovate-oblong sessile pellucid-punctate with glandular dots near the margin, *sep. lanceolate acute fringed with shortly stalked glands,* pet. elliptic entire without dots or glands, styles half the length of the capsule.—*E. B.* 371. *R.* vi. 347.—Fragrant. St. 2 feet high, simple, smooth, slender. Fl. in terminal dense panicles.—Bushy limestone hills. P. VII. VIII. E.

11. *H. pul'chrum* (L.); *st.* erect round *glabrous, l. cordate amplexicaul* pellucid-punctate *glabrous, sep. broadly ovate blunt fringed with sessile glands,* pet. ovate-lanceolate fringed with glands.—*E. B.* 1227.—St. 12—18 in. high, nearly simple. Fl. in loose, axillary, opposite, and terminal cymes. Buds tipped with red. Anth. red. A nearly prostrate form in Shetland (var. *procumbens* Rostr.).—Dry heaths, banks, woods. P. VI. VII. E. S. I.

**** *Styles* 3; *stam. united throughout their lower half in* 3 *sets, a scale between each set; pet. equal-sided.* ELODES.

12. *H. elódes* (L.); *st.* ascending round *shaggy rooting below l. roundish-ovate sessile* pellucid-punctate *shaggy*, sep. ovate bluntish glabrous fringed with shortly stalked (reddish) glands, pet. ovate entire, styles nearly as long as the capsules.—*E. B.* 109. *E. palustris R.* vi. 342.—St. prostrate below, then ascending and leafy, 6—8 in. long. Fl. in terminal and axillary few-flowered cymes.—Spongy bogs. P. VII. VIII. E. S. I.

Order XVIII. ACERACEÆ.

Cal. 5-, rarely 5—9-parted, imbricate. Pet. the same number, inserted round an hypogynous disk. Stam. generally 8, inserted on the hypogynous disk. Ovary 2-lobed, 2-celled. Style 1. Stigmas 2. Fruit winged, separating into two indehiscent nuts (samaras) each with 1 cell and 1—2 seeds. Embryo curved, albumen 0.—Trees with opposite leaves. Stip. 0.

1. ACER. Fl. some imperfect. Calyx 5-parted. Pet. 5. Stam. usually 8, longer in the male flowers.

1. A'CER *Linn.* Maple.

1. *A. campes'tre* (L.); l. 5-lobed, lobes entirely or slightly cut, *corymbs erect*, sep. and pet. linear hairy, wings of the fruit horizontally diverging, ovary downy [var. *hebecarpum* DC., or glabrous, var. *leiocarpon* Wallr.], stam. of the male flowers as long as the corolla.—*E. B.* 304.—A small tree with corky fissured bark.—Woods and hedges. T. V. VI. *Maple.* E. S.? I.

‡2. *A. Pseudo-plat'anus* (L.); l. 4-lobed, unequally serrate, *racemes pendulous*, ovary downy with spreading wings, stam. of the male flowers twice as long as the corolla.—*E. B.* 303. *R.* v. 164.—A large handsome tree.—In hedges and plantations even in the North of Scotland. T. V. VI. *Sycamore.* E. S. I.

Order XIX. GERANIACEÆ.

Sep. 5, persistent, imbricate. Pet. 5, clawed, twisted in the bud. Stam. generally monadelphous, 2 or 3 times as many as the petals, some often abortive. Fruit of 5 carpels cohering round a long beaked axis, each terminated by an indurated style which finally twists up, separating from the axis and carrying with it the carpel. Seeds solitary, without albumen. Cotyledons convolute, plaited.—L. with stipules.

1. Geranium. Sep. 5. Pet. 5. Stam. 10, monadelphous, alternately larger and with glands at their base, all perfect. Carp. rounded at the top; the long ultimately recurved beak glabrous internally.

2. Erodium. Sep. 5. Pet. 5. Stam. monadelphous, 5 sterile, 5 fertile with glands at their base. Carp. with 2 lateral depressions at the top; the long ultimately spirally twisted beak bearded internally.

1. Gera'nium *Linn.* Cranesbill.

* *Root consisting of long fibres springing from a short thick rhizome, perennial.*

†1. *G. phæ'um* (L.); peduncles 2-flowered, pet. roundish wedgeshaped rather longer than the mucronate sepals, carp. hairy below *transversely wrinkled* above, *seeds punctate-striate.*—*E. B.* 322. *R.* v. 197.—St. erect, 2 feet high. L. 5-lobed; lobes acute, cut, serrate. Fl. purplish black, very rarely white.—In woods and thickets, rare. P. V. VI. E. S.

[*G. nodósum* (L.); peduncles 2-flowered, *pet.* obcordate *long, sepals awned, carp.* downy *even,* l. 3—5-lobed, lobes ovate acuminate serrate.—*E. B.* 1091.—St. 18 in. high, slender, erect. Fl. pale purple.—In Cumberland, Herts, and Yorksh.—*G. striátum,* L., which resembles this but has a hairy stem, was found near Filby, Cumb.; Chepstow, and near Plymouth. Colonists.] E.

2. *G. sylvat'icum* (L.); peduncles 2-flowered, pet. obovate slightly notched long, sepals awned, carp. hairy even, hairs spreading glandular, *seeds dotted,* l. palmate 7-lobed, lobes cut and serrate, st. erect glandular hairy above, *filaments* of stam. *subulate, fruitstalks erect.*—*E. B.* 121.—St. erect, 2—3 feet high. Fl. purplish blue, claws of the petals bearded, lower half of filaments hairy. Sometimes the fl. are pale rose-coloured, pet. smaller and nearly entire, and st. more decidedly hairy.—Woods and thickets in the North, rare. P. VI. VII. E. S. I.

3. *G. praten'se* (L.); ped. 2-flowered, pet. obovate entire or slightly notched long, sepals awned, *carp. hairy* even, *hairs spreading glandular, seeds minutely netted,* l. palmate 7-lobed, lobes cut and serrate, st. diffuse glandular hairy above, hairs deflexed, *filaments* of stam. filiform *with a triangular-ovate base, fr.-st. deflexed.*—*E. B.* 404.—St. 1—3 feet high. Fl. large, purple, claw of pet. ciliate. Base of filaments slightly hairy.—Moist pastures. P. VI.—VIII. E. S. I.

4. *G. sanguin'eum* (L.); *peduncles mostly single-flowered*, pet. obcordate long, sepals awned, *carp. smooth crowned with a few bristles*, seeds minutely wrinkled and dotted, l. nearly round 7-lobed, lobes deeply 3-fid and cut, st. diffuse hairy, *hairs spreading* horizontally.—*E. B.* 272.—Fl. large, purple; filaments dilated at the base.—β. *G. prostratum* (Cav.); st. dwarf tufted nearly simple decumbent, fl. flesh-coloured. *G. lancastriense* With.—[A small-flowered procumbent form with crowded l. and l.-segments more tapering is var. *micranthum* B. White].—In dry places, rare. β. Sands in Walney Island, Lancashire. P. VII. E. S. I.

** *Roots fusiform, rhizome wanting, perennial; ped. 2-flowered.*

†5. *G. pyrenäicum* (Burm. f.); fruitstalks deflexed, *pet.* obcordate *twice as long as the* mucronate *sep., claws densely ciliate*, carp. smooth with adpressed hairs, seeds smooth, l. reniform 7—9-lobed, lobes of lower l. oblong blunt trifid and toothed at the end, *t.* erect *villose.*—*E. B.* 405. *R.* v. 191. *G. perenne* (Hook.).—Fl. light purple or nearly white. Claws of the pet. with a dense tuft of hairs on each side. Fertile anth. 10. Segments of the upper leaves more acute. St. spreading; 1—3 feet high, clothed with dense short down and long hairs intermixed.—Roadsides and pastures, rare. P. VI. VII. E. S.? I.

*** *Root fibrous, rhizome* 0, *annual; ped. 2-flowered.*

† *Sep. spreading.*

6. *G. mol'le* (L.); *pet.* oblong deeply *bifid* ½ as long as or ½ longer than the mucronate sepals, *claws ciliate, carp. transversely wrinkled glabrous, seeds smooth*, l. roundish-reniform in 7—9 deep wedgeshaped segments trifid at the end, st. diffuse pubescent.—*E. B.* 778. *R.* v. 191.—Fl. small, purple, or white with lilac claws. Softly pubescent, glandular above.—Dry places. A. IV.—VIII. E. S. I.

7. *G. rotundifólium* (L.); *pet.* spathulate *entire* blunt rather longer than the shortly awned sepals, *claws glabrous, carp.* not wrinkled *with spreading hairs, seeds netted*, l. reniform in 5—7 broadly wedgeshaped incise-crenate segments, st. diffuse pubescent.—*E. B.* 157. *R.* v. 190. *G. viscidum* Ehrh.—Fl. small, flesh-coloured. Peduncles shorter than the leaves.—Old walls and waste places, rare. A. VI. VII. E. I.

8. *G. pusil'lum* (L.); pedicels deflexed after flowering, *pet. notched hardly exceeding the* mucronate *sepals, claws* slightly ciliate, *carp.* not wrinkled *with adpressed hairs, seeds smooth*,

l. reniform palmate with 5—7 trifid lobes, *st.* diffuse *downy.*—*E. B.* 385. *R.* v. 190. *G. rotundifolium* Fries.—St. usually prostrate, clothed only with short down. Fl. small, bluish purple. Claws of the pet. only slightly ciliate. Fertile anth. 5. Styles pale flesh-coloured. Ped. shorter than the leaves.—Waste places. A. VI.—IX. E. S. I.

9. *G. dissec'tum* (L.); pet. bifid about equalling the awned sepals, claws slightly ciliate, *carp.* not wrinkled *with erect hairs, seeds netted,* l. in 5—7 deep laciniate segments with linear lobes, *st.* diffuse *hairy.*—*E. B.* 753. *R.* v. 189.—Fl. small, bluish purple. *L. divided almost to the base, longer than the peduncles.*—Waste places. A. VI.—VIII. *Dove's-foot.* E. S. I.

10. *G. columbinum* (L.); *pet.* obovate *emarginate* with a short blunt tooth in the notch about equalling the awned sepals, *claws ciliate, carp.* not wrinkled *with a few minute scattered hairs,* seeds netted, l. in 5—7 deep laciniate segments, st. diffuse with adpressed hairs.—*E. B.* 259. *R.* v. 198.—Fl. small rose-coloured. L. divided almost to their base. *Peduncles longer than the leaves; pedicels very long.*—On gravelly and limestone soils. A. VI. VII. E. S. I.

†† *Sep. appressed with flower and fruit.*

11. *G. lúcidum* (L.); *pet.* obovate *entire, claws glabrous* very long near equalling the *transversely wrinkled pyramidal calyx, carp. netted triply keeled* glandular-hairy at the summit, seeds smooth, l. reniform in 5 blunt incise-crenate mucronate segments, st. spreading ascending —*E. B.* 75. *R.* v. 187.—Fl. small, rose-coloured. St. and l. glabrous and shining, often strongly tinged with red.—Lindley considered his *G. Raii* most allied to this species, differing by its "shaggy calyx and simply keeled fruit."—Walls and hedgebanks. A. V.—VIII. E. S. I.

12. *G. Robertiánum* (L.); pet. obovate entire or slightly emarginate, claws glabrous very long nearly equalling the long-awned hairy and slightly glandular sepals, *carp. transversely wrinkled* downy, seeds smooth, l. ternate or quinate, leaflets stalked trifid incise-pinnatifid, st. spreading erect.—*E. B.* 1486. *R.* v. 187.—Fl. purple, sometimes white. Cal. with a very few glandular hairs, not transversely wrinkled.—β. *G. purpureum* (Forst.); pet. narrower, sep. glandular-hairy, carp. glabrous and more wrinkled, l. in narrower segments *E. B. S.* 2648. [A less extreme form is *G. modestum* (Jord.)].—Hedgebanks. β. Southern sea-coast. A. V.—IX. E. S. I.

2. Ero'dium *L'Hérit.* Storksbill.

1. *E. cicutárium* (L'Hér.); st. procumbent hairy, peduncles many-flowered, perfect *stam. dilated* not toothed below glabrous, beak hairy, a concentric furrow below the circular glandless depression on the carpel, *l. pinnate, leaflets* sessile *pinnatifid* cut, *stip. lanceolate.*—*E. B.* 1768.—St. diffuse, leafy, with scattered hairs. Fl. purplish or white. Leaflets very deeply divided, their segments lanceolate or linear, acute. In Jersey specimens the l. are ovate and short, and their segments short broad and bluntish.—*a. E. pimpinellifolium* (Cav.); 2 pet. with a spot, lts. ovate incise-pinnatifid with bluntish lobes, furrow on the carp. conspicuous. *E. commixtum* Jord.—*β. E. triviale* (Jord.); pet. not spotted, lts. incise-pinnatifid, carp.-furrow faint.—*γ. E. pilosum* (Bor.); pet. not spotted, lts. almost pinnate, carp.-furrow obsolete. (*β.* & *γ. E. chærophyllum* Cav.?).—Waste ground. A. VI.—IX. E. S. I.

2. *E. moschátum* (L'Hér.); st. procumbent hairy, peduncles many-flowered, perfect stam. toothed at the base glabrous, beak downy, a concentric furrow below the circular *glandular depression* on the carpel, *l. pinnate, leaflets* nearly sessile, *ovate unequally cut, stip. oval.*—*E. B.* 902.—Much larger than the preceding, and diffusing a strong musky scent when handled. Leaflets less deeply cut.—Waste places, rather rare. A. VI. VII. E. I.

3. *E. marit'imum* (L'Hér.); st. prostrate slightly hairy, peduncles 1—2-flowered, pet. very minute, a transverse furrow below the semicircular depression on the carpel, *l. simple ovate-cordate* stalked lobed and crenate.—*E. B.* 646.—St. often very fleshy. Fl. very small. Pet. pale red, very minute, often wanting.—Sandy and gravelly places, particularly near the sea, rare. P. V.—IX. E. I.

Order XX. BALSAMINACEÆ.

Sep. 5 (2 upper usually wanting), irregular, deciduous, lower spurred, imbricate in the bud. Pet. 5, irregular, the lateral united in pairs. Stam. 5. Anth. 2-celled, opening at the apex by a longitudinal fissure, more or less cohering. Ovary 5-celled. Fr. capsular with 5 elastic valves. Seeds solitary or many, pendulous; albumen 0.—L. with glands in place of stipules. Plants succulent. Many raphides. Minute but fertile fl. which do not open are found.

1. IMPATIENS. Sep. 3, the lower one hoodlike with a spur. Pet. 3, upper one symmetrical, lateral unequally 2-lobed or each formed of 2 combined. Anth. cohering.

1. IMPA'TIENS *Linn.* Balsam.

1. *I. Noli-tan'gere* (L.); l. ovate coarsely serrate, peduncles many-flowered solitary, *spur loosely recurved not emarginate.*—*E. B.* 937. *R.* v. 198. b. *St.* 5. 15.—Fl. large, yellow spotted with orange. St. 1—2 feet high, tumid at the joinings.—Damp woody places in mountainous districts, rare. A. VI.—IX. *Yellow Balsam.* E. S.

*2. *I. biflóra* (Walt.); l. ovate coarsely serrate, peduncles about 4-flowered solitary, *spur closely reflexed emarginate.*—*I. fulva* (Nutt.) ed. viii. *E. B. S.* 2794.—Fl. orange-yellow spotted with red. Each serrature of the l. with a reflexed glandular tooth. St. 2—3 feet high.—An American plant quite naturalized by the Wey and other rivers in Surrey. A. VIII. E.

*3. *I. parviflóra* (DC.); l. elliptic serrate, ped. erect 3—12-flowered, spur short straight.—*Sy. E. B.* 315.—Fl. small, yellowish. St. 6—18 in. high. L. very acute at both ends.—Waste places. A Russian plant now quite naturalized. A. VII.—IX. E.

Order XXI. OXALIDACEÆ.

Sep. 5, equal, persistent, imbricate in the bud. Pet. 5, equal, often cohering at the base, twisted in the bud. Stam. 10, more or less monadelphous, those opposite to the pet. longer than the others. Anth. 2-celled. Ovary 5-celled. Styles 5. Caps. 5—10-valved. Seeds several. Embryo straight, in cartilaginous albumen.—With stipules.

1. OXALIS. Sep. 5, connected below. Pet. 5, often connected below. Stam. 10, monadelphous, 5 outer ones shorter. Styles 5. Seed with an elastic coat.

1. OX'ALIS *Linn.* Wood-Sorrel.

1. *O. Acetosel'la* (L.); *stemless*, rhizome creeping toothed, l. ternate, leaflets obcordate hairy, peduncles with 2 scaly bracts at about the middle 1-flowered, caps. ovoid, seed longitudinally ribbed.—*E. B.* 762. *R.* v. 199.—Fl. white with purple veins or rarely purple or blue. Cor. about 4 times as long as the calyx.—Woods and shady places. P. V. *Wood-Sorrel.* E. S. I.

†2. *O. corniculáta* (L.); *st. diffuse. with procumbent branches* pubescent, l. ternate, leaflets obcordate, *stipules oblong united to the base of the petioles*, peduncles about 2-flowered shorter than the leaves, partial fruitstalks reflexed, caps. narrowly oblong, seeds transversely ribbed, root fibrous.—*E. B.* 1726. *R.* v. 199. —Fl. yellow. L. mostly in pairs.—Waste ground in Devon and Cornwall. A. VI.—IX. E.

*3. *O. stric'ta* (L.); st. erect, stipules 0, ped. 2—8-fl. longer than the l., fl. cymose, fruitstalks erect, stoloniferous.—*Sy. E. B.* 312. *R.* v. 199.—Cymes close, terminal; fl. yellow. L. in imperfect whorls.—Naturalized in Cornwall, Devon, Glamorgan, &c. B. VII. VIII. E.

Order XXII. LINACEÆ.

Sep. 4—5 persistent, imbricate. Pet. 4—5, twisted in the bud, clawed, deciduous. Stam. as many as the pet., connected into an hypogynous ring with intermediate teeth (abortive stamens). Ovary with about as many cells and styles as the sepals, stigmas capitate. Caps. generally tipped with the hardened base of the styles, with 3—5 complete and 4—5 incomplete dissepiments, and no central axis. Seeds 1 in each spurious cell, pendulous, with albumen.—L. without stipules, alternate.

1. Linum. Sep. 5, entire. Pet. 5. Stam. 5. Styles 5. Caps. with 10 cells and 10 valves.

2. Radiola. Sep. 4, connected below, deeply trifid. Pet. 4. Stam. 4. Styles 4. Caps. with 8 cells and 8 valves.

1. Li'num *Linn.* Flax.

* *Leaves scattered.*

1. *L. angustifólium* (Huds.); caps. downy within, *sep. elliptic* pointed ciliate, l. linear-lanceolate, *st. many.*—*E. B.* 381.—Fl. pale blue. St. 1—2 feet long, lax, diffuse, branching irregularly. —Sandy and chalky places in the South. P. VII. E. I.

[*L. usitatis'simum* (L.); caps. glabrous within, *sep. ovate* pointed ciliate, l. lanceolate, *st. solitary.*—*E. B.* 1357. *St.* 26. 12.—Fl. blue. St. 1—1½ foot high. Sep. 3-veined.—*β. crepitans* (Boenn.); smaller and more branched, caps. opening with elasticity, seeds paler.—Escaped from cultivation. A. VII. *Common Flax.*]

2. *L. peren'ne* (L.); *sep. obovate* obscurely 5-veined glabrous, inner sep. very blunt, l. linear-lanceolate, st. many, fruitstalks erect.—*E. B.* 40.—Fl. blue. St. 1—2 ft. long, erect or decumbent.—Chalky [and limestone] places, rare. P. VI. VII. E. S.

** *Leaves opposite. Flowers white.*

3. *L. cathar'ticum* (L.); sep. elliptic pointed, l. obovate, upper l. lanceolate.—*E. B.* 382.—Fl. small; sep. serrate; pet. acute. St. one or more, slender, 2—6 in. high. Panicle forked, spreading.—In dry pastures. A. VI.—VIII. E. S. I.

2. Radi'ola *Hill*. Flax-seed. All-seed.

1. *R. linoïdes* (Roth), *R. millegrana* (Sm.) ed. viii.—*E. B.* 893.—St. 1—2 in. high, repeatedly forked, with solitary minute white fl. in the forks as well as at the ends of the branches. Sep. deeply and acutely 3-cleft, connected below into a tube.—Damp sandy places. A. VII. VIII. E. S. I.

Division II. CALYCIFLORÆ.

(Orders XXIII.—XXXIX.)

Petals distinct, and stamens perigynous or epigynous.

A. Petals and stamens perigynous (except in Orders XXVIII. XXIX., XXXI. and XXXV.).

Order XXIII. CELASTRACEÆ.

Sep. 4—5, imbricate in the bud. Pet. 4—5, inserted in the margin of a fleshy disk surrounding the ovary. Stam. alternate with the petals, inserted in the disk. Ovary sunk in the disk, more or less connected with it, 3—5-celled; cells 1—2-seeded, ovules erect. Embryo straight.—Trees or shrubs. Stip. minute, deciduous.

1. Euonymus. Cal. flat, 4—5-lobed; disk peltate. Pet. 4—5, inserted in the margin of the disk. Stam. 4—5, inserted in the disk. Style 1. Caps. 3—5-celled, 3—5-angled, loculicidal. *Seeds* solitary in each cell, *with a fleshy aril, not truncate* at the hile.—Leaves simple.

1. Euon'ymus *Linn*. Spindle-tree. Prickwood.

1. *E. europæ'us* (L.); pet. oblong, fl. mostly 4-cleft and 4-androus, branches tetragonous smooth and even, l. ovate-

lanceolate minutely serrate, caps. obtusely angular not winged. —*E. B.* 362. *R.* vi. 309. *St.* 27. 3.—Orange-coloured aril enclosing the seed. Bark green. L. glabrous. Fl. forming small umbels, greenish white. Fruit rose-coloured.—Hedges and woods. Sh. V. VI. E. S. I.

Order XXIV. RHAMNACEÆ.

Cal. 4—5-cleft, valvate in the bud. Pet. distinct, inserted in the throat of the calyx. Stam. opposite to the pet. and equalling them in number. Ovary wholly or in part superior, 2—3—4-celled, surrounded by a glandular disk. Seeds solitary, erect. Embryo straight. Fr. fleshy or dry.—Shrubs. Stip. minute.

1. RHAMNUS. Cal. pitcher-shaped, 4—5-cleft. Pet. 4 or 5, or sometimes 0, inserted with the stam. on the margin of the tube of the calyx. Fruit fleshy, with 2—4 cells and as many seeds.

1. RHAM'NUS *Linn.* Buckthorn.

1. *R. cathar'ticus* (L.); thorns terminal, *fl. 4-cleft* diœcious, petioles much longer than the stipules, *l. roundish-oval sharply toothed,* fr. with 4 seeds.—*E. B.* 1629.—*Branches opposite.* Serratures of the l. incurved, glandular. Notch in the seeds closed. *Styles* 4, united halfway up.—Hedges and thickets. Sh. V.—VII. *Buckthorn.* E. S. I.

2. *R. Fran'gula* (L.); spineless, *fl. 5-cleft* perfect, *l. elliptic-obovate acuminate* narrowed below *entire,* fr. with 2 seeds, style simple.—*Sy. E. B.* 319.—*Branches alternate.* Fl. in small clusters greenish white, small.—Hedges and thickets. Sh. V. VI. *Black Alder.* E. S. ? I.

Order XXV. LEGUMINOSÆ.

Cal. inferior. Sep. 5, more or less combined, odd one inferior. Cor. papilionaceous (in our plants), inserted into the base of the calyx. Pet. 5, odd one superior and external. Stam. 10 (in our plants), monadelphous or diadelphous. Ovary free, 1-celled. Fruit a legume; placenta on the upper suture; style from the upper suture. Embryo bent over the edge of the cotyledons, or straight.—All our plants have papilionaceous flowers and 10 stamens in one bundle or in two bundles of 9 and 1 L. mostly stipulate; lts. often stipulate.

Tribe I. *LOTEÆ.* Pod continuous. Cotyledons rising above the ground and becoming green leaves. *Leaves of* 1 *or* 3 *leaflets or pinnate with a terminal leaflet* [or digitate].

* *Leaflet solitary. Stamens monadelphous.*

1. Ulex. *Cal. of* 2 *parts; the upper with* 2, *the lower with* 3 *minute teeth*, a bract on each side at the base. *Pod thick*, few-seeded, scarcely longer than calyx which nearly equals the corolla.

2. Genista. *Cal.* 2-*lipped; upper bifid, lower* 3-*toothed.* Style subulate, ascending. Stigma terminal, oblique. Cor. much exceeding calyx.—Many foreign species have ternate leaves.

** *Leaf of* 3 *leaflets. Stamens monadelphous.*

3. Sarothamnus. *Cal.* 2-*lipped; the upper with* 2, *the lower with* 3 *teeth.* Style long, curved, thickened upwards, channelled within. Stigma terminal, capitate, small. *Pod flat.*

4. Ononis. *Cal.* 5-*cleft*; segments narrow, the lower ones longer. Keel beaked. Style filiform, ascending. Stigma terminal, subcapitate. Pod thick.

*** *Leaf digitate. Stamens monadelphous.*

[5. Lupinus. Cal. deeply 2-lipped. Keel rostrate. Style filiform curved. Stigma capitate. Pod flattened.]

**** *Leaf of* 3 *leaflets. Stamens diadelphous; one free.*

6. Medicago. Cal. with 5 nearly equal teeth. Keel blunt. Filaments of the stamens filiform. Ovaries curved. Pod 1-celled, hooded or spirally twisted. Seeds 1 or many.

7. Melilotus. Cal. with 5 nearly equal teeth. Keel blunt. Filaments filiform, not adhering to the claws of the petals. *Ovary straight.* Pod subglobose or oblong, 1-celled, 1—4-seeded, longer than the calyx. Pet. distinct, deciduous. Fl. in long loose racemes.

8. Trifolium. Cal. with 5 unequal teeth. Keel blunt. *Filaments slightly enlarged upwards and more or less adhering to the claws of the petals.* Pod oval, 1—4-seeded, included in the calyx or slightly protruding. *Pet. slightly combined, persistent.*—Fl. in close racemes.

9. TRIGONELLA. Cal. of 5 nearly equal teeth. Keel blunt. *Filam. filiform, not adhering to the claws of the petals.* Pod compressed, truncate, straight, 6—8-seeded, protruding from the calyx. *Pet. distinct, deciduous*; wings and keel nearly equally long.—Fl. 1, 2 or 3 together. Common peduncle shorter than the petiole.

10. LOTUS. Cal. with 5 nearly equal teeth. *Keel* ascending, *with a narrowed point* (beak). Wings connivent at their upper margin. Longer filaments dilated upwards. Style kneed at the base, filiform-subulate. *Pod linear many-seeded,* 2-valved, imperfectly divided by transverse partitions.

***** *Leaf imparipinnate.* † Stamens monadelphous.

11. ANTHYLLIS. *Cal. tubular inflated,* 5-cleft, segments unequal. *Keel not beaked.* Style filiform. Stigma capitate.

†† Stamens diadelphous; one free.

12. OXYTROPIS. Cal. with 5 teeth. *Keel with a narrow straight point.* Pod imperfectly 2-celled, *cells formed by the inflexed margin of the upper suture.*

13. ASTRAGALUS. Cal. with 5 teeth. *Keel blunt.* Pods imperfectly 2-celled, *cells formed by the inflexed margin of the lower suture.*

Tr. II. *VICIEÆ.* Pod continuous. *Stam. diadelphous*; one free. Cotyledons remaining under ground. *L. pinnate without the term. odd leaflet* or apparently simple.

* *Tube of stam. very obliquely truncate.*

14. VICIA. Cal. 5-fid or 5-toothed. *Style filiform*; *its upper part hairy all over, or bearded on the underside* and at the same time hairy or glabrous. Pods 1-celled, 2-valved.

** *Tube of stam. transversely truncate.*

15. LATHYRUS. Cal. 5-fid or 5-toothed. *Style flattened upwards, hairy beneath the stigma.* Pods 1-celled, 2-valved.

Tr. III. *HEDYSAREÆ.* Pod dividing transversely into 1-seeded joints. *L. imparipinnate.*

16. ORNITHOPUS. Cal. long, tubular, with 5 nearly equal teeth, 2 upper ones slightly combined and converging. Keel blunt. *Pod* long, compressed, *of many* 1-*seeded indehiscent joints equally narrowed on both sides at the joinings.*—Apex of the common peduncle bearing a small pinnate leaf just below the flowers.

17. ARTHROLOBIUM. Cal. long, tubular, with 5 nearly equal teeth, 2 upper ones combined up to their middle and straight. Keel blunt. *Pod* long, terete, *of many* 1-seeded indehiscent *joints scarcely narrowed at the joinings.—No leaf at the apex of the peduncles.*

18. HIPPOCREPIS. Cal. short, bellshaped, with 5 nearly equal teeth, 2 upper ones combined up to their middle. Keel narrowed into a beak. *Pod* long, compressed, of *many* 1-seeded *crescent-shaped joints, so that each pod has many notches on one side.*

19. ONOBRYCHIS. Cal. with 5 nearly equal subulate teeth. Keel obliquely truncate, longer than the wings. *Pod* 1-*celled,* compressed, indehiscent, 1-*seeded, upper suture straight, lower curved toothed winged or crested.*

Tribe I. *Loteæ.*

1. U'LEX *Linn.* Furze. Whin. Gorse.

1. *U. europæ'us* (L.); young l. shaggy beneath furrowed, primary spines strong terete-polygonal furrowed rough, st. hairy, fl. lateral, *bracts ovate lax, cal. shaggy,* wings longer than keel.—*E. B.* 742.—St. shrubby, 4—6 feet high, much branched, spreading. Fl. bright yellow, from both the primary and secondary spines. Cal.-teeth converging, finely downy. Spines branching at their base and up to about their middle, not exceeding the flowers.—*U. strictus* (Mack.); primary spines small slender tetragonal, plant 1—2 ft. high, with upright branches.—Does not come true from seed. *E. B. S.* 2988.—Heaths. Sh. II.—VI. E. S. I.

2. *U. Gal'lii* (Planch.); young l. glabrous ciliate furrowed, primary spines strong deflexed subterete striate smooth, st. hairy, fl. lateral and terminal, *bracts minute* adpressed, *cal. finely downy, wings longer than keel.*—*E. B. S.* 2987.—St. 2—5 ft. high. Fl. orange, pet. more or less divaricate; wings not straight and thus often seeming shorter than keel. Cal.-teeth

diverging, pubescent. Pods bursting in spring. A dwarf form is often taken for *U. minor*, a tall one for *U. europæus*.—Heaths. Sh. VIII.—XI. E. I.

3. *U. minor* (Roth); young *l. glabrous* ciliate furrowed, primary *spines* slender *terete striate smooth*, st. hairy, fl. lateral and terminal, *bracts very minute adpressed, cal. finely downy*, wings shorter than keel.—*E. B.* 743. *U. nanus* (Forst.) ed. viii. —St. shrubby procumbent. Primary spines short, spreading, branched at their base only. Fl. half the size of those of *U. europæus*, from the primary spines and exceeded by them, pale; pet. scarcely separated when full-blown. Cal.-teeth diverging. Pod persistent for nearly a year.—Heaths. Sh. VIII. IX. E.

2. Genis'ta *Linn.*

1. *G. pilósa* (L.); st. procumbent without thorns, l. obovate-lanceolate blunt, *stipules ovate blunt*, branches peduncles calyx standard keel and underside of the l. silky, *peduncles lateral accompanied by a tuft of leaves, pods hairy.*—*E. B.* 208.—Fl. small, yellow, collected towards the end of the branches. St. much branched, furrowed, woody.—Dry sandy and gravelly heaths, rare. Sh. V. E.

2. *G. tinctória* (L.); st. with erect branches without thorns, l. lanceolate or elliptic hairy at the edges, *stipules minute subulate, fl. racemose*, cor. and *pods glabrous.*—*E. B.* 44.—Branches erect, 1—2 feet high, striate, glabrous, downy above. Fl. yellow. Keel as long as the standard.—*β. G. humifusa* (Dicks.); st. and branches procumbent, l. ovate or oblong, *pods hairy on the back* of each valve. St. angular, 6—10 in. long.—In pastures and thickets. *β.* Lizard district, Cornwall, and St. David's Head, Pemb. Sh. VII.—IX. *Dyer's-weed.* E. S.

3. *G. ang'lica* (L.); st. *spinous* leafless below, flowering branches glabrous without thorns, l. ovate-lanceolate, *stip.* 0, *fl. solitary* in the axils of the l., cor. and pods glabrous.—*E. B.* 132—St. 1 foot high, round, leafless, sometimes quite prostrate, with short leafy branches bearing the yellow flowers. Keel longer than the standard.—Moist peaty heaths. Sh. V. VI. *Needle Whin.* E. S.

3. Sarotham'nus *Wimm.* Broom.

1. *S. vulgáris* (Wimm.)—*Spartium* L., *E. B.* 1339. *Cytisus* (Hook.); *S. scoparius* (Koch) ed. viii.—St. 2—3 feet high, an-

gular, glabrous; or (*β prostratus* Bailey) prostrate and spreading, at Kynance Cove, Cornwall, and Channel I. L. ternate or simple, obovate. Fl. axillary, solitary or in pairs, shortly stalked, large, bright yellow. Pods dark brown, hairy at the edges; seeds many.—Dry hills and heaths. Sh. V. VI. E. S. I.

4. ONO'NIS *Linn.* Rest-harrow.

1. *O. répens* (L.); stoloniferous, st. procumbent uniformly hairy, fl. axillary solitary stalked, leaflets broadly oblong, *pods* ovate erect *falling short of the calyx.—E. B. S.* 2659.—Shrubby. Usually without spines. St. rooting at their base. Wings equalling the keel; standard a little longer. Seeds tubercular. —*α.* glandular, fl.-l. equalling or surpassing cal., pod shorter than calyx.—*β. O. maritima* (Dum.); more or less spinose glandular-villose, fl.-l. falling short of cal., pod as long or longer than calyx [= var. *horrida* Lange].—Barren sandy places. P. VI.—IX. E. S. I.

2. *O. spinósa* (L.); not stoloniferous, st. erect or ascending bifariously hairy, fl. axillary solitary stalked, leaflets oblong, *pods* ovate erect *exceeding the calyx.—E. B.* 682.—Shrubby. Usually spinous [or almost unarmed, var. *mitis* L.]. St. mostly erect. Wings falling short of the keel, which falls short of standard. Seeds tubercular.—Barren places. P. VI.—IX. E. S.

3. *O. reclináta* (L.); viscid, pubescent, st. ascending, fl. axillary, pedicels 1-flowered shorter than the l. fl. or pod without bracts, cor. about equal to the calyx, leaflets obovate-cuneate serrate at the tip, stipules ovate, *pods cylindrical reflexed*, seeds 14—18 tubercular.—*E. B. S.* 2838.—St. 5—6 in. high, much branched.—Sandy places. With *Bupleurum aristatum* at Berry Head, Devon. *Mr. E. M. Holmes* ! (*J. of B.* vi. 58). Channel Isles. *Galloway. A. VII. E.

5. LUPI'NUS *Linn.*

*1. *L. Nootkátensis* (Sims); st. stout leafy, l. of 6—8 cuneate oblong somewhat mucronate lts., petioles about equalling lts., stip. linear-acuminate, raceme long partially whorled, bracts long exceeding the buds, upper lip of cal. bifid, lower broad 3-toothed, cor. blue or purple.—*Bot. Mag.* 1311.—More or less densely villous. A large showy plant, naturalized in many parts of Scotland, whence it has been reported as *L. perennis* L. P. V.—VIII. S.

6. MEDICA'GO *Linn.*

* *Pods without spines.*

(Sp. 1, 2, and 3 are said to be sometimes monadelphous.)

*1. *M. sativa* (L.); st. erect, *racemes* many-flowered, *pods* compressed *loosely spiral with 2 or 3 turns downy with adpressed hairs*, pedicels shorter than the calyx or bract, leaflets obovate-oblong dentate above emarginate mucronate.—*E. B.* 1749.—St. angular when young. Pods twisted into a loose open spiral. Fl. large, yellow or violet.—Hedgebanks and borders of fields, scarcely naturalized. P. VI. VII. *Lucerne.* E S.

2. *M. sylves'tris* (Fries); st. rather quadrangular pithy ascending, racemes many-flowered, *pods forming one complete flat ring*, pedicels shorter than the cal. longer than the bract, lts. obovate-oblong dentate above emarginate mucronate—*E. B. S.* 2980.—Fl. large, yellow or blackish green with darker streaks.—Sandy and gravelly places in Norfolk and Suffolk. P. VI. VII. E. [‡ I.]

3. *M. falcáta* (L.); st. usually terete nearly solid prostrate, racemes many-flowered, *pods straightly sickle-shaped twisted not forming a ring*, pedicels shorter than the cal. longer than the bract, lts. obovate-oblong dentate emarginate mucronate.—*E. B.* 1016.—Fl. large, yellow.—Sandy and gravelly places in Norfolk and Suffolk. P. VI. VII. *Yellow Medick.* E.

4. *M. lupulína* (L.); *spikes* many-flowered *dense ovoid*, *pods* compressed *kidney-shaped with a spiral point* with longitudinal branched prominent veins, stip. obliquely ovate slightly toothed, leaflets roundish-ovate denticulate emarginate mucronate.—*E. B.* 971.—St. procumbent or ascending, spreading widely. Pods 1-seeded, glabrous or slightly hairy. Fl. small, yellow.—[Var. *Willdenowiana* (Koch); pods with yellowish spreading glandular hairs.] Waste ground. A. or B. V.—VIII. *Black Medick.* E. S. I.

** *Pods edged with spines, compactly spiral.*

5. *M. arab'ica* (Huds.); *peduncles* 1—4-*flowered*, *pods* compressed of 2—6 turns *veined with 4 ridges and a central furrow on the edge*, spines in 2 rows divergent subulate curved, leaflets triangular-obcordate, stip. toothed.—*E. B.* 1616. *M. maculata* (Sibth.) ed. viii.—Lts. with a purple spot in the centre. Edge of pods broad; spines springing from the margin and the ridge next to it on each side, compressed and furrowed on both sides, variable in length.—On a gravelly soil. A. V.—VIII. E. S. I.

6. *M. min'ima* (L.); peduncles 1—6-flowered, *pods* of 4 turns *smooth with a thin edge*, spines in 2 rows divergent subulate hooked, leaflets obovate, stip. nearly entire.—*E. B. S.* 2635.—Edge of the pods with 3 ridges, the central one so prominent as to be easily taken for the true margin, no central furrow but the central ridge common to the 2 rows of spines. Sides of the pods smooth. Spines varying considerably in length and the whole plant in hairiness.—In sandy fields in the South-east, rare. A. V. E.

7. *M. denticuláta* (Willd.); peduncles 1—5-flowered, *pods* of 2 or 3 turns *deeply netted* with a thin edge, spines in 2 rows divergent subulate hooked, leaflets obcordate, *stip. laciniate* [seeds narrowly reniform].—*E. B. S.* 2634.—Edge of the pods as in *M. minima.* Spines about equalling the diameter of the pod. Glabrous.—β. *M. apiculata* (Willd.); spines very short without hooks, often scarcely longer than their own breadth so as to appear little more than tubercles, ped. 3—10-flowered.—On sandy ground near the sea, rare. A. V.—VIII. E.

[*M. lappacea* (Desr.); ped. 1—4-flowered, pods of 3—5 turns with a thin edge, spines hooked in 2 rows spreading horizontally, stip. laciniate, seeds reniform.—*R.* vol. 22. t. 70.—Heads more than double as large as in Sp. 7, spines stronger, pods less strongly veined, seeds much broader.—Waste and cultivated ground, introduced. A. V.—VIII.] E. S.

7. Melilo'tus *Hill.* Melilot.

1. *M. officinális* (Lam.); *wings keeled and standard equal, pods* ovoid *acute* compressed transversely wrinkled *hairy,* leaflets serrate truncate narrowly ovate, stip. setaceous entire.—*E. B.* 1340.—*M. macrorrhiza* Pers., Koch.—St. erect, 2—3 feet high. Fl. in lateral racemes, yellow.—Waste places. B.? VI.—VIII. *Common Melilot.* E. S. I.

2. *M. arven'sis* (Wallr.); *wings and standard equal longer than the keel, pods* ovoid blunt mucronate *rounded* and slightly keeled *on the back* transversely plicate glabrous, lts. obcordate or oblong serrate uppermost lanceolate, stip. awlshaped entire.—*E. B. S.* 2960. M. officinalis *Koch.*—St. erect. Fl. yellowish, in long racemes. Pods brown.—Waste places. Cambridge, Thetford, &c. B. VI.—VIII. E. I.

†3. *M. al'ba* (Desr.); *wings and keel equal but shorter than the standard, pods ovate blunt* mucronate netted *glabrous,* leaflets obovate the upper ones oblong serrate blunt, stip. awlshaped entire.—*M. leucantha* Koch, *E. B. S.* 2689.—St. erect. Fl. white.—Sandy and gravelly places near the sea, rare. B. VII. VIII. E. S. I.

*4. *M. ind'ica* (All.); wings and keel equal but shorter than the standard, *pods globular-ovoid blunt* mucronate netted glabrous, lts. obovate serrate at the end, stip. awlshaped entire. —*Sy. E. B.* 344. *M. parviflora* (Desf.) ed. viii.—Slender. Smaller in all its parts than the other species, 6—15 in. high. Fl. very small, pale yellow; cal.-teeth triangular. Pods olive-green.—Waste places. A. VII. VIII. E.

8. Trifo'lium *Linn.* Clover.

**Fl. sessile; cal. with an elevated thickened often hairy line or ring of hairs in its throat, not inflated.*

1. *T. praten'se* (L.); *heads* ovate dense *sessile*, cal. 10-veined hairy not half so long as the corolla, teeth setaceous ciliate, free part of *stip. blunt ovate* abruptly bristle-pointed, adpressed, leaflets oval emarginate upper ones entire apiculate.—*E. B.* 1770. *St.* 15. 11.—Lts. of cultivated plant [var. *sativum*, Schreb.] usually all quite entire. Stip. adpressed to petiole; veins much branched and anastomosing. Upper part of cal. usually hairy; teeth 5, 4 nearly equal in length to the tube, the lower one twice as long. Heads of fl. sometimes slightly stalked. Fl. purplish, sometimes white. St. erect.—β. *parviflorum*[1]; head stalked, calyx-teeth as long as or longer than the corolla.—*Fl. Dan.* 2782.—Mountainous pastures, fields. β in dry places. P. V.—IX. *Purple Clover.* E. S. I.

2. *T. médium* (L.); *heads* subglobose lax *stalked*, calyx 10-veined glabrous not half as long as the corolla, teeth setaceous hairy, free part of *stip. lanceolate* acuminate spreading, leaflets elliptic or lanceolate apiculate.—*E. B.* 190.—Stip. spreading; veins branching, parallel, scarcely at all joining. Cal. glabrous; teeth ciliate, 4 of them as long or rather longer than the tube, the fifth $\frac{1}{3}$ longer. Heads of fl. large. Fl. purplish. St. ascending, zigzag.—Dry elevated pastures. P. VI.—IX. E. S. I.

3. *T. ochroleúcon* (Huds.); heads subglobose dense stalked solitary terminal, *cal.* 10-veined *pubescent* about half as long as the corolla, *teeth erect subulate lower one rather longer than the tube the others two-thirds shorter,* stip. lanceolate-subulate, leaflets elliptic-oblong the lower one emarginate or cordate.—*E. B.* 1224. *St.* 15. 15.—Cal. with acute teeth having 1 strong vein. St. 1½ foot high, erect. Lower l. on very long stalks. Fl. cream-coloured, at length turning brown.—Dry gravelly soils and clays of the East of England. P. VI. VII. E.

---

[1] Mr. Burkill (J. of B. xxxix. 1901, p. 235) considers this an abnormal state.—H. & J. G.

4. *T. incarnátum* (L.) ; *heads* ovate *at length cylindrical* stalked solitary terminal, cal. 10-veined hairy, *teeth patent in fruit lanceolate-subulate nearly equal* rather longer than their tube and *falling short of the cor., stip. ovate blunt,* leaflets obovate retuse or obcordate, st. erect and together with the l. and stip. villose. —Mouth of fr.-cal. hairy. Stip. sometimes rather acute especially the lower ones.—*a* ; hairs of stem patent, heads oblong, fl. reddish purple.—*β*. *T. Molinerii* (Balb.) ; st. with adpressed hairs, heads conical, fl. nearly white. *E. B. S.* 2950.—Var. *a* is cultivated. *β* native near Lizard Point, Cornwall. A. VI. VII. E.

*5. *T. stellátum* (L.) ; *heads globose* stalked terminal, *calyx* 10-veined hairy, *teeth subulate from a broad base equal exceeding the corolla patent in fruit 3-veined and netted,* throat closed with hairs, stip. ovate rather acute denticulate, *leaflets obcordate,* st. spreading and together with the l. and stip. villose. -*E. B.*1545. *St.* 16. 5.—Fl. cream-coloured, small. Calyx of the fruit remarkably large, its teeth spreading like a star. St. short.—Shingly beach near Shoreham. A. VI. VII. E.

6. *T. arven'se* (L.) ; *heads nearly cylindrical* stalked *very hairy,* cal. 10-veined, *teeth subulate-setaceous* hairy nearly equal exceeding the cor. at length slightly spreading, stip. ovate or lanceolate acuminate, leaflets linear-oblong.—*E. B.* 944. *St.* 16. 3.—St. erect or, in a maritime form [var. *perpusillum* DC.] (Ray Syn. t. 14. f. 2), procumbent with globose heads ; st. and l. finely hairy. Points of the lower stip. sometimes very slender. Fl. small, almost concealed by the very hairy calyx. Seeds oval, greenish yellow ; radicle not prominent.—Sandy fields. A. VII.– IX. *Hare'sfoot Trefoil.* E. S. I.

7. *T. striátum* (L.) ; heads ovate or oblong sessile terminal and lateral solitary or the terminal in pairs, calyx 10-veined hairy, *teeth subulate* unequal *straight* mucronate about equalling the cor., *tube ventricose in fruit,* stip. ovate cuspidate, leaflets obcordate or obovate, veins equal and straight at the margins. —*E. B.* 1843. *St.* 16. 6, 7.—St. procumbent, 4—10 in. long and as well as the l. silky. Stip. with reddish veins. Fl. small. Seeds oval, brownish yellow; radicle not prominent.—*β*. *erectum* (Leight.) ; st. erect, heads long subconical lateral shortly stalked, cor. longer than calyx.—Dry and sandy places. A. VI. VII. E. S. I.

8. *T. scábrum* (L.) ; heads ovate sessile terminal and lateral solitary, cal. 10-veined hairy, *teeth lanceolate* mucronate about equalling the cor. *with* 1 *strong prominent vein* at length patent, *tube cylindrical in fruit,* stip. ovate cuspidate, leaflets obovate,

veins thickened and curved at the margins.—*E. B.* 903.—St. procumbent, spreading. Fl. small. Cal. of the fruit very rigid Seeds oblong, reddish yellow; radicle not prominent.—Dry sandy places. A. V.—VII. E. S. I.

9. *T. Boccóni* (Savi); *heads oblong-ovate* sessile terminal usually 2 together, cal. 10-veined hairy, teeth lanceolate-subulate mucronate about as long as the cor. with 1 strong prominent *vein* erect, tube cylindrical in fruit, *stip. oblong with a long subulate point*, leaflets oblong obovate or roundish obovate in the lower leaves, veins equal and straight at the margins, seeds with the radicle slightly prominent.—*E. B. S.* 2868.—St. 2—6 in. high, erect. Fl. small, pale yellow. Seeds oval, brownish yellow. Heads dense, somewhat conical.—In dry places, near the Lizard Point, Cornwall. A. VII. E.

10. *T. squamósum* (L.); *heads ovate-globose* stalked terminal, cal. strongly veined, *teeth ciliate at first subulate erect* falling short of the cor. *afterwards broad leaflike acute spreading 1-veined the lower one longest and 3-veined*, tube hairy above obconical in fruit, *stip. broadly subulate very long*, leaflets oblong-obovate.—*T. maritimum* (Huds.) ed viii. *E. B.* 220.—St. spreading, usually procumbent. Fl. pale red, small.—Muddy salt marshes. A. VI. VII. E.

** *Fl. sessile; throat of the calyx naked within; heads few-flowered, at length producing thick stellate fibres (abortive calyces) from their centre which ultimately fold over the fruit.*

11. *T. subterráneum* (L.); heads 2—5-flowered erect deflexed in fruit, calyx glabrous, teeth filiform hairy nearly equal shorter than the corolla, tube inflated in fruit and at length split longitudinally, *abortive calyces many slender with 5 points*, stip. ovate pointed, leaflets obcordate.—*E. B.* 1048.—St. prostrate and as well as the l. hairy. Fl. white, considerably longer than the calyx. The abortive calyces are remarkably characteristic of this species.—Dry gravelly places. A. V. VI. E. I.

*** *Fl. sessile or stalked; throat of the calyx naked within, not inflated; pods 2—4-seeded.*

12. *T. glomerátum* (L.); *heads* globose *sessile* terminal and axillary, *calyx sessile* 10-veined, *teeth ovate very acute veiny* nearly equal reflexed, stip. ovate taper-pointed, leaflets obcordate the upper ones obovate, seeds 2.—*E. B.* 1063.—St. procumbent. Fl. rose-coloured; standard persistent, striate. Seeds transversely ovate-reniform, radicle prominent.—Gravelly places East and South of England, rare. Wicklow, I. A. VI. E. I.

13. *T. stric'tum* (L.); heads globose axillary, calyx sessile 13-veined, teeth subulate unequal somewhat spreading, stip. broad pointed serrate, leaflets obovate the upper ones oblong, seeds 2.—*E. B. S.* 2949.—St. diffuse, short. Ped. short. Pods slightly projecting; seeds ovate, radicle slightly prominent.—Jersey. Near the Lizard Point, Cornwall. A. VI.—VIII. E.

14. *T. suffocátum* (L.); heads roundish sessile axillary, *cal.* sessile, *teeth lanceolate* acute *recurved* longer than the corolla, stip. ovate pointed, leaflets obcordate, seeds 2.—*E. B.* 1049.—St. short, usually buried in the sand. Fl. small, erect. Cal. scarcely striate. Seeds roundish, radicle prominent.—Sandy sea-shores, rare. A. VI. E.

15. *T. répens* (L.); heads roundish, *ped. axillary longer than the leaves, fl. stalked at length deflexed*, calyx glabrous half as long as the corolla, teeth lanceolate unequal erect, stip. ovate abruptly cuspidate, leaflets obovate or obcordate, *seeds* 4, *stems creeping.—E. B.* 1769.—Fl. white, sometimes pink, rarely lilac-purple (*T. Townsendii*); standard striate. L. often with a dark spot at their base. Pod covered by the faded corolla.—Meadows and pastures. P. V.—IX. *Dutch or White Clover.* E. S. I.

*16. *T. hyb'ridum* (L.); *heads globular depressed*, ped. axillary longer than the leaves, fl. stalked at length deflexed, cal. glabrous half as long as the (pinkish) corolla, *teeth nearly equal subulate* erect, *stip. ovate-lanceolate attenuate*, leaflets obovate-lanceolate, seeds 2—4, *st. erect or ascending not rooting.—Sy. E. B.* 361.—Fl. white or pale pink, turning brown. St. 1—2 ft. high, branched, wavy.—[β. *T. elegans* (Savi); st. decumbent solid (hollow in the type), heads smaller, stip. narrower.] Introduced. P. VII.—IX. *Alsike Clover.* E. S. I.

**** *Fl. sessile: throat of the calyx naked within, inflated after flowering and arched above.*

17. *T. fragíferum* (L.); heads globose, peduncles axillary exceeding the leaves, *involucre multifid equalling the calyx*, calyx of the fruit membranous netted downy, *stip. ovate with a long slender point*, leaflets obovate emarginate minutely serrate, seeds 2.—*E. B.* 1050. *St.* 16. 8.—St. creeping. Fl. purplish red. Heads large, remarkable when in fruit for their curious calyces enclosing the pods.—Damp pastures. P. VII. VIII. E. S. I.

[*T. resupinátum* (L.); heads hemispherical at length globose, peduncles axillary short, *bracts minute*, calyx of the fruit membranous netted woolly, *stip. subulate-lanceolate from an ovate base*, leaflets obovate minutely serrate, seeds 2.—*Sy. E. B.* 364. *St.* 16. 9.—St. prostrate or ascending. Fl. small, reversed in position. Pod included.—Introduced. A. VII.] E.

***** *Fl. stalked, yellow; throat of the calyx naked within, not inflated; cor. persistent; standard covering the pod.*

18. *T. procum'bens* (L.); *heads oval dense with about* 40 *fl.*, ped. axillary equalling or exceeding the leaves, fl. at length reflexed, *standard dilated and deflexed in front* (*not folded*) striate much exceeding the pod, *style much shorter than the pod, stip.* ½ *ovate* acute entire, lts. obovate emarginate, *central petiole longest*, seeds oval, radicle scarcely prominent.—*E. B.* 945. *St.* 15. 15. *T. agrarium*, Huds. (not L.).—Primary stem erect, branches procumbent or ascending. Fl. turning tawny. Radicle causing a slight irregularity in the otherwise regularly oval seeds.—Dry pastures. A. VI.—VIII. E. S. I.

*19. *T. agrárium* (L.); *heads oval dense of about* 50 *fl.*, ped. axillary equalling or exceeding the l., fl. at length reflexed, *standard dilated* striate much exceeding the pod, style nearly as long as the pod, stip. oblong-lanceolate not enlarged at the base, lts. narrowly obovate *the central petiolule not longer than the lateral.*—*St.* 16.—St. erect. Fl. deep yellow turning brown.—Fields and roadsides, established in some places. P. VI.—VIII. E. S.

20. *T. dúbium* (Sibth.); *heads close about* 12 *fl.*, ped. axillary straight, pedicels very short, fl. at length reflexed, *standard folded* furrowed *truncate* covering the pod, style much shorter than the pod, stip. ovate falling short of petioles, lts. obcordate intermediate one stalked.—*E. B.* 1256. *T. minus* (Sm.) ed. viii. *T. filiforme* Koch.—St. wiry, ascending or prostrate. Fl. turning dark brown.—Dry places. A. VI.—VIII. E. S I.

21. *T. filifor'me* (L.); *fl. few* (2—7) *in lax racemes*, ped. axillary flexuose, pedicels as long as the calyx-tube, fl. at length reflexed, *standard* folded *not furrowed deeply notched* scarcely covering the pod, lts. sessile.—*E. B.* 1257. *T. micranthum* Koch.—St. prostrate, very slender. Standard much narrower in proportion than in *T. minus.*—Dry places. A. VI. VII. E. S.? I.

9. Trigonel'la *Linn.*

1. *T. purpuras'cens* (Lam.); *clusters stalked* axillary *of* 1–3 *stalked fl.*, calyx glabrous, teeth slender acute nearly equal erect, stip. ovate with long taper points, leaflets obcordate, *seeds* 6—8.—*E. B.* 1047. *Falcatula ornithopodioides* (Bab.) ed. viii.—St. prostrate. Fl. small, pet. all distinct. Pod. compressed,

blunt, transversely furrowed, slightly hairy, curved, longer than the calyx, opening with 2 valves.—Dry gravelly places, rare. A. VI. VII. E. I.

10. Lo'tus *Linn.* Bird's-foot Trefoil.

1. *L. corniculátus* (L.); claw of the standard obovate transversely vaulted, *calyx-teeth adpressed in the bud* subulate from a triangular base, *points of the 2 upper teeth converging, heads* 5—10-*flowered.—E. B.* 2090.—Glabrous or slightly hairy. St. ascending. Angle between the 2 upper calyx-teeth rounded.—*α*; st. short, lts. obovate thin, stip. rather narrowly ovate.—*β. villosus* (Ser.); upper part of st. l. and cal. hairy with long spreading hairs.—*γ. crassifolius* (Pers.); pilose, st. cæspitose, leaflets obovate fleshy, stip. ovate.—Pastures, dry banks, &c. P. VI.—VIII. E. S. I.

2. *L. ten'uis* (W. & K.); cal.-teeth [short] subulate adpressed in bud, points of upper 2 teeth converging [*wings oblong-obovate lower margin abruptly curved to the apex,* head 2—5-flowered] *leaflets linear* [*acute*] *or linear-obovate,* stip. ½-ovate.—*E. B. S.* 2615.—Glabrous or slightly hairy, st. filiform long procumbent or ascending. [Pet. turning green in drying.—Meadows and moist banks.] P. VI.—VIII. E. S.

3. *L. uliginósus* (Schk.); claw of the standard linear, *calyx-teeth spreading like a star in the bud* subulate from a triangular base, *two upper teeth diverging, heads* 8—12-*flowered,* leaflets obovate, stip. roundish-ovate.—*E. B.* 2091. *L. major* Scop. ed. viii.—Hairy. St. usually erect, 1—3 feet high. Angle between the 2 upper calyx-teeth acute. Sometimes glabrous, the margins and veins of the l. stip. bracts and sep. ciliate, st. erect or procumbent.—In damp places. P. VII. VIII. E. S. I.

4. *L. angustis'simus* (L.); claw of the standard linear, calyx-teeth straight in the bud subulate, *pod linear straight* 6 *times as long as the calyx, heads about* 2-*flowered.—E. B.* 925. *L. diffusus* (Ser.).—Standard broader than long, not exceeding the wings, fading to a greenish colour. Peduncle of the fl. as long as the l., of the fr. twice as long; leaflets and stip. ovate-lanceolate acute; st. procumbent.—*β. Seringianus* (Bab.); peduncle of the fl. and fr. as long as the leaves, leaflets obovate-oblong, stip. ovate acute, st. ascending. *L. angustissimus* (Ser.).—South of England near the sea. A. VII. VIII. E.

5. *L. his'pidus* (Desf.); claw of the standard subulate, calyx-teeth straight in the bud subulate, *pod rugose terete twice as long*

as the calyx, heads few (3—4)-*flowered,* leaflets obovate-lanceolate, stip. half cordate, st. procumbent.—*E. B. S.* 2823.—Standard longer than broad, exceeding the wings, not turning green. —Near the sea in Devon, Dorset, and Cornwall.—A. VII. VIII.
E.

11. ANTHYL'LIS *Linn.* Lady's Finger.

1. *A. Vulnerária* (L.); herbaceous, l. pinnate, leaflets unequal, heads of fl. in pairs, calyx of 5 ovate pointed teeth.—*E. B.* 104. *St.* 49. 4. 5.—Pod semiorbicular, long-stalked, upper suture arched outwards, 1-seeded. St. 6—12 in. high, silky. Root-l. simple, oval-oblong. Fl. yellow, in terminal pairs of crowded many-flowered heads.—β. *coccinea,* L. (*A. Dillenii,* Schult.): invol.-l. nearly equalling fl., plant smaller, fl. red-tipped. *Dill. Elth.* 320. [The following vars. have also been recorded:—*ovata* (Bab.) having a large broadly-ovate terminal lt.; *Allionii* (DC.) having the st. clothed with many spreading hairs; *maritima* (Koch) also very hairy, with tall erect branched st. and many flowering heads.]—Dry pastures. P. VI.—VIII. E. S. I.

12. OXY'TROPIS *Cand.*

1. *O. uralensis* (DC.); stemless, leaflets ovate acute in about 12 pairs, *peduncles* exceeding the leaves *erect silky,* bracts equalling the calyx, *pods erect* ovate-oblong inflated *silky* imperfectly 2-celled.—*Astragalus uralensis* Sm. *E. B.* 466.—Rootstock woody, branched. Pods abrupt with a very oblique acute point. *Fl. bluish purple.* More silky and hairy than Sp. 2.—Dry hilly pastures in Scotland. P. VII. S.

2. *O. campes'tris* (DC.); stemless, leaflets lanceolate in about 12 pairs, *peduncles* rather exceeding the l. *ascending hairy,* bracts equalling the calyx, *pods* erect ovate inflated *hairy* imperfectly 2-celled.—*E. B.* 2522. *St.* 19. 12.—Rootstock woody, prostrate. Pods narrowed upwards with a slightly oblique point. *Fl. yellowish* tinged with purple.—Clova mountains. [Perthsh.]. P. VII.
S.

13. ASTRAG'ALUS *Linn.* Milk-vetch.

1. *A. dan'icus* (Retz.); st. prostrate, *stip. united,* leaflets blunt in 8—10 pairs, racemes ovate, *peduncles exceeding the leaves, pods* ovate hairy *erect* stalked in the calyx.—*E. B.* 274. *A. hypoglottis* (L.) ed. viii.—Stip. opposite to the leaves. St. a few inches long, slender. Leaflets small. Fl. in rather large heads, ascending, purple. Ovary twice as long as its stalk.—Chalky and gravelly places, rare. P. VI. VII. E. S. I.

2. *A. alpinus* (L.); st. prostrate, *stip.* ovate *free,* leaflets elliptic blunt in 10—12 pairs, racemes short close, peduncles equalling the leaves, *pods* oblong hairy narrowed at both ends *pendulous* stalked in the calyx.—*E. B. S.* 2717. *St.* 19. 13.—Stip. sometimes slightly connected at the base. St. a few inches long, slender. Fl. few, drooping, white tipped with purple.—Lofty mountains. Aberdsh., Forf., Perthsh. P. VII. S.

3. *A. glycyphyl'los* (L.); st. prostrate, stip. ovate-lanceolate free, leaflets ovate in 5—6 pairs, racemes ovate, peduncles falling much short of the leaves, *pods linear incurved erect glabrous.*—*E. B.* 203.—*St.* 2—3 feet long, scarcely branched, nearly glabrous. Fl. in short dense racemes, dull yellow. Pods an inch long.—Hedges and thickets on a chalky or gravelly soil, rare. P. VI. *Wild Liquorice.* E. S.

Tribe II. *Viciea.*

14. Vi'cia, *Linn.* Vetch.

* *Upper part of the style equally hairy all over.*

† Peduncles long, few-flowered; cal. not gibbous at the base on the upperside.—Annuals. Ervum. *Tares.*

1. *V. hirsúta* (S. F. Gray); *ped.* 1—6-*fl.* about equalling the leaves, leaflets in 6—8 pairs linear-oblong truncate mucronate, calyx-teeth equal as long as their tube the 2 upper ones converging, *pods sessile oblong 2-seeded hairy, hile long linear.*—*Ervum* Sm. *E. B.* 970. *Cracca minor* Godr.—Stip. 2-lobed, outer lobe trifid with setaceous segments, inner lanceolate. Fl. small, pale blue, standard entire. Calyx-teeth subulate. Pod obliquely truncate; its upper suture nearly straight and prominent at the end. Seeds globose, compressed, red with darker spots, smooth. Ped. rarely 1—2-flowered and pods glabrous.—Corn-fields and hedges. A. VI.—VIII. *Hairy Tare.* E. S. I.

2. *V. gemel'la* (Crantz); *ped.* 1—2-*fl.* about equalling the leaves, *leaflets* linear-oblong *blunt* mucronate in 4—6 pairs, calyx-teeth unequal shorter than the tube the 2 upper ones shortest "diverging," *pods shortly stalked linear-oblong about 4-seeded glabrous,* hile oblong.—*Ervum* Sm. *E. B.* 1223. *V. tetrasperma* (Moench) ed viii. *St.* 32. 14.—Stipules half-arrowshaped. Fl. small, pale blue; standard with blue streaks, emarginate. Calyx-teeth long triangular. Pod rounded; its

upper suture decurved at the end. Seeds 3—5, globose, dull brown, slightly rough. [A narrow and acute leaved form, var. *tenuissima* (Druce), has been mistaken for Sp. 3.]—Fields and hedges. A. VI.—VIII. *Smooth Tare.* E. S. I.

3. *V. grácilis* (Lois.); ped. 1—4-fl. becoming twice as long as the leaves, *leaflets* linear *acute* in 3—4 pairs, calyx-teeth unequal shorter than their tube, the 2 upper ones shortest, *pods* linear 5—8-*seeded* glabrous, hile roundish-oval.—*E. B S.* 2904. —Stip. half-arrowshaped. *Fl. twice as large as those of V. tetrasperma*, pale blue; standard emarginate. Calyx-teeth long-triangular, "two upper slightly converging." Upper suture of pod slightly decurved at the end. Seeds globose, variegated with dark brown and yellow, smooth, *half as large and hile half as long* as those of Sp. 2.—Fields and hedges in the South. A. VI.—VIII. E.

†† Peduncles long, many-flowered; calyx gibbous at the base on the upperside.—Perennial.

4. *V. sylvat'ica* (L.); ped. exceeding the leaves, leaflets elliptic blunt mucronate in about 8 pairs, *stip.* lunate *deeply toothed at the base, teeth setaceous, calyx-teeth* shorter than their tube, *subulate.*—*E. B.* 79. *St.* 31. 3.—St. many feet long, climbing by *branched tendrils.* Fl. many, cream-coloured and streaked with blue or purple. Hile extending about half round the seed. —Woods and thickets.—P. VII. VIII. *Wood Vetch.* E. S. I.

5. *V. Or'obus* (DC.); leaflets ovate-oblong or ovate-lanceolate mucronate in many (7—10) pairs, *stip.* half-arrowshaped *slightly toothed at the base, calyx-teeth* longer than their tube, 2 *upper ones triangular*, the others triangular-subulate.—*Orobus sylvaticus* L. *E. B.* 518.—St. ascending, 1—1½ foot long. Tendrils reduced to a short slender point. Fl. many, cream-coloured streaked with purple. Pods linear-oblong. Hile extending about ½ round the seed.—Northern rocky woods. P. V. VI. *Bitter Vetch.* E. S. I.

** *Upper part of the style hairy all over, the hairs rather longer below the stigma but scarcely bearded; calyx gibbous at the base.*

6. *V. Crac'ca* (L.); ped. long many-flowered, lts. lanceolate mucronate silky in about 10 pairs, *stip.* half-arrowshaped *entire*, calyx-teeth shorter than their tube, upper pair minute, others subulate, standard sinuate at about the middle of each side, its limb and claw equally long, pods linear-oblong smooth. *E. B.*

1168. *St.* 31. 6.—St. 3—4 feet long. Tendrils branched. Fl. blue varied with purple. Seeds subglobose, black. Hile linear, extending ½ round the seed. [A dwarf greyer and more pubescent form is var. *argentea* (C. & G.).] Hedges. P. VI.—VIII. E. S. I.

*** *Style bearded below the stigma, in other respects glabrous or uniformly hairy all over its upper part; calyx gibbous at the base on the upperside.*

7. *V. bithyn'ica* (L.); ped. falling short of the leaves 1—2-flowered, *lts. of upper l. in 2 pairs elliptic-lanceolate mucronate or linear-lanceolate acute,* stip. half-arrowshaped toothed, calyx-teeth longer than their tube lanceolate-subulate, *pods* linear-oblong *hairy.—E. B.* 1842. St. 32. 5.—St. 12—18 in. long. Fl. almost always solitary, purple. "Seeds globose, speckled with black and grey; hile oval." Upper part of the style hairy all over. The inland plants have broader leaflets and more cut stip. than those [var. *angustifolia* Sy.] found near the sea.—Bushy places on a gravelly soil, rare. P. VII. VIII. E.

8. *V. sépium* (L.); *fl.* 4—6 *in small axillary nearly sessile clusters,* leaflets in 4—8 pairs ovate blunt mucronate gradually smaller upwards on the petiole, stip. half-arrowshaped undivided or lobed, *calyx-teeth unequal* shorter than their tube, 2 upper ones curved upwards, pods linear-oblong glabrous.—*E. B.* 1515. *St.* 31. 16.—St. about 2 feet high. Fl. purplish. Calyx hairy. Hile linear, extending about ⅔ round the seed. Upper part of the style nearly or quite glabrous, bearded. L. more or less hairy, leaflets sometimes ovate-lanceolate truncate.—Woods and hedges. P. VI.—VIII. E. S. I.

[*V. hybrida* (L.) and *V. lævigata* (Sm.) are now lost.]

9. *V. lútea* (L.); fl. solitary axillary, leaflets elliptic-lanceolate acute or rounded at the end apiculate in 5—8 pairs, *calyx-teeth unequal,* upper ones very short and curved upwards, lower one longer than the tube, *standard glabrous,* pods elliptic-oblong hairy.—*E. B.* 481. *St.* 31. 13.—St. procumbent, 1—2 feet long. Fl. sulphur-coloured. Hairs on the pod bulbous. Seeds round. compressed, with a short hile. L. varying greatly in hairiness. —Pebbly and sandy ground near the sea. P. VI.—VIII. E. S.

10. *V. sativa* (L.); *fl. axillary solitary or in pairs,* leaflets in 5—7 pairs elliptic-oblong retuse or obcordate mucronate, upper ones narrower or linear truncate mucronate, *calyx-teeth equal* lanceolate-subulate long equalling their tube, standard glabrous, *pods linear slightly silky,* seeds globose smooth.—Seeds slightly

compressed; hile linear, occupying about $\frac{1}{4}$ of the circumference.—*a. V. sativa* (Sm.); *leaflets all elliptic- or ovate-oblong*, the lower ones shorter and broader, fl. usually in pairs, *pods mostly parallel to the st.*, st. 1—1½ foot high. *Sy. E. B.* 392. *St.* 31. 10.—*β. V. angustifolia* (L.); leaflets of the upper l. linear-lanceolate, lower ones obovate retuse or obcordate, fl. solitary or in pairs, pods mostly patent, st. slender. *E. B. S.* 2614. *St.* 31. 11.—*γ. V. Bobartii* (Forst.); leaflets of the upper l. linear, fl. solitary, *pods patent*, st. prostrate. *E. B. S.* 2708. *V. angustifolia* Sm.—*a* is only known in cultivation. *β* and *γ* in dry places. A. V. VI. *Common Vetch.* E. S. I.

**** *Style bearded below the stigma; calyx not gibbous.*

11. *V. lathyroi'des* (L.); fl. axillary solitary, leaflets in 1—3 pairs obovate or oblong retuse mucronate, calyx-teeth subulate straight as long as their tube, pods linear glabrous, *seeds nearly cubical tubercular*, hile short oblong.—*E. B.* 30. *St.* 31. 12.—St. procumbent, 3—5 in. long. Fl. small purple.—Dry gravelly and sandy places. A. V. VI. E. S. I.

15. LATH'YRUS *Linn.* Vetchling.

* *No true leaflets.*

1. *L. Aph'aca* (L.); ped. 1-fl., *petioles leafless forming tendrils*, *stip. very large leaflike* hastate-ovate.—*E. B.* 1167.—St. weak, climbing. Fl. yellow, rarely 2 together. Remarkable for its want of l., which are replaced by the stipules, rarely 1 or 2 elliptic leaflets occur. Pods subfalcate; seed smooth, compressed.—Sandy and gravelly fields, chiefly in the South, rare. A. V.—VIII. E.

2. *L. Nissólia* (L.); ped. long 1—2-flowered, *petioles leaflike* but leafless linear lanceolate, *no tendrils*, *stip. minute* subulate. —*E. B.* 112.—St. mostly erect. Petioles grasslike. Fl. purple on long stalks. Pods cylindrical; seeds tubercled, round; hile small, oval.—Bushy grassy places, rare. P. VI. E.

** *Petioles with one pair of leaflets and a tendril.*

[*L. sphær'icus* (Retz.): ped. 1-fl. with a long point shorter than the petiole, lts. linear-lanceolate, seeds globose; has been found in Hertfordshire.]

3. *L. hirsútus* (L.); ped. 2-flowered, *lts. linear-lanceolate*, pods hairy, *seeds* globose tubercled.—*E. B.* 1255.—St. winged, climb-

ing to the height of 1 or 2 feet. Pods linear-oblong, covered with bulbous hairs. Fl. 2 or 1, pale blue with a crimson standard. Hile oblong.—Rare. Essex. Surrey. A. VI. VII. E.

4. *L. praten'sis* (L.); *st. angular, ped. many-flowered,* lts. lanceolate mucronate slightly silky, *calyx-teeth subulate,* pods obliquely veined, seeds globose smooth.—*E. B.* 670.—Creeping. St. 2—3 feet high, climbing, not winged. Pods linear-oblong, compressed. Fl. racemose, drooping, bright yellow. *Hile small,* oblong.—Moist meadows. P. VII. VIII. E. S. I.

†5. *L. tuberósus* (L.); *st. angular,* ped. long many-flowered, *lts. obovate-oblong* mucronate, upper *cal.-teeth triangular,* pod net-veined, seeds globose smooth.—*Sy. E. B.* 401.—Creeping and tuberous. St. about 2 ft. high, not winged. Fl. 2-5 together, purple. Pods linear-oblong, subcylindric. Hile small.—Hedges and fields. Abundant at Fyfield, Essex. P. VIII. E.

6. *L. sylves'tris* (L.); *st. winged,* ped. many-flowered, *lts. linear-lanceolate* or lanceolate, *calyx-teeth triangular-subulate,* 2 upper ones short, pods net-veined, *seeds* compressed roughish nearly *half surrounded by the hile.—E. B.* 805.—St. climbing to the height of 5—6 feet. Pods linear-oblong curved. Fl. greenish yellow variegated with purple. Broader-leaved varieties often pass for *L. latifolius,* which differs by its elliptic pointed lts. and rougher seeds with a shorter hile, but is not a native. —Woods and thickets. P. VII.—IX. E. S.

*** *Petioles with 2 or more pairs of leaflets and a tendril.*

7. *L. palus'tris* (L.); st. winged, ped. long many-flowered, *lts. in 2 or 3 pairs linear-lanceolate acute,* stip. half-arrowshaped lanceolate, pods linear-oblong compressed, seeds globose smooth ¼ surrounded by the hile.—*E. B.* 169.—St. 2—3 feet high. Fl. bluish purple.—Boggy meadows, rare. P. VI. VII. E. S. I.

8. *L. marit'imus* (Big.); *st. angular not winged,* ped. short many-flowered, lts. in 3 or 4 pairs oval, *stip. large oval cordate-hastate,* pods oblong obliquely net-veined, seeds globose ⅓ surrounded by the hile.—*Pisum* L. *E. B.* 1046.—St. prostrate. Leaflets large, blunt but apiculate; petioles often recurved. Fl. purple variegated.—*β. acutifolius* (Bab.); leaflets elliptic-lanceolate acute, petioles straight, stems slender straggling.—Pebbly sea-shores, rare. *β.* Burrafirth, Unst, Shetland. P. VII. VIII. *Sea Pea.* E. S. I.

**** *L. pinnate, without tendrils.* OROBUS.

9. *L. monta'nus* (Bernh.); *st.* simple *winged*, peduncles 2—4-flowered, l. of 2—3 pairs of oblong or lanceolate blunt apiculate leaflets *without tendrils*, stip. half-arrowshaped broad, *pods cylindrical*, seeds globose $\frac{1}{3}$ surrounded by the hile.—*L. macrorrhizus* (Wimm.) ed. viii. *O. tuberosus* L. *E. B.* 1153.—Root tuberous. L. glaucous beneath. Fl. purple, variegated with red and blue.—*β. O tenuifolius* (Roth): lts. linear.—Woods and thickets in hilly countries. P. V.—VII. E. S. I.

10. *L. niger* (Wimm.); *st. branched not winged*, ped. many-flowered, l. of 3—6 pairs of lanceolate or oblong leaflets *without tendrils*, *stip. linear-subulate* the lower ones half-arrowshaped, pods slightly compressed, seeds oval $\frac{1}{3}$ surrounded by the hile. —*E. B. S.* 2788.—Turns black in drying. Fl. variegated with red, blue, and purple. Pods linear. Seed dark brown, perfectly smooth.—Rocky woods in Scotland, rare. P. VI. VII. S.

Tribe III. *Hedysareæ.*

16. ORNI'THOPUS *Linn.* Bird's Foot.

1. *O. perpusil'lus* (L.); ped. exceeding the leaves, calyx-teeth triangular acute $\frac{1}{3}$ the length of their tube, beak scarcely as long as a joint of the pod.—*E. B.* 369.—A small prostrate plant, 3—12 in. long. L. with $5\frac{1}{2}$—$12\frac{1}{2}$ pairs of elliptic downy leaflets. Fl. small; calyx hairy, cor. white with crimson veins. Pods curved, joints beadlike wrinkled lengthwise.—Dry sandy and gravelly places. A. V.—VII. E. S. I.

17. ARTHROLO'BIUM *Desv.*

1. *A. ebracteátum* (DC.); ped. about equalling the l. 2—4-flowered, stip. minute distinct, l. pinnate with many pairs of elliptic-oblong leaflets, the lowest pair remote from the stem.—*E. B. S.* 2844.—St. prostrate, filiform. Fl. small, yellow, standard red externally. Pod. curved upwards, joints cylindrical rugose.—Channel and Scilly Islands. A. VI. VII. E.

18. HIPPOCRE'PIS *Linn.* Horseshoe Vetch.

1. *H. comósa* (L.); pods umbellate, their joints rough curved neither dilated nor bordered, joinings glabrous, peduncles longer than the leaves.—*E. B.* 31.—St. procumbent, often a foot long. Fl. yellow. Leaflets 7—13, obovate, blunt or emarginate, apiculate.—Dry calcareous banks. P. V.—VIII. E.

19. ONOBRY'CHIS *Mill.* Sainfoin. Cock's-head.

1. *O. viciæfólia* (Scop.); wings shorter than the calyx, keel about as long as the standard, st. ascending, pods with netted spinous elevations on the disk and short sharp flat teeth on the lower suture.—*E. B.* 96. *St.* 19. 10. *O. sativa* (Lam.) ed. viii. —St. often 2 feet long. Fl. in long dense racemes, crimson, variegated. Tube of the calyx silky, short; teeth very long. Leaflets elliptic-oblong, mucronate, entire glabrous above, in about 12½ pairs.—On chalky and limestone hills. P. VI. VII. E.

Order XXVI. ROSACEÆ.

Cal. 4—5-parted, or 8—10-parted in 2 rows, free or adhering to and enclosing the ovary; odd lobe superior. Pet. usually 5, equal, perigynous. Stam. perigynous, usually indefinite. Carp. several or solitary, distinct or connate or adnate to the calyx. Styles distinct, often lateral. Fruit various. Seeds nearly without albumen, embryo straight.—L. alternate, usually compound, with stipules.

Suborder I. AMYGDALEÆ.

Fruit a drupe. Calyx deciduous, inferior, quite free from the solitary ovary. Stipules free.

1. PRUNUS. Drupe fleshy, indehiscent; its nut smooth or furrowed.

Suborder II. ROSEÆ.

Carpels several, distinct from each other and from the calyx. Stipules united to the petiole.

Tribe I. *SPIRÆIDÆ. Carpels* (follicles) several, *not included in calyx-tube*; seeds 1—6, suspended from the inner edges of the follicles. Sep. persistent, in one row.

2. SPIRÆA. Cal. 5-cleft. Stam. many, inserted with the pet. on a disk adhering to the calyx. Follicles 1 or more, usually distinct. Seeds 2—6.

Tr. II. *SANGUISORBEÆ. Carpels* 1—3, *enclosed in the dry calyx-tube,* which is narrowed at the top.

3. SANGUISORBA. *Fl. perfect.* Cal. 4-cleft, with 2 or 3 scales at its base, tube quadrangular. Pet. 0. *Stam.* 4,

opposite to the segments of the calyx. Nuts 1—2. Style terminal. Stigma capitate, covered with oblong spreading prominences. Seeds suspended.

4. POTERIUM. *Fl. monœcious or polygamous.* Cal. 4-cleft, with 3 scales at its base, tube quadrangular. Pet. 0. *Stam.* 20—30. Nuts 2—3. Style terminal. Stigma brush-like, with filiform divisions. Seed suspended.

5. AGRIMONIA. Fl. perfect. *Calyx 5-cleft, without scales; tube turbinate, armed with hooked bristles above. Pet.* 5. *Stam.* 12—20, inserted with the pet. into a glandular ring in the throat of the calyx. Nuts 1—3. Style terminal. Seed suspended.

6. ALCHEMILLA. Fl. perfect. *Cal.* 8-*parted*, the alternate parts smaller, contracted at the throat, unarmed. *Pet.* 0. *Stam.* 1—4, inserted into a ring in the throat of the calyx and *opposite to the smaller segments.* Nuts 1—5. *Style basal.* Seeds ascending.

Tr. III. *DRYADEÆ.* Fruit not included in calyx-tube, of 5 or more small dry nuts (in *Rubus* drupes) inserted on a dry or succulent receptacle. Calyx persistent, open, nearly flat.

* *Attachment of the seed near to that of the style, radicle superior.*

7. SIBBALDIA. *Cal. concave*, 10-parted, in 2 series, 5 exterior parts smaller. *Pet.* 5. *Stam.* 5. Style lateral. *Fr. of* 5—10 *small nuts* placed on a dry receptacle. Seed ascending.

8. POTENTILLA. Cal. concave, 8—10-parted, in two series, 5 exterior parts smaller. Pet. 4—5. *Stam. many.* Style lateral or nearly terminal. *Fr. of many small nuts placed upon a flattish dry receptacle.* Seed pendulous or ascending.

9. COMARUM. Cal., cor., stam., and pistils as in *Potentilla. Receptacle ultimately large, spongy, persistent.* Style lateral near the top of the nut. Seed pendulous.

10. FRAGARIA. Cal., cor., stam., and pistils as in *Potentilla. Receptacle large, succulent, pulpy, deciduous.* Style lateral near the base of the nut. Seed ascending.

11. Rubus. *Cal.* concave or flattish, *5-parted.* Pet. 5. Stam. many. Styles nearly terminal. *Carp. many, succulent* (drupes), *placed upon a hemispherical or conical spongy receptacle.* Seed pendulous.

** *Attachment of the seed distant from that of the style, radicle inferior. Nuts with long awnlike styles.*

12. Dryas. *Cal.* 8—10-cleft, *in one row.* Pet. 8—9. Stam. many. Fr. of many small nuts, tipped with the persistent hairy *styles,* which are *not jointed.* Receptacle flat, dry. Seed ascending.

13. Geum. *Cal.* 10-cleft, *in 2 rows,* the outer parts smaller. Pet. 5. Stam. many. Fr. of many small nuts, tipped with the persistent *jointed styles hooked at the joining.* Receptacle elongated, dry. Seed ascending.

Tr. IV. *ROSIDÆ.* *Fruit* formed of many small dry nuts *enclosed in the fleshy calyx-tube.*

14. Rosa. *Cal.* urceolate, contracted at the mouth, *ultimately fleshy,* 5-fid. Pet. 5. Stam. many, inserted with the petals on the rim of the tube of the calyx.

Suborder III. POMEÆ.

Calyx-tube thick and fleshy, in fruit adhering to the carpels and forming a 1—5-celled pome: thus fl. appearing superior.

15. Cratægus. Calyx-segments 5, acute. Pet. 5. Styles 1—5. Fr. oval or round, *concealing the upper end of the* 1—5 *bony* 1—2-seeded *carpels.*

16. Cotoneaster. Calyx-segments 5. Pet. 5. Styles 2—5. Fr. turbinate, *its nuts* adhering to the sides of the calyx but *not cohering at the centre.*—Stam. erect, as long as the teeth of the calyx.

17. Mespilus. Calyx-segments 5, leaflike. Pet. 5. Styles 2—5. Fr. turbinate with *the upper end of the bony carpels exposed*; disk dilated, almost as broad as the fruit.

18. Pyrus. Cal. 5-toothed. Pet. 5. Styles 2—5. *Fr.* fleshy *with 5 cartilaginous or membranous distinct* 2-seeded *cells.* Testa cartilaginous.

Suborder I. *Amygdaleæ*.

1. Pru'nus *Linn.*

* *Young leaves convolute. Drupe glaucous.*

1. *P. spinósa* (L.); ped. solitary or in pairs, l. elliptic or ovate-lanceolate rather downy beneath. *P. communis* (Huds.) ed. viii. —*α*; branches spinous, ped. solitary glabrous, l. usually glabrous, fr. globose. *Sy. E. B.* 408.—[*β. macrocarpa* (Wallr.) *P. fruticans* (Weihe); branches spinous, ped. glabrous usually in pairs, l. pubescent on the veins beneath, fl. and fr. larger than in *α*, fr. globose.]—*γ. P. insititia* (L.); branches slightly spinous, ped. and underside of l. usually downy, fr. globose. *E. B.* 841.—**δ. P. domestica* (L.); branches without spines, ped. glabrous, l. hairy about the midrib beneath, fr. oblong. *E. B.* 1783.—*α* is a shrub with crooked and much armed black branches and *fl. before the leaves*; *β* and *γ* taller shrubs with *fl. and l. usually together*, *γ* with straight and slightly armed brown branches; *δ* a small tree with straightish branches.—Thickets. *δ* not indigenous. Sh. IV. V. *α. Sloe. Blackthorn. γ. Bullace. δ. Plum.* E. S. I.

** *Young leaves conduplicate. Drupe not glaucous.* Cerasus.

2. *P. Pádus* (L.); arborescent, l. obovate-lanceolate finely serrate glabrous, *fl. in pendulous racemes*, fr. roundish-oblong.—*E. B.* 1383.—A small tree. L. minutely doubly serrate. Fl. white, many, in a lax raceme. Fr. black, harsh, bitter; nut wrinkled.—Woods and hedges. T. V. *Bird-Cherry.* E. S. I.

3. *P. A'vium* (L.); arborescent, *l. long-stalked drooping* oblong-obovate *suddenly cuspidate incise-serrate downy beneath, calyx-tube constricted* below the entire sepals, pet. flaccid, "fr. heart-shaped."—*E. B.* 706.—A tree 20—30 feet high. Outer scales of the leaf-buds deflexed. Flower-buds not leafy. Fl. in umbels. Pet. bifid, with a minute claw. "Fr. firm, bitter, black or red with staining juice."—Woods. T. V. *Wild Cherry. Gean.* E. S. I.

4. *P. Cer'asus* (L.); fruticose, *l. short-stalked, not drooping* oblong-obovate or ovate-lanceolate *doubly crenate-serrate glabrous, calyx-tube not constricted*, "pet. firm," fruit round.—*E. B. S.* 2863.—An erect bushy shrub 3—8 feet high. Umbels scattered. Outer scales of the leaf-buds erect. Inner scales of the flower-buds leaflike. Sep. crenate-serrate. Pet. subemarginate, with a claw. "Fr. juicy, acid, always red, not staining." —Hedges. Sh. V. *Dwarf Cherry.* E. I.

Suborder II. *Roseæ*. Tribe I. *Spiræidæ*.

2. SPIRÆ'A *Linn.*

‡1. *S. salicifólia* (L.); *shrubby, stip.* 0, l. elliptic-lanceolate unequally serrate glabrous, racemes terminal compound, stam. exceeding the petals.—*E. B.* 1468.—A shrub 4—5 feet high with smooth round wandlike branches. Fl. flesh-coloured, in dense erect racemes.—Damp places in the North and in Wales, rare. Sh. VII. E. S.

2. *S. Ulmária* (L.); *herbaceous, stip. rounded, toothed*, l. interruptedly pinnate, *leaflets* ovate *undivided, the terminal one larger palmately* 3—5-lobed, fl. in compound cymes, caps. glabrous twisted together.—*E. B.* 960. *St.* 18. 8.—St. about 3 feet high, angular, branched. L. with a few large serrate leaflets and very minute intermediate ones; st. l. usually downy beneath. Cymes with long side branches. Fl. yellowish, sweet-scented. Pet. roundish. [β. *denudáta* (Bœnn.); st.-l. green and glabrous on under surface.]—Meadows and by water. P. VI.—VIII. *Meadow-sweet.* E. S. I.

3. *S. Filipen'dula* (L.); herbaceous, stip of the root-leaves linear acute entire, those of the stem rounded and cut, l. interruptedly pinnate, *leaflets all oblong deeply cut and serrate*, fl. in a panicled cyme, caps. hairy, straight but adpressed.—*E. B.* 284. *St.* 18. 7.—Root producing knobs. St. 1—1½ foot high, round, simple, panicled at the top. L. mostly radical, spreading; leaflets small, many, intermediate ones small. Fl. yellowish-white tinged with red. Pet. obovate.—Dry chalky and limestone pastures. P. V. VII. *Dropwort.* E. S. I.

Tribe II. *Sanguisorbeæ*.

3. SANGUISOR'BA *Linn.* Great Burnet.

1. *S. officinális* (L.); spikes ovate-oblong, stam. about equalling calyx, leaflets cordate-oblong.—*E. B.* 1312.—L. pinnate, glabrous; leaflets about 13, stalked, opposite, blunt, coarsely serrate. Spikes rarely long, cylindrical. Fr. oblong, winged chiefly in its upper half, transverse section terete, 4 wings thin. —Damp meadows. P. VI.—VIII. E. S. I.

4. POTE'RIUM *Linn.* Lesser Burnet.

1. *P. Sanguisor'ba* (L.); herbaceous, st. slightly angular, cal. of the *fruit* hardened *quadrangular with 4 thin entire wings and netted veins.*—*E. B.* 860.—L. pinnate, with many small ovate coarsely serrate subsessile leaflets glabrous or slightly hairy beneath. Lower part of the stems and petioles often downy.—On a dry calcareous soil. P. VI.—VIII. E. S. I.

‡2. *P. polyg'amum* (W. & K.); herbaceous, st. slightly angular, calyx of *fruit* hardened *4-winged pitted,* pits with elevated and *denticulate* edges.—*E. B. S.* 2989. *P. muricatum* (Spach) ed. viii.—*α. P. platylophum* (Jord.); fr.-cal. with rather blunt denticulate wings, pits with sharply denticulate edges.—*β. P. stenolophum* (Jord.); fr.-cal. with sharp entire wings, pits with rather bluntly denticulate edges.—Both of these closely resemble Sp. 1, but are usually larger in all their parts. Leaflets usually oval.—Dry places. P. VI. VII. E.

5. AGRIMO'NIA *Linn.* Agrimony.

1. *A. Eupatória* (L.); *cal.-tube of fr. obconic* furrowed to the base, *exterior spines spreading,* l. interruptedly pinnate serrate shaggy beneath.—*E. B.* 1335. *St.* 59. 4.—St. erect, about 2 feet high. Spikes long with distant yellow flowers. Leaflets deeply serrate.—Fields and roadsides. P. VI. VII. E. S. I.

2. *A. odoráta* (Mill.); *cal.-tube of fr. bellshaped* not furrowed, *exterior spines declining,* l. interruptedly pinnate hairy and with minute glands beneath.—*E. B. S.* 2982.—Scented. Spikes long. Fl. yellow; pet. obovate-lanceolate, wedgeshaped below, distant, spreading, flat, turning up at the end. Styles spreading. Leaflets deeply and sharply serrate throughout. Cal.-tube of the large fruits rarely with shallow furrows in its upper half. Taller than Sp. 1, usually more branched; l. and lts., fl. and fr. larger.—Waste places, rare. P. VI. VII. E. S. I.

6. ALCHEMIL'LA *Linn.* Lady's Mantle.

1. *A. vulgáris* (L.); *l. reniform* or nearly orbicular plaited 7—9-lobed, lobes rounded serrate throughout green beneath, fl. in terminal corymbs, or rather racemose cymes.—*E. B.* 597.—Slightly hairy. Fl. yellowish green. L. large on long stalks, st.-l. sessile with large notched connate stipules.—*β. A. filicaulis* (Buser) [1]; st.-l. and petioles silky, old l. wavy with broad waves.

[1] Rev. E. F. Linton in a paper on the segregates of *A. vulgaris* (J. of B. 1895, p. 110) states that *A. montana* Willd. to which this var. was referred in Ed. viii. is not British.

[γ. *A. alpestris* (Schmidt); whole plant nearly glabrous, l. slightly hairy on the nerves beneath and with tufts of silky hairs on the teeth.] —Moist pastures. P. VI.—VIII. *Common Lady's Mantle.* E. S. I.

2. *A. alpina* (L.); *radical l. digitate, divisions* 5—7 *separated* to their base oblong blunt adpressed serrate at the end white and silky beneath, fl. in interrupted spikes of small lateral and terminal corymbs, achene oblong-ovoid suddenly acuminate, st. slightly branched simple below.—*Sy. E. B.* 425. *St.* 51. 2.—St., cal. and underside of the l. beautifully silky. Leaflets rarely slightly combined, *outer ones* of the radical l. usually nearly *opposite* to each other. Branches usually undivided, ascending.—Mountains. P. VI. VII. *Alpine Lady's Mantle.* E. S. I.

3. *A. conjunc'ta* (Bab.); *radical l. suborbicular peltate-palmate, divisions* 5—7 much connected below oblong blunt adpressed serrate at the end white and very silky beneath, fl. in small lateral and terminal corymbs, achene ovate-ovoid gradually acuminate, st. much branched.—*E. B. S.* 2983. *A. argentea* (Don); not Lam.—Often taken for *A. alpina* but its lts. are much larger, their lobes broader and placed in the radical leaves so that *the* 2 *external ones* almost if not quite *touch each other so as to present the appearance of a peltate leaf*; st.-branches long alternate spreading and often again subdivided; fl. in small nearly simple distant corymbs. [Faroe Isles. *A. fissa* Fl. Dan. 2101. Dauphine.] Clova! Glen Sannox, Arran! P. VI. VII. S.

4. *A. arven'sis* (Scop.); *l. palmate* 3-*fid wedgeshaped below* hairy, lobes with 3—6 teeth at the end, fl. sessile axillary.—*Aphanes* Linn. *E. B.* 1011.—St. prostrate or ascending, 4—5 in. long. Fl. very small, greenish, in small hairy inconspicuous tufts.—Dry fields on sand and gravel. A. V.—VIII. *Parsley Piert.* E. S. I.

Tribe III. *Dryadeæ.*

7. SIBBAL'DIA *Linn.*

1. *S. procum'bens* (L.); l. ternate, leaflets wedgeshaped with 3 teeth at the end, fl. corymbose, pet. lanceolate.—*E. B.* 897. *St.* 17. 5. *Pot. Sibbaldii*, Hall.—L. pilose on both sides. St. woody, procumbent. Pet. very small, yellow. Pistils and stam. very variable in number.—Dry summits of Scottish mountains. Above Highcup Scar, Teesdale. P. VII. E. S.

8. POTENTIL'LA *Linn.* Cinque-foil.

* *Hairs on the receptacle shorter than the glabrous carpels.*

† Leaves pinnate.

1. *P. rupes'tris* (L.); *st. erect* dichotomous, leaflets roundish-ovate unequally cut and serrate 5—7 on the lower leaves, on the uppermost 3.—*E. B.* 2058.—Fl.-shoots annual. *Pet. white,* much longer than the calyx. Base of l. wedgeshaped. St. 1—2 feet high.—Craig Breidden, Montgomeryshire. Radnorshire. P. V. VI. E.

2. *P. Anserina* (L.); *st. creeping,* l. interruptedly pinnate, leaflets many oblong acutely serrate silky beneath, peduncles solitary.—*E. B.* 861. *St.* 4. 7.—*Fl. yellow,* large. L. green above, white and silky beneath; or white and densely silky on both sides.—Roadsides. P. VI. VII. *Silver Weed.* E. S. I.

†† Leaves digitate.

3. *P. argen'tea* (L.); *st. ascending, l. quinate,* leaflets obovate-cuneate incise-serrate *white and downy beneath* their margins revolute.—*E. B.* 89. *St.* 17. 7.—Fl. yellow, small, in terminal corymbs.—Dry gravelly places. P. VI. VII. E. S.

4. *P. ver'na* (L.); *st. prostrate,* lower l. of 5—7 *obovate leaflets* serrate towards the end bristly on the margin and ribs beneath, teeth 2—4 on each side, *lower stipules narrowly linear.*—*E. B.* 37. *St.* 17. 8.—St. woody, about 5 in. long. Fl. yellow, solitary or 2 or 3 together. The terminal tooth of the l. usually smallest and shortest.—Dry pastures. P. IV. V. E. S.

5. *P. rúbens* (Vill.); st. *ascending,* lower l. quinate, leaflets obovate-cuneate somewhat hairy deeply cut in the upper half, teeth about 4 on each side, *stip. all ovate.*—*P. maculata* (Pourr.) ed. viii. *P. alpestris* (Hall.). *P. salisburgensis* (Haenke). *St.* 17. 10. *Sy. E. B.* 429.—Larger than the preceding. All the teeth of l. equal. Outer sep. oblong, blunt; inner twice as broad ovate, acute. [A weak form with short stems and deeply cut lts. has been referred to var. *debilis* (Koch).]—Mountains. P. VI. VII. E. S.

6. *P. rep'tans* (L.); *st. filiform* prostrate *rooting, l.* quinate *stalked,* leaflets obovate bluntly serrate, peduncles solitary, pet. obcordate, *carpels asperous.*—*E. B.* 862.—L. on long stalks, often with a bunch of small l. in their axils, sometimes solitary, usually in pairs. Leaflets blunt, rough or hairy on their ribs and margins. Fl. on long stalks, yellow.—Sometimes the l. and cal. are covered with long silky hairs on both sides. [A small tufted form is var. *microphylla* (Tratt.).]—Roadsides and banks. P. VI.—IX. E. S. I.

7. *P. silves'tris* (Neck.); st. procumbent or ascending, l. ternate sessile or shortly stalked, lower l. quinate on long stalks, lts. lanceolate or obovate-cuneate incise-serrate, pet. obcordate, *carpels longitudinally wrinkled.*—*P. Tormentilla* (Nestl.) ed. viii. *Tormentilla officinalis* L. *E. B.* 863. *St.* 34. 12.—Rootstock large, woody. L. all nearly sessile, except the lowest, which are often ternate; lts. acute, rather hairy. Stip. deeply cut. Fl. small, yellow, usually with 8 sep. and 4 pet.—*P. mixta* (Nolte)[1]; l. stalked, lts. obovate-oblong acutely serrate, stip. entire. A hybrid between Sp. 6 and 7.—*β. P. procumbens* (Sibth.); leaflets obovate-cuneate deeply cut, l. 5-nate or 3-nate stalked, stip. entire or trifid. Fl. usually larger. *Sy. E. B.* 431.—In dry places. *β.* Woods and hedgebanks. P. VI.—VIII. E. S. I.

††† Leaves ternate.

*8. *P. norvégica* (L.); st. erect, lts. obovate coarsely toothed, pet. falling short of cal. obovate, carp. glabrous longitudinally wrinkled.—*E. B.* iii. Suppl. 435 A.—St. 6—12 in. Fl. crowded in term. cymes. Naturalized in Yorks. and elsewhere. A. or P. VI.—VIII. E.

*** Hairs on the receptacle long, carpels hairy at the scar or all over.*

9. *P. fruticósa* (L.); *shrubby, l. pinnate*, leaflets mostly 5 oblong acute entire hairy with revolute margins.—*E. B.* 88.—St. 3—4 feet high. Fl. large, yellow, terminal.—Teesdale. Wastdale. Galway. Clare. Sh. VI. VII. E. I.

10. *P. Fragarias'trum* (Ehrh.); st. procumbent, *l. ternate* greyish green, *leaflets roundish obovate serrate* silky on both sides, pet. narrowly obcordate not contiguous, carp. glabrous except at the scar smooth or wrinkled transversely.—*E. B.* 1785.—Hairs on the upperside of l. bulbous-based. Fl. small, white.—Woods, banks. P. IV. V. *Barren Strawberry.* E. S. I.

9. COM'ARUM *Linn.* Marsh Cinque-foil.

1. *C. palus'tre* (L.).—*E. B.* 172. *Potentilla* (Scop.).—St. ascending, 1 foot high, reddish. L. pinnate. Leaflets 5—7, elliptic oblong, acute, sharply serrate. Fl. several, dark purple; cal. purple within; pet. small. Differing from *Potentilla* by its enlarged spongy receptacle.—Marshes and peaty bogs. P. VII. E. S. I.

---

[1] This has been regarded as a hybrid between *P. reptans* and *P. procumbens*, taking the latter as a species. Two other hybrids have been recorded—*P. suberecta* (Zimm.) = *P. procumbens* × *P. silvestris*, and *P. italica* (Lehm.) = *P. reptans* × *silvestris*. See J. of B. 1888, p. 78, and 1893, p. 325.—H. & J. G.

10. FRAGA'RIA *Linn.* Strawberry.

1. *F. ves'ca* (L.); cal. of the fruit spreading or reflexed, *hairs on* the peduncles spreading those of the *pedicels adpressed upwards* silky.—*E. B.* 1524. *E. B. S.* 2742.—Stoloniferous. Stole continued by an axillary shoot at each rosette (a sympode); one scale between each two rosettes. L. bright green. Flowering stems short, erect, mostly simple, few-flowered. Lts. sessile. Hairs on pedicel of first fl. spreading, on the underside of l. adpressed. Carp. smooth, glabrous, on all parts of the receptacle, superficial. Pet. about as long as broad, white throughout with 2 slight notches at the end, contiguous; claw indistinct. —Woods and thickets. P. V. VI. *Wood Strawberry.* E. S. I.

†2. *F. moscháta* (Duchesne); cal. of the fruit spreading or reflexed, *hairs on* the peduncles and *pedicels spreading and somewhat deflexed.*—*E. B.* 2197. *F. elatior* (Ehrh.) ed. viii.—Fl. imperfectly diœcious. Pet. $\frac{1}{3}$ broader than long, white, entire; claw distinct, bright yellow. "Base of receptacle without carpels." Larger and more hairy than *F. vesca.*—Woods, rare. P. VI.–IX. *Hautboy Strawberry.* E. S.

11. RU'BUS *Linn.*[1 2] Bramble.

A. FRUTESCENTES.

Stem shrubby. Leaves subquinate. Stipules linear, affixed to the petioles. Flowers subpanicled. Succulent carpels forming a compound many-seeded berry. Receptacle conical.

---

[1] In the descriptions by *stem* is meant the *barren stem* of the year; the prickles are called *patent* when they spread at right angles to the st. and *subpatent* when a little declining; the shape of the leaflets, unless otherwise stated, is taken from those of the barren stem. The term *seta* is used to express a *hair or bristle tipped with a gland*; *aciculi* are *strong bristles.*—By *R. G.* the plates in Weihe and Nees's *Rubi Germanici* are intended.—See Babington's *British Rubi.*

When the Continental plants are better known it is feared that considerable changes of nomenclature will be necessary.

[2] From the Author's notes it was his evident intention to entirely rewrite the account of this genus, but this intention was not fulfilled. Since the last edition appeared a very large amount of work has been done and a full account of the genus has been published by the Rev. W. Moyle Rogers in his valuable '*Handbook of British Rubi.*' With his kind permission we have added as an appendix the 'Conspectus of Species' from that work, giving here the account as in ed. viii. with the exception of a few alterations actually made by the Author.—H. & J. G.

i. *Idæi.* Stems suberect, biennial. Ripe fruit separating from its receptacle.—Leaves often pinnate.

1. *R. Idæ'us* (L.).; st. erect round pruinose, prickles setaceous straight, l. 5-pinnate or ternate white beneath, term. lt. long-stalked, interm. lts. sessile not imbricate, prickles of fl.-shoot and ped. deflexed, fl. axillary and terminal corymbose.—*E. B.* 2442. *R. G.* 47.—Creeping. St. 4—6 ft. high; prickles small, usually many. L. usually pinnate, rarely ternate. Fr. red or amber-coloured.—*β. R. Leesii* (Bab.); l. ternate, lts. all roundly ovate subsessile imbricate. *E. B. S.* 2981.—*γ. rotundifolius*; l. like those of *β* but term. lt. long-stalked, l. of flowering shoot similar but upper ones simple.—Damp edges of woods and heaths. Sh. VI. E. S. I.

ii. *Fruticosi.* Stems biennial or subperennial. Ripe fruit not separating from its receptacle. Leaves digitate, pedate or rarely subpinnate.—R. fruticosus *Hook.*

(1) *Suberecti.* Stems usually suberect, glabrous or slightly pilose, not setose nor felted. Prickles equal.—*Sepals* densely white-felted within, pilose externally *with a narrow border of white felt.*

2. *R. suberec'tus* (Anders.); st. erect obtuse-angled at the top, *prickles few* small uniform from a large compressed base confined to the angles of the stem, l. 3—5—7-nate, *lts. flexible flat*, term. lt. cordate-acuminate, basal lts. subsessile *those of fl.-shoot narrowed to the base*, fl. racemose or subpanicled, sep. reflexed.—*E. B.* 2572.—St. 3—6 ft. high. Prickles conical, scarcely longer than the longitudinal extent of their base. L. rarely ternate; lts. thin, unequally serrate. Stam. pale, exceeding styles. Fr. dark red.—Boggy woods and heaths. Sh. VI. VII. E. S. I.

3. *R. fis'sus* (Lindl.); st. erect or subarcuate obtuse-angled, *prickles* many straight or deflexed *from an oblong slightly dilated base not confined to the angles*, l. 5—7-nate, lts. coriaceous plicate, term. lt. cordate-ovate, basal lts. sessile, lateral lts. of fl.-shoot often gibbous at the base, pan. simple racemose-corymbose, *fr.-cal. erect-patent.*—Creeping extensively. St. 1—2 ft. high. Prickles much longer than the longitudinal extent of their base. Lts. unequally serrate. Stam. slightly exceeding styles. Fr. dark red.—Wet places. Sh. VI.—VIII. E. S. I.

4. *R. plicátus* (W. & N.); st. suberect obtuse-angled, prickles straight or deflexed from an oblong dilated base confined to the angles, l. quinate, *lts.* more or less *plicate* thin *pilose* not felted *beneath*, term. lt. cordate-acuminate, basal lts. usually subsessile,

lateral lts. of fl.-shoot rhomboidal-ovate dilated at the base, fl. racemose or corymbose, rachis and ped. pilose not felted, fr.-cal. reflexed.—*E. B. S.* 2714. *R. G.* 1. *R. fruticosus* (Arrh.)—St. rarely 4 ft. high. Prickles unequal, usually conical, much longer than the longitudinal extent of their base. Upper l. sometimes pinnate-septenate. Cal. bright green externally. Stam. falling short of styles.—Heaths. Sh. VI. VII. E. S. I.

5. *R. affi'nis* (W. & N.); st. suberect or subarcuate angular smooth, prickles strong slightly deflexed or declining from a dilated compressed base confined to the angles, l. quinate, *lts.* coriaceous *wavy towards the end* green and opaque on both sides subpilose above paler and *silky-pubescent beneath,* term. lt. cordate-oval cuspidate, basal lts. stalked *those of the fl.-shoot narrowed at the base, branches of the* compound leafy *pan. corymbose* erect-patent often long, sep. acuminate greenish white-felted with pale edges externally patent.—*R. G.* 3.—St. often arching, but apparently it does not root at the end. Stam. exceeding styles.—*β. R. lentiginosus* (Lees); prickles declining, lts. narrower and nearly glabrous; pet. very small, cal. erect-patent.—Heaths and open woods. Sh. VII. VIII. E. S. I.

6. *R. hemistémon* (Müll.); st. arcuate angular nearly or quite glabrous with subsessile glands, prickles slender short declining from a short oblong base, l. quinate, lts. coriaceous irregularly dentate green on both sides paler with many short hairs beneath, term. lt. oblong-ovate acuminate, *basal lts.* very shortly stalked those *of the flowering shoot rather dilated at the base,* pan. oblong abrupt often nearly subracemose with long 1—2-fl. branches, term. fl. nearly sessile, rachis and peduncles hairy not felted, *sep.* ovate-acuminate greenish subtomentose *clasping the fruit.*—St. often-arching. Prickles rather unequal. Stam. white falling short of "pale green" styles.—Hedges and thickets. Warw., Card., Aberd. Sh. VII. VIII. E. S.

(2) *Rhamnifolii.* Stems arching more or less, rooting at the end, slightly pilose, not setose nor felted nor glaucous. Prickles usually confined to the angles of the stem, nearly equal, from a depressed and compressed base.

7. *R. Lindleiánus* (Lees); st. erect-arcuate smooth shining, prickles strong declining compressed dilated below, l. quinate, lts. subcoriaceous shining above pale green pilose (often slightly felted) beneath, term. lt. obovate or roundish acuminate, basal lts. stalked not overlapping the interm. lts., *branches* of the compound leafy pan. *patent* or divaricate short corymbose, *rachis hairy* most prickly in the middle its top and the pedicels felted

its prickles strong declining.—*R. nitidus* Bell-Salt., Bab. (not W. & N.).—St. angular throughout, appearing as if varnished. Term. lt. often much narrowed below. Pan. often very compound, close, usually long, blunt and convex at the end. Pet. not contiguous, oblong, white. Stam. slightly exceeding styles; both greenish. . Fr. small.—Hedges and borders of thickets. Sh. VII. VIII. E. S.

8. *R. rhamnifólius* (W. & N.); st. arcuate angular furrowed upwards, prickles strong patent or declining, l. quinate, lts. coriaceous *flat* opaque above greenish-white-felted beneath, term. lt. ovate or cordate subcuspidate, *basal lts.* stalked *not imbricate*, pan. felted often dense and blunt at the end with axillary racemose few-flowered distant branches and strong declining prickles. —*E. B. S.* 2604. *R. G.* 6. R. cordifolius *R. G.* 5.—St. usually bright red. Prickles yellow or tipped with red. Lts. hard but felted beneath, finely serrate. Sep. dull green externally. Pet. roundish, white. Stam. exceeding styles; both greenish. Primordial fr. oblong.—Hedges and thickets. Sh. VII. VIII. E. I.

9. *R. incurvátus* (Bab.); st. arcuate-prostrate angular, prickles strong patent or declining, l. quinate concave, *lts.* coriaceous acuminate *with incurved wavy edges* shining and subglabrous above *greenish-white-felted beneath*, term. lt. cordate-ovate, pan. narrow leafy below with short approximate patent *corymbose branches its top and pedicels* hairy and *felted* its prickles slender deflexed, sepals ovate acuminate.—St. purple, strong. Lts. flat except at the edges, doubly dentate, soft beneath. Pet. roundly obovate, pink. Stam. exceeding styles; both pink. Primordial fr. almost hemispherical.—Heaths and open woods. Sh. VII. E. S.

10. *R. imbricátus* (Hort); st. arcuate-prostrate angular, prickles small strong declining from a long compressed base, *l. convex* quinate, *lts. convex* coriaceous opaque and subglabrous above paler with scattered hairs beneath *cuspidate imbricate*, term. lt. roundly cordate-obovate, *pan.* narrow leafy below with ascending distant *long racemose branches its top and pedicels hairy scarcely felted* its prickles slender declining, sep. abruptly cuspidate.—St. purplish red. "Basal lts. overlap the interm., interm. the term. leaflet." Pet. obovate, white. Styles greenish yellow. Primordial fr. subglobose.—*β. R. ramosus* (Blox.); st. erect-arcuate, l. not imbricate broadly ovate shining.—Top of pan. felted; pet. pink. Fil. white, styles brownish pink. *J. of B.* ix. 330.—By the Wye below Monmouth. *β.* Devon, Cornw., and Warw. Sh. VI. VII. E.

11. *R. latifólius* (Bab.); *st.* arcuate-prostrate angular *furrowed*, prickles small slender compressed slightly declining from a very long compressed base, l. quinate, lts. thin pilose on both sides coarsely and doubly dentate not felted beneath, *term. lt. cordate acuminate*, basal lts. sessile imbricate, pan. short leafy hairy with ascending few-flowered corymbose branches its top and pedicels felted and hairy its prickles slender short declining.—St. strong, green. Lts. very broad. Pet. ovate, clawed. Stam. exceeding styles and incurved. Primordial fruit apparently about hemispherical.—Open woods. Cramond Bridge and Colinton near Edinburgh. Acharn, Perthshire. Sh. VII. VIII. S.

(3) *Villicaules.* Stems arching more or less, rooting, pilose or bald, often felted, having subsessile glands and rarely a few setæ and aciculi. Prickles confined to the angles of the stem, nearly equal, or a few smaller scattered on the faces. Basal leaflets stalked, not overlapping the intermediate leaflets.

a. *Discolores.* St. with strong equal prickles and closely adpressed pubescence. Leaves white-felted beneath.

12. *R. dis'color* (W. & N.); *st. arcuate-prostrate* angular furrowed *stellately downy*, prickles declining or deflexed from a much dilated compressed base, l. quinate, lts. coriaceous convex rugose above *finely white-felted* beneath, term. lt. obovate-cuspidate, pan. long narrow felted its few lower branches axillary many-flowered its prickles strong hooked, *cal. finely white-felted.*—*R. G.* 20. *R. fruticosus* Sm. *E. B.* 715.—St. nearly prostrate unless supported, often nearly glabrous and glaucous. Lts. usually hairless but with fine hard felt beneath, usually with decurved edges. Pet. pink. Fil. and styles purple; anth. greenish; stam. exceeding styles. Drupes many, small, acid.—*β. pubigerus*; st. angular stellately hairy with spreading hairs, prickles slender from a dilated oblong depressed scarcely compressed base patent or deflexed.—Hedges and thickets. Common. Sh. VII. VIII. E. S. I.

13. *R. thyrsoïdeus* (Wimm.?); *st. erect-arcuate* angular furrowed *subglabrous*, prickles declining or deflexed from a much-dilated compressed base, l. quinate, *lts.* flat rather coriaceous glabrous above *hairy and greenish-white-felted beneath*, term. lt. cordate-ovate-acuminate, pan. long narrow its lower branches many axillary few-flowered its prickles strong hooked, *cal. hairy and felted.*—R. fruticosus *R. G.* 7.—St. arching highly, self-supporting. Lts. wavy and a little turned up at the edge, hairy

and softly but often finely felted beneath. Pet. white. Fil. white; anth. rather fuscous. Drupes few, subacid.—Hedges and thickets. Sh. VII. VIII. E. I.

b. *Sylvatici.* Stem with moderate mostly equal prickles and (often dense) hairy-woolly patent (but often deciduous) pubescence. Leaves usually green but sometimes white-felted beneath.

14. *R. leucóstachys* (Sm.); *st. arcuate-prostrate* angular *hairy woolly* and felted, prickles many subpatent slender from a dilated compressed base, l. quinate, *lts.* flat hairy and shining and *softly yellowish or whitely felted beneath,* term. lt. obovate ovate or roundish cuspidate, *pàn. long* narrow felted hairy setose its branches short few-flowered its prickles slender declining or angularly deflexed, cal. green-felted hairy setose aciculate.—*a*; st. arcuate-prostrate, prickles mostly on the angles of the stem equal, lts. coriaceous obovate sublobate-serrate yellowish-white-felted and shining beneath.—*E. B. S.* 2631.—St. nearly prostrate, covered with loose spreading clustered hairs. Pan. very narrow, long. Pet. pinkish. Stam. with pink base upright, exceeding the pink-based styles. Fr. purplish black.—*β. R. vestitus* (Weihe); st. arcuate, prickles unequal scattered, lts. cordate-roundish cuspidate irregularly dentate pale green beneath. *R. G.* 33. St. often nearly round and bearing a few aciculi and setæ. Lts. rather thin.—Hedges, thickets and woods. Sh. VII. VIII. E. I.

15. *R. Hirtifólius* (Müll.); st. arcuate-prostrate angular *pilose*, prickles nearly equal small declining from a dilated compressed base, l. quinate, *lts.* flat *pale green and pilose only on the veins beneath, term. lt. obovate acuminate,* panicle narrowly pyramidal its branches few-flowered its prickles few slender, sepals greenish felted and with pale edges externally, stam. exceeding the styles.—Pan.-branches often 1-fl. Sep. ovate, leaf-pointed. —*R. pyramidalis* Kaltenb..—Bushy and heathy places. Near Plymouth. Isle of Wight. Sh. VII. VIII. E.

16. *R. Grabows'kii* (Weihe); *st. arcuate* angular *subglabrous,* prickles many equal declining or deflexed much dilated and compressed below, l. quinate, *lts.* plicate opaque and glabrous above *ashy-felted beneath* irregularly dentate imbricate, *term lt. cordate* abruptly cuspidate (on the fl.-shoot much dilated below), pan. long leafy below its branches ascending racemose-corymbose its prickles many deflexed, cal. ashy-felted and hairy.—*R. carpinifolius* Borr. *E. B. S.* 2664? *R. montanus* Focke.—St. with

a few often clustered hairs. Pan. hairy but not felted. "Stam. white scarcely exceeding the green styles." Fr. sparingly produced.—Hedges and thickets. Sh. VII. VIII. E. I.

17. *R. Coleman'i* (Blox.); st. arcuate angular subglabrous, prickles many nearly equal declining compressed and dilated below, l. quinate, *lts. convex* opaque above *green and hairy on the veins beneath* irregularly dentate imbricate, term. lt. roundly cordate acuminate (on the fl.-shoot roundish or broadly oval), pan. long leafy below its branches ascending corymbose or the axillary ones racemose its prickles many slender deflexed or declining with many setæ and aciculi, sep. ashy-felted hairy. —*R. fusco-ater β. Colemani* Bab.—St. with a few aciculi and setæ and scattered hairs. Pan. not felted. Pet. white. The position of this plant is doubtful; but it seems to be most allied to *R. Grabowskii* of our species.—Hedges near Coventry and Packington. Sh. VII. VIII. E.

18. *R. Sal'teri* (Bab.); *st. arcuate-prostrate angular furrowed subglabrous, prickles subpatent slender* compressed from a dilated compressed base, l. quinate, *lts.* thin *coarsely and doubly patently dentate* green on both sides hairy only on the veins beneath, term. lt. broadly obovate cuspidate-acuminate subcordate, pan. long lax hairy its ultra-axillary branches short few-flowered corymbose patent its rachis wavy its prickles slender declining, sep. hairy felted erect-patent.—St. green, becoming quite glabrous, with a few aciculi and setæ occasionally. Lts. hard and rough beneath.—*a. Salteri*; lts. lobed or doubly serrate, rachis of cylindrical panicle nearly straight, branches corymbose patent, sep. erect-patent. A few sunken setæ on the pan. and calyx. Pet. white.—*β. R. calvatus* (Blox.); lts. coarsely dentate the teeth distant with recurved tips interstices denticulate, rachis of lax pan. wavy, branches subracemose and ascending the uppermost corymbose and patent, sep. loosely reflexed. Many sunken setæ on the panicle. Pet. pinkish or deep rose-coloured.—These plants seem to be the extremes of one species.—Open woods and hedges, rare. Sh. VII. VIII. E. I.

19. *R. carpinifólius* (W. & N.); *st. erect-arcuate* angular striate with patent clustered hairs, prickles slender conical compressed declining from a dilated compressed base, l. quinate, lts. thin irregularly but finely serrate pilose above rather paler or whitish and densely hairy beneath, term. lt. obovate-acuminate or cuspidate, *pan. narrow* racemose hairy setose its lower branches axillary short few-flowered its prickles deflexed or declining, sep. hairy loosely reflexed from the fruit.—*R. G.* 13.—St. forming a large lofty arch. *L. with very fine* but irregular

acute *teeth* remarkably directed forwards. Pan. often nearly simple, cylindrical. Pet. white or reddish.—Open places in hilly districts. Sh. VII. VIII. E. S. I.

20. *R. villicaúlis* (W. & N.); *st. arcuate* angular patently pilose, prickles strong conical-compressed subpatent from a dilated compressed base, l. quinate, lts. thin dentate-serrate pale green and shining and often densely hairy but the hairs only on the veins beneath, term. lt. obovate or roundly cordate-obovate subacuminate, *pan. open compound* leafy hairy subsetose felted its branches corymbose its prickles declining or deflexed, sep. hairy setose aciculate loosely reflexed from the fruit.—*R. G.* 17. *R. villicaulis* and *R. pampinosus* Bab.—St. becoming nearly naked, rarely a few setæ and aciculi.—*β. R. adscitus* (Genev.); st. aciculate and setose. *R. vulgaris* (Lindl.). Heads of setæ very deciduous.—Woods and hedges. Sh. VII. VIII. E. S. I.

21. *R. macrophyl'lus* (Weihe); *st. arcuate-prostrate* angular patently pilose, prickles short slender conical-compressed declining from a large dilated compressed base, l. quinate, *lts. doubly and patently dentate* or irregularly dentate-serrate pilose above pale green felted or velvety or hairy only on the veins beneath, term. lt. elliptic or roundly obovate or obovate cuspidate or acuminate more or less cordate, pan. hairy felted setose its branches short few-flowered corymbose the lower axillary subracemose and ascending its prickles declining, sep. ovate-attenuate hairy felted setose loosely reflexed from the fruit.—The following forms have been considered species.—*a. R. umbrosus* (Arrh.); prickles slender from a large base, l. quinate, *lts. doubly and patently dentate* velvety or slightly felted beneath, term. lt. broadly obovate cuspidate, pan. with slender prickles, *tip of sep. linear*, cor. rose-coloured. Stam. white, scarcely exceeding the dull styles. *R. carpinifolius* (Blox.). Term. lt. sometimes divided into three.—*β. R. macrophyllus* (W. & N.); prickles small short from a very large base, l. 5—3-nate, lts. irregularly dentate-serrate hairy only on the veins or rarely felted beneath, term. lt. elliptic or broadly obovate, pan. with slender prickles, *sep. leaf-pointed*, cor. white. *E. B. S.* 2625. *R. G.* 12. Lts. often very large. St. often with a few aciculi and setæ. Stam. pinkish, exceeding the pink styles. A very variable plant.—*γ. R. Schlechtendalii* (W. & N.); prickles short small from a very large base, l. usually quinate, *lts. doubly and patently dentate* usually hairy only on the veins not felted beneath, *term. lt. long obovate* acuminate wedgeshaped or subcordate at the base, pan. with strong prickles, sep. with a linear point, cor. white. *R. G.* 11. Hardly distinguishable from var. *β*.—*δ. R. amplificatus* (Lees); prickles short from a very large base,

l. usually quinate, *lts.* somewhat doubly patently dentate *hairy only on the veins beneath* not felted, term. lt. broadly obovate acuminate, pan. very large its prickles strong compressed from a very large base, sep. usually leaf-pointed, cor. white or pinkish. Remarkable for the very strong prickles with exceedingly long compressed bases on its panicle.—*ε. glabratus* (Bab.); prickles short from a very large base, l. quinate, *lts.* irregularly or rather doubly dentate *slightly hairy only on the veins beneath, term. lt. roundly cordate* or broadly obovate subcordate, pan. with slender prickles, sep. (apparently) leaf-pointed. L. nearly glabrous beneath. Term. lt. very round.—Woods and thickets. Sh. VII. VIII. E. S. I.

c. *Spectabiles.* Stem with rather unequal prickles and scattered aciculi and setæ and usually dense pubescence.

22. *R. mucronulátus* (Bor.); *st.* arcuate *subterete* patently pilose, *prickles* few conical *slender* declining from an oblong dilated base, l. quinate, *lts. thick finely dentate-serrate* rough and pilose above hairy only on the (reddish) veins beneath, *term. lt.* broadly obovate *cuspidate* cordate below, pan. narrow lax leafy hairy setose felted its branches long 1—3-flowered its prickles small slender declining, sep. ovate-attenuate hairy setose felted with a linear point.—*Sy. E. B.* 451. *R. mucronatus* Blox. (not Ser.)—St. becoming nearly naked; a few setæ and aciculi. Pet. pale pink. Pink stam. exceeding green styles.—*R. festivus* (Warr.) seems to differ only by angular st. and much thinner doubly dentate leaves.—Banks and hedges. Sh. VII. VIII. E. S. I.

23. *R. Sprengel'ii* (Weihe); st. prostrate terete pilose, *prickles unequal deflexed* from a large compressed base, l. 3—5-nate pedate, lts. thin green with scattered hairs only on the veins beneath, term. lt. elliptic-acuminate, pan. lax hairy felted setose its axillary branches patent few-flowered the ultra-axillary divaricate its prickles few slender deflexed, sep. ovate-acuminate erect-patent often leaf-pointed.—*α. R. Borreri* (Bell-Salt.); st. arcuate-prostrate thick with scattered aciculi and setæ, prickles unequal, l. usually quinate, pan. rather thyrsoid or with a subcorymbose top, dull stam. falling short of styles.—*R. rubicolor* (Blox.) seems an extreme form of this.—*β. R. Sprengelii* (Weihe); st. usually quite prostrate slender, prickles small, aciculi and setæ very few, l. usually ternate, lts. flexible, pan. lax few-flowered pyramidally subcorymbose; rose-coloured stam. exceeding styles. *R. G.* 10.—Heaths and woods. Sh. VI. VII. E.

d. *Radulæ.* Stem rough with small elevated rigid points on which the deciduous setæ and aciculi were seated; prickles nearly equal.

24. *R. Bloxámii* (Lees); st. arcuate-prostrate angular slightly furrowed, *prickles small subpatent* unequal, aciculi and setæ short many, hairs many, l. 5- or 3-nate, *lts. coarsely* doubly *dentate green and pilose on both sides, term. lt. roundly obovate cuspidate, pan. long leafy to the top* felted *its short branches and top corymbose* its prickles slender declining, sep. ovate-acuminate loosely reflexed from the fruit.—St. thick, rarely furrowed. L. subpedate. Upper floral l. simple. Pet. white. Pale stam. falling short of styles.—Woods. Sh. VII. VIII. E.

25. *R. Hys'trix* (Weihe); st. arcuate-prostrate angular slightly furrowed, prickles slender declining from a dilated compressed base, aciculi and hairs few short equalling the many setæ and much shorter than the prickles, l. quinate pedate, *lts. coarsely and rather doubly and patently dentate green and pilose on both sides, term. lt. oblong-obovate* acuminate, pan. long leafy its branches long racemose ascending but the uppermost and ultra-axillary patent or divaricate its rachis wavy its prickles strong declining the uppermost slender, *sep.* lanceolate-attenuate *loosely adpressed to the fruit.—R. G.* 41.—Lts. flat, but wavy at the edges, hairy but not felted beneath. Pet. pink. Pale stam. exceeding pink styles.—Hedges and thickets. Sh. VII. VIII. E. I.

26. *R. rosáceus*(Weihe); st. arcuate-prostrate angular, *prickles* slender nearly equal declining from a dilated compressed base *a few shorter slightly exceeding the nearly equal aciculi setæ and hairs,* l. quinate-pedate or ternate, lts. doubly-dentate-serrate pilose above paler and pilose only on the veins beneath, term. lt. obovate- or oblong-acuminate usually subcordate below, *pan. subpyramidal truncate* leafy below its branches racemose ascending or the ultra-axillary patent corymbose or simple its rachis more or less wavy its prickles slender declining, sep. lanceolate-attenuate loosely adpressed to the fruit.—*R. G.* 36.—Perhaps not distinct from *R. Hystrix*. Its more elegant pyramidal and abrupt panicle and the more unequal prickles less markedly separated from the aciculi are the chief differences. "Pet. pale pink. Pale stam. exceeding pink-based styles."—Woods and hedges. Sh. VII. VIII. E. I.

27. *R. præ'ruptorum* (Boul.); st. arcuate-prostrate subterete, *prickles* many slender unequal declining *slightly dilated at their base,* setæ hairs and very slender aciculi unequal many, l. quinate-pedate or ternate, *lts. coarsely unequally and doubly serrate*

pilose above paler and *pilose only on the veins beneath*, term. lt. obovate acuminate, pan. long narrow leafy below its branches corymbose its rachis straight its prickles slender declining its aciculi setæ and hairs many and unequal, *sep. ovate-attenuate aciculate with long setæ* felted loosely reflexed from the fruit.—*R. G.* 42. *R. hirtus β. Menkii* Bab. *R. pygmæus* Bab. not Weihe.—St. perhaps prostrate. Pan. rather long with distant short axillary branches, uppermost very short and often divaricate. Pet. white or pinkish. "Pale stam. exceeding pink-based styles."—Hedges. South-eastern counties. Sh. VII. VIII. E.

28. *R. scáber* (Weihe); st. arcuate-prostrate subangular subsulcate, *prickles strong short* nearly equal *declining or deflexed* from a long compressed base, aciculi setæ and hairs few very short, l. 3—5-nate, lts. doubly dentate opaque and pilose above pale green and pilose beneath, term. lt. broadly obovate cuspidate or acuminate subcordate below, pan. subpyramidal leafy felted truncate or blunt at the end its axillary branches erect-patent racemose the ultra-axillary racemose-corymbose or simple its prickles short declining or deflexed from a long base its aciculi strong its setæ and hairs nearly equal, sep. ovate-acuminate loosely reflexed from the fruit.—*R. G.* 32.—Stam. inflexed exceeding styles.—The typical plant is slender and elegant, pan. often nearly simple with long peduncles. The large form (*R. Babingtonii* Bell-Salt.) is often an enormous plant with very rough long prostrate stems, a very large panicle with very large and long branches. Pet. white.—Open woods. Sh. VII. VIII. E.

29. *R. rúdis* (Weihe); st. arcuate angular subsulcate, prickles strong conical compressed nearly equal subpatent exceeding the *nearly equal and short aciculi setæ and hairs*, l. quinate, *lts. coarsely and doubly serrate* (or lobate-serrate) *greenish-white-felted* beneath, term. lt. elliptic or broadly oblong-obovate acuminate, pan. long leafy its branches ascending corymbose-racemose the uppermost and ultra-axillary divaricate its rachis straight its prickles strong declining or deflexed from a long base the uppermost slender, sep. ovate-attenuate strongly reflexed.—*R. G.* 40.—The nearly equal not scattered prickles, short aciculi setæ and hairs, jagged lts. felted beneath, and strongly reflexed sepals are marks of this species. Pet. white. Stam. exceeding styles.—Hedges and thickets. Sh. VII. VIII. E. S. I.

30. *R. Rádula* (Weihe); st. arcuate angular, prickles slender declining from a dilated compressed base exceeding the many short *unequal aciculi setæ and hairs*, l. quinate-pedate, *lts. finely* but doubly and patently dentate *greenish-white-felted* beneath, term. lt. obovate acuminate or subcuspidate, pan. long leafy its

branches short corymbose ascending its prickles strong declining from a long base the uppermost slender, sep. ovate loosely reflexed from the fruit.—*a. R. Radula* (Weihe); prickles on the st. unequal, term. lt. obovate acuminate. Connivent stam. much exceeding styles. *E. B. S.* 452. *R. G.* 39.—*β. R. Leightonii* (Lees): prickles on the st. nearly equal, term. lt. obovate cuspidate.—*γ. denticulatus* (Bab.); term. lt. broadly quadrangular-obovate cuspidate subcordate below broadly but faintly dentate, the teeth denticulate.—Hedges. Sh. VII. VIII. E. S. I.

(4) *Glandulosi.* Stems arcuate-prostrate or prostrate, rooting, hairy. Prickles abundant, very unequal, scattered, passing gradually into abundant aciculi and setæ.

a. *Koehleriani.* Leaves quinate or rarely ternate. Prickles and setæ thickened at their base.

31. *R. Koehl'eri* (Weihe); st. arcuate-prostrate roundish or angular pilose, prickles very unequal slightly declining from a compressed base, aciculi and setæ very unequal, *lts.* unequally or rather doubly dentate *even above* pale green *hairy on the veins beneath*, term. lt. cordate-ovate, *basal lts. not imbricate* stalked, pan. open leafy its branches short patent corymbose or the axillary branches racemose its prickles many long slender declining its aciculi setæ and hairs many unequal, *sep.* ovate-attenuate *patent or reflexed from the fruit.* Pale stam. exceeding pink styles.—*a. R. Koehleri* (Weihe); st. with many prickles aciculi and setæ, *lts. hairy only on the veins and rough beneath, pan. open* truncate often broad-topped its branches long corymbose patent or the axillary branches racemose and ascending, *term. ped. of pan. and branches shorter than the lateral ones. R. G.* 25. *E.B.S.* 2605.—*β. R. infestus* (Weihe); st. with many strong prickles aciculi and setæ, *lts. hairy only on the veins and soft beneath, pan.* broad but compact *rounded at the end* its branches moderate rather corymbose erect-patent, *term. ped. of pan. and branches shorter than the lateral ones,* prickles of pan. strong deflexed.—*γ. R. pallidus* (Weihe); st. with fewer but strong prickles aciculi and setæ, *lts. often slightly felted* hairy on the veins *and soft beneath,* pan. narrow its branches short corymbose-racemose patent or the axillary branches ascending, *term. ped. of branches often longer than the lateral ones. R. G.* 29.—*δ. R. cavatifolius* (Mull.); st. subglabrous, aciculi and setæ unequal inconspicuous, hairs very few, *lts. not felted* but hairy on the veins beneath, *term. lt. cordate cuspidate-attenuate, pan.* abrupt *with very short terminal peduncles.*—All are usually very prickly; their very strong prickles pass very gradually into aciculi. and those into setæ; st. often quite covered by their broad bases. Pet. white.—Hedges and thickets. Sh. VII. VIII. E. S. I.

32. *R. fusco-âter* (Weihe); *st.* arcuate-prostrate angular *hairy*, prickles unequal slightly declining from a very large compressed base, setæ and strong unequal aciculi many, *lts.* irregularly or rather doubly dentate *even above* green and hairy beneath, term. lt. broadly cordate-ovate acuminate or subcuspidate, *basal lts. stalked*, pan. long subpyramidal leafy below its branches patent corymbose or the axillary branches erect-patent racemose its prickles many unequal longest at about the middle of the fl.-shoot its hairs setæ and aciculi many unequal, *sep.* ovate-attenuate setose aciculate *patent or adpressed to the fruit.*—*R. G.* 26.—The lts. are all imbricate and rather rough beneath. Pet. pink. Stam. incurved exceeding styles.—Heaths. Sh. VII. VIII. E.

33. *R. emersisty'lus* (Müll.); st. arcuate-prostrate angular hairy, prickles slightly declining unequal from a compressed base a few smaller, *setæ aciculi and hairs very short nearly equal*, *lts.* rather irregularly and rather doubly dentate *thick* even above green and hairy on the veins beneath, *term. lt. orbicular-ovate* attenuate *with a cordate base*, basal lts. very shortly stalked imbricate, pan. often leafy nearly to the top its *axillary branches corymbose* its prickles small unequal declining its setæ and hairs many unequal, sep. ovate-attenuate with a rather leaflike point setose aciculate patent or adpressed to the fruit.—*R. Briggsii* (Blox.) *J. of B.* vii. t. 88.—Stam. exceeding styles.—Heaths in the South. Sh. VII. VIII. E.

34. *R. diversifólius* (Lindl.); *st.* arcuate prostrate angular *sparingly pilose*, prickles unequal subpatent from a compressed base, aciculi and setæ many unequal, *lts.* often irregularly or towards their tip rather doubly dentate wavy at the edge *rugose above* pale green hairy and felted beneath. term. lt. broadly cordate obovate-acuminate *basal lts. subsessile imbricate, pan. long leafy nearly to the top* its branches erect-patent subracemose its prickles longest at about the middle of the fl.-shoot its hairs setæ and aciculi short equal, *sep. ovate acute* felted setose *patent or loosely adpressed to the fruit.*—R. fusco-ater *Bab.* formerly. *J. of B.* viii. t. 107.—A very prickly plant which differs much in appearance from *R. fusco-ater*. Lts. usually soft beneath. Pet. white. Stam. much exceeding styles.—Hedges. Sh. VII. VIII. E.

35. *R. mutab'ilis* (Genev.); stem arcuate-prostrate angular sparingly pilose and setose, prickles moderate unequal subpatent from a long compressed base, aciculi strong very unequal mostly short, l. quinate, lts. doubly or mostly lobate-dentate pilose

above hairy or felted *beneath, term. lt. obovate acuminate, basal lts. stalked not imbricate, pan.* long *narrow pyramidal* leafy nearly to the top its *branches and branchlets subcorymbose subpatent its rachis very prickly* aciculate and setose its prickles many very strong from large compressed bases, sep. ovate felted setose reflexed with rather leaflike points.—Pet. dull white. "Filam. pink. Styles greenish."—*R. obliquus* (Blox.) seems to belong here.—Hedgebanks in the South. Sh. VII. VIII. E.

36. *R. Lejeun'ei* (Weihe); st. arcuate-prostrate subangular with scattered hairs and setæ, prickles mostly small a few longer declining from a long compressed base, *aciculi very short,* l. quinate-pedate, lts. opaque and pilose above paler and hairy only on the veins beneath lobate-serrate towards the tip, *basal lts. stalked* not imbricate, term. lt. obovate-acuminate, pan. broad hairy leafy setose its top corymbose its axillary *branches ascending* subracemose its prickles slender declining its setæ many unequal, *sep. ovate* felted setose loosely adpressed to the fruit.—*R. G.* 31.—Sep. often having a short linear point; its allies when furnished with an appendage to the sepals have it leaflike.—Hedges. Sh. VII. VIII. E.

b. *Bellardiani.* Leaves ternate or quinate pedate; basal leaflets not imbricate, stalked. Prickles mostly confined to the angles of the very hairy aciculate and setose stems.

37. *R. pyramidális* (Bab. not Kaltenb.); st. very nearly prostrate roundly angular, *prickles* many *short strong* much declining or deflexed from a very large compressed base, hairs few, aciculi and setæ many nearly equal, l. ternate or rarely quinate-pedate, *lts. convex* irregularly dentate-serrate nearly equal opaque and pilose above paler and *pilose beneath,* term. lt. obovate-cuspidate, *pan. pyramidal* leafy below its top and branches racemose felted its *rachis straight rigid* its prickles slender declining its hairs and setæ many unequal, sep. lanceolate-attenuate loosely adpressed to the fruit.—St. usually quite prostrate. Pan. very pyramidal. Pet. very narrow, greenish white. Stam. white exceeding pinkish styles.—Edges of woods, rare. Abundant at Llanberis, N. Wales. Sh. VII. VIII. E.

38. *R. Gun'theri* (Weihe); st. arcuate-prostrate terete, *prickles slender unequal* declining from a large subcompressed base, aciculi setæ and hairs short nearly equal, l. ternate or rarely quinate-pedate, *lts. flat* irregularly or doubly dentate-serrate nearly equal opaque and pilose above pilose greenish white or slightly felted beneath, term. lt. obovate-acuminate, pan. narrow leafy its branches distant ascending racemose few-flowered its

rachis wavy its prickles few slender declining its hairs and setæ many equal, sep. ovate-lanceolate loosely reflexed from the fruit. —*R. G.* 21. *R. saltuum* Focke.—Only slightly prickly. Pan. very loose, its rachis forming an angle at each leaf. Pet. narrow, pale pink. Stam. equalling styles.—Shady places. Sh. VII. VIII. E. I.

39. *R. humifúsus* (Weihe); st. arcuate-prostrate nearly round, prickles many slender very unequal declining from a long compressed base, *hairs setæ and very slender aciculi unequal* many patent, l. quinate-pedate, *lts. finely but doubly patently dentate* pilose and opaque above pilose greenish white and shining beneath, term. lt. obovate-oblong subcuspidate, pan. broad leafy below its branches corymbose its rachis rather wavy its prickles few slender declining its hairs and setæ many unequal, sep. ovate slightly aciculate shortly setose loosely reflexed from the fruit. —*R. G.* 35. *R. hirtus a.* Bab.—Pan. sometimes nearly simple. Pet. white, large. Stam. exceeding styles. Sometimes the underside of the lts. is felted.—Woods and thickets. Sh. VII. VIII. E. S. I.

40. *R. foliósus* (Weihe); st. arcuate-prostrate angular, prickles many slender unequal declining from a long compressed base, setæ and very slender aciculi scattered unequal, hairs few, l. quinate-pedate, *lts. unequally dentate-serrate* pilose and opaque above *paler and pilose beneath*, term. lt. roundly cordate acuminate, pan. long narrow leafy to the top its branches short erect-patent corymbose its rachis rather wavy its prickles very slender many declining its hairs and setæ many unequal, *sep. ovate-attenuate aciculate setose* hairy laxly reflexed from the fruit.—*R. exsecatus* Müll.—*β. R. adornatus* (Müll.), has term. lts. obovate-acuminate or even narrowed below.—Hartshill Wood, Warw. Plymouth. Sh. VII. VIII. E.

41. *R. glandulósus* (Bell.); st. arcuate-prostrate nearly round, prickles small declining from a long compressed base, aciculi setæ and hairs many nearly equal, l. ternate or rarely quinate, lts. nearly equal oblong cuspidate hairy only on the veins beneath, term. lt. subcordate-ovate-acuminate, pan. felted aciculate very setose its branches erect-patent axillary its top racemose its prickles slender declining, sep. ovate-attenuate aciculate setose felted loosely adpressed to the fruit or patent.—*a. R. Bellardi* (Weihe); l. ternate, lts. nearly equal oblong doubly dentate-serrate shortly pilose on the veins beneath, lateral lts. divaricate, pan. with axillary distant corymbose branches its rachis usually wavy, stam. slightly exceeding styles. *R. G.* 44. R. glandulosus *E. B. S.* 2883. The divaricate lts. are remarkable, and the

very open panicle. Sometimes the l. are quinate. A subvariety, *R. dentatus* Blox., has a slightly angular stem, lts. ovate-acuminate-cuspidate with a cordate base ashy green beneath, lateral lts. patent or ascending.—β. *R. hirtus* (W. & K.); l. quinate coarsely and unequally serrate with long and dense hairs on the veins and shining beneath, term. lt. subcordate-ovate-acuminate, pan. often long its branches racemose or corymbose shortly setose its rachis nearly straight. *R. G.* 43. *R. fuscus* Lees. A subvariety, *R. rotundifolius* Blox., *R. deflexidens* Boul ?, has a slightly angular stem, l. ternate or rarely quinate, lts. doubly dentate cuspidate, term. lt. nearly round with a subcordate base.—γ. *R. Reuteri* (Merc.): l. quinate coarsely and rather doubly dentate-serrate with few hairs on the veins beneath, term. lt. obovate-rhomboidal acuminate, pan. truncate its branches short subcorymbose few-flowered upper 1—3-flowered very aciculate setose and hairy, rachis nearly straight. Much more hairy and setose than the other forms and with some stronger declining or deflexed prickles on the stem.—Woods. Sh. VII. VIII. E. I.

(5) *Cæsii.* Stems most often arcuate-prostrate, terete or slightly angular, usually with a glaucous bloom; aciculi and setæ few or none; prickles unequal.

42. *R. Balfouriánus* (Blox.); st. arcuate-prostrate nearly round patently pilose, aciculi and setæ few, prickles slender unequal scattered patent from an oblong subcompressed base, l. quinate, lts. dentate-serrate green on both sides rugose and pilose above *hairy* (not felted) *beneath,* term. lt. cordate or ovate acute, basal lts. subsessile imbricate, pan. loose leafy hairy rather setose its branches long distant few-flowered racemose-corymbose erect-patent, *sep. ovate-acuminate erect-patent,* styles pale flesh-coloured, fr. oblong its torus oblong stalked.—A very variable plant, approaching *R. corylifolius* in some states. Pet. pale pink. Stam. exceeding styles.—Hedges. Sh. VII. VIII. E. I.

43. *R. corylifólius* (Sm.): st. arcuate-prostrate nearly round or obtuse-angled nearly glabrous, aciculi setæ and subsessile glands very few, prickles subulate slender nearly equal subpatent or rarely deflexed from a long base, l. quinate, *lts.* doubly serrate green on both sides rugose with scattered hairs above paler and *felted beneath,* term. lt. roundly cordate or ovate cuspidate or acuminate, basal lts. subsessile imbricate, pan. and its branches subcorymbose, *sep ovate cuspidate reflexed from the fruit,* pet. roundly ovate, styles greenish, torus oblong stalked.—St. slightly glaucous. Pan. usually with 2 or 3 long axillary branches,

felted, more or less prominently setose at the top.—*a. R. sublustris* (Lees); st. nearly round reddish green, *prickles slender* subpatent *from a long base*, lts. ashy-felted beneath, *term. lt. often subtrilobed* roundish cordate, rachis straightish with few prickles. *R. corylifolius* Sm. *Sy. E. B.* 455. St. thick. Pet. white. Stam. slightly exceeding styles.—*β. conjungens* (Bab.); st. rather angular, reddish green, *prickles* slender strong subpatent or deflexed from a compressed *very long base* often with slightly deflexed points, lts. ashy-felted beneath, term. lt. cordate-ovate or broadly obovate with a subcordate base, rachis straightish with few prickles. *R. Wahlbergii* Bell-Salt. St. round at the base, angular but flat-sided above. Pet. white or pink.—*γ. purpureus* (Bab.); st. angular purple often a little floccose, prickles strong subpatent or deflexed from a long compressed base, lts. pale-green- or white-felted beneath, term. lt. roundly- or subcordate-obovate, rachis rather wavy with many prickles. *R. Wahlbergii* Arrh. St. round at the base, angular and often furrowed above. Pet white or pink.—Hedges and thickets. Sh. VI.—VIII. E. S. I.

44. *R. althæifólius* (Host); st. prostrate slightly angular with scattered hairs and setæ, prickles many unequal slender patent from an oblong compressed base, l. quinate or ternate, *lts. crenately lobed* pale green with hairs on the veins loosely white-felted beneath, *basal lts. of the ternate l. retrorsely bipartite of the quinate l.* sessile *imbricate, term. lt. rhomboidal-obovate* subcordate below, pan. leafy its axillary branches and top racemose-corymbose with very few short setæ, prickles on the middle of the rachis longest slender, *sep.* ovate-subacuminate setose *loosely adpressed* to the (black-blue) fruit, pet. obovate, styles flesh-coloured at their base. *R. deltoideus* Müll. ?—St. round at the base, above bluntly angular or even furrowed. Pan. rather long, open; top formed of irregular corymbs of nearly simple peduncles. Pet. nearly white. Pale stam. about equalling pinkish styles.—Hedges. Sh. VI.—VIII. E.

45. *R. tuberculátus* (Bab.); st. arcuate-prostrate slightly angular with scattered short hairs and setæ, *prickles* many unequal slender patent *from an oblong tuberculiform base*, l. ternate or quinate, lts. rather doubly dentate hairy on the veins beneath green on both sides, *basal lts.* of the ternate l. bilobate *of the quinate l.* nearly sessile *imbricate*, term. lt. roundly cordate subcuspidate, pan. leafy its axillary branches racemose its top corymbose, prickles from the middle to the top of the pan. and ped. slender and longest, sep. ovate acuminate aciculate setose loosely adpressed to the fruit. *J. of B.* viii. t. 106. *R. nemorosus δ. ferox* Leight. St. very bluntly angular. Pan. short

its branches few-flowered. Pet. pinkish. Stam. yellow. Styles greenish.—Mr. Warren combines my *R. diversifolius* and *R. Briggsii* with this under the name of *R. dumetorum*. His *R. concinnus* is near *R. corylifolius*, but has not felted l. on my authentic specimen. It may be distinct.—Hedges. Sh. VII. VIII. E. I.

46. *R. cæ'sius* (L.); st. prostrate terete glaucous, prickles small unequal subulate declining or deflexed from a long compressed base, l. ternate or rarely pinnate, *lts. unequally cut or coarsely serrate*, term. lt. ovate rhomboidal-ovate or 3-lobed, lateral lts. rather bilobed subsessile, pan. nearly simple often very small, sep. ovate acuminate with a long linear point adpressed to the glaucous fruit, pet. obovate notched, styles greenish.—α. *agrestis*; st. very slender, prickles few small, lts. flat lobate-serrate rather pilose on both sides, term. lt. rhomboidal-ovate acuminate rounded below. *R. cæsius α. aquaticus* and β. *agrestis* R. G. 46. A. Pan. often nearly simple, or its branches rarely more than once divided but often very long.—β. *R. tenuis* (Bell-Salt.); st. very slender, prickles many strong but small nearly equal deflexed, lts. flat (?) doubly serrate pilose on both sides or villose beneath, term. lt. obovate or cordate-obovate acuminate, fr. black "not glaucous." St. sometimes having a very few aciculi and setæ. Bases of prickles much enlarged.—γ. *ulmifolius*; st. slender purplish, prickles many small deflexed or declining, aciculi setæ and hairs few short, lts. rather rugose lobate-serrate pilose only on the veins or hairy or slightly ashy-felted beneath, *term. lt.* roundly cordate acuminate often 3-*lobed* or rarely divided into 3 sessile lts. of which the term. is narrowed below. *R. cæsius β. pseudo-cæsius* R. G. 46. B. f. 1. Often much larger than var. α and β. St. thicker. L. very broad.—δ. *intermedius*; st. thicker greenish purple, prickles many slender very unequal subpatent, aciculi and setæ few strong very short, lts. lobate-serrate pilose on the veins beneath, *term. lt. triangular-cordate* acuminate 3-lobed or 3-partite *or divided into* 3 *sessile lts.* of which the term. is narrowed to the base.—ε. *R. pseudo-Idæus* (Lej.); st. thick green slightly glaucous, prickles slender violet-coloured subpatent, aciculi and setæ few very short, l. ternate or *quinate-pinnate*, lts. doubly-serrate ashy-felted beneath, lateral lts. all sessile, *term. lt. stalked roundly-cordate. R. G.* 46. B. f. 2. I have only seen one specimen of this var., found at Hunsdon, Herts.—ζ. *serpens*; *st.* slender *green*, l. ternate, lts. lobate serrate pilose on the veins beneath, lateral lts. with a large backward lobe, term. lt. obovate acuminate subcordate below, ped. and sep. very setose felted scarcely hairy.—Hedges and thickets. common. Sh. VI. VII. *Dewberry.* E. S. I.

B. HERBACEI.

Stem nearly or quite herbaceous. Leaves ternate or simple. Stipules usually attached to the stem. Fl. umbellate or nearly solitary. Receptacle flat.

i. *Saxatiles.* Stems slender, prostrate. Flowers umbellate or nearly solitary. Succulent carpels large, few, distinct.

47. *R. saxat'ilis* (L.); st. rooting annual, prickles none or very few minute weak, *l. ternate,* fl.-shoot erect with a *terminal few-flowered corymb,* pet. lanceolate about equalling the calyx.—*E. B.* 2233. *R. G.* 9.—St. very slender. Fl.-shoots radical, about a span high. Fl. white. Fr. of 1—4 large drupes.—Stony mountains or high hills. P. VII. VIII. E. S I.

ii. *Arctici.* No sterile stems but a long subterranean rhizome. Flowers terminal, nearly solitary. Succulent carpels adhering together.

48. *R. Chamæmórus* (L.); fl.-shoot erect unarmed 1-flowered herbaceous, fl. diœcious, l. simple lobed plicate.—*E. B.* 716. *R. G.* 49.—Rhizome woody. Fl. large, white. Fr. large, red, afterwards orange-yellow.—Alpine turf-bogs. P. VI. *Cloudberry. Knotberry.* E. S. I.

[*R. arc'ticus* (L.); st. erect unarmed nearly 1-flowered herbaceous, l. ternate, pet. obovate twice as long as the calyx, stam. connivent, succulent carpels many.—*E. B.* 1585. *R. G.* 48.—Rhizome subterranean. Fl.-shoot 6—10 in. high. Lts. nearly equal. Fl. rose-coloured.—Isle of Mull and Ben y Gloe. Probably a mistake. P. VI.] S.

Index to the Rubi.

Index to the Rubi (cont.).

12. DRY'AS *Linn.*

1. *D. octopet'ala* (L.); l. crenate-serrate blunt, sep. 3 or 4 times as long as broad more or less pointed, base of the cal. hemispherical.—*E. B.* 451. *St.* 20. 3.—Fl. large, white. Pet. 8. L. simple, white with fine dense woolly pubescence beneath. St. prostrate, woody. Seeds obovate-oblong apiculate. —Alpine situations, particularly on limestone. P. VI. VII.
E. S. I.

13. GE'UM *Linn.* Avens.

1. *G. urbánum* (L.); fl. erect, *pet. obovate*, cal. of the fruit reflexed, *carpophore 0, lower joint of the style much longer than the glabrous upper joint*, radical l. interruptedly pinnate and lyrate, stem-l. ternate, stip. large rounded lobed and cut.—*E. B.* 1400. *St.* 5. 7.—St. 2 feet high. Fl. small, bright yellow, calyx green. Upper joint of the style with a few minute hairs at its base.—Hedges and thickets. P. VI.—VIII. *Wood-Avens.*
E. S. I.

[*G. intermédium* (Ehrh.); fl. erect or nodding, *pet. roundish with a wedgeshaped claw*, cal. of the fruit patent, *carpophore* 0 or short, *lower joint of the style longer than the hairy upper joint*, radical l. interruptedly pinnate and lyrate, stem-l. 3-lobed, stipules round toothed.—*Sy. E. B.* 458.—St. 1—2 feet high. Fl. larger than in Sp. 1, less than in Sp. 2, yellow, calyx purplish. Upper joint of style clothed with long hairs but with a rather long glabrous point. · Perhaps a hybrid; or rather it consists of extreme forms of Sp. 1 and 2.—Damp woods. P. VI. VII.] E. S. I.

2. *G. rivále* (L.); fl. nodding, pet. broadly obovate emarginate or obcordate with a long wedgeshaped claw, cal. of the fruit erect, *carpophore long, lower joint of the style equalling the long hairy upper joint*, radical l. interruptedly pinnate and lyrate, stem-l. ternate, stip. small ovate toothed.—*E. B.* 106. *St* 3.—St. about 1 foot high. Fl. large, purplish brown with darker veins, calyx purplish. Carpophore nearly equalling the calyx. Upper joint of the style with a short glabrous point.—Damp woods. P. VI. VII. *Water-Avens.* E. S. I.

Tribe IV. *Rosidæ.*

14. Ro'sa *Linn.*[1] Rose.

i. Spinosissimæ. Styles free, scarcely protruding. Sep. mostly persistent. St. short, erect, with many slender unequal prickles lessening gradually into aciculi and setæ. [Lts. usually 9 roundish. Fr. without or with but a small disk.]

1. *R. spinosis'sima* (L.); prickles crowded very unequal mostly straight subulate, *sep. simple* acuminate, *fr. nearly globular with no disk.*—*E. B.* 187.—[A dwarf shrub.] St. erect with short compact branches. Fl. solitary, white. Fr. dark purple or black, ripe in Sept. [*α.* Lts. simply serrate glabrous and glandless. *β. R. pimpinellifolia* (L.); lts. simply serrate, ped. glandular. *γ. R. Ripartii* (Déségl.); lts. doubly serrate somewhat glandular.] Sandy and chalky heaths and sea-shores. Sh. V. VI. E. S. I.

---

[1] From the Author's MS. it is evident that he intended to thoroughly revise the account of this genus; we have therefore made some slight alterations, preserving as far as possible the actual wording of the last edition, and have added, from the works of Baker, Crépin, and Déséglise, brief notices of most of the forms in the London Catalogue, ed. ix., but have not attempted to verify the authorities for the varietal names.

H. & J. G.

2. *R. hiber'nica* (Sm.); prickles scattered, larger slightly falcate, lts. simply [or irregularly] serrate without glands, sep. [usually] pinnate not glandular on the back, *fr. with a small disk.*—*E. B.* 2196.—[Usually a compact bush.] Fl. 1–3 rarely more, pale pink. Fr. urceolate blood-red with subpersistent sep., ripe in Oct. [α. Lts. slightly hairy on the midrib above and on the veins beneath or (var. *glabra*, Bak.) quite glabrous, ped. naked. β. *Grovesii* (Bak.); robust, very glaucous, sep. almost always simple, fr. ovate-urceolate. γ. *cordifolia* (Bak.); ped. hispid glandular.—Generally considered a hybrid between Sp. 1 and 10.] Chiefly in the N. of England and N. of Ireland. β. Surrey. Sh. VI.—VII. E. S. I.

3. *R. involúta* (Sm.); prickles crowded, *lts. doubly serrate downy and often glandular beneath,* sep. simple or pinnate glandular on the back, fr. subglobose with no disk.—*E. B. S.* 2594.—St. erect [or arching]. Fl. 1–3. Fr. red or pink, with persistent sep., ripe in Sept. and Oct. [Prof. Crépin considers this to be a series of hybrids of *R. spinosissima* with *R. villosa, R. tomentosa,* and possibly *R. rubiginosa.* The following are the principal forms described as British.—α. *Smithii* (Bak.); lts. naked above hairy principally on the midrib beneath, serration but slightly compound, ped. and cal.-tube densely aciculate, sep. simple. β. *Robertsoni* (Bak.); lts. glabrous above when mature, hairy and inconspicuously glandular beneath, serration compound, cal.-tube usually naked, sep. compound. γ. *Webbii* (Bak.); lts. glabrous above hairy on the midrib beneath, ped. and fr. naked, sep. more compound than in the other forms. δ. *Moorei* (Bak.); lts. nearly glabrous above, thinly hairy and densely glandular beneath, serration nearly simple, ped. and cal.-tube densely aciculate. ε. *Wilsoni* (Borr.); lts. thinly hairy on the ribs below, almost eglandular, serration simple, sep. nearly simple, fr. ovate-urceolate, ped. aciculate and setose. ζ. *occidentalis* (Bak.); near *R. Wilsoni*, lts. smaller slightly hairy, petiole glandular, ped. densely aciculate, fr. naked. η. *Nicholsonii* (Crép.); lts. glabrous above slightly hairy and glandular beneath, serration compound, fr. and ped. densely setose. θ. *gracilescens* (Bak.); a robust Irish form, lts. thinly hairy, serration copiously compound, ped. aciculate, cal.-tube naked. ι. *R. Sabini* (Woods); lts. thinly hairy above more hairy below, serration copiously compound, ped. and cal.-tube densely aciculate, sep. with 2—4 small pinnæ. κ. *lævigata* (Bak.); lts. like *R. Sabini*, ped. and cal.-tube naked, sep. simple not glandular. λ. *R. Doniana* (Woods); a small form with lts. densely hairy, ped. and cal.-tube densely aciculate, sep. almost simple.] Chiefly in the North. Sh. VI. E. S. I.

ii. VILLOSÆ. [St. free.] Sep. mostly persistent, densely glandular on the back. St. tall suberect with scattered uniform nearly straight prickles. Lts. [usually 7, terminal one ovate, generally] very hairy and glandular beneath.

[*R. pomif'era* (Herrm.); lts. large *oblong-lanceolate* doubly serrate downy, *sep. copiously pinnate,* fr. globose with no disk. Not a native. Sh. VI. ?]

4. *R. villósa* (L.); prickles slender nearly straight, *lts. ovate* doubly serrate [usually] softly downy, *sep. quite persistent slightly pinnate*, fr. subglobose with no disk.—*E. B.* 2459.—*R. mollis* (Sm.) ed. viii.—[An erect bush with straight branches.] Fl. 1—3 deep rose colour [rarely white] ped. very short. Fr. bright red with *connivent sep.*, ripe in Aug. [β. *cærulea* (Woods); l. softer and greyer with very few glands on the petiole, ped. with very few glands or quite naked, fr. pendent. γ. *R. pseudo-rubiginosa* (Lej.); lts. nearly glabrous but very glandular beneath, ped. and cal.-tube densely aciculate.] Northern Counties. Sh. VI. VII. E. S. I.

5. *R. tomentósa* (Sm.); prickles mostly uniform slender straight or slightly curved, *lts. elliptic* or subovate [usually] doubly serrate [and glandular and hairy or] downy especially beneath, *sep. subpersistent copiously pinnate*, fr. urceolate [or subglobose] *with a slight disk* [usually setose].—*E. B.* 990—St. erect with arching branches. Fl. 1—3 [—10] pink or white; ped. rather long. Fr. bright red, ripe in Sept. [The following are some of the principal forms described as British:—β. *R. cinerascens* (Dum.); lts. hairy on both sides simply serrate, fr. subglobose, styles hairy. γ. *R. subglobosa* (Sm.); lts. densely hairy almost eglandular irregularly or doubly serrate, fr. subglobose. δ. *R. farinosa* (Rau.); lts. densely grey-downy, somewhat glandular beneath, serration compound, fr. naked, styles woolly. ε. *R. cuspidatoides* (Crép.); lts. broadly oval hairy glandular beneath, fr. subglobose, ped. and fr. aciculate, styles hairy. ζ. *fœtida* (Ser.); lts. slightly hairy very glandular doubly serrate, styles glabrous. η. *R. scabriuscula* (Sm.); lts. almost glabrous above hairy principally on the veins beneath and sparsely glandular, fr. ovate aciculate, sep. reflexed. θ. *Woodsiana* (Groves); nearly allied to *scabriuscula* but lts. narrower with more compound serration, sep. decidedly erect persistent, styles hairy. ι. *sylvestris* (Woods); lts. naked above when mature thinly hairy and very glandular beneath, ped. and fr. densely aciculate and glandular. κ. *obovata* (Bak.); prickles strongly curved slender, lts. obovate thinly hairy and eglandular beneath with very sharp double serration, ped. very short quite naked.]—Hedges and thickets. Sh. VI. VII. E. S. I.

iii. RUBIGINOSÆ. [Styles free.] Sep. subpersistent. St. suberect with scattered nearly uniform hooked prickles becoming more slender upwards [and sometimes also with smaller straight ones]. Lts. [usually 7] very glandular beneath. Fr. with a small disk.

6. *R. Eglantéria* (L.); *prickles many the larger hooked the smaller subulate unequal straight*, lts. doubly serrate [thinly hairy and very glandular beneath], sep. persistent copiously pinnate, primordial fr. pear-shaped, disk very small.—*E. B.* 991. *R. rubiginosa* (L.) ed. viii.—St. erect [or somewhat arching]. Fl. 1—4 rose-coloured, ped. [usually] very setose, styles hairy. Fr. subglobose (except the first) ripe in Oct. Foliage scented.

[*R. comosa* (Rip.) has been described as having more ovate fr., and *R. apricorum* (Rip.) more globose fr. The name *R. umbellata* (Leers) has been applied to plants with several fl. together, and *R. rotundifolia* (Reichb.) to a small form with nearly straight prickles and very small lts. *β*. *R. echinocarpa* (Rip.); more glandular with setæ on the fr. and the branches often setigerous. *γ*. *jenensis* (M. Schultze under *R. rubiginosa*); "ped. recept. and back of the sep. glandular" Hanb. & Marsh., *Flor. of Kent*.—Bushy places. Sh. V.—VII. *Sweet Briar*. E. S. I.

7. *R. micran'tha* (Sm.); *prickles uniform* hooked, lts. doubly serrate [hairy principally on the veins] and glandular beneath, sep. subpersistent with long leaflike points and minute lanceolate simple pinnæ, primordial fr. rounded at the base, disk moderate.—*E. B.* 2490.—St. suberect with long lax branches. Fl. few pale; ped. [usually] very setose; styles glabrous. Fr. ovate-urceolate, bright scarlet, ripe in Sept. or Oct. Foliage scarcely scented, lts. rounded at the base. [*β*. *R. permixta* (Déségl.); lts. glabrous but very glandular below, fr. setose. *γ*. *Briggsii* (Bak.); more luxuriant, lts. less glandular beneath, fr. and ped. quite naked. *δ*. *R. hystrix* (Lem.); lts. narrow, sharply serrate glabrous but densely glandular beneath, ped. densely aciculate, fr. naked.—Hedges and] Thickets. Sh. VI., VII. E. S. I.

8. *R. agres'tis* (Savi); prickles rather unequal hooked with *a few aciculi*, lts. [copiously] doubly serrate [usually glabrous or slightly hairy on the veins and usually very] glandular beneath, *sep. subpersistent closely pinnate*, primordial fr. ovate or subglobular, disk moderate.—*E. B. S.* 2653. *R. inodora* (Fr.) ed. viii. *R. sepium* (Thuill.)—St. suberect with long lax branches. Fl. 1—3, pink; *ped. naked*; *styles hairy*. Foliage scarcely scented. [*β*. *R. cryptopoda* (Bak.); lts. glabrous above, hairy on the main veins and thinly glandular beneath, ped. very short, fr. subglobose. *γ*. *R. Billietii* (Pug.); lts. small hairy on both sides, petioles very hairy eglandular, sep. erect after flowering, fr. ovoid. *δ*. *R. inodora* (Fr.); lts. broadly oval, less attenuate at the base becoming glabrous above, petioles glabrous, sep. reflexed after flowering, fr. ovoid. Prof. Crépin is not satisfied that either of the last two occur in Britain.]—Local. Sh. VI. E. I.

iv. CANINÆ. [Styles free. Sep. subpersistent or deciduous.] St. arching, with scattered uniform hooked prickles becoming more slender upwards. [Lts. usually 7 rarely glandular beneath. Fr. with a disk.]

9. *R. obtusifólia* (Desv); prickles equal very strongly hooked, lts. ovate-oblong simply or doubly serrate usually hairy with inodorous glands on the veins and often on the surface beneath, sep. reflexed deciduous pinnate, fr. ovate or subglobose, disk flat, styles densely hairy.—Hook. *Fl. Lond.* t. 117. *R. canina* (L.) vars. *tomentella* &c. ed. viii.—*α*. Lts. simply serrate hairy on both sides. *β*. *R. frondosa* (Spreng.); lts. small simply serrate much rounded at the base, fr. small subglobose. When

the ped. is glandular it is *R. canina* var. *concinna* (Bak.). γ. *R. tomentella* (Lem.); lts. hairy on both sides doubly serrate, ped. short. When the ped. is glandular it is var. *decipiens* (Dum.). δ. *R. Borreri* (Woods); lts. almost glabrous above hairy beneath with very compound serration, ped. slightly glandular, styles thinly hairy.—Heaths and hedges. Sh. VI. VII. E. S. I.

10. *R. canína* (L.); prickles equal hooked, lts simply or doubly serrate glabrous or hairy, *sep. reflexed deciduous* pinnate, fr. ovate-urceolate or roundish, disk flat.—St. erect with long arching branches. Lts. flat or keeled. Fl. white or pale rose-pink; ped. mostly naked; styles distinct hairy. [* *Lts. glabrous on both sides.*—α. *R. lutetiana* (Lem.); a large bush, lts. ovate often glaucous sharply simply serrate, fl. 1—4, sep. naked on the back, fr. ovate-urceolate. *R. Andegavensis* (Bast.) has ped. glandular. β. *R. surculosa* (Woods); a robust form, lts. openly simply serrate rounded at the base, fl. 10—30 together. γ. *R. sphærica* (Gren.); lts. broader simply serrate, petioles slightly pubescent, fr. globose. δ. *R. senticosa* (Ach.); a small slender form with quite globular fr. ε. *R. dumalis* (Bechst.); lts. doubly serrate, petioles glandular, sep. gland-ciliated. *R. inconspicua* (Déségl.) has ped. glandular; when fr. also glandular it is *R. aspernata* (Déségl.). ζ. *R. vinacea* (Bak.); serration of lts. very open and compound, fr. oblong. *R. latebrosa* (Déségl.) has ped. and floral branches glandular. η. *R. Blondeana* (Rip.); lts. with very compound serration, glandular on the veins beneath, ped. slightly glandular. ** *Lts. glabrous above hairy beneath.*—θ. *R. urbica* (Lem.); lts. simply serrate. *R. semiglabra* (Rip.) has lts. hairy only on the veins beneath. ι. *R. arvatica* (Bak.); lts. doubly serrate hairy and glandular on the veins beneath. *** *Lts. hairy on both sides.*—κ. *R. dumetorum* (Thuil.); lts. simply serrate soft grey-green, fl. often deeper coloured. λ. *R. cæsia* (Sm.); lts. softly hairy beneath slightly so above, teeth slightly compound, ped. glandular. μ. *incana* (Bak.); lts. very glaucous with a few glands beneath, serration very compound.]—Hedges and thickets. Sh. VI. VII. E. S. I.

11. *R. glau'ca* (Vill.); prickles equal hooked, lts. simply or doubly serrate glabrous or hairy, *sep. ascending after flowering subpersistent*, fr. ovate-urceolate or roundish, disk narrow, *ped. short* often almost hidden by the large bracts, styles woolly.—Usually a compact bush. Fl. bright rose-pink. Fr. ripening in Sept. α. *R. Reuteri* (Godet); lts. glabrous simply serrate, fr. ovate-urceolate or roundish. β. *R. subcristata* (Bak.); lts. glabrous doubly serrate. *R. Hailstoni* (Bak.) is a form with aciculate stem and less persistent sep. γ. *subcanina* (Christ under *R. Reuteri*); lts. slightly hairy on the veins beneath, sep. spreading, ped. somewhat longer, fl. paler. δ. *R. implexa* (Gren.); lts. glabrous above hairy principally on the veins beneath simply serrate. ε. *R. coriifolia* (Fr.); lts. hairy on both sides simply serrate, fr. subglobose. ζ. *R. Lintoni* (Scheutz under *R. coriifolia*); lts. hairy glandular beneath doubly serrate, sep. eglandular on the back, ped. eglandular. η. *R. Bakeri* (Déségl.); lts. naked above when mature thinly hairy beneath and glandular on the midrib and main veins, serration copiously compound, sep. thinly glandular on the back, fr. oblong or turbinate. ι. *R. Watsoni* (Bak.); lts. hairy on both sides doubly serrate, ped. sometimes glandular. κ. *R. celerata* (Bak.); "habit and l. of *tomentella* with the fr. and sep. of this section." *Baker.*—Hedges and thickets, chiefly in the N. Sh. VI. VII. E. S. I.

v. SYSTYLÆ. Sep. deciduous. *Styles* [more or less] *united into a protruding column*. St. arching or trailing, with uniform strong hooked prickles. [Fr. with a conspicuous disk.]

12. *R. stylósa* (Desv.); prickles equal hooked, lts. usually simply serrate slightly hairy beneath [rarely also above], sep. reflexed deciduous pinnate with a tapering point, fr. ovate *with a prominent disk*.—*E. B.* 1895. *R. bibracteata* (Bast.) ed. viii.—*St. erect-arching* to 8—12 ft. in height. Fl. pale pink [or white] 3—6; ped. aciculate and setose; column of styles falling short of stam. glabrous. [*α. Desvauxii* (Bak.); lts. slightly hairy above decidedly so beneath, fl. white, ped. glandular. *β. opaca* (Bak.); lts. much more hairy, fl. white, ped. naked. *γ. R. systyla* (Bast.); lts. glabrous above hairy principally on the veins beneath, fl. pink, ped. glandular. *δ. R. leucochroa* (Desv.); like *R. systyla* but fl. white, ped. glandular shorter, disk less prominent, styles less agglutinated. *ε. R. pseudo-rusticana* (Crep.); lts. glabrous or with a few hairs on the midrib beneath, fl. white, ped. glandular, disk very prominent. *ζ. R. virginea* (Rip.); lts. glabrous simply serrate, fl. white, ped. naked, fr. globose. *η. R. evanida* (Christ); lts. almost glabrous doubly serrate, ped. glandular.] Chiefly in the S. of England. Sh. VI. VII. E.

13. *R. arven'sis* (Huds.); prickles equal hooked, lts. [usually] simply serrate and quite glabrous, sep. reflexed deciduous with a short point simply pinnate, fr. subglobose with a convex disk.—*E. B.* 188.—*St. trailing*. Lts. nearly flat. Fl. white usually 4—6; ped. glandular; column of styles equalling the stam. glabrous. [*β. R. bibracteata* (Bast.); more robust, sep. rather more compound, fr. obovoid. *γ. R. gallicoides* (Déségl.); upper part of st. very glandular, fr. ovoid. *δ. R. ovata* (Lej.); prickles dilated circular at the base, lts. hairy on the midrib beneath, fl. larger, fr. ovoid. *ε. reptans* (Crép.); lts. doubly serrate.]—Hedges and thickets. Sh. VI. VII. E. S. I.

13*. *R. sempervirens* (L.); l. evergreen shining, those on the fl. branches often of only 5 lts., column of styles hairy. *R. Melvini* (Towndrow) from Madresfield, Worc., with glabrous styles has been referred to this species.

Index to Rosa.

Suborder III. *Pomeæ.*

15. CRATÆ'GUS *Linn.* Hawthorn.

1. *C. Oxyacan'tha* (L.); spinose, l. obovate 3—4-lobed cut and serrate cuneate at the base, fl. corymbose, cal. not glandular, styles 1—3.—*a. C. oxyacanthoides* (Thuill.); l. lobed, *ped. and cal. usually glabrous*, cal.-lobes triangular-acuminate, styles 1—3, fruit oval, nuts 1—3. *Sy. E. B.* 479.—*β. C. monogyna* (Jacq.); l. deeply lobed usually acute, *ped. and cal. villose*, cal.-lobes lanceolate acuminate, style 1 bent, fruit subglobose, rarely style straight and cal. glabrous. *E. B.* 2504.—[*C. kyrtostyla* Fing.]—Hedges and thickets; *β* the more common form and flowering later. [L. sometimes more deeply cut, var. *laciniata* (Auct.).] T. V. VI. E. S. I.

16. COTONEAS'TER *Medic.*

1. *C. integerrimus* (Medic.); l. roundish-ovate rounded at the base, flowerstalks and margins of the calyx downy.—*E. B. S.* 2713. *C. vulgaris* (Lindl.) ed. viii.—Pet. rose-coloured. Fr. small pendulous, red.—Great Orme's Head, Caernarvonshire. Sh. V. E.

17. MES'PILUS *Linn.* Medlar.

1. *M. german'ica* (L.); l. lanceolate undivided downy beneath, fl. solitary.—*E. B.* 1523. [*Pyrus* Hook. fil.]—L. entire simply or doubly serrate. In the wild state it is spinous.—Hedges and thickets in Surrey, Sussex, [and Cheshire ?] T. V. VI. E.

18. PY'RUS *Linn.*

1. *P. commúnis* (L.); l. ovate serrate, fl. corymbose, *fruit turbinate, styles distinct.*—*E. B.* 1784.—Branches rather spinous. Germen woolly. Leaves sometimes obovate suddenly contracted into a long very acute point.—*a. Pyraster* (L.); l. acuminate downy beneath when young ultimately glabrous, fr. elongate-pyriform obconical at the base.—*β. P. Achras* (Gaert.); mature l. acute or cuspidate slightly downy beneath, fr. globose-pyriform rounded below.—*γ. P. cordata* (Desv.), *P. Briggsii* (Syme); l. cordate-ovate nearly glabrous, fr. small globose or pyriform.—Hedges and woods. T. IV. V. *Wild Pear-tree.* E.

2. *P. Málus* (L.); l. ovate acute serrate, fl. in a sessile umbel, *fr. globose, styles combined below.*—*a. P. acerba* (DC.); young branches calyx-tube and underside of the l. glabrous. *E. B.* 179.—*β. mitis* (Wallr.); the same parts pubescent or woolly. *Sy. E. B.* 490.—Woods and hedges. T. V. *Crab-tree.* E. S. I.

[*P. domes'tica* (Ehrh.); *l. pinnate* downy beneath serrate, *fl. panicled, fr. obovate.*—Fr. resembling a small pear.—One tree in Wyre Forest; now lost.—T. V. *Service-tree.*] E.

3. *P. Aucupária* (Ehrh.); *l. pinnate* downy beneath serrate, *fl. corymbose, fr.* (small red) *globose.*—Corymb 4—6 in. across. Lts. 12—16.—Hilly woods. Mountains.—T. V. VI. *Rowan tree. Mountain-Ash.* E. S I.

[*P. pinnatifida* (Ehrh.), *P. semipinnata* (Roth), E.B. iii. Suppl. 485B, with oblong-lanceolate bluntly-lobed l. pinnate below, occurs in shrubberies. *P. Aucuparia × Aria* ?]

4. *P. fen'nica* (Bab.); l. oblong serrate usually pinnate below with 1—4 pairs of free decurrent lts. and 1—2 pairs of nearly free lobes, *lts. and lobes oblong or narrowly elliptical,* underside grey-webbed, fl. corymbose.—*Sorbus fennica* Fr.—Lobes blunt or acute as the term. tooth is more or less prominent. Hybrid (?) between Sp. 3 and 5.—Glen Catacol, Arran, S., in small quantity. S.

5. *P. intermédia* (Ehrh.); *l. oblong doubly serrate near the apex pinnatifid below, lobes oblong-lanceolate* serrate, underside

white and downy, fl. corymbose.—*E. B.* iii. Suppl. 484 *a* & *b*. *P. scandica* (Syme) ed. viii.[1]—Fr. scarlet. Lower part of the l. often subpinnate, the lobes becoming more and more combined as they approach the extremity of the leaf which is only deeply and doubly serrate. [β. *P. minima* (Ley); l. narrower with shallower lobes, fr. small globose. J. of B. 1897, t. 372.]—[West of England, Wales.] Arran. [β. Brecon.] T. V. E. S.

6. *P. A'ria* (Ehrh.); *l. of fl.-shoot roundly oval* or elliptic, entire below unequally and doubly serrate or with many small broad lobes especially towards the end, *lateral veins* 9—14 on each side, underside of l. uniformly snowy-white-felted, fl. corymbose.—*E. B.* 1858.—Lobes deepest a little below the end of the leaf.—Chalky banks and limestone rocks. T. V. E. I.

7. *P. rupic'ola* (Syme); *l. of fl.-shoot obovate* narrow and entire below with many small broad lobes on the upper half, deepest near the end, *lateral veins* 5—8 on each side, *underside of l. uniformly snowy-white-felted*, fl. corymbose.—*Sy. E. B.* 483.—Often the basal half of the edge of l. is entire.—Exposed rocks. T. V. E. I.

8. *P. latifólia* (Syme); *l. of fl.-shoot elliptic* with triangular-oval acutely serrate lobes deepest near the middle of the sides of leaf, lateral veins 5—9 on each side, *underside of l. ashy-felted flocculent*, fl. corymbose.—*Sy. E. B.* 484. *P. scandica* Bab. [*P. rotundifolia* Bechst. (non Mœnch).] L. often very acute with very acute lobes which are longer than broad, basal ¼ finely serrate. Fr. dark or reddish brown.—Hilly woods. T. V. E.

9. *P. torminális* (Ehrh.); *l.* ovate or cordate *lobed glabrous, lobes triangular acute* serrate the lower ones larger and spreading, fl. corymbose.—*E. B.* 298.—Fr oval brown.—Woods and hedges in the South. T. IV. V. *Wild Service-tree.* E.

Order XXVII. LYTHRACEÆ.

Cal. tubular, lobed; valves valvate or distant in the bud, sometimes with intermediate teeth. Pet. between the lobes of the calyx, very deciduous. Stam. inserted in the tube of the cal., below the pet. and equalling them or 2, 3, or 4 times as many. Ovary free, 2—4-celled. Style 1. Caps. membranaceous,

---

[1] The Rev. A. Ley in a paper on this group in 'Science Gossip,' 1895, p. 113, considers the broader-leaved plant with larger fruit from the West of England distinct from the Arran plant, using the name of "*P. scandica* Asch." for the latter.—H. & J. G.

usually (by abortion) 1-celled, with many seeds and a central placenta, covered by the calyx. Embryo straight.—Stip. 0.

1. LYTHRUM. *Cal. tubular, cylindrical,* with 8—12 teeth; 4—6 broader, erect; alternate teeth subulate. Pet. 4—6. Stam. as many as or twice the number of the petals. *Style filiform.* Caps. 2-celled many-seeded.

2. PEPLIS. *Cal. bell-shaped* with 12 teeth, of which 6 are broader and erect, 6 subulate. Pet. 6, minute, fugacious. Stam. 6. *Style very short.* Caps. 2-celled, many-seeded.

1. LY'THRUM *Linn.* Loosestrife.

1. *L. Salicária* (L.); *l. lanceolate from a cordate base opposite* or whorled, fl. in whorled leafy spikes, bracts 0, subulate calyx-teeth twice as long as the others, stam. 12.—*E. B.* 1061.—Upper l. usually falling short of the flowers, or so large as totally to destroy the spiked appearance of the plant. St. 2—4 feet high and l. nearly glabrous, or downy with crisped hairs. Fl. large, purple or crimson.—Ditch-banks and damp places. P. VII. VIII. *Purple Loosestrife.* E. S. I.

2. *L. Hyssopifólia* (L.) ; *l. alternate linear-lanceolate* blunt, fl. axillary solitary, *bracts* 2 *minute* subulate, calyx-teeth all short, stam. 6.—*E. B.* 292.—St. mostly procumbent, spreading, simple or branched. Fl. small, light purple, Glabrous.—Damp places where water has stagnated, rare. A. VI.—X. E.

2. PEP'LIS *Linn.* Water-Purslane.

1. *P. Por'tula* (L.); l. opposite obovate stalked, fl. axillary solitary sessile.—*E. B.* 1211. *St.* 1. 7.—Pet. often wanting. Cal.-tube short, shortly bell-shaped, shorter than the capsule. St. 4—6 in. long, prostrate, creeping.—Damp places. A. VII. VIII. E. S. I.

Order XXVIII. TAMARISCACEÆ.

Cal. 4—5-parted, persistent, imbricate in the bud. Pet. 4—5, withering, from the base of the calyx. Stam. equal to or twice as many as the pet., from the margin of a shieldlike disk. Caps. 1-celled, 3-valved, many-seeded, loculicidal. Placentas often only at the base. Seeds ascending, crowned with a tuft of hairs.

1. TAMARIX. Styles 3, patent. Seeds from base of capsule; crowned with a tuft of simple papilliform hairs.

1. TAM'ARIX *Linn.* Tamarisk.

[1. *T. ang'lica* (Webb); l. glabrous rather narrowed at the base spurred, fl.-buds ovoid, *hypogynous ring* with 5 angles *narrowed into the filaments* of the cordate shortly apiculate anthers, caps. roundish-trigonous at the base abruptly narrowed towards the apex.—*T. gallica* Sm. (*T. gallica* L. is a Mediterranean tree.) *E. B.* 1318. Webb in Hook. J. of B. iii. 422 t. xv.—St. shrubby, with slender leafy branches. L. minute. Spikes lateral, rather panicled, slender. Fl. small, pink.—South-west coast of England. Planted. Sh. VII.] E.

Order XXIX. ONAGRACEÆ.

Cal. tubular, adnate to the ovary wholly or in part, with 2—4 lobes valvate in the bud. Pet. as many as the calyx-lobes, twisted in the bud, inserted at the top of the tube. Stam. 2, 4, or 8, inserted with the petals. Ovary of several cells, with a central placenta. Style 1, filiform, stigma capitate or lobed. Fr. a berry or capsule with 4 cells. Albumen 0.—L. alternate or opposite, not dotted. Exstipulate. Many raphides.

1. EPILOBIUM. Cal. 4-cleft, deciduous. Pet. 4. Stam. 8. Style filiform, stigma clavate or cruciform. Caps. linear, 4-celled, 4-valved. Seeds many, bearded.—Fl. not yellow.

2. ŒNOTHERA. Cal.-limb tubular below. Seeds not bearded. In other respects like *Epilobium*.—Pet. yellow.

3. LUDWIGIA. Limb of the cal. 4-cleft, persistent. Pet. 4, or 0 in our plant. *Stam.* 4. Style filiform, deciduous; *stigma capitate. Caps. obovate,* 4-valved, 4-celled, many-seeded, with a loculicidal dehiscence.

4. CIRCÆA. *Limb of the cal. 2-cleft*, deciduous, its tube closed by a cupshaped disk. *Pet.* 2, obcordate. Stam. 2, alternate with the petals. Style simple; stigma emarginate. *Caps.* 1- or 2-*celled*, cells 1-seeded, seeds erect.—Fl. white.

1. EPILO'BIUM *Linn.*[1] Willow Herb.

A. *Fl. irregular, cal. without a free tube, stam. and style, ultimately declining, l. scattered.*

1. *E. angustifolium* (L.); st. erect round, l. lanceolate veined,

[1] In Rev. E. S. Marshall's list of *Epilobia* in Lond. Cat. ed. 9, p. 20, thirty-three hybrids are reported. See also Mr. Marshall's papers in J. of B. 1890, p. 2, & 1891, p. 7.—H. & J. G.

pet. obovate shortly clawed, style exceeding the stamens.—Rhizome far-creeping. St. 3—6 ft. high. Fl. crimson.—α. *E. macrocarpum* (Steph.); l. lanceolate, fl.-buds obovate cuspidate, caps. long ($2\frac{1}{2}$ in.) erect. *Sy. E. B.* 495.—β. *E. brachycarpum* (Leight.); l. lanceolate-attenuate (broadest near their base), fl.-buds narrowly obovate obliquely acute, caps. short (1 in.) spreading. *Sy. E. B.* 496.—Damp shady places. β. The cultivated form. P. VII. *Rose-Bay.* E. S. I.

[*E. Dodonæi* (Vill.), *E. rosmarinifolium* (Haenke), *Sy. E. B.* 494, reported from Glen Tilt, was probably a mistake, a small alpine form of Sp. 1 being taken for it.]

B. *Fl. regular, cal. with a free tube, stam. and style erect, l. opposite, upper l. lanceolate alternate.*

a. Turionate, *i. e.*, producing radical suckers.

2. *E. hirsútum* (L.); st. woolly glandular, *l. clasping* slightly decurrent oblong-lanceolate denticulate-serrate, buds erect mucronate, sep. lanceolate, *seeds* tubercular oblong *acute below.* —*E. B.* 838.—Suckers thick, fleshy, with distant scales; others leafy and ending in rosettes. St. 4—5 ft. high, terete, branched. L. most hairy upon the ribs. Fl. large.—Wet places by ditches and streams. P. VII. VIII. *Great Willow-Herb.* E. S. I.

B. Stoloniferous. * *Stoles autumnal rosulate, st. erect.*

† Stem mostly terete, stigmas 4-cleft.

3. *E. parviflórum* (Schreb.); st. downy, *l. sessile* lanceolate from a rounded base denticulate, lowest l. shortly stalked, buds erect ovoid, sep. lanceolate, *seeds* tubercular obovate-oblong *rounded below.*—*E. B.* 795.—Autumnal rosettes on short stalks or sessile. St. 1—2 ft. high, nearly simple. *L. uniformly hairy.* Fl. small.—*E. rivulare* (Wahl.) is a subglabrous state.—Damp places. P. VII. VIII. E. S. I.

4. *E. montánum* (L.); st. downy, l. shortly stalked ovate-lanceolate from a rounded base denticulate, buds nodding ovoid, sep. lanceolate, seeds tubercular oblong blunt at both ends or rather narrowed below.—*E. B.* 1177.—Root truncate. Rosettes fleshy, oblong, sessile. St. 6—24 in. high. *Edges and veins of l. pubescent.* L. sometimes in threes.—Dry places. P. VI. VII. E. S. I.

5. *E. lanceolátum* (S. & M.); st. slightly angular downy, *l.* stalked lanceolate denticulate-serrate *narrowed to an entire base,*

buds nodding ovoid, sep. broadly linear acuminate, seeds tubercular obovate-oblong subacute below.—*E. B. S.* 2935.—Root not truncate. Rosettes loose, sessile. St. 1—2 ft. high, obscurely or very bluntly angled especially below. L. flaccid, mostly pendulous.—South of England, rare. P. VII.—IX. E.

†† Stem with raised lines, stigma entire.

6. *E. róseum* (Schreb.); rosettes loose, *l. stalked ovate narrowed and acute at both ends* serrulate, buds usually nodding ovoid acuminate, sep. lanceolate acuminate, seeds obovate-oblong narrowed to a rounded base.—*E. B.* 693.—St. 1—2 ft. high, branching, with two sharp and two blunt angles. Petioles long.—Damp places. P. VII. VIII. E. S. ? I.

7. *E. tetragónum* (L.); rosettes dense subsessile, *l. strap-shaped much denticulate-serrate*, limb of interm. l. decurrent, buds erect, seeds oblong-obovate tubercular.—*Sy. E. B.* 502. *Curt.* i. 66.—St. 1—2 ft. high, with 2—4 raised lines, usually much branched.—Damp places. P. VII. VIII. E. S.

8. *E. Lam'yi* (F. Schultz); rosettes dense, *l. lanceolate* denticulate-serrate *slightly rounded at the base with sometimes a short petiole*, buds erect.—*R.* 23. 13.—St. 1—3 ft. high. Nearly allied to Sp. 7 but differing by the glaucous lanceolate more acute less strongly dentate l., close pubescence and larger fl.—Woods and roadsides. P. VII.—VIII. E.

** *Stoles æstival long-jointed with small leaves, primary stem erect, stigma usually entire.*

9. *E. obscúrum* (Schreb.); *l. tapering* from a rounded base sessile *remotely denticulate*, lower l. oblong blunt, buds erect, seeds oblong-obovate.—*E. virgatum* G. & G., *Sy. E. B.* 503.—Resembling *E. tetragonum*, but the caps. much shorter and l. very different.—Wet places. P. VII. VIII. E. S. I.

*** *Stoles æstival long-jointed with very small leaves ending in autumnal bulbs which become detached, base of stem cordlike.*

10. *E. palus'tre* (L.); stoles long slender, st. rooting near the base, *l. narrowly lanceolate from a wedgeshaped base* sessile not decurrent, *top of raceme nodding*, sep. lanceolate, *seeds subfusiform* attenuate and acute below and the testa prolonged above smooth.—*Sy. E. B.* 504.—St. round, often with two rows of down, 6—18 in. high. L. entire or denticulate.—*E. ligulatum* (Baker) is a broad-leaved form.—Bogs. P. VII. VIII. E. S. I.

**** *Stoles or barren stems æstival rosulate.*

11. *E. alpinum* (L.); *barren stems short their upper l. closely placed,* fl.-st. erect from a short rooting base, l. oval or oblong blunt narrowed below not acuminate, upper l. lanceolate, buds nodding, *sep. linear-lanceolate acute,* seeds lanceolate-obovate pointed below apiculate.—*E. B.* 506, 2001. *E. anagallidifolium* (Lam.) ed. viii.—St. filiform, simple, with two raised lines, 3—4 in. long. L. pale green. Fl. pale.—Higher mountains. P. VII. S.

***** *Stoles æstival scale-bearing not rosulate.*

12. *E. alsinefólium* (Vill.); stoles (yellowish) with small roundish distant scales, st. erect from a long rooting base, *l. ovate-acuminate repand-dentate* shortly stalked, buds nodding, *sep. linear-oblong, seeds subfusiform.*—*Sy. E. B.* 505.—St. mostly simple, rather thick, with 2 raised lines, 3—12 in. long. L. shining, subpellucid. Fl. large, purplish. Caps. upright, long, long-stalked.—Higher mountains. P. VII. E. S. I.

2. ŒNOTHE'RA *Linn.* Evening Primrose.

*1. *Œ. bien'nis* (L.); l. obovate-lanceolate flat toothed, pet. longer than the stamens, st.-l. elliptic or ovate-lanceolate, caps. roundly tetragonal *tapering upwards.*—*E. B.* 1534. *St.* 5. 5. —Fl. large, many, bright yellow. St. 2—3 ft. high, leafy. —Sandy coast of Lancashire. (American.) B. VII.—IX. E.

*2. *Œ. odoráta* (Jacq.); l. subsessile, radical l. linear-lanceolate toothed, st.-l. ovate-lanceolate attenuate, pet. longer than the stamens, caps. clavate.—*Sy. E. B.* 509.—Fl. large, bright yellow. St. 2—3 ft. high. Foliage brighter green and fl. deeper yellow than in Sp. 1.—Channel Isles. Coasts of Som. and at Plymouth. (Patagonian.) B. VII.—IX. E.

3. LUDWIG'IA *Linn.*

1. *L. apet'ala* (Walt.); st. procumbent rooting glabrous, l. opposite ovate acute narrowed into a petiole, fl. axillary solitary sessile without petals.—*E. B. S.* 2593. *St.* 22. 3. *L. palustris* (Ell.) ed. viii.—St. 6—8 in. long, round, branching, often reddish. Fl. with 2 small bracts at the base. Caps. ovate obtusely quadrangular, with the persistent calyx spreading horizontally.—Pools and marshes. Suss., Hants, and Jersey. A. VI. E.

4. CIRCÆ'A *Linn.*

1. *C. lutetiána* (L.); *l. ovate* or slightly cordate below repand denticulate *opaque, petioles subterete, bracteoles* 0, pet. deeply

emarginate, lobes broadly obovate, cal. hairy, ovary 2-celled, fr. broadly obovate.—*E. B.* 1056. *St.* 23. 1.—Pet. as long as the herbaceous sepals, broad below. Fr. persistent.—Woods and hedge-banks. P. VI.—VIII. *Enchanter's Nightshade.* E. S. I.

2. *C. alpina* (L.); *l. cordate* acuminate repand-dentate shining, *petioles flat* with membranous wings, *bracteoles setaceous*, pet. bifid, lobes oblong, cal. glabrous, ovary 1-celled.—*E. B.* 1057. *St.* 23. 2.—Usually glabrous. Pet. shorter than the membranous sepals, narrow below. Fr. soon falling. Bracteoles deciduous.—β. *C. intermedia* (Ehrh.); fl. larger, sep. less membranous equalling the petals. Whole plant much larger.—Woods and thickets in hilly districts. P. VII. VIII. E. S. I.

Order XXX. HALORAGACEÆ.

Cal. adnate to the ovary, limb minute. Pet. minute, from the throat of the calyx. Stam. 1—8, inserted with the petals. Ovary of 1 or more cells. Styles equal in number to the cells. Fr. dry, not bursting, usually crowned with rim of calyx. Seeds pendulous; albumen fleshy.—Stip. 0.

1. MYRIOPHYLLUM. Monœcious. Cal. 4-parted. Pet. 4, soon falling, exceeding the calyx in the male, small and reflexed or 0 in the female. Stam. 8. Styles 4, villose. Fr. tetragonal, separable into 4 hard nuts.

2. HIPPURIS. Calyx-limb very minute, obsoletely 2-lobed. Pet. 0. Stam. 1. Style filiform. Stigma simple, acute. Fruit a 1-celled nut.

1. MYRIOPHYL'LUM *Linn.* Water-Milfoil.

1. *M. verticillátum* (L.); *fl. all axillary whorled*, bracts pinnatifid.—*E. B.* 218.—L. whorled, pinnatifidly divided into setaceous segments. Bracts more or less longer than the flowers, pinnatifid or (*M. pectinatum* DC.) pectinate.—Ponds and ditches, rare. P. VII. VIII. E. I.

2. *M. spicátum* (L.); *fl.* whorled *forming a leafless spike*, bracts small entire, spike erect when in bud.—*E B.* 83.—L. 4 in a whorl, submersed. Lower bracts often pectinate.—Ponds and ditches. P. VI. VII. E. S. I.

3. *M. alterniflórum* (DC.); sterile fl. alternate about 6 forming a leafless spike, *spike nodding when in bud* afterwards erect, *fertile* fl. about 3 together *in axillary whorls* at the base of the

spike.—*E. B. S.* 2854.—L. 3 or 4 in a whorl, submersed.—Ponds and ditches. P. V.—VIII. E. S. I.

2. Hippu'ris *Linn.* Mare's-tail.

1. *H. vulgáris* (L.); l. linear 6—12 in a whorl with a hard point.—*E. B.* 763. *St.* 44. 1.—St. simple, or sometimes branching at the base, erect. Fl. in the axil of each of the upper leaves, often without stamens. In deep water the submersed leaves are long flaccid pellucid and not hard at the end; in running water often wholly submersed flaccid and barren.—In stagnant water and slow streams. P. VI. VII. E. S. I.

Order XXXI. CUCURBITACEÆ.

Cal. 5-toothed, tube adnate to the ovary. Cor. 5-cleft, often scarcely distinguishable from the calyx, netted with veins. Stam. 5, more or less cohering. Anth. sinuous. Ovary 3—5-celled or spuriously 1-celled; placentas parietal. Style short. Stigmas lobed. Fr. more or less succulent. Seeds flat, in an aril; embryo flat; albumen 0.—Succulent, climbing with extra-axillary tendrils; often monœcious or diœcious. Stip. 0.

1. Bryonia. Cal. 5-toothed. Cor. 5-cleft. Male: Stam. 5 in 3 bundles. Fem: Style 3-fid. Fruit a globose few-seeded berry. Seeds oval, compressed, more or less bordered.

1. Bryo'nia *Linn.* Red Bryony.

1. *B. dioi'ca* (Jacq.); l. palmate 5-lobed dentate rough on both sides with hard points, fl. diœcious, cal. of the fertile fl. half as long as the corolla.—*Sy. E. B.* 517.—St. climbing. Tendrils simple. Fr. red.—[*B. alba* (L.) has the fertile cal. as long as the cor. and black fruit. It is said to be monœcious.]—Hedges and thickets. P. V.—IX. E.

Order XXXII. PORTULACEÆ.

Sep. 2, rarely 3 or 5, cohering at the base; imbricate in the bud. Pet. usually 5, from the base of the calyx. Stam. indefinite, inserted with the petals, often opposite to the petals; filaments distinct, Ovary 1-celled. Style 0. Stigmas several. Caps. opening transversely or by 3 valves; placenta central. Embryo curved round the albumen.—Stip. scarious.

1. MONTIA. Cal. of 2 sepals, persistent. Cor. 5-parted, with 3 segments smaller than the others, tube split to the base in front. Stam. 3, inserted in the throat and opposite to the smaller segments of the corolla. Ovary turbinate. Style very short. Stigmas 3. Caps. 3-valved, 3-seeded.

2. CLAYTONIA. Cal. of two sepals, persistent. Cor. of 5 free petals. Stam. 5, opposite to and adhering to the petals. Style 3-cleft. Caps. globose, 3-valved, 3-seeded.

1. MON'TIA *Linn.* Blinks.

1. *M. fontána* (L.).—*E. B.* 1206. *St.* 11. 1.—L. opposite, spathulate, entire. Valves of the caps. rolled longitudinally inwards after the seeds have fallen.—*a. minor* (All.); st. short ascending rigid, ped. axillary and terminal, seeds netted-asperous. —*β. major* (All.), *M. rivularis* (Gm.); st. flaccid ascending, ped. axillary, seeds netted.—Watery places. *β* in water. A. IV.—VIII. E. S. I.

2. CLAYTO'NIA *Linn.*

*1. *C. perfóliata* (Donn); radical l. fleshy rhomboidal-spathulate, upper st.-l. 2 connate, raceme subverticillate with 1 or 2 basal flowers, pet. nearly or quite entire.—*Sy. E. B.* 260.—St. tufted, ascending, 4—12 in. high. Pet. white, just overtopping the calyx.—Naturalized. (N.W. American.) A. IV.-VI. E. S.

*2. *C. sibir'ica* (L.); radical l. ovate-acuminate, upper l. opposite sessile, fl. mostly solitary in raceme, pet. bifid.—*E. B.* iii. Suppl. 260 A. *C. alsinoïdes* (Sims) ed. viii.—St. often 1 ft. high. Fl. more and larger than in Sp. 1.—Naturalized. (N.W. American.) A. IV.—VII. E. S.

[*Portulaca oleracea* (L.) is a common weed near Richmond, Surrey.]

Order XXXIII. PARONYCHIACEÆ[1].

Cal. 5-parted, imbricate in the bud. Pet. usually minute and resembling abortive stam., or 0. Stam. indefinite, opposite the sep. (when the same number), somewhat hypogynous. Ovary free. Styles 2 or 3. Fr. dry, 3-valved or indehiscent. Embryo more or less curved. Albumen mealy.—With stipules.

---

[1] Benth. & Hook. Gen. Plant. include this in the Order *Illecebraceæ* among the *Monochlamydeæ*.—H. & J. G.

1. CORRIGIOLA. Sep. 5. *Pet.* 5, *oblong, equalling the sepals.* Stam. 5. *Stigmas* 3, sessile. Fr. 1-seeded, indehiscent. Seed suspended by its seedstalk, which arises from the base of the capsule.—L. alternate.

2. HERNIARIA. Sep. 5. *Pet.* 5, *filiform*, inserted with the 5 stam. on a perigynous ring. *Stigmas* 2, nearly sessile. Fr. 1-seeded, indehiscent, membranous.—L. opposite.

3. ILLECEBRUM. *Sep.* 5, thickened, *horned at the back.* Pet. 0 or 5, subulate, inserted with the 5 stam. on a perigynous ring. Stigmas 2. *Fr.* 1-seeded, *furrowed, bursting along* the 5 furrows.

1. CORRIGI'OLA *Linn.* Strapwort.

1. *C. littorális* (L.); st. leafy amongst the flowers.—*E. B.* 668.—Fl. stalked, white, small, in small clusters. Stem-l. oblong, narrow below. St. many from the crown of the root, prostrate, slender.—Sandy shores. Slapton Sands and near the Start Point, Devon. Helston, Cornwall. A. VII. VIII. E.

2. HERNIA RIA *Linn.* Rupturewort.

1. *H. glábra* (L.); st. herbaceous prostrate clothed all round with minute decurved hairs, l. narrowed below, clusters of sessile fl. axillary collected on the lateral branches into a slightly leafy spike.—*E. B.* 206.—Pale yellowish green. St. procumbent or subterranean; autumnal shoots ascending, irregularly branched with spreading not imbricate leaves. Lateral branches resembling leafy spikes from the dense aggregation of the clusters. Cal. glabrous or with a very few hairs. L. glabrous.—Rare. Suffolk. Wilsford, Lincoln. Six-mile Bottom, Camb. Finchley Common. A. or B. VII. E.

2. *H. ciliáta* (Bab.); st. suffrutescent prostrate clothed on its upperside alone with minute decurved hairs, *l. orbicular-ovate* ciliate, clusters of sessile fl. axillary upon the lateral branches and distinct.—*E. B. S.* 2857.—Dark green. Root strong, woody. St. spreading extensively from the crown of the root but scarcely rooting until the autumn, when they are prostrate. and regularly alternately branched and the leaves are imbricate in two rows. Clusters 1—3 together in small distinct bunches. Fl. larger than in Sp. 1. Sep. tipped with a strong bristle.—Very rare. Lizard Point. Guernsey. P. VII. VIII. E.

3. *H. hirsúta* (L.); st. herbaceous prostrate *clothed with straight spreading hairs*, l. oblong narrowed at both ends, cal.

hairy resembling a minute bur. *E. B.* 1379.—Sandy ground at Christchurch, Hants. *Mr. Townsend.* The Finchley-Common plant was *H. glabra*!

3. Ille'cebrum *Linn.*

1. *I. verticillátum* (L.).—*E. B.* 895.—St. procumbent, filiform, glabrous. Fl. whorled, axillary, minute, white. L. roundish, variable in size, shorter or longer than the flowers.—Boggy places in Devon and Cornwall. P. VII. E.

Order XXXIV. CRASSULACEÆ.

Sep. 3—30, more or less united at the base. . Pet. the same number, free or slightly connected, inserted at the base of the calyx. Stam. inserted with the pet. and the same or twice their number. Hypogynous scales 1 at the base of each carpel or inconspicuous. Carpels the same nnmber as and opposite to the pet., free or slightly connected, 1-celled. Fr. of several follicles opening on their face, with slightly albuminous seeds on the inner suture.—Exstipulate.

1. Tillæa. Sep., pet., and stam. 3—4. Carp. 3—4, constricted in the middle and 2-seeded. Hypogynous scales 0.

2. Sedum. *Sep.* and *pet.* 5, rarely 4 or 6. Stam. 10 or 12. Hypogynous scales entire. Carp. 5 or 6 many-seeded.—*S. Rhodiola* has 4 sep., 4 pet., 8 stam., 4 carp., and is subdiœcious.

[3. Sempervivum. *Sep.* 6—10. Pet. the same. Stam. twice as many as the pet. *Hypogynous scales laciniate.* Carp. as many as the petals.]

4. Cotyledon. Sep. 5. *Pet. cohering into a tubular 5-cleft corolla. Stam.* 10, *inserted on the corolla.* Hypogynous scales 5. Carp. 5.

1. Tillæ'a *Linn.*

1. *T. muscósa* (L.); st. branched and decumbent at the base, fl. axillary sessile trifid.—*E. B.* 116. *R. I.* t. 191.—Very minute, about 1 in. long, reddish. L. opposite, oblong, blunt, concave above, connate. Sep. ovate or lanceolate, acute, bristle-pointed. Pet. nearly subulate, white tipped with red.—On barren sandy heaths in the South and East. A. VI. VII. E.

2. Sedum *Linn.* Stonecrop.

* *Rootstock thick, many-headed. Leaves flat. Stems annual.*

† Flowers yellow, subdiœcious, 4-parted. Rhodiola Linn.

1. *S. róseum* (Scop.); root fleshy, l. oblong smooth.—*S. Rhodiola* (DC.) ed. viii. *Rhodiola rosea* L. *E. B.* 508.—Fl. in a compact terminal cyme, subdiœcious. Stam. 8. L. alternate, acuminate, usually dentate in their upper half. Rhizome large, having a remarkable smell. St. 6—8 in. high, simple.—Wet alpine rocks. P. VI. VII. *Roseroot.* E. S. I.

†† Flowers white or purple, perfect, 5-parted.

2. *S. Teléphium* (L.); *uppermost l. long-oval* dentate *rounded at the base and sessile,* lower l. obovate or oblong narrowed below, corymb dense, *ovaries* flattened and slightly *furrowed on the back.*—*R. I.* 968. *S. purpurascens* Koch, *Sy. E. B.* 526.—St. 1—2 feet high. L. large, often orbicular. Pet. recurved from the middle, faintly channelled at the end. Stam. 10.—[*S. maximum* Sut., *R. I.* 969, has lower l. broad, uppermost cordate at the base, back of the ovaries convex.]—Hedgebanks and thickets. P. VII. VIII. *Orpine. Live-long.* E. S. I.

3. *S. Fabária* (Koch); l. dentate oblong-*lanceolate* or lanceolate *all narrowed to a slight petiole,* corymb dense, *ovaries not furrowed.*—*Sy. E. B.* 527. *St.* 83. 9.—Resembling *S. Telephium.* Pet. spreading. Stam. 10.—Hedges and thickets. P. VIII. *Orpine. Live-long.* E. I.

** *Root small, weak, without any rooting shoots. L. subterete.*

4. *S. villósum* (L.); st. erect, l. linear blunt flat above not produced at the base, pet. ovate acute, *l. st. and panicle glandular-pubescent.*—*E. B.* 394. *St.* 6. 12.—Pet. rose-coloured with a purple streak. St. 3—4. in. high. Seedlings with a rosette.—Wet mountain pastures. B. VI. VII. E. S.

*** *Root small, producing rooting shoots. Leaves subterete.*

† Flowers white. Leaves blunt.

5. *S. al'bum* (L.); *flowering st. curved at base then erect,* barren st. procumbent rooting, *l. oblong* subcylindrical flattened above spreading, pet. lanceolate, *panicle* much branched *glabrous.*—*E. B.* 1578.—Pet. white. St. 4—5 in. high, purplish leafy.—[*β. S. micranthum* (Bast.); l. oblong-obovate flattened

on both sides blunt. *Sy. E. B.* 529.]—Rocks and walls, not common. [*β.* Arundel.] P. VII. VIII. E.

6. *S. dasyphyl'lum* (L.); *flowering* and barren *st. procumbent, l. ovoid fleshy* gibbous, pet. and sep. ovate blunt, *panicle* small *glandular-pubescent.—E. B.* 656.—Pet. white. Sep., pet., and carp. often in sixes. L. very short and thick, glaucous, often tinged with red, opposite on the barren shoots. Flowering st. 3—4 in. long.—Rocks and walls, rare. P. VI. VII. E. I.

7. *S. ang'licum* (Huds.); st. procumbent at the base ascending, *l.* ovoid fleshy gibbous *spurred at the base,* pet. lanceolate acute, sep. ovate bluntish, *cyme bifid smooth.—E. B.* 171.—Pet. white spotted with red. L. mostly alternate. Flowering st. 3—6 in. long.—Sandy and rocky places. A. VII. VIII. E. S. I.

†† Flowers yellow. ‡ *Leaves blunt.*

8. *S. ácre* (L.); l. ovoid thick tumid spurred at the base, pet. lanceolate acute, *sep.* ovate *blunt gibbous at the base,* cyme trifid smooth.—*E. B.* 839.—L. closely imbricate on the barren shoots, very acrid.—Walls and dry places. P. VI. VII. *Wall-Pepper.* E. S. I.

[*S. sexanguláre* (L.); *l. linear* terete spurred at the base, pet. lanceolate acute, *sep.* lanceolate *acute not gibbous,* cyme trifid smooth.—*E. B.* 1946.—Old walls; a doubtful native. P. VII.] E.

‡‡ *Leaves acute.*

9. *S. reflex'um* (L.); *l. subulate* scattered spurred at the base *convex on both sides* the lowermost recurved, fl. cymose, sep. ovate rather acute, pet. lanceolate blunt.—*E. B.* 695.—Barren st. long. L. many, patent or reflexed. Cyme nearly level-topped, its outer branches spreading or recurved. Base of filam. and lateral edges of carp. with glandular hairs.—*β. S. albescens* (Haw.) l. more slender glaucous not recurved. *S. glaucum* Sm. *E. B.* 2477.—Walls and rocks, common. *β.* Dry hills rare. P. VII. VIII. E. S. I.

10. *S. rupes'tre* (L.); *l. linear-lanceolate* spurred at the base *flattened,* fl. imperfectly cymose, sep. elliptic, pet. lanceolate.—*E. B.* 170. *S. elegans* Lej.—Barren st. short with densely imbricate adpressed *glaucous* leaves. *Fl. corymbose* rather than cymose. Filam. and ovaries glabrous.—On limestone rocks. Bristol. Cheddar. Orme's Head, &c. P. VI. VII. E. I.

11. *S. Forsteriánum* (Sm.); l. lanceolate spurred at the base flattened, *cyme round-topped* compact, sep. ovate, pet. lanceolate.—*E. B.* 1802.—Barren st. short, erect, densely leafy. *L. forming small rose-like tufts, bright green. Fl. truly cymose.* Filam. and ovaries glabrous. Very different from the last in appearance; but there are 2 forms, (1) with short obconic barren stems and green leaves, (2) with very short globular barren stems and glaucous leaves.—Damp rocks in Wales, Shropshire, and West Somerset. P. VI. VII. E.

3. Sempervi'vum *Linn.* Houseleek.

[*S. tectórum* (L.); l. glabrous ciliate obovate-oblong, pet. 12 or more entire at the margins, hypogynous scales short convex resembling glands.—*E. B.* 1320.—L. succulent, green with purple tips, forming large roselike tufts.—Walls and roofs (planted). P. VII.] E. I.

4. Cotyle'don *Linn.* Navelwort.

1. *C. Umbilícus* (L.); lower l. peltate concave, bracts entire, fl. pendulous.—*E. B.* 325.—Raceme usually simple. St. 6—12 in. high, leaves mostly collected at its base. Fl. greenish yellow. —Chiefly on rocks and walls in the West. P. VI.—VIII. E. S. I.

Order XXXV. RIBESIACEÆ.

Cal. superior, 4—5-cleft, regular. Pet. 4—5, small, inserted at the mouth of the tube and alternating with the stamens. Ovary 1-celled, with 2 opposite parietal placentas. Style 2—4-cleft. Berry many-seeded; cell filled with pulp. Albumen horny. Stip. 0.

1. Ribes. Cal. 5-cleft. Berry many-seeded, crowned with the persistent calyx.

1. Ri'bes *Linn.*

* *Peduncles 1—3-flowered. Stems spinous.* Gooseberry.

†1. *R. Grossulária* (L.); ped. with 2 minute bracts, cal. bell-shaped, sep. reflexed oblong, pet. ovate.—*E. B.* 1292 and 2057. —Thorns 1—3, at the base of the young branches. Germen and fruit smooth (*R. Uva-crispa* L.) or pubescent or glandular. L. rounded, 3—5-lobed and cut, glabrous or hairy.—Hedges and thickets, doubtfully native. Sh. IV. V. E. S.

** *Flowers racemose. Spines none.* Currants.

2. *R. alpinum* (L.); diœcious, *racemes upright* both in flower and fruit glandular-pilose, *bracts* exceeding the pedicels *lanceolate, cal. glabrous,* l. shining beneath.—*E. B.* 704. *St.* 51. 6.—Male raceme dense, of many flowers; fem. of 2—5 fl. Limb of the calyx nearly flat. Berries scarlet. L. with 3 acute deeply serrate lobes.—Woods. North of E., rare. Sh. IV. V. E.

3. *R nigrum* (L.); *racemes pendulous* downy with a separate simple pedicel at the base of each, *bracts subulate* falling short of the pedicels, *cal. pubescent,* l. glandular-punctate beneath.—*E. B.* 1291.—Racemes lax. Calyx-limb bell-shaped. Berries large, black. L. with 3—5 acute serrate lobes.—In damp and swampy places. Sh. IV. V. *Black Currant.* E.

4. *R. rúbrum* (L.); racemes mostly glabrous and pendulous, *bracts* shorter than the pedicels *ovate, cal.* nearly flat *glabrous,* l. bluntly 5-lobed.—*E. B.* 1289.—*a. sativum* (R.); l. glabrous, raceme glabrous.—*β. R. petræum* (Sm. *not* Wulf.); racemes slightly downy, in flower upright, in fruit pendulous. *E. B.* 705.—*R. spicatum* (Robs.); racemes erect with fl. and fr., "fl. nearly subsessile," was perhaps a sport from this. *E. B.* 1290.—Woods. *β.* North of E., S. Sh. IV. V. *Red Currant.* E. S.

Order XXXVI. SAXIFRAGACEÆ.

Cal. 4—5-cleft, superior or inferior. Pet. 4—5, rarely 0. Stam. 5—10, free, perigynous or hypogynous. Glandular disk present or wanting. Ovary of 2 carpels cohering by the inflexed sides or margins. Styles 2, persistent, usually diverging. Seeds many. Albumen fleshy.—Stip. 0.

1. SAXIFRAGA. Cal. 5-fid or 5-parted, more or less adnate to the ovary or free. Cor. of 5 petals. Stam. 10, rarely 5. Styles 2, persistent. Caps. 2-celled, with 2 beaks, opening by a pore between the beaks.

2. CHRYSOSPLENIUM. Cal. 4-fid, half superior. Cor. 0. Stam. 8, rarely 10. Styles 2. Caps. 1-celled, with 2 beaks, opening into the form of a cup.

3. PARNASSIA. Cal. 5-cleft, inferior. Pet. 5. Stam. 5, perigynous, with 5 scales bearing glandular bristles interposed. Stigm. 4, sessile. Caps. 1-celled, with 4 valves.

1. Saxifraga *Linn.* Saxifrage.

Sec. 1. With decumbent barren shoots at the base.

* *Cal. reflexed inferior, fl. panicled, flowering stems leafless.*

† Filaments enlarged upwards. Robertsonia Haw.

1. *S. umbrósa* (L.); *l. obovate* with cartilaginous crenatures or sharp notches tapering at the base into *dilated footstalks with flat edges*, panicles racemose.—*a*; l. crenate or dentate spreading. *E. B.* 663.—*β. S. punctata* (Haw.); l. nearly round acutely serrate erect. *R. I.* t. 622, 623.—*γ. S. serratifolia* (Mack.); l. oblong acutely serrate erect. *E. B. S.* 2891.—West of Ireland. Yorkshire (var. a), a doubtful native. P. VI. *London Pride. St. Patrick's Cabbage.* [E.] I.

2. *S. el'egans* (Mack.); *l. round* smooth shining acutely serrate, *footstalks broad convex beneath with flat edges,* panicle racemose.—*E. B. S.* 2892.—L. not tapering into the footstalks.—Very rare. Top of Turk Mountain, Killarney, and Connor Hill, Kerry. P. VI. I.

3. *S. hirsúta* (L.); *l. oval* sharply serrate, *footstalks linear semicylindrical with raised edges* hairy, panicles racemose.—*E. B.* 2322.—*Footstalks* slightly *tapering* upwards. L. longer than broad. Perhaps not distinct from Sp. 4?—Very rare. Gap of Dunloe and Connor Hill, Kerry. Hungry Hill, Co. Cork. P. VI. I.

4. *S. Géum* (L.); *l. transversely oval* or reniform crenate or dentate, *footstalks semicylindrical with raised edges* hairy, panicle racemose.—*E. B. S.* 2893.—L. always broader than long, usually hairy, sometimes glabrous, often beautifully netted with purple beneath.—West of Ireland. P. VI. I.

[*S. Andrews'ii* (Harv.). *Sy. E. B.* 549. *R. Guthriana* Hort. Is a garden hybrid. P. VI.]

†† Filaments subulate. Spatularia Haw.

5. *S. stelláris* (L.); *l. oblong wedgeshaped* dentate-serrate narrow below *scarcely stalked*, panicle corymbose of few flowers.—*E. B.* 167. *St.* 35. 3.—Stems tufted. Fl.-stalks 1—5 in. high, naked. Pet. ovate, clawed, white with 2 transverse yellow spots on their lower half.—L. rarely quite entire.—Damp rocks on mountains. P. VII. E. S. I.

** *Cal. at length reflexed inferior, st. leafy, l. undivided.*

6. *S. Hir'culus* (L.); st. erect, barren shoots prostrate filiform, l. alternate lanceolate flat entire, root-l. narrowed into a footstalk, sep. blunt fringed at the margin, pet. blunt with 2 hard points near the base.—*E. B.* 1009. *St.* 35. 8.—Fl. few or solitary, terminal. Pet. obovate, spreading, yellow dotted with red. St. 4—8 in. high, downy in the upper part.—Wet moors, rare. P. VIII. E. S. I.

*** *Cal. erect or spreading half inferior, st. leafy, l. simple toothed or with rigid jointless cilia, also a transparent gland at the tip.*

7. *S. aïzoï'des* (L.); st. decumbent below, l. alternate linear-oblong mucronate ciliate entire flat above convex beneath, sep. blunt.—*E. B.* 39. *St.* 35. 9.—Fl. in a leafy panicle, with glutinous downy stalks. Pet. bright yellow often spotted with scarlet. St. 3—6 in. long. L. rigidly ciliate or with several strong teeth.—Wet places on mountains. P. VII.—IX. E. S. I.

**** *Cal. erect or spreading superior or half inferior, st. more or less leafy, l. lobed, cilia all jointed. S. hypnoides,* Hook.—See Baker in *J. of B.* viii. 280, 355.

8. *S. cæspitósa* (L.); *root-l.* crowded 3—5-cleft *blunt* veined fringed, fl. 1—5, pet. rounded 3-veined, *cal. half inferior,* sep. blunt.—Barren st. very short. Fl. white. Germen broad and rounded below. Common base of the l. not furrowed. Anth. long, cordate with an open notch. Horns of caps. spreading.—*β. S. incurvifolia* (D. Don); stem-l. more numerous with incurved lobes. *E. B. S.* 2909.—Very rare. Caernarvonshire. Westmoreland. Aberdeenshire. Kerry. P. V. VI. E. S. I.

9. *S. decip'iens* (Ehrh.)[1]; *l.* crowded 3—4-cleft, *lobes lanceolate pointed* fringed, fl. few (usually 3) loosely panicled, *pet. obovate* 3-veined, cal. half inferior, *sep. bluntish.*—*S. hirta* Sm. *E. B.* 2291 (not Haw.). *S. Sternbergii* (Willd.) ed. viii.—L. on the barren shoots as well as the radical l. 3-cleft, the lateral lobes often divided halfway down, lobes diverging. Barren shoots 3—6 or 8 in. long, weak. Fl.-st. hairy glandular, bearing 2 or 3 deeply 3-lobed l. and a few simple linear bracts;

---

[1] Prof. Engler (Monogr. Saxif.) includes as vars. under this name *S. Sternbergii,* Willd., *S. quinquefida,* Haw. (=*S. sponhemica,* Gmel.), and *S. grœnlandica,* L., and has referred some British specimens to the last named.—H. & J. G.

lateral fr.-ped. much overtopping the intermediate one. Fl. white. Anth. large, round, with a deep parallel-sided notch. Horns of caps. slightly spreading.—Llyn y Cwm, N. Wales. Summit of Brandon Mountain, Kerry; Galty More, Tipperary; Black Head, Co. Clare. West of Scotland. *Mr. G. Don.* P. VII. E. S. I.

10. *S. sponhem'ica* (Gm.); l. of the barren shoots 3—5-lobed with a very broad base fringed, lobes linear acute, fl. few (2—4), *pet. oblong* 3-veined, cal. half inferior deeply divided into triangular *subulate sepals.*—*S. affinis* Don. *E. B. S.* 2903. *S. pinnatifida* Haw.—The 5-cleft l. few. St. 1—3 in. high, erect, with a few linear simple leaves, glandular. Fl. white. Anth. ½ as large as in Sp. 8 and 9, cordate-ovate. Horns of caps. spreading.—The original *S. affinis* has pet. with inflexed sides: *S. platypetala* (*E. B.* 2276) has them broad and flat.—Mountains. P. VII. E. S. I.

11. *S. hypnoi'des* (L.); root-l. 3—5-cleft *those of the trailing shoots undivided* or *3-cleft*, lobes all *acute bristle-pointed* and fringed, calyx half inferior, *sep. triangular-subulate or ovate acute.*—*E. B.* 454.—Fl. white. Anth. very small, oblong-ovate. Horns of caps. divaricate. Differs from the preceding by having usually *buds in the axils of the barren shoots.*—Mountains. P. V.—VII. E. S. I.

[*S. pedatifida* (Sm.).—*E. B.* 2278.—Probably of garden origin. See J. of B. xxi. p. 152.]

Sec. 2. Without barren shoots at the base.

* *Stem leafy.*

12. *S. tridactylites* (L.); *st.* panicled erect *leafy*, l. wedge-shaped 3—5-fid with a flat petiole, lowermost l. often simple and spathulate, *peduncles* 1-flowered much longer than the fruit *with* 2 *bracts at the base*, cal. superior.—*E. B.* 501. *St.* 33. 15.—Whole plant viscid, 2—4 in. high. Fl. scattered, many, small, white.—Walls and dry banks. A. IV.—VII. E. S. I.

13. *S. granuláta* (L.); st. erect slightly leafy, *radical l. reniform crenately lobed* with channelled petioles, stem-l. nearly sessile 3—5-fid, fl. in a cymose panicle, *cal. half inferior*, pet. obovate-oblong 2 or 3 times as long as the sepals, *roots bearing* many small round downy *bulbs.*—*E. B.* 500.—St. 6—12 in. high. Fl. large, white.—Gravelly banks. P. V. E. S. I.

14. *S. cer'nua* (L.); *st.* erect *simple* 1-*flowered* leafy, radical l. reniform palmately lobed stalked, upper l. nearly sessile sub-trifid, uppermost entire, *axils bearing bulbs, cal. quite inferior.*—*E. B.* 664.—Rarely flowering; fl. replaced by reddish bulbs. St. 3—6 in. high.—Rocks on the top of Ben Lawers, very rare. P. VI.—VIII. S.

15. *S. rivuláris* (L.); *st.* ascending *branched* few-flowered leafy, radical l. subreniform stalked with 3-5 rounded lobes, uppermost l. lanceolate entire, cal. half inferior.—*E. B.* 2275.—St. 1—2 in. long. Fl. few, stalked, white.—Wet places on the coldest parts of mountains. P. VIII. S.

** *Stem leafless.*

16. *S. nivális* (L.); st. erect leafless, *l. all radical roundish-obovate* dentate-serrate narrowed into a footstalk, *fl. in capitate cymes*, cal. half inferior, pet. longer than the calyx.—*E. B.* 440. *St.* 35. 4.—St. 3—6 in. high, usually simple, sometimes with 1 branch. Fl. in a dense cluster, white.—Alpine rocks. P. VII. E. S. I.

Sec. 3. Stems procumbent with opposite l. and terminal flowers.

17. *S. oppositifólia* (L.); st. procumbent, *l. opposite in* 4 *rows* oblong blunt fringed, sep. ciliate without glands, pet. ovate.—*E. B.* 9.—Very different from our other species. L. with a pore at tip. Fl. large, purple.—Damp alpine rocks. P. IV. V. E. S I.

2. Chrysosple'nium *Linn.* Golden Saxifrage.

1. *C. alternifólium* (L.); *l. alternate,* lower l. subreniform hairy crenate upon long stalks.—*Sy. E. B.* 564. *St.* 12.—Crenatures of the lower l. emarginate, upper l. glabrous with the crenatures often rather acute. St. erect, 4—5 in. high, *branching only near the top.* Fl. umbellate, nearly sessile, deep yellow. Stam. usually 8.—Boggy places. P. IV. E. S. I.

2. *C. oppositifólium* (L.); *l. opposite,* lower l. roundish-cordate shortly stalked wavy.—*E. B.* 450. *St.* 4. 6.—*St. branching from the base,* 4—6 in. long, decumbent, straggling, rooting. Fl. paler and more scattered than in the last. L. usually glabrous, sometimes slightly hairy. Stam. usually 8.—Damp shady places. P. IV. V. E. S. I.

3. Parnas'sia *Linn.*

1. *P. palus'tris* (L.); petal-like scales 9—13, pet. with a short claw, radical l. cordate stalked, st.-l. clasping.—*E. B.* 82.—Pet. white, veined. Glands of the scales yellow. L. mostly radical. St. 8—10 in. high.—Wet and boggy ground. P. VIII.—X. E. S. I.

B. Petals and stamens epigynous, inserted round an epigynous disk. Cal.-tube adnate to the ovary.

Order XXXVII. UMBELLIFERÆ.

Cal. 5-toothed or entire, adherent to the ovary, limb often scarcely visible. Pet. 5, usually flexed at the point. Stam. 5, inserted with the pet. round the stylopode. Ovary 2-celled, crowned with a double fleshy disk (stylopode). Styles 2. Fr. consisting of 2 carpels (mericarps) adhering by their face (commissure) to a common axis from which they ultimately separate and become pendulous. Seed solitary, pendulous. Albumen horny.—Inflorescence umbellate. Æstivation imbricate, except in *Hydrocotyle* and *Crithmum.*—Each carpel has 5 primary, and often 4 intermediate secondary ridges; and in the substance of the pericarp are usually linear receptacles of oil (vittæ or stripes) under the ridges or the spaces between them. These parts are sometimes either wanting or only slightly apparent. The stripes are "solitary" when there is only one in each space between the primary ridges, and "2, 3, &c. together" when 2, 3, or more occur in each space. They and the ridges are best seen by making a horizontal section of the fruit.

Suborder I. ORTHOSPERMÆ.

Sutural side of seed flat. Umbels various.

* *Umbels imperfect or simple; no fr.-stripes. "No carpophore."*

Tribe I. *HYDROCOTYLEÆ.* Fr. laterally flattened, its back even or acute. Umbel irregular or imperfect.

1. Hydrocotyle. Cal.[1] inconspicuous. Pet. ovate entire. Fr. of 2 flat nearly orbicular carpels, each with 5 filiform ridges, of which the dorsal and 2 lateral are often inconspicuous; the 2 others arched. Stripes 0. Commissure linear.

---

[1] By calyx, throughout this Order, the free margin is intended.

Tr. II. *SANICULEÆ.* Fr. ovoid; transverse section nearly round. Umbel simple or imperfect.

2. SANICULA. *Cal. of* 5 *leaflike teeth. Pet.* erect, obovate, *with a long inflexed connivent point. Fr.* subglobose, *covered with hooked spines*; ridges 0; stripes many.

3. ASTRANTIA. Cal. of 5 leaflike teeth. Pet. erect, with a long inflexed point. *Fr. with* 5 *plaited dentate ridges*; stripes 0.

Tr. III. *ERYNGIEÆ.* Fr. ovoid; transverse section nearly round. Fl. in a head-like umbel.

4. ERYNGIUM. Cal. of 5 leaflike teeth. Pet. erect, oblong, with a long inflexed point. Fr. *covered with chaffy scales, without ridges* or stripes.

** *Umbels perfect or compound.*

† Carpels with 5 primary ridges only.

Tr. IV. *AMMINEÆ.* Fr. of two pentagonal carpels with 5 prominent ridges; the commissure about as broad as either of the other four sides.

A. *Leaves compound.*

5. CICUTA. *Cal. of* 5 *leaflike teeth. Pet. obcordate* with an inflexed point. Fr. subdidymous. Carp. with 5 equal broad flattened ridges, the lateral marginal; stripes solitary.

6. APIUM. *Cal. inconspicuous. Pet.* roundish *entire* with an oblique acute or small involute point. Fr. roundish ovoid, didymous. Carp. with 5 filiform equal prominent ridges and solitary stripes.—*Inv.* 0.

7. PETROSELINUM. Cal. minute or inconspicuous. Pet. roundish entire with a narrow incurved point. Fr. ovoid. Carp. with 5 filiform equal ridges and solitary stripes. Carpophore bipartite.—Partial inv. of many; general of few leaves.

8. SISON. Cal. inconspicuous. *Pet. broadly obcordate, deeply notched* with an inflexed point. Fr. ovoid. Carp. with 5 filiform prominent equal ridges and *solitary clavate stripes.*

9. APINELLA. Diœcious. Cal. inconspicuous. *Pet. of the barren fl. lanceolate* with the point inflexed, *of the fertile fl. ovate* with a short inflexed point. Fr. ovoid. Carp. with 5 filiform prominent equal *ridges with a single stripe beneath each of them, but none between.*

10. ÆGOPODIUM. Cal. inconspicuous. Pet. obovate, notched with an inflexed point. Fr. oblong. Carp. with 5 filiform ridges; the *interstices without stripes.* Stylopode conical.

11. CARUM. Cal. inconspicuous. *Pet.* obcordate *with an inflexed point.* Fr. oblong. Carp. with 5 filiform ridges, the stripes solitary or 1—3 together. *Stylopode depressed.*

12. PIMPINELLA. Cal. inconspicuous. Pet. obcordate with an inflexed point. Fr. ovoid. Carp. with 5 filiform equal ridges; *stripes 3 or more together. Stylopode tumid.* Styles of the fr. divaricate or recurved. *Inv.* 0 (rarely of 1 leaf).

13. SIUM. Cal. of 5 small teeth. Pet. obcordate with an inflexed point. Fr. ovoid or subdidymous. *Carp. with* 5 filiform equal *blunt ridges*; stripes 3 or more together. *Stylopode depressed or shortly conical.* Styles of the fr. divaricate or recurved.—Inv. general and partial.

B. *Leaves simple.*

14. BUPLEURUM. Cal. inconspicuous. *Pet. roundish entire with a closely involute broad retuse point.* Fr. subdidymous. Carp. with equal, winged, or filiform and sharp, or inconspicuous ridges. Stylopode depressed.

Tr. V. *SESELINEÆ.* Fr. of two 5-ribbed or pentagonal carpels. Commissure much the broadest side of the carpel.

A. *Stripes solitary, between the ribs.*

15. ŒNANTHE. Cal. of 5 lanceolate teeth. *Pet. obcordate* with an inflexed point. Fr. ovoid cylindrical or subturbinate, crowned with the long *suberect styles.* Carpels more or less corky, with 5 blunt convex ridges.

16. ÆTHUSA. Cal. inconspicuous. *Pet. obcordate with an acute inflexed point.* Fr. shortly ovoid, crowned with the *reflexed styles.* Carp. with 5 thick acutely keeled ridges.

17. FŒNICULUM. *Cal. inconspicuous. Pet. roundish entire with a broad blunt inflexed lobe.* Fr. oblong. Carp. with 5 prominent bluntly keeled ridges. Stylopode conical.

18. SESELI. *Cal. with acute teeth.* Pet. obcordate with a broad inflexed lobe. Fr. ovoid or oblong, slightly dorsally compressed. Carp. with 5 thick blunt ridges. Stylopodes conical.

B. *Stripes 2 or more together, between the ribs.*

19. HALOSCIAS. Cal. of 5 small persistent teeth. *Pet. ovate with* an inflexed lobe and *short claw.* Fr. elliptic, terete or slightly dorsally compressed. *Carp. with* 5 *sharp somewhat winged ridges.* Interstices and commissure with many stripes. *Seeds not adhering to the carpel, without stripes.*

20. SILAUS. Cal. inconspicuous. *Pet.* ovate-oblong *entire or slightly emarginate* with an inflexed lobe, *sessile, truncate or appendaged at the base.* Fr. oblong, terete or slightly dorsally compressed. Carp. with 5 sharp somewhat winged ridges. Stripes many, inconspicuous.

21. MEUM. Cal. inconspicuous. *Pet.* entire, elliptic, *acute at both ends,* with an incurved point. Fr. as in *Silaus.*

22. CRITHMUM. Cal. inconspicuous. *Pet. elliptic* with a broad base, *entire, involute.* Fr. oblong, terete. Carp. with 5 elevated sharp slightly winged ridges. *Seed free*; *with many stripes.*

Tr. VI. *ANGELICEÆ.* Fr. of two much and dorsally compressed carpels, with a double wing on each side.

23. ANGELICA. Cal. inconspicuous. Pet. lanceolate, entire, acuminate, incurved. *Carp. with* 3 dorsal elevated *filiform ridges* and 2 marginal ridges dilated into broad wings; *interstices with solitary stripes. Seed adhering to the carpel.*

[24. ARCHANGELICA. Cal. minutely 5-toothed. Pet. elliptic, entire, acuminate, incurved. *Carp. with* 3 dorsal elevated *thick ridges* and two marginal ridges dilated into broad wings. *Stripes* 0. *Seed free, with many fine lines.*]

25. SELINUM. Cal. inconspicuous. *Pet. oblong, notched,* with an incurved lobe. Carp. with 3 dorsal elevated thin ridges, and 2 marginal dilated rather broad wings. *Interstices with solitary stripes. Seed adhering to the carpel.*

Tr. VII. *PEUCEDANEÆ.* Fr. of two much and dorsally compressed carpels, with a single wing on each side; wing flat or thickened towards the edge, formed of those of the two carpels combined.

26. PEUCEDANUM. Cal. of 5 teeth or inconspicuous. Pet. obovate or obcordate with an inflexed point. Fr. with a dilated thin flat margin. Carp. with equidistant ridges, 3 dorsal filiform, 2 *lateral close to the base of the dilated margin* inconspicuous; stripes solitary.

27. PASTINACA. Cal. of 5 very small or inconspicuous teeth. *Pet. roundish, entire,* involute with an acute point. Fr. with a dilated flat margin. Carp. with slender ridges, 3 dorsal equidistant, 2 *lateral distant near the outer edge of the dilated margin; stripes linear, solitary.*—Fl. yellow.

28. HERACLEUM. Cal. of 5 minute teeth. *Pet. obcordate* with an inflexed point, *outer ones radiant.* Fr. as in *Pastinaca, but the stripes short clubshaped.*—Fl. white.

29. TORDYLIUM. Cal. of 5 awlshaped teeth. Pet. obcordate with an inflexed lobe, outer ones radiant. *Fr. with a thickened wrinkled margin.* Carp. with slender ridges, 3 dorsal equidistant, 2 *lateral distant close to the thickened margin*; stripes 1—3 together.

†† Carpels with primary and secondary ridges.

[Tr. VIII. *SILERINEÆ.* Fr. dorsally compressed. Carp. with 5 primary ridges, the lateral marginal, and 4 less prominent secondary ridges.]

[30. SILER. Cal. of 5 teeth. Pet. obovate with an inflexed lobe, regular. Fr. dorsally compressed. Carp. with elevated filiform blunt ridges; one stripe under each secondary ridge and 2 on the commissure.]

Tr. IX. *DAUCINEÆ.* Fr. somewhat dorsally compressed. Carp. with 5 primary ridges, the lateral ones on the inner face; and 4 secondary, forming *rows of prickles.*

31. DAUCUS. Cal. of 5 teeth. Pet. obcordate with an inflexed lobe, exterior usually radiant and bifid. Fr. dorsally compressed. Carp. with bristly primary ridges; secondary ridges equal winged with 1 row of spines.

Suborder II. CAMPYLOSPERMÆ.

Sutural side of seed with inflexed edges or deeply furrowed lengthwise. Umbels compound or perfect.

Tr. X. *CAUCALINEÆ.* Fr. contracted or rounded. Carp. with the lateral primary ridges on the inner face; 4 *secondary more prominent, prickly*; stripes solitary.

32. Caucalis. Cal. of 5 teeth. Pet. obcordate with an inflexed point, outer ones radiant and bifid. Fr. slightly laterally compressed. Carp. with filiform bristly *primary, and* more or less prominent *secondary ridges, all bearing* 1—3 rows of *prickles*.—Umbels sometimes simple.

33. Torilis. Cal. of 5 teeth. Pet obcordate with an inflexed point, outer ones radiant and bifid. Fr. slightly laterally compresssd. *Carp. with* bristly primary ridges, with *many prickles on the spaces* between them.

Tr. XI. *SCANDICINEÆ.* Fr. compressed or contracted on the sides, often beaked or narrowed at the top. *Carpels with primary ridges only.*

34. Scandix. Cal. inconspicuous. Pet. obovate with an inflexed point. *Fr. with a very long beak. Carp. with* 5 *blunt ridges*; stripes none.

35. Chærophyllum. Cal. inconspicuous. Pet. obcordate with an inflexed point. *Fr. hardly beaked. Carp. with* 5 *equal blunt ridges,* often only apparent at the top; stripes solitary or wanting.

36. Myrrhis. Cal. inconspicuous. Pet. obcordate with an inflexed point. Fr. not beaked. *Carp. formed of a double membrane*; the outer with elevated keeled ridges hollow within, the inner close to the seed; stripes 0.

Tr. XII. *SMYRNIEÆ.* Fr. ovoid or didymous compressed or contracted on the sides. Carpels with primary ridges only, not beaked nor remarkably narrowed at the top.

[37. Echinophora. Cal. of 5 teeth. Pet. obcordate with an inflexed point, the exterior larger and bifid. Fl. of the ray, sterile on long stalks, fertile central and solitary. Fr. ovoid terete, imbedded in the enlarged prickly receptacle. Carp. with 5 depressed equal striate wavy ridges; interstices with single stripes and covered by a cobweblike membrane.]

38. Conium. Cal. inconspicuous. Pet. obcordate with a short inflexed point. Fr. ovoid, laterally compressed. *Carp. with 5 prominent or wavy crenate ridges,* the lateral marginal; interstices striate; stripes 0[1]

---

[1] I have never seen any stripes, which are said to be sometimes present.

39. PHYSOSPERMUM. Cal. of 5 teeth. Pet. obcordate with an inflexed point. Fr. laterally compressed, didymous. *Carp.* reniform-globose, *with 5 filiform slender equal ridges, the lateral within the margin*; *stripes solitary.*

40. SMYRNIUM. Cal. inconspicuous. Pet. lanceolate or elliptic, entire, with an inflexed point. Fr. laterally compressed, didymous. *Carp.* reniform-oblong *with 3 dorsal prominent sharp ridges, the 2 lateral marginal and inconspicuous*; *stripes many.*

Suborder III. CŒLOSPERMÆ.

Seed with the base and apex curved inwards.

Tr. XIII. *CORIANDREÆ. Fr. globose* or didymous. Primary ridges of the carpels often inconspicuous, secondary more prominent, all without wings.

[41. CORIANDRUM. Cal. of 5 teeth. Pet. obcordate with an inflexed point, outer ones radiant and bifid. *Fr. globose.* Carp. scarcely separating, the primary ridges inconspicuous, the 4 secondary conspicuous prominent keeled; interstices without stripes; commissure with 2 stripes.]

Suborder I. *Orthospermæ.* Tribe I. *Hydrocotyleæ.*

1. HYDROCOTYLE *Linn.* Pennywort.

1. *H. vulgáris* (L.); l. peltate nearly circular 9-veined doubly crenate, petioles pilose, umbels 3—6-fl., fr. emarginate below.—*E. B.* 751.—Fl. and fr. almost sessile. Umbels or rather heads often proliferous in the centre and bearing a second head. St. creeping extensively. L. upon stalks which considerably exceed the peduncles. Ripe fr. with purplish dots.—Bogs and marshy places. P. V.—VII. E. S. I.

Tribe II. *Saniculeæ.*

2. SANIC'ULA *Linn.* Sanicle.

1. *S. europæ'a* (L.); lower l. palmate 3—5-lobed, lobes trifid unequally serrate, fertile fl. sessile, barren fl. slightly stalked.—*E. B.* 98.—Umbels many, capitate, in an irregular slightly umbellate panicle. Styles persistent, reflexed. Fr. covered with hooked prickles. St. ascending about a foot high.—Woods and thickets. P. VI. VII. E. S. I.

3. ASTRAN'TIA *Linn.*

†1. *A. májor* (L.); lower l. palmately 5—7-fid, lobes oblong acutely unequally incise-serrate, inv.-l. entire, cal.-teeth ovate-lanceolate narrowed to an acute point.—*E. B. S.* 2990. *St.* 29. 8.—Inv. equalling the umbel, usually straw-coloured. Cal.-teeth exceeding petals.—Woods in hilly districts. Above Stokesay Castle, Shropshire, probably introduced. P. VI.—VIII. E.

Tribe III. *Eryngieæ.*

4. ERYN'GIUM *Linn.* Eryngo.

1. *E. marit'imum* (L.); *radical l. suborbicular plaited coriaceous* spinous stalked, upper l. amplexicaul palmately lobed, *inv.-l. 3-lobed* spinous exceeding the heads, scales of the receptacle 3-lobed.—*E. B.* 718.—St. 1 foot or more in height, much branched, leafy, rigid, *glaucous.* Fl. in heads rather than umbels, blue.—Sandy sea-shores. P. VII. VIII. *Sea-Holly.* E. S. I.

2. *E. campes'tre* (L.); *radical l.* 2 or 3 *times pinnatifid* spinous stalked, st.-l. amplexicaul pinnatifid, inv.-l. lanceolate spinous longer than the heads, scales of the receptacle undivided.—*E. B.* 57.—More bushy and slender than the last. Pet. purplish or white. Petioles thick, semiterete, channelled.—On waste ground, very rare. P. VII. VIII. E.

Tribe IV. *Ammineæ.*

5. CICU'TA *Linn.* Water-Hemlock.

1. *C. virósa* (L.); fibres of the root filiform, l. tripartite, leaflets linear-lanceolate acute serrate decurrent.—*E. B.* 479.—St. 3—4 ft. high, very thick terete and hollow below. Lower l. on long stalks; leaflets 1—2 in. long. Umbels large; general inv. 0 or of 1 or 2 slender leaflets, partial of many subulate leaflets. Fl. white. Herb poisonous.—Ponds and ditches, not common. P. VII. VIII. *Cowbane.* E. S. I.

6. A'PIUM *Linn.*

* *Pet. with involute point, cordate base, minute claw. Partial inv.* 0.

1. *A. gravéolens* (L.); glabrous, l. pinnate or ternate, leaflets of the upper l. wedgeshaped.—*E. B.* 1210.—St. 1—2 feet high,

branched, furrowed, leafy. Umbels terminal or lateral, frequently almost sessile, accompanied by one or two ternate leaves. Fl. small, whitish.—Marshes and ditches, especially near the sea. P. VI.—VIII. *Celery.* E. S. I.

** *Pet. with oblique acute point and wedgeshaped base. Partial inv. present.*—HELOSCIADIUM Koch.

2. *A. nodiflórum* (R. fil.); st. procumbent at the base and rooting, l. pinnnate, *leaflets* ovate or ovate-lanceolate unequally *bluntly serrate*, umbels opposite to the l. longer than their peduncles or nearly sessile.—*E. B.* 639.—St. 1—2 feet long.—*α*; general inv. soon falling of 1–3 lts., anth. purple.—*β*. *A. repens* (R. fil.); creeping, lts. unequally and actually incise-serrate, umbels shorter than the peduncles, anthers yellow.—*γ*. *ocreatum*; creeping, l. roundish ovate small bluntly toothed, umbels very shortly stalked.—*E. B.* 1431.—Banks of ditches and brooks. *γ*. Barnes, Surrey. P. VII. VIII. E. S. I.

3. *A. inundátum* (R. fil.); st. creeping or floating, l. pinnate, *lts. of lower l. in capillary segments*, of upper l. wedgeshaped and trifid, umbels generally with 2 rays.—*E. B.* 227.—Usually submersed, a few of the upper l. and the fl. rising above the water. Partial umbels very small.—[*β*. *Moorei* (Sy.); lts. of lower l. linear or strap-shaped, of upper broader.]—Ponds. P. VI. VII. E. S. I.

7. PETROSELI NUM *Hill.*

*1. *P. sativum* (Hoffm.); *l. tripinnate* shining, leaflets of the lower l. ovate-cuneate trifid and toothed of the upper l. ternate lanceolate entire.—*E. B. S.* 2793. *Carum Petroselinum* Benth.—Partial inv. filiform. Fl. yellow. L. greenish.—Rocks and old walls. B. VI.—VIII. *Parsley.* E.

2. *P. seg'etum* (Koch); lower *l. pinnate*, leaflets nearly sessile ovate lobed and serrate, upper l. entire or trifid.—*Sison* Sm. *E. B.* 228. *Carum* Benth.—Umbels very irregular. General inv. of 1—2 leaves. *Fl. whitish.* St. erect, terete, nearly leafless above, 1—1½ foot high, wiry.—Damp calcareous fields and near the sea. B. VIII. IX. E.

8. SI'SON *Linn.* Stonewort.

1. *S. Amómum* (L.).—*E. B.* 954.—St. erect, panicled, 2—3 feet high. Lower l. pinnate; lts. oblong lobed cut and serrate. Upper l. divided into narrow segments. Partial umbels and fl.

small. Much like *Petr. segetum*.—Dampish places on a calcareous soil. B. VIII. E.

[*Falcaria vulgaris* Bernh., *R.* 1862, having digitate l. with linear-lanceolate sharply serrate lts., and *Ammi majus* L., *R.* 1864, having a conspicuous general invol. of long slender trifid bracts, occur as cornfield aliens.]

9. APINEL'LA *Necker*. (*Trinia* Hoffm. ed. viii.) Honewort.

1. *A. glau'ca* (O. Kuntze); glabrous, inv. 0 or of 1 leaf, ridges of the fr. blunt.—*T. glaberrima* Hoffm. *Pimpinella dioica* Sm. *E. B.* 1209, *Sy. E. B.* 579.—L. tripinnate, glaucous green; lts. linear or filiform. Root crowned with the remnants of former leaves. St. branched, erect, 6—8 in. high. Diœcious.—Dry limestone hills, rare. P. V. VI. E.

10. ÆGOPO'DIUM *Linn.* Gout-weed.

1. *Æ. Podagraria* (L.).—*E. B.* 940.—St. 1—2 feet high, erect, furrowed. L. 2 or 3 times ternate; leaflets ovate-acuminate, unequal at the base, acutely serrate. Creeping.—Damp places. P. VI. VII. E. S. I.

11. CA'RUM *Linn.*

* *Root fusiform or fascicled. Stripes* 1—2 *in each interstice.*

[*C. Car'vi* (L.); *partial involucre* 0, general 0 or of 1 leaf, l. bipinnate, leaflets cut into linear segments.—*E. B.* 1503.—St. 1—2 feet high, branched. Root fusiform. Carp. aromatic.—Meadows and pastures. B. VI. *Caraway.*] E. S. I.

1. *C. verticillátum* (Koch); *general and partial involucres of many leaves* small, l. pinnate, leaflets divided to the base into capillary spreading segments.—*E. B.* 395.—St. 1—1½ foot high. Root fasciculate. *Segments of the leaflets* spreading so as to appear *whorled* and quite surrounding the petiole. L. mostly long-stalked, radical.—Marshy places in hilly districts, rare. P. VIII. E. S. I.

** *Root a tuber.*—BUNIUM *Linn.*

† Stripes 3 in each interstice.

2. *C. flexuósum* (Fr.); general involucre of 1—3 leaves, partial more numerous, *fr. oval* narrowing upwards crowned with the *long stylopode and erect styles.*—*E. B.* 988. *B. denudatum* (DC.). *Conopodium denudatum* (Koch). — Involucres sometimes altogether wanting. Root a solitary tuber. St. a foot or more

high, very slender below, bearing a few l. with linear segments. Radical l. triternate with long footstalks tapering downwards.—Sandy and gravelly pastures. P. V. VI. *Pig-nut.* E. S. I.

†† Stripes 1—2 in each interstice.

3. *C. Bulbocas'tanum* (Koch); *general and partial involucres of many leaves, fr. oblong* crowned with the *short stylopode and reflexed styles,* interstices with single stripes.—*E. B. S.* 2862.—Involucres always present. Root a solitary tuber. St. about 2 ft. high. Lower l. bipinnate, with a triangular outline, rather many near the base of the stem, rarely 1 or 2 radical upon long footstalks tapering downwards.—Chalky fields in Camb., Bucks., Beds., and Herts. P. VI. VII. E.

12. PIMPINEL'LA *Linn.* Burnet-Saxifrage.

1. *P. májor* (Huds.); l. pinnate, *leaflets all ovate serrate* somewhat cut the terminal one 3-lobed, *st. angularly striate.—E. B.* 408. *P. magna* (L.) ed. viii.—St. 1—2 feet high, leafy. Lateral leaflets sometimes 3-lobed. Styles longer than the ovary, as long as or longer than the oval fruit.—*β. P. dissecta* (Retz); l. all divided into long linear cut segments.—Shady hills. P. VII. VIII. E. S. I.

2. *P. Saxífraga* (L.); l. pinnate, leaflets of the lower l. roundish-ovate serrate somewhat cut, *those of the stem-l. bipinnatifid with linear segments, st. terete,* ped. glabrous.—*E. B.* 407.—St. 1—2 feet high, naked above. Styles shorter than the ovary. L. sometimes all pinnatifid. *β. dissecta* (With.); l. finely cut.—[A large hairy form with lts. of lower l. cordate is var. *nigra*, Mill.] Dry pastures. P. VII.—IX. E. S. I.

13. SI'UM *Linn.*

* *Stripes superficial. Lateral ridges marginal. Stylopode depressed.* SIUM Koch.

1. *S. latifólium* (L.); l. pinnate, *lts.* oblong-lanceolate *evenly serrate* pointed, umbels terminal, inv.-l. many lanceolate.—*E. B.* 204.—St. 3—5 feet high, angular, furrowed, erect. L. of 9—13 very large distant lts.—Ditches and rivers, rare. P. VII. VIII. E. S. I.

** *Stripes deeply seated. Lateral ridges not marginal. Stylopode shortly conical.* BERULA Koch.

2. *S. erectúm* (Huds.); l. pinnate, *leaflets unequally lobed and cut* ovate, of the stem-l. lanceolate, umbels lateral, inv. of many

lanceolate entire or cut leaves.—*E. B.* 139. *S. angustifolium* (L.) ed. viii.—St. 1—3 feet high, round, striate, erect. Lts. often very deeply cut and lobed.—Ditches. P. VIII. E. S. I.

14. Bupleu'rum *Linn.* Hare's-ear.

* *Fruit granulate.*

1. *B. tenuis'simum* (L.); st. branched, l. linear acute, umbels lateral and terminal minute, partial umbels of 3—5 flowers usually overtopped by their involucres, carp. granular between the 5 ridges.—*E. B.* 478.—St. very slender, wiry, 6—12 in. long.—Chiefly in pastures near the sea. A. VIII. IX. E.

** *Fruit not granulate.*

2. *B. aristátum* (Bartl.)[1]; st. branched, l. linear-lanceolate acuminate 3-veined, l. of the partial involucres elliptic-lanceolate cuspidate somewhat awned with branching veins, pedicels short equal.—*E. B.* 2468.—St. 1—6 in. high.—Park Hill, Torquay, and Berry Head, Devon. Cow Gap near Eastbourne, Suss. Channel Islands. A. VI. E.

3. *B. falcátum* (L.); st. branched, l. 5—7-veined, *lower l. elliptic-oblong* on long stalks, *upper l. linear-lanceolate* acute sessile 5—7-veined, partial involucre of 5 lanceolate pointed l. as long as the flowers.—*E. B. S.* 2763.—Pedicels as long as the fruit. St. 1—4 feet high.—Near Ongar, Essex. P. VIII. E.

4. *B. rotundifólium* (L.); st. branched above, *l. oval perfoliate,* fr. with striate interstices.—*E. B.* 99.—St. 12—18 in. high. General inv. 0. Partial inv. connivent.—Corn-fields on a calcareous soil. A. VII. *Thorough-wax.* E.

Tribe V. *Seselineæ.*

15. Œnan'the *Linn.* Water-Dropwort.

* *Root fascicled, fibres more or less thickened or tuberous.*

1. *Œ. fistulósa* (L.); stoloniferous, *st. and petioles hollow,* root-l. 2—3-pinnate with 3-fid leaflets, *stem-l.* simply pinnate *shorter than their petioles,* leaflets linear, fr. angular turbinate.—*E. B.* 363.—St. 1—3 feet high, remarkably hollow. Stem-l.

---

[1] 2=*B. opacum,* Lange, Prod. Fl. Hisp. iii. 71, Bull. Soc. Bot. Fr. xxxvii. xv. Differs from *B. aristatum,* umbel 2-4 rays (not 4-6), l. of partial invol. imbricate, adpressed much exceeding the fl., strongly net-veined.—Author's MS.

distant, with very long stalks. Stoles with simply pinnate leaves. Umbels small, globose in fruit; general involucre 0. Fruit tipped with the long slightly diverging rigid styles.—By ponds and ditches. P. VII.—IX. E. S. I.

2. *Œ. pimpinelloïdes* (L.); root o long fibres bearing round or ovoid knobs beyond their middle, root-l. bipinnate with obovate-wedgeshaped 3-lobed lts., st.-l. pinnate with linear acute lts., uppermost l. simple, radiant pet. obcordate divided to the middle, *fr.* subcylindrical *with an enlarged corky base* but not narrowed at the top.—*E. B. S.* 2991. *Jacq. Aust.* t. 394.—St. ½—3 ft. high, alternately branched. General inv. 0—6-leaved; partial of many l., about as long as the barren fl. *Partial umbels close together forming one compact flat-topped compound umbel.* Fr. nearly cylindrical; cal. erect-patent.—Southern counties. P. VI. VII. E. I.

3. *Œ. Lachenalii* (Gmel.); root of long subclavate fleshy fibres tapering at both ends, *root-l.* bipinnate *with oblong entire or wedgeshaped and* bluntly 2—3-lobed *lts.*, lower st.-l. 2—3-pinnate with *linear* acute lts., upper l. simply pinnate, radiant pet. divided to the middle, *fr. oblong not corky below rounded and contracted at the top.*—*Œ. pimpinelloïdes* Sm. *E. B.* 347.—St. 1—3 feet high, slightly branched. General inv. of many leaves, sometimes wanting; partial of many leaves, shorter than the barren flowers. Outer fl. on long stalks, mostly barren; inner fl. fertile, nearly sessile. Radiant pet. roundish obcordate with a short narrow claw. *Partial umbels distinct, spherical.* Fr. crowned with the inflexed calyx. Root-leaves soon vanishing.—Marshes. P. VII.—IX. E. S. I.

4. *Œ. peucedanifólia* (Poll.); root of elliptic-oblong knobs, radical l. bipinnate, st.-l. pinnate, *lts. all linear acute,* external fr. nearly cylindrical with a corky base, but not narrowed at the top.—*E. B.* 348. *Œ. silaïfolia* (Bieb.?) ed. viii.—St. 2—3 feet high, branched. Outer fl. stalked, mostly barren; cal. very unequal; radiant pet. small, obcordate with an attenuate base, notch ⅓ their length. Fr. usually slightly narrower downwards, in the middle of the umbel much narrower and appearing to want the corky base; cal. erect or inflexed. General involucre 0; partial of many leaves shorter than the flowers. Distinguished from the preceding by its pet., uniform leaflets, want of a general involucre, and fruit.—The true name of this plant is very doubtful.—In freshwater marshes. P. VI. E.

5. *Œ. crocáta* (L.); root of large fusiform tubers, radical l. 2—3-pinnate, stem-l. pinnatifid, *leaflets stalked roundish or*

oblong-wedgeshaped variously cut those of the upper l. narrower, *fr. cylindrical* oblong striate longer than its pedicel.—*E. B.* 2313. —Poisonous. St. 3—5 feet high, much branched. L. large, lts. broad. Inv.-l. various in number and shape.—Wet places. P. VII. E. S. I.

** *Root of whorls of slender fibres.* PHELLANDRIUM L.

6. *Œ. Phellan'drium* (Lam.); st. erect fusiform below, l. tripinnate, *lts.* ovate pinnatifid cut spreading, *of the submersed l. multifid with capillary diverging segments,* umbels lateral opposite to the leaves, fr. *ovate.*—*E. B.* 684.—St. 2—3 feet high, very thick below, stoloniferous. Segments of the l. many, fine, acute, pale green; submersed l. dark green. The flowering root dies each year; but the plant is continued by the offsets.—In wet ditches and ponds. B.? VII.—IX. *Horsebane.* E. S. I.

7. *Œ. fluviat'ilis* (Colem.); st. floating, l. bipinnate, *lts.* simple or pinnatifid, *of the submersed l. cuneate cut pellucid with many parallel veins,* umbels lateral opposite to the l., fr. broadly oblong.—*A. N. H.* xi. 188. *E. B. S.* 2944.—A decumbent floating plant well marked by the submersed lts. being divided into finger-like acute broadly linear parallel *segments deeply cut at the end.* Lts. of the upper l. broader than those of *Œ. Phellandrium.*—Streams. B. or P. VII.—IX. E. I.

16. ÆTHU'SA *Linn.* Fool's Parsley.

1. *Æ. Cynâpium* (L); partial involucre of 3 leaves longer than their umbel, l. all doubly pinnate, leaflets lanceolate decurrent pinnatifid.—*E. B.* 1192.—St. 4—18 in. high. L. deltoid, dark green, lurid, stinking. General inv. 0; partial long, narrow, pendulous, all on one side. Herb poisonous.—Cultivated land. A. VII. VIII. E. S. I.

17. FŒNIC'ULUM *Mill.* Fennel.

1. *F. vulgâre* (Mill.); st. terete below, l. 3—4 times pinnate, *segments all capillary long in the upper l. flaccid,* umbels of many rays concave.—*E. B.* 1208. *F. officinale* (All.) ed. viii.—Involucre 0. St. 3—4 feet high, usually filled with pith, branching. Umbels large. Fl. yellow. Whole herb aromatic. Segments of l. channelled, usually capillary in the wild plant; but the cultivated plant with awlshaped segments is probably the same species.—Rocks and walls, particularly near the sea. P. VII. VIII. E. I.

18. Ses'eli *Linn.*

1. *S. Libanótis* (Koch); l. doubly pinnate cut, segments lanceolate mucronate, the lowermost leaflets crossing, general involucre of many leaves, fr. hairy.—*Athamanta* L. *E. B.* 138. —St. 1—3 feet high, covered at the base with the fibrous remains of decayed petioles. Umbels terminal, convex, with many downy rays.—Chalk hills of Cambr., Herts, and Sussex. P. VII. VIII. E.

19. Halos'cias *Fries.* Scottish Lovage.

1. *H. scot'icum* (Fr.); l. twice ternate, leaflets ovate somewhat rhomboidal dentate-serrate opaque, involucre of 5—7 linear-lanceolate leaves, cal. 5-toothed.—*Ligusticum* L. *E. B.* 1207.—St. herbaceous, nearly simple, striate, tinged with red, 1—1½ foot high. Lts. large, lobed and cut. Interstices with 3, commissure with 6 stripes. *Seed quite free in the carpel.*—Rocks on the Northern sea-coast. P. VII. E. S. I.

20. Sila'us *Bernh.* Sulphur-wort.

1. *S. flaves'cens* (Bernh.); st. angular, radical l. 3—4 times pinnate, leaflets lanceolate entire or bifid, terminal lt. tripartite, general involucre of 1—2 leaves, partial of many leaves.—*E. B.* 2142. *S. pratensis* (Bess.) ed. viii.—St. 1—2 feet high. L. mostly radical, stem-l. decreasing upwards. Fl. pale yellow.—Damp meadows and pastures. P. VI.—IX. E. S. I.

21. Me'um *Mill.* Bald-money.

1. *M. athaman'ticum* (Jacq.); l. bipinnate, leaflets in many threadshaped acute segments.—*E. B.* 2249.—St. 1—2 feet high, round, clothed at the base with the fibrous remains of the decayed petioles. Fl. many, whitish yellow. General involucre of 2 or 3 leaves, partial more numerous. Highly aromatic.—Dry mountainous pastures. P. VI. VII. E. S.

22. Crith'mum *Linn.* Samphire.

1. *C. marit'imum* (L.).—*E. B.* 819.—St. 6—12 in. long. L. fleshy, 2—3-pinnate; leaflets lanceolate, narrowed at both ends, few. Involucre of many lanceolate acute leaves. Fl. whitish. —On rocky sea-coasts. P. VIII. E. S. I.

Tribe VI. *Angeliceæ.*

23. Angel'ica *Linn.*

1. *A. sylves'tris* (L.); leaflets equal ovate-lanceolate or ovate incise-serrate not decurrent, lateral lts. rather unequal at the base.—*E. B.* 1128.—St. 2—3 feet high, slightly downy above, purplish. Fl. pinkish white. Inv. deciduous. Lts. often subcordate at the base.—Wet places. P. VII. VIII. E. S. I.

24. Archangel'ica *Hoffm.*

[*A. officinális* (Hoffm.); leaflets all sessile partly decurrent, terminal lt. trifid, foliage stalks and even fl. bright green, l. 2—3 feet wide (*E. B.* 2516), is not a native. Watery places. P. VII.—IX.]

25. Selinum *Linn.*

1. *S. Carvifólia* (L.); *st. angular furrowed*, rays of umbel scabrous.—*J. of B.* xx. 129. t. 229. *R.* xxi. 101.—L. 2—3-pinnate; lts. with linear-lanceolate mucronate serrulate lobes. No general invol., partial of many linear acuminate scales.—Broughton wood, Lincolnshire. Between Fordham and Chippenham, Cambridgeshire. P. VII. VIII. E.

Tribe VII. *Peucedaneæ.*

26. Peuced'anum *Linn.*

1. *P. officinále* (L.); *l. 5 times ternate, leaflets linear* very long acute *flaccid*, general involucre 3-leaved deciduous, pedicels much longer than the fruit.—*E. B.* 1767.—Fl. yellow. St. terete, striate, 2—3 feet high. Stripes of the commissure superficial.—Salt marshes. Kent. Essex. P. VII.—IX. E.

2. *P. palus'tre* (Moench); *l. 3-pinnate, leaflets pinnatifid* with linear-lanceolate acuminate segments, *general involucre of many persistent* lanceolate deflexed *leaves*, st. furrowed.—*Selinum* Sm. *E. B.* 229.—St. erect, 3—5 feet high. Fl. white. Stripes of the commissure deeply seated.—Marshy and fenny places, rare. P. VII. VIII. E. S.?

†3. *P. Ostrúthium* (Koch); *l. biternate, leaflets broadly ovate lobed cut and serrate*, sheath very large, *general involucre* 0, cal.-segments inconspicuous. Fr. broadly winged.—*E. B.* 1380.—St. 1—2 feet high. Fl. white.—Moist meadows, rare. P. VI. *Masterwort.* S.

27. PASTINA'CA *Linn.* Parsnep.

1. *P. sativa* (L.); st. angular furrowed, l. pinnate downy beneath, leaflets ovate-oblong crenate-serrate often with a lateral lobe at the base, inv. 0, fr. oval.—*E. B.* 556. *Peucedanum* (Benth.)—St. 2—3 feet high. Fl. yellow. L. generally shining above, downy beneath.—Hedgebanks on a calcareous soil. B. VII. E. I.

28. HERAC'LEUM *Linn.* Cow-parsnep.

1. *H. Sphondyl'ium* (L.); l. pinnate, leaflets lobed or pinnatifid cut and serrate, fr. at length glabrous.—*E. B.* 939.—St. 4 feet high. Lower l. very large. L. sometimes narrow (var. *angustifolium* Huds.). Umbels large flattish. Fl. white or reddish, outer fl. radiant.—Hedge-banks. P. VII. *Hog-weed.* E. S. I.

29. TORDYL'IUM *Linn.* Hartwort.

†1. *T. max'imum* (L.); outermost pet. radiant with 2 *equal lobes, partial involucres linear* shorter than the umbel, fr. hispid the thickened margin slightly crenate.—*E. B.* 1173.—St. 2—4 feet high. Fl. reddish.—Waste ground, very rare A. VII. E.

[*T. officinále* (L.), *E. B.* 2440, was a mistake.]

Tribe VIII. *Silerineæ.*

30. SI'LER *Crantz.*

[*S. trílobum* (Crantz).—*R.* xxi. 153. *J. of B.* ix. t. 118.—St. solid, glaucous when young. Radical l. triternate, irregularly and coarsely serrate, long-stalked, dark green, glabrous. St.-l. ternate. Umbels terminal, large. Inv. of 1—3, partial of 5—8 bracts. Fr. large, crowned with persistent reflexed styles.—Naturalized at Cherry Hinton, Camb. P. V. VI.] E.

Tribe IX. *Daucineæ.*

31. DAU'CUS *Linn.* Carrot.

1. *D. Caróta* (L.); *radical l. with an oblong narrow outline* bipinnate with incise-dentate lts. and *acute segments, upper l. broader* below with lanceolate segments.—*E. B.* 1174.—St. 2—3 ft. high, hairy; branches ascending. Umbel of fr. usually concave [sometimes convex with mature fr., f. *convexa*, Linton]

Prickles of fr. slender, mostly distinct, about equalling its breadth, spreading, tipped with 1—3 recurved minute bristles. —Pastures. B. VI.—VIII. E. S. I.

2. *D. gum'mifer* (Lam.); *radical l. triangular* broad 2—3-pinnate with ovate cut or pinnatifid lts. and *blunt mucronate segments,* upper l. narrower below.—*D. maritimus* With. (not Lam.).—*E. B.* 2560.—St. short, very hispid below; branches divaricate from the base. Prickles of fr. usually flattened and often united below and shorter than its breadth, incurved, tipped with one bristle. L. shining above, rather fleshy. Umbel of fr. usually convex.—The forms of *Daucus* on the South coast deserve further study.—Sea-coasts in the South-west, rare. B. VII. VIII. E. I.

Suborder II. *Campylospermæ.* Tribe X. *Caucalineæ.*

32. Cau'calis *Linn.* Hen's-foot.

1. *C. daucoïdes* (L.); l. bipinnate, leaflets pinnatifid with linear-acute segments, general involucre 0, partial umbels of few fl. with involucres of 3—5 leaves, secondary ridges of the fr. each with *one row of glabrous hooked prickles.—E. B.* 197.—St. 6—12 in. high, furrowed, hairy at the joints. General umbels 3-cleft; partial bearing about 3 large oblong very prickly fruits. Fl. small, reddish.—Cornfields on a chalky soil. A. VI. E.

†2. *C. latifólia* (L.); *l. pinnate, leaflets lanceolate* decurrent *coarsely serrate,* inv.-l. oblong membranous, secondary ridges of the fr. with 2 *or* 3 *rows of retrorsely scabrous* prickles.—*E. B.* 198. *Turgenia* Hoffm.—St. 1—2 feet high, rough. General umbels about 3-cleft; partial bearing about 5 large oblong very prickly fruits. Fl. large, pink.—Cornfields, mostly on a chalky soil, very rare. Formerly abundant in Cambridgeshire. A. VII. E.

33. Tori'lis *Adans.* Hedge-Parsley.

1. *T. Anthris'cus* (Gaert.); l. bipinnate, leaflets ovate-oblong incise-serrate, umbels long-stalked terminal, *general involucre of many leaves,* fr. with subulate incurved *prickles not hooked at the tip.—E. B.* 987.—St. erect, 1—3 feet high. Umbels on long stalks. Fr. densely prickly. Fl. small, white or reddish.—Hedges and banks. A. VII. VIII. E. S. I.

2. *T. infes'ta* (Spr.); l. bipinnate, lts. ovate-lanceolate incise-serrate, *umbels long-stalked terminal, general inv. of one leaf or* 0, *fr. with* spreading asperous *prickles hooked at the tip.* - *E. B.* 1314. *T. helvetica* Gm. *Caucalis arvensis* Huds.—St. erect, usually much and densely branched, 6—18 in. high. Umbels on long stalks. Fr. densely prickly, primary ridges with adpressed prickles. Fl. small, white or reddish. Styles scarcely twice as long as the stylopode.—Fields. A. VII. VIII. E.

3. *T. nodósa* (Gaert.); lower l. bipinnate, upper l pinnate, lts. deeply narrowly and uniformly pinnate, *umbels nearly sessile* dense *lateral, no involucres, outer carpels with bristles* hooked at the tip, *inner often warted.*—*E. B.* 199.—St. diffuse, often prostrate. Umbels very small, nearly globular.—Banks and dry places.—A. V.—VII. E. S. I.

Tribe XI. *Scandicineæ.*

34. SCAN'DIX *Linn.* Shepherd's Needle.

1. *S. Pec'ten* (L.); fr. rough dorsally compressed glabrous with bristly edges, beak three times as long as fruit, lts. of partial inv. entire or bifid longer than the pedicels.—*E. B.* 1397.—St. often a foot high. L. light green, triply pinnate; segments short linear. Umbels 1—2 together, small. Fl. often slightly radiant. Styles always straight. Stylopode purple. Fr. and beak nearly 2 in. long. Partial involucres sometimes much divided.—Fields. A. VI.—IX. E. S. I.

35. CHÆROPHYL'LUM *Linn.* Chervil.

* *Ridges only apparent on the beak of the fr.; no stripes.*
ANTHRISCUS Bernh.

1. *C. sylves'tre* (L.); st. hairy below glabrous upwards swollen below the joinings, *umbels terminal stalked,* l. bipinnate, leaflets pinnatifid, *fr. smooth and shining narrow to its tip.*—*Anthriscus* Hoffm. *E. B.* 752—.*St.* 3 feet high, erect, leafy, *furrowed,* hollow, branched. Partial involucre of several ovate-lanceolate cilate leaflets. Umbels at first drooping. Pet. oblong-obovate, scarcely emarginate, with a short inflexed point.—Hedges and banks. P. IV.—VI. *Wild Chervil.* E. S. I.

[*C. satívum* (Lam.); st. hairy above the joinings only, *umbels lateral sessile,* l. tripinnate, leaflets ovate pinnatifid, *fr. linear smooth about twice as long as its beak.*—*Anth. Cerefolium* Hoffm.

E. B. 1268.—St. 1—3 feet high, slender, striate, much branched. Partial involucre of 3 unilateral linear-lanceolate leaflets. Peduncles downy.—Waste ground. Probably an escape from cultivation. A. V. VI. *Garden-Chervil.*] E. S. I.

2. *C. Anthris'cus* (Lam.); st. glabrous, *umbels lateral stalked*, l. tripinnate, leaflets pinnatifid, *fr. ovate with hooked bristles* about twice as long as its glabrous beak.—*E. B.* 818. *A. vulgaris* Bernh.—St. erect, 2 feet high, branched. L. slightly hairy. Umbels on rather short stalks. Partial involucre of few ciliate leaflets.—Waste places. A. V. VI. E. S. I.

** *Ridges 5, equal, blunt; stripes solitary.*

3. *C. tem'ulum* (L.); st. thickened beneath the joinings rough, l. bipinnate, *leaflets* ovate-oblong pinnatifid with rather acute mucronate segments, pet. glabrous, *styles equalling the stylopode.*—*E. B.* 1521.—*St.* 3—4 *feet high, round, solid, spotted,* rough below, hairy near the top. Umbels at first nodding. Pet. deeply obcordate.—Hedge-banks. P. VI. VII. *Rough Chervil.* E. S. I.

36. Myr'rhis *Scop.* Sweet Cicely.

1. *M. Odoráta* (Scop.); l. downy beneath, leaflets of the partial involucres lanceolate-acuminate.—*E. B.* 697.—St. 2—3 feet high, round, leafy, hollow. L. very large, tripinnate. Leaflets ovate-lanceolate pinnatifid. Umbels terminal. Fl. many, white. Fr. large, nearly an inch long, dark brown. Whole plant highly aromatic. Pastures in hilly districts. P. V. VI. E. S. I.

Tribe XII. *Smyrnieæ.*

37. Echinoph'ora *Linn.*

[1. *E. spinósa* (L.); l. pinnate, leaflets pinnatifid with spinous awlshaped entire segments.—*E. B.* 2413.—Formerly found on the Lancashire and Kentish shores. P. VII.] E.

38. Coni'um *Linn.* Hemlock.

1. *C. maculátum* (L.); leaflets of the partial involucres unilateral ovate-lanceolate with an attenuate point shorter than the umbels.—*E. B.* 1191.—*St.* 3—5 feet high, erect, round, hollow, glaucous, *spotted with purple*, branched. L. tripinnate; leaflets lanceolate, pinnatifid with acute cut segments. Readily

distinguished by its fœtid smell, spotted stem, unilateral partial involucres, and wavy crenate ridges of the fruit. Highly poisonous.—Hedge-banks and waste places. B. VI. VII. E. S. I.

39. PHYSOSPER'MUM *Cusson.*

1. *P. commutátum* (Spreng.); radical l. triternate, leaflets wedge-shaped cut or deeply 3-lobed with acute segments, stem-l. ternate lanceolate entire.—*E. B.* 683. *P. cornubien'se* (DC.) ed. viii.—St. 1—3 feet high, erect, round, striate, bearing a few small ternate leaves with linear-lanceolate segments, the uppermost represented by a barren lanceolate acute sheath. Umbels terminal and axillary, long-stalked. Carp longer than broad; the coat loose. Seed free.—Devon and Cornwall, rare. P. VII. VIII. E.

40. SMYR'NIUM *Linn.* Alexanders.

1. *S. Olusátrum* (L.); st. terete, stem-l. ternate stalked serrate.—*E. B.* 230.—St. 3—4 feet high, stout, branched, leafy, furrowed. Radical l. very large, 3—4 times ternate; all with large membranous sheaths and large ovate shining cut and serrate leaflets. Fl. greenish yellow in rounded often dense umbels. Fr. nearly black, aromatic.—Waste ground and near ruins. B. V. VI. E. S. I.

Suborder III. *Cœlospermæ.* Tribe XIII. *Coriandreæ.*

41. CORIAN'DRUM *Linn.* Coriander.

[1. *C. satívum* (L.).—*E. B.* 67.—St. 12—18 in. high, leafy, round, striate. L. bipinnate, cut, with broad roundish or wedge-shaped segments; upper l. more divided with linear segments. Fl. white.—Waste places, scarcely naturalized. A. VI.] E. S.

Order XXXVIII. HEDERACEÆ.

Cal. 4—5-toothed, adnate to the ovary. Pet. 5—16, rarely wanting; valvate in the bud. Stam. as many as the pet. and alternate with them or twice as many, inserted below the margin of an epigynous disk. Ovary with 2 or more cells. Styles as many as the cells. Fr. succulent or dry, of several cells each with 1 pendulous seed. Perisperm fleshy. Embryo minute (not so in our *Hedera*).—L. alternate without stipules.

1. Hedera. Cal. superior, limb of 5 teeth. Pet. 5—10, not cohering at the apex. Stam. 5—10. Styles 5—10, connivent or combined into one. Berry 5-celled and 5-seeded, crowned with the calyx.

1. Hed'era *Linn.* Ivy.

1. *H. Hélix* (L.); l. coriaceous ovate or cordate and 5-lobed, lobes angular, umbels simple downy erect.—*E. B.* 1267.—Climbing by means of rootlike fibres. L. of the flowering branches ovate-oblong, acute, entire. Berries black. Embryo like that of *Cornaceæ*.—Rocks, old walls, hedges. Sh. X XI. E. S. I.

Order XXXIX. CORNACEÆ.

Cal. 4-lobed, adnate to the ovary. Pet. 4, oblong, broad at the base, inserted at the top of the calyx-tube; valvate in the bud. Stam. 4. Ovary 2-celled. Style filiform. Fruit a drupe, crowned with the remains of the calyx. Seed pendulous, solitary. Embryo in the axis of fleshy perisperm and as long as it. —Leaves opposite, exstipulate.

1. Cornus. Calyx-limb superior, of 4 teeth. Pet. 4. Stam. 4. Style 1. Drupe with a 2-celled and 2-seeded nut.

1. Cor'nus *Linn.*

1. *C. sanguin'ea* (L.); *arborescent*, branches straight, l. ovate cuspidate green on both sides, cymes flat without an involucre. —*E. B.* 249. *St.* 52. 3.—Shrub 5—6 feet high. Old bark reddish. Fl. many, white, in terminal cymes. Fr. dark purple. L. mostly opposite, strongly veined, acutely cuspidate, rounded below.—Hedges and thickets. Sh. VI. *Dog-wood.* E. I.

2. *C. suecica* (L.); *herbaceous*, l. all opposite sessile ovate, fl. umbellate shorter than the *4-leaved petal-like involucre.*—*E. B.* 310. *St.* 52. 1.—Flowering shoots about 6 in. high, annual, springing from the procumbent or subterranean creeping woody leafless stems. Fl. dark purple with yellow stamens, in a small solitary terminal umbel with an inv. of 4 ovate white l. tipped with purple. Fr. red.—Moist alpine moors. P. VII. E. S.

Division III. COROLLIFLORÆ.

(Orders XL.—LXIII.[1])

Pet. united. Stamens epipetalous; except in *Ericaceæ* and part of *Plantaginaceæ*, which have hypogynous, and *Campanulaceæ*, which has epigynous stamens.

Order XL. LORANTHACEÆ.

Cal. adnate to the ovary, with two bracts at its base; limb entire or lobed. Cor. of 4—8 more or less united petals. Stam. as many as and opposite to the petals, with which the filaments more or less combine; anth. sometimes adnate to the petals. Ovary 1, 1-celled with 1 pendulous ovule. Style filiform or 0. Stigma capitate. Fr. succulent. Perisperm fleshy.—Parasitical plants with entire mostly opposite leaves and no stipules. Connects this Division with Div. Calycifloræ and Monochlamydeæ.

I. Viscum. Diœcious. Male: Cal. 0. Pet. 4, ovate, fleshy, united at the base. Anth. adnate to the petals, many-celled. Fem.: Cal. an obscure entire superior margin. Pet. 4, erect, somewhat triangular, minute. Stigma sessile, blunt. Berry 1-seeded, crowned with the calyx.

1. Vis'cum *Linn.* Mistletoe.

1. *V. al'bum* (L.); st. repeatedly forked, branches terete, l. ovate-lanceolate blunt, fl. in the forks of the stem sessile clustered.—*E. B.* 1470.—Evergreen, yellow, succulent. Spreads by runners under living bark of tree. Male fl. about 3 together, female about 5, yellowish. Berries white, pellucid, globular, viscid. L. of male plant broader.—Parasitical on various trees. P. III. IV. E

Order XLI. CAPRIFOLIACEÆ.

Cal. adnate to the stamen ovary, usually with bracts at the base; limb 4—5-lobed. Cor. regular or irregular, 4—5-cleft. Stam. 4 or 5, free, on the corolla, and alternate with the lobes. Ovary

---

[1] Orders XL.—XLVI. and Tribe 3 of XLVII. are often regarded as Calycifloral. Their ovary is inferior. Order XL. is sometimes placed in Monochlamydeæ.

3—5-celled. Stigmas 1—3. Fruit not bursting, 1- or many-celled, usually fleshy. Perisperm fleshy.—L. opposite. Stip. rare.

1. Adoxa. Cal. ½-inferior, 2—3-cleft. *Cor. rotate*, 4—5-lobed. *Stam.* 8—10, *in pairs alternate with the lobes* of cor.; *anth.* 1-*celled.* Styles 5—10. Fr. 4—5-celled; cells 1-seeded.—L. ternate, lobed.

2. Sambucus. Cal.-limb 5-cleft. *Cor. rotate*, 5-lobed. Stam. 5. Stigmas 3, sessile. Fr. 3—4-seeded.—L. pinnate.

3. Viburnum. Cal.-limb 5-cleft. *Cor. bell- or funnel-shaped*, 5-lobed. Stam. 5. Stigmas 3, sessile. Fruit 1-seeded.—L. simple.

4. Lonicera. Cal.-limb small, 5-cleft. *Cor. tubular* or funnelshaped, limb 5-fid or irregular. Stam. 5. *Style filiform. Stigma capitate.* Fruit 1—3-celled, few-seeded.

5. Linnæa. *Cal.-limb* 5-cleft, *with lanceolate subulate equal deciduous segments.* Cor. turbinate, bell-shaped, 5-lobed. *Stam.* 4, rarely 5, 2 longer. Style filiform; stigma capitate. Fr. dry, 3-celled; 2 cells barren, 1 single-seeded.—Two large and 2 minute bracts at the base of the fruit.

1. Adox'a *Linn.* Moschatel.

1. *A. Moschatel'lina* (L.).—*E. B.* 453.—Rhizome white, fleshy, toothed, soboliferous. St. solitary, erect, simple, 3—4 in. high, with 2 opposite leaves, and a head of 4 whorled and 1 terminal flowers. Root-l. ternate, 3-lobed. Stam. often more or less united in pairs, showing their number to be normally 4. Odour musky. Terminal fl. usually divided in fours, the others in fives, but the numbers vary.—Woods and shady hedge-banks. P. IV. V. E. S. I.

2. Sambu'cus *Linn.* Elder.

1. *S. Eb'ulus* (L.); *herbaceous*, st. furrowed, *stip. leaflike* ovate serrate, l. pinnate, leaflets lanceolate serrate, cyme with 3 principal branches.—*E. B.* 475.—St. 2—4 feet high. Cymes terminal, compact. Fl. white, reddish externally; anth. purple. Fr. reddish black.—Hedge-banks. P. VIII. *Dwarf Elder. Danewort.* E. S. I.

2. *S. nigra* (L.); *arborescent, stip. inconspicuous* or wanting, l. pinnate, leaflets ovate cuspidate serrate, cymes with 5 principal branches.—*E. B.* 476.—A small tree. Cymes large, terminal.

Fl. cream-coloured. Fr. black, rarely green or white.—*β. *laciniata* (Lam.); l. 2–3-pinnate, lts. laciniate.—γ. *rotundifolia* (DC.); lts. usually 3 orbicular.—Woods and hedges. γ. Isle of Wight. T. VI. *Elder.* E. S. I.

3. Vibur'num *Linn.* Guelder-rose.

1. *V. Lantána* (L.); *l. oblong* with a cordate base *finely denticulate-serrate* downy beneath, stip. 0, pubescence stellate. —*E. B.* 331.—A shrub with round mealy branches. Young shoots, petioles, and undersides of the l. densely, upperside more sparingly, covered with stellate down. Cymes terminal. Fl. white, not radiant, all perfect; fr. black.—Hedges and thickets on a calcareous soil. T. V. *Mealy Guelder-rose. Wayfaring tree.* E.

2. *V. Op'ulus* (L.); *l.* 3–5-*lobed,* lobes acuminate and dentate, stip. (?) linear, petioles with glands.—*E. B.* 332. *St.* 27. 6.—Branches glabrous, tetragonal when young. L. slightly downy beneath, with linear adnate stipules (?). Cymes large, with linear bracts; fl. white, inner ones fertile, outer barren and radiant. Fr. red.—Hedges and thickets. T. VI. VII. *Common Guelder-rose.* E. S. I.

4. Lonice'ra *Linn.* Honeysuckle.

‡1. *L. Caprifólium* (L.); *fl.* ringent whorled terminal *sessile,* l. deciduous glabrous on both sides blunt, *upper l. connate-perfoliate,* style glabrous.—*E. B.* 799.—St. twining. Fl. white or purplish. Fr. orange. Pericarp and placenta becoming fleshy. Upper pairs of leaves connate, the rest distinct.—Thickets. Sh. V. VI. E.

2. *L. Periclym'enum* (L.); *fl.* ringent in terminal stalked clusters, *l. all distinct* deciduous oval, st. twining.—*E. B.* 800. —Fl. pale yellow, externally red. Fr. red. Pericarp, placenta, bracts, and axis becoming fleshy. L. sometimes downy beneath, often lobed when young.—Woods and hedges. Sh. VI.—IX. *Honeysuckle. Woodbine.* E. S. I.

3. *L. Xylos'teum* (L.); *peduncles 2-flowered* downy as long as the flowers, calyx-limb deciduous, berries slightly connected at the base, l. oval downy, st. erect.—*E. B.* 916.—Shrub. Fl. pale yellow. L., bracts, cal., cor. externally, filaments, and style downy. Fr. scarlet.—Thickets. Native in Sussex (*Borrer*) Sh. V. E.

[*Symphoricarp'os racemosus* (Michx.), a North American shrub, with roundish entire or sinuate-lobed l., fl. in small terminal and axillary racemes, the cor. bellshaped pink bearded within and with large opaque greenish-white 4-celled 2-seeded berries, is often planted in hedges. *Snowberry*.]

5. LINNÆ'A *Linn.*

1. *L. boreālis* (L.).—*E. B.* 433.—St. trailing and creeping. L. opposite, broadly ovate, stalked, dark green above, paler beneath. Peduncles long, erect, 2-flowered, from short lateral branches with 2—4 leaves. Fl. drooping, flesh-coloured, purple within.—Woods, chiefly of fir, in the North. P. VII. E. S.

Order XLII. RUBIACEÆ.

Cal. superior, entire or lobed. Cor. regular, 4—6-lobed. Stam. 4—5, alternate with the lobes of the corolla. Ovary 1, 2-celled, with solitary erect ovules. Style 1, often bifid. Stigmas 2. Fr. a didymous indehiscent pericarp. Embryo straight in horny albumen.—St. herbaceous, square. Many raphides.

1. SHERARDIA. Cor. funnelshaped. Fr. crowned with the deeply 6-toothed calyx, dry.—Fl. umbellate, involucrate.
2. ASPERULA. Cor. funnelshaped. Fr. dry. Limb of the calyx inconspicuous.
3. GALIUM. Cor. rotate, 4-fid. Fr. dry. Limb of cal. inconspicuous.
4. RUBIA. Cor. rotate, 5- (or rarely 4-)fid. *Fr. succulent*, 2-lobed. Limb of cal. inconspicuous.

1. SHERAR'DIA *Linn.* Field-Madder.

1. *S. arven'sis* (L.).—*E. B.* 891.—St. mostly decumbent, branched, leafy. L. 6 in a whorl, obovate-lanceolate, acute. Fl. lilac, in a small sessile terminal umbel with 7—8 inv.-leaves. Cal.-segm. 4, 2 bifid. [*α.* cal.-teeth lanceolate aciculate accrescent; *β. mutica* (Wirtg.) cal.-teeth triangular not aciculate not accrescent.] —Fields. A. V.—VII. E. S. I.

2. ASPER'ULA *Linn.*

1. *A. cynan'chica* (L.); *l.* 4 *in a whorl* linear, upper pair much smaller, fl. corymbose, bracts lanceolate mucronate, cor. rough.—*E. B.* 33.—Root fusiform. Stems many, diffuse or ascending, branched. Lowest l. obovate; interm. obovate-lanceolate; uppermost lanceolate-attenuate. Fl. generally lilac.

Fr. wrinkled and tubercled.—Dry banks in limestone districts. P. VI. VII. *Quinancy-wort.* E. I.

2. *A. odoráta* (L.); *l.* 6—9 *in a whorl* lanceolate, margins rough with forward prickles, fl. in stalked terminal cymose corymbs, *fr. hispid.*—*E. B.* 755.—St. erect, about 6 in. high. Fl. white. L. broad, lower usually in sixes. Whole plant fragrant.—Woods. P. V. VI. *Woodruff.* E. S. I.

[*A. arven'sis* (L.); *l.* 6—10 *in a whorl* linear-lanceolate blunt, fl. clustered terminal surrounded by long ciliate bracts, *fr. glabrous.*—*E. B. S.* 2792.—Like *Sherardia arvensis.* Fl. bright blue.—Introduced. A. VI.] E.

[*A. taurína* (L.); l. 4 in a whorl elliptic acuminate 3-veined, fl. corymbose, cor.-tube very long, fr. rather rough.—*Sy. E. B.* 662.—Cadeby, Leices. Casterton, Westm.] E.

3. GA'LIUM *Linn.* Bedstraw.

A. Leaves 3-veined.

* *Fl. in a terminal panicle, perfect, white; fr.-stalks erect.*

1. *G. boreále* (L.); *l.* 4 *in a whorl* lanceolate, st. erect panicled, fruitstalks patent, *fr. covered with hooked bristles.*—*E. B.* 105.—St. about 18 in. high; branches many, leafy. Fl. in compact panicles.—Moist rocky places. P. VII. VIII. E. S. I.

** *Fl. axillary, yellow; lateral fl. imperfect; fr.-stalks deflexed.*

2. *G. Cruciáta* (Scop.); *l.* 4 *in a whorl* elliptic-oblong hairy, fl. corymbose bracteate, *terminal fl. fertile,* lateral fl. mostly male, fr. smooth.—*E. B.* 143.—St. simple above, 1—2 feet high, hairy. Fl. small, about 8 together in small corymbs, falling short of the leaves.—Hedges and thickets. P. V. VI. *Crosswort.* E. S.

B. Leaves 1-veined. Root annual.

* *Flowers axillary; lateral fl. imperfect.*

[*G. saccharátum* (All.).—*E. B.* 2173.—Accidental. A. VI.—VIII.]

** *Fl. in axillary cymes, all perfect, white or greenish.*

3. *G. tricor'ne* (Stokes); *l.* 6—8 *in a whorl* linear-lanceolate with marginal *backward prickles,* st. rough with deflexed prickles,

cymes 3-flowered, *fr. granular* on reflexed peduncles.—*E. B.* 1641.—St. procumbent, spreading. Fl. small, all 3 appearing perfect, the middle one usually alone fertile, cream-coloured. Fr. large, a double globe, covered with small granulations.—Dry calcareous fields. A. VI.—IX. E.

4. *G. Aparíne* (L.); l. 6—8 in a whorl linear-lanceolate with marginal backward prickles, st. rough with deflexed prickles, cymes few- (about 3-) flowered, fruitstalks divaricate straight, *fruit covered with short hooked bristles.*—*E. B.* 816.—St. straggling amongst bushes, 3—4 feet long. The marginal prickles near the extremity of the l. point forwards, the rest backwards. Fl. small, pale. Ped., or rather flowering branches, with several l. at the primary divisions. Fr. large.—Common. A. VI.—VIII. *Goose-grass. Cleavers.* E. S. I.

5. *G. spúrium* (L.); l. 6—8 in a whorl linear-lanceolate with marginal backward prickles, st. rough with deflexed prickles, cymes with 3—9 *flowers*, fruitstalks divaricate straight.—Closely resembling the preceding; distinguished by its more numerous green flowers, floral leaves solitary ("or in pairs"), fruit of about half the size.—*α*; fr. smooth. *E. B.* 1871.—*β*. *G. Vaillantii* (DC.); fr. hispid. *E. B. S.* 2943.—Fields. *α*. Forfar; *β*. Saffron Walden and Chesterford, Essex. Oxon. Dorset. Warwicksh. A. VII. E. S.

6. *G. ang'licum* (Huds.); *l. about* 6 *in a whorl linear-lanceolate* bristle-pointed with marginal *forward prickles, st.* rough *with deflexed prickles*, cymes small forked with divaricate bifid branches.—*E. B.* 384.—St. 6—8 in. high, spreading, slender, brittle. L. usually 6 in a whorl, the lowermost sometimes in fours. Branches of the small panicles often spreading nearly at right angles with their stalk. Fr. granular, nearly black.—Old walls and dry sandy places. A. VI. VII. E.

C. Leaves 1-veined. Root perennial. Fl. in terminal panicles, *white* (*except in G. verum*).

* *Fruit not granular. No downward prickles on the stem.*

7. *G. erec'tum* (Huds.); *l. about* 8 *in a whorl lanceolate* mucronate the margins rough with forward prickles, midrib slender, *branches of the pyramidal panicle all ascending*, fruitstalks divaricate, fr. oval smooth, pet. taper-pointed.—*E. B.* 2067.—St. glabrous or hairy, erect, not much branched. L. lanceolate, scarcely at all obovate, those of the main st. erect, patent; veins not translucent; margins with 2 rows of prickles pointing

forwards. Fl. white.—α; l. lanceolate.—[β. *G. cinereum* (Sm.): l. 6—8 in a whorl linear. *E. B. S.* 2783. Perhaps a distinct species, *G. diffusum* (Hook.), but a doubtful native.]—*G. aristatum* (Sm.) has l. in sixes, but is probably a state of this species. *E. B. S.* 2784.—Banks and pastures. P. VI. and IX. E. I.

8. *G. Mollúgo* (L.); *l. about* 8 *in a whorl lanceolate-obovate or obovate-oblong* cuspidate the margins rough with forward prickles, branches of the broad pan. spreading *lower ones horizontal* or deflexed, fruitstalks divaricate, fr. glabrous.—*E. B.* 1673.—St. ascending or diffuse, square, thickened at the joinings, glabrous or hairy. L. slightly translucent, veined, hardly separated at the base, those of the main st. horizontal or declining. Pan. large. Fl. small, white. Styles nearly free.—β. *G. insubricum* (Gaud.); l. about 6 in a whorl obovate abruptly cuspidate, pan.-branches few-flowered terminating in trichotomous umbels, floral l. large, bracts large usually solitary.—γ. *Bakeri* (Syme); l. 6—8 in a whorl linear-lanceolate, pan. with few-flowered ascending branches. It may be distinct.—Hedges and thickets. P. VII. VIII. E. S. I.

9. *G. vérum* (L.); *l. about* 8 *in a whorl linear-setaceous with revolute margins* channelled above *downy beneath,* panicles many small densely flowered subterminal, fruitstalks patent, *fruit smooth,* pet. blunt apiculate.—*E. B.* 660.—St. erect, slightly branched, somewhat woody, with many whorls of narrow deflexed leaves. *Fl. golden yellow,* rarely green or straw-coloured, (*ochroleucum*[1] Sy.) usually in many small dense panicles collected into a kind of terminal spike. St. and upper surface of the l. sometimes downy or rough. On loose sands the st. are much more branched and the fl. sometimes solitary [Var. *maritimum* DC.].—Dry and sandy places. P. VII. VIII. E. S. I.

** *Fruit granular, not hairy. St. without downward prickles.*

10. *G. saxat'ile* (L.); *l. about* 6 *in a whorl obovate* pointed flat, midrib slender, panicles corymbose small, fl.- and fr.-stalks erect-patent, pet. acute. *E. B.* 815.—St. many, *procumbent,* much branched. L. suddenly narrowed to a point, smooth with a few marginal forward prickles; lower l. roundly obovate. It turns black in drying.—Heaths. P. VII. VIII. E. S. I.

11. *G. umbellátum* (Lam.); *l.* 6—8 *in a whorl linear or linear-lanceolate* mucronate with revolute edges, *midrib slender*

---

[1] Apparently a hybrid between sp. 8 & 9.—H. & J. G.

prominent, panicles few-flowered, *fl.- and fr.-st. erect-patent,* fr. faintly granular, pet. acute.—*G. pusillum* Sm. *E. B.* 74. *G. sylvestre* (Poll.) ed. viii.—St. many, slender, square, diffuse, *ascending.* L. often nearly glabrous or with *marginal hairs* (*not prickles*) spreading or backward. Lower part of stem and leaves sometimes densely covered with patent hairs. Panicle very variable in size. Fr. very minutely granular.—*a. G. montanum* (Vill.); l. linear-lanceolate, pan. with short ascending branches and few-flowered cymes.—*β. G. nitidulum* (Thuill.); l. linear, pan. with somewhat spreading branches and compact cymes.—Limestone hills. P. VI. VII. E. S. I.

*** *Stem rough with downward prickles.*

12. *G. uliginósum* (L.); *l.* 6—8 *in a whorl linear-lanceolate bristle-pointed the margins rough like the angles of the stem with backward prickles,* panicles small axillary few-flowered trichotomous, the branches patent 3-fid, fruitstalks divaricate straight, fruit granular.—*E. B.* 1972.—Turning blackish when dry. Stems slender, brittle, about a foot high, weak. L. usually 6 in a whorl, discoloured at the tip, shortly acuminate. Fr. dark brown.—Wet places. P. VII. VIII. E. S. I.

13. *G. palus'tre* (L.); *l.* 4 *in a whorl broadly linear* broader upwards *blunt, midrib slender,* panicle diffuse, *fr.-st.* straight, *spreading at right angles,* fr. smooth.—Continuing green. St. 1—2 feet high, slender, usually branched. Fl. small, white. St. and branches nearly smooth. L. narrow, lowest usually in sixes, upper in fours of which 2 are smaller.—On dry ground the l. are broader, those of the barren shoots often obovate: then known by its blunt l. and downward prickly stems.—*G. Witheringii* (Sm.) differs only by having rough edges to the leaves.—A very strong form, much larger in all respects, is the *G. elongatum* (Presl). Var. *microphyllum* (Lange) has narrower and shorter l. *E. B.* 1857.—Wet places by ditches and rivers P. VI. VII. E. S. I.

4. Ru'bia *Linn.* Madder.

1. *R. peregrína* (L.); l. 4—6 in a whorl elliptic or lanceolate shining smooth above without veins the margin and keel rough with reflexed bristles.—*E. B.* 851.—Old st. terete; shoots spreading, square. L. rigid persistent. Fl. in panicled cymes. Cor. rotate, 5-cleft; lobes oval, suddenly narrowed into a slender point.—Stony and sandy thickets chiefly in the South. P. VI.—VIII. E. I.

Order XLIII. VALERIANACEÆ.

Cal. superior; limb various, toothed, or inconspicuous, or involute and ultimately resembling a pappus. Cor. tubular, 3—5-lobed, unequal or irregular often spurred or gibbous at the base. Stam. 1—3, inserted in the tube, free, fewer than the cor.-lobes. Ovary with 1 perfect fertile cell and often 2 abortive cells; ovule solitary, pendulous. Fr. dry.—No stipules.

1. KENTRANTHUS. *Cor.* 5-lobed, *with a spur*. *Stam.* 1. Fr. 1-celled, indehiscent, crowned with the limb of the calyx expanded into a feathery pappus.

2. VALERIANA. *Cor.* 5-lobed, gibbous but *without a spur*. *Stam.* 3. *Fr.* 1-*celled*, indehiscent, crowned with the limb of the calyx expanded into a feathery pappus.

3. VALERIANELLA. Cor. 5-lobed, without a spur. Stam. 3. *Fr.* 3-*celled*, indehiscent, *crowned with the erect unequally toothed limb of the calyx*, 2 of the cells usually empty inflated or filiform.

1. KENTRAN'THUS *Neck.*

1. *K. rúber* (DC.); l. lanceolate stalked, upper l. ovate-lanceolate sessile, spur much shorter than the cor.-tube twice as long as the germen.—*Valeriana* Sm. *E. B.* 1531.—St. 1—2 feet high. Fl. purple or white.—Chalk-pits and old walls. P. VI.—IX. *Red Valerian.* E. I.

*2. *K. Calcitrápa* (Dufr.); radical l. ovate entire, stem-l. pinnatifid, spur very short.—*Sy. E. B.* 665.—Eltham, Kent [apparently now extinct]. E.

2. VALERIANA *Linn.* Valerian.

1. *V. officinális* (L.); *l. all pinnate*, lts. 9—21 lanceolate dentate-serrate terminal one not larger than the others, st. sulcate solitary, fr. glabrous ovate-oblong.—*R.* xii. 727. *St.* 9.—St. 2—4 ft. high. Fl. flesh-coloured. Radical l. on long stalks. —*α*; lts. usually 9—11 near together, their anterior edge nearly entire, the posterior edge strongly toothed. *With suckers*, not stoles.—*β. V. sambucifólia* (Mikan); lts. of rt.-l. ovate-acute, of st.-l. oblong-lanceolate, all toothed on both edges. *E. B.* 698. Term. lt. of rt.-l. often slightly the largest. Stoles long.—Ditches, marshes and damp places. P. VI. VII. E. S. I.

*2. *V. pyrenáica* (L.); *l. heartshaped* serrate stalked, upper l. with 1—2 pairs of small lanceolate leaflets.—*E. B.* 1591.—St. 2—3 feet high, furrowed. Fl. light rose-coloured.—Woods, rare. P. VI. VII. E. S. I.

3. *V. dioíca* (L); fl. imperfectly diœcious, *root-l. ovate* stalked, *stem-l. pinnatifid* with a large terminal lobe, fr. glabrous.—*E. B.* 628. *St.* 9.—St. a foot or more in height, simple. Fl. flesh-coloured; rather large with protruded stam. when barren; or small with included stamens and forming a closer corymb when fertile. Stoloniferous.—Boggy places. P. V. VI. E. S.

3. Valerianel'la *Mill.*

* *Fruit with 2 barren cells, fertile cell corky on the back.*

1. *V. olitória* (Poll.); fr. compressed oblique, barren cells without furrows, dissepiment incomplete, bracts ciliate.—*E. B.* 811. *St.* 2. 3. *R.* xii. 708.—About 6 in. high. L. ovate-spathulate, upper ones narrower. Fl. in *terminal dense cymes* with oblong linear opposite bracts. Fr. 3-celled; 1 cell fertile with its back formed of a thick gibbous spongy mass usually traversed by one furrow, separated by a groove on each side from 2 barren slightly confluent cells each having a slender rib on its side and their junction marked by a slight furrow, [glabrous or (var. *lasiocarpa* Reich.) pubescent].—Corn-fields and banks. A. V. VI. *Corn-Salad.* E. S. I.

** *Fruit with 2 barren conspicuous cells, fertile cell not corky.*

‡2. *V. carináta* (Loisel.); *fr. oblong boatshaped* crowned with 1 straight tooth, *cells nearly equal* each with a single rib on the back, barren cells contiguous in their whole length and with a deep furrow between them, fl. in dense cymes.—*E. B. S.* 2810. *R.* xii. 708.—About 6 in. high. Fl. pale blue. Root-l. spathulate, st.-l. oblong. Bracts ciliate. Fr.-section crescent-shaped.—Hedge-banks, rare. A. IV.—VI. *Lambs' Lettuce.* E. S. I.

3. *V. rimósa* (Bast.); fr. subglobose crowned with 1 erect membranous leaf, *barren cells larger than the fertile ones inflated* contiguous having a narrow furrow between them, fl. scattered.—*E. B. S.* 2809. *R.* xii. 709. *V. Auricula* (DC.) ed. viii.—About a foot high. Fl. distant, in the forks of a repeatedly forked cyme. Lower l. obovate attenuate downwards, upper l. oblong. Bracts ciliate. Section of the fruit nearly round. Crown of one oblong-blunt obliquely truncate tooth, sometimes with a minute tooth on each side; or of 3 acute teeth, of which one is much the longest and often 3-pointed.—Cultivated land. A. VII. VIII. E. S. I.

*** *Barren cells* 0, *or reduced to a rib.*

4. *V. dentáta* (Poll.); *fr.* oval *crowned with the small* oblique unequally 4-toothed *calyx flat in front with a space* enclosed between 2 elevated curved ribs convex behind, cyme lax spreading its branches long divaricate.—*E. B.* 1370. *R.* xii. 710. *V. Morisonii* DC.—Teeth of the crown spreading or all incurved except the largest. Fr. smooth or [var. *mixta* Dufr.] hairy Fl. corymbose.—Corn-fields and banks. A. VI. VII. E. S. I.

5. *V. eriocar'pa* (Desv.); fr. pilose crowned with the large toothed open *nearly regular rather obliquely truncate net-veined calyx*; otherwise like *V. dentata.*—*Sy. E. B.* 673. *R.* xii. 712. *Coss. Atl.* 24 E.—Worc., Dorset, Cornwall. The Welsh plant is *V. dentata.* A. VI. E.

Order XLIV. DIPSACACEÆ.

Fl. capitate, involucrate. Cal. superior, surrounded by an involucel (or free outer calyx) which closely invests the ripe fruit. Cor. 4—5-fid with unequal lobes. Stam. 4, inserted in the tube, free; anth. not. cohering. Style 1. Stigma simple. Ovary 1-celled, with a pendulous ovule. Fr. crowned with the pappus-like calyx. Embryo in fleshy albumen.—No stipules.

1. DIPSACUS. Involucel forming a thickened margin to the ovary. Calyx cupshaped. Cor. 4-fid. *Receptacle with spinous exserted scales* shorter than the involucre. Fr. with 4 sides and 8 little depressions.

2. KNAUTIA. Involucel terminating in 4 small teeth. Calyx cupshaped with radiant teeth. *Receptacle hairy hemispherical*; *scales* 0. Fr. with 4 sides and 4 little depressions.

3. SCABIOSA. Involucel membranous or minute. *Calyx of 4 or 5 bristles. Receptacle cylindrical with sunken scales.* Fr. nearly cylindrical with 8 excavations.

1. DIP'SACUS *Linn.* Teasel.

1. *D. sylves'tris* (Huds.); l. opposite simple sessile, stem-l. connate, *scales of the receptacle straight at the end exceeding the flowers, involucres curved upwards.*—*E. B.* 1032.—St. 5—6 feet high, prickly, leafy, branched. Heads of pale lilac fl. large, conical, overtopped by the slender ascending involucre.—Hedges and roadsides. B. VIII. IX. *Wild Teasel.* E. S. ? I.

[*D. Fullónum* (L.); *scales of the receptacle hooked at the end equalling the fl., invol. reflexed,* otherwise like *D. sylvestris.*—*E. B.* 2080.—Scarcely naturalized. B. VIII. IX. *Teasel.*] E.

2. *D. pilosus* (L.); *l. stalked with a lt. at the base on each side,* scales of the receptacle obovate-cuspidate straight, involucres deflexed.—*E. B.* 877.—St. 3—4 feet high, slender, branched, rough, leafy. *Heads* of white fl. small, *globose,* exceeding the involucres.—Moist shady places. B. VIII. *Shepherd's Rod.* E.

2. Knaut'ia *Coult.*

1. *K. arven'sis* (Coult.); lower l. simple, stem-l. pinnatifid, st. bristly, calyx with about 8 awned teeth.—*E. B.* 659.—St. 2—3 feet high, slightly branched, with a few leaves. Radical l. many, sometimes pinnately lobed. Fl. purple, in large convex long-stalked heads; *outer usually unequal and radiant.* Inv. bluntish.—Sometimes l. all simple narrowly lanceolate entire or superficially crenate.—Fields. P. VII.—IX. *Field Scabious.* E. S I.

3. Scabio'sa *Linn.* Scabious.

1. *S. Succísa* (L.); root-stock abrupt, heads of fl. and fr. nearly globose, *involucel hairy* 4-fid *herbaceous,* cor. 4-cleft, l. oblong entire, upper l. narrower mostly entire.—*E. B.* 878.—St. 1—3 feet high, rarely branched. Radical l. many, stem-l. usually few. *Fl. all alike,* purplish blue, rarely white. St. and both sides of the l. hairy or glabrous.—Meadows and pastures. P. VII.—X. *Devil's bit.* E. S. I.

2. *S. columbária* (L.); heads of fr. globose, *involucel membranous notched furrowed throughout,* no distinct base, cor. 5-cleft radiant, radical l. oblong stalked crenate entire or lyrate, uppermost l. pinnatifid with linear segments.—*E. B.* 1311.—St. 12—18 in. high. Radical l. blunt, or, rarely, lanceolate and acute, on long stalks; upper l. rarely entire, linear. Fl. purplish, anth. yellow.—On a calcareous soil. P. VII. VIII. E. S.

[*S. marítima* (L.); involucel and its base furrowed, cor. 5-cleft; has been found near St. Ouen's Bay, Jersey; and *S. atro-purpúrea* (L.) with large dark purple fl., which is probably a var. of it, is naturalized at Folkestone.]

Order XLV. COMPOSITÆ.

Fl. surrounded by an involucre formed of scales (phyllaries), collected together in a head looking like a single flower. Cal. superior; limb inconspicuous or forming a toothed bristly or feathery pappus. Cor. tubular or ligulate; both kinds in the same head or only one of them. Stam. 5, inserted in the tube; filaments free; anthers united into a tube surrounding the style, often with tails at their base. Fr. indehiscent, dry, with an erect seed without perisperm.—No stipules.

The following arrangement of the genera is different from that used in the arrangement of the species. In cases of difficulty they may both be used with advantage.

Suborder I. CORYMBIFERÆ.

Flowers of the disk tubular and perfect; marginal flowers often ligulate and female or neuter. Style not swollen below its branches.—Juice watery.

A. *Pappus more or less hairlike.*

* Anthers without bristles at their base.

† *Leaves opposite, cauline.*

(1.) 1. Eupatorium. Heads few-flowered. Phyll. imbricate, oblong. Receptacle naked. *Fl. all tubular*-funnel-shaped, *perfect* (reddish purple). Anth. included. Branches of the style exserted, cylindrical, blunt.

†† *Leaves radical.*

(2.) 2. Petasites. *Heads* many-flowered. Fem. fl. filiform, obliquely truncate or shortly ligulate, in many rows in the fem. heads, none or in 1 row in the male heads. Male fl. tubular, few and central in the fem. heads, occupying the whole disk in the male heads. Receptacle naked. Phyll. in one row.—*Subdiœcious.* Heads panicled

(3.) 3. Tussilago. *Heads* many-flowered. Fl. of ray narrowly ligulate, fem. in many rows; of disk male, tubular, 5-cleft. Receptacle naked. Phyll. in 1 row, with membranous margins.—Heads solitary.

††† *Leaves alternate, cauline.*

(4.) 22. SENECIO. Fl. of ray in 1 row, ligulate, fem., rarely 0; of disk perfect, tubular. Inv. cylindrical or conical, of 1 *row of equal phyll. not membranous at the margin*, with or without smaller scales at its base.

(5.) 21. DORONICUM. Fl. of ray in 1 row, ligulate, fem.; of disk perfect, tubular. *Inv. hemispherical, of 2 or 3 rows of equal phyll.* Pappus wanting in the ray.

(6.) 8. LINOSYRIS. *Heads not radiant.* Fl. all perfect, tubular (yellow). Receptacle naked, pitted; pits with elevated dentate margins in our plants. Phyll. imbricate. Fr. compressed, silky, without a beak.

(7.) 4. ASTER. *Fl. of ray* fem., ligulate, *in* 1 *row*; of the disk perfect, tubular, Receptacle naked, pitted. Phyll. imbricate and a few scales on the peduncle. Pappus in many rows. Fr. compressed, without a beak.

(8.) 5. ERIGERON. *Fl. of ray* fem., ligulate, *in many rows*; of the disk mostly perfect, tubular. Receptacle naked. Phyll. imbricate. Pappus in many rows. Fr. compressed, without a beak.

(9.) 7. SOLIDAGO. Pappus in 1 row. Fr. terete. (Fl. all yellow.) Otherwise like *Aster*.

** Anthers with 2 bristles at their base.

† *Receptacle without scales. Fr. cylindrical or tetragonal.*

(10.) 9. INULA. Fl. of ray fem., ligulate, rarely subtubular; of disk perfect, tubular. Receptacle naked. Phyll. imbricate in many rows. *Pappus hair-like, uniform, in* 1 *row*.

(11.) 10. PULICARIA. Phyll. laxly imbricate in a few rows. Pappus in 2 rows, outer short cuplike membranous toothed rarely wanting, inner hairlike. Otherwise like *Inula*.

†† *Receptacle without scales or scaly only at the margin. Fr. cylindrical or compressed. Pappus hairlike.*

(12.) 12. GNAPHALIUM. *Fl. all tubular*; outer fem.; central perfect. Receptacle flat, not scaly. Inv. hemispherical, imbricate; phyll. equalling the fl. but not intermixed with them.—Cor. of the fem. fl. often inconspicuous.

(13.) 11. FILAGO. Outer fl. fem., filiform, in several rows; *outermost ones intermixed with the inner phyllaries*; central fl. few, perfect, tubular. *Receptacle conical, scaly at the margin.* Inv. subconical, imbricate; phyll. lanceolate, longer than the flowers.

(14.) 13. ANTENNARIA. Heads subdiœcious. *Male* fl. tubular; style almost simple; *pappus clavate.* Fem. fl. filiform. Receptacle convex, not scaly. Inv. hemispherical, imbricate; phyll. coloured at the end.

B. *Pappus* 0, *or membranous.*

† Receptacle without scales. Heads radiant. Fl. of the ray fem., ligulate, in 1 row; of the disk herm., tubular.

(15.) 6. BELLIS. Phyll. in 2 rows, equal, blunt. Receptacle conical. *Fr. compressed.* Pappus 0.

(16.) 17. CHRYSANTHEMUM. Receptacle flat or concave. *Fr. of disk terete,* without wings; of the ray slightly angular or somewhat winged. Pappus 0 or of 3 minute teeth.

(17.) 16. MATRICARIA. Receptacle at length conical. Fr. angular, not winged. Pappus 0, or a slight membranous border.

†† Receptacle without scales. Heads discoidal.

(18.) 19. ARTEMISIA. Fl. of disk perfect; of the ray fem., slender, in 1 row; or all herm. and tubular. Involucre roundish. Phyll. imbricate. Receptacle naked or hairy. Fr. obovate, with *a small epigynous disk,* without pappus.

(19.) 20. TANACETUM. Fl. as in *Artemisia.* Involucre hemispherical. Phyll. imbricate. Receptacle naked. Fr. oblong, angular, with *a large epigynous disk* (as broad as the fruit), crowned with a slight membranous border.

††† Receptacle scaly. Pappus scale-like.

(20.) 24. GALINSOGA. Heads radiant. Fl. of ray fem., ligulate, in one row; of the disk perfect, tubular. Receptacle conical. Phyll. 4—5, broad, ciliate, in one row. Fr. prismatic, with a pappus of oblong scales.

†††† Receptacle scaly throughout. Pappus 0.

(21.) 15. ANTHEMIS. Heads radiant. Fl. of the ray fem., or neuter, ligulate, in 1 row; of the disk perfect, tubular.

Receptacle convex or conical. Phyll. imbricate, of few rows. Fr. terete, or bluntly tetragonal, without pappus, but with a more or less prominent margin.

[ANACYCLUS. Fr. compressed, winged at the edges. Otherwise like *Anthemis.*]

(22.) 14. ACHILLEA. Heads radiant. Fl. of the ray fem., ligulate, short; of the disk perfect, tubular; *tube plane-compressed, 2-winged.* Receptacle nearly flat, afterwards often narrow and lengthened. Inv. ovate or oblong. Phyll. imbricate. *Fr. compressed,* without pappus.

(23.) 18. DIOTIS. Heads discoidal. Fl. all perfect, tubular; tube compressed, with 2 auricles at the base. Receptacle convex, with concave downy-topped scales. Inv. bellshaped. Phyll. imbricate. *Fr.* compressed, *crowned with the persistent auricled tube of the cor.*; pappus 0.

C. *Pappus of 2—5 stiff bristles. Receptacle scaly throughout.*

(24.) 23. BIDENS. Heads discoidal, sometimes radiant. Fl. (of the ray neuter, ligulate;) of the disk herm., tubular. Receptacle flat. Phyll. in 2 rows, outer row spreading. Branches of the style surmounted by short cones. Fr. compressed, angular, rough at the edges; the angles ending in 2—5 hispid bristles.

Suborder II. CYNAROCEPHALEÆ.

Flowers all tubular. Style usually thickish below its branches, which often combine into a more or less perfect cylinder. Involucre imbricate.

a. Basal scar of fr. transverse, rounded. Anthers tailed.

25. SAUSSUREA. Fl. all perfect. *Phyll. unarmed.* Receptacle scaly. Pappus in two rows, outer of short rough bristles, inner feathery.

26. CARLINA. Fl. all perfect. *Outer phyll.* lax, leaflike *spinous*; inner linear, membranous, coloured and resembling a ray. Receptacle with cleft scales. Pappus in one row branched and feathery, united in a ring below.

27. ARCTIUM. Involucre globose. *Phyll.* ending in *hooked* points. Receptacle flat, with rigid subulate scales. Fr. compressed, oblong. Pappus short, hairlike, distinct.

b. Fr. attached obliquely. Anthers scarcely tailed.

* *Pappus in many rows of different lengths; inner row longest, longer than the fruit.*

28. SERRATULA. Heads diœcious by abortion. *Phyll.* sharp, *unarmed.* Scales of the receptacle split longitudinally into linear bristles. Fr. compressed, not beaked; basal scar oblique. Pappus persistent.

** *Pappus in many rows, unequal; second row longest, equal to or shorter than the fruit; rarely none.*

29. CENTAUREA. Anthers with papillose filaments. Receptacle chaffy. Fr. attached laterally above the base to the receptacle. Pappus hairlike, rarely 0.

c. Fr. with a transverse basal scar. Pappus in many rows, equal, long. Anth. scarcely tailed.

30. ONOPORDUM. Receptacle honeycombed. Fr. 4-ribbed. Pappus rough. Otherwise like *Carduus*.

31. CARDUUS. *Phyll.* simple, *spinous*, pointed. Receptacle with fimbriate scales. Fr. compressed, oblong. *Pappus* long, hairlike or feathery, *united into a ring at the base and deciduous.*—Includes *Cnicus* Linn. and *Cirsium* DC.

d. Fr. with a transverse basal scar. Pappus in many rows. Filaments monadelphous, glabrous. Anth. scarcely tailed.

32. MARIANA. Phyll. leaflike at the base, narrowed into a long spreading spinous point. Receptacle scaly. Fr. compressed, *its terminal scar surrounded by a papillose ring.* Pappus hair- or scale-like, united into a ring at the base, deciduous.

Suborder III. CICHORIACEÆ or LIGULIFLORÆ.

Flowers all ligulate and perfect.—Juice milky.

* *Pappus* 0. *Receptacle without scales.*

33. LAPSANA. Heads 8—12-flowered. Phyll. in 1 row, erect, with 4—5 short bracts at their base. Fr. compressed, striate, deciduous, not enveloped in the phyllaries.

** *Pappus like a crown, or of many entire broad scales. Receptacle without scales.*

34. ARNOSERIS. Heads many-flowered. Phyll. in 1 row, about 12, keeled, linear-lanceolate, *at length converging*, a few small bracts at their base. Fr. angular, crowned with a short elevated entire margin.

35. CICHORIUM. Head many-flowered. Phyll. in 2 rows, outer of about 5, lax, shortish; inner of 8—10, longer, converging, *at length reflexed.* Receptacle sometimes slightly pilose. Fr. obovate, compressed striate. Pappus of 2 rows of minute erect chaffy scales.

*** *Pappus feathery. Receptacle scaly.*

36. HYPOCHŒRIS. Heads many-flowered. Phyll. oblong, imbricate. Fr. glabrous, muricate, often beaked. Pappus in 2 rows, outer short and setaceous, inner long and feathery; or in 1 row and feathery.

**** *Pappus feathery, or on the exterior fruit scaly. Receptacle without scales.*

37. THRINCIA. Inv. oblong. Phyll. in 1 row, with a few additional at the base. Fr. beaked. Pappus in 2 rows; outer setaceous, deciduous; inner longer, feathery, dilated at the base. *Marginal row of fruits* enveloped in the phyllaries, scarcely beaked, and *with a short crown-like pappus.*

38. LEONTODON. Inv. subimbricate; exterior phyll. much smaller, in 1—3 rows. *Fr. uniform,* slightly beaked. *Pappus of all the fr. in* 2 *rows*; outer setaceous, persistent; inner longer, *feathery,* dilated at the base; *or in* 1 *row feathery.*

39. TRAGOPOGON. Inv. simple, of 8—10 equal phyll. connected at the base. Fr. longitudinally striate, with a long beak: basal scar lateral. *Pappus in many rows, feathery, interwoven in the ray.*

40. PICRIS. Phyll. in 1 row, equal, *with unequal linear* often spreading *ones at the base. Fr.* terete, *transversely striate, constricted or slightly beaked* above. *Pappus in* 2 *rows, feathery*; *external row rather hairlike.*

41. HELMINTHIA. Phyll. in 1 row, equal, with equal subulate adpressed ones at the base, and *surrounded by 3—5 leaflike loose bracts. Fr.* compressed, transversely rugose, *rounded at the end and with a slender beak* longer than the fruit. Pappus in several rows, feathery.

***** *Pappus filiform, deciduous, never feathery nor dilated at the base. Receptacle generally without scales. Fruit compressed.*

42. LACTUCA. Heads few-flowered. Phyll. with a membranous margin, imbricate in 2—4 rows; outer row shorter. Fr. plane compressed, contracted and prolonged into a filiform beak which is neither crowned nor muricate.

43. TARAXACUM. Heads many-flowered. *Inv. double*; inner phyll. in 1 row, erect; outer few, short, lax or adpressed, imbricate. *Fr.* subcompressed, *muricate and suddenly contracted above,* prolonged into a filiform beak.

44. SONCHUS. Heads many-flowered. Phyll. imbricate in 2 or 3 rows, unequal. *Fr. plane compressed, truncate, not beaked.*

45. MULGEDIUM. Heads many-flowered. Inv. double; inner phyll. in 1 row; outer short, lax, imbricate. *Fr. compressed, constricted above, and ending in a ciliate disk.* Outer rows of the pappus rigid and brittle.

****** *Fruit terete, ribbed. Otherwise like the preceding section.*

46. CREPIS. Heads many-flowered. Inv. double; inner phyll. in 1 row; outer short, lax. *Fr.* terete, *narrowed upwards or beaked.* Pappus soft.

47. HIERACIUM. Heads many-flowered. Phyll. imbricate, many, oblong. Fr. truncate, not beaked, with a very short crenulate margin. Pappus brittle.

[*Anomalous Genus.* Order AMBROSIACEÆ Link.]

[48. XANTHIUM. Heads monœcious.—Male: inv. of 1 row of free phyll., many-flowered. Receptacle scaly. Cor. funnelshaped, 5-cleft. Anth. free. Stigma blunt, entire. —Fem. fl. 2, enclosed within the inv. which ends in 1—2 beaks, is covered with hooked spines, and hardens over the fruit. Cor. 0. Stam. 0. Stigmas 2, diverging, linear. Fr. compressed, each occupying a cell in the involucre.]

Suborder I. *TUBULIFLORÆ.* Flowers all tubular, regular, with 4—5 teeth; or the outer ligulate.

Tribe I. *Eupatorieæ.*

Fl. all tubular, regular, perfect. Anth. not tailed. Style-branches blunt, terete, subclavate.

1. Eupato'rium *Linn.* Hemp-Agrimony.

1. *E. cannab'inum* (L.); l. in 3 or 5 deep lanceolate serrate segments the middle one longest.—*E. B.* 428.—St. herbaceous, erect, striate scabrous, 2—3 feet high. Heads in a fastigiate corymb, 5—6-flowered. Phyll. about 10; 5 exterior short, blunt. Florets reddish purple. L. downy. Herb slightly aromatic.—Banks of streams. P. VIII. IX. E. S. I.

Tribe II. *Tussilagineæ.*

Fl. of ray female, filiform or ligulate; of disk male, tubular. Anth. not tailed. Style-branches connate or short, with conical tips.

2. Petasi'tes *Mill.* Butterbur.

1. *P. officinális* (Moench); l. roundish-cordate unequally toothed downy beneath with approximate basal lobes, stigmas of the submale fl. short ovate, female fl. truncate obliquely.—*E. B.* 430, 431. *R.* xvi. 901. *Tussilago Petasites* (L.), *P. vulgaris* Desf.—Soboliferous. Panicle long and lax in the female plant, ovoid and dense in the male. Fl. appearing before the l., on stout erect stalks which are clothed with concave tumid petioles either leafless or with a small limb. L. very large, radical, ultimately often 3 feet broad, glabrous above.—Swamps. P. IV. E. S. I.

[*P. frágrans* (Presl), *Sy. E. B.* 781, which has shortly ligulate female fl., is established in some places in the South.—*P. albus* (Gaert.), *Sy. E. B.* 782, with white or cream-coloured fl., and much smaller deeply scolloped l., is established in Scotland.]

3. Tussila'go *Linn.* Coltsfoot.

1. *T. Far'fara* (L).—*E. B.* 429. *R.* xvi. 904.—Soboliferous. Fl. appearing before the l., in bright yellow solitary heads, erect in blossom and seed, drooping before and after flowering; their stalks clothed with scalelike smooth bracts. L. roundish-cordate, angular, toothed, downy beneath.—Moist chalky and clay soils. P. III. IV. E. S. I.

Tribe III. *Asteroideæ.*

Fl. of ray female or neuter, ligulate (or in *Linosyris* 0); of disk tubular. Anth. not tailed. Style-branches flattened with a subconical tip. Leaves alternate.

4. (7.) As'TER *Linn.* Starwort.

1. *A. Tripólium* (L.); st. glabrous corymbose, l. linear-lanceolate fleshy smooth, involucre imbricate, phyll. blunt membranous the inner ones longer.—*E. B.* 87. *R.* xvi. 907.—St. 1—2 feet high, erect, leafy, many-flowered. Heads large; disk yellow; rays bright blue, often wanting. [A dwarf form branched from the base, from N. of Scotl., is referred to var. *arcticum* (Fr.)].—Muddy salt marshes. P. VIII. IX. E. S. I.

[*A. salignus* (W.) is established in Wicken Fen, Cambridgeshire, and *A. longifolius*[1] at and about Seggieden, Perth; but they are probably both escapes from gardens. *A. Novi-Belgii* (L.) and other species are occasionally found.]

5. (8.) ERIG'ERON *Linn.* Fleabane.

*1. *E. canaden'se* (L.); *st.* much branched hairy *panicled many-headed*, l. linear-lanceolate ciliate.—*E. B.* 2019.—St. erect, 1—2 feet high. Heads many, small, yellowish. Inv. cylindrical, scarcely shorter than the fl. of the ray, finally spreading.—Waste ground, rare (American). A. VIII. IX. E.

2. *E. ácre* (L.); *st. corymbose, branches* alternate, usually 1-headed, l. linear-lanceolate entire spreading, lower l. narrowed below, *ray erect scarcely longer than the disk*, inner female fl. filiform many.—*E. B.* 1158. *R.* xvi. 917.—St. erect, 6—18 in. high, simple below, corymbosely branched above, often several from one root. Fl. yellow, the ray pale blue.—Dry gravelly places and walls. B. VII. VIII. *Blue Fleabane.* E. S. I.

3. *E. alpínum* (L.); st. mostly with a single head, l. lanceolate, lower l. narrowed below, *ray spreading twice as long as the disk*, inner female fl. tubular filiform many.—*E. B.* 464. *R.* xvi. 914.—St. 4—8 in. high, usually ending in a solitary head with a yellow disk and light-purple ray. Inv. hairy.—

---

[1] Prof. Asa Gray referred the Perthsh. plant to *A. Novi-Belgii.*—H. & J. G.

β. *E. uniflorum* (Sm. *not* L.) has a shorter and more erect ray and a rather more hairy involucre. *E. B.* 2416.—Highland mountains. P. VII. VIII. S.

6. (15.) Bel'lis *Linn.* Daisy.

1. *B. peren'nis* (L.); l. obovate-spathulate crenate-dentate. —*E. B.* 424.—St. a short procumbent rhizome producing l. only at its end. Stalks simple, each bearing a single head. Sometimes all the fl. are ligulate; rarely all are tubular.—Banks and pastures. P. III.—X. E. S. I.

7. (9.) Solida'go *Linn.* Golden Rod.

1. *S. Virgaúrea* (L.); st. erect slightly angular, l. lanceolate narrowed at both ends, lower l. elliptic stalked serrate, raceme erect simple or compound, phyll. lanceolate acute, fr. downy.— *E. B.* 301. *R.* xvi. 911.—St. usually 1—3 feet high, leafy, nearly simple, ending in a long cluster of yellow heads.—β. *angustifolia* (Gaud.); l. all lanceolate [entire or obscurely serrate]. —γ. *S. cambrica* (Huds.); st. 2—6 in. high, l. ovate-lanceolate, heads larger.—Woods and thickets. γ on mountains. P. VII. —IX. E. S. I.

8. (6.) Linosy'ris *Cand.* Goldilocks.

1. *L. vulgáris* (Cass.); herbaceous, l. linear glabrous entire, heads corymbose, inv. lax.—*Chrysocoma Linosyris* L. *E. B.* 2505. *Aster Linosyris* (Bernh.).—St. 12—18 in. high, simple, leafy. L. single-ribbed, smooth or rough, very many, more or less dotted. Fl. yellow.—Limestone cliffs, rare. P. VIII. IX. E.

Tribe IV. *Inuleæ.*

Heads never diœcious. Fl. of ray female or neuter, ligulate, in one row, or wanting; of disk perfect, tubular. Anth. with slender tails. Style-branches broadened upwards and rounded. L. alternate.

9. (10.) I'nula *Linn.*

‡1. *I. Helénium* (L.); *outer phyll. ovate, inner obovate,* l. unequally dentate downy beneath cordate-ovate acute clasping, root-l. stalked elliptic-oblong, fr. quadrangular glabrous.—*Sw. E. B.* 766.—St. 3—4 feet high, round, furrowed, solid, leafy, branched above. Heads few together or solitary, terminal, very

large; fl. bright yellow, those of the ray ligulate. Phyll. reflexed.—Moist pastures. P. VII. VIII. *Elecampane.* E. I.

2. *I. salicína* (L.); outer phyll. oblong-lanceolate, inner linear, l. lanceolate ½-clasping tuberculate-serrate scabrous-ciliate glabrous above the edge revolute, st. and underside of l. pilose-hairy.—*Sy. E. B.* 768. *J. of B.* iv. t. 43.—L. with crisped hairs on the veins beneath as also the st., disk of l. beneath pilose. Heads terminal, solitary or 3—5 in a corymb. Rays yellow.—Shore of Lough Derg, Co. Galway. P. VII. VIII.? I.

3. *I. Cony'za* (DC.); *outer phyll. lanceolate, inner linear acute,* l. ovate-lanceolate downy denticulate, lower l. narrowed into a haft, fl. of the ray tubular-ligulate, fr. terete.—*R.* xvi. 923. *Conyza squarrosa* L. *E. B.* 1195.—St. 1—2 feet high, leafy. Heads corymbose. Phyll. reflexed, leaflike. Fl. yellow, those of the ray deeply divided on the inner side.—Calcareous soils. P. VII.—IX. *Ploughman's Spikenard.* E.

4. *I. crithmoïdes* (L.); phyll. linear *taper-pointed, l. fleshy* linear blunt or with 3 points, fr. terete.—*E. B.* 68.—St. about a foot high, slightly branched near the top, each branch ending in a solitary head with an orange-coloured disk and yellow rays. —On rocks and in muddy salt marshes by the sea. P. VII. VIII. *Golden Samphire.* E. S. I.

10. (11.) PULICA'RIA *Gaert.*

1. *P. vulgáris* (Gaert.); *l. lanceolate* wavy *narrow at the base* and somewhat clasping, st. much branched downy, heads lateral and terminal hemispherical with *very short rays.*—*Inula Pulicaria* L. *E. B.* 1196.—St. 6—12 in. high, leafy. Heads small; fl. yellow. Fr. terete. Outer pappus of small distinct scales.—Moist sandy heaths. A. VIII. IX. E.

2. *P. dysenter'ica* (Gaert.); *l. oblong* cordate at the base clasping downy beneath, st. panicled woolly, heads axillary and terminal corymbose, *rays much exceeding the disk.*—*Inula* L. *E. B.* 1115.—Creeping, floccose. St. 12—18 in. high, leafy. Heads larger than in Sp. 1, bright yellow; fr. angular. Outer pappus cuplike, crenulate.—Damp places. P. VIII. IX. E. S. I.

Tribe V. *Gnaphalieæ.*

Heads with fem. marginal fl., or diœcious. Fl. all tubular. Anth. with slender tails. Style-branches broadened upwards and rounded. Pappus hairlike, rarely 0.

11. (13.) FILA'GO *Linn.* Cudweed.

* *Phyll. not spreading with the ripe fruit, in 2 rows.*

1. *F. german'ica* (L.); cottony, st. proliferous at the summit, l. lanceolate wavy acute, *heads obscurely 5-angled* half-sunk in wool forming axillary and terminal clusters not surrounded and overtopped by l., phyll. longitudinally folded linear cuspidate with glabrous points.—*E. B.* 946. *F. canescens* Jord.—Grey. St. erect or ascending, 4—12 in. long, usually simple below, bearing a solitary terminal cluster of heads, afterwards producing from just below it 2 erect branches which are again proliferous. Heads 20—40 in each cluster; furrows obscure; l. nearest to the cluster much narrowed upwards and acute. Tips of phyll. yellow, rarely reddish.—Dry fields. A. VII. VIII. E. S. I.

2. *F. apiculáta* (G. E. Sm.); cottony, st. proliferous at the summit, *l. all* oblong *blunt* apiculate, *heads prominently 5-angled* half-sunk in wool forming lateral axillary and terminal *clusters* surrounded and *overtopped by* 1—2 *blunt l.,* phyll. boatshaped cuspidate with glabrous points.—*Sy. E. B.* 737. *F. lutescens* Jord.! Pl. nov. Fr. iii. t. 7.—Greenish. Smelling like Tansy. St. mostly erect, with short erect branches below. Heads larger than in Sp. 1, 10—20 in a cluster; furrows deep; cluster often seeming lateral from only 1 branch being produced just below it; l. nearest to the cluster scarcely narrowed upwards, blunt apiculate. Tips of phyll. purple.—Sandy places. A. VII. VIII. E.

3. *F. spathuláta* (Presl); silky, st. proliferous, l. oblong-obovate, *heads prominently 5-angled not deeply sunk in wool* forming axillary and terminal *clusters overtopped by* 2—3 *acute l.,* phyll. cuspidate boatshaped with glabrous points.—*F. Jussiæi* C. & G. Atl. Fl. Par. t. 26. *Sy. E. B.* 738.—Whitish. St. usually branched from near its base; branches mostly horizontal. Heads larger than those of Sp. 1, 8—15 in each cluster; furrows very deep. Tips of phyll. yellow.—Dry fields. A. VII. VIII. E.

** *Phyll. at length spreading.*

4. *F. min'ima* (Fr.); *st. forked,* l. linear-lanceolate acute *flat* adpressed, heads pyramidal in lateral and terminal clusters longer than the leaves, *phyll. bluntish* cottony with glabrous points.—*E. B.* 1157. *Gnaphalium arvense* Willd.—St. slender, erect, 2—6 in. high, branched, the branches forked; or prostrate and spreading. Fl. yellowish in very small heads. Whole plant cottony, greyish.—Dry sandy and gravelly places. A. VI.—IX. E. S. I.

5. *F. gal'lica* (L.); st. forked, *l. linear* acute, *heads* conical in axillary and terminal clusters *shorter than the leaves*, outer phyll. cottony with bluntish glabrous points gibbous at the base and enclosing the marginal fr.—*E. B.* 2369.—St. 6—8 in. high slender. L. narrowing upwards from the base, upright, afterwards revolute.—Dry gravelly places, very rare. Berechurch, Essex. Bayford, Herts. A. VII.—IX. E.

12. (12.) GNAPHA'LIUM *Linn.*

‡1. *G. lúteo-album* (L.); *st. simple branched at the base* slightly corymbose above, *heads* densely clustered *leafless*, l. linear-oblong wavy woolly on both sides half clasping, lower l. broader at the end and blunt, upper l. narrowing and acute. —*E. B.* 1002.—Woolly. St. 3—12 in. high, decumbent below, then erect or ascending. Heads collected at the extremity of the stem; *inv. straw-coloured*; fl. tinged with red.—Sandy fields. [Eastern Counties, doubtfully native.] Jersey and Guernsey. A. VI. VIII. E.

2. *G. uliginósum* (L.); *st. diffuse much branched, heads* in terminal dense clusters *shorter than the leaves*, l. linear-lanceolate cottony on both sides.—*E. B.* 1194.—St 3—5 in. high, much branched, decumbent or ascending. Heads collected at the extremity of the st. and branches; *inv. yellowish brown*. Fr. glabrous or hairy.—*β. G. pilulare* (Wahl.); fr. papillose.— Wet sandy places. A. VII. VIII. E. S. I.

3. *G. sylvat'icum* (L.); *st. simple* nearly erect, *heads in axillary clusters forming an interrupted* leafy *spike*, l. acute linear-lanceolate, st.-l. narrower.—*R.* xvi. 58. *G. rectum* Sm. *E. B.* 124. —St. 3—24 in. high, upper half constituting the spike. Upper l. very narrow.—Woods and heaths. P. VII.—IX. E. S. I.

4. *G. norvégicum* (Gunn.); st. simple nearly erect, *heads in a close terminal* leafy *spike*, l. silky or cottony on both sides lanceolate, st.-l. acuminate mucronate broad.—*R.* xvi. 58. *G. sylvaticum* Sm. *E. B.* 913.—St. 6—12 in. high, spike distinctly terminal. St.-l. broad. Fl. longer in proportion to the inv. Phyll. dark brown.—Highland mountains. P. VIII. S.

5. *G. supínum* (L.); *cæspitose*, st. decumbent, flowering st. erect, *heads* 1—5 *distant*, l. linear downy on both sides mostly radical.—*E. B.* 1193.—Height 2—3 in. Cæspitose, very leafy at the root. Flowering st. with few leaves, which are downy on both sides. Heads sessile forming a sort of capitate spike, or stalked subracemose, or solitary.—Highland mountains. P. VII. S.

13. (14.) Antenna'ria *Gaert.*

1. *A. dioica* (Gaert.); *shoots procumbent,* flowering st. simple erect, corymb dense terminal, phyll. oblong dilated upwards blunt coloured, root-*l. obovate spathulate* glabrous above cottony beneath, stem-l. nearly equal linear-lanceolate adpressed.—*Gnaphalium* Sm. *E. B.* 267.—St. prostrate, woody, ending in a tuft of many l. and producing prostrate leafy stoles. Flowering st. 4—8 in. high, quite simple, cottony. Heads 4—5, erect, slightly stalked. Phyll. white or rose-colour.—*β. A. hyperborea* (D. Don); l. cottony on both sides. *E. B. S.* 2640.—Mountain heaths. P. VI. VII. *Cat's-foot.* E. S. I.

*2. *A. margaritácea* (R. Br.); *st. erect* corymbosely branched above leafy, *l. linear-lanceolate* acute cottony below, heads in level-topped corymbs.—*E. B.* 2018. *Anaphalis* (Benth. & Hook.).—St. 2—3 feet high, cottony. L. alternate, slightly cottony above, densely beneath. Inv. white. Fl. yellowish.—Moist meadows, rare. Established by rivers in Monm., Glam., and Merioneth. P. VIII. E.

Tribe VI. *Anthemideæ.*

Heads usually radiant, fl. of ray fem. or neuter, ligulate or slender and tubular; of disk tubular. Anth. not tailed. Style-branches truncate. Pappus often wanting or crownlike, rarely formed of scales or slender hairs.

14. (22.) Achille'a *Linn.* Yarrow.

1. *A. Ptar'mica* (L.); *l. shining linear-lanceolate* attenuate *acute glabrous* smooth *uniformly and finely serrate,* teeth adpressed mucronate minutely scabrous at the margin, ray 8—12-flowered equalling the involucre, corymb compound.—*E. B.* 757. *R.* xvi. 1024.—St. about 2 feet high, slightly branched above, erect, leafy, angular, smooth. Phyll. with a dark-brown membranous margin. Limb of the radiant florets longer than broad, white. Disk broad, white. L. sometimes very narrow; lower teeth not deeper than the others.—Moist meadows and thickets. P. VII. VIII. *Sneezewort.* E. S. I.

[*A. decólorans* (Schrad.); l. opaque bluntish downy thickly dotted coarsely and doubly serrate with spreading teeth laciniate and radiating at the base, ray 5- or 6-flowered about equalling the involucre.—*A. serrata* Sm. *E. B.* 2531.—Not known except in gardens. P. IX.]

[*A. tomentósa* (L.); *l. with a linear-lanceolate outline pinnatifid* woolly *lobes* crowded linear acute, *trifid* in the lowermost leaves, 2—3-fid in the intermediate, *uppermost simple*, corymb repeatedly compound, ray equalling about half the involucre.—*E. B.* 2532.—St. 10—12 in. high, decumbent at the base, woolly, simple. Phyll. woolly, edged with brown. *Disk and rays golden yellow.*—Scarcely naturalized. P. VII. VIII.]

2. *A. Millefólium* (L.); *l. with a lanceolate outline* bipinnatifid woolly or nearly glabrous, *lobes cut with linear segments, rachis entire* or subdentate with entire teeth, corymb dense, rays equalling about half the involucre.—*E. B.* 758. *R.* xvi. 1024. —St. erect, 6—18 in. high, nearly glabrous or woolly. Phyll. nearly glabrous with a brown margin. Heads small. *Fl. white*, occasionally reddish or purple. [Extreme forms are var. *lanata* (Koch), whole plant densely villous; and var. *alpestris* (Wimm. & Grab.), l. more divided, phyll. with broad dark margins.]—Pastures and waste ground. P. VI.—VIII. *Yarrow. Millefoil.* E. S. I.

[*A. tanacetifólia* (All.).—*Sy. E. B.* 728. *R.* xvi. 1027.—An escape.]

15. (21.) AN'THEMIS *Linn.* Chamomile.

* *Scales of the receptacle lanceolate or oblong abruptly ending in an acute rigid point.*

[*A. tinctória* (L.); *receptacle hemispherical*, fr. tetragonal crowned, with a membranous undivided border, l. bipinnatifid downy beneath, segments parallel decurrent serrate.—*E. B.* 1472.—St. 1—2 feet high, much branched, cottony. Heads on long stalks, solitary, terminal; *disk and rays bright yellow.* Scales not protruding.—Fields.—Not a native. B.? VII. VIII.]

1. *A. arven'sis* (L.); *receptacle conical*, fr. tetragonal, l. bipinnatifid hairy, segments linear-lanceolate.—*E. B.* 602. *R.* xvi. 1004.—St. 1—2 feet high, striate, downy, much branched. Segments of the l. parallel and at length converging. Heads on long stalks, solitary, terminal; disk convex, bright yellow; *ray white, always having styles.* Scales just appearing above the fl. of the disk, lanceolate. Outer fr. crowned with a tumid plicate-rugose ring, inner with an acute margin.—Borders of cultivated fields, rare. A. VI. VII. *Corn-Chamomile.* E. S. I.

[*A. ang'lica* (Spr.); "*receptacle flat,*" fr. crowned with a very narrow entire border, l. pinnatifid somewhat hairy, lobes incise-serrate acute bristle-pointed rather fleshy.—*A. maritima* Sm. *E. B.* 2370.—Probably a maritime form of *A. arvensis.* —Sea-shore. Sunderland. Not recently found. A. VII.] E.

** *Scales of the receptacle linear setaceous acute.* MARUTA Cass.

2. *A. Cot'ula* (L.); receptacle long conical, fr. terete tubercular-striate crowned with a crenulate margin surrounding a slightly convex disk, l. bipinnatifid nearly glabrous, lobes linear acute mostly entire.—*E. B.* 1772. *R.* xvi. 1000.—St. 1—2 feet high, branched, angular, furrowed. Heads solitary on long terminal stalks; scales confined to the central part of the receptacle; disk yellow; *ray white, usually without styles.* Cor.-tube 2-winged. Phyll. blunt, with white membranous margins. Fetid and acrid. [A procumbent sea-shore form with fleshy l. is var. *maritima* Bromf.]—Fields and waste places. A. VII.—IX. E. S. I.

*** *Scales of the receptacle thin membranous blunt.*

3. *A. nob ilis* (L.); receptacle conical, fr. subtrigonous, smooth, l. bipinnate, leaflets linear-subulate slightly downy rather fleshy acute.—*E. B.* 980. *St.* 27. 15.—*St. procumbent,* 1 foot long, much branched. Heads solitary, terminal; disk yellow; ray white; cor.-tube cylindric. Pleasantly aromatic. Gravelly and sandy places. P. VII. VIII. *Chamomile.* E. I.

[*Anacy'clus radiátus* (Lois.). Berehaven, Co. Cork. Accidental.]

16. (17.) MATRICA'RIA *Linn.* Feverfew.

‡1. *M. Parthénium* (L.); *l. stalked pinnate, lts. ovate or oblong* pinnatifid, lobes cut, st. branched, *heads corymbose,* phyll. linear blunt, receptacle convex, fr. crowned, with a short jagged membrane.—*Pyrethrum* Sm. *E. B.* 1231.—St. erect, 2 feet high, branched, furrowed, panicled. Heads in small corymbs terminating the stem and branches; disk yellow; ray white.—Waste places, not very common. P. VII. VIII. *Feverfew.* E. S.

2. *M. inodóra* (L.); *l. sessile pinnatifid with many capillary* pointed *segments,* st. branched, *heads solitary,* phyll. lanceolate blunt, receptacle ovate, *fr.* rugose and *with 2 glandular spots on the external face* just below the elevated entire border.—*Pyrethrum* Sm. *E. B.* 676.—St. erect, 12—18 in. high, smooth, angular. L. in very narrow mostly alternate segments. Heads solitary, ending the branches; margin of phyll. cut and fuscous; ray white; disk yellow. Base of the invol. turbinate afterwards truncate; recept. hemispherical afterwards conical, much longer than broad. Fr. with 3 prominent smooth ribs.—

β. salina; l.-segm. short fleshy linear bluntish convex above, principal ribs keeled beneath, inv. and recept. as in typical plant. *E. B.* 979. L. with short crowded mostly opposite segments.—*γ. phæocephala* (Rupr.); l.-segm. long, base of inv. subumbilicate, recept. as broad as long, fr. larger, margin of phyll. broadly and darkly coloured.—Fields and waste places. *β.* Sea-shore. *γ.* Near sea-shore in the north of Scotland. A. VII. VIII. E. S. I.

3. *M. Chamomil'la* (L.); l. bipinnate smooth, segments capillary simple or divided, heads solitary or subcorymbose, *receptacle hollow conical,* phyll. linear blunt.—*E. B.* 1232.—St. erect, 1 foot high, branched. Heads on long naked stalks or forming an irregular corymb; disk yellow; ray white.—Cultivated and waste ground. A. VI. VII. *Wild Chamomile.* E.

[*M. discoïdea* (DC.) of low stature with densely leafy st., short-stalked rayless heads with broadly-membranous phyll., a native of N. America, is established in many parts of Ireland, Cornwall and elsewhere. VI. VII.] E. S. I.

17. (16.) CHRYSAN'THEMUM *Linn.*

1. *C. Leucan'themum* (L.); lower l. obovate stalked, stem-l. oblong blunt cut sessile pinnatifid at the base, *phyll.* lanceolate blunt *with a narrow dark purple membranous margin,* fl. of *ray white,* fr. without a border.—*E. B.* 601. *St.* 2. 11.—St. erect, 1—2 feet high, simple, striate. Lower l. narrowing into a winged and auricled stalk. Heads solitary, terminal, large; disk yellow.—Fields. P. VI.—VIII. *Ox-eye.* E. S. I.

†2. *C. seg'etum* (L.); l. glabrous toothed dilated outwards and lobed, upper l. clasping, *phyll.* ovate blunt, *with a broad membranous margin,* fl. of *ray yellow.*—*E. B.* 540.—St. a foot high, alternately branched, angular. L. incise-serrate or lobed in the upper part, simply toothed below. Heads solitary, terminal.—Corn-fields. A. VI.—VIII. *Corn-Marigold.* E. S. I.

18. (23.) DIO'TIS *Desf.* Cotton-weed.

1. *D. candidis'sima* (Desf.).—*E. B.* 141. *D. maritima* (Cass.) ed. viii.—Densely cottony and white. St. about a foot long, decumbent below, densely leafy, corymbose above. L. sessile, oblong, blunt, flat, crenate, persistent. Heads in terminal corymbose tufts. Inv. cottony. Fl. yellow.—Sandy sea-shores, rare. [Extinct in England?] P. VIII. IX. E. I.

19. (18.) ARTEMIS'IA *Linn.* Wormwood.

* *Receptacle pilose.*

1. *A. Absin'thium* (L.); heads drooping hemispherical fl. not all perfect, l. silky in many deep lanceolate blunt segments, outer phyll. linear silky, inner roundish scarious.—*E. B.* 1230. —St. bushy, 1—2 feet high. Heads in erect leafy panicles. Floral l. simple. Fl. dull yellow, the outer row female.—Waste ground. P. VII. VIII. *Wormwood.* E. S. I.

** *Receptacle naked.*

2. *A. campes'tris* (L.); heads drooping ovate glabrous, fl. not all perfect, *l. silky* with many linear-lanceolate mucronate segments, stem-l. once or twice pinnate with linear segments, st. wandlike procumbent before flowering, *phyll. ovate glabrous with a scarious margin.*—*E. B.* 338.—Barren st. cæspitose. Flowering st. slender, 1—2 feet long, ascending when the fl. appear, leafy, smooth. Fl. yellow, those of disk sterile; inv. purplish. —Sandy heaths in Norf. and Suff., rare. P. VIII. IX. E.

3. *A. vulgáris* (L.); heads ovate, fl. not all perfect, *l. woolly* and white beneath pinnatifid *with lanceolate acuminate cut and serrate segments, phyll. woolly.*—*E. B.* 978.—St. 2—3 feet high, erect, leafy. Clusters leafy, nearly simple, erect. Fl. few, reddish or brownish yellow.—α. l.-segm. oblong, racemes open.—β. *A. coarctata* (Forsell); l.-segm. linear-lanceolate, racemes subspicate condensed.—Waste ground. P. VII.—IX. *Mugwort.* E. S. I.

4. *A. marit'ima* (L.); heads oblong, *florets few all perfect, l.* downy *pinnatifid with linear blunt segments,* phyll. oblong outer woolly inner scarious.—*E. B.* 1706.—St. procumbent or ascending, woolly, much branched. Fl. reddish yellow. Racemes drooping.—β. *A. gallica* (Willd.); heads erect (inconstant). *E. B.* 1001.—Salt marshes. P. VIII. IX. E. S. I.

[*A. Stelleriana* (Besser), with a nearly simple raceme, large erect globose-campanulate heads, pinnatifid l. with broad blunt lobes and with st., l., and phyll. densely white-felted, is naturalized on the coast of Cornwall and Co. Dublin. P. VIII.] E. I.

[*A. cærules'cens* (L.), *E. B.* 2426, is not a native. P. VIII. IX.]

[*Cotula coronopifólia* (L.), R. xvi. 998, with small button-like yellow heads without ray-florets, and with succulent lanceolate or acutely lobed clasping l., occurs as an alien in Cheshire and elsewhere. A. VIII.] E.

20. (19.) Tanace'tum *Linn.* Tansy.

1. *T. vulgáre* (L.) ; l. bipinnatifid, lts. serrate.—*E. B.* 1229. —St. 2—3 feet high, leafy. Heads in a terminal corymb. Fl. golden yellow. Fr. with an entire crown. St. 2—3 feet high. —Waysides. P. VIII. E. S. I.

Tribe VII. *Senecioneæ.*

Fl. of ray fem. ligulate; of disk tubular. Anth. tailless. Style-branches truncate or slightly rounded. Pappus hairlike.

21. (5.) Doroni'cum *Linn.* Leopard's-bane.

†1. *D. Pardalian'ches* (L.) ; *l. cordate* denticulate, *lowermost l. on long stalks, intermediate with clasping auricles at the base of the stalk,* uppermost sessile clasping.—*E. B. S.* 2654.—St. 2—3 feet high, erect, solitary, hollow, hairy. L. hairy, minutely toothed, soft, blunt, the uppermost acute. Lowest petioles not auricled. Heads several, phyll. lanceolate-subulate. Fl. yellow. The earlier heads overtopped by the later ones. Fr. oblong, furrowed, of disk hairy, of ray glabrous.—Damp and hilly woods and pastures, rare. P. V.—VII. E. S.

†2. *D. plantagin'eum* (L.?); *l. ovate* denticulate, radical on long stalks rounded or subcordate produced at the base, *stem-l. sessile* clasping the lowermost with a winged and auricled stalk. —*Sy. E. B.* 762.—Crown of the root woolly. St. 2—3 feet high. Stem-l. narrowed in their lower half but sessile, uppermost with a long taper point. Heads usually solitary, or, if more, the lateral ones not overtopping the terminal one. Phyll. subulate. Fr. of ray glabrous. Fl. yellow.—Damp places, rare. P. VI. VII. E. S.

22. (4.) Sene'cio *Linn.* Ragwort.

A. Involucre with small scales at its base.

* *Fl. all tubular, or marginal ones ligulate but mostly revolute.*

1. *S. vulgáris* (L.); l. half-clasping pinnatifid, segments distant oblong blunt and together with the rachis and auricles acutely and unequally toothed, lower l. narrowed into a stalk, heads in clustered racemes, *outer phyll. very short* adpressed with black points, ray 0.—*E. B.* 747.—Smooth or woolly. *Not viscid.* St. 6—12 in high, branching. Heads small,

involucre oblong-conical, glabrous; fl. yellow; fr. silky.—Rarely [var. *radiatus*, Koch] there is a single row of ligulate minute revolute marginal flowers.—Common. A. I.—XII. *Groundsel.* E. S. I.

2. *S. sylvat'icus* (L.); l. deeply pinnatifid downy, segments oblong unequally toothed, *heads corymbose,* involucre downy, *outer phyll. very short glabrous* (or slightly downy), ray small revolute, *fr. silky.—E. B.* 748.—*Slightly viscid.* St. 1—2 feet high, erect, more or less branched, hairy. L. narrower than in Sp. 3. Inv. conical; fl. yellow.—*S. lividus* (Sm.) is a slight var. with the upper l. more distinctly auricled and clasping.—Dry and gravelly places. A. VII.—IX. E. S. I.

3. *S. viscósus* (L.); *l.* deeply pinnatifid *viscid glandular-hairy,* segments oblong unequally toothed and lobed, heads in an irregular corymb, *involucre viscid, outer phyll. half the length of the inner hairy,* ray small, *fr. glabrous.—E. B.* 32.—*Very viscid.* St. 1—2 feet high, much branched, spreading. Heads on long stalks; inv. cylindrical; fl. yellow.—Waste ground, rare. A. VII.—IX. E. S. I.

** *Heads with spreading rays. Leaves pinnatifid.*

[**S. squal'idus* (L.); *l. pinnatifid* glabrous, *segments linear or oblong distant* toothed *irregular,* heads loosely corymbose, involucre glabrous, *outer phyll. few small,* fr. silky.—*E. B.* 600. *S. chrysanthemifolius* DC.—St. much branched, leafy, smooth. L. sessile, often auricled, deeply and irregularly lobed. Heads few, broad. Outer phyll. very small, sometimes very few. Many awlshaped scattered bracts below the heads. Fl. yellow. —Walls. Oxford. Bideford, Devon. Cork. [The hybrid with Sp. 1 has been found at Oxford and Cork.] A. VI.—X.] E. I.

4. *S. erucifólius* (L.); *l. pinnatifid* margins somewhat revolute cottony beneath, lower l. stalked, *segments linear* the lowermost smallest entire and clasping the stem, *outer phyll. half as long as the inner, ribs of all the fr. silky.—S. tenuifolius* Sm. *E. B.* 574. —Creeping slightly. St. erect, 2 feet high, angular, furrowed, somewhat cottony, simple. Lower l. oblong-ovate, deeply pinnatifid, cottony especially beneath; segments often linear. Fr. each having a persistent pappus. Fl. yellow. When the l. are divided into very narrow segments it is *S. tenuifolius* Jacq. —Calcareous soils. P. VII. VIII. E. S. I.

5. *S. Jacobæ'a* (L.); *l. glabrous, lower l. oblong-ovate attenuate below lyrate-pinnatifid* stalked, stem-l. sessile bipinnatifid, segments spreading oblong deeply and irregularly toothed and cut,

lowermost much divided clasping, outer phyll. scattered few, *fr. hairy those of the ray glabrous.*—*E. B.* 1130.—Root fleshy. St. 2—3 feet high, smooth, striate, branched, leafy. Corymb with erect branches. Fr. of the ray with deciduous pappus. Fl. yellow. Ray sometimes wanting [var. *discoideus* Koch].—Waste ground. P. VII.—IX. *Ragwort.* E. S. I.

6. *S. aquat'icus* (Hill); *l. glabrous, lower l.* stalked *crenate or dentate obovate or oblong* slightly prolonged at the base *undivided or sublyrate* blunt, upper l. lyrate or pinnately cut, segments oblong or linear, st. round, corymbosely branched, *fr. all glabrous.*—*E. B.* 1131.—St. erect, 1—4 feet high, simple or branched in the upper half, branches ascending. Terminal lobe of the lower l. rounded below and narrowed into its stalk. Fl. yellow.—A larger much branched form has l. all lyrate, term. lobe truncate or subcordate below, seg. subspathulate, and was supposed to be *S. erraticus* (Bert.) [1].—In marshy places. P. VII. VIII. E. S. I.

*** *Heads with spreading rays. Leaves undivided.*

7. *S. paludósus* (L.); l. sessile long lanceolate tapering *sharply serrate cottony beneath,* st. straight hollow, corymbs terminal, ray of 13—16 flowers.—*E. B.* 650.—St. 4—6 feet high, somewhat woolly. Fl. yellow: those of the ray narrow.—Fen ditches, very rare. P. V.—VII. E.

8. *S. saracen'icus* (L.); *l.* sessile lanceolate acute *glabrous irregularly serrate* especially the uppermost, st. straight solid, corymbs terminal, ray of 6—7 flowers.—*E. B.* 2211. *S. salicetorum* Godr.—St. 3—5 feet high, smooth. L. broad. Corymb many-headed. Fl. yellow.—There may be a second plant here with l. broadest below their middle very finely toothed, uppermost nearly entire. Its l. are much prolonged and rather glaucous. It closely resembles *S. Doria* Jacq.—Watery places, local. P. VIII. E. S. I.

B. Involucre without scales at its base. Heads with a spreading ray. Leaves nearly entire. CINERARIA L.

9. *S. palus'tris* (Hook.); shaggy, *st. much branched* and corymbose above, *l. broadly lanceolate half-clasping,* lower l. *sinuate-dentate.*—*E. B.* 151.—St. 3 feet high, thick, hollow, leafy. Heads erect. Fl. bright yellow.—Fen ditches, now become very scarce. B. VI. VII. E.

---

[1] The form with lyrate lower l. is var. *pennatifidus* G. & G.—H. & J. G.

10. *S. campes'tris* (DC.); shaggy, *st. simple, rt.-l. oblong* nearly entire narrowed below, *stem.-l lanceolate*, heads corymbose, involucre woolly below nearly glabrous in the upper half, fr. hispid.—*E. B.* 152.—St. 6—8 in. high, with small st.-leaves. Heads erect, 1—6, in a simple corymb. Involucre often almost glabrous, pale. Fl. yellow. In very wet seasons it is often thrice as large throughout.—Chalk Downs. P.? VI. E.

11. *S. spathulæfólius* (DC.); shaggy, st. simple, rt.-l. ovate-spathulate arachnoid above more woolly beneath, st.-l. ovate-oblong narrowed into broadly winged petioles, upper sessile linear or lanceolate clasping, invol. woolly, fr. hispid.—*J. of B.* xx. t. 226. *R.* xvi. 978. *Schultz, Herb. Norm.* 690!—St. 1—3 ft. high, with large clasping st.-leaves often much enlarged at the base. Fl. yellow.—Near Holyhead. South part of Mickle Fell range, Yorkshire. *Mr. J. Backhouse*! P. or B. VI. VII. E.

[*S. Cinerária* (DC.); a Mediterranean species, with l. densely white-felted beneath and deeply pinnatifid, the segm. broadening and lobed at the extremities, is naturalized and hybridizes with *S. Jacobæa* at Dalkey, Co. Dublin. See *J. of B.* xl. (1902) p. 401, t. 444.]

Tribe VIII. *Helianthoideæ.*

Heads discoid with all the fl. similar and perfect; or with fem. radiant ligulate flowers. Anth. without tails or scarcely tailed. Style-branches truncate. Pappus of 2—4 bristles or of oblong scales.

23. (24.) Bi'dens *Linn.*

1. *B. tripartíta* (L.); *l. stalked 3-partite*, segments lanceolate serrate, fr. obovate-cuneate usually with 2 bristles.—*E. B.* 1113. *R.* xvi. 941.—St. 1—3 feet high, with opposite branches. L. narrowed into winged footstalks, sometimes undivided, sometimes pinnate-5-fid. *Heads* terminal, solitary, *suberect.* Fl. brownish yellow.—Marshy places. A. VIII. IX. E. S. I.

2. *B. cer'nua* (L.); *l. sessile* connate lanceolate *undivided* serrate, fr. cuneate usually with 3—4 bristles.—*E. B.* 1114. *R.* xvi. 941.—St. 1—3 feet high, with opposite branches. L. simple narrowed below but not stalked. *Heads* terminal, solitary, *drooping.* Fl. brownish yellow.—Sometimes [var. *radiata* Gray] radiant marginal fl. are found.—Watery places. A. VIII. IX. E. S. I.

24. (20.) GALINSO'GA *R. & P.*

*1. *G. parviflora* (Cav.); subglabrous, receptacle conical, pappus of 8—16 scales.—*Sy. E. B.* 765. *R.* xvi. 983.—St. 1—2 feet high; branches opposite. L. opposite, ovate, stalked. Fl. of ray few, broadly ligulate, short, white; of the disk about as long as the phyllaries, yellow.—A South-American plant escaped from Kew Gardens, and quite naturalized in many places. A. VII.—IX. E.

Tribe IX. *Cynarocephaleæ.*

Fl. all tubular. Style of the perfect fl. thickened and often with a tuft of hairs below the branches, which are united or free and downy externally.

Section 1. *CARLINEÆ.*—Heads many-flowered, never diœcious. Phyllaries in many rows, distinct, often spinous. Filaments distinct, naked. Fr. mostly villose. *Pappus in* 1—2 *rows, not surrounded by an elevated margin.*

25. SAUSSU'REA *Cand.*

1. *S. alpina* (DC.); l. nearly glabrous above cottony beneath, lower ones ovate-lanceolate, upper sessile lanceolate, all distantly toothed, heads few in a dense corymb, involucre subcylindrical, phyll. adpressed hairy.—*Serratula* L. *E. B.* 599.—St. 3—12 in. high, erect, downy, simple, ending in a small corymb of heads with pinkish fl. and purple anthers. Fl. scented like Heliotrope. Fr. glabrous.—In alpine situations. P. VIII. E. S. I.

26. CARLI'NA *Linn.*

1. *C. vulgáris* (L.); st. corymbose one- or many-headed, l. oblong-lanceolate sinuate spinous, outer phyll. bipinnatifid spinous, inner linear-lanceolate attenuate acute ciliate in the lower half, bracts shorter than the heads.—*E. B.* 1144.—St. 6—12 in. high, usually cottony, leafy. Spines many, short. Root-l. lanceolate or linear-lanceolate. Underside of the l. and phyll. often cottony. Heads large; inner phyll. cream-coloured (a false ray); anth. yellow.—Dry sandy heaths. B. VII.—X. E. S. I.

27. ARC'TIUM *Linn.* Burdock.

1. *A. május* (Bernh.); *heads loosely subcorymbose long-stalked hemispherical* and open in fr. *glabrous* (green), phyll. equalling

or exceeding fl. subulate inner row shorter than the others, subcylindrical upper part of fl. more than ½ as long as the lower part. —*Sy. E. B.* 699.—St. 3—4 ft. high, centre and usually most of the branches ending in corymbs. L. broadly cordate-ovate, blunt; *petioles solid* with prominent angles, deeply furrowed. *Heads very large*, a few of the lower sometimes with short stalks. Fl.-heads not umbilicate. Fr. yellowish, ultimately dark brown, irregularly rugose.—A form with more spherical and webbed heads (*A. tomentosum* Bab. [1]) is common near Cambridge and is the *L. major* v. *subtomentosa* Lange!—Waste places. B. VIII. E. S. I.

2. *A. nemorósum* (Lej.); *heads racemose subsessile ovate* and contracted at the mouth in fr. slightly webbed, phyll. equalling or exceeding the fl. subulate inner row lanceolate shorter than the others, subcylindrical upper part of fl. as long as the lower part.—*Sy. E. B.* 701. *A. intermedium* Bab.—St. 2—4 feet high. L. convolute, cordate, oblong-ovate, petioles hollow, rather angular, nearly flat above. Heads all nearly sessile, less than in Sp. 1, three usually placed close together at the end of each branch, ovate-prolonged when young, not umbilicate. Most of the phyll. ascending.—Local? B. VIII. E. S. I.

3. *A. minus* (Bernh.); *heads racemose shortly stalked globular* slightly contracted at the mouth in fr. slightly webbed (greenish), *phyll. falling short of the fl.* subulate inner row equalling the others and gradually subulate, subcylindrical upper part of fl. about as long as the lower part.—*Sy. E. B.* 702. *Fl. Dan.* 2662. *R.* xv. 811.—Smaller than either of the preceding. Central st. mostly nodding and as well as the branches having scattered small heads; term. head solitary. L. deeply cordate-prolonged; petioles hollow, slightly angular, nearly round, scarcely furrowed. Fl.-heads not umbilicate. Fr. fuscous with black blotches.—Waste places. B. VIII. E. S. I.

4. *A. púbens* (Bab.); *heads subracemose stalked hemispherical* and open in fr. much webbed (greenish), phyll. equalling the fl. subulate inner row about equalling the others and gradually subulate, subcylindrical upper part of fl. as long as the lower part.—*Sy. E. B.* 700. *Fl. Dan.* 2663. *R.* xv. 812. *A. intermedium* (Lange) ed. viii.—St. about 3 feet high. L. deeply cordate-prolonged; petioles hollow, scarcely angular, slightly furrowed. Stalks of the heads rather long, those of the lower heads longest. Heads usually with much wool, twice as large

---

[1] The true *A. tomentosum* is not a native of England.

as those of *A. minus*. Fr. dark brown with a few paler spots towards the top. [A form with densely tomentose heads is *A. interm.* var. *subtomentosum* Ar. Benn.]—Waste places. B. VIII. E. I.

Section 2. *SERRATULEÆ*.—Heads many-flowered; flowers all tubular, perfect or diœcious, the external row sometimes female. Involucre of many rows of distinct phyllaries. Filaments distinct. *Pappus in many rows of different lengths, inner row longest, hairlike or feathery, surrounded by a margin.*

28. SERRAT'ULA *Linn.* Saw-wort.

1. *S. tinctória* (L.); l. with bristly serratures pinnatifid or lyrate, heads oblong, phyll. ovate adpressed, inner ones linear coloured.—*Sy. E. B.* 704. *St.* 3. 16.—St. 2—3 feet high, straight, erect, angular, branched above. L. rarely entire. Fl. purple.—*a*; heads stalked in a lax corymb.—*β. S. monticola* (Bor.); heads subsessile few large close together. *E. B.* 38. A curious dwarf form in Lizard district.—Groves and thickets. P. VIII. E. S.

Section 3. *CENTAUREÆ*. Heads many-flowered, discoidal; outer row of fl. usually barren, enlarged, irregular. Phyllaries in many rows. Filaments distinct. *Pappus in many rows of different lengths, second row longest, setaceo-pilose, placed within the margin which surrounds the epigynous disk, rarely* 0.

29. CENTAU'REA *Linn.* Knapweed.

* *Phyll. with a scarious pectinate not decurrent appendage.*

‡1. *C. Jácea* (L.); *phyll.-appendages* erect *rounded*, pappus 0, l. linear-lanceolate lower l. broader and toothed.—*E. B.* 1678. —Outermost phyll. with deeply pinnatifid appendages, few innermost entire, the rest irregularly jagged. Heads radiant.—Near Hastings, *E. N. Bloomfield.* J. of Bot. xxii. 248. P. VIII. IX. E.

2. *C. nígra* (L.); phyll.-appendages patent or erect lanceolate or ovate-lanceolate pectinate their teeth subulate, pappus 0 or short deciduous, l. lanceolate or ovate-lanceolate acute.—St. ½—2 ft. high. L. green; lower narrowed into long stalks, entire or sinuate-dentate; upper sessile. Heads globose. Few innermost phyll.-appendages torn. Fr. oblong, downy.—*a*; phyll.-appendages erect or patent ovate-lanceolate usually quite covering the phyll. their teeth long. *E. B.* 278. *Mart. Fl. Rust.* 130. *R.* xv. 761. St. usually with long 1-headed

branches. Pappus 0 or short. Heads sometimes radiant. Autumnal forms have erect-patent branches, ending in solitary heads; vernal have almost divaricate branches.—β. *C. decipiens* (Thuill.); phyll.-appendages erect lanceolate or ovate-lanceolate usually not wholly covering the phyll. their teeth short. St. usually simple, 1-headed. L. broader than those of α. Teeth often scarcely longer than the breadth of the brown appendage; 3 inner rows of phyll. usually protruding. Pappus 0. Heads usually (perhaps always) radiant. *Sy.E.B.* 707. *C. nigrescens* (Bab.). The plant when seen is easily distinguishable from the radiant form of *C. nigra*, although hardly to be separated by characters.—Meadows and pastures. β. South of England, rare. P. VI.—IX. E. S. I.

** *Phyll. lanceolate, their upper half with a somewhat scarious deeply toothed or fringed decurrent margin.*

3. *C. Cy'anus* (L.); phyll. erect adpressed deeply toothed, pappus rather shorter than the fruit, *l. linear-lanceolate*, the lowermost toothed or pinnatifid.—*E. B.* 277.—St. 1—3 feet high, loosely cottony, leafy. L. slightly cottony above, densely beneath. Involucre greenish yellow; phyll. often tinged with purple in their upper half, margins brown decurrent with whitish teeth. Heads with large radiant *blue flowers*, disk purple.—Corn-fields. A. VI.—VIII. *Corn Bluebottle.* E. S. I.

4. *C. Scabiósa* (L.); phyll. erect adpressed, the triangular-ovate black pectinate appendages not covering the inv., teeth ascending setaceous short, pappus as long as the fruit, *l. pinnatifid* roughish, segments lobed with hard points.—*E. B.* 56.—St. 2—3 feet high, rough, furrowed. L. hispid, lobes of the upper ones entire. Heads on long naked stalks, solitary. Involucres usually rather woolly; phyll. pale, with dark acute membranous pectinate decurrent appendages; teeth paler, short, not longer than ½ the width of the phyllary. Fl. purple, outer row radiant or 0. Rarely the inv. is quite covered by the appendages. [β. *succisæfólia* (Marsh.); root-l. entire, upper st.-l. entire, lower sometimes slightly lobed at base.]—Fields and hedges. [β. Sutherland, *E. S. Marshall & W. A. Shoolbred.*] P. VII.—IX. *Great Knapweed. Matfellon.* E. S. I.

[*C. paniculáta* (L.); phyll. erect adpressed rigid with subulate teeth and a short term. rigid point innermost narrow long toothed at the end, pappus much shorter than the fruit, lower l. pinnatifid with linear segments.—*R.* xv. 780.—St. about a foot high, panicled above, rough, rather cottony. Heads cylindric-oblong. Fl. purplish.—Quenvais and St. Ouen's Bay, Jersey. B. VII.]

*** *Phyll. horny at the end, spines palmate or pinnate.*

‡5. *C. solstitiális* (L.); *phyll. woolly* palmately spinous, *central spine* of the intermediate ones *very long* needle-shaped, inner phyll. with a roundish scarious appendage, *heads terminal* solitary, st. winged with the decurrent bases of the linear-lanceolate *entire* hoary *leaves*, root.-l. lyrate.—*E. B.* 243.—St. 1—2 feet high, branched, spreading. Involucres sometimes glabrous. Fl. yellow.—Cultivated land, probably introduced. A. VII.—IX. *Yellow Star-Thistle.* E.

6. *C. Calcitrápa* (L.); *phyll. glabrous* palmately spinous, *central spine strong channelled*, innermost phyll. with a scarious blunt appendage, *heads lateral* sessile solitary, pappus 0, l. deeply *pinnatifid*, lobes of the root-l. lanceolate toothed, of the stem-l. linear.—*E. B.* 125.—St. furrowed, slightly hairy, branched, spreading, about a foot high. Fl. purplish.—Gravelly and sandy places. A. VII. VIII. *Common Star-Thistle.* E.

[*C. as'pera* (L.); phyll. palmately spinous, *spines nearly equal* 3—5, innermost phyll. with a scarious blunt lanceolate or slightly spathulate appendage, *heads terminal* solitary, pappus of all the fr. in several rows, *l. linear* coarsely toothed narrowed below sessile rough, lower l. (and those of the primary stem?) broader incise-dentate with clasping auricles.—*C. Isnardi* L. *E. B.* 2256.—St. procumbent; branches long slender simple leafy. L. slightly toothed or entire. Fl. purple.—Typical *C. aspera* has its upper leaves sessile but not clasping.—Channel I. P. VII. VIII.]

[*C. Salaman'tica* (L.) and *C. leucophæa* (Jord.) are said to have been found in Jersey with *C. paniculata*; but I have not seen specimens.]

Section 4. *CARDUINEÆ.*—Heads many-flowered; flowers all tubular. Involucre in many rows of distinct spinous phyllaries. *Filaments distinct. Pappus in many rows, not surrounded by a prominent margin.*

30. ONOPOR'DUM *Linn.* Cotton Thistle.

1. *O. Acan'thium* (L.); st. erect many-headed, l. elliptic-oblong woolly on both sides sinuate spinous decurrent, outer phyll. lanceolate-subulate recurved and spreading.—*E. B.* 977.—St. 4—5 feet high, woolly, with broad spinous wings, branched. Inv. nearly globose, large, somewhat cottony; phyll. fringed with minute spinous teeth. Fl. purple.—Waste ground in South-east. B. VIII. E.

31. CAR'DUUS *Linn.* Thistle.

[Many hybrids apparently occur in this genus.]

* *Pappus rough.* CARDUUS Sm., DC., Koch.

1. *C. nútans* (L.); l. decurrent spinous lanceolate sinuate, *heads solitary drooping* hemispherical, *phyll. lanceolate* cottony *outer ones reflexed.*—*E. B.* 1112.—St. 2 feet high, erect, angular, furrowed, cottony, interruptedly winged. L. hairy on both sides, with woolly veins beneath, pinnatifid with 3-lobed wavy spinous-ciliate segments ending in strong spines. Heads large; fl. crimson; unopened anth. purple. Inv. hemispherical, internal phyll. contracted above the base and then lanceolate.—[A hybrid with Sp. 2 (*Newbouldi*, H. C. Wats.) occurs.]—Waste ground. B. V.—VIII. *Musk-Thistle.* E. S. I.

2. *C. cris'pus* (L.); l. decurrent spinous-ciliate lanceolate glabrous or cottony beneath deeply pinnatifid, lobes trifid and dentate, *heads roundish, phyll. linear-subulate* erect or ascending. —*E. B.* 973.—St. about 3 feet high, continuously winged. Phyll. not contracted above their base.—*α*; l. lanceolate usually cottony beneath, heads small clustered subglobular (or ovoid *C. polyanthemos* Koch), phyll. ending in a weak spine, ped. winged to the top, central tubercle of fr. not angular.—*β*. *A. acanthoides* (L.); l. broadly lanceolate less downy beneath, head twice as large solitary or rarely 2 or 3 together subglobular, phyll. strongly spinous erect, ped. often naked at the top, central tubercle of fr. 5-angled.—Dry banks and waste places. *β* is the less common plant. B.? VI.—VIII. E. S. I.

3. *C. pycnoceph'alus* (L.)[1]; l. decurrent sinuate spinous broadly lanceolate cottony beneath, segments ovate lobed, heads many crowded sessile subcylindrical, phyll. ovate-lanceolate attenuate.—*E. B.* 412.—St. about 3 feet high, slightly branched, with broad deeply lobed continuous spinous wings. L. deeply sinuate or pinnatifid. Involucres nearly glabrous. Fl. pink.—Sandy places near the sea. B.? VI.—VIII. E. S. I.

** *Pappus feathery.* CNICUS Linn., Sm. CIRSIUM Adans.

† Leaves spinous-hairy above, flowers purple.

4. *C. lanceolátus* (L.); *l. decurrent* white and cottony beneath

---

[1] This description refers to the common British form, *C. tenuiflorus* (Curt.). The more typical *C. pycnocephalus* from Plymouth has but slightly winged long branches, bearing 1—3 larger heads, with more spreading phyll. and the l. densely cottony beneath.—H. & J. G.

pinnatifid, lobes bifid with lanceolate entire segments each terminated by a strong spine, *involucres ovate* shaggy, phyll. lanceolate spinous spreading.—*E. B.* 107.—St. 3—4 feet high, erect, furrowed, hairy, with strong spinous wings. Heads terminal, solitary or 2 or 3 together, large.—Waste ground. B. VII. VIII. *Spear Thistle.* E. S. I.

5. *C. erioph'orus* (L.); *l. half-clasping not decurrent* white and cottony beneath deeply pinnatifid, lobes bifid the segments lanceolate entire alternately pointing upwards and downwards and each terminated by a strong spine, *involucres globose* shaggy, phyll. lanceolate with a long spinous-tipped reflexed point.—*E. B.* 386.—St. 3—4 feet high, much branched, furrowed, hairy. Root-l. 1—2 feet long, linear with long divergent lobes which form double rows in a very regular manner. Stem-l. similar but smaller. Heads very large; inv. covered with a dense white web. A remarkably conspicuous plant.—Waste ground on a limestone soil. B. VIII. *Woolly-headed Thistle.* E.

†† Leaves not spinous-hairy above.

a. *Limb of the cor. 5-parted to its base.* BREEA Less.

6. *C. arven'sis* (Robs.); heads subdiœcious, *l. subsessile oblong-lanceolate* pinnatifid spinous wavy, inv. ovoid subglabrous, *phyll. broadly lanceolate adpressed* terminating in a short spreading spine, rhizome creeping.—*E. B.* 975.—St. erect, 3—4 feet high, leafy, angular, corymbose above. L. very spinous, sessile or very slightly decurrent, varying greatly in width.—†β. *setosus*; l. lanceolate flat entire or slightly lobed. *Cir. setosum* M. B.—[A less spinous form with some of the upper l. entire, intermediate between the type (*horridus*) and β, is var. *mitis* (Koch, under *Cirsium*). γ. *argen'teus* (Buch.-White); l. densely white-tomentose beneath.]—Fields and roads. β. Culross and Kirkwall, S. [and many other places]. P. VII. β. IX. *Creeping Thistle.* E. S. I.

b. *Limb of the cor. 5-parted to its middle.*

7. *C. palus'tris* (L.); *l. decurrent* lanceolate deeply pinnatifid spinous, inv. ovoid crowded, *phyll. ovate-lanceolate* adpressed mucronate.—*E. B.* 974.—St. solitary, erect, 3—5 feet high, wand-like, with wavy spinous wings throughout, slightly branched. *Heads in a terminal cluster, small.* Fl. purple or white. Underside of the l. usually cottony. Inv. with a slight web.—Wet meadows. A. VII. VIII. E. S. I.

8. *C. praten'sis* (Huds.); *l. mostly radical lanceolate wavy or lobed* pilose above cottony beneath fringed with minute prickles,

stem-l. not decurrent few clasping, *inv.* globose *solitary* terminal slightly cottony, phyll. lanceolate-attenuate adpressed mucronate, rhizome creeping.—*E. B.* 177. *Cir. anglicum* DC.—St. 1—2 feet high, cottony, usually quite simple and single-headed, leafless in the upper half with a few scaly bracts, springing singly from the rhizome. L. broad, soft, sinuate-dentate, rarely with small 2—3-fid lobes, not pinnatifid, fringed with small but unequal prickles, lower l. stalked. Occasionally there are 2 or 3 fl. on a stem.—*C. Forsteri* (Sm.) is probably a hybrid between this and *C. palustris*. It has l. slightly decurrent lanceolate all pinnatifid spinous cottony beneath, st. panicled, inv. ovoid slightly cottony, root cæspitose producing several stems. *Sy. E. B.* 695.—Boggy meadows. P. VI.—VIII. E. I.

9. *C. tuberósus* (L.); l. lanceolate deeply pinnatifid pilose above hairy or slightly cottony beneath fringed with minute prickles, *stem-l. sessile not decurrent,* lobes 2—3-fid, inv. ovoid terminal 1—3 together slightly cottony, phyll. *lanceolate mucronate* adpressed, *root of elliptic tapering fleshy fibres.*—*E. B.* 2562. *Cir. bulbosum* DC., Koch.—Not stoloniferous. St. 2 feet high, erect, round, hairy, leafless above the middle with a few minute bracts. Lower l. stalked, stem-l. nearly or quite sessile. May be a hybrid between Sp. 2 and 10.—*C. Woodwardii* (Wats.) much resembles this, and may be a hybrid between *C. acaulis* and *C. pratensis*. *Sy. E. B.* 696.—Greatridge Wood near Boyton, and at Avebury, Wilts. P. VIII. IX. E.

10. *C. acaúlis* (L.); *l. glabrous* radical lanceolate pinnatifid, lobes subtrifid spinous, inv. ovoid glabrous nearly sessile mostly solitary, outer phyll. ovate inner ones gradually longer adpressed, root *with filiform fibres.*—*E. B.* 161. *St.* 24. 16.—St. usually almost wanting; sometimes 3—12 in. long, leafy, woolly. L. all stalked, glabrous except a few hairs on the ribs beneath. Heads very large, fl. crimson.—*β. C. dubius* (Willd.); st. much branched woolly a foot or more in height. *Willd. Fl. Berol.* f. 11. Perhaps a hybrid between this and *C. arvensis.*—Dry calcareous pastures. *β.* Saffron Walden, Essex. *Mr. G. S. Gibson.* P. VII.—IX. *Ground-Thistle.* E.

11. *C. heterophýllus* (L.); *l. clasping not decurrent* glabrous above *white and downy beneath lanceolate serrate fringed with minute prickles,* root-l. with long stalks clasping at the base, heads ovoid truncate below slightly downy, phyll. ovate or lanceolate acuminate adpressed.—*E. B.* 675.—*Creeping.* St. 3—4 feet high, furrowed, cottony, slightly branched above. Heads large and handsome. L. very large, undivided or laciniate.—*C. Carolorum* (Jenn. in Edin. Bot. Tr. ix. 257) seems to

be a hybrid of *C. heterophyllus* and *C. palustris*.—Moist mountain-pastures. P. VII. VIII. E. S.

Section 5. *SILYBEÆ.*—*Filaments monadelphous. Pappus in many rows.*

32. Maria'na *Hill* [*Silybum* Gaert. ed. viii.]. Milk-Thistle.

‡1. *M. lac'tea* (Hill).—*Carduus* L. *E. B.* 976.—St. 3—4 feet high, ribbed and furrowed. L. very large, oblong-lanceolate, wavy, clasping; radical l. pinnatifid, usually variegated with green and milk-white. Heads large, globose. Phyll. closely adpressed below, leaflike, with a long terminal recurved spine. Fl. purple; tube very long.—Waste places. B. VI. VII. E. S. I.

Suborder II. *LIGULIFLORÆ.* Fl. all perfect, ligulate.—Style cylindrical above; branches long, blunt, equally pubescent. Stigmatic lines prominent, narrow, terminating below the middle of the branches.—Juice milky.

A. *Receptacle naked. Pappus* 0.

33. Lap'sana *Linn.* Nipplewort.

1. *L. communis* (L.); l. dentate or lobed stalked, lower l. lyrate, involucres glabrous angular, st. panicled.—*E. B.* 844.—St. and l. hispid or nearly glabrous. St. 1—3 feet high, branched above. Heads small, with yellow fl. in terminal panicles with small subulate bracts at the subdivisions. Inv. of fr. erect.—Waste and cultivated ground. A. VII. VIII. E. S. I.

B. *Receptacle naked. Pappus like a crown, of many entire broad scales.*

34. Arnos eris *Gaert.* Swine's Succory.

1. *A. pusil'la* (Gaert.).—*E. B.* 95.—St. 3—8 in. high, swelling and hollow upwards, leafless, with a minute bract at the base of each branch. Each branch overtopping its predecessor and gradually thickening up to the solitary small terminal head of yellow flowers. Inv. connivent over the fr. when its phyll. become remarkably keeled. Receptacle honycombed towards the margins. Fr. small, obovate, attenuate below, 5-angled. L. radical, oblong, toothed.—Gravelly and sandy fields, rare. A. VI.—VIII. E. S.

35. Cicho'rium *Linn.* Succory. Chicory.

1. *C. In'tybus* (L.); lower l. runcinate hispid on the keel, upper l. oblong or lanceolate clasping entire heads axillary in pairs nearly sessile.—*E. B.* 539. *St.* 6. 15.—St. 2—3 feet high, bristly, alternately branched. Heads many, fl. bright blue, handsome. Floral l. lanceolate from a broad clasping base.—Waste places on a gravelly or chalky soil. P. VII. VIII. E. S. I.

C. *Receptacle scaly. Pappus feathery.*

36. Hypochæ'ris *Linn.* Cat's-ear.

* *Pappus with an outer row of shorter bristles.*

1. *H. glábra* (L.); st. branched leafless glabrous, *l. oblong, inv.* glabrous *equalling the flowers.*—*E. B.* 575.—St. 3—10 in. high, scaly; primary stem simple, leafless, but lateral branched stems prostrate leafy. L. spreading in a circle on the ground, glabrous, except a few scattered hairs. Outer row of fruits destitute of a beak; the rest with a long beak.—β. *H. Balbisii* (Lois.); all the fruits with long beaks.—[γ. *erostris* (Coss. & Germ.). Fr. without beaks.]—Sandy and gravelly places. β. In Kent and Salop. A. VII. VIII. E. S.

2. *H. radicáta* (L.); st. branched leafless glabrous, l. runcinate *blunt, inv. falling short of the flowers.*—*E. B.* 831.—St. about a foot high, scaly, each branch terminating in a rather large solitary head. L. spreading upon the ground, rough. Stalks slightly thickened beneath the heads. *Fr. all beaked.*—Waste ground. P.? VII. E. S. I.

** *Pappus in one row.* Achyrophorus Scop.

3. *H. maculáta* (L.); st. simple or slightly branched almost leafless, l. obovate-oblong undivided toothed pilose, phyll. bristly on the back.—*E. B.* 225.—St. about a foot high, stout, slightly hairy. L. often all radical. Heads large, fl. deep yellow.—Chalky and limestone hills, rare. P. VII. VIII. E.

D. *Receptacle without scales. Pappus feathery or on the exterior fruits scaly.*

37. Thrin'cia *Roth.*

1. *T. hir'ta* (Roth); l. lanceolate sinuate-dentate or entire hispid or hairy with forked or simple hairs, stalks simple pilose below.—*E. B.* 555. *Leontodon* (L.).—L. all radical, sometimes

nearly or quite entire, occasionally runcinate. Stalks often purplish, quite simple, longer than the leaves, somewhat hairy in their lower half. Phyll. downy on the margins at the apex or hairy.—Gravelly places and fields; also in the Fens. P. VI.—IX. E. S. I.

38. LEON'TODON *Linn.* Hawkbit.

* *Pappus feathery and with an outer row of bristles.*

1. *L. his'pidus* (L.); l. radical oblong-lanceolate runcinate hispid with forked hairs, stalks simple naked or with 1 or 2 minute scales thickened upwards hispid.—*E. B.* 554.—L. with regular spreading or reflexed narrow teeth. Stalks green, erect, longer than the leaves. Head drooping in bud, afterwards erect. Inv. nearly always hairy. Fl. glandular at the end. Fr. muricate.—*β. *L. hastilis* (L.). [Almost glabrous throughout.]—Meadows and pastures. β. Diptford, Devon. *J. of B.* xix. p. 312. P. VI.—IX. E. S. I.

** *Pappus in one row, feathery.* OPORINIA Don, DC.

2. *L. autumnális* (L.); l. radical linear-lanceolate toothed or pinnatifid nearly glabrous, stalk branched scaly and thickened upwards.—*E. B.* 830.—L. all radical tapering at the base, often with long linear spreading segments, usually somewhat hairy particularly on the midrib beneath. Inv. nearly always hairy. Pappus brownish.—β. *O. pratensis* (Less.); l. glabrous, stalk mostly simple, inv. shaggy with greenish black hairs. *Apargia Taraxaci* Sm. *E.B.*1109.—γ. *sordidus*; l. hairy, st. branched, inv. as in β, plant very large.—Meadows and pastures. β. Mountains. γ. Highland glens. P. VIII. E. S. I.

39. TRAGOPO'GON *Linn.* Goat's-beard.

1. *T. minus* (Mill.); *inv. about twice as long as the flowers*, ped. slightly thickened at the very top, l. tapering from a dilated base to a long slender acute point.—*Sy. E. B.* 799. *T. minor* (Fr.) ed. viii.—St. 2 feet high, branched, erect. L. clasping the stem. Phyll. 8, in 2 rows. *Fl. yellow*, truncate, 5-toothed. Anth. dark brown. Marginal fr. angular, striate; angles squamously toothed; interstices tubercled.—Meadows and pastures. B. ? VI. VII. E. S. I.

2. *T. praten'se* (L.); *inv. equalling or shorter than the flowers*, ped. slightly thickened at the very top, l. tapering from a dilated base to a long linear acute point keeled.—St. 1½—2 feet high, branched, erect. L. clasping the stem. Phyll. 8, in 2

rows. *Fl. yellow*, truncate, 5-toothed. Anth. yellow. There are 2 forms of this plant:—(*a*) inv. equalling the fl., marginal fr. obscurely striate and rough throughout (*Sy. E. B.* 798); (*b*) [var. *Symei* Ar. Benn.] inv. rather shorter than the fl., marginal fr. (in my specimens) yellow slightly furrowed and quite smooth (*E. B.* 434).—Meadows and pastures, less frequent than the preceding. B.? VI. E. S. I.

*3. *T. porrifólium* (L.); *inv. longer than the flowers, ped. much thickened upwards*, l. tapering slightly dilated just above the base.—*E. B.* 638.—St. 3—4 feet high, erect, branched. L. slightly broader just above the base, then gradually narrowing to an acute point. Heads twice as large as in the two preceding. Inv. usually ⅓ longer than the fl., sometimes only equalling or even falling short of them. *Fl. purple.* Marginal fr. with scalelike tubercles throughout but particularly on the ribs.—Moist meadows. B. VI. *Salsify.* E. I.

40. Pi'cris *Linn.*

1. *P. hieracioïdes* (L.); st. rough with forked and hooked bristles, l. linear or lanceolate dentate or sinuate, upper l. somewhat clasping, heads solitary terminating the stem and branches, outer phyll. lax oblong bristly on the keel glabrous on the margin, fr. constricted just below the pappus.—*E. B.* 196.—St. 1—3 feet high, divaricately branched above, irregularly corymbose, very rough. Florets yellow.—*P. arvalis* (Jord.) with branches ascending and heads in an umbellate corymb passes gradually into the type.—Dry banks. B. VII.—IX. E.

41. Helmin'thia *Juss.* Ox-tongue.

1. *H. echioïdes* (Gaert.); st. erect hispid with rigid 3-fid and hooked hairs from tubercular bases, phyll. 5 ovate-cordate.—*E. B.* 972. *Picris* (L.).—St. 2—3 feet high, branched, covered as well as the leaves and involucre, with strong prickles springing from white tubercles and with 3 minute hooks at the apex (glochidate). L. clasping.—Dry banks. A. VII.—IX. E. I.

E. *Receptacle generally without scales. Pappus filiform, deciduous, never feathery, nor dilated at the base. Fruit compressed.*

42. Lactu'ca *Linn.* Lettuce.

* *Beak long, white. Leaves with a bristly keel.*

1. *L. salig'na* (L.); upper *l. linear entire acuminate* with a sagittate base, lower l. pinnatifid, *beak twice as long as* the fruit.

—*E. B.* 707.—St. 2 feet high, slender, wavy, slightly branched. Heads in small alternate tufts forming long clusters. Flowers yellow.—Chalky places and near the sea. B. VII. VIII. E.

2. *L. virósa* (L.); *upper l. horizontal oblong* auricled and clasping *mucronate-dentate* or sinuate, *beak equalling the black fruit.*—*E. B.* 1957.—St. scabrous, 2—4 feet high, leafy, branched above, panicled. Heads scattered, with many heartshaped acute bracts, rarely runcinate. Plant full of acrid milky juice.—Dry banks. B. VII. VIII. *Acrid Lettuce.* E. S.

3. *L. Serríola* (L.); *upper l. upright* arrowshaped at the base and clasping *sinuate, beak equalling the pale fruit.*—*E. B.* 268.—*St. slightly scabrous below,* 2—5 feet high, leafy, panicled. Heads scattered, with many heartshaped acute bracts. Juice rather less acrid than in Sp. 2.—Waste places, rare.—B. VII. VIII. *Prickly Lettuce.* E.

** *Beak short. Leaves with a smooth keel.*

4. *L. murális* (Fresen.); florets 5, l. lyrate-runcinate angled and toothed clasping terminal lobe largest, *beak much shorter than the fruit,* heads panicled.—*Prenanthes* L. *E. B.* 457.—St. erect, a foot high, smooth, round, hollow. Flowers bright yellow. Fruit black.—Banks and old walls. P. VII. E. I.

43. TARAX'ACUM *Juss.* Dandelion.

1. *T. officinále* (Weber); l. runcinate toothed, fr. linear-obovate blunt and muricate at the top longitudinally striate with a long beak.—Stalks single-headed, radical, hollow. Fl. yellow. L. all radical, very variable, glabrous or slightly hispid.—*a. Leontodon Taraxacum* (L.); *outer phyll. linear deflexed inner ones simple at tip,* fr. yellow its upper half muricate, crown of the root glabrous or woolly, l. runcinate broad.—*E. B.* 510.—*β. T. lævigatum* (DC.); *outer phyll. erect-patent ovate, inner* gibbous or *appendaged at tip,* fr. reddish yellow muricate at the top, beak with a thickened and coloured base, l. runcinate-pinnatifid with unequal teeth.—*γ. T. erythrospermum* (DC.); *outer phyll. lanceolate, adpressed or patent, inner* gibbous or *appendaged at tip,* fr. bright red muricate at the top, beak with a thickened and coloured base, l. runcinate-pinnatifid with unequal teeth and intermediate smaller ones. Lowermost l. sometimes obovate and dentate (or runcinate when it becomes *T. obovatum* DC.). *Sy. E. B.* 803.—[*δ. T. udum* (Jord.); outer phyll. at first adpressed ultimately spreading, inner almost simple at tip, styles bright yellow, fr. yellowish green, muricate-aculeate at the top.]—*ε. L. palustre* (Sm.);

outer phyll. ovate-acuminate, adpressed, inner simple at tip, fr. pale yellow or brown muricate at the top, l. oblong and entire sinuate-dentate or runcinate; or outer phyll. ovate-lanceolate, or (*L. leptocephalum* R.) lanceolate. *E. B.* 553.—Very common, γ in dry places, δ in bogs or damp places. P. III.—X. E S. I.

44. SON'CHUS *Linn.* Sowthistle.

1. *S. oleráceus* (L.); l. undivided or pinnatifid toothed clasping, *auricles spreading arrowshaped, fr. transversely rugose and longitudinally ribbed,* st. branched, heads subumbellate, inv. usually glabrous.—*E. B.* 843.—St. 2—3 feet high. L. flattish, lower stalked. Fl. yellow.—Common. A. VI.—VIII. *Sowthistle.* E. S. I.

2. *S. as'per* (Hill); l. undivided or pinnatifid sharply toothed clasping, *auricles rounded, fr. longitudinally ribbed not transversely rugose,* st. branched, heads subumbellate, inv. usually glabrous.—*E. B. S.* 2765, 2766.—St. 2—3 feet high. L. crisped, lower stalked. Fl. yellow.—Common. A. VI.—VIII. *Sowthistle.* E. S. I.

3. *S. arven'sis* (L.); l. lanceolate runcinate sharply toothed cordate at the base, uppermost l. entire, *st. simple, heads corymbose, inv. and ped. glandular-hairy,* fr with transversely rugose ribs, creeping.—*E. B.* 674.—St. 3—4 feet high, leafy. L. long, acute. Heads large, fl. yellow. Glandular pubescence sometimes wanting (var. *glabrescens* Guenth. G. & W.). [A narrow-leaved shore form is var. *angustifolius* Meyer.]—Fields and waste ground, also in fens. P. VIII. IX. *Corn-Sowthistle.* E. S. I.

4. *S. palus'tris* (L.); l. linear-lanceolate all acutely arrowshaped denticulate, lower l. long with 2—4 linear-lanceolate lobes, *st. simple,* heads corymbose, inv. and ped. glandular-hairy, fr. with finely rugose ribs, *no stoles.*—*E. B.* 935.—St. 4—6 feet high, leafy. Fl. lemon-coloured.—Marshes, very rare. P. VII. VIII. E.

45. MULGE'DIUM *Cass.*

1. *M. alpinum* (Less.); l. glabrous lyrate arrowshaped at the base, terminal lobe large triangular-hastate acute, st. simple, heads racemose, bracts ped. and inv. glandular-hairy, fr. oblong not attenuate with many ribs.—*Sonchus cæruleus* Sm. *E. B.* 2425. *Lactuca* Benth.—St. 3 feet high, glabrous below, leafy. L. gradually smaller upwards, cordate-acute on the barren shoots. Heads small, many. Fl. blue.—Clova Mountains. P. VIII. S.

F. *Receptacle generally without scales. Pappus filiform, never feathery nor dilated at the base. Fruit terete, ribbed.*

46. CRE'PIS *Linn.* Hawk's-beard.

* *Fruit with a long subulate beak.* BARKHAUSIA Moench.

1. *C. taraxacifólia* (Thuil.); l. rough runcinate-pinnatifid, buds erect, *inv.* bristly and downy *covering half the pappus,* outer phyll. lanceolate with a membranous margin, bracts herbaceous, *fr. all equally beaked.—E. B. S.* 2929.—St. 1—2 feet high, hispid, angular, furrowed, purple below, at length branched, corymbose. *L. mostly radical* lyrate-runcinate with backward teeth or deeply pinnatifid with the terminal lobe large. Stem-l. few, sessile, clasping, deeply pinnatifid and toothed. Fl. yellow, purple beneath. Fr. narrowing very gradually into a setaceous beak of about its own length, ribs rough.—Limestone districts. B. VI. VII. E. I.

2. *C. fœ'tida* (L.); l. hairy runcinate-pinnatifid, unopened buds nodding, inv. hairy and downy, outer phyll. lanceolate acute downy, *marginal fr. slightly beaked shorter than the inv., central fr. with long beaks equalling inv., pappus protruding.— E. B.* 406.—St. 6—12. in., high, hairy, round, branched. L. mostly radical; stem-l. few, small, lanceolate, deeply toothed at the base, sessile. Heads solitary, terminal, on long simple stalks. Midrib of the phyll. at length much thickened and hardened. Ribs of the fr. rough.—Chalky places, rare. B. VI. VII. E.

[*C. setósa* (Hall.); l. runcinate-dentate or lyrate-runcinate, st.-l. sagittate entire or incise-dentate below, buds erect, *inv. not quite covering the pappus,* margin of the outer lanceolate acute phyll. and the bracts and the back of the inner phyll. and the ped. hispid with rigid simple bristles.—*E. B. S.* 2945.—Plant 1—2 feet high. Stem-l. large, rather strapshaped, clasping. —Fields. Introduced with seed. A. VII. VIII.] E. I.

** *Fruit narrowed upwards or obscurely beaked. Pappus silky.*

[*C. pul'chra* (L.), *E. B.* 2325, was probably an error.]

3. *C. virens* (L.); *outer phyll. adpressed linear, inner ones glabrous within,* l. lanceolate remotely dentate runcinate or pinnatifid, *uppermost l.* linear-arrowshaped clasping *with flat margins,* st. subcorymbose, *fr. shorter than the pappus* oblong slightly

narrowed upwards with smooth ribs.—*C. tectorum* Sm., *E. B.* 1111 (*not* Linn.).—Very variable. St. 1—3 feet high, or diffuse. Fl. yellow. [*C. tectorum* (L.) has revolute-margined upper l., phyll. downy within, fr. somewhat beaked with scarious ribs; Mr. Brotherston found it as an escape near Kelso.]—Common. A. VI.—IX. E. S. I.

[*C. nicæensis* (Balb.); *outer phyll. adpressed linear, inner glabrous within*, l. lyrate-pinnatifid hispid, stem.-l. sagittate clasping, uppermost *lanceolate entire*, ped. and inv. usually glandular-hairy, fr. oblong scarcely narrowing upwards, strongly-ribbed *scabrous* shorter than the pappus.—R. 1440.—Habit of Sp. 1. St. 1—3 feet high nearly leafless above, corymbosely branched. Heads rather large, fl. yellow.—Fields, introduced with seed. B. VI. VII. E. S. I.]

4. *C. bien'nis* (L.); *outer phyll. oblong-linear lax, inner downy within*, l. runcinate-pinnatifid hispid, uppermost l. lanceolate clasping dentate-pinnatifid, st. subcorymbose, fr. oblong slightly narrowed upwards with nearly smooth ribs and about as long as the pappus.—*E. B.* 149.—St. 1—3 feet high, hispid, nearly leafless above, corymbosely branched. Heads large; fl. yellow. *L. radical and extending halfway up the stem.*—Chalky places, rare? B. VI. VII. E. I.

5. *C. succisæfólia* (Tausch); *phyll. lanceolate-attenuate, outer ones very short adpressed, l. entire* nearly glabrous oblong blunt, lower ones narrowed into a footstalk, *upper l.* sessile and *somewhat clasping*, st. corymbose, ped. and inv. glandular-hairy, fr. much striate slightly narrowed upwards as long as the pappus which is shorter than the involucre.—*C. hieracioides* (W. & K.) ed. viii., *Hieracium molle* Sm. *E. B.* 2210.—St. 2—3 feet high, erect, simple below; l. few. Heads few; fl. yellow.—Woods in the North. P. VII. VIII. E. S.

*** *Fruit not beaked, cylindrical. Pappus stiff, brittle.*
ARACIUM Monn.

6. *C. paludósa* (Moench); phyll. lanceolate much attenuate glandular-pilose, outer ones short, *l. ovate-oblong taper-pointed* runcinate-dentate narrowed into a footstalk glabrous, *upper l.* ovate-lanceolate *cordate* and clasping acute entire or dentate, st. subcorymbose, fr. striate scarcely narrowed upwards.—*Hieracium* L. *E. B.* 1094.—St. 2 feet high, leafy, simple, angular. L. large. Fl. yellow.—Damp woods and shady places. P. VII. —IX. E. S. I.

47. HIERA'CIUM *Linn.* Hawkweed.

[The following entirely new account of the genus has been drawn up under the direction of Mr. F. J. Hanbury, from his notes and specimens, by Miss R. F. Thompson. Where possible Professor Babington's descriptions have been retained.]

i. *PILOSELLOIDEA.* Stoloniferous. St. scapelike. Fr. minute, crenulate at top, striate. Hairs of pappus equal, very slender.

1. *H. Pilosel'la* (L.); stoles slender leafy rooting, *scape 1-headed leafless,* l. oblong or lanceolate hairy on both sides whitish and densely floccose beneath, inv. ovate below ultimately conical, inner phyll. acute, styles yellow.—*E. B.* 1093; *Mon. Brit. Hierac.*[1] Pl. 1; *Exsicc.*[2] Fasc. ii. 26.—Stoles many sometimes flowering. Fl. pale yellow; outer striped with red or purple externally.—β. *pilosissimum* (Wallr.); stoles short thick, l. st. and inv. with long silky hairs, heads large, phyll. all lanceolate.—*R.* xix. 1468; *Mon. Brit. Hierac.* Pl. 2; *Exsicc.* Fasc. iv. 76.—γ. *nigrescens* (Fr.); stoles long straight, scape long, inv. densely setose, l. less hairy.—*Exsicc.* Fasc. v. 101.—δ. *concinnatum* (F. J. Hanb.); very dwarf, scape and inv. densely floccose setose, *not hairy,* outer ligules striped dark crimson.—Dry banks, common. δ. Ben Macdhui. P. V.—VIII. E. S. I.

2. *H. aurantiacum* (L.); stoles often wanting, st. slightly hairy densely corymbose at top more hairy setose and floccose above, *l.* obovate-lanceolate *green* and hairy on both sides *not floccose beneath,* phyll. blunt, styles brown.—*E. B.* 1469; *Mon. Brit. Hierac.* Pl. 3.—St. 15—20 in. high. Phyll. dark. Fl. dark orange.—Woods, pastures and waste places, naturalized. P. VI.—VIII. E. S. I.

*3. *H. praten'se* (Fr.); stoles leafy, st. pilose floccose and setose above, corymbose at top, l. narrowly obovate-lanceolate green on both sides pilose slightly floccose beneath, phyll. linear blunt, styles bright yellow.—*H. collinum* Fr. *J. of B.* vi. t. 86; *Edin. Bot. Tr.* x. t. i.; *Mon. Brit. Hierac.* Pl. 4; *Exsicc.* Fasc. ii. 27.—Stoles not very long. Corymbs dense. Phyll. dark green hairy setose, with a pale margin and tip. Fl. yellow.—Near Selkirk and Edinburgh. P. VI. VII. S.

---

[1] An Illustrated Monograph of the British Hieracia, by F. J. Hanbury.
[2] Set of British Hieracia, E. F. and W. R. Linton.

ii. *PULMONAREA.* Radical rosettes in autumn which produce the persistent root-l. of the next year. Phylls. interruptedly or irregularly imbricate. Fr. short, truncate, not crenulate, striate. Hairs of pappus unequal, rigid.

* *Alpiniformes.* Inv. shaggy or silky. Outer phyll. lax; inner acuminate or acute. Fl. hairy externally, more or less pilose at the tips. St.-l. 1 or few, or leaflike bracts. —*H. alpinum* L.

4. *H. alpinum* (L.); deep green, *st.* 1-*headed* hairy floccose, *rt.-l.* lanceolate or *ovate spathulate* narrowed into petioles, st.-l. 1 or few sessile, buds slightly nodding, *inv. hemispherical* shaggy with long soft black-based hairs setose, *phyll. few broad acuminate* lax, styles yellow.—*H. alpinum* Backh., *Sy. E. B.* 827; *Mon. Brit. Hierac.* Pl. 5; *Exsicc.* Fasc. iii. 51.—St. 4—8 in. high. Usually 1 st.-leaf. Head large, always solitary; fl. bright yellow, with short hairs at the top.—*a*; rt.-l. ovate-spathulate.—*β. insigne*; rt.-l. lanceolate with a few large teeth, heads very large.—Lofty mountains of Scotland. P. VII. VIII. S.

5. *H. holoseric'eum* (Backh.); *green*; st. 1-headed shaggy silky floccose, rt.-l. lanceolate-*spathulate* or linear-lanceolate *blunt* hairy on both sides narrowed into winged petioles, st.-l. few small sessile, buds nodding, *inv. turbinate shaggy with long silky* white black-based *hairs, outer phyll. very lax leaflike blunt, inner phyll. adpressed linear acute*, styles yellow.—*H. alpinum* Sm. *E. B.* 1110; *Mon. Brit. Hierac.* Pl. 6; *Exsicc.* Fasc. ii. 28.—St. 3—9 in. high. Usually 1—2 st.-leaves. Head solitary. L. entire or sometimes slightly denticulate. Fl. densely hairy at the tips, bright yellow.—Lofty mountains. Cumberland. Grampian and Breadalbane Mountains. P. VII. VIII. E. S.

6. *H. exim'ium* (Backh.); green, *st. usually* 1-*headed* hairy floccose, *rt.-l. lanceolate acute sharply toothed* hairy on both sides narrowed into *broadly winged petioles*, st.-l. small slender, buds nodding, *inv. truncate below* shaggy with rather silky black-based hairs, *phyll. many linear-attenuate*, outer phyll. small lax. —*Sy. E. B.* 825; *H. villosum* Sm., *E. B.* 2379, not L.; *Mon. Brit. Hierac.* Pl. 7; *Exsicc.* Fasc. i. 1.—St. 6—15 in. high; l. few. Rarely more than 1 large head. L. sometimes entire, usually with large teeth. Flowers twice in cultivation, (1) with one head, (2) with many nearly parallel branches and many heads.—*a.* St. long, rt.-l. lanceolate dentate, *styles livid.*— *β. tenellum* (Backh.); st. short slender, rt.-l. lanceolate or linear-lanceolate, *styles yellow.* Considered a distinct species

by Dr. Grenier.—*Mon. Brit. Hierac.* Pl. 8; *Exsicc.* Fasc. iii. 52.—Lofty mountains. Clova. Braemar. Breadalbane Mountains. P. VII. VIII. S.

7. *H. calenduliflórum* (Backh.); deep green, st. usually 1-headed hairy floccose, *rt.-l. broadly ovate-spathulate blunt* apiculate sharply toothed hairy on both sides narrowing into broadly winged petioles, st.-l. very small slender, buds nodding, inv. truncate below shaggy with rather silky black-based hairs, phyll. many linear-attenuate, outer phyll. lax, *styles livid.—Sy. E. B.* 824; *Mon. Brit. Hierac.* Pl. 9; *Exsicc.* Fasc. ii. 29.—St. 6—14 in. high. St.-l. few. Head most usually solitary, very large. Youngest l. sometimes rather acute. Original l. nearly circular. Nearly allied to *H. eximium.*—Lofty mountains. Lochnagar. Clova. P. VII. VIII. S.

8. *H. granitic'olum* (W. R. Linton); st. 4—8 in. high, 1-3-headed, floccose setose with black-based hairs, rt.-l. rosulate ovate-spathulate cuneate at the base toothed in the lower half, inner l. lanceolate coarsely toothed rounded or blunt at the apex, narrowed into winged petioles, somewhat hairy on both surfaces and on the margins, primary and outer l. coriaceous more or less glabrous. St.-l. solitary linear with one or two bract-like l. above, *inv. rounded thickly shaggy with black-based hairs,* setose at the base, phyll. outer broad adpressed, inner slightly acute, all white-tipped. Styles nearly pure yellow. Fl. light yellow strongly pilose at the tips.—Lofty mountains. Clova. P. VII. VIII. S.

9. *H. gracilen'tum* (Backh.); green, st. usually 1-headed hairy floccose, *rt.-l. lanceolate or oblong*-lanceolate attenuate below, inv. ventricose rounded below shaggy with black soft hairs setose, *phyll. few broad acuminate adpressed with floccose tips* outermost lax, *styles livid.—Sy. E. B.* 828; *Mon. Brit. Hierac.* Pl. 10; *Exsicc.* Fasc. iv. 77. *H. alpinum* var. *melanocephalum* Fries, not Bab., nor *H. melanocephalum* Tausch.—St. 6—10 in. high, hairy and floccose, with straight simple diverging branches and many heads when under cultivation; st.-l. usually 2 or more. L. broad, narrowed gradually below; original l. roundish. Petioles sometimes slightly winged. Heads usually solitary, rather large, fl. bright yellow, nearly or quite glabrous at the back.—Lofty mountains. P. VII. VIII. S.

10. *H. petiolátum* (Elfst.); green, st. 4—7 in. high hairy and usually 1-headed, with many heads under cultivation, rt.-l. obovate or lanceolate subacute with few coarse teeth towards the base, narrowed into long shaggy somewhat winged petioles,

scattered hairs on both surfaces, *very metallic-looking*, st.-l. usually 2 linear lanceolate, entire or with few small teeth towards the base, inv. rounded to campanulate rather shaggy with white black-based hairs, phyll. dark green outer narrow somewhat lax unequal in length tips recurved slightly white in bud. *Styles very dark.*—Ben-na-muic-dhui. P. VII. VIII. S.

11. *H. globósum* (Backh.); *glaucous* or green, st. few-headed floccose, *rt.-l. obovate* or ovate-lanceolate subacute glabrous above entire or dentate narrowed into petioles, *inv. rounded below ultimately globose* greenish-black with short black-based hairs, *phyll. many attenuate acute adpressed, styles yellow.*—*Sy. E. B.* 829; *Mon. Brit. Hierac.* Pl. 11; *Exsicc.* Fasc. iii. 54. —Stem 6—12 in. high, usually 1-headed, rarely branching even from near the base, leafless or with few narrow bract-like l. Original l. blunt. Innermost l. sometimes acute. Petioles short sometimes slightly winged. *Buds globose.* Fl. bright yellow.—Cairngorm Mountains. P. VII. VIII. S.

** *Nigricantes.* Inv. villose or hairy. Phyll. adpressed or few outermost lax. Fl. nearly hairless externally, minutely pilose or subglabrous at the tips.—*H. nigrescens* Hook.

12. *H. nigres'cens* (Willd.); green, st. 4—6 in. high 1—2-headed floccose setose with few white hairs towards the base, rt.-l. ovate inner longer lanceolate finely toothed hairy on both surfaces, st.-l. usually 1 lanceolate finely toothed or bract-like entire *sessile* floccose with scattered hairs, inv. dark, phyll. subacute outer rather lax *very setose* slightly floccose *not hairy.* Fl. golden yellow glabrous. Styles dark.—*β. commutatum* (Lindeb.); st. 6—10 in. high, 1—3 large heads, hairy throughout, l. larger broader *more deeply toothed* floccose and hairy beneath, phyll. longer more attenuate acute *hairy with few setæ. Fl. faintly pilose.*—*γ. gracilifolium* (F. J. Hanb.); *l. much longer narrower very acutely toothed* especially towards the base decurrent into slightly winged petiole, inv. setose not hairy slightly floccose. Styles not so dark.—*Exsicc.* Fasc. iv. 78.—*a.* Breadalbane Mountains. *β.* Cairn Toul. *γ.* Ben Lawers. P. VII. VIII. S.

13. *H. atrátum* (Fr.) *f.*; st. 6—12 in. high simple or branched floccose with few white hairs below setose above, rt.-l. primary subrotund almost entire, inner ovate to ovate-lanceolate dentate somewhat hairy on both surfaces all on long hairy petioles, st.-l. usually 1 large lanceolate acute *petiolate sharply toothed at the*

base almost glabrous, inv. large *always very dark*, phyll. long narrow acuminate *somewhat floccose* with setæ *microglands* and few hairs. Fl. almost glabrous. Styles dusky.—*Exsicc.* Fasc. v. 102.—High mountains. Inverness, Ross, and Perthshire. P. VII. VIII. S.

14. *H. curvátum* (Elfstr.); green, *rt.-l.* ovate or lanceolate *coarsely and irregularly dentate* in their lower half hairy with slender petioles, st. simple or branched with few heads and few lanceolate or oblong shortly stalked l., *inv. ovate below,* dark green with soft black-based hairs setose, ped. densely floccose, phyll. acuminate, outer ones blunt, *styles brownish.*—*Sy. E. B.* 832; *Mon. Brit. Hierac.* Pl. 12; *Exsicc.* Fasc. iv. 79; *H. nigrescens* Willd. pt.—Variable in height. Fl. deep brilliant yellow, pilose at the tips. Phyll. overtopping the buds. L. rarely nearly glabrous above. St.-l. usually 1, narrowed at both ends. St. floccose, branching much in cultivation.—Highland mountains. P. VII. VIII. S.

15. *H. Backhous'ei* (F. J. Hanb.); dark green, st. 5—15 in. high with scattered black-based hairs throughout floccose above with numerous hairs and setæ, *rt.-l. very erect* ovate and ovate-lanceolate with *long curved forward-pointing teeth* or almost entire, decurrent to winged petioles *sub-coriaceous glossy* glabrous or with few scattered hairs on both surfaces, *st.-l.* 2—4, *semi-amplexicaul* toothed, upper bract-like entire, inv. large rounded below greenish-black with rather short black-based hairs and few setæ. Phyll. adpressed attenuate sub-acute, outer short rather lax. Fl bright yellow somewhat pilose. Buds globose. Styles rather livid. Pappus distinctly tawny.—*Mon. Brit. Hierac.* Pl. 13; *Exsicc.* Fasc. iii. 55.—Lofty mountains around Braemar. Breadalbane Mountains. P. VII. VIII. S.

16. *H. lingulátum* (Backh.); green, *rt.-l.* lanceolate or oblong apiculate *denticulate* or dentate hairy with short petioles, st. simple or branched with few heads and few lanceolate or lanceolate-attenuate acute sessile l., *inv. broad becoming truncate below very dark* with soft black-based hairs slightly floccose, phyll. attenuate acute incumbent, *style livid.*—*Sy. E. B.* 834; *Mon. Brit. Hierac.* pl. 14; *Exsicc.* Fasc. i. 2; *H. saxifragum* Bab.—St. 15—24 in. high. Fl. bright yellow. Phyll. greenish black, overtopping the buds. *L. coarsely hairy above,* entire towards the end. Original l. broad and rounded. St. with scattered black-based hairs throughout, a little floccose. St. and inv. nearly without setæ.—Mountain glens. Clova. Braemar. Breadalbane Mountains. P. VII. VIII. S.

17. *H. senes cens* (Backh.); green, rt.-l. elliptic-lanceolate denticulate or remotely dentate hairy with slender petioles, st. simple with few heads and 1 or 2 *linear-lanceolate l. with* slender *stalks*, inv. ovate or turbinate below with short black-based hairs and many setæ slightly floccose, phyll. acute adpressed (?), styles yellow.—*Sy. E. B.* 833; *Mon. Brit. Hierac.* pl 15; *Exsicc.* Fasc. i. 3.—St. 15—18 in. high. Fl. golden. Phyll. of the buds with a tuft of white down at the tips. L. with scattered hairs on both sides, entire at both ends. Original l. broader. Lowest ped. above the middle of the stem.—Grassy slopes and edges of streams in mountain districts. Clova. Braemar. Ben Voirlich, Dumbartonshire. Breadalbane Mountains. Co. Down. P. VII. VIII. S. I.

18. *H. Marshall'i* (Linton); st. erect 7—16 in. high few-headed with scattered white hairs, sparsely floccose and setose above, peduncles thick, floccose with black-based hairs and setæ interspersed, rt.-l. few rather large outer roundly-ovate to ovate nearly entire, inner ovate-acuminate with *large forward-pointing teeth especially towards the base*, yellowish-green fleshy upper surface dull nearly glabrous with rough white hairs beneath on the margin and midrib, narrowed into shaggy winged petiole, st.-l. usually solitary short-stalked or sessile toothed, or bract-like and almost entire if above, heads 2—3 rather large; young buds ovoid and white-tipped. Inv. rounded, phyll. very broad subacute, outer short and rather lax, dark olive-green densely hairy setose tips very senescent. Fl. bright golden yellow pilose externally and at the tips. Styles darkened.—*Mon. Brit. Hierac.* pl. 16; *Exsicc.* Fasc. i. 4.—High mountains. P. VII. VIII. S.

19. *H. chrysan'thum* (Backh.); green, rt.-l. ovate acute at both ends *sharply irregularly and deeply toothed* hairy with long slender petioles, st. simple or branched with few heads and linear-lanceolate or subulate stalked st.-l., inv. rounded below floccose with short black-based hairs and setæ, phyll. many *linear-attenuate* outermost small and lax, styles dull yellow.—*Sy. E. B.* 830; *Mon. Brit. Hierac.* pl. 17; *Exsicc.* Fasc. i. 5.—St. 9—15 in. high. Fl. golden. Phyll. greenish-black, overtopping the drooping buds, incumbent. L. with scattered hairs on both sides, entire towards the end. Original l. blunt, roundish. St. often with only 1 head, with subplumose hairs below, floccose setose and with black-based hairs above, sometimes branching throughout. Hairs on st. and inv. with dull-reddish tips. Buds sometimes pilose at the tip.—*β. microcephalum* (Backh.); st. simple with 1 or few nearly erect heads, l. dentate or nearly entire, inv. urceolate, styles rather livid.—

Mon. Brit. Hierac. pl. 18; *H. atratum* Bab.—γ. *gracilentiforme* (F. J. Hanb.); l. all more evenly dentate or almost entire, st. sometimes many 3—9 rising from the base 1-headed. Fl. strongly ciliate-tipped. Styles faintly livid.—α. High Scotch mountains. β. Lochnagar. γ. Cumberland and Westmoreland about Helvellyn. P. VII. VIII. E. S.

20. *H. sin'uans* (F. J. Hanb.); st. 6—15 in. high simple or branched fistular slightly floccose, hairy and setose, heads 1—4 large, *rt.-l.* narrow ovate to lanceolate outer blunt inner longer narrower acuminate *all with rounded sinuous forward-pointing teeth* decurrent into shaggy somewhat winged petioles, green sometimes reddish paler beneath firm glabrous above with scattered white hairs beneath *with ciliate wavy margins*, st.-l. 1 or none bract-like or lanceolate acute with short petiole or sessile, *toothed wavy* hairy slightly floccose. Buds short round, inv. urceolate-campanulate, phyll. outer short rather lax inner longer acute with setæ and black-based hairs. Fl. deep golden yellow pilose. Styles *very dark.*—*Mon. Brit. Hierac.* Pl. 19; *Exsicc.* Fasc. ii. 30. High mountain glens. P. VII. VIII. S.

21. *H. centripet'ale* (F. J. Hanb.); st. erect solid striate 14—16 in. high usually branched *peduncles arcuate* densely setose floccose and pilose, rt.-l. ovate to ovate-lanceolate narrowed at the base all dentate usually bi-dentate with long acute teeth or *rather evenly scalloped and bearing glandular teeth*, upper surface with scattered deciduous white hairs, margin and under surface with long silky white hairs, st.-l. usually 1 often bract-like entire or large near the base and toothed like the rt.-l., heads numerous, buds very dark floccose-tipped, inv. dark velvety campanulate, phyll. long narrow acute, outer short rather lax sub-acute, with long black setæ black-based hairs and pilose tips. Fl. bright yellow shortly pilose-tipped, styles livid.—*Mon. Brit. Hierac.* Pl. 20; *Exsicc.* Fasc. i. 6.—Rocky mountain glens. P. VII.—IX. S.

22. *H. submurórum* (Lindeb.); st. 10—18 in. high usually simple setose hairy and sparsely floccose above, rt.-l. few, outer broadly ovate truncate coarsely toothed towards the base almost entire towards the blunt rounded apex, inner narrower lanceolate acuminate unequal at the base decurrent with few large irregular teeth, light green, glabrous above hairy beneath; st.-l 1—2, upper bract-like slightly toothed floccose hairy, lower large petiolate ovate-lanceolate acuminate cuneiform decurrent irregularly toothed nearly entire towards apex almost glabrous above or hairy both surfaces and margin. Heads usually 2—3, buds very long cylindric, inv. campanulate, phyll.

unequal outer dark rather lax inner narrow acute much paler, somewhat pilose-tipped densely setose with few hairs. Fl. golden yellow, ciliate-tipped, styles dark.—*Mon. Brit. Hierac.* Pl. 21.—High mountains. Argyleshire, Perthshire. P. VII. VIII. S.

23. *H. hyparct'icum* (Almq.) *f*; st. 12—20 in. high, branched striate floccose throughout somewhat hairy setose above, ped. long arcuate densely floccose setose with few hairs; rt.-l. 3—7, outer elliptic oval apiculate nearly entire, inner narrow ovate or elliptic, finely toothed in lower part almost entire towards the apex decurrent into hairy stiff recurved petioles; pale yellowish green firm glabrous above very floccose with few hairs beneath. St.-l. 1 or none small narrow sessile floccose beneath, heads 2—5, buds cylindric, inv. long *gradually narrowed into the peduncle* dark, outer phyll. short rather blunt, inner paler narrow gradually attenuate subacute, hairy setose the margins and tips very floccose. Fl. pilose-tipped, styles almost pure yellow.—*Mon. Brit. Hierac.* Pl. 22.—Ben More of Assynt, Sutherland. P. VII. S.

Amplexicaulia. Plant yellowish green, rarely glaucous, viscid glandulose. Rt.-l. rosulate persistent until after flowering; st. with few large l., all clothed with yellow-headed setæ. Phyll. acuminate setose. Fl. ciliate.

[*H. amplexicaule* (L.); yellowish green, st. 6—18 in. high, paniculately and corymbosely branched floccose setose with few white hairs towards the base, *st.-l.* 1—6 *large,* lower oblanceolate or oblong, upper and bracts at base of corymbs broadly ovate acuminate *amplexicaul with large rounded auricles,* rt.-l. rather rigid oblanceolate gradually decurrent into short petioles subobtuse coarsely dentate; inv. ovate at the base, phyll. numerous acuminate lax densely setose with senescent tips. Fl. pale yellow, very ciliate-tipped, styles yellow. The whole plant densely clothed with short yellow-headed setæ. *Exsicc.* Fasc. i. 7.—Naturalized on old walls and rocks at Oxford, Hawes, Cleish Castle Kinross, and other places. P. VII. VIII. E. S.]

Cerinthoidea. Plant glaucous not glandular with simple or denticulate hairs. Rt.-l. rosulate persistent until after flowering; st. with few leaves. Phyll. with simple or gland-tipped hairs. Heads large. Fl. ciliate.

24. *H. callistophyl'lum* (F. J. Hanb.); st. erect 1—2 ft. high often much branched hairy somewhat floccose, ped. spreading hairy setose densely floccose, *rt.-l.* outer *almost balloon-shaped*

slightly apiculate almost entire at the apex decurrent into long shaggy petiole, *with coarse outward-pointing teeth*; inner more acute *sharply toothed*, all light green glaucous often purplish nearly glabrous above ciliate margins hairy beneath, st.-l. usually 1 sessile acute sharply toothed hairy floccose beneath. Heads 2—6 large, buds short cylindric, inv. oblong campanulate, phyll. dark, outer short rather blunt, inner long narrow, *all ciliate-tipped with few setæ and many black-based hairs with very long fine white tips*, sparingly floccose. Fl. bright yellow faintly pilose or glabrous, styles almost pure yellow.—*Mon. Brit. Hierac.* Pl. 23; *Exsicc.* Fasc. iii. 56.—*β. cremnanthes* (F. J. Hanb.); rt.-l. narrower more acute with *very long acute triangular frequently hooked teeth* entire or nearly so towards the apex, heads few, phyll. hairy setose *very floccose* with less ciliate tips. *Exsicc.* Fasc. v. 103.—*γ. glandulosum* (F. J. Hanb.); ped. *very setose*, inv. floccose with *many long yellow-headed setæ and few black-based hairs*, heads few with *frequently an adnate tendency*.—*a.* Rocky margins of high mountain streams in Central Scotland. *β.* Aberdeenshire, Argyleshire, Inverness-shire, Yorkshire. *γ.* Argyleshire, Perthshire. P. VII. VIII. S.

25. *H. ang'licum* (Fr.); glaucous green, rt.-l. ovate-lanceolate apiculate or acuminate denticulate with *long* shaggy winged *petioles*, st. with few l. and few heads, *st.-l. clasping* ovate acuminate, inv. ventricose hairy setose, phyll. acuminate, *styles livid*.—*Sy. E. B.* 836; *Mon. Brit. Hierac.* Pl. 24; *Exsicc.* Fasc. iii. 57; *H. cerinthoides* (Backh.).—St. 12—18 in. high, branching in cultivation. Original rt.-l. roundish. Rhizome short. Fl. pale yellow. St. with *long arcuate* ascending *lateral ped.*, st.-l. usually solitary sessile scarcely clasping, phyll. all adpressed.—*β. acutifolium* (Backh.) *very glaucous*, st. much branched, l. *very acute* or acuminate *sharply and coarsely toothed* nearly glabrous broadly winged, inner phyll. acute attenuate. *Exsicc.* Fasc. iii. 58.—*γ. jaculifolium* (F. J. Hanb.); l. often *javelin-shaped*, long petioled, phyll. shorter floccose margined with senescent tips, *fl. undeveloped*, styles prominent. *Exsicc.* Fasc. v. 104.—*δ. longibracteatum* (F. J. Hanb.); l. blue-green extremely glaucous firm glabrous above slightly hairy and floccose beneath, *phyll. extraordinarily attenuate very floccose*. *Exsicc.* Fasc. ii. 31.—*ε. cerinthiforme* (Backh.); st. usually 15—24 in. high, st.-l. usually 2, lower oblong-lanceolate acuminate *constricted below the middle and enlarged again into a round amplexicaul base* denticulate, upper usually small entire not so clasping, rt.-l. broadly ovate to ovate-lanceolate acute, phyll. slender acute. *Exsicc.* Fasc. iv. 80.— . *Hartii* (F. J. Hanb.); rt.-l. obovate not persistent,

st.-l. *3—5* large ovate or obovate acute entire or minutely and acutely toothed sessile strongly amplexicaul, phyll. broad rather obtuse lax.—η. ×*hypochæroides* (Gibs.); st. red erect branched *each branch bearing* 2 *heads* floccose, ped. very floccose, rt.-l. ovate subacute petiolate firm fleshy *beautifully spotted purpled and floccose beneath,* st.-l. usually 1 *rather large broadly clasping.* Heads very truncate, phyll. short broad sub-obtuse, fl. *golden yellow* subglabrous or slightly pilose before expansion, *styles pure yellow.*—Mountains. *a.* Teesdale. Craven. Lake District. Scotland. Ireland. *β.* Cairn Toul, Aberdeen Links, and other localities in Scotland. Co. Donegal and Antrim in Ireland. *γ.* Lake District. *δ.* Abundant in Sutherlandshire; Skye. Co. Antrim. *ε.* Widely distributed over the British Isles. *ζ.* Slieve League, Co. Donegal. *η.* Limestone scars near Settle, Yorkshire. P. VII. VIII. E. S. I.

26. *H. iricum* (Fr.); glaucous green, rt.-l. ovate or oblong-lanceolate acute with *short* shaggy winged *petioles,* st. leafy corymbose at top, *st.-l. clasping* broadly ovate taper-pointed denticulate or dentate, lower ones narrowed below, ped. and inv. hairy setose floccose, inv. truncate below, phyll. blunt, *styles livid.*—*E. B. S.* 2915; *Exsicc.* Fasc. iii. 59; *H. Lapeyrousii* Bab.—St. 1—3 ft. high, very leafy rigid not branching 1-headed on barren soil. Pet. rigid diverging. Original rt.-l. blunt. L. often purplish at the end. St.-l. decreasing successively upwards, upper l. broad and rounded below. Phyll. broad, narrowed upwards.—Chiefly in mountain districts. P. VII. VIII. E. S. I.

27. *H. flocculósum* (Backh.); ashy-green, rt.-l. ovate or oval blunt or acutish with small teeth below *floccose on both sides* narrowed into long petioles, *st.-l. few large* ½-clasping, *st.* sub-corymbose *floccose* throughout, ped. long straight-based, inv. ovate below floccose setose hairy, phyll. acuminate incumbent, styles rather livid.—*Sy. E. B.* 848; *Exsicc.* Fasc. iv. 81; *H. stelligerum* (Backh.) (not Froel.).—St. 1½—2 ft. high. Fl. bright yellow. L. rosulate, persistent; original spathulate; innermost acutish. Petioles rather slender and woolly. St.-l. stalked; uppermost sessile.—Margins of high Alpine streams. P. VII. VIII. E. S. I.

28. *H. breadál'banense* (F. J. Hanb.); st. 12—18 in. high, simple or branched many-headed, singularly devoid of setæ, ped. thick densely floccose, rt.-l. yellow-green paler beneath, rather thick roughly hairy on both sides or almost glabrous, outer oval apiculate almost entire, inner ovate subacute denticulate, all abruptly narrowed to short shaggy petiole, st.-l.

usually 1, large shortly-stalked acute and sharply toothed, sometimes another bract-like. Heads 3—9, inv. rounded at the base constricted above dark green hoary with white hairs and floccose down, with very few setæ, phyll. rather broad moderately acute. Fl. slightly pilose behind the tips, styles pure yellow.—*Exsicc.* Fasc. v. 105.—Frequent in the Breadalbane Mountains. P. VII. VIII. S.

29. *H. lang'wellense* (F. J. Hanb.); st. 1—2 ft. high, branched, many-headed hairy floccose, peds. arcuate slender hairy floccose setose, rt.-l. spreading, outer ovate or obovate rounded at apex almost entire, inner broadly lanceolate acute with large sharp teeth in the lower half; decurrent into long shaggy winged petioles, nearly glabrous above with white hairs beneath firm rather pale green. St.-l. usually 1 large sometimes another high up bract-like; lanceolate acuminate somewhat coarsely toothed towards the base glabrous above sparsely floccose and hairy beneath. Inv. dark conical in fruit floccose setose hairy, outer phyll. short lax rather blunt, inner longer more attenuate sub-acute, floccose-margined ciliate-tipped, fl. few, bright yellow ciliate below scarcely ciliate at the tips, styles fuliginous.—*Exsicc.* Fasc. iii. 60.—Mountain gorges and cliffs. P. VII. VIII. S.

30. *H. li'ma* (F. J. Hanb); glaucous, st. 8—14 in. high, many-headed branched or simple floccose hairy, ped. arcuate hairy setose floccose, rt.-l. glaucous green purplish below ovate nearly entire or coarsely and irregularly toothed near the base abruptly narrowed into long shaggy petioles, inner more lanceolate acute toothed, felted on both sides with stiff curved hairs, st.-l. sometimes absent or bract-like or 1 large lanceolate acute sharply toothed with few hairs on both sides. Inv. grey densely hairy setose sparsely floccose, truncate, phyll. long slender acute porrect in bud, outer shorter more obtuse, fl. orange-yellow, glabrous, styles pure yellow.—*Exsicc.* Fasc. iv. 82.—β. *Brigantum* (F. J. Hanb.); plant more robust, heads fewer in number singularly globose and truncate at the base shaggy with grey-tipped hairs, l. dark green acute gradually decurrent into short hairy petioles, the bulbous-based hairs on the l. less rigid. Fl. pilose.—Limestone Cliffs. *a.* Cheddar. Great Orme's Head. β. Settle. P. V.—VIII. E.

31. *H. cloven'se* (Linton); st. 8—16 in. high, subglabrous above floccose below, ped. straight or arcuate floccose setose not hairy, rt.-l. ovate to ovate-acuminate dentate often with large spreading teeth near the base the lowest sometimes reflexed, hairy on both surfaces purple-blotched, st.-l. absent or lanceolate short-

stalked, entire or dentate. Heads in a lax irregular corymb, inv. dark green velvety, phyll. broad-based, attenuate acute floccose at the base, porrect in bud. Fl. orange-yellow, glabrous-tipped, styles usually pure yellow.—*Exsicc.* Fasc. i 8. Highland mountains. P. VII. VIII. S.

[*H. villósum* (L.), specimens marked from Clova district are found in several Herbaria, notably in the Royal Herb. Kew, the Nat. Hist. Mus., Barras Bridge Newcastle-on-Tyne, and that of Sir J. E. Smith. Some of the plants are correctly named, but others are *H. alpinum*, *H. eximium*, or *H. senescens*. It is extremely probable that cultivated specimens of *H. villosum* were circulated by mistake, instead of some of these very hairy alpine forms that abound in the Clova district. Nothing approaching *H. villosum* L. has been found during the past fifty years.]

Oreadea. Plant glaucous, st. with long simple rarely denticulate hairs, rt.-l. not persistent green bluish green or paler beneath margins very ciliate, inv. large broad hairy floccose with many minute yellow-headed setæ, and few dark larger setæ sometimes. Styles yellow. Fl. ciliate or glabrous.

32. *H. Griffith'ii* (F. J. Hanb.); st. 6—12 in. high, simple or branched furrowed hairy floccose, ped. spreading floccose minutely setose with few hairs, rt.-l. rosulate ovate-lanceolate to lanceolate acute with very sharp teeth in the lower half gradually narrowing into shaggy petioles, hairy on both sides slightly spotted floccose beneath with very ciliate margins, st.-l. 1—3 long lanceolate or linear-lanceolate very acute with long narrow sharp teeth sessile hairy. Heads 1—8, inv. rather dark, hairy minutely setose somewhat floccose, phyll. long narrow acute inner pale-margined. Fl. golden yellow subglabrous at the tips, styles slightly livid.—*Exsicc.* Fasc. vi. 127.—Nant Francon, Carnarvonshire. P. VII. VIII. E.

33. *H. Leyi* (F. J. Hanb.); st. reddish 6—15 in. high almost glabrous or scabrid striate, ped. green sparsely floccose setose with few hairs arcuate thickened upwards 1—5-headed, rt.-l. dark bluish-green often with black spots or clouded with purple, outer ovate or ovate-lanceolate sub-entire, inner lanceolate acute petiolate with small acute teeth in the upper part large teeth towards the base thin but firm coriaceous glabrous above with soft white hairs beneath, st.-l. 1 or absent shortly stalked lanceolate acute glabrous above scattered hairs beneath, inv. green warted setose sparsely floccose minutely setose with

short black-based hairs, phyll. very long attenuate subacute outer short rather lax. Fl. bright yellow almost glabrous-tipped, styles darkened.—*Exsicc.* Fasc. vi. 128, 129.—Mountain cliffs. P. VI.—VIII. E. S.

34. *H. Careno'rum* (F. J. Hanb.); st. 10—12 in. high 1—3-headed floccose with few scattered white hairs, ped. floccose minutely setose with scattered hairs, rt.-l. ovate-lanceolate decurrent into longish shaggy petioles acutely dentate sparsely hairy, st.-l. 1 or absent lanceolate acute slightly toothed sessile almost glabrous above floccose beneath margin and midrib ciliate, inv. narrow, phyll. pale sub-glaucous green with lighter margins (tips purplish in cultivation) outer appressed small floccose margined all with black-based hairs and minute setæ. Buds long narrow. Fl. yellow glabrous-tipped. Styles yellow. —*Exsicc.* Fasc. iv. 83.—Ben Hope. P. VII. VIII. S.

35. *H. Schmidt'ii* (Tausch); cæsius green, st. 5—12 in. high simple or branched 1—4-headed fistulose, ped. minutely setose floccose with long white hairs with short black bases, rt.-l. bluish green, outer ovate apiculate almost entire inner ovate-lanceolate or lanceolate-acute toothed decurrent into long shaggy petioles hairy above sometimes glabrous very hairy beneath and on margins, st.-l. 1 or absent almost sessile linear bract-like or lanceolate acute entire hairy slightly floccose. Inv. ovate at the base hairy slightly floccose with minute setæ, phyll. long narrow sub-acute somewhat pilose-tipped and floccose-margined outer lax. Fl. bright yellow glabrous or sub-ciliate, styles pure yellow.—*Exsicc.* Fasc. i. 9; v. 106.—*β. crinigerum* (Fr.); glaucous green, l. hairy both sides, st.-l. 1—2 sub-petiolate, head showy, inv. with simple hairs eglandular. Fl. glabrous.—*γ. eustomon* (Linton); st. more solid, l. ovate-acuminate or ovate-lanceolate rather fleshy very glaucous glabrous above. Fl. large lemon-yellow, slightly pilose-tipped.—*Exsicc.* Fasc. vi. 130.—*δ. devoniense* (F. J. Hanb.); 12—14 in. high scarcely floccose with few setæ, st.-l. large toothed narrowing into semi-amplexicaul petiole, rt.-l. broadly ovate glabrous above less hairy below than type. Fl. ciliate externally glabrous-tipped.—*Exsicc.* Fasc. iv. 84.—Mountain glens. γ. Penard Castle, Glamorgan. δ. Countisbury, North Devon. P. VII. VIII. E. S. I.

36. *H. lasiophyl'lum* (Koch); glaucous, st. 6—16 in. high, simple or branched 1—6-headed hairy, ped. densely floccose setose slender long rather arcuate, rt.-l. *broadly oval* or ovate-lanceolate entire or remotely denticulate apiculate innermost

acute original subrotund, coarsely hairy on both sides and on the margins or destitute of hairs above, floccose with shaggy petioles dilated at the base; st.-l. lanceolate very acute narrowed into short petiole almost entire hairy floccose. Inv. subglobose slightly hairy floccose minutely setose, phyll. attenuate rather acuminate pilose-tipped outer subobtuse. Fl. glabrous-tipped, styles yellow.—*Exsicc.* Fasc. iii. 61.—*β. planifolium* (F. J. Hanb.); ped. short erect, heads small, *rt.-l. very broad flat ovate very truncate-based* softly hairy with long shaggy petioles.—*Exsicc.* Fasc. v. 107.—*γ. euryodon* (F. J. Hanb.); rt.-l. long narrow subentire towards the apex, cut towards the base into *extraordinarily broad triangular irregular teeth*, almost glabrous above slightly hairy and floccose beneath, phyll. floccose-margined not pilose-tipped, st.-l. large coarsely toothed glabrous above.—Mountain rocks. *γ.* Herefordshire; Gloucestershire. P. V.—VIII. E. S. I.

37. *H. far'rense* (F. J. Hanb.); dark green, st. 1—2 ft. high, slender few-headed few-l. with scattered hairs throughout, ped. long subarcuate minutely setose, densely floccose with few hairs, rt.-l. grass-green paler beneath *lanceolate* acute decurrent into long winged shaggy petiole long almost entire or only denticulate hairy on both surfaces frequently purpled, st.-l. 2—3 upper small sessile lower large lanceolate acute or subacute petiolate floccose hairy on both sides. Inv. ventricose rounded minutely setose *densely floccose* with very few hairs, outer and middle phyll. dark floccose-margined, inner paler less floccose, subacute pilose-tipped. Fl. orange glabrous-tipped, styles olive-yellow. —*Exsicc.* Fasc. i. 10.—Rocks. P. VII. VIII. E. S. I.

38. *H. eustáles* (Linton); st. 12—18 in. high somewhat hairy floccose above branching little, st.-l. usually 1 sometimes another bract-like linear-lanceolate high up, rt.-l. light green narrow ovate-oblong acute long-petioled decurrent thinly hairy below glabrescent or hairy above, margin slightly crenate denticulate or subentire, st.-l. similar to rt.-l. petiolate denticulate. Heads few, ped. straight long very floccose with few simple hairs setose. Inv. very floccose, with short black-based hairs large and small setæ, phyll. narrowly acuminate markedly pilose-tipped floccose-margined. Fl. pilose-tipped, styles livid. —Breadalbane and Glen Shee. P. VII. VIII. S.

39. *H. prox'imum* (F. J. Hanb.); st. 1—2 ft. high simple or branched hairy floccose, rt.-l. yellowish green often purpled with pink round the margins thick leathery rough on both sides with bulbous-based hairs petiolate sinuate ovate or ovate-lanceolate almost entire, st.-l. 2—4 rapidly decreasing in size

upwards, shortly petioled or sessile lanceolate acute or subacute irregularly toothed. Inv. floccose hairy, phyll. broad obtuse adpressed margins densely floccose hairy with few setæ. Fl. orange-yellow glabrous-tipped, styles dusky. — *Exsicc.* Fasc. i. 11.—Sandy links and cliffs. P. VII. VIII. S. I.

40. *H. caledon'icum* (F. J. Hanb.); st. 6—14 in. high, erect simple or branched hairy somewhat floccose, ped. long slender subarcuate floccose minutely setose with few hairs, rt.-l. large grass-green ovate to ovate-lanceolate acute or apiculate outer rounded at apex, decurrent into broadly winged hairy petioles coriaceous glabrous above softly hairy beneath and on margin with long glandular teeth, st.-l. 1—2 large lanceolate acute upper sessile or semi-amplexicaul, lower decurrent into hairy winged petiole toothed towards the base hairy beneath glabrous above. Heads 2—4 or more, inv. dark green truncate hairy minutely setose, outer phyll. dark blunt pilose-tipped floccose-margined inner pale almost naked. Fl. large rather orange glabrous-tipped, styles olive-brown.—*Exsicc.* Fasc. iv. 85; vi. 131.—*β. platyphyllum* (Ley); st. 9—24 in. high, branched, ped. long erect hairy setose very floccose, rt.-l. broadly ovate to broadly ovate-lanceolate broad-based with irregular coarse somewhat sagittate teeth thick firm floccose beneath, phyll. strongly floccose-margined incurved in bud. Fl. mostly stylose, styles dark olive-green.—*H. pollinarium*, var. *platyphyllum* (Ley). — *Exsicc.* Fasc. iv. 90.—*a.* Sandy cliffs, Scotland. *β.* Mountain rocks, South Wales. P. VI.—VIII. E. S.

41. *H. rubicun'dum* (F. J. Hanb.); st. 10—20 in. high robust erect simple or branched slightly floccose hairy, ped. long straight thickened upwards flattened floccose hairy setose, rt.-l. bluish-green much purple-blotched on both sides broad lanceolate apiculate gradually narrowing into hairy petiole entire or denticulate sometimes coarsely dentate coriaceous glabrous above softly hairy beneath, st.-l. large 1—2 broadly lanceolate dentate apiculate slightly stalked or sessile glabrous above hairy beneath upper bract-like. Heads 1–6 large showy, inv. rounded in fl. truncate in fr. hairy minutely setose floccose, phyll. broad adpressed rather blunt outer dark inner much paler, fl. yellow glabrous or sparingly ciliate, styles slightly livid.—*Exsicc.* Fasc. i. 32.—*β. Boswelli* (Linton); st. 6—16 in. high, *sinuous*, rt.-l. *thin* ovate-oblong or narrow ovate acuminate the margins ciliate waved, st.-l. 1 or absent ovate acuminate or lanceolate shortly petioled, inv. rather globose, phyll. broad subulate very obtuse floccose hairy with few setæ. Fl. glabrous, styles livid yellow.—*Exsicc.* Fasc. vi. 132.—Rocks and mountains. P. VI.-VIII. E. S. I.

42. *H. Oreades* (Fr.) var. *subglabratum* (F. J. Hanb.); *intensely glaucous*, st. 12—18 in. high, simple or branched *nearly glabrous*, ped. floccose setose with few hairs, rt.-l. oblong or ovate-lanceolate apiculate denticulate or almost entire narrowly decurrent into long petioles glabrous above ciliate beneath and on the margins, st.-l. 1—3 large lanceolate acute denticulate or almost entire decreasing in size upwards narrowing into winged semi-amplexicaul petiole or sessile glabrous or slightly floccose with very few hairs. Heads 2—5, inv. truncate slightly floccose minutely setose with few hairs, phyll. linear attenuate, outer and middle adpressed obtuse, inner acuminate incumbent white-tipped. Fl. yellow ciliate-tipped, styles yellow.—*Exsicc.* Fasc. ii. 33.—Rocks by the sea, North of Scotland. P. VII. S.

43. *H. argenteum* (Fr.); *very glaucous, rt.-l.* lanceolate acute *denticulate towards the middle* or entire glabrous above with short petioles, st. simple or branched with few sessile or stalked l., branches long straight, inv. broadly ventricose becoming subtruncate below slightly setose hairy and floccose, *phyll. blunt* adpressed, styles yellow.—*Sy. E. B.* 843; *Exsicc.* Fasc. i. 13; *H. pallidum β. persicifolium* Bab.—St. 1—2 ft. high. Buds cylindrical. Inner phyll. subacute pale-edged. Fl. bright yellow. L. with scattered hairs beneath and at the edges. Petioles usually short. Original l. bluntish. Upper st.-l. small narrow. β. *septentrionale* (F. J. Hanb.); inv. truncate more hairy and setose, rt.-l. more erect often broadly lanceolate strongly toothed.—*Exsicc.* Fasc. vi. 133.—Mountain districts of England, Scotland, and Wales. β. Rocks by the coast and stream-sides, Sutherlandshire. P. VII. VIII. E. S.

44. *H. nitidum* (Backh.); *dark green, rt.-l.* lanceolate acute *coarsely and sharply toothed* glabrous above with short petioles, st.-l. 1 or 0, st. few-headed, ped. scaly, inv. ventricose setose and hairy slightly floccose, phyll. acuminate or bluntish, styles yellow.—*Sy. E. B.* 844; *Exsicc.* Fasc. iv. 86.—St. 15—24 in. high, nearly hairless. Heads large; fl. bright yellow. L. slightly hairy beneath; teeth very large not directed downwards. Petioles usually short. Original l. roundish.—β. *siluriense* (F. J. Hanb.); ped. slender *very floccose* hairy hardly setose, inv. *less setose but more hairy* than type, outer phyll. sub-obtuse, inner long attenuate acute. Fl. glabrous.—By mountain streams. β. Cwm Tarrell, Breconshire. P. VII. VIII. E. S.

45. *H. Sommerfelt'ii* (Lindeb.); st. 8—12 in. high, erect slender simple or branched almost glabrous, ped. slightly floccose setose and hairy, rt.-l. green purple-blotched, outer

ovate, inner oblong-lanceolate apiculate almost entire or with sharp glandular teeth glabrous above hairy and floccose beneath with ciliate margins, decurrent into long petioles, st.-l. 1 or 0, sessile or shortly petiolate linear-lanceolate acute entire or slightly denticulate. Inv. truncate dark slightly floccose hairy setose, phyll. outer short broad subobtuse ciliate-margined, inner attenuate subacute reflexed almost naked ciliate-tipped. Fl. yellow glabrous-tipped, styles slightly livid.—*Exsicc.* Fasc. ii. 34.—*β. tactum* (F. J. Hanb.); l. deeply blotched, with very long acute hooked forward-pointing teeth; st.-l. large toothed, phyll. few very broad obtuse minutely setose with ciliate margins. Fl. slightly pilose.—*Exsicc.* Fasc. v. 108.—*γ. splendens* (F. J. Hanb.); much stronger and more hairy than type; st. hairy floccose, ped. long spreading very setose hairy floccose, rt.-l. bluish green much blotched coriaceous deeply toothed glaucous almost glabrous above hairy somewhat floccose beneath, st.-l. large deeply toothed with very ciliate-margins. Fl. pilose-tipped, styles rather livid.—*Exsicc.* Fasc. vi. 134.—Mountains. *β.* Scotland. *γ.* Wales. P. VII. VIII. E. S. I.

46. *H. scot'icum* (F. J. Hanb.); st. 1—2 ft high, erect simple or branched hairy floccose, ped. somewhat arcuate hairy with few setæ, rt.-l. large coriaceous tinged with purple ovate to ovate-lanceolate apiculate hairy beneath and at the margins glabrous above almost entire or irregularly toothed, st.-l. 1—6 large decreasing upwards broadly lanceolate acuminate hairy beneath subglabrous above denticulate or coarsely dentate, lower petioled, upper sessile or clasping linear entire. Heads 1—10, inv. truncate finely setose hairy, phyll. broad obtuse outer short floccose-margined, inner pale more attenuate ciliate-tipped. Fl. orange-yellow glabrous-tipped.—*Exsicc.* Fasc. i. 14.—*β. occidentale* (F. J. Hanb.); st. 1—3-headed, st.-l. usually 2 shortly stalked, rt.-l. few broadly ovate subacute very wide towards the truncate base abruptly narrowed to long petiole almost entire. L. bright green glabrous above rather glaucous with few long hairs beneath, styles pure yellow.—Mountain and sea cliffs. *β.* Banks of Carrick River, Co. Donegal. P. VII. VIII. S. I.

47. *H. onosmoïdes* (Fr.), var. *buglossoïdes* (Arv. Touv.); green, st. 1½—4 ft. high, furrowed branched hairy at the base, ped. slender striate floccose with hairs and setæ much thickened upwards bearing scaly bracts, rt.-l. yellow-green lighter beneath glaucous firm hairy both surfaces lanceolate decurrent into long winged petioles toothed veins prominent below, st.-l. 3—6 decreasing upwards firm glaucous almost or quite glabrous

above, sessile or shortly stalked lanceolate apiculate coarsely and sharply toothed. Inv. very truncate and angular, outer phyll. commencing near thickened part of ped., phyll. adpressed subacute or bluntish very slightly floccose with few hairs and setæ. Fl. large pale yellow glabrous-tipped, styles almost pure yellow.—*Exsicc.* Fasc. iv. 87.—Sandhills and railway banks in East Ross and East Sutherland. P. VII. S.

48. *H. saxifrágum* (Fr.) *β. pseudonosmoïdes* (Dahlst.); glaucous, st. 8—14 in. high 1—4-headed hairy below floccose, ped. floccose with few hairs and setæ, rt.-l. ovate-lanceolate or lanceolate apiculate irregularly dentate or almost entire with long winged petioles roughly hairy on both sides, st.-l. 1—3 decreasing upwards lanceolate almost entire towards the apex somewhat coarsely dentate towards the base slightly hairy on both surfaces sessile or with winged petiole. Inv. sub-cylindric hairy setose slightly floccose, phyll. outer short subacute inner long narrow acute. *Fl. olive-yellow almost always stylose* glabrous-tipped, styles yellow.—*Exsicc.* Fasc. i. 12.—*γ. orimeles* (F. J. Hanb.); green slightly glaucous, st. erect wiry purple below scabrid 12—16 in. high, hairy slightly floccose, ped. spreading thickened upwards densely floccose setose with some hairs bracteate, rt.-l. firm rigid glaucous somewhat hairy both surfaces ovate to ovate-lanceolate denticulate or almost entire with long petiole, st.-l. ovate-lanceolate to lanceolate often dentate entire towards apex sessile or short-stalked semi-amplexicaul almost glabrous above hairy sparsely floccose beneath. Inv. truncate hairy floccose some setæ, phyll. deep green broad subobtuse porrect in bud. Fl. yellow ciliate, styles yellow or slightly livid.—*Exsicc.* Fasc. vi. 135.—*β.* Exposed banks and cliffs, Scotland. *γ.* Mountain crags, Wales, Scotland. P. VII. VIII. E. S.

49. *H. hiber'nicum* (F. J. Hanb.); st. 12—22 in. high, reddish purple green above, simple 1—3-headed recurving hairy floccose, ped. thickened upwards bracteate very floccose. rt-l. ovate-lanceolate obtuse not persistent entire glabrous above sparingly hairy below, st.-l. 5—9 decreasing upwards ovate-lanceolate to linear-lanceolate acute sometimes sharply toothed in the lower half, grey-green almost glabrous above slightly hairy and floccose beneath. Inv. truncate angular humpy, phyll. all rather broad obtuse floccose minutely setose with few short hairs, outer very small lax extending into the ped. Fl. yellow glabrous, styles slightly livid.— Cliffs of the Mourne Mountains, Co. Down; Moynalt, Co. Donegal. P. VII. I.

Vulgata. Plant green or glaucous with soft simple eglandular hairs. St. few or many-leaved. Phyll. irregularly imbricated; with short hairs and floccose down, with or without setæ. Styles yellow-livid or dusky. Fl. glabrous or sub-ciliate.

* *Scapigera.* St. scapiform, bare, or 1—2-leaved.

50. *H. stenolèpis* (Lindeb.); st. 6—14 in. high simple or branched furrowed almost glabrous hairy towards the base, ped. long ascending bracteate setose densely floccose hardly hairy, rt.-l. cæsius-green purpled beneath, ovate or oblong-lanceolate obtuse, or lanceolate acute or acuminate unequally broadly based sometimes truncate or cuneate or cordate almost entire towards the apex, irregularly and deeply dentate the teeth descending into the long floccose very hairy petioles, glabrous above somewhat floccose beneath, st.-l. 0—1 linear or lanceolate acute long-stalked entire or sharply dentate. Heads small 1—5, inv. sub-truncate floccose setose hairy, phyll. long narrow acute dark green, inner paler. Fl. bright yellow pilose externally, sub-ciliate at the tips, styles yellow or very slightly livid.—*Exsicc.* Fasc. vi. 136.—*β. anguinum* (W. R. Linton); rt.-l. rosulate nearly erect ovate-oblong denticulate or almost entire with few large teeth, subglabrous, inv. darker less hairy setose and floccose than type, long snake-like in bud, styles livid.—*Exsicc.* Fasc. i. 15.—Mountains and cliffs. *β.* High hills above Moffat, Scotland. P. VI. VII. E. S. I.

51. *H. hypochæroides* (Gibs.); glaucous green, rt.-l. ovate blunt apiculate rather truncate or cordate below denticulate, petioles slender, st. *leafless* rigid simple or forked with *straight-based ped., inv. truncate below* floccose hairy setose, *phyll. broad and blunt* incumbent, styles yellow.—*Sy. E. B.* 842; *Exsicc.* Fasc. ii. 35; *H. Gibsoni* Backh.—St. 6—18 in. high. Fl. bright yellow. Phyll. not cuspidate. L. rosulate, persistent, very broad, blotched with purple above.—*β. saxorum* (F. J. Hanb.); *rt.-l. narrower more acute* paler green purpled when young losing this when older, ped. longer less straight and rigid, *phyll.* porrect in bud, *more acute* darker less white-margined.—*γ. Cyathis* (Ley.); rt.-l. green or slightly blotched, original obovate retuse at the tip decurrent into broadly winged hairy petioles hairy beneath and on the margins, inner long oval or narrow elliptic acute often toothed, *bud very short forming from the first an open cup*, phyll. *recurved in bud*, outer lax.—*Exsicc.* Fasc. v. 109.—Mountains. *α.* England; Wales; Ireland. *β.* Red sandstone rocks, Breconshire, Montgomeryshire. *γ.* Limestone rocks near Merthyr Tydfil, Breconshire. P. VII. VIII. E. I.

52. *H. aggregatum* (Backh.); deep green, rt.-l. ovate blunt coarsely dentate below floccose beneath when young, st.-l. 1 lanceolate narrowed at both ends, *st. corymbose* and floccose at the top, *ped. aggregate*, inv. subtruncate below loosely floccose setose nearly hairless, *phyll. blunt.*—*Sy. E. B.* 845; *Exsicc.* Fasc. iii. 62; *H. bifidum* Koch?—St. 12—20 in. high, usually simple. Ped. forming a close corymb as in *H. umbellatum*. Heads many. Buds cylindrical. Fl. bright yellow or orange. Styles yellow. L. rosulate, persistent; original l. nearly round; innermost acutish. Petioles rather winged, slightly hairy. St.-l. subsessile. Phyll. adpressed.—*β. prolongatum* (F. J. Hanb.); *l. longer* ovate-lanceolate acute not floccose beneath, *ped. much elongated, inv. longer* and more hairy.—*Exsicc.* Fasc. v. 110.—By rocky streams in the Highlands. *β.* Glen Lochay, Perthshire. P. VII. VIII. S.

53. *H. Pictórum* (Linton); st. 1½—2 ft. high, subglabrous 1—4-headed, ped. somewhat floccose setose with few hairs, rt.-l. pale green glabrous above subglabrous paler and turning purple beneath, *nerves very conspicuous*, ovate or ovate-acuminate denticulate often sharply dentate at the base, suddenly narrowed into slightly winged hairy petioles, st.-l. 1 or absent petiolate lanceolate acuminate dentate, floccose beneath glabrous above. Inv. ovoid ventricose hairy setose somewhat floccose, phyll. adpressed, outer dark green subulate, inner paler with purpled and senescent tips, acuminate. Fl. gamboge-yellow glabrous upwards, styles olive-yellow.—*Exsicc.* Fasc. iv. 88.—*β. dasythrix* (Linton); l. dull green hairy beneath with densely ciliate margins, nerves inconspicuous, slightly and regularly dentate towards the base, inv. shaggy with white hairs.—*Exsicc.* Fasc. v. 111.—Central Highlands of Scotland. P. VII. VIII. S.

54. *H. britan'nicum* (F. J. Hanb.); st. 12—18 in. high, branched striate with scattered hairs, ped. long somewhat arcuate very floccose with few hairs setose, rt.-l. firm glaucous glabrous above hairy beneath and on the margins *much furrowed by deep parallel veining*, outer broadly ovate apiculate subentire near the apex, *very coarsely and irregularly toothed* towards the *remarkably truncate base*, inner narrower more acute less truncate more deeply toothed, *the teeth or appendages in l. often extending a long* way down the petiole, st.-l. large near the base, or bract-like at the point of branching. Heads 3—6, inv. truncate at the base ultimately conical, grey floccose with long hairs and few setæ, phyll. long narrow acute porrect in bud, inner almost naked, outer broad lax. Fl. bright yellow glabrous-tipped, styles yellow or slightly dusky.—*Exsicc.* Fasc. ii. 36.—Limestone dales and scars. Yorkshire, Derbyshire, Stafford. P. VII. VIII. E.

55. *H. riväle* (F. J. Hanb.); st. 12—20 in. high, branched floccose with few hairs, ped. long slender *densely floccose setose* scarcely hairy, rt.-l. bright green much paler beneath often purpled, outer oval apiculate, inner lanceolate acute all somewhat toothed especially towards the base, *very unequally based,* truncate or cuneate or almost cordate, abruptly narrowed into very long slender shaggy petioles, roughly hairy on both sides or glabrous above, st.-l. 1—2 bract-like linear lanceolate denticulate or entire, or if low on the st. long lanceolate acute long-stalked toothed, floccose beneath. Heads 4—10, inv. conical *grey with floccose down* setose with few hairs, phyll. long acute *floccose-margined.* Fl. large yellow pilose-tipped, styles yellow or dingy yellow.—*Exsicc.* Fasc. i. 16.—β. *subhirtum* (F. J. Hanb.); rt.-l. more or less dentate or bidentate, st. floccose with long spreading hairs, inv. dark hairy but little floccose and with few setæ, *phyll. ciliate-tipped not floccose-margined,* the whole plant much greyer than type.—*Exsicc.* Fasc. vi. 137.—Rocky streams. P. VI.—VIII. E. S.

56. *H. pollinárium* (F. J. Hanb.); st. 12—18 in. high, furrowed simple or branched, heads 2—10 corymbose closely aggregated, *ped. short thick* straight-based *mealy with floccose down* setose not hairy, rt.-l. grass green coriaceous nearly glabrous above softly hairy beneath fleshy brittle, boat-shaped elliptic, outer retuse or blunt, inner longer more acute, with few minute apiculate patent teeth, narrowing into a long slender winged petiole, st.-l. frequently small placed high linear bract-like floccose, or if low on the stem large lanceolate acuminate almost glabrous petiolate slightly clasping. Inv. truncate below very grey floccose setose not hairy, phyll. few short broad blunt floccose-margined ciliate-tipped. Fl. yellow glabrous or sub-pilose, styles light olive-brown.—*Exsicc.* Fasc. iv. 89.—North coast of Scotland. P. VII. S.

57. *H. murórum* (L. pt.); green, rt.-l. ovate often cordate and with large patent or descending teeth below, petioles shaggy, st.-l. 0 or 1 stalked and placed high, heads subcorymbose, *ped. arcuate* ascending, inv. thinly clothed floccose setose with few hairs, phyll. acuminate erect overtopping the glabrous-tipped buds, styles livid.—*Sy. E. B.* 846.—St. 12—18 in. high. Young heads cylindrical. L. usually thin. St.-l. ovate-acuminate, often rounded or cordate below. Styles rarely yellowish.—β. *pulcherrimum* (F. J. Hanb.); st. simple or branched corymbose or panicled, *ped. straight slender* hairy setose densely floccose, rt.-l. primary oval blunt, inner ovate-lanceolate obtuse nearly entire at apex dentate abruptly decurrent towards the base, shortly pilose on both sides, st.-l.

1—2 short-petioled. *Inv.* small dark *urceolate* phyll. long narrow subacute sparingly floccose *densely setose.* Heads 9—15. *Fl.* deep golden yellow *markedly pilose-tipped,* styles livid.—*Exsicc.* Fasc. vi. 138.—Limestone cliffs, Wales, Yorkshire.—γ. *micracladium* (Dahlst.); *ped. long spreading not corymbose* sparsely floccose, rt.-l. small, inner lanceolate acute usually dentate or with few irregular teeth at base, st.-l. *large* almost entire, *phyll.* few long narrow *acute* setose with few hairs, outer slightly floccose-margined. Fl. sub-ciliate, styles *very livid.*—*Exsicc.* Fasc. i. 17.—Rocky sides of mountain-streams in the British Isles.—δ. *camptopetalum* (F. J. Hanb.); ped. spreading *almost straight,* rt.-l. oval *sharply and finely dentate glabrous above,* hairy beneath, st.-l. *irregularly coarsely toothed.* Fl. dirty yellow, stylose glabrous-tipped, styles dirty yellow.—Above the falls of Allt-na-Caillich, Ben Hope, Sutherlandshire.—ε. *ciliatum* (Almq.); *ped. arcuate* densely floccose setose, rt.-l. large, outer elliptic denticulate or almost entire, inner lanceolate acute dentate with large irregular spreading teeth *subglabrous above hairy below and on margins,* st.-l. large broadly lanceolate acuminate coarsely toothed hairy *floccose beneath,* phyll. subacute *all ciliate-tipped and floccose-margined. Fl. pilose, styles yellow.*—Stream sides.—ζ. *pachyphyllum* (Purchas); rt.-l. outer very blunt *almost retuse,* inner more acute broadly ovate all *mucronate, cordate or hastate at base* with descending patent teeth, glaucous and glabrous above, under surface and margins hairy, *deeply stained with purple and purple marginal line,* st.-l. usually 0 or bract-like. Heads 3—10 broad blunt, phyll. outer *lax short* broad blunt, inner narrower subacute, woolly at the tips floccose-margined with black-based hairs and minute setæ. Fl. deep yellow pilose, styles rather dusky.—*Exsicc.* Fasc. iii. 64.—Limestone rocks in the Wye Valley.—η. *sagittatum* (Lindeb.); l. *sagittate at the base, hairy on both sides and margin,* inv. broad with rounded base minutely setose with short hairs, phyll. acute floccose-margined and tipped. Fl. ciliate, styles very dark.—Scotland.—θ. *subulatidens* (Dahlst.); l. ovate or ovate-lanceolate acute *with truncate base* toothed with large prominent teeth, yellowish green veins conspicuous on under surface. St.-l. 0—1 deeply coarsely toothed, peds. and phyll. setose floccose with few hairs, styles yellow.—*Exsicc.* Fasc. vi. 141. Forfar. Wales.—*sanguineum* (Ley); rt.-l. oval or elliptic nearly glabrous, *older turning blood-red beneath* unequally based finely denticulate, st. branched furrowed 0—1-leaved, ped. and phyll. floccose setose, the latter with many grey hairs. Heads large semi-globose.—*Exsicc.* Fasc. vi. 140.—Yorkshire. Breconshire.—Many other varieties have been distinguished that cannot be mentioned here.—Woods and rocks. P. VI.—VIII. E. S. I.

** *Caulescentia.* Stem usually long, more or less leafy.

58. *H. eúprepes* (F. J. Hanb.); st. 10—18 in. high, simple or with long straight branches hairy or bristly, ped. *very short straight* floccose sparingly hairy and setose, rt.-l. ovate-lanceolate obtuse or acute denticulate or almost entire (Scottish plants coarsely toothed) gradually narrowed into long petioles, grass-green lighter and purpled beneath, softly hairy on both sides, st.-l. usually 1 large lanceolate acute or acuminate denticulate petiolate hairy. Heads often numerous, *crowded very adnate,* inv. very truncate and conical dark green, phyll. few dark with lighter margins and purplish tips, adpressed rather blunt floccose-tipped when young hairy sparingly floccose and setose. Fl. pale yellow glabrous-tipped, styles faintly livid beneath.—*Exsicc.* Fasc. iii. 66.—*β. glabratum* (Linton); l. narrower often strongly dentate glabrous above, ped. and inv. less hairy setose and floccose.—*Exsicc.* Fasc. i. 19.—Mountain cliffs. P. VI.—VIII. E. S.

59. *H. orcaden'se* (W. R. Linton); st. 1—2 ft. high, hairy floccose, ped. thickened upwards almost straight panicled floccose hairy setose, rt.-l. rosulate outer broad ovate narrowing to petiole rounded or apiculate at the apex dentate or denticulate or almost entire, inner lanceolate acute dentate with medium-sized patent teeth hairy or subglabrous margins ciliate, st.-l. 1—4 with short winged petiole acutely dentate in the lower part, ovate-lanceolate nearly glabrous above. Heads 2—9 small, inv. small dark green velvety rounded at the base constricted above, hairy setose slightly floccose, phyll. few broad blunt, inner subacute with paler margins all adpressed with white tips and margins. Fl. deep golden yellow glabrous or faintly pilose before expansion, styles greenish yellow.—*Exsicc.* Fasc. vi. 142.—Cliffs at Hoy, Orkney. P. VII. VIII. S.

60. *H. rubiginósum* (F. J. Hanb.); st. 1—2 ft. high, simple or branched robust *purplish-red,* hairy below, ped. somewhat floccose setose hairy, rt.-l. numerous *large broadly ovate* subacuminate sharply irregularly toothed thickly hairy on both sides *blotched with rusty purple,* st.-l. 1—4, *subsessile or slightly clasping* ovate-lanceolate acute coarsely irregularly toothed hairy on both surfaces and margins. Heads 2 or many large, inv. large broad-based, phyll. broad subacuminate slightly floccose, hairy with few setæ. Fl. deep golden yellow glabrous-tipped, styles nearly pure yellow.—*Exsicc.* Fasc. iv. 91.—*H. vulgatum* (Fr.) var. *rubescens* Backh.—Limestone scars. Derbyshire, West Yorkshire, Westmoreland. P. VII.—IX. E.

61. *H.* [illegible] (Almq.); st. 1—2½ ft. high, simple or branched corymbose purpled at the base hairy floccose, ped. long spreading floccose setose, rt.-l. ovate lanceolate apiculate denticulate narrowing into petiole, st.-l. 2—5 large, broadly ovate-lanceolate to lanceolate acuminate with *large irregular spreading teeth* entire towards the apex, hairy or glabrous above hairy and floccose beneath, lower petiolate upper sessile, *purpled round the margins.* Heads 2—12, inv. dark, outer phyll. short broad subacute, inner long narrow acute setose scarcely floccose with few hairs. Fl. golden yellow, glabrous-tipped, styles livid.—*Exsicc.* Fasc. iii. 67.—River-banks in Wales. [illegible] E.

62. *H. cæsium* (Fr.); cæsius or dull green, rt.-l. ovate or lanceolate rounded or narrowed below irregularly dentate-serrate with patent or ascending teeth, petioles slender, st.-l. 0 or 1 placed low, st. few-headed with *straight-based ped., inv. rounded below* floccose hairy slightly setose, *phyll. bluntish,* inner ones acute incumbent, styles slightly livid.—*Exsicc.* Fasc. v. 115; *H. murorum Sm. E. B.* 2082; *Fl. Dan.* 2598.—St. 12—18 in. high. Young heads roundish. Fl. bright yellow. Phyll. not cuspidate; inner very slender, pointed. Rootstock long. L. nearly glabrous above, coriaceous, not fringed with coarse hairs; innermost acute. St.-l. not stalked, narrowed below.—β. *Smithii* (Baker); *l. and st. purpled,* outer rt.-l. oblong ovate denticulate with sheathing shaggy petioles, inner longer narrower acute or acuminate dentate sometimes coarsely towards the unequal bases, st.-l. wanting or bract-like, heads 3—4, phyll. linear acute hairy scarcely floccose with few setæ, ped. long arcuate.—γ. *coracinum* (Ley); l. yellow-green paler beneath, elliptic or elliptic-lanceolate acute or acuminate decurrent deeply toothed at the base. Heads 4—8 in compact corymb, ped. short spreading ascending densely floccose with few setæ, phyll. subobtuse densely hairy scarcely setose, outer few short. Fl. rather broad sub-glabrous, styles rather dark.—Mountains. β. Yorkshire; Derbyshire. γ. Brecon Beacons. F. VII. VIII. E. S. I.

63. *H.* [illegible] (F. J. Hanb.); st. 6—12 in. high, stiff erect, ped. setose *densely floccose,* rt.-l. few, outer oval or ovate blunt apiculate almost entire, inner ovate-lanceolate very acute *laterally curved from the apex to the base of long shaggy petiole,* dentate especially towards the base, green coriaceous, veins prominent, roughly hairy on both surfaces or subglabrous above, st.-l. 1 lanceolate acute dentate petiolate *curved like rt.-l.* Heads 1—2, inv. *cylindric-campanulate* hairy setose, phyll. adpressed with floccose tips and margins, inner sub-acute, *buds*

long cylindrical. Fl. deep golden yellow, glabrous-tipped, styles rather livid.—*β. petrocharis* (Linton); l. elliptic oblong, *denticulate. Styles uniformly livid,* st.-l. often wanting *sub-entire* or denticulate oblong narrowed to both ends.—*Exsicc.* Fasc. i. 20.—Alpine rocks. P. VII. VIII. S.

64. *H. cam'bricum* ((Baker) F. J. Hanb.); glaucous, st. 6—16 in. high, scarcely hairy, ped. long floccose with few hairs, rt.-l. rosulate 4—5 thin glaucous lanceolate acute narrowed to the base with long petioles sharply dentate nearly glabrous, st.-l. 0 or 1 small near the base. Heads 3—4, inv. small, outer phyll. dark short broad floccose-margined, inner paler narrower, all acute floccose-tipped hairy somewhat setose. Fl. bright yellow, *styles yellow.*—*Exsicc.* Fasc. iv. 92.—Limestone rocks. Grt. Orme's Head, Wales. P. VII. VIII. E.

65. *H. vagen'se* (Ley); st. 12—18 in. high, slender drooping branched or simple, sparsely hairy floccose above, ped. long ascending floccose setose, rt.-l. light glaucous green, long narrowly ovate or elliptic *narrowed at the base into long petioles* or truncate, acuminate with *many long acuminate very unequal teeth* which are often *continued down the petiole nearly to its base,* glabrous above hairy beneath, st.-l. similar, or narrower linear; all the l. with *pellucid veins and midrib.* Heads few large, inv. ovate after flowering, phyll. at first loosely incurved, *erect in bud* hairy setose, styles yellow.—*Exsicc.* Fasc. iii. 63; *H. britannicum* (F. J. Hanb.) var. *vagense* (F. J. Hanb.).—Riverside rocks in the Wye Valley, Wales. P. VI. E.

66. *H. holophyl'lum* (W. R. Linton); st. 10—12 in. high, corymbose-paniculate hairy below floccose above, ped. arcuate ascending floccose, rt.-l. persistent deep green somewhat coriaceous oblong-ovate *entire* or somewhat denticulate *rounded and blunt at each end,* inner pointed, subglabrous above slightly hairy below, st.-l. 1—3 ovate-lanceolate acuminate narrowing into petiole. Inv. pale green, constricted in fl., truncate below in fr., phyll. bluntish, the margins and tips pale green floccose with few hairs and setæ, the outer with lax tips. Fl. yellow glabrous-tipped, styles long pure yellow, dusky when old.—*Exsicc.* Fasc. ii. 39.—Limestone cliffs. Derbyshire. P. VII. VIII. E.

67. *H. cæsiomurórum* (Lindeb.); st. 1—2½ ft. high wiry fistular simple or branched, ped. spreading floccose setose, rt.-l. dull deep green rather glaucous with prominent veins beneath, outer blunt apiculate, inner long lanceolate acuminate almost entire towards the apex with large spreading teeth in the lower half, petioles long slender, st.-l. petiolate lanceolate acute

coarsely and sharply toothed. Heads 2—7, inv. *broadly campanulate in fl.* subglobose in fr. green hairy somewhat setose floccose, phyll. adpressed, outer few broad gibbous somewhat floccose-margined. Fl. golden yellow pilose-tipped when young styles slightly darkened.—*Exsicc.* Fasc. i. 21.—Subalpine glens in Wales and Scotland. P. VI.—VIII. E. S.

68. *H. Orárium* (Lindeb.) *f.*; st. 1—3 ft. high, simple or branched, panicled or few-headed, ped. hairy setose floccose, rt.-l. few not persistent, ovate-lanceolate denticulate or dentate in the lower half almost entire towards the apex hairy on both surfaces *somewhat floccose beneath*, st.-l. 2—5 petiolate lanceolate acute decurrent dentate. Inv. broad at the base rounded, buds rather narrow cylindric, phyll. obtuse, inner acute, hairy setose floccose-margined. Fl. especially the inner very pilose-tipped, styles light yellow.—*β. fulvum* (F. J. Hanb.); st. few-leaved scabrid, rt.-l. *acutely dentate* with *long petioles.* Fl. small *reddish yellow or orange*, phyll. adpressed bristly densely setose, *styles very dark long and scarcely cleft.*—*Exsicc.* Fasc. iv. 93.—Banks of streams in Scotland. *β.* Caithness and Sutherland. P. VII. VIII. S.

69. *H. dúriceps* (F. J. Hanb.); st. 15—20 in. high wiry, reddish-purple hairy below almost glabrous above, ped. straight slender rigid floccose, with *dark and minute yellow-headed setæ*, rt.-l. small lanceolate subacute, primary oval apiculate, dull green often blotched with purple, almost glabrous above with scattered white hairs beneath, st.-l. narrow lanceolate acute petiolate if low down, sessile if springing from the point of branching. Inv. *small hard compact,* deep green conical when mature, buds short and stout, phyll. narrow acute densely setose sparingly floccose and pilose-tipped. Fl. small rather numerous *often stylose,* ciliate-tipped especially when young, styles livid.—*Exsicc.* Fasc. vi. 144.—*β. cravoniense* (F. J. Hanb.); rt. and st.-l. more numerous pilose on both surfaces, deeply irregularly dentate, ped. less setose, more hairy and floccose, inv. hairy, sparsely setose and floccose. Fl. always stylose. The whole plant more robust than type. Rocks and cliffs by streams in Scotland. *β.* Stream-sides, Craven, Yorkshire; Lancashire. P. VI.—VIII. E. S.

70. *H. dissim'ile* (Lindeb.) *f.*; st. 10—18 in. high, hairy below, ped. floccose with few hairs and setæ. Rt.-l. rosulate narrowly decurrent at the base with long slender very hairy red petioles, *sharply dentate* or denticulate, outer ovate, inner lanceolate acute, yellowish green thin but firm not glaucous, sub-glabrous above hairy beneath, st.-l. 1—3 lanceolate acute

petiolate *very deeply irregularly toothed* hairy *slightly floccose beneath.* Heads 1—3, inv. broad-based, phyll. outer lax, inner long attenuate, somewhat floccose-margined. Fl. golden yellow pilose-tipped, styles very livid.—*Exsicc.* Fasc. iv. 94.—*β. poliænum* (Dahlst.); l. broader less sharply toothed, *ped. only floccose,* heads numerous, inv. *very floccose* with fine hairs and few minute yellow-headed setæ, tips of phyll. purple-brown.—*γ. porrigens* (Almq.); st.-l. more numerous *broadly lanceolate* petiolate *upper sessile,* all with rather *coarse teeth* in the lower half, ped. densely floccose and setose, phyll. floccose-margined, *not hairy* but with *many long minute yellow-headed* setæ.—Mountain glens. *α.* Frequent in the Breadalbanes. *β.* Caithness, Inverness. *γ.* Nr. Watersmeet, Countisbury, N. Devon. P. VII. VIII. E. S.

71. *H. vulgátum* (Fr.); green or glaucous, l. oblong or lanceolate often with patent or *forward teeth* on lower half, rt.-l. narrowed into petiole, st.-l. often many all or uppermost sessile, head panicled or subcorymbose, inv. and *straight* ascending *peds.* floccose setose with few hairs, *phyll. equally attenuate* acutish incumbent, styles livid.—*Exsicc.* Fasc. i. 40; *H. sylvaticum,* Sm. *E. B.* 2031.—Very variable, 1—3 ft. high. The more common plant has panicled heads cylindrical when young, glaucescent l. purplish beneath, persistent rt.-l., few st.-l., greenish inv.; but sometimes the st. is very leafy, heads panicled, rt.-l. evanescent.—*β. maculatum* (Sm.); has few broad purpled-based st.-l., and imperfectly corymbose heads. The spots on the l. are not constant; in shade they are absent.—*Exsicc.* Fasc. iii. 68; *Sy. E. B.* 849.—*γ. dædalolepium* (Dahlst.); st.-l. 2—3 lower slightly petiolate, heads numerous small narrow, phyll. narrow equally broad with few fine small hairs small setæ and scarcely floccose.—*Exsicc.* Fasc. v. 115.—*δ. cacuminum* (Ley); upper st.-l. long *large lingulate,* nearly glabrous entire or *with shallow irregular teeth,* ped. *bracteate,* phyll. broad obtuse, *styles pure yellow.*—*Exsicc.* Fasc. v. 116.—*ε. sejunctum* (W. R. Linton); st.-l. 2—4 rarely 9, yellowish-green. hairy above floccose beneath, *firm sharply dentate with several large cusped teeth.* heads 4—12 sub-umbellate, floccose hairy thinly setose, fl. glabrous-tipped, styles livid.—*Exsicc.* Fasc. vi. 26. Many other forms are named but they defy definition.—Woods, banks, walls. *δ.* Teesdale, North Wales. *ε.* Scotland. P. VI.—IX. E. S. I.

72. *H. surreiánum* (F. J. Hanb.); st. 1—2 ft. high, usually simple, hairy below, ped. usually straight ascending floccose with fine hairs not setose, rt.-l. not persistent elliptic almost entire, st.-l. 3—5 lower ovate decurrent into shaggy petiole

blunt at apex, upper longer apiculate uppermost sessile, hairy on both surfaces and margins floccose beneath denticulate or dentate. Heads 2—5, inv. cylindric, outer phyll. short blunt, inner longer paler subacute almost naked, *very sparingly hairy* floccose and few small setæ. Fl. orange-yellow, slightly ciliate-tipped, styles almost pure yellow.—*Exsicc.* Fasc. vi. 147.—*β. megalodon* (Linton); st.-l. broader, more coarsely toothed, rt.-l. rounder with longer petioles, inv. greener when dry, not so dark.—*Exsicc.* Fasc. vi. 148.—Banks near Witley, Surrey. P. VI. E.

73. *H. stenophy'es* (W. R. Linton); st. 1½—2 ft. high, hairy below, *ped. arcuate* floccose, rt.-l. rosulate speading, outer ovate-oblong with few blunt teeth, inner lanceolate-oblong acute, with cuneate base gradually decurrent into long petiole coarsely toothed, hairy both surfaces, st.-l. usually 1 *petiolate lanceolate acuminate* with few large acute patent teeth. Heads 3—8, inv. dark cylindric, phyll. broad dark greenish, inner paler-margined, acute hairy floccose at the base with few setæ. Fl. cup-shaped, rich yellow, glabrous-tipped, styles livid.—*Exsicc.* Fasc. vi. 149.—Mountains. P. VI. VII. S.

74. *H. subanfrac'tum* (Marshall); st. 9—20 in. high, rigid fistular simple or branching glabrous below floccose upwards with few setæ, ped. *erect* ascending floccose setose, rt.-l. oblong lanceolate obtuse or linear-lanceolate to lanceolate acute or apiculate, blade from 2—4 in. long gradually narrowing into long slender hairy petioles, strongly dentate with large forward-pointing teeth, gland-tipped, bright green above with prominent pellucid veins sometimes glaucous, paler beneath, firm glabrous above slightly hairy beneath and at the margins, st.-l. usually 1 lanceolate petiolate slightly or deeply toothed, sometimes linear bract-like. Heads 1—4, inv. campanulate very dark, buds cylindric, phyll. very dark obtuse narrowing upwards hairy glandular floccose-tipped. Fl. golden yellow, ciliate-tipped, styles rather livid.—*Exsicc.* Fasc. ii. 41.—Rocky subalpine streamlets. P. VI.—VIII. S.

75. *H. angusta'tum* (Lindeb.); st. 10—16 in. high simple sometimes branched, almost glabrous, ped. slightly floccose setose with few hairs, rt.-l. small ovate-lanceolate or lanceolate acuminate or acute denticulate or almost entire, st.-l. 0—3 lanceolate acute or linear bractlike, entire or slightly dentate towards the base, somewhat fleshy yellow-green, paler somewhat floccose beneath. Heads 1—3 small, phyll. long narrow blunt or acuminate, slightly hairy with little floccose and few setæ, inner acute almost naked. Fl. small, yellow, styles somewhat livid.—*Exsicc.* Fasc. vi. 150, 151.—Ravines, Lake District, Scotland. P. VII. VIII. E. S.

76. *H. subramo'sum* (Lönn.); st. 1—2½ ft. high branched slightly floccose with few hairs, ped. erect spreading densely floccose slightly hairy setose bracteate, rt.-l. 2—3 large outer ovate-oblong blunt almost entire, inner lanceolate acute, irregularly deeply toothed almost glabrous above, hairy beneath and on margins, st.-l. 1—3, deeply irregularly acutely toothed hairy floccose below. Heads large numerous, inv. hairy floccose with few setæ. Fl. ciliate.—Shore between Burntisland and Pettycue, Fife. P. VI. S.

77. *H. diaph'anum* (Fr.); st. 1—1½ ft. high fistular with few hairs, ped. long spreading very floccose setose with few hairs, rt.-l. large ovate-elliptical denticulate hairy both surfaces, st.-l. 3—5 lower lanceolate acute dentate or denticulate decurrent into long winged almost amplexicaul white-haired petioles, almost glabrous above hairy below, upper sub-sessile floccose with few hairs. Inv. *large truncate dark blackish green comparatively glabrous* or with scattered hairs and numerous setæ, phyll. outer broad blunt, inner paler slightly floccose-tipped. Fl. yellow glabrous-tipped, styles yellow or slightly livid.—*β. stenolepis* (Lindeb.); st.-l. rather coriaceous, *not floccose*, heads numerous 2—6, *buds extraordinarily long and narrow*, setose, phyll. *very narrow more acute* slightly floccose *with floccose-tips, minutely setose* with few hairs. Fl. golden yellow, slightly pilose.—Rocky cliffs. P. VII. VIII. E.

78. *H. diaphanoïdes* (Lindeb.); st. 1½ ft. high, many-headed hairy slightly floccose, ped. corymbose spreading erect or rarely incurved setose floccose with some hairs, rt.-l. dull often cæsius green sometimes crimsoned, fleshy rosulate elliptic or oblong or lanceolate decurrent into short hairy winged petioles deeply toothed entire towards the apex, glabrous on both sides or hairy beneath, st.-l. 1—3 far apart often sessile lanceolate acute or acuminate coarsely dentate towards the base. Heads corymbose dark, outer phyll. broad subacute, inner long narrow acute thickly setose, not hairy nor floccose. Fl. slightly pilose, styles bright yellow or often dusky.—*Exsicc.* Fasc. ii. 42.—*β. apiculatum* (Linton); l. fresh green with more cuneate base, blunter more apiculate, phyll. broad acuminate obtuse, white-tipped, heads in lax irregular subcorymbose panicle.—*Exsicc.* Fasc. ii. 43. Somerset. Cheshire. Wales. Perthsh. Forfar. P. VII. VIII. E. S.

79. *H. sciaph'ilum* (Uechtr.); st. 1—2½ ft. high, branched, hairy slightly floccose, ped. long wiry spreading erect or arcuate bracteate densely setose floccose not hairy, rt.-l. rigid light green often crimsoned broadly lanceolate blunt or subacute dentate denticulate or almost entire hairy on both surfaces, st.-l. 1—8

broadly lanceolate acuminate apiculate entire towards the apex toothed below, upper sessile lower petiolate, somewhat coriaceous glabrous above hairy beneath. Heads many, inv. narrow cuneate very setose, buds cylindric, outer phyllaries short rather lax, inner long attenuate, outer and middle floccose-margined. Fl. rich yellow sub-pilose, styles nearly pure yellow.—*Exsicc.* Fasc. i. 22.—β. *pulchrius* (Ley); st. shorter stouter less leafy, rt.-l. obovate obtuse coarsely toothed with shallow teeth, st.-l. elliptic or ovate somewhat acute, coarsely toothed. Heads sub-umbellate with dark phyll. and longer setæ.—*Exsicc.* Fasc. v. 117.—Woods, rocks, and banks. β. Mountain cliffs in Wales. P. VI.—IX. E. S. I.

iii. *ACCIPITRINA.* Forming closed buds at the base of the st. in autumn. No true rt.-leaves with the flowers. St. leafy. Phyll. in many rows. Hairs of pappus rigid, unequal.

* *Rigida.* St.-l. numerous, lower sub-petiolate, upper sessile, narrow or broad lanceolate more or less toothed, outer phyll. slightly lax with straight margins, inv. nearly glabrous or with minute yellow setæ, few hairs, sometimes slightly floccose.

80. *H. goth'icum* (Backh.); st. $1\frac{1}{2}$—4 ft. high erect rigid simple or branched nearly glabrous throughout, ped. erect or spreading rigid floccose, l. 7—20 dark green paler and rather glaucous beneath firm, lanceolate or ovate, acute or subobtuse denticulate or dentate in the middle, entire towards apex, aggregate towards the base of st. sometimes stalked, upper smaller sessile almost bractlike. Heads 1—5, inv. dark green globose-ventricose, cylindric in bud, rounded and sometimes truncate at the base after flowering, nearly glabrous or with few hairs, phyll. dark green broad obtuse rather lax with incumbent tips. Fl. bright not deep yellow glabrous-tipped, styles yellow with minute dark hairs.—*Sy. E. B.* 851; *Exsicc.* Fasc. iii. 69; iv. 96.—β. *latifolium* (Backh.); plant much more robust, l. large very broad obtuse entire or denticulate rarely dentate, ped. elongated, styles pure yellow.—*Exsicc.* Fasc. iii. 70.—γ. *Stewartii* (F. J. Hanb.); l. very broadly lanceolate acute with broadly winged petiole, almost entire towards the apex, with sharp narrow curved forward-pointing teeth often $\frac{3}{4}$ in. long towards the base, lowest ovoid blunt, upper and middle sessile, ped. very long bracteate bearing large deep golden yellow fl., outer phyll. lax, styles very livid.—*Exsicc.* Fasc. vi. 152.—δ. *basifolium* (Lindeb.); rt.-l. rosulate 4—8, ovate-oblong or ovate-lanceolate narrowed at both ends, dentate, st.-l. fewer in number decreasing immediately in

size, almost bract-like at the apex of st., styles somewhat dusky.—Heathy or grassy places in subalpine districts. α. England; Scotland. β. Scotland. γ. Ireland. δ. Scotland. P. VII.—IX. E. S. I.

81. *H. sparsifólium* (Lindeb.); st. ½—2 ft. high, simple or branched erect hairy below, ped. long spreading floccose, rt.-l. small spathulate not persistent, st.-l. many deeply blotched with brownish-purple pale green beneath, nearly entire or denticulate in the middle, glabrous above somewhat floccose beneath, lower ovate entire decurrent into winged petioles, upper long narrow acute with somewhat clasping base, uppermost linear. Heads few, inv. large broad dark with long white hairs and minute setæ, phyll. imbricated with recurved tips and curved margins, outer small lax obtuse, inner pale narrower more acute. Fl. bright yellow, almost glabrous, styles pure yellow. The l. of the Welsh specimens are always more toothed, broader and blunter, and the inv. less clothed than in those from Scotland.—*Exsicc.* Fasc. i. 24.—Cliffs and sides of streams in subalpine districts. P. VII. VIII. E. S .I.

82. *H. rig'idum* (Hartm.) β. *pullatum* (Dahlst.); st. 1—2½ feet high, simple or branched hairy floccose, ped. spreading bracteate, l. ovate-lanceolate or lanceolate apiculate lower almost entire, denticulate or coarsely toothed in the middle, almost glabrous above, hairy beneath purpled. Heads 1 or many, outer phyll. lax short, inner paler subobtuse, very sparingly clothed with few hairs and scattered yellow setæ, *styles livid.—Exsicc.* Fasc. v. 118.—γ. *acrifolium* (Dahlst.), l. long narrow mostly coarsely toothed below the middle, inv. minutely setose with few hairs very slightly floccose at the base, *styles yellow.—Exsicc.* Fasc. iii. 71.—δ. *Friesii* (Hartm.); ped. long straight *densely floccose*, l. denticulate with fine sharp teeth, almost glabrous above *hairy and floccose beneath*, inv. broad unequally based, with few hairs and slightly floccose, outer phyll. short dark subobtuse, inner paler more attenuate glabrous.—ε. *tridentatum* (Fr.); l. sessile or sub-petiolate upper only bractlike *hairy below not floccose* dentate in the middle, *ped. slender hairy floccose*, inv. truncate below, constricted after flowering, phyll. pale-margined, hairy hardly floccose, *styles livid.*—ζ. *nidense* (F. J. Hanb.); l. very numerous, upper sessile rounded at the base, lower long-petioled decurrent to the stem acutely pointed *with long acuminate teeth*, buds cylindric, styles somewhat darkened.—*Exsicc.* Fasc. v. 119.—Other vars. or forms have been described.—Mountain glens and banks of streams. P. VII.—IX. E. S.

83. *H. cantianum* (F. J Hanb.); 18—20 in. high, usually branched hairy floccose, ped. long spreading arcuate bracteate or leafy felted with floccose down, l. 10—17 upper short broad curiously sharply toothed somewhat amplexicaul with broad bases, lower petiolate decurrent with large irregular coarse teeth, hairy on both surfaces somewhat floccose beneath. Heads many small, outer phyll rather lax, sparsely clothed with yellow setæ, few hairs and slightly floccose. Fl. glabrous, styles pure yellow. *Exsicc.* Fasc. ii. 45.—*β. subrigidum* (Linton); peduncles more densely floccose, inv. more hairy, styles somewhat livid, narrower less numerous and less coarsely-toothed l.—*Exsicc.* Fasc. vi. 154.—Hangley Woods, Kent. *β.* Near Witley, Surrey. P. VI.—VII. E.

** [illegible] plant green or glaucous. Stem always leafy. Heads large few, ped. usually springing from the axils of the leaves. Fl. often ciliate.

84. *H. [illegible]* (Lindeb.); st. 9—16 in. high, hard wiry simple or branched about the middle, hairy below and sparsely floccose, ped. erect slender bracteate densely floccose with numerous very minute setæ, l. green, paler and often purpled beneath, rad. l. few not persistent, st.-l. 3—6 ovate lanceolate to lanceolate, upper sessile, lower decurrent into winged petioles, subglabrous above hairy and floccose beneath, with few minute distinct teeth on the margin. Heads 1—5 rather small, inv. truncate, cylindric in bud, turbinate later, outer phyll. dark-green short lax, inner paler; all somewhat recurved later rather broad blunt slightly floccose-tipped with few hairs and setæ. Fl. medium yellow, glabrous-tipped, styles pure yellow.—Cliffs. Unst, Shetland. P. VII. VIII. S.

85. *H. [illegible]* (Beeby); st. 3½—9 in. high simple or branched hairy and purpled below floccose above, ped. densely floccose, rt. l. rosulate persistent oval-elliptic with few forward-pointing teeth on each side, olive-green often purpled, subglabrous above paler more hairy below with ciliate margins, petiole short hairy, st.-l. 0—2, clasping broadly lanceolate acute or acuminate sharply toothed. Heads rather small, inv. somewhat truncate at the base, outer phyll. dark short broad obtuse, inner longer paler less obtuse, all somewhat recurved sparingly ciliate-tipped setose sparsely floccose. Fl. small orange-yellow, glabrous-tipped, styles nearly pure yellow.—A form of this plant is issued in *Exsicc.* Fasc. vi. 155.—Pastures and sheltered rocks. Shetland, West Sutherland. P. VII. VIII. S.

86. *H. truncátum* (Lindeb.) *f.*; st. 10—24 in. high, hairy floccose, simple or branched, ped. long somewhat arcuate floccose with very few hairs, rt.-l. ovate not persistent, st.-l. 7—11 lanceolate acuminate or acute rounded at the base sessile denticulate in the middle entire towards the apex hairy or subglabrous above, hairy and floccose beneath, deep green paler and bluish-green beneath. Heads 1—6, inv. large dark truncate nearly glabrous with minute setæ slightly floccose, phyll. short broad blunt inner paler less obtuse. Fl. orange-yellow glabrous-tipped, styles yellow.—Among ferns and long herbage. Northmaven, Shetland. P. VIII. S.

87. *H. protrac'tum* (Fr.) *f.*; st. 1—2 ft. high, simple or branched leafy hairy and reddish below floccose above, ped. straight ascending thickened upwards floccose hairy with few small setæ, rt.-l. persistent elliptic blunt apiculate somewhat hairy, st.-l. long narrow apiculate sessile amplexicaul or decurrent into winged sheathing hairy petioles, all nearly entire or with very minute teeth, deeply blotched dark purplish brown bluish green beneath, glabrous above densely ciliate on margins and below. Heads few or many frequently adnate, inv. rather narrow rapidly tapering, buds cylindric, phyll. green lax, outer short, blunt with recurved floccose tips hairy with minute yellow setæ. Fl. bright yellow glabrous-tipped, styles pure yellow.—Cliffs. Shetland Islands. P. VII. VIII. S.

88. *H. dovren'se* (Fr.) *f.*; st. 1—2 ft. high, simple or branched hairy throughout, ped. curved ascending bracteate much thickened floccose with scattered hairs and setæ, rt.-l. oblong or lanceolate-oblong stalked small bluntish, st.-l. lanceolate sessile cordate or broad-based half-clasping green often purpled paler beneath, roughly hairy on both surfaces somewhat floccose, leathery slightly denticulate, with short marginal hairs. Heads 3—many, inv. *remarkably truncate* even in the cylindrical bud, outer phyll. dark green short slightly floccose-tipped, inner paler, all very blunt hairy and minutely setose. Fl. orange-yellow glabrous-tipped, styles livid.—β. *hethlandiæ* (F. J. Hanb.); ped. less floccose, l. almost entire very acute, inv. floccose, phyll. not floccose-tipped.—γ. *spectabile* (Marshall); more robust, st.-l. more numerous tougher longer *broader more dentate* glabrous above midrib and margins hairy. Heads 3—11, phyll. very broad *floccose-margined* sparingly hairy with few setæ. Fl. pilose-tipped before expansion.—*Exsicc.* Fasc. iii. 72.—Rocks and Highland glens. α and β. Shetland. γ. Glen Shee and the Clova district. P. VII. VIII. S.

89. *H. Dew'ari* (Sy.); *l.* oblong-lanceolate denticulate, lowest oval with winged stalks, *upper l. ovate-lanceolate ½ clasping*, ped. slightly floccose and setose scarcely hairy, heads in lax panicle, inv. subcylindrical obconic slightly hairy and setose, phyll. dark green few, the outer short adpressed inner with pale edges, *styles fuscous*.—Edin. Bot. Trans. xiii. 211, t. 5; *Exsicc.* Fasc. ii. 47.—Bright green. St. 1—3 ft. high. Root-l. rare and persistent.—Mountain glens. P. VII.—IX. S.

*** *Prenanthoidea.* St. leafy without basal rosette. Lower l. petiolate, upper and middle amplexicaul with rounded auricles, glaucous and reticulate beneath. Phyll. few irregularly imbricated. Fl. ciliate.

90. *H. prenanthoïdes* (Vill.); st. leafy, *l. all clasping* net-veined and glaucous beneath *hairy on both sides*, lowest narrowed into winged auricled petioles, interm. pinched above their base, uppermost l. cordate-lanceolate, heads in a corymbose panicle, *ped.* short lax very floccose and *setose*, inv. cylindrical rather floccose *very setose, outer phyll. few much the shortest, inner all about equal* and blunt, styles with dark hairs, fr. pale.—*E. B.* 2235; *Exsicc.* Fasc. ii. 48.—St. 1—3 ft. high rather rigid, usually hairy. Pan. leafy below. Heads small. Pappus whitish.—River-sides in the North. P. VII.—IX. E. S. I.

[*H. Bor'reri* (Sy.); like the preceding but l. fewer, the lower abruptly contracted into long petioles, interm. l. regularly oval not pinched above the base, styles yellow.—*E. B.* 859; *H. juranum*, Fr.; *H. denticulatum* Borr. MS. not Sm.—Harehead wood near Selkirk. P. VII. VIII.]

**** *Foliosa.* Stem leafy without basal rosette. L. all more or less amplexicaul, or sessile, paler sometimes rather glaucous and reticulate beneath. Phyll. obtuse adpressed. Fl. glabrous-tipped.

91. *H. stric'tum* (Fr.); st. leafy, *l. sessile* oblong-lanceolate denticulate glaucous beneath, lowest lanceolate much narrowed below *uppermost l.* rounded below *scarcely clasping*, heads irregularly corymbose, *ped.* straight floccose *scarcely setose*, inv. truncate *thinly* floccose *setose* and hairy, *phyll. irregularly imbricate* blunt, outermost rather acute, styles with dark hairs, fr. fuscous.—*H. denticulatum* Sm. *E. B.* 2122; *Exsicc.* Fasc. iii. 73.—St. 1—3 ft high slightly hairy. L. broad, not auricled, mostly with bulbous hairs beneath. Heads thicker and paler than those of *H. prenanthoides*. Phyll. pale-edged, not in two

distinct ranks. Pappus reddish.—β. *reticulatum* (Lindeb.); st. taller more branched many-fl., *ped. widely spreading*, *l. reticulate* floccose hairy, inv. small minutely setose, phyll. broad narrowing upwards with pale margins. Fl. minutely ciliate, *styles yellow.*—*Exsicc.* Fasc. iv. 98.—γ. *angustum* (Lindeb.); st. glabrous few or many-leaved, *l. slender narrow small* reticulate entire or finely denticulate deep green and mostly glabrous above, reddish or blue-green *floccose* and short-haired beneath, inv. floccose with few setæ not hairy, phyll. more equally broad, not narrowing to tips. *Fl. glabrous*, styles yellowish.—*Exsicc.* Fasc. vi. 157.—δ. *opsianthum* (Dahlst.); much more robust, with larger broader leaves sometimes coarsely dentate subglabrous above, upper l. floccose, lower floccose and hairy, inv. hairy with few setæ.—*Exsicc.* Fasc. vi. 158.—ε. *subcrocatum* (Linton); l. broadly ovate-acuminate. Inv. very dark, nearly glabrous or with few setæ and sparsely floccose. Fl. small glabrous, *styles dark.*—*Exsicc.* Fasc. v. 120.—ζ. *amplidentatum* (F. J. Hanb.); l. nearly glabrous with strongly ciliate margins *floccose* especially above, upper sessile, lower with winged clasping petioles very acute sharply irregularly toothed. Fl. glabrous, styles yellow.—*Exsicc.* Fasc. v. 121.—Mountain glens and river banks. P. VII. VIII. E. S. I.

92. *H. corymbósum* (Fr.); st. 2—4 ft. high rigid nearly glabrous very leafy, ped. floccose, heads 6—40 in a *spreading leafy branched corymbose panicle*, l. ovate or lanceolate acute dentate towards the middle narrowing from a broad somewhat clasping base, lowest l. narrowed into petioles, deep green glaucous reticulate-veined sometimes floccose beneath more or less hairy on both sides; inv. ovate at the base not constricted with scattered hairs and setæ, phyll. greenish-black with paler margins adpressed blunt outer rather lax. Fl. glabrous-tipped, styles yellowish.—*Exsicc.* Fasc. iii. 74. *H. eupatorium* (Griseb.). —β. *prælongum* (Lindeb.); l. very long narrow acute reticulate, deeply sharply toothed, lower subentire with winged petioles not persistent, middle l. broad to narrow lanceolate somewhat clasping, upper sessile with rounded base. Heads small, styles yellow.—γ. *salicifolium* (Lindeb.); *l. always floccose* somewhat hairy, lower almost entire, middle broadly lanceolate acute very slightly clasping denticulate in the middle sometimes sharply dentate, inv. truncate at the base, setose slightly floccose with few or no hairs, phyll. rather narrow.—*Exsicc.* Fasc. v. 122.—Mountain districts. P. VII.—IX. E. S. I.

93. *H. aurátum* (Fr.); st. 2—3 ft. high erect rigid usually with long spreading leafy branches, ped. bracteate floccose, l.

bluish-green, lower oblong-lanceolate obtuse apiculate petiolate almost entire not persistent, middle l. broadly lanceolate narrowing at both ends semi-amplexicaul, upper lanceolate acute sessile with rounded base, denticulate or dentate, very reticulate-veined, almost glabrous above somewhat floccose and hairy beneath. Heads many corymbose, inv. oblong conical after flowering dark green, with yellow setæ scarcely floccose, phyll. obtuse or sub-obtuse outer very short. Fl. golden yellow glabrous, styles pure yellow.—*Exsicc.* Fasc. iv. 99.—*β. thulense* (F. J. Hanb.); st.-l. fewer almost entire or with few minute teeth ovate-lanceolate apiculate, clasping somewhat constricted towards the base at times, phyll. broader more floccose, purpled tips.—Mountain districts. *β.* Shetland. P. VII.—IX. E. S. I.

94. *H. crocátum* (Fr.); st. 1½—4 ft. high rigid smooth sub-corymbose, *branches nearly simple,* ped. thickened upwards floccose with few setæ, l. oblong or narrow lanceolate sessile with a *broad base* falsely 3-veined semi-amplexicaul dentate or nearly entire, lower l. often narrowed very gradually downwards but slightly enlarged again at the base, glabrous or with hairy margins and below. Inv. very broad-based, phyll. *never lax adpressed broad obtuse black* minutely setose or almost glabrous sometimes pale-margined. Fl. glabrous, styles olive. —*Exsicc.* Fasc. v. 123.—Mountain districts. P. VII. VIII. E. S. I.

95. *H. marit'imum* (F. J. Hanb.); st. 1—3 ft. high erect reddish-purple corymbose the branches ascending from the axils of upper leaves, ped. long floccose with short bristly hairs, *l. long narrow acute or subacute* practically entire or minutely serrate or with short sharp prickles at the margins, fleshy upper sessile sometimes a little clasping the lowest narrowing into short petiole with blunter apex bluish green below with short bristles on both surfaces and margins, some slightly floccose. Inv. truncate angular below, *phyll. extraordinarily broad obtuse* imbricated almost glabrous very dark, the inner paler. Fl. orange yellow glabrous, styles smoky yellow.—North coast of Sutherlandshire. P. VII. S.

***** *Sabauda.* Stem leafy robust branched. Lower st.-l. narrowed at the base into petioles; upper broad-based sessile. Phyll. broad obtuse unicoloured sparingly hairy with few or no setæ.

96. *H. boreále* (Fr.); st. leafy, l. ovate or lanceolate dentate below falsely 3-veined, upper broad sessile scarcely clasping,

lowest l. much narrowed below but scarcely stalked, heads in a rather leafy corymb or panicle, top of *ped. floccose,* ovate-based *inv. uniformly blackish-green nearly glabrous* or pilose, phyll. blunt adpressed, *styles livid blackish.—H. sabaudum* Sm. *E. B.* 349. *Exsicc.* Fasc. ii. 49.—St. 2—4 ft. high, rigid usually hairy and often very leafy below. Base of corymb or panicle leafy; branches nearly erect. Interm. l. narrowed to a rounded base; upper with a broad rounded or subcordate base. Phyll. turning black, scarcely at all setose or floccose, rarely with spreading tips. Pappus whitish.—β. *calvatum* (F. J. Hanb.); st. 15 in.—2 ft. high, *very glabrous,* l. few with a tendency to crowding at the base. Heads few large, phyll. broader more obtuse less clothed and very dark, styles very dark.—Banks. β. Carnarvonshire. P. VIII. IX. E. S. I.

****** *Umbellata.* Stem rigid leafy without a basal rosette, sub-umbellate or corymbose at the top. St.-l. narrowed at the base sessile, the nerves on the under surface loosely anastomosing not reticulate-veined. Phyllaries broad obtuse. Ligules glabrous.

97. *H. umbellátum* (L.); st. leafy, *l. all linear* or oblong-lanceolate *narrowed below sessile* net-veined, heads in an umbellate corymb, top of ped. floccose, turbinate-based *inv. uniformly dark green glabrous, phyll. blunt with recurved points, styles yellow.—E. B.* 1771; *Exsicc.* Fasc. ii. 50.—Stem 1—4 ft. high, rigid, hairy below. L. usually all alike, or upper rather broad and rounded at the base. Outer phyll. often very small, slender, acute. Inv. very rarely straw-coloured, rarely with a few white hairs. Pappus whitish.—β. *coronopifolium* (Bernh.); st. 1—3 ft. high corymbose or paniculate, ped. rigid erect or suberect, l. very long narrow sharply and acutely toothed, subglabrous.—*Exsicc.* Fasc. v. 124.—γ. *littorale* (Lindeb.); st. 6—12 in. high hairy floccose, l. long narrow almost entire especially crowded together at the base of the stem, floccose on both surfaces not hairy, veins prominent.—δ. *curtum* (Linton); 8—16 in. high, l. very short almost entire, fl. few rather large on short spreading ped., outer phyll. broad obtuse reflexed.—ε. *Ogweni* (Linton); l. fewer, upper sessile, lower narrowing into petioles, panicle sub-umbellate, tips of phyll. not recurved, nearly glabrous, not hairy slightly floccose with few scattered setæ.—Sandy and stony places. γ. Channel Islands. δ. Sandbanks, Carnarvonshire. ε. Banks of the Ogwen, Carnarvonshire. P. VII.—IX. E. S. I.

Index to the Hieracia.

Index to the Hieracia (cont.).

Anomalous Genus. Order AMBROSIACEÆ *Link.*

48. Xanthium *Linn.*

[*X. strumárium* (L.); st. without spines, lower l. heartshaped 3-lobed at the base coarsely dentate, beaks of the fr. 2 straight. —*E. B.* 2544.—Involucre of the fr. oval, downy.—Rich waste land, not naturalized. A. VIII. IX.] E.

[*X. spinósum* (L.); has also been found.]

Order XLVI. CAMPANULACEÆ.

Cal. superior, 5-fid or entire. Cor. gamopetalous, inserted on the calyx, 5-lobed, regular or irregular. Stam. inserted with, but not adhering to the cor., alternate with its lobes; anth. distinct or cohering, 2-celled, opening longitudinally. Fr. dry, capsular, opening by lateral fissures or valves at the top, many-seeded. Embryo straight, in the axis of fleshy perisperm.—No stipules.

Tribe I. *LOBELIEÆ.* *Cor. irregular.* *Anth. cohering.* Style glabrous with a fringe of hairs below the stigma.

1. LOBELIA. Cal 5-fid. Cor. irregular; tube split to the base at the back; limb 2-lipped, 5-parted. Anth. 5, cohering. Stigma blunt, surrounded by a cup-shaped fringe. Caps. 2—3 celled, opening at the end by 2—3 valves.

Tr. II. *CAMPANULEÆ.* *Cor. regular.* Anth. usually free. Style pubescent.

2. JASIONE. Cal. 5-fid. Cor. rotate, with 5 long linear segments. *Anth. cohering* at their base. Style hairy, bifid. *Caps.* 2-celled, *opening broadly at the end by short teeth.*

3. PHYTEUMA. Cal. 5-parted. *Cor. 2-lipped, with 5 long linear segments. Anth. free*; filaments dilated at the base. Style hairy, 2—3-fid. Caps. 2—3-celled, bursting at the sides.

4. CAMPANULA. Cal. 5-parted. *Cor.* mostly bellshaped, *with 5 broad* and shallow *segments.* Anth. free; filaments dilated at the base. Stigma 3—5-fid. *Caps.* not long, 3—5-celled, *opening by lateral pores below the cal.-limb.*

5. SPECULARIA. Cor. rotate. Caps. linear-oblong, prismatic, opening by lateral pores between the segments of the calyx. Otherwise like *Campanula.*

6. CERVICINA. Caps. half superior, 3-celled, opening by 3—5 valves above the segments of the calyx. Otherwise like *Campanula.*

Tribe I. *Lobelieæ.*

1. LOBE'LIA *Linn.*

1. *L. Dortman na* (L.); l. radical linear entire of 2 parallel tubes, st. simple nearly leafless.—*E. B.* 140.—Rootstock fleshy, with filiform runners. L. blunt, 1—2 in. long, submerged. St. 12—18 in. high, with or without 2—3 small bractlike leaves. Fl. pale lilac, distant, in a raceme, slightly raised above the water.—In lakes with a gravelly bottom. P. VII. E. S. I.

2. *L. ŭrens* (L.); st. nearly upright leafy, lower l. obovate or oblong slightly toothed, upper lanceolate serrate, fl in long

terminal racemes.—*E. B.* 953.—St. 1—2 ft. high, branched, angular, roughish. Racemes erect, simple, lax. Fl. light blue. —Heath near Axminster, Devon. Between Lostwithiel and St. Veep, Cornw. P. VIII. IX. E.

Tribe II. *Campanuleæ.*

2. JASIO'NE *Linn.* Sheep's Scabious.

1. *J. montána* (L.); root simple, l. oblong bluntish wavy, fl. in long stalked heads.—*E. B.* 882.—St. several from the crown of the root, 6—12 in. long, simple or branched, pilose, leafy below, bare and usually glabrous above. Radical l. in a rosette. Fl. small, in terminal heads with involucres. Bracts glabrous or hairy. Cal.-segm. subulate, glabrous. Cor. light blue.—[Extreme forms are:—var. *major* (M. & K.); root thick, stems many tall, heads large, and var. *littoralis* (Fr.); cæspitose, stems prostrate, simple, l. glabrous, heads small. A dwarf form with large l. and very large heads occurs in Shetland.]—Dry places. B. VII. E. S. I.

3. PHYTEU'MA *Linn.*

1. *P. orbiculáre* (L.); *heads of fl. globose, of fr. oblong,* l. crenate-serrate, lowermost cordate-ovate stalked, upper l. linear-lanceolate sessile, outer bracts ovate-lanceolate attenuate, *stigmas* 3.—*E. B.* 142.—St. 4—18 in. high, each with 1 terminal head of blue flowers.—Chalky downs. P. VII. E.

2. *P. spicátum* (L.); *heads of fl. oblong, of fr. elongate cylindrical,* lower l. cordate-ovate somewhat doubly serrate stalked, upper l. linear-lanceolate sessile, bracts linear, *stigmas* 2.—*E. B. S.* 2598.—St. 1—2 feet high, each with a solitary terminal head of cream-coloured flowers. Spike of fruit often 2—3 in. long.—Woods and thickets about Waldron, Sussex. P. VII. E.

4. CAMPAN'ULA *Linn.* Bell-flower.

* *Caps. sessile, erect; pores at the base.*

1. *C. glomeráta* (L.); l. minutely crenate-serrate, lowermost stalked ovate-lanceolate generally cordate at the base, upper l. half-clasping sessile ovate acute, *fl. sessile* in terminal and axillary clusters.—*E. B.* 90.—St. 6—18 in. high. Bracts ovate-acuminate, shorter than the large erect flowers. Cal. hoary,

with lanceolate segments. Cor. funnelshaped, large, deep blue, downy. L. often hoary beneath.—Dry calcareous pastures. P. VII. VIII. *Clustered Bell-flower.* E. S.

** *Caps. stalked, nodding; pores at the base.*

2. *C. latifólia* (L.); *l. ovate-lanceolate* acuminate *doubly serrate* hairy, lower ones stalked, upper l. nearly sessile, fl. racemose, peduncles 1-flowered, cal.-segments lanceolate-acuminate glabrous finely serrate, st. erect slightly angular.—*E. B.* 302. *St.* 72. 3.—St. 3—4 feet high, simple, leafy. Cor. very large, blue, glabrous, hairy within.—Woods and thickets chiefly in the North. P. VII. VIII. *Giant Bell-flower.* E. S. I.

3. *C. Trachélium* (L.); *l. coarsely doubly serrate* hispid, *lower ones cordate* with long stalks, upper l. nearly sessile ovate or lanceolate-acuminate, fl. racemose, peduncles 2—3-flowered, cal.-segments triangular-lanceolate entire erect, st. erect angular.—*E. B.* 12.—St. 2—3 feet high, mostly simple, leafy. Cor. truly bellshaped, large, deep blue.—Hedges and thickets chiefly in the South. P. VII. VIII. *Nettle-leaved Bell-flower.* E. I.

†4. *C. rapunculoïdes* (L.); *l. unequally crenate serrate* scabrous, lower ones cordate with long stalks, upper l. sessile lanceolate, *fl.* racemose *unilateral*, peduncles 1-flowered, cal.-segments linear-lanceolate entire *at length reflexed*, st. erect slightly angular, root creeping.—*E. B.* 1369.—St. 2 feet high, simple, leafy. Cor. pale blue.—Hedges, very rare. Near Kirkcaldy, Fifeshire. *Boswell* (*Syme*). P. VII. VIII. E. S.

5. *C. rotundifólia* (L.); *radical l. cordate or reniform* shorter than their stalks, *stem-l. linear*, the lower ones lanceolate, *fl.* 1 *or more* racemose, cor. turbinate-campanulate.—*E. B.* 866.—St. 6—12 in. high. Radical l. soon vanishing. Cor. blue. Cal.-segments linear subulate equalling ⅓-corolla.—β. *lancifolia* (M. & K.); lower st.-l rather broadly lanceolate, upper l. gradually smaller, fl. often solitary.—γ. *speciosa* (More); as β, but fl. much larger and more erect, cor.-lobes short. *Fl. Dan.* 2711.—[δ. *hirta* (M. & K.) lower part of st. with rigid hairs.]—Dry and hilly places. β on mountains. γ Innis Boffin, I.; Hebrides. P. VII. VIII. *Hairbell.* E. S. I.

*** *Caps. stalked, erect; pores just below cal.-segments.*

[*C. persicifólia* (L.); l. smooth slightly serrate, root-l. obovate narrowed into a petiole, stem-l. linear-lanceolate sessile, raceme few-flowered, *cal.-segments lanceolate*.—*E. B. S.* 2773. —St. 1—2 feet high. L. long, narrow, with very narrow serratures. Fl. very large, often solitary. Cal -segments entire. —" Woods near Cullen, Banffshire, and Thorpe Arch, Yorkshire." Not a native. P. VII.]

‡6. *C. Rapun'culus* (L.); l. crenate, root-l. elliptic-lanceolate narrowed into a petiole, stem-l. linear-lanceolate, panicle erect racemose, *cal.-segments subulate*.—*E. B.* 283.—St. 3 feet high, angular, rough. Fl. small, pale blue. Cal.-segments entire.—Sandy soil in the South. B. VII. VIII. *Rampion.* E.

7. *C. pat'ula* (L.); l. crenate, root-l. oblong-elliptic narrowed into a petiole, stem-l. linear-lanceolate, *panicles very lax, fl. on long stalks erect, cal.-segments toothed at the base* subulate.—*E. B.* 42.—St. 2 feet high, terminating in a very loose spreading panicle. Fl. purplish blue, funnelshaped, open.—Hedges and thickets. B. VII. VIII. E.

5. SPECULA'RIA *Heist.*

1. *S. hyb'rida* (A. DC.); st. simple or branched, l. slightly crenate wavy oblong sessile, lower l. spathulate, cal. rough, its segments lanceolate longer than the cor. shorter than the ovary. —*Campanula* L. *E. B.* 375.—St. 3—12 in. high, rough with rigid minute hairs. Fl. few, terminal, solitary, small, lilac.—Corn-fields. A. VI.—IX. E. S.

6. CER'VICINA *Del.* [*Wahlenbergia* Schrad. ed. viii.].

1. *C. hederácea* (Druce); l. roundish cordate angularly 5-lobed stalked alternate, st. filiform prostrate, peduncles solitary. —*Campanula* L. *E. B.* 73.—St. branched slender, creeping greatly. Peduncles longer than the leaves. Fl. pale blue, bellshaped, narrow, at first nodding, afterwards erect. Cal.-segments subulate. Caps. nearly globose.—Damp peaty places in the South and West. P. VII. VIII. *Ivy-leaved Bell-flower.* E. S. I.

Order XLVII. ERICACEÆ [1].

Cal. 4—5-parted, persistent. Cor. gamopetalous, 4—5-parted, usually regular and marcescent; or sometimes only slightly cohering below. Stam. 8—10, hypogynous. Anth. 2-celled,

[1] Monotropeæ, Vaccinieæ, and Pyrolaceæ are considered distinct orders by Maout and Decaisne.

opening by 2 pores or slits and often with spur-like appendages at the base. Ovary surrounded by a disk or scales, free or adhering to the corolla. Fr. capsular or baccate, with several cells, many-seeded.—No stipules.

* *Anthers opening by pores.*

Tribe I. *ARBUTEÆ. Fr. baccate, fleshy, superior.* Disk and stam. hypogynous. Petals cohering.—L. usually broad.

1. Arbutus. Cal. 5-parted. Cor. globose or ovoid with a small contracted 5-cleft reflexed border, deciduous. Stam. 10, with flattened filaments. Anth. compressed, with 2 pores at the apex, fixed at the back beneath the apex and there furnished with 2 reflexed appendages. *Berry* globose, *tubercled; cells* 5, *many-seeded.*

2. Arctostaphylos. Fr. with 5 1-seeded cells, smooth. Otherwise like *Arbutus.*

Tr. II. *ERICEÆ. Fr. capsular, dry, superior.* Anth. 2-celled. Disk and stam. hypogynous. Testa close. Petals cohering.

3. Andromeda. Caps. of 5 cells and 5 valves, dry. Otherwise like *Arbutus.*—L. usually broad.

4. Calluna. *Cal.* 4-*parted,* membranous, coloured, *longer than* the 4-cleft bellshaped persistent but fading corolla, surrounded by 4 green bracts. Stam. 8, with dilated filaments. Caps. 4-celled; *dissepiments adhering to the axis; valves opening at the dissepiments and separate from them.*

5. Erica. Cal. 4-parted. Cor. bellshaped or ovoid, often ventricose, 4-toothed, persistent, fading. Stam. 8. Caps. 4-celled; *valves opening between the dissepiments and carrying a part with them.*

6. Bryanthus. *Cal.* 5-*parted.* Cor. ovoid, deciduous mouth contracted, 5-toothed. Stam. 10, included; *filaments slender, longer than the anthers; cells short, truncate,* opening by pores at the apex. *Stigma peltate, with* 5 *tubercles.* Caps. 5-celled with 5 valves opening at the dissepiments.

7. Boreita. Cal. 4-cleft. Cor. ovoid, ventricose, limb 4-toothed. Stam. 8, included; *filaments flattened shorter than*

the linear anthers, which are sagittate below; cells loosened and opening by oblique pores at the apex. *Stigma simple, truncate.* Caps. 4-celled, with 4 valves opening at the dissepiments.

8. Azalea. Cal. 5-parted. Cor. bellshaped, 5-cleft. Stam. 5, equal, snorter than the corolla; anth. roundish; *cells opening by a longitudinal fissure.* Stigma capitate. *Caps.* 2-3-celled *with 2 or 3 bifid valves whose inflexed edges form the double partitions.*

Tr. III. *VACCINIEÆ. Fr. baccate, fleshy, inferior.* Disk and stam. epigynous. Petals cohering.

9. Vaccinium. Cal. entire or 4—5-toothed or lobed. Cor. 4—5-cleft or toothed. Stam. 8—10; anth. oblong, bifid at the summit. *Berry globose,* crowned by the persistent limb of the calyx, 4—5-celled, *many-seeded.*

Tr. IV. *PYROLEÆ.* Fr. capsular, dry, superior. *Seeds with a loose testa.* Disk 0. Stam. hypogynous. Anth. opening by pores. Petals scarcely cohering.

10. Pyrola. Cal. 5-parted. Cor. of 5 connivent petals. Stam. 10; anth. inverted, with 2 *cells each opening by a round pore at the base. Style 5-lobed.* Caps. 5-valved, opening from near the base to the top; margins of the valves connected by a web.

11. Moneses. Cal. 5-parted. Cor. of 5 petals connected below. Stam. 10; anth. inverted, with 2 *cells each furnished with a tubular horn opening at the end. Stigma 5-parted, radiant.* Caps. 2-celled, 5-valved, opening from the top to the base with glabrous sutures.

** *Anthers opening by a transverse fissure.*

Tr. V. *MONOTROPEÆ.* Fr. capsular, dry, superior. Seeds with a loose testa. Disk 0. Petals scarcely cohering.

12. Monotropa. Cal. 4—5-parted. Cor. of 4—5 *petals each with a hooded honey-bearing base.* Stam. 8—10. Anth. kidney-shaped, 1-celled, 2-valved. Stigma peltate. Caps. 5-celled, 5-valved, many-seeded.

Tribe I. *Arbuteæ.*

1. Ar'butus *Linn.* Strawberry-tree.

1. *A. Unédo* (L.); bark rough, l. elliptic-lanceolate serrate coriaceous glabrous, panicle terminal nodding, pedicels glabrous. —*E. B.* 2377.—An evergreen tree. Fl. whitish, pendulous. Fr. red.—Killarney where it is truly wild. T. IX. X. I.

2. Arctostaph'ylos *Adans.*

1. *A. alpina* (Spr.); procumbent, *l. thin wrinkled serrate* fading but persistent, clusters terminal.—*Arbutus* L. *E. B.* 2030. *St.* 6. 8.—St. woody, trailing, long. L. obovate, netted. Fl. white, hairy about the mouth. *Berry* smooth, *black.*—Dry barren spots on the Highland mountains. Sh. V. S.

2. *A. Uva-ur'si* (Spr.); procumbent, *l. coriaceous obovate entire shining* evergeen, clusters terminal.—*Arbutus* L. *E. B.* 714. *St.* 6. 8.—St. woody, trailing, long. L. blunt, quite entire, rigid. Fl. rose-coloured, smooth. *Berry* globose, *scarlet,* superior.—Dry stony mountain heaths. Sh. VI. E. S. I.

Tribe II. *Ericeæ.*

3. Androm'eda *Linn.*

1. *A. polifólia* (L.); l. alternate lanceolate with revolute margins glaucous beneath, fl. clustered terminal.—*E. B.* 713.—St. slender, woody, prostrate below. Ped. variable in length. Fl. drooping, ovate, pink, occasionally 4-fid and 8-androus. L. evergreen, acute. Peduncles 2 or 3 times as long as the flowers. —Peat bogs. Sh. V.—IX. E. S. I.

4. Callu'na *Salisb.* Ling.

1. *C. Erica* (DC.)—*E. B.* 1013. *C. vulgaris* (Salisb.) ed. viii. —A low tufted shrub. L. small, sessile, closely imbricate, patent, in 4 rows, keeled, each with 2 small spurs at the base, nearly or quite smooth. Fl. small, shortly stalked, drooping, lilac-rose-coloured or white, with the lower ped. leafy; sep. and pet. oblong erect; fl.-raceme ending in a leafy shoot. L. sometimes hoary (var. *pubescens,* Koch).—The supposed *C. atlantica* is not constant to the characters recorded, even on the same bush.—Dry heaths. Sh. VI.—VIII. E. S. I.

5. ERI'CA *Linn.* Heath.

* *Cor. globose or urceolate, stam. included, filaments filiform flattened, stigma peltate.* ERICA D. Don.

1. *E. Tet'ralix* (L.); *l.* 4 in a whorl lanceolate or linear ciliate *downy above and on the midrib beneath* with revolute edges, fl. in an umbellate head, *sep. linear downy* ciliate, anth. spurred, *ovary downy.*—*E. B.* 1014.—St. branched below and often especially about the middle, simple in the upper part, densely leafy below, the whorls more distant towards the top and usually leaving a leafless space next to the flowers. Tips of young shoots green. Young. l. always downy above, old l. sometimes glabrous. Sep. downy and mealy. Fl. rose-coloured. Style usually included.—β. *Watsoni*; cor. ventricose, fl. more or less racemose. *Sy. E. B.* 888. It may be a hybrid.—Boggy heaths. β. Truro, Cornwall. Sh. VII. VIII. E. S. I.

2. *E. Mackaii* (Hook.); *l.* 4 in a whorl ovate ciliate *the midrib beneath and upper surface glabrous* with revolute edges, fl. in an umbellate head, *sep. ovate lanceolate glabrous*, anth. spurred, *ovary glabrous.*—*E. B. S.* 2900. *E. Mackaiana* (Bab.) ed. viii.—St. irregularly branched throughout, particularly above, densely and equally leafy quite up to the flowers. Tips of young shoots pink. L. and sepals quite without down; l. mealy beneath but the midrib bare; sep. with a small portion of meal near the apex beneath, otherwise quite bare. Fl. purplish. Style protruded.—I have seen forms of this taken for *E. ciliaris.*—Very wet moors between Roundstone and Clifden, and between Carna and Lough Sheedah, Co. Galway. Sh. VIII. IX. *Mackay's Heath.* I.

3. *E. cinérea* (L.); *l.* 3 *in a whorl* linear-lanceolate acute *keeled beneath with a central furrow glabrous, fl. in dense whorled racemes*, sep. linear-lanceolate smooth acute keeled, anth. spurred, ovary glabrous.—*E. B.* 1015.—St. with many upright branches. L. flat above, minutely serrulate. Fl. reddish purple.—Dry heaths. Sh. VII. VIII. *Fine-leaved Heath.* E. S. I.

4. *E. ciliáris* (L.); *l.* 4 in a whorl *ovate ciliate* with revolute edges. *fl. in terminal unilateral racemes*, anth. not spurred, mouth of the cor. oblique.—*E. B. S.* 2618.—St. long, straggling, ending in a long raceme of large oblong purple flowers and producing many short barren branches. Style protruded. Ovary glabrous.—Sandy heaths. Wareham, Dorset. Edgecome Downs near Carclew, Cornwall. Mr. More, the author, and others have failed to find it near Clifden, Co. Galway. Sh. VII. VIII. E.

** *Cor. bellshaped or shortly tubular, stam. exserted, filaments flattened, style capitate.* GYPSOCALLIS D. Don.

5. *E. mediterránea* (L.); *l. 4 in a whorl linear glabrous* flat above *convex* with a central furrow *beneath*, decurrent line from the l. reaching but not extending beyond the next whorl, fl. axillary drooping racemose, *cor. cylindrical-urceolate* twice as long as the coloured calyx, *anth. terminal* not spurred opening throughout nearly their whole length.—*E. B. S.* 2774. *E. hibernica* Syme.—St. 2—5 feet high, with many upright rigid branches terminating in leafy racemes of flesh-coloured flowers but afterwards prolonged. L. many, erect-patent. Bracts above the middle of the pedicels. Stam. and style slightly exserted, style afterwards elongated. Ovary glabrous.—Mountain bogs. West of Mayo and Galway. Sh. IV. I.

6. *E. vágans* (L.); l. 4—5 in a whorl linear glabrous, fl. axillary crowded, *cor. short* bellshaped, sep. small ovate blunt, *anth. lateral* ovate of 2 distinct cells gibbous at the base.—*E. B.* 3.—St. 1—2 feet high, copiously branched. Fl. usually collected in large numbers considerably below the top of the branches, cor. red or white. Anth. dark purple, not spurred. Ovary glabrous.—West of Cornwall. Sh. VII. VIII. *Cornish Heath.* E.

6. BRYAN'THUS *S. G. Gmel.* [*Phyllodoce* Salisb. ed. viii.].

1. *B. taxifólius* (A. Gray); l. linear denticulate, pet. glandular-hairy, calycine segments lanceolate acute, anth. two-thirds shorter than the glabrous filaments.—*Menziesia* Sm. *E. B.* 2469.—St. 4—5 in. high, branched, naked below, densely hairy above. Ped. terminal, all together, simple. Fl. large, pale bluish red.—Sow of Athol, Perthshire. Sh. VI. VII. S.

7. BORET'TA *Neck.* [*Dabeocia* D. Don, ed. viii.].

1. *B. cantab'rica* (O. Kuntze).—*Menziesia* Sm. *E. B.* 35.—St. bushy, 1—2 feet long, ultimately decumbent. L. ovate or elliptic, flat, with revolute edges, white and cottony beneath. Fl. large, purple, sometimes white, drooping, on short stalks, in terminal simple unilateral clusters. Anth. very large.—Western Galway and Mayo. Sh. VIII. *St. Dabeoc's Heath.* I.

8. AZA'LEA *Linn.*

1. *A. procum'bens* (L.).—*E. B.* 865. *Loiseleuria* Desv. The original and only *Azalea* of Linn —St. woody, spreading pro-

cumbent. L. small, opposite, revolute. Fl. small, on simple stalks, terminal, collected together.—Summits of the Highland mountains. Sh. V. VI. S.

[*Lédum palus'tre* (L.), Fl. Dan. t. 1031, having linear obtuse l. with strongly recurved margins, the lower surface as well as the young shoots covered with reddish-brown felt, and terminal clusters of white fl. with rotate cor. and conspicuously exserted stam., has been known for many years in Perthsh. Introduced ?—See J. of B. 1894, p. 274.]

Tribe III. *Vaccinieæ.*

9. VACCIN'IUM *Linn.*

* *Anth. with 2 dorsal horns. L. deciduous. St. erect.*

1. *V. Myrtil'lus* (L.) ; *l. ovate serrate* glabrous, fl. solitary, st. acutely angular.—*E. B.* 456.—St. woody, about a foot high, branching. Fl. subglobular, greenish tinged with red, nodding. Berries nearly black.—A small form with st. buried and l. like *Salix herbacea* is *V. microphylla* Lange.—Stony woods and heaths. Sh. V. *Bilberry.* E. S. I.

2. *V. intermédium* (Ruthe) ; *l. elliptical apiculate* slightly narrowed below *denticulate* pale green with minute stalked glands and veined beneath, fl. in small terminal drooping racemes, st. subterete.—*Linn. Journ.* xxiv. t. 3.—St. woody, slightly angular, minutely downy above. Fl. urceolate-campanulate, pale pink, scented. Berries globose dark violet, rarely found. L. darker green than in Sp. 1. [Considered a hybrid between Sp. 1 & 4.]—Moors, Cannock Chase. *Bonney.* Sh. VIII. E.

3. *V. uliginósum* (L.) ; *l. obovate entire glaucous and veined beneath,* fl. several together, st. terete.—*E. B.* 581. *St.* 12.—St. woody. Fl. ovoid, flesh-coloured, nodding.—Berries nearly black.—Mountain bogs. Sh. V. *Bog-Whortleberry.* E. S.

** *Anth. without horns on the back. L. evergreen.*

4. *V. Vitis-idæ'a* (L.) ; *l. obovate dotted beneath,* margins revolute and *somewhat crenate,* fl. racemose terminal, *cor. bell-shaped.*—*E. B.* 598.—Evergreen. St. woody, 6—8 in. high, straggling. L. like those of Box, dark green above. Fl. pink, 4-cleft. Berries red, acid.—Mountain heaths. Sh. VI. VII. *Red Whortleberry. Cowberry. Cranberry* of the North. E. S. I.

5. *V. Oxycoc'cos* (L.); *l. ovate entire* with revolute margins *glaucous beneath*, fl. terminal on long simple peduncles, *cor. rotate* with reflexed segments.—*E. B.* 319. *Oxycoccus quadripetala* (Gilib.).—St. procumbent, filiform, rooting. L. small. Fl. bright rose-colour. Cor. deeply divided, remarkably reflexed. Berries crimson.—Wet bogs. Sh. VI. VII. *Cranberry.* E. S. I.

[*V. macrocarp'um* (Ait.); l. oblong with flat margins, fl. lateral on long simple peduncles.—Soughton Bog, Mold, Flintshire. An American plant, probably sown there.]

Tribe IV. *Pyroleæ.*

10. Pyr'ola *Linn.* Winter-green.

1. *P. rotundifólia* (L.); l. nearly round entire or slightly crenate, fl. racemose, cal.-segments lanceolate acute, *style bent down and curved upwards at the end longer than the ascending stam.*, stigma annular with 5 erect blunt points.—*E. B.* 213.—Fl. white, rather many, expanded. Style longer than the petals. Stam. all turned upwards. L. many.—*β. arenaria* (Koch); l. smaller, st. with bracts throughout, cal.-segments shorter and broader. *Sy. E.B.*896.—Damp bushy places and reedy marshes. *β.* Sand-hills near Lytham, Lancashire. P. VIII. E. S. I.

2. *P. média* (Sw.); l. nearly round or roundish-oval slightly crenate, fl. racemose, cal.-segments ovate-acute, *stam. all. regularly inflexed shorter than the nearly straight declining style, stigma annular with 5 erect points.*—*E. B.* 1945.—Fl. milk-white tinged with pink, rather many, less expanded than in the preceding. Style projecting a little beyond the corolla, longer than the ovary, always nearly straight. L. many.—Woods in the North. P. VII. VIII. E. S. I.

3. *P. minor* (L.); l. roundish-oval crenate, fl. racemose, cal.-segments ovate-prolonged acute, *stam.* regularly inflexed *equalling the straight style, stigma without a ring* 5-lobed pointless.—*E. B.* 2543.—*St.* 13. 12.—Fl. pale pink, many, on very short pedicels, nearly closed. Style shorter than the ovary, included. L. many.—Mossy woods and thickets. P. VI. VII. E. S. I.

4. *P. secun'da* (L.); l. ovate acute serrate, *fl. in a secund raceme*, cal.-segments triangular rounded notched, stam. regularly inflexed equalling the *long straight style*, stigma 5-lobed without

a ring or points.—*E. B.* 517. *St.* 13. 13.—Fl. white drooping, oval-oblong, nearly closed. Style very long, exserted. L. many.—Mossy alpine woods. P. VII. E. S. I.

11. Mone'ses *Salisb.*

1. *M. grandiflóra* (Gray).—*E. B.* 146. *Pyrola uniflora* L.—L. few, roundish, serrate. Fl. solitary, terminal, large, drooping, white, open, nearly an inch broad. Stam. shorter than the pet. and closely adpressed to them. Stigmas very large.—Woods in the north and west of Scotland, rare. P. VI. VII. S.

Tribe V. *Monotropeæ.*

12. Monot'ropa *Linn.*

1. *M. Hypop'itys* (L.); fl. in a drooping cluster, 8 stam. in terminal, 10 in lateral fl., fr. erect, bracts and fl. glabrous externally.—*E. B.* 69. *Hypopitys Monotropa* (Crantz).—Inner side of the pet., filaments, germen and style glabrous (*H. glabra* Bernh., DC.) or hairy (*M. multiflora* Scop., DC.).—Plant 6—8 in. high, succulent, simple, clothed with ovate scales, terminating in a short cluster, dingy yellow, turning nearly black. Fl. with large scale-like bracts. Fr. ovoid erect. Not parasitical. Clusters sometimes erect.—Woods. P. VII. VIII. *Yellow Bird's nest.* E. S. I.

Order XLVIII. AQUIFOLIACEÆ.

Sep. inferior, 4—9, imbricate. Cor. regular, 4—6-parted, imbricate. Stam. inserted upon the base of the corolla and alternate with its lobes. Disk 0. Ovary 2—6-celled; ovules solitary, pendulous, with a cup-shaped seed-stalk. Fr. fleshy, not bursting; seeds stony, 2—6.—Stipules small, deciduous.

1. Ilex. Cal. 4—5-fid, persistent. Cor. rotate, 4—5-fid. Stam. 4—5. Stigmas 4—5, nearly sessile. Fr. fleshy, containing 4—5 seeds.

1. I'lex *Linn.* Holly.

1. *I. Aquifólium* (L.); l. ovate acute spinous wavy shining, peduncles axillary short many-flowered, fl. somewhat umbellate. —*E. B.* 496. *St.* 7. 4.—A small tree. L. evergreen, often

quite entire on the upper branches, edged with strong spinous teeth and terminated by a spine on the lower ones. Fl. white. Berries scarlet.—Woods and hedges. T. VI.—VIII. E. S. I.

Order XLIX. OLEACEÆ.

Cal. gamosepalous, divided, persistent; or none. Cor. with 4—8 divisions, valvate, rarely 0. Stam. 2. Ovary free, 2-celled; ovules pendulous. Stigma entire or bifid. Fr. a berry, drupe, or capsule, often 1-seeded.—No stipules.

1. LIGUSTRUM. Fr. fleshy, a berry containing 2 seeds. Cal. cupshaped with 4 minute teeth. Cor. funnelshaped; limb 4-cleft, spreading. Stam. 2.

2. FRAXINUS. Fr. (a samara) dry, of 1 or 2 single-seeded cells, compressed and leaflike at the end, pendulous. Cal. 0 or 4-cleft. Cor. 0, or of 4 petals. Cal. and cor. wanting in our plant.—Fl. sometimes with only stam. or with pistils only.

1. LIGUS'TRUM *Linn.* Privet.

1. *L. vulgáre* (L.); l. elliptic-lanceolate entire glabrous, panicles terminal compound dense.—*E. B.* 764. *St.* 14. 1.—A bushy shrub, 6—8 feet high, with straight smooth branches and opposite leaves. Fl. white. Berries globose, black, rarely yellow.—Thickets in the South. Sh. VI. VII. E. I.

2. FRAX'INUS *Linn.* Ash.

1. *F. excel'sior* (L.); l. pinnate with 4—8 pairs of nearly sessile ovate-lanceolate acuminate serrate leaflets, cal. wanting.—*E. B.* 1692. *St.* 44. 7.—A handsome tree. Usually diœcious. Fl. appearing before the l., in axillary clusters with no perianth. —[β. *E. heterophyllus* (Vahl); l. simple and pinnate. *E. B.* 2476.]—Woods and hedges. T. IV. V. E. S. I.

Order L. APOCYNACEÆ.

Cal. in 4 or 5 persistent divisions. Cor. regular, 4—5-lobed, deciduous, twisted in the bud. Stam. 5, filaments distinct. Anth. 2-celled, basifixed; pollen granular. Ovaries 2, 1-celled; or 1 of 2 cells. Stigmas 1. Fr. 2 follicles. Seed with fleshy perisperm.—Rudimentary stipules rarely seen.

1. VINCA. Cor. salvershaped; tube long, with 5 angles at the mouth, closed by speading hairs and the connivent stamens; limb flat, 5-lobed. Stigma capitate with a ring at its base. Fr. of 2 erect long slender follicles or rarely one, 2 glands alternating with them.

1. VIN'CA *Linn.* Periwinkle.

†1. *V. mínor* (L.); st. procumbent, l. lanceolate-elliptic, their margins as well as those of the small lanceolate calyx-segments glabrous.—*E. B.* 917.—St. rooting. Flowering branches erect. Fl. smaller than those of the next, blue, rarely white.—Woods and thickets. P. V. VI. *Lesser Periwinkle.* E. I.

[2. *V. májor* (L.); st. somewhat ascending, l. ovate acute or subcordate, their margins as well as those of the long subulate calyx-segments ciliate.—*E. B.* 514.—St. at first ascending, afterwards prostrate and rooting. Flowering shoots erect. Fl. large, purplish blue.—Hedges and thickets, introduced. P. IV. V. *Greater Periwinkle.*] E. I.

Order LI. GENTIANACEÆ.

Cal. inferior, divided, persistent. Cor. regular, 4—8-fid, hypogynous, marcescent; imbricate and twisted, rarely induplicate in the bud. Stam. inserted on the cor., as many as the segments. Anth. versatile. Ovary of 2 carpels with the edges slightly inflexed or meeting. Caps. many-seeded, generally two-valved. —No stipules.

Subord. I. GENTIANEÆ. Corolla twisted in the bud. L. opposite.

* *Style deciduous.*

Tribe I. *CHLOREÆ.* Corolla rotate.

1. BLACKSTONIA. Cal. 8-parted. Cor. nearly rotate, 8-parted. Stam. 8. Style 1. Stigma 2—4-cleft. Caps. 1-celled placentas on the inflexed margin of the valves.

Tr. II. *ERYTHRÆEÆ.* Corolla funnel- or salver-shaped.

2. ERYTHRÆA. Cal. 4—5-fid. Cor. funnel-shaped, limb short, 4—5-fid. *Stam.* 4—5. Anth. erect, at length spirally twisted. Style simple; stigmas 2. Caps. imperfectly 2-celled from the inflexed margins of the valves.

3. Cicendia. Cal. 4-partite or lobed. Cor. salver-shaped, limb short, 4-fid. *Stam.* 4. Anth. erect, not twisted. Stigma capitate, undivided. Caps. 1- or imperfectly 2-celled.

** *Style persistent or stigma sessile.*

Tr. III. *SWERTIEÆ.* Style often wanting, stigma persistent.

4. Gentiana. Cal. 4—5-cleft. Cor. funnel- or salver-shaped, limb 4—5-cleft. Stam. 4—5, straight. Stigmas 2. Caps. 1-celled, seeds on the inflexed margins of the valves.

Subord. II. MENYANTHEÆ. Corolla induplicate in the bud. L. alternate, or opposite to the fl.-stems.

5. Limnanthemum. Cal. 5-parted. Cor. rotate, thin; limb 5-parted, smooth on the disk, hairy or scaly at the base within. Stam. 5. Stigma with two toothed lobes. Caps. 1-celled, not bursting.—*L. simple*, floating.

6. Menyanthes. Cal. 5-parted. Cor. funnelshaped, fleshy; limb 5-parted hairy within. Stam. 5. Stigma capitate, notched. Caps. 1-celled, 2-valved; valves bearing the seeds along their middle.—*L. ternate.*

Suborder I. *Gentianeæ.* Tribe I. *Chloreæ.*

1. Blackstonia *Huds.* [*Chlora* Linn. ed. viii.]. Yellow wort.

1. *B. perfoliáta* (Huds.); lowermost l. elliptic-oblong narrowed below, stem-l. broadly connate.—*E. B.* 60. *R.* xvii. 1060.—St. 12—18 in. high, simple perfoliate. Stem-l. triangular-ovate, connected by their whole breadth in rather distant pairs, glaucous. Panicle forked, many-flowered. Cal. divided to its base into linear-subulate segments. Cor. bright yellow. Stigmas yellow.—Damp chalky places. A. VII.—IX. E. I.

Tribe II. *Erythrææ.*

2. Erythræ'a *Necker.* Centaury.

* *Stamens from top of cor.-tube. Caps. included.*

1. *E. ramosiss'ima* (Pers.); st. much branched acutely quadrangular, l. ovate the uppermost oblong lanceolate, *fl. all stalked* axillary and terminal, *cal. rather shorter than the tube of*

the opening corolla, cor.-lobes elliptic-oblong blunt.—*E. B.* 458. *Sy. E. B.* 910 b. *E. pulchella* (Fr.) ed. viii.—St. quite simple, 1 in. high and 1-flowered; or very much branched, 6—8 in. high, with very many flowers. Radical l. very few. Panicles forked, a fl. in the fork; *lateral fl. distant from the floral leaves.* Fl. rose-coloured.—"β. *E. tenuiflora* (Link); branches erect forming a term. fastigiate lengthened corymb." *Towns. MS.*—In each species the length of the tube must be observed exactly when the flower is opening.—Sandy ground. β. Low ground between Cowes and Newport, Isle of Wight. *Townsend.* A. VII.—IX. E. S. I.

2. *E. Centaúrium* (Pers.); st. branched above quadrangular, l. elliptic-oblong, the upper ones acute, *fl. nearly sessile* corymbosely panicled, *cal. not half as long as the tube of the opening corolla, cor.-lobes oval.—Sy. E. B.* 909.—St. 6—18 in. high, usually simple below. Panicles of fl. lax. *Lateral fl* apparently stalked, but *sessile between the two small floral leaves.* Radical l. many. β. *capitata* (Koch); pan. compact, plant dwarf. *E. latifolia, E. B. S.* 2719.—Dry pastures. A. VII. VIII. E. S. I.

3. *E. latifólia* (Sm.); st. short simple 3-cleft at the top, *l. broadly oval blunt* 5—7*-veined, fl. in compact* round dense forked *term. tufts subsessile,* cal. about equalling cor.-tube, cor.-lobes lanceolate.—*Sy. E. B.* 907.—St. about 3 in. high. Fl. ½ as large as in Sp. 2. Lowest l. sometimes almost orbicular.—Sands by sea near Liverpool, very rare. A. VII. E.

4. *E. littorális* (Fries); st. simple solitary or several from the crown of the root, l. oblong-linear blunt narrowed below, radical l. crowded spathulate, *fl. sessile* between the floral l. densely corymbose, *calyx as long as the tube of the opening corolla,* cor.-lobes oval blunt.—*Sy. E. B.* 908. *E. chloödes* Gren., not *E. linariifolia.*—St. 2—6 in. high. Corymb usually trichotomous, dense; branches sometimes long. Fl. rose-coloured. Radical l. narrow, many.—Sandy sea-shores. A. VII. VIII. E. S.

** *Stamens from base of cor.-tube. Ripe caps.* ⅓ *protruded, unripe slightly so.*

5. *E. capitáta* (W.); st. short simple, st.-l. ovate or subspathulate blunt 3-veined, fl. sessile in compact term. tufts, cal. about equalling *cylindrical cor.-tube,* cor.-lobes oval blunt.—*Linn. Journ.* xviii. t. 15.—St. less than 3 in. high. Fl.-tufts sessile; there are often from outer bracts a few long-stalked tufts over-

topping primary tuft. Root-l. in rosette, with 3 long and often 2 short veins. *Cor.-tube not narrowed at top*, not lengthening after flowering. Stam. from quite base of cor.-tube, otherwise free.—Downs at Cornwall, Isle of Wight, Sussex, Northumb. &c. A. or B. VII. VIII. E.

3. CICEN'DIA *Adans.*

1. *C. filifor'mis* (Delarb.); cal. bellshaped with 4 ovate acute lobes, st. threadshaped forked.—*Exacum* Sm. *E. B.* 235. *Microcala* Hoffm. & Link.—St. 1—4. in. high. Radical l. linear-lanceolate, stem.-l. subulate, all sessile. Fl. yellow, solitary, on long stalks. Stigma capitate. Damp sandy places. A. VII. VIII. E. I.

[*C. pusil'la* (Griseb.); *cal.* 4-*parted* with linear segments, st. slender branching from its base.—*E. B. S.* 2994.—St. much branched throughout, 1—4 in. high. L. all narrowly linear-lanceolate. Fl. pink. Stigma 2-lobed.—On spots sometimes flooded. Paradis, Guernsey. *Capt. Gosselin.* A. VI. VII.]

Tribe III. *Swertieæ.*

4. GENTIA'NA *Linn.* Gentian.

* *Cor. funnel- or somewhat salver-shaped.*

1. *G. Amarel'la* (L.); cor. 5-cleft hairy in the throat, *cal.-lobes* 5 *nearly equal* lanceolate, l. sessile ovate-lanceolate, radical l. oval-spathulate.—*E. B.* 236. *Sy. E. B.* 917.—Very variable in size and in the number of the flowers, 3—12 in. high, erect. St. square, much branched, sometimes from the base. Fl. rarely 4-cleft. Caps. stalked or sessile. Cor.-tube obconical or subcylindrical.—*α*. *G. Amarella* (Sm.); branches of st. erect, cal.-segm. nearly equal, cor.-tube cylindrical a little exceeding calyx, fl. lurid purple. [A small annual form with basal l. ovate-lanceolate is *G. uliginosa* Willd.]—*β*. *G. germanica* (Willd.); st. much branched ascending, cal.-segm. unequal, cor.-tube obconical much exceeding calyx, fl. bluish-lilac. *J. of B.* ii. t. 15.—Dry calcareous fields. A. or B. VIII. IX. *Felwort.* E. S. I.

2. *G. campes'tris* (L.); cor. 4-cleft hairy in the throat, *cal.-lobes* 4, 2 *outer ones very large* ovate, l. ovate-lanceolate.—*E. B.* 237.—St. 3—10 in. high. Fl. pale lilac; cor.-tube slightly

thicker upwards. Caps. nearly sessile. Upper l. and sepals pointed. [An annual form with ovate-lanceolate lower l. and rather smaller fl. is *G. baltica* Murb.]—Dry limestone hills. A. or B. VIII. IX. E. S. I.

3. *G. nivális* (L.); cor. 5-cleft with minute intermediate bifid lobes, *throat naked, cal. cylindrical with* 5 *equal lobes and keeled angles,* l. ovate lowermost broadly elliptic.—*E. B.* 896.—St. erect, slightly branched, 1—6 in. high. Fl. bright blue.—Top of Highland mountains, very rare. A. VIII. S.

4. *G. ver'na* (L.); cor. 5-cleft with small intermediate bifid lobes, throat naked, cal. with prominent angles and sharp teeth, l. ovate lower ones crowded, *st. cæspitose single-flowered* with 1 or 2 pairs of leaves.—*E. B.* 493. *St.* 40. 12.—St. prostrate, rooting, each ending in a roselike tuft of l. and a single short flowering shoot. Fl. rather large, vivid blue.—Barren limestone districts. Teesdale, Durham. Burrin, Co Clare; Gort, Galway, Tuam, &c. P. IV.—VI. E. I.

** *Cor. bellshaped, its throat naked.*

5. *G. Pneumonan'the* (L.); *cor.* 5-cleft, cal. entire with linear blunt lobes, *fl. mostly solitary* slightly stalked, *l. linear* blunt.—*E. B.* 20.—St. 4—10 in. high, leafy, simple, erect or ascending. Fl. very large, deep blue within and with a broad greenish band down the middle of each segment.—Moist turfy heaths. A. VIII. IX. E.

Suborder II. *Menyantheæ.*

5. LIMNAN'THEMUM *S. P. Gm.*

1. *L. peltátum* (Gm.); l. roundly heartshaped floating wavy at the edges, ped. clustered 1-fld., cor. ciliate.—*Villarsia* Vent. *E. B.* 217. *R.* xvii. 1042. *L. nymphæoides* (Link) ed. viii.—Floating. St. long, round, branched. L. resembling those of *Castalia speciosa,* but much smaller. Fl. yellow. Caps. sometimes 3-valved.—Still places in rivers, rare. P. VII. VIII. E.

6. MENYAN'THES *Linn.* Buckbean.

1. *M. trifoliáta* (L.).—*E. B.* 495.—St. ascending, round, leafy. L. ternate. Leaflets equal, obovate, wavy. Raceme long-stalked, opposite to a leaf, many-flowered. Cor. flesh-coloured, densely fringed within.—Boggy places. P. V.—VII. E. S. I.

Order LII. POLEMONIACEÆ.

Cal. inferior, 5-parted, persistent. Cor. hypogynous, regular, 5-lobed. Stam. 5, unequal, on the tube of the corolla. Ovary 3-celled. Stigmas 3-fid. Caps. 3-celled, 3-valved; valves separating at the axis.—No stipules.

1. POLEMONIUM. Cal. 5-fid. Cor. rotate, with a short tube and 5-lobed limb; throat nearly closed by the dilated bases of the filaments.

1. POLEMO'NIUM *Linn.* Jacob's Ladder.

1. *P. cærúleum* (L.); st. angular, l. glabrous pinnate, leaflets ovate-lanceolate pointed, panicle downy glandular.—*E. B.* 14.—St. 1—2 feet high, simple, hollow. L. alternate, leaflets many. Fl. many, somewhat drooping, bright blue or white.—Bushy hilly places, rare. P. VII. E. S.

Order LIII. CONVOLVULACEÆ.

Cal. inferior, of 5 persistent imbricate often unequal sepals. Cor. hypogynous, regular, deciduous. Stam. 4—5, from near the base of the corolla. Ovary of 2—4 cells, few-seeded, surrounded by an angular hypogynous disk. Style 1, rarely 2. Caps. with the valves separating from the edges of the dissepiments or bursting transversely.—No stipules.

* *With leaves and leaflike cotyledons. Æstivation plaited.*

1. CONVOLVULUS. Cor. bellshaped, with 5 prominent plaits and 5 shallow lobes. Style simple; stigmas 2. Caps. 2—4-celled; cells 2-seeded.

** *Without leaves or cotyledons. Æstivation imbricate.*

2. CUSCUTA. Cal. 4—5-cleft. Cor. roundish-urceolate or bellshaped, 4—5-parted, with as many scales alternating with the segments at the base within. Stam. 4—5. Styles 2, rarely 1. Caps. bursting transversely, 2-celled, 4-seeded.

1. CONVOL'VULUS *Linn.* Bindweed.

* *Bracts minute, distant from the flower.*

1. *C. arven'sis* (L.); l. ovate- or strapshaped-hastate, peduncles mostly 1-flowered.—*E. B.* 312.—St. many, angular,

twining or prostrate, leafy, branched. Peduncles sometimes 2-flowered. Cor. beautifully variegated with pink and white. Caps. 2-celled. Roots descending remarkably deep. Plant glabrous, or st. and l. downy.—Fields and hedges. P. VI.—VIII. E. S. I.

** *Bracts* 2 *large, close to the flower.* VOLVULUS *Medic.* CALYSTEGIA *R. Br.*

2. *C. sépium* (L.); l. ovate- or triangular-hastate, peduncles 1-flowered square.—*E. B.* 313.—St. twining, many feet long, with large rather distant leaves. Fl. solitary, axillary, large, white, rarely pink. Bracts quite enclosing the calyx. Fr. imperfectly 2-celled through the shortness of the dissepiment.—Hedges and thickets. P. VII. VIII. E. S. I.

3. *C. Soldanel'la* (L.); l. reniform slightly angular fleshy, peduncles 1-flowered with 4 membranous angles.—*E. B.* 314.—*St. short, procumbent.* Fl. large, solitary, axillary, very handsome, pink with yellow bands. Bracts rather shorter than the calyx. Rootstock long creeping.—Sandy sea-shores. P. VI. VIII. *Sea-side Bindweed.* E. S. I.

2. CUS'CUTA *Linn.* Dodder.

Clusters sessile in all our species.

1. *C. europæ'a* (L.); cor.-tube cylindrical afterwards ventricose, *scales adpressed* to inside of tube bifid distant below with rounded spaces between them, *cal. much shorter than corolla.*—*E. B.* 378.—St. threadshaped, branching, reddish. Fl. in rather large clusters, yellowish.—Parasitical upon herbaceous plants. A. VIII. IX. *Greater Dodder.* E. S.

[*C. Epilinum* (Weihe); cor.-tube ventricose, *scales adpressed* fringed with teeth distant below with rounded spaces, *cal. with fleshy segments deltoid below nearly as long as the cor.-tube.*—*E. B. S.* 2150. *C. densiflora* Soy.-Willm.—St. slender, nearly simple, pale green. Fl. in rather small distant clusters, whitish. Scales bifid, with 4—8 teeth on each lobe. Ventricose cor.-tube with 5 longitudinal prominences; segm. ventricose. Styles at first erect, soon bowing outwards; stigmas converging.—Parasitical upon Flax and very injurious to the crop. A. VIII. *Flax Dodder.*] E. S. I.

2. *C. Epíthymum* (Murr.); *st. twining irregularly,* cor.-tube cylindrical, *scales converging usually equalling the tube of the*

cor. fringed with teeth and rounded at the end *close together below with narrow acute spaces,* cal. bellshaped *shorter than the cor.-tube.*—*E. B.* 55.—St. slender, red. Fl. small, with a reddish thin cal., and white cor. with spreading ovate-acute segments. Sep. broad, ovate-apiculate, longer than their tube, with patent tips. Anth. blunt or notched at the end. Scales broad; the connecting membrane adpressed throughout.—[See *C. Ulicis* (Godr.); scales less than in Sp. 2 or 3, less deeply fringed converging. *Bull. Bot. Fr.* xxiii.]—Parasitical upon small shrubby plants. A. VII.—IX. *Lesser Dodder.* E. S. I.

†3. *C. Trifólii* (Bab.); *st. clasping like a ring,* cor.-tube cylindrical, *scales converging usually equalling half the tube of the cor.* fringed with teeth and rounded at the end *distant below with rounded spaces,* cal. narrowed below about as long as the cor.-tube.—*E. B. S.* 2898. Not *C. Epithymum β. Trifolii R.* xviii. 1343.—*St.* slender, branching, reddish yellow, forming dense broad circular patches. Fl. small, white. Cal. fleshy, usually tipped with red; sep. lanceolate, about as long as their tube, adpressed. Anth. apiculate. Scales narrow; connecting membrane not adpressed, forming cuplike spaces between itself and the corolla.—Parasitical upon Clover chiefly. A. VII.—IX. *Clover-Dodder.* E.

[*C. approximata* (Bab.); like Sp. 2 but scales divided by acute spaces and truncate, was found on Bockhara Clover. *A. N. H.* xvi. t. i.; and *C. hassiaca* (Pfeiff.); *fl. stalked,* cor.-tube bellshaped closed with converging scales, stig. capitate, Heliotrope-scented, on Lucerne. Both introduced. A. VIII. IX.]

Order LIV. BORAGINACEÆ.

Fl. mostly in scorpioidal cymes, symmetrical. Cal. inferior, 5-parted, persistent. Cor. hypogynous, usually regular. Stam. 5, inserted on the corolla. Ovary of two 2-parted carp., with each lobe 1-seeded; ovules pendulous. Style simple, from near the base between the lobes of the ovary. Fr. separating in 4 nutlets or 2 bilocular portions. Seeds without albumen.—L. alternate.—Stip. 0.

Tribe I. *CYNOGLOSSEÆ.* Nutlets 4, on the persistent base of the style.—Stam. included.

1. Asperugo. Cal. 5-cleft with alternate smaller teeth, enlarged and compressed in fruit. Cor. funnelshaped with

rounded scales in the throat. Filaments of stam. short. *Nutlets* tubercled, *compressed*, attached by their narrow side, covered by the compressed calyx.

2. Cynoglossum. Cal. 5-cleft. Cor. funnelshaped, the mouth closed with prominent blunt scales. Filaments very short. *Nutlets roundish-ovate*, depressed, muricate, attached by the upper part of their inner edge.

Tr. II. *ANCHUSEÆ.* Nutlets 4, on an hypogynous disk, with an excavated space surrounded by a tumid ring at their base.

3. Borago. Cal. in 5 deep segments. *Cor. rotate*; tube very short; throat with short erect emarginate scales. *Stam. exserted*; filaments bifid, the inner fork bearing the anther; anthers linear-lanceolate, connivent in the form of a cone.

4. Anchusa. Cal. 5-fid. *Cor. funnelshaped with a straight tube*; throat closed by prominent blunt scales. *Stam. included, subsessile.* Nutlets depressed.

5. Lycopsis. Cal. in 5 deep segments. Tube of the cor. curved; limb oblique. Otherwise like *Anchusa.*

6. Symphytum. Cal. 5-cleft or 5-parted. *Cor. cylindrical-bellshaped, throat closed by* a prominent cone of connivent *lanceolate-subulate scales.* Stam. exserted from the tube but covered by the scales; filaments short. Nutlets ovate.

Tr. III. *LITHOSPERMEÆ.* Nutlets 4, affixed to an hypogynous disk, their base not excavated but attached by a flat or rather convex suface.

7. Echium. Cal. in 5 deep segments. *Cor. sub-bellshaped; throat dilated, naked*; limb irregular. *Stam. exserted; filaments very long, unequal. Style bifid.* Nutlets wrinkled, attached by a flat triangular base.

8. Pulmonaria. Cal. tubular, 5-fid. Cor. funnelshaped, its throat naked. *Stam. included* in the tube; *filaments very short. Style simple. Nutlets* smooth, *attached by their truncate base, which has a central tubercle.*

9. Pneumaria. Cal. in 5 deep segments. Cor. bellshaped, with a short thick cylindrical tube with 5 minute protuberances in its throat. *Stam. protruded* beyond the throat;

filaments rather long. Style simple. *Nutlets* smooth, *inflated, rather drupaceous, attached laterally near their base by a flat surface; seeds free.*

10. LITHOSPERMUM. *Cal. in 5 deep segments.* Cor. funnel-shaped; throat naked or with 5 minute scales. Stam. included in tube; filaments very short. Style simple. *Nutlets* smooth or tubercular, *stony, attached by a truncate flat base.*

11. MYOSOTIS. Cal. 5-parted. *Cor. convolute in the bud, salvershaped*; throat closed with scales; limb 5-fid, blunt. Stam. included; filaments very short. Style simple. Nutlets smooth, convex externally, keeled within, attached by a minute lateral spot near their base.—Distinguished from all the other genera by the convolute corolla.

Tribe I. *Cynoglosseæ.*

1. ASPERU'GO *Linn.* Madwort.

†1. *A. procum'bens* (L.).—*E. B.* 661.—St. procumbent, angular, rough, with short deflexed bristles. L oblong, rough, hispid, lower ones stalked, upper sessile. Fl. small, axillary, solitary, blue, upon short peduncles. Col. of the fr. much enlarged.—Rich waste ground, rare. A. VI. VII. E. S.

2. CYNOGLOS'SUM *Linn.* Hound's-tongue.

1. *C. officinále* (L.); l. downy acute, lower l. elliptic contracting into a petiole, *upper l. lanceolate narrowed below* subcordate half-clasping, nutlets with thickened prominent margin.—*E. B.* 921.—About 2 feet high. Covered with soft adpressed hairs. Cor. dull crimson, veiny; veins disappearing in drying. Nutlets flat in front. Fetid, rarely [var. *subglabrum* Merat] subglabrous and nearly scentless.—Waste ground. B. VI. VII. E. S. I.

2. *C. montánum* (L.); *l. slightly hairy* acute nearly glabrous and shining above rough beneath, inferior oblong narrowed into a long petiole, *upper l. lanceolate slightly narrowed* below clasping, nutlets without thickened edge.—*C. sylvaticum* Sm. *E. B.* 1642.—Clothed with straight spreading hairs. Cor. reddish, changing to blue. L. sometimes very rough.—Shady situations. B. VI. VII. E. I.

Tribe II. *Anchuseæ.*

3. Bora'go *Linn.* Borage.

*1. *B. officinális* (L.); lower l. obovate blunt attenuated below, segments of the cor. ovate acute flat spreading.—*E. B.* 36. —Fl. blue. Anth. very prominent. Stem.-l. much narrowed below so as to appeared stalked, eared at the base. Whole plant hispid with bulbous hairs. St. spreading, 12—18 in. high.—On rubbish and waste ground. B. VI. VII. E. S. I.

4. Anchu'sa *Linn.* Alkanet.

‡1. *A. officinális* (L.); *l. lanceolate* hispid, spikes crowded unilateral, bracts ovate-lanceolate, cal.-segm. bluntish hairy on both sides, scales of cor. hairy.—*E. B.* 662.—Fl. deep purple. Cal.-segm. narrow, longer than tube. St. 1—2 feet high, rough with deflexed hairs.—Waste ground, rare. P. VI. VII. E. S.

‡2. *A. sempervírens* (L.); *l. ovate*, lower l. on long stalks, peduncles axillary each bearing 2 dense spikes with an intermediate flower, cal.-segments hairy on the outside only, bracts minute lanceolate, scales of the cor. downy.—*E. B.* 45.—Fl. blue, rather salver- than funnel-shaped. Cal.-segments narrow. St. 1½—2 feet high, rough with spreading somewhat deflexed hairs.—Waste ground near ruins, rare. P. V.—VIII. E. I.

5. Lycop'sis *Linn.* Bugloss.

1. *L. arven'sis* (L.); l. lanceolate repand-dentate very hispid, cal. of fr. bellshaped erect.—*E. B.* 938.—Fl. small, blue. Whole plant very hispid with strong hairs each rising from a scaly tubercle.—Fields and hedges. A. VI. VII. E. S. I.

6. Sym'phytum *Linn.* Comfrey.

1. *S. officinále* (L.); l. ovate-lanceolate attenuate below, *stem-l. very decurrent* lanceolate, *st. winged* in the upper part.—*E. B.* 817.—Height 1—2 feet, branching. Racemes in pairs, drooping. Fl. yellowish white or purple. Cal.-segments somewhat spreading and pubescence rougher in the usually purple-flowered variety, *S. patens* Sibth. *Sy. E. B.* 1116.—Common in damp places. P. V. VI. E. S. I.

2. *S. tuberósum* (L.); l. ovate-oblong attenuate below, stem-l. lanceolate, *uppermost slightly decurrent, st. scarcely winged* nearly simple.—*E. B.* 1502.—St. 12—18 in. high. Fl. yellowish white,

whole plant smaller and slenderer than the preceding. Anth. twice as long as their filaments.—Damp woods and river-banks, rare. P. VI. VII. E. S.

[*S. peregrinum* Ledeb. (*asperrimum* Bab. not Bieb.) Bot. Mag. 6466, *S. tauricum* Willd., and *S. orientale* L. have been noticed in England, but are not natives.]

Tribe III. *Lithospermeæ.*

7. E'CHIUM *Linn.* Viper's Bugloss.

1. *E. vulgáre* (L.); tubercular hispid, st. erect simple, l. lanceolate 1-ribbed, stem-*l. narrowed below* sessile, *fl. in four lateral scorpioidal cymes*, stam. exceeding the corolla.—*E. B.* 181.—St. 1—2 feet high. Lower l. narrowing into a footstalk. Fl. reddish, then bright blue.—Dry places. B. VI. VII. E. S. I.

2. *E. plantagin'eum* (L.); pilose-hispid, st. erect branched diffuse, lower branches prostrate, radical l. oblong-ovate stalked, *stem.-l. oblong narrowed from a cordate half-clasping base* with lateral ribs, *spikes panicled* long simple, stam. scarcely exceeding the corolla.—*E. B. S.* 2798.—Stam. very unequal, 1 short, 2 intermediate, and 2 longer. Fl. violet-blue. Root reddish.—Jersey. Land's-end, Cornwall. B. VI.—IX. E.

8. PULMONA'RIA *Linn.* Lungwort.

[*P. officinális* (L.); l. ovate-lanceolate with a long stalk, upper l. oblong sessile.—*Sy. E. B.* 1098.—Fl. pale purple. St. 1 ft. high. L. spotted.—Woods and thickets, scarcely naturalized. P. IV. V.] E.

1. *P. angustifólia* (L.); l. narrow-lanceolate narrowed to the base, upper l. sessile.—*E. B.* 1628.—Fl. pink, then blue. St. 1 ft. high. L. less frequently spotted.—Thickets, rare. P. IV. V. E.

9. PNEUMA'RIA *Hill.* (*Mertensia* Roth, ed. viii.)

1. *P. marit'ima* (Hill); st. procumbent branched, l. ovate acute rough with hard dots glabrous fleshy glaucous, nutlets smooth.—*E. B.* 368.—Spreading, very glaucous. Fl. in cymes, purplish blue. Protuberances in throat of cor. yellow. L. tasting like oysters. Nutlets free, forming a pyramid, exceeding the calyx. Pericarp membranous; seeds smaller than the cavity.—Northern sea-shores. P. V.—VIII. E. S. I.

10. LITHOSPER'MUM *Linn.* Gromwell.

1. *L. officinále* (L.); *st. erect* much branched, l. lanceolate acute veined hispid with bulbous adpressed bristles above hairy beneath, *throat of the cor. with a crown of scales, nutlets smooth.* —*E. B.* 134.—St. 12—18 in. high. Fl. greenish yellow, nutlets grey, highly polished, stony, 2 or 3 ripening in each calyx. Root whitish.—Dry and stony places. P. VI.—VIII. *Gromwell.* E. S. I.

2. *L. purpureo-cærúleum* (L.); barren shoots prostrate creeping, fl.-st. erect forked, l. lanceolate acute hispid, hairs on the upperside adpressed bulbous, cor. much exceeding the calyx, nutlets smooth.—*E. B.* 117. Fl. large, bright blue. Nutlets silvery white, highly polished.—This species and the following have no scales in the throat of the cor., but merely 5 longitudinal downy folds (*Rhytispermum* Link).—Thickets on a limestone soil, rare. P. V.—VII. E.

3. *L. arven'se* (L.); st. erect branched, l. lanceolate rather acute hispid with adpressed bulbous hairs on both sides, cor. rather exceeding the calyx, *nutlets tubercled.*—*E. B.* 123.—St. 1 ft. high. Fl. small, white. Nutlets pale brown, polished, pitted, 3 or 4 perfected in each calyx. Root bright red.—Cornfields. A. V.—VII. *Corn Gromwell.* E. S. I.

11. MYOSO'TIS *Linn.* Scorpion-grass.

i. Persistentes. *Cal. of fruit shorter than its stalk.*

* Hairs on cal. all straight and adpressed.

1. *M. scorpioïdes* (L.); fr.-cal. open, its *teeth short* triangular, cor.-limb flat longer than the tube, its *lobes slightly emarginate, style about equalling cal., pubescence of st.* (*usually*) *spreading.*—*E. B.* 1973. *M. palustris* (With.) ed. viii.—L. bluntish, apiculate. Raceme quite leafless. Cor. large, bright blue with a yellow eye. Cal. divided about $\frac{1}{3}$ down.—*α*; rhizome long leafy stoloniferous, st. angular with decurrent prominent lines, pubescence spreading.—*β*. *M. strigulosa* (R.); rhizome short leafless, st. more slender and erect not angular its pubescence adpressed, l. less blunt, style equalling cor.-tube. *M. coronaria* Dum. Base of st. sometimes patently hairy.—Ditches and watery places. P. VI.—VIII. *β*. VI. *Forget-me-not.* E. S. I.

2. *M. répens* (Don); fr.-cal. open, its *teeth narrow lanceolate* acute, *style falling short of cal.,* cor.-limb flat longer than the tube, its *lobes slightly emarginate, pubescence of the st. spreading*

—*E. B. S.* 2703.—L. rather acute. Stoloniferous. St. slightly angular. Raceme usually slightly leafy (1—4 leaves) below. Cor. pale blue. Cal. divided fully halfway down.—Boggy places. P. VI.—VIII. E. S. I.

3. *M. cæspitósa* (Schultz); fr.-cal. open, its *teeth narrow lanceolate* bluntish, *cor.-limb equalling the tube, its lobes entire, style very short, pubescence of the st. adpressed.—E. B. S.* 2661. M. lingulata *Lehm.* (name only).—L. usually blunt or even emarginate. St. round, with a decurrent line. Raceme usually slightly leafy below. Cor. smaller than in the preceding, bright blue, segments narrower and rounded at the end. Style about equalling the cal.-tube.—Watery places. P. VI.—VIII. E. S. I.

** Hairs on cal. not all straight, but some or all hooked.

4. *M. alpes'tris* (Schm.); *cal attenuate below* deeply 5-cleft open with fruit with straight and a few curved adpressed bristles, pedicels ascending, *cor -limb longer than the tube flat, style equalling ½ cal.*, nutlet keeled not rounded at the end, *root-l. on long stalks pointed.—E. B.* 2559. *M. rupicola*, Sm.—L. oblong-lanceolate, stalks of the lower ones slender. Fl. large, handsome, blue, sweet-scented in the evening.—Breadalbane mountains. Micklefell, Teesdale. P. VII. VIII. E. S.

5. *M. sylvat'ica* (Hoffm.); *cal. rounded below ¾-5-cleft closed with fruit*, its tube with spreading hooked bristles, pedicels divergent, *cor.-limb longer than tube flat, style nearly equalling cal.*, nutlet keeled on one side upwards bluntish, *root.-l.* on short dilated stalks *bluntish.—E. B. S.* 2630.—L. oblong-lanceolate; stalks of the oblong-ovate lower l. dilated. Fl. large, handsome, blue. Cal. divided more than halfway down.—Shady places, rare. P. V. VI. E. S.

6. *M. arven'sis* (Hill); *cal. half-5-cleft* closed with fruit, tube with spreading hooked bristles, pedicels divergent, *cor.-limb equalling tube concave*, cor.-lobes entire, *style very short*, racemes stalked.—*E. B. S.* 2629. *M. intermedia* Link.—L. oblong, acute; lower l. oblong-obovate, blunt. Fl. usually small. A large-flowered plant found in shade is often taken for *M. sylvatica*.—Cultivated land and thickets. A. VI.—VIII. E. S. I.

ii. Fugaces. *Cal. of fruit not shorter than its stalk, its tube with spreading hooked bristles.*

7. *M. collína* (Hoffm.); *fr.-cal. open and ventricose* as long as the diverging pedicels, cor.-limb shorter than exserted tube con-

cave, *style about equalling* ½ *cal.*, racemes stalked usually with 1 distant flower, fr.-pedicels spreading, hairs on the l. straight.—*E. B.* 2558. *St.* 42. 11. *M. hispida* Koch.—Usually slender, erect; or cæspitose with prostrate branches. L. oblong, blunt; lower obovate. *Fl.* small, unchangeably *blue.*—β. *Mittenii* (Baker). Fl. pale, lowest bracteate.—[*M. stricta* Link, *M. arvensis* (R.) *St.* 42. 14, has its fr-cal closed, very short adpressed fr.-pedicels, sessile racemes leafy below. Lowest fr.-ped. not ½ as long as calyx. Probably a native of Britain.]—Dry banks. A. IV. V. E. S. I.

8. *M. versic'olor* (Sm.); *fr.-cal. closed oblong bellshaped* longer than the ascending pedicels, cor.-limb much shorter than tube concave, *style equalling cal.*, racemes stalked.—*E. B.* 480. —St. erect, simple at the base. L. narrow, oblong, acutish; upper ones frequently opposite. *Fl.* small, at first *pale yellow, afterwards blue.* Lobes of fr.-cal. often erect, therefore cal. not truly closed.—On plants in damp places the fl. are at first white and the cal. is "less deeply divided;" sometimes the fl. are all yellow, *M. Balbisiana* (Jord.).—Meadows and banks. A. V. VI. E. S. I.

Order LV. SOLANACEÆ.

Cal. inferior, 5-parted, persistent. Cor. hypogynous, regular or slightly irregular, 5-cleft, deciduous, plicate in bud; the lobes imbricate or imbricate-plicate (in *Solanum* valvate). Stam. 4—5, inserted on the cor., alternate with the lobes. Ovary 1—2- or imperfectly 4-celled. Fr. a berry or capsule. Seeds many. Embryo usually curved, in fleshy albumen, often not in the axis.—No stipules.

Tribe I. *SOLANEÆ.* Cor. rotate; lobes nearly regular and equal, valvate in the bud. Anth. opening by pores.

1. SOLANUM. *Cor.-limb* 5-cleft, *reflexed.* Anth. erect connivent. Berry roundish with 2 or more cells.

Tr. II. *ATROPEÆ.* Cor. tubular; tube plicate in bud; lobes slightly unequal, imbricate in the bud. Anth. opening longitudinally at the margin.

2. ATROPA. *Cor. bellshaped* with 5 equal lobes. Cal. 5-parted, patent and dilated with fruit. Stam. included. Fr. a globose 2-celled berry.

3. HYOSCYAMUS. *Cor. funnelshaped with* a short tube and 5 *unequal* blunt *lobes*. Stigma capitate. Fr. a dry 2-celled caps., ventricose below, furrowed, *opening transversely* by a convex lid.

[LYCIUM. Cal. small and adpressed to base of fruit. *Cor. funnelshaped* with a short tube and 5 equal patent lobes. Stam. exserted. Fr. a 2-celled berry.]

4. DATURA. Cor. funnelshaped, angular, 5-lobed. Cal. deciduous. Stigma 2-lobed. *Caps. 4-valved, with* 2 *partially bipartite cells.*

Tribe I. *Solaneæ.*

1. SOLA'NUM *Linn.* Nightshade.

1. *S. nigrum* (L.); *st. herbaceous* with tubercled angles, l. ovate bluntly dentate or wavy, fl. drooping.—*St.* 1. 4.—Umbel from the intermediate spaces between the leaves. L. attenuate below. Fl. white.—Fr.-stalks thickened upwards. Berries globular. St. a foot or more high.—*α*; hairs incurved upwards, l. sinuate, berry black. *Sy. E. B.* 931.—[*β*. *S. ochroleucum* (Bast.); l. sinuate-dentate, berries yellowish-green.]—*γ*. *S. miniatum* (Bernh.); angles of the st. with prominent tubercles, l. sinuate-dentate more deeply toothed, pubescence patent, berries scarlet. *Sy. E. B.* 932. It may be distinct.—Waste ground. *γ*. Kent, Jersey, [&c.]. A. VII.—X. *Black Nightshade.* E. S. I.

2. *S. Dulcamára* (L.); *st. shrubby*, zigzag, l. cordate-ovate, *upper l. hastate* auricled, fl. drooping.—*E. B.* 565. *St.* 18. 3.—Corymb opposite to leaves. Fl. purple with 2 green spots at the base of each segment. Berries ovate, red [rarely greenish-yellow]. St. climbing to the height of 12—14 feet, nearly round, almost glabrous throughout, or (*S. littorale* Raab) st. and l. downy with patent hairs.—*β*. *marinum* (Bab.); branches of the present year and l. fleshy and usually clothed with hairs incurved upwards, st. angular prostrate diffuse much branched, l. nearly all cordate not hastate. *S. lignosum seu Dulcamara marina* Ray 265.—Woods and hedges, common. *β*. Pebbly sea-beach. Sh. VI. VII. *Bittersweet.* E. S. I.

Tribe II. *Atropeæ.*

2. AT'ROPA *Linn.* Deadly Nightshade. Dwale.

1. *A. Belladon'na* (L.); st. herbaceous, l. broadly ovate-

acuminate entire, fl. solitary axillary on short stalks.—*E. B.* 592.—St. 3 ft. high. Fl. lurid purple, drooping. Berry violet-black, highly poisonous.—Waste places, rare. P. VI.—VIII. E. S. I.

3. Hyoscy'amus *Linn.* Henbane.

1. *H. niger* (L.); l. oblong pinnatifid or sinuate sessile and clasping, lower l. stalked, fl. nearly sessile axillary unilateral.—*Sy. E. B.* 936. *St.* 3. 4.—St. 1—2 feet high. Fl. lurid yellow with dark veins, drooping. Fr. erect. Whole herbage downy glandular, viscid, fetid. Fl. sometimes without dark veins.—Waste places, preferring a calcareous soil. B. V.—VII. E. S. I.

Lyc'ium *Linn.*

*1. *L. chinen'se* (Mill.); l. narrowly lanceolate narrowed at the base, cal. 2-lipped, cor.-tube as long as the limb, berry oblong.—*Sy. E. B.* 933. *L. barbarum* ed. viii.—A straggling shrub with long pendulous spinous branches. Fl. bluish. Filaments woolly at the base. Berry red.—Seems quite naturalized on north Norfolk coast and elsewhere. Sh. VI.—VIII. E.

4. Datu'ra *Linn.* Thorn-apple.

[1. *D. Stramónium* (L.); l. ovate unequally sinuate-dentate glabrous, caps erect spinose.—*E. B.* 1288.—St. 1—2 ft. high. Fl. white (purplish in *D. Tatula*), large, erect. Fr. densely spinose. Caps with 4 dissepiments below, only 2 at the top.—Waste ground, rare. A. VI. VII.] E.

Order LVI. OROBANCHACEÆ.

Cal. variously divided, persistent. Cor. irregular, usually 2-lipped, persistent, imbricate in the bud. Stam. on the cor., 4, didynamous. Anth. 2-celled; cells distinct, parallel. Ovary in a fleshy disk, 1-celled, with 2 or more parietal placentas. Stigma 2-lobed. Fr. capsular, 2-valved, with many minute seeds.—Leafless root-parasites.

1. Orobanche. Cal. 4-cleft or of 2 usually bifid sepals. Cor. ringent, 4—5-cleft, deciduous, its base persistent. Bracts 1—3.
2. Lathræa. Cor. 2-lipped; the upper lip concave, deciduous, entire; its base persistent. Otherwise like *Orobanche.*

1. OROBAN'CHE *Linn.* Broom-rape.

* *Sepals 2, entire or bifid, separate or connected below in front. Bract 1. Valves of caps. cohering at each end.*

1. *O. major* (L.); sep. 2-veined equally bifid nearly equalling the cor.-tube, cor. bellshaped ventricose at the base in front, its back curved, lips wavy obscurely denticulate (not fringed), upper lip concave nearly entire its sides patent, middle lobe of lower lip much longer than lateral lobes, *stam. inserted at the base of the cor.* glabrous below but their upper part and the style glandular-pubescent.—*R. I.* f. 900 & 923. *E. B.* 421. *O. Rapum-genistæ* (Thuill.)[1].—*Stig. of 2 distant yellow lobes.* Anth. white when dry.—Parasitical upon Broom, Furze, and other shrubby leguminous plants. P. V.—VII. E. S. I.

2. *O. rúbra* (Sm.); sep. 1-veined lanceolate attenuate exceeding the cor.-tube, undivided, *cor.* bellshaped its back curved *glandular-pubescent externally and the upper lip internally, lips acutely denticulate* and crisped, upper lip emarginate its sides patent, lateral lobes of lower lip nearly equal, intermediate lobe rather long, *stam. inserted near to the base of the corolla* slightly pilose within below but their top and the upper part of the style slightly glandular-pilose.—*Sy. E. B.* 1011. *R. I.* f. 885. Probably *O. Epithymum* DC.—Scarcely a foot high. Purplish red. *Stigma 2-lobed, pale pink.* Anth. fuscous when dry. Sep. with a second faint vein near their anterior margin; and in the dry plant there is an appearance of several more. L. few. Sweet-scented.—Parasitical upon *Thymus Serpyllum.* P. VI.—VIII. E. S. I.

3. *O. caryophyllácea* (Sm.); sep. many-veined lanceolate equally bifid falling short of the cor.-tube touching or connate in front, *cor. tubular-bellshaped curved on the back,* lips spreading, upper one 2-lobed its lobes porrect, lobes of lower lip nearly equal rounded wavy, *stam. inserted above the base of the cor. hairy within below* but their upper part and the style glandular-hairy.—*E. B. S.* 2639.—*O. Galii* Duby.—Scarcely a foot high. *Stigma of 2 nearly separate dark purple lobes.* Anth. fuscous, yellow when dry. Sep. with crisp glandular hairs externally, each lobe with 1 strong vein and several slender ones. Cor. similarly hairy on both sides.—On *Galium Mollugo* in Kent. P. VI. VII. E.

---

[1] We have kept the Linnean name for this sp., *O. elatior* Sutt. having been differentiated earlier than *O. Rapum-genistæ* Thuill.—H. & J. G.

4. *O. elátior* (Sutt.); sep. many-veined equally bifid equalling the cor.-tube connate in front, *cor. curved tubular slightly compressed above, upper lip 2-lobed toothed its lobes inflexed*, lobes of lower lip 3 nearly equal acute toothed, stam. inserted above the base of the cor. glandular-hairy in the lower half within.—*E. B.* 568.—Stem 2—3 ft. high. *Stigma bilobed, yellow.* Upper lip of the cor. usually with an elevated point between the lobes. Cor. glandular externally. Stam. sometimes slightly hairy above. Anth. whitish when dry.—Parasitical upon *Centaurea Scabiosa*, rare. P.? VI. VII. E.

5. *O. Pic'ridis* (F. W. Schultz); *sep.* 1—3-veined entire or toothed below in front narrowed into 1 or 2 subulate points, cor. tubular-bellshaped its back *nearly straight* and compresed *slightly curved at each end*, lips denticulate wavy *upper not notched* its sides porrect, *stam.* inserted below the middle of the cor-tube *hairy in their lower half within*, style slightly glandular-hairy below in front and above throughout, stigma bilobed.—*E. B. S.* 2956.—Height 6—18 inches. *Stigm.-lobes just touching, purple.* Anth. fuscous, pale purple or yellowish. — Parasitical upon *Picris*. Cambridgeshire. Kent. Pemb. Hunts. Isle of Wight. P.? VII. E.

6. *O. Hed'eræ* (Duby); sep. 1-veined ovate below narrowed into 1 or 2 subulate points about equalling the cor-tube, *cor.* tubular *arcuate*, lips denticulate wavy, upper one bilobed porrect (straight when dry), lobes of lower lip nearly equal the *middle one longest*, stam. inserted below the middle of the cor.-tube glabrous with a few scattered hairs on their lower part, style glabrous with a few hairs on the upper part, stigma scarcely bilobed.—*O. barbata* E. B. S. 2859, not Poir.—St. purplish, about a foot high. *Lobes of stigma attached together by at least* $\frac{1}{3}$ *of their circumference, yellow.* Anth. fuscous, rather paler when dry. St. purplish.—Parasitical upon Ivy. P. VI. VII. E. I.

7. *O. minor* (Sm.); *sep. many-veined* ovate below suddenly narrowed into 1 or 2 subulate points equalling or exceeding the cor.-tube. cor. tubular arcuate, [usually tinged with purple,] lips bluntly denticulate wavy, upper lip porrect (inflexed when dry) notched, lobes of lower lip nearly equal, stam. inserted below middle of the cor.-tube glabrous with a few scattered hairs below, style glabrous with a line of distant hairs on its anterior side, *stigma bilobed.*—*E. B.* 422.—*Lobes of stig. not much connected, purple.* Anth. yellow when dry. [*β. flavescens* (Reut.); st. and fl. yellow.]—Parasitical chiefly upon *Trifolium pratense.* A.? VI. VII. E. I.

8. *O. amethys'tea* (Thuill.); sep. many-veined ovate below narrowed into 1 or 2 subulate points, cor. tubular its *back curved immediately from the base otherwise straight*, lips unequally acutely denticulate wavy, upper lip hooded porrect notched, lobes of lower lip unequal the middle one larger, stam. inserted in the curvature of the cor. glabrous with many hairs at the base within, stigma bilobed.—*Sy. E. B.* 1017. *Atl. Fl. Par.* t. 19. E. *O. Eryngii* Duby.—*Lobes of stig. attached by* ¼, *purple.* Anth. dusky brown when dry. Perhaps a form of *O. minor.*—Parasitical upon *Daucus gummifer.* Whitsand Bay, Cornwall. Rock End, Torquay. *Mr. Townsend.* Dorset. A.? VI. E.

** *Sepals 4 or 5, connected below. Bracts 3. Valves of capsule separating at the top.* PHELIPÆA Desf.

[*O. arenaria* (Bork.)[1]; cal. of 5 sep. tubular with triangular-subulate teeth falling short of the cor.-tube, cor. tubular slightly curved in front, middle of the tube compressed on the back, throat slightly inflated externally glandular, *lobes of lips blunt* with reflexed edges, the lower lip hairy within, *suture of anth. hairy.*—*Atl. Fl. Par.* t. 19. L.—Height 12—18 inches. Stigma scarcely 2-lobed; style pale yellow (?), glandular. Filaments glabrous with a few hairs at their base. Lateral bracts linear-subulate, intermediate lanceolate attenuate above. St., scales and cal. glandular-pubescent.—Parasitical upon *Achillea Millefolium*? Alderney! (the Jersey plant is Sp. 9). P. VII. VIII.]

9. *O. purpúrea* (Jacq.); cal. of 5 sep. tubular with lanceolate acute teeth falling short of the cor.-tube, cor. tubular curved in front, middle of tube compressed, *lobes of lips acute* with reflexed edges, lower lip hairy within, *anth. glabrous.*—*E. B.* 423. *Atl. Fl. Par.* t. 19. K. *O. cærulea* (Vill.) ed. viii.—About a foot high. Fl. bluish purple. Stig. scarcely 2-lobed, white. St., scales, bracts, cal. and cor. glandular-pubescent.—Grassy pastures. Herts. Norf. Isle of Wight. Chepstow. P. VI. VII. E.

[*O. ramósa* (L.); cal. of 4 sep. tubular with triangular ovate acuminate teeth, anth. glabrous, st. usually branched.—*E. B.* 184.—Sown with Hemp.]

2. LATHRÆ'A *Linn.* Toothwort.

1. *L. Squamária* (L.); st. simple, fl. pendulous secund, lower lip of the cor. 3-cleft.—*E. B.* 50. *G. E. Smith S. Kent.* t. 3.—

---

1 Mr. Ar. Bennett refers the Alderney plant to *O. Millefolii* (*O. cærulea* var. *Millefolii*, Reich.), see Marquand, Guernsey Flora, p. 137.—H. & J. G.

Rootstock fleshy, with thick scales. St. 3—8 in. high. Young raceme decurved. Bracts ovate or lanceolate. Style straight or curved. Upper lip nearly entire, or bifid.—Woods and thickets, parasitical upon Hazels &c. P. IV. V. E. S. I.

Order LVII. SCROPHULARIACEÆ.

Cal. 4—5-cleft, persistent. Cor. irregular or 2-lipped or personate (subrotate in *Verbascum*), deciduous, imbricate in the bud. Stam. on the cor., usually 4, didynamous, or 2 or 5. Ovary free, 2-celled. Style simple; stigma 2-lobed. Fr. capsular, 2-celled; placentas attached to the dissepiment or ultimately central. Embryo straight, in axis of fleshy albumen. —No stipules.

* *Stamens 5.*

1. VERBASCUM. Cal. of 5 sepals. Cor. rotate; segments unequal, spreading. Stam. unequal, 2 or more, hairy at the base.

** *Stamens 4, didynamous.*

[ERINUS. Cal. in 5 deep segments. Cor. 5-parted with nearly equal emarginate segments and a short tube. Caps. 2-celled.]

2. DIGITALIS. Cal. in 5 deep segments. *Cor. bellshaped,* oblique, 4—5-fid. Caps. septicidal, 2-celled.

3. ANTIRRHINUM. Cal. 5-parted. *Cor. personate,* gibbous at the base (*no distinct spur*); lower lip 3-fid; a prominent palate closing the mouth. Caps. opening by 2 or 3 pores at the top, 2-celled.

4. LINARIA. Cal. 5-parted. *Cor. personate, spurred*; lower lip 3-fid; a prominent palate closing the mouth. Caps. with valves or teeth at the top, 2-celled.

5. SCROPHULARIA. Cal. 5-lobed (in *S. vernalis* 5-cleft). *Cor. globose*; limb minute, of 2 short lips, upper 2-lobed often with a scale (the rudiment of a fifth stam.) within, lower 3-lobed. Caps. 2-valved, the edges inflexed, 2-celled.

6. LIMOSELLA. Cal. 5-cleft. *Cor.* 5-fid, *bellshaped,* equal. Caps. globose, 2-valved, 1-celled; placenta central, free or connected with a short dissepiment below.—L. radical.

7. MELAMPYRUM. *Cal. tubular*, 4-toothed. Cor. ringent; upper lip compressed laterally with reflexed edges; lower furrowed 3-fid. Caps. oblong, obliquely acuminate, compressed. *Seeds* 1—2 *in each cell, smooth.*

8. MIMULUS. Cal. prismatical, 5-toothed. Cor. ringent; upper lip folded back at the sides. Seeds many.

9. PEDICULARIS. *Cal. inflated*, 5-toothed. Cor. ringent; upper lip compressed laterally; lower plane, 3-lobed. *Caps.* compressed, *acute. Seeds many*, angular.

10 RHINANTHUS. *Cal.* inflated, 4-*toothed.* Cor. ringent; upper lip compressed laterally; lower plane, 3-lobed. *Caps.* compressed, *blunt. Seeds* many, *compressed, with an orbicular margin.—Alecterolophus* (Hall.).

11. BARTSIA. *Cal. bellshaped*, 4-fid. Cor. tubular, ringent; upper lip much arched, not compressed. *Caps. pointed*; cells many-seeded. *Seeds compressed at the hile and with winged ribs at the back* (large).

12. EUFRAGIA. Cal. tubular, 4-cleft. Cor. tubular, 2-lipped. Caps. pointed; cells many-seeded. *Seeds slightly angular, very minute, crenate-ribbed*; *hile basal.*

13. EUPHRASIA. Cal. tubular or bellshaped, 4-fid or 4-toothed. Cor. tubular, 2-lipped; lower lip with 3 notched or emarginate lobes. Anth. unequally pointed. *Caps. blunt or emarginate*; cells many-seeded. *Seeds* rather angular, *longitudinally ribbed*; *hile subapical.*

14. ODONTITES. Lower cor.-lip with 3 entire lobes. Anth. with 2 equal points; otherwise like *Euphrasia.*

15. SIBTHORPIA. Cal. in 5 deep spreading segments. *Cor. rotate, irregularly* 5-*cleft.* Caps. compressed, orbicular, 2-seeded, 2-valved.

*** *Stamens* 2.

16. VERONICA. Cal. 4—5-parted. Cor. rotate, unequally 4-lobed, lower lobe the smallest. Caps. compressed, 2-celled

1. VERBAS'CUM *Linn.*[1] Mullein.

* *Leaves strongly decurrent. Raceme dense, nearly simple.*

1. *V. Thap'sus* (L.); l. ovate-oblong crenate densely woolly on both sides all decurrent, st. simple, spike dense, pedicels shorter than the calyx, cor. rotate, segments oblong blunt, filaments woolly 2 longer nearly glabrous, anth. all nearly equal.—*E. B.* 549. *V. Schraderi* Koch, *R.* xx. 1637.—St. 4—5 feet high. Cor. about twice as long as the calyx. Filaments with white wool: the 2 glabrous ones about 4 times as long as their slightly decurrent anthers.—Waste ground. B. VII. VIII. *High-taper.* E. S. I.

** *Leaves not decurrent. Racemes branched, panicled. Anth. all reniform, not decurrent.*

† Flowers yellow or whitish; hairs on the filaments white.

2. *V. Lychnitis* (L.); *l.* crenate *nearly glabrous above* woolly and powdery beneath, lower l. elliptic-oblong wedgeshaped below scarcely stalked, upper l. sessile ovate-acuminate with a rounded base, st. angular panicled above with ascending branches, stam. equal, filaments all with white hairs.—*E. B.* 58. *R.* xx. 1648. —Height 2—3 feet. Fl. on short stalks, small, many, [yellow (rare in Britain) or var. *album*, Mill.] *cream-coloured* or white.—Roadsides and waste places. B. VI.—VIII. *White Mullein.* E.

3. *V. pulverulen'tum* (Vill.); *l.* obscurely crenate clothed *with mealy deciduous wool on both sides,* lower l. oblong-elliptic attenuated into a stalk, upper l. sessile acuminate, st. terete panicled above with patent branches, stam. nearly equal scarlet with white hairs.—*E. B.* 487. *R.* xx. 1667.—Height about 3 feet. Fl. on very short stalks, which, as well as the calyx, are densely covered with wool, *bright yellow.* Cal.-teeth often glabrous.—Roadsides in Norfolk and Suffolk. B. VII. *Hoary Mullein.* E.

†† Flowers yellow; hairs on the filaments purple.

4. *V. nigrum* (L.); l. doubly crenate nearly glabrous above subpubescent beneath, lower l. cordate or ovate-oblong with long stalks, upper l. cordate-ovate nearly sessile, st. angular, raceme elongated, pedicels twice as long as the calyx, stam. equal with

---

[1] Hybrids occur in this genus. Four are figured in *Sy. E. B.* t. 843, 844, 845, 846.

purple hairs.—*E. B.* 59. *R.* xx. 1649.—Fl. in clusters in a nearly simple long spike, small, bright yellow.—[*β. tomentosum* (Bab.); l. subpubescent above woolly beneath, fl. smaller.]—Banks and waysides. [*β.* Alderney.] P. VII. VIII. *Dark Mullein.* E. S.

*** *Leaves not, or very slightly decurrent.*

5. *V. Blattária* (L.); l. crenate glabrous, lower l. ovate-oblong blunt sinuate, upper l. oblong or subcordate semiamplexicaul, *pedicels solitary nearly twice as long* as the bract, stam. and anth. unequal.—*E. B.* 393. *R.* xx. 1653.—Height 5—6 feet. Raceme glandular-pilose. *Fl. cream-coloured.* Filaments with purple hairs, the 2 longer hairy only on the inside.—On gravelly banks, rare. B. VIII. *Moth-Mullein.* E. *I.

6. *V. virgátum* (Stokes); l. doubly serrate slightly glandular-hairy, lower l. oblong-lanceolate sublyrate lobate-crenate-serrate, upper l. oblong-acuminate semiamplexicaul, *pedicels* 1—5 *together shorter than the bracts*, stam. nearly equal.—*E. B.* 550. *R.* xx. 1655.—Usually much stouter than Sp. 5. Height 3—4 feet. Raceme glandular-pilose. Fl. yellow. Filaments with purple hairs, 2 rather longer and hairy only within.—On gravelly banks, rare. B. VIII. E. I.

[*Erínus alpínus* (L.); l. spathulate deeply serrate smoothish, ped. terminal subcorymbose.—Subcæspitose. Fl.-shoots 3—8 in. long, ascending.—In abundance on the old bed of the river near Tanfield, Yorkshire. Hexham, Northumberland. P. VII.]

2. Digita'lis *Linn.* Foxglove.

1. *D. purpurea* (L.); l. ovate-lanceolate crenate downy beneath, lower l. narrowed into footstalks, sep. ovate-oblong acute 3-veined downy, cor. blunt glabrous externally, upper lip scarcely cloven, segments of the lower lip ovate rounded.—*E. B.* 1297. —L. often crenate-dentate or -serrate. Fl. sometimes white or flesh-coloured. St. 3—4 feet high.—Hedgebanks and woods. P. VI.—VIII. E. S. I.

3. Antirrhi'num *Linn.* Snapdragon.

*1. *A. május* (L.); l. lanceolate opposite or alternate glabrous, fl. racemose, *sep. ovate* blunt much shorter than the cor., upper cor.-lip bifid.—*E. B.* 129.—Height 1—2 feet. Cor. 1½ in. long, purplish-red or white.—Old walls and calcareous cliffs. P. VII.—IX. *Great Snapdragon.* E.

2. *A. Oron'tium* (L.); l. linear-lanceolate opposite or alternate, fl. loosely spiked distant, *sep. linear longer than the corolla.*—*E. B.* 1155. *St.* 27.—About a foot high. Fl. purple.—Dry sandy and gravelly fields. A. VII. VIII. E. I.

4. Lina'ria *Hill.* Toadflax.

* *Stems trailing. Fl. axillary.*

*1. *L. Cymbalária* (Mill.); l. roundish heartshaped 5-lobed glabrous, st. procumbent.—*E. B.* 502.—Stems slender, rooting. Fl. solitary, axillary, upon long stalks, pale blue.—Old walls. P. V.—X. *Ivy-leaved Toadflax.* E. S. I.

2. *L. Elat'ina* (Mill.); *l. ovate-hastate,* lower l. ovate, *cor.-spur straight,* peduncles glabrous.—*E. B.* 692.—Fl. solitary, axillary, upon long slender stalks, small, yellow, with the upper lip purple.—Corn-fields. A. VII.—IX. *Fluellin.* E. I.

3. *L. spuria* (Mill.); *l. roundish-ovate, spur curved* upwards, peduncles hairy.—*E. B.* 691.—Fl. similar to the last but larger. L. with here and there a small tooth.—In this species and the two preceding some of the fl. are often regular with 5 spurs or partially so with 2, 3, or 4.—Gravelly and sandy corn-fields. A. VII.—IX. E.

** *Stems erect or rarely diffuse.* † Fl. solitary.

4. *L. vis'cida* (Moench); *l. linear-lanceolate* blunt glandular-pubescent mostly alternate, fl. axillary, peduncles 3 times as long as the calyx, segments of upper cor.-lip diverging, seeds oblong sulcate.—*E. B.* 2014. *L. minor* (Desf.) ed. viii.—Fl. small; the tube, upper lip, and spur of the cor. purplish; lower lip yellowish. St. erect, 4—10 in. high, branched, glandular-pubescent. Lower l. nearly spathulate.—Sandy and gravelly fields. A. VI.—VIII. E. S. I.

†† Fl. racemose.

[*L. Pelisseriána* (Mill.); glabrous, l. linear, the lower ternate or quaternate, upper alternate, sterile branches radical prostrate with ternate lanceolate or ovate l., *fl. racemose,* peduncles as long as the bracts, sepals linear acute twice as long as the capsule, seeds nearly flat with a fringed wing one side smooth the other tubercular.—*E. B. S.* 2832.—Fl. purple with darker veins. St. one or more from each root, erect, about a foot high. Caps. bilobed.—Jersey. A. V.]

*5. *L. supína* (Desf.); glabrous, rachis, ped. and *sep. glandular-hairy*, l. linear blunt mostly whorled, *sep. linear-spathulate* shorter than the caps. or spurs, seeds smooth nearly flat with a striate wing.—*Sy. E. B.* 858. *L. maritima* DC. *Icon. Gall.* 12. —St. diffuse or ascending. Fl. capitate-racemose, yellow; throat and spur with slender purple lines. Styles entire.—Plymouth and Poole; a ballast plant. Perhaps a native at Hayle, and St. Blazey's Bay, Cornwall. A. VII. VIII. E.

*6. *L. purpúrea* (Mill.); glabrous, l. linear-lanceolate scattered, lower l. irregularly in fours, fl. narrowly racemose, sep. linear shorter than the caps. and long incurved spur, seeds angular with a network of elevated lines.—*Sy. E. B.* 960.—Fl. purple or yellow with the lips purple; spur two or three times as long as the ped. which is usually shorter than the bract. St. erect, leafy.—Old walls. Naturalized. P. VII. VIII. E.

7. *L. répens* (Mill.); glabrous, l. linear scattered or partly whorled, fl. racemose, *sep. lanceolate* as long as the spur but shorter than the caps., *seeds angular with transverse elevated lines. E. B.* 1253. *L. striata* DC.—Fl. white with blue veins. St. erect, branched, leafy, 1—1½ foot high, slender. Seeds much smaller than those of *L. vulgaris*.—*L. italica* and *L. sepium* may be hybrids between this and *L. vulgaris*.—Calcareous soils, particularly near the sea, rare. P. VII.—IX. E. I.

8. *L. vulgáris* (Mill.); glabrous, rachis and peduncles glandular-hairy, l. linear-lanceolate scattered crowded, fl. racemose imbricate, *sep. ovate acute glabrous* shorter than the caps. or spur, seeds tubercular-asperous with a smooth orbicular margin. —*E. B.* 658.—Fl. large, yellow, rarely milk-white with an orange palate. St. erect, 2 feet high, as well as the l. glabrous. Common and partial flower-stalks occasionally glabrous.—The state called *Peloria* with 5 spurs and an equal and regular cor. is rarely found. *E. B.* 260.—β. *latifolia* (Bab.); l. narrowly lanceolate, fl.-l. often lanceolate very glaucous, fl. twice as large in a few-fl. lax raceme, ped. glabrous, spur directed perpendicularly downwards. *Sy. E. B.* 964. *L. speciosa* Ten. ?—Hedges on a gravelly soil. P. VI. VII. *Yellow Toadflax.* E. S. I.

5. Scrophula'ria *Linn.* Figwort.

* *Cal. of* 5 *rounded lobes. Cor. purplish; upper lip with a scale (staminode) on its inner side.*

1. *S. nodósa* (L.); *l.* ovate acute subcordate glabrous *doubly and acutely serrate, lower teeth largest*, st. acutely 4-angular,

cymes lax, sep. roundish-ovate with a narrow membranous margin, *staminode wedgeshaped slightly emarginate* [rarely entire].—*E. B.* 1544.—Root tuberous, thick, knotty. St. 2—3 feet high. Bracts small, lanceolate, acute. Fl. greenish purple, lurid, sometimes green or pale. Caps. ovate.—Moist hedges and thickets. P. VI. VII. *Knotted Figwort.* E. S. I.

2. *S. umbrósa* (Dum.)[1]; *l.* ovate-lanceolate acute subcordate glabrous sharply *serrate, lower teeth smaller,* st. and petioles winged, *cymes lax few-* (4—8-) *flowered,* sep. roundish with a broad membranous margin, *staminode bilobed with diverging lobes.*—*S. Ehrharti,* C. A. Stev. *E. B. S.* 2875. Not *S. alata* Gil.—St. tall. *Bracts leaflike,* lanceolate, acute. Fl. dark purple. Caps. subglobose, blunt.—Wet places. P. VIII. IX. E. S. I.

3. *S. aquat'ica* (L.!); *l.* cordate-oblong roundly blunt glabrous *crenate-serrate,* st. and petioles winged, *cymes dense corymbose many-* (1—15-) *flowered,* sep. roundish blunt with a broad membranous margin, *staminode roundish-reniform.*—*E. B.* 854. *S. Balbisii* Horn., Koch.—St. 2—5 feet high. *Bracts linear, blunt.* Fl. dark purple, occasionally milk-white. Caps. ovoid, pointed.—In wet places. P. VII. VIII. E. S. I.

4. *S. Scorodónia* (L.); *l.* cordate-triangular *with large double teeth downy on both sides,* st. bluntly quadrangular downy, cymes lax few-flowered, sep. roundish downy with a membranous margin, staminode roundish entire.—*E. B.* 2209.—St. 2—3 feet high. L. wrinkled. *Bracts leaflike,* lower exactly like the leaves. Fl. purple. Caps. ovoid, acute.—In moist places. West of Cornwall. Tralee, Kerry? Jersey. P. VII. E. I. ?

*** Cal. of 5 deep acute segments. Cor. yellow; no staminode.*

‡5. *S. vernális* (L.); *l. downy* cordate-acute doubly serrate, st. winged hairy, *cymes axillary* corymbose with leaflike bracts, sep. oblong with a recurved apex.—*E. B.* 567.—St. about 2 feet high. *Fl. yellow, inflated*; the mouth much contracted. Caps. ovoid, acute. Differing greatly from the other species and allied in appearance to some of the *Calceolariæ.*—Waste places, rare. P. IV. V. E. S. ?

6. LIMOSEL'LA *Linn.* Mudwort.

1. *L. aquat'ica* (L.); l. lanceolate-spathulate on long stalks, ped. axillary crowded shorter than the petioles.—*E. B.* 357.

---

[1] See Du Mortier in *Bull. Belg.* vii. 36.

St. 30. 15.—Very small; st. 0, except the naked stoles. Fl. small, white or rose-coloured. Caps. minute, ovoid. [β. *tenuifolia* (Hook. f.); smaller, l. linear.]—Muddy places where water has stagnated. A. VII.—IX. E. S. I.

7. Melampy'rum *Linn.* Cow-wheat.

1. *M. cristátum* (L.); *spikes densely imbricate 4-sided, bracts heartshaped* acuminate pectinate-dentate lower ones with a long leaflike recurved point.—*E. B.* 41.—Bracts rose-coloured at the base. Fl. yellow, tinged with purple. L. linear-lanceolate, acute, entire, with netted veins beneath. St. 8—12 in. high.—Woods and thickets in the Eastern Counties. A. VII. *Crested Cow-wheat.* E.

2. *M. arven'se* (L.); *spikes lax conical, bracts ovate-lanceolate-attenuate* pinnatifid with subulate segments and with a few large glandular points beneath, cal. hispid equalling the cor.-tube with long lanceolate-attenuate teeth from an ovate base, cor. closed.—*E. B.* 53.—Bracts purple-rose-colour. Fl. variegated with yellow, rose-colour, and purple. L. linear-lanceolate, acute, rough-edged, slightly downy on both sides, entire. St. 8—18 in. high.—Corn-fields and dry banks in the Eastern Counties and the Isle of Wight. A. VII. *Purple Cow-wheat.* E.

3. *M. praten'se* (L.); *fl. axillary secund in distant pairs,* upper bracts lanceolate with 1 or 2 teeth at the base, *cor.* 4 times as long as the glabrous calyx *closed, lower lip projecting.*—*E. B.* 113.—Teeth and tube of the cal. about equal in length. L. lanceolate, or [var. *latifolium* Schueb. & Mart.] ovate-lanceolate, or with a cordate base, entire, varying greatly in size. Fl. large, pale yellow, horizontally patent. St. 6—12 in. high.—β. *ericetorum* (D. Oliv.); hispid, l. lanceolate or linear-lanceolate, bracts toothed, fl. near together, cor.-tube whitish.—γ. *M. montanum* (Johnst.); smaller in all its parts, bracts quite entire, l. linear-lanceolate hispid, the 2 lower obovate-lanceolate blunt.—δ. *hians* (Druce); "cor. deep yellow, with open lips, palate not closing tube."—Woods and thickets. β. West of Ireland. γ. Mountains. A. VI.—VIII. E. S. I.

4. *M. sylvat'icum* (L.); fl. axillary secund in distant pairs, bracts mostly all entire linear-lanceolate, *cor.* about twice as long as the glabrous calyx, *open lips equal* in length.—*E. B.* 804.—Teeth of the cal. longer than the tube. L. linear-lanceolate, entire. Fl. very small, deep yellow, erect. St. 12 in. high.—Alpine woods. A. VII. E. S. I.

8. MIM'ULUS *Linn.*

*1. *M. Langsdorffii* (Donn); l. roundish ovate veined, lower ones stalked, uppermost l. clasping, st. creeping, fl. yellow.—*Sy. E. B.* 967.—*M. luteus* ed. viii.—An American plant, naturalized in many boggy places. P. VI.—IX. E. S. I.

9. PEDICULA'RIS *Linn.*

1. *P. palus'tris* (L.); *st. solitary erect branched throughout,* l. pinnatifid, segments oblong blunt lobed, *cal. ovoid pubescent, 2-lobed,* lobes incise-dentate crisped.—*E. B.* 399.—Upper lip of the cor. with a short truncate beak with a triangular tooth on each side. Fl. large, crimson. St. 12—18 in. high, angular, with alternate branches. Whorl of ovate-acute scales at root-crown.—Marshy and boggy places. A.? V.—VII. *Lousewort.* E. S. I.

2. *P. sylvat'ica* (L.); *st. branched at the base erect, branches long spreading* prostrate, l. pinnatifid, segments ovate lobed, *cal. oblong glabrous* irregularly 5-lobed, upper lobe lanceolate, other lobes with 3 leaflike divisions.—*E. B.* 400.—Upper lip of the cor. as in the last. Fl. large, rose-colour. Summit of the ped. with a loose membranous cuticle enclosing the base of the calyx. Primary st. erect, often very short; branches prostrate. Whorl of ovate-lanceolate crenate undivided reflexed leaves at root-crown.—Wet heathy and rather hilly pastures. P.? V.—VIII. *Red Rattle.* E. S. I.

10. RHINAN'THUS *Linn.* Yellow Rattle.

1. *R. Crista-gal'li* (L.); l. oblong-lanceolate serrate, fl. in lax spikes, cal. glabrous, *lobes of the upper cor.-lip short roundish,* bracts ovate incise-serrate, seeds with a membranous border.—*E. B.* 657. *R. minor* (Ehrh.).—Lateral lobes of the upper cor.-lip very blunt, shorter than broad, bluish. *Bracts green throughout.* Style downy near the top. Caps. as long as broad. Cor.-tube straight. St. 1—2 feet high, nearly simple. Rarely (var. *Drummond-Hayi* White) cal. pubescent with short hairs. [A taller form with black-dotted st. is var. *fallax* (Druce); a bushy much-branched form with narrow l. from N. of Scotl. is apparently the var. *angustifolius* G. & G.]—Meadows and pastures. A. VI. E. S. I.

2. *R. májor* (Ehrh.); l. linear-lanceolate serrate, fl. in crowded spikes, cal. glabrous, *lobes of the upper cor.-lip oblong,* bracts with an attenuate point incise serrate.—Cal. often slightly downy on

its edges. Lateral lobes of the upper cor.-lip longer than broad purple; the central part truncate. Cor.-tube slightly curved Style glabrous. Caps. often longer than broad. Anth. very villose. *Bracts yellowish with green points.*—*a. platypterus* (Fries); seed not twice as broad as its wing. *R. major* Koch, *R. I.* f. 975.—*β. stenopterus* (Fries); seed quite twice as broad as its wing. *R. major* E. B. S. 2737.—*γ. apterus* (Fries); seed not winged but rounded and longitudinally ribbed or furrowed on the back. *R. Reichenbachii* Drej.—Cultivated land. *α.* Hastings. *β.* North of England and Scotland. *γ.* Arbroath and Monifief, Forfarshire. A. VII. VIII. E. S.

11. Bart'sia *Linn.* Red Eye-bright.

1. *B. alpina* (L.); l. opposite ovate slightly clasping bluntly serrate.—*E. B.* 361.—Creeping. St. square, 4—8 in. high, simple. Fl. forming a short dense leafy spike, purplish blue, downy. Cal. purplish, viscid. Anth. hairy.—Alpine pastures, rare. P. VI. VII. E. S.

12. Eufrag'ia *Griseb.* Marsh Eye-bright.

1. *E. viscósa* (Benth.); l. opposite, upper l. alternate ovate-lanceolate sessile acutely serrate.—*Bartsia* L. *E. B.* 1045.—L. sometimes linear-lanceolate. St. round, 3—12 in. high, simple. Root fibrous. Fl. distant, axillary, upper ones crowded, yellow. Anth. hairy. St., l., and cal. viscid.—Damp places in the West of E., South-west of S., and South of I. A. VII.—IX. E. S. I.

13. Euphra'sia *Linn.* Eye-bright.

1. *E. officinális* (L.); l. ovate or oblong-lanceolate nearly sessile serrate (3—5 teeth on each side), lobes of the lower cor.-lip emarginate, of the upper lip patent sinuate-dentate, anth. hairy, seeds with ribs.—*E. B.* 1416.—St. 1—8 in. high. Fl. axillary, solitary, sessile, crowded towards the ends of the branches.—*α*; glandular-pubescent above and on the calyx, *caps. oblong-obovate,* seeds ovoid greyish. L. usually large and broad, sometimes densely imbricate (*E. ericetorum* Jord.?) —*β. E. nemorosa* (Pers.), pubescent not glandular, caps. linear-oblong, *seeds fusiform* yellowish. L. usually narrow, sometimes (*E. Salisburgensis* Funk.?) with very long teeth.—Some authors divide this into many species; but even the above are scarcely distinguishable at all times.—Pastures, woods, heaths. A. VI. —VIII. *Common Eye-bright.* E. S. I.

By permission of Mr. Fredk. Townsend we have compiled the following short account of the British segregates from his '*Monograph of the British Species of Euphrasia*' and we are further indebted to him for kindly revising the MS. H. & J. G.

* *Parviforæ.* L. not more than twice as long as broad, usually much less. Cor.-tube not lengthening after the fl. opens.

† Fl. from 6–10 mm. in length.

1. *E. strict'a* (Host); st. usually branched below with few branches eglandular, l. nearly or quite glabrous, cauline ovate or ovate-lanceolate about double as long as broad with 6–10 awned teeth, floral suberect ovate very acute with shortly-cuneate base broader than the cauline with 8–14 long-awned or sometimes only acute teeth, lower l. deciduous, spike much lengthening in fr., cal. glabrous or with minute hairs with lanceolate-acuminate awned teeth not accrescent, cor. large usually pale violet rarely blue or white, lobes of upper lip denticulate rarely bilobate, fr. cuneate-obovate truncate or subemarginate ciliate with surface hairy or nearly glabrous falling short of the cal.-teeth.—Widely distributed. E.S.I.

2. *E. boreális* (Towns.); st. stout simple or branched below eglandular, l. erect-patent subglabrous or more or less setose, cauline ovate obtuse with 6–10 obtuse (or the lowest somewhat acute) teeth, floral broadly ovate with 6–10 acute sometimes shortly-awned or occasionally obtuse teeth, spike usually dense, cal. subglabrous with triangular-lanceolate acuminate teeth more or less accrescent, cor. large white or violet or with upper lip violet and under white, lobes of upper lip reflexed emarginate or denticulate, fr. elliptic or oblong attenuate below emarginate ciliate with the surface almost glabrous about equalling the cal.-teeth.—Widely distributed. E. S. I.

3. *E. brevipíla* (Burn. & Grem.); st. usually branched below eglandular, l. with short glandular hairs, cauline ovate or ovate-oblong acute or obtuse with 6–10 obtuse acute or even awned teeth, floral ovate with shortly-cuneate base broader and shorter than the cauline with 8–14 awned or acuminate teeth, spike becoming much elongated, cal. with short glandular hairs and triangular lanceolate teeth not or but slightly accrescent, cor. large pale violet bluish or white, lobes of upper lip emarginate or entire and denticulate, fr. oblong or cuneate-obovate truncate or emarginate strongly ciliate with surface hairy or glabrous equalling or exceeding the cal.-teeth.—Widely distributed, principally in grazing pastures. E. S. I.

†† Fl. from 4–7 mm. in length.

4. *E. nemorósa* (H. v. Mart.); st. stout eglandular much branched, branches often again branched, l. dull green plicate beneath glabrous patent arcuate, cauline ovate or ovate-lanceolate acute with 8–14 acute scarcely awned teeth, floral patent or recurved ovate broader and shorter than the cauline with 8–12 very acute or shortly-awned teeth, spike becoming elongated, cal. nearly glabrous with triangular or triangular-lanceolate teeth becoming somewhat inflated, cor. small white or bluish

upper lobe minutely-denticulate, fr. cuneate-obovate emarginate long-ciliate with surface hairy or glabrous, equalling or often exceeding the cal.-teeth.—Common. E. S. I.

5. *E. cur'ta* (Wettst.); st. usually stout branched below, l. greyish-green rugose beneath usually more or less densely clothed with comparatively long bristles, cauline ovate acute with 8–14 acute not awned teeth, floral often nearly orbicular with 8–14 acute or shortly-awned teeth, cal. clothed wholly or on the nerves and margin with short white hairs, becoming somewhat inflated, teeth short, cor. small usually whitish or pale lilac rarely wholly blue, lobes of upper lip emarginate or denticulate, fr. cuneate-obovate truncate or emarginate strongly ciliate with hairy or rarely glabrous surface equalling or exceeding cal.-teeth.—β. *glabres'cens* (Wettst.); l. nearly glabrous, fr. often much exceeding cal.-teeth. Approaching closely to Sp. 4.—γ. *pic cola* (Towns.); "plant smaller in all its parts, teeth of cauline- and floral-l. 8–12."—Widely distributed. E. S.

6. *E. occidentális* (Wettst.); st. stout ascending branched below, l. clothed with small stiff bristles and short glandular hairs, cauline ovate acute or subacute with 6–10 acute or subacute teeth, floral broadly ovate acute with 8–14 acute teeth, spike usually dense, cal. with small stiff bristles and glandular hairs, teeth lanceolate-acuminate, cor. small whitish, lobes of upper lip entire or emarginate, fr. elliptic emarginate ciliate with surface nearly glabrous equalling or exceeding cal.-teeth.—Maritime. E I.

7. *E. latifólia* (Pursh); st. straight simple or slightly branched at or below the middle, internodes long, l. more or less densely clothed with stout white bristles sometimes with glandular hairs, cauline few ovate or cuneate-obovate obtuse with 4–10 broad obtuse teeth, floral broadly-oval or nearly orbicular subobtuse or acute with cuneate base with 6–12 broad subobtuse or acute but not awned teeth, spike dense, cal. with stout white bristles and sometimes glandular hairs accrescent, teeth broad acute, cor. rather small usually whitish, lobes of upper lip reflexed denticulate, fr. elliptic emarginate ciliate with surface hairy, equalling or exceeding cal.-teeth.—N. of Scotland. S.

8. *E. foulaen'sis* (Towns.); st. stout simple or slightly branched, l. almost glabrous, cauline few distant ovate obtuse with 4–6 obtuse teeth, floral similar sometimes acute with acute but not acuminate nor awned teeth, spike lengthening slightly, cal. glabrous or slightly setose accrescent, teeth triangular-lanceolate, cor. small usually purple, lobes of upper lip entire, fr. elliptic-elongate emarginate ciliate usually exceeding cal.-teeth.—N. of Scotland. S.

9. *E. grac'ilis* (Fries); st. straight slender simple or branched about the middle, l. small conspicuously shorter than the internodes green or reddish with few very short hairs on the upper surface and on the nerves beneath, cauline ovate often cuneate-based acute with 6–8 acute teeth, floral ovate often cuneate-based with 6–10 cuspidate-acuminate or shortly-awned teeth, spike much lengthening, cal. glabrous becoming somewhat inflated, teeth lanceolate-acuminate, cor. small white with blue lines bluish or violet, lobes of upper lip entire or faintly denticulate, fr. linear-elliptic truncate or somewhat emarginate ciliate with glabrous surface equalling or more usually exceeding cal.-teeth and floral l.—Common. E. S. I

10. *E. scot'ica* (Wettst.); st. firm simple or slightly branched at or below the middle, l. rigid almost glabrous with a few short hairs on the margin ovate or ovate-oblong 6–8-toothed, cauline and lower floral with obtuse teeth, upper floral with cuneate base and shortly awned teeth the lower incurved, spike interrupted below, cal.-teeth broadly-triangular acute and as well as the nerves clothed with minute bristles, cor. small whitish or violet and white, lobes of upper lip emarginate, lower lip equalling the upper and slightly exceeding the tube, fr. oblong narrowed below, upper part ciliate and pilose, equalling or exceeding cal.-teeth and equalling the floral l.—On mountains and near the sea, Devon, Somerset, York, & N. of Scotland. E. S.

** *Grandifloræ.* L. not more than twice as long as broad, usually much less. Cor.-tube lengthening after the flower opens.

11. *E. Rostkóviana* (Hayne); st. tall usually branched below with long scattered glandular hairs, l. plicate striate more or less densely clothed with white bristles and long more or less wavy glandular hairs, cauline ovate acute or shortly-acuminate with 6-12 acute (not awned) teeth, floral broadly-ovate with 6–12 acute teeth, spike lengthening, cal. clothed with bristles and glandular hairs, teeth triangular-lanceolate, cor. very large ultimately much exceeding cal. white or more or less violet, lobes of upper lip reflexed emarginate or bilobed, fr. elliptic emarginate strongly ciliate with shortly-pilose surface equalling or but little exceeding cal-teeth.—Widely distributed. E. S. I.

[*E. campes'tris* (Jord.); closely related to Sp. 11, but differing by its shorter st. branching higher up, smaller fl. and narrower l. with shorter glandular hairs, is perhaps British.]

12. *E. Ker'neri* (Wettst.); st. eglandular usually branched below, l. eglandular with minute bristles especially on the margins and nerves, cauline ovate-elliptic acute with 8–14 triangular-acute teeth, floral oval acute with 6–12 acuminate or mucronate teeth, spike much lengthening, cal. eglandular with lanceolate acuminate scabrid teeth, cor. usually very large ultimately much exceeding cal. usually whitish with violet stripes, the upper lip often violet, lobes of upper lip reflexed bilobed, fr. oblong-obovate emarginate ciliate with shortly-pilose surface falling short of the cal.-teeth.—On limestone in the Southern and Midland Counties. E.

*** *Angustifoliæ.* L. narrow, usually more than twice as long as broad.

13. *E. salisburgen'sis* (Funck); st. simple or branched below, l. quite glabrous or with very few minute bristles below plane beneath, cauline lanceolate 2–5 times as long as broad usually very acute with 4–6 distant elongate patent awned teeth, floral ovate-lanceolate with 4–10 (usually 6) similar teeth, spike ultimately much elongated, cal. glabrous or with minute bristles, teeth triangular-lanceolate, cor. rather small usually whitish sometimes bluish purple or violet, lobes of upper lip reflexed emarginate or denticulate, fr. cuneate-elongate truncate-emarginate quite glabrous or very slightly hairy on the upper part, equalling or exceeding cal.-teeth.—On limestone W. of Ireland. I.

The following hybrids are recognised as British by Mr. Townsend:—

E. stricta×*brevipila.*
E. brevipila×*scotica.*
E. curta v. glabrescens×*brevipila.*
E. gracilis×*brevipila.*
E. scotica×*gracilis.*
E. Rostkoviana×*brevipila.*
E. Rostkoviana×*nemorosa.*

14. ODONTI'TES *Hall.*

1. *O. rúbra* (Gilib.); *l. narrowed from near the base* opposite linear-lanceolate-attenuate *remotely serrate, floral l.* usually longer than the fl. *with* 2—4 *teeth* and an entire end, cal.-segments as long as their tube lanceolate acute, *cor. pubescent* open, lobes of the lower lip oblong, style protruded even before the fl. opens, caps. oblong.—*Bartsia Odontites* Huds. *E. B.* 1415.—St. about a foot high, much branched. Fl. many, pink, in leafy unilateral spikes. Filaments, anth., and stam. hairy. —*a. O. verna* (Dum.); branches ascending straightish, l. rounded below, cal.-teeth narrowed below.—*β. O. serotina* (Dum.); branches spreading and curving up, l. narrowed below, cal.-teeth narrowed below.—[*O. simplex* (Krok) = *Euphr. Odontites* var. *litoralis* (Fr.), a small form with stout usually unbranched st., broader more obtuse l., and exserted fr., is recorded from Scotland.]—Corn-fields and waste places. A. VII. VIII. E. S. I.

15. SIBTHORP'IA *Linn.*

1. *S. europæ'a* (L.).—*E. B.* 649.—A slender trailing plant with filiform creeping stems, and alternate long-stalked roundish reniform leaves with a few large crenatures. Fl. very small, axillary, solitary, on short stalks, pinkish, inconspicuous.—Damp shady places in the South and South-west. P. VI.—IX. E. I.

16. VERON'ICA *Linn.* Speedwell.

* *Racemes axillary.* (*Root perennial.*)

1. *V. scutelláta* (L.); *l. linear-lanceolate* acute sessile minutely denticulate, racemes alternate, fruitstalks slender reflexed, *caps. of* 2 *flattish orbicular lobes*, st. erect.—*E. B.* 782.—Stoloniferous. St. weak, 1 ft. high, glabrous or [var. *villosa*, Schum.] hairy. Fl. pale flesh-coloured, with darker lines. Sep. small, lanceolate, acute, shorter than the capsule.—Boggy places. P. VI.—VIII. E. S. I.

2. *V. Anagal'lis* (L.); *l. lanceolate serrate* acute sessile, racemes opposite, fruitstalks spreading, *caps. oval slightly notched*,

st erect.—*E. B.* 781.—Stoloniferous. St. glabrous, thick, succulent, hollow, 12—24 in. high. Fl. pale blue. Sep. lanceolate. Racemes sometimes glandular (*V. anagalliformis*, Bor.). Whole plant usually glabrous. [A small annual form is var. *montioides* (Boiss.), see Hiern, J. of B. 1898, p. 321.]—In water. P. VI.—VIII. *Water Speedwell.* E. S. I.

3. *V. Beccabun'ga* (L.); *l. stalked oval* crenate-serrate, racemes opposite, fruitstalks spreading, *caps. roundish* tumid slightly notched, *st. procumbent* at the base rooting.—*E. B.* 655.—Glabrous, succulent. Fl. bright blue, rarely pink or flesh-coloured.—Ditches and streams. P. V.—VIII. *Brooklime.* E. S. I.

4. *V. Chamæ'drys* (L.); l. nearly sessile cordate-ovate incise-serrate, racemes usually opposite, fruitstalks ascending, *caps. flat obcordate deeply notched* ciliate shorter than the cal., *st. bifariously hairy* ascending.—*E. B.* 623.—St. about a foot long. Fl. large, many, handsome, blue. Sep. lanceolate, acute. L. on autumnal shoots slightly stalked. Hedgebanks. P. V. VI. *Germander Speedwell.* E. S. I.

5. *V. montána* (L.); *l. stalked* broadly ovate serrate, fruitstalks ascending, *caps. orbicular* notched their margins crenulate and ciliate *longer than the cal., st. diffuse hairy all round* procumbent.—*E. B.* 766.—St. often above a foot long. Racemes lax. Fl. few, pale blue. Caps. very large, quite flat. Sep. ovate-lanceolate, acute.—Woods and thickets. P. V. VI. E. S. I.

6. *V. officinális* (L.); l. shortly stalked oval serrate, *racemes spikelike* many-flowered, fruitstalks erect, *caps. obcordate truncate* bluntly notched longer than the cal., st. prostrate creeping hairy.—*E. B.* 765.—St. 6—12. in. long. Racemes erect, much longer than the leaves. Caps. not always notched. St., l., and calyx sometimes smooth.—*β. V. hirsuta* (Hopk.); l. ovate-lanceolate, caps. abrupt undivided. *E. B. S.* 2673. Much smaller than the true *V. officinalis.* I have not seen wild specimens.—Dry banks and heaths. P. VI.—VIII. *Common Speedwell.* E. S. I.

** *Racemes terminal; cor.-tube longer than broad.*

7. *V. spicáta* (L.); l. ovate or lanceolate crenate-serrate entire at the end, lower l. blunt stalked, *raceme spikelike long dense,* bracts longer than the pedicels, caps. ovate emarginate with a very long style.—*E. B.* 2.—St. erect, branching at the base, about 6 in. high, or in *V. hybrida* taller. Fl. blue. Lower l. oval with

a wedgeshaped base, or ovate with a rounded or slightly cordate base (*V. hybrida* L. *E. B.* 673).—Rare. On chalky heaths near Newmarket and Bury; and on limestone cliffs. P. VII. VIII. *Spiked Speedwell.* E.

*** *Racemes terminal; cor.-tube very short.* † Seeds flat.

8. *V. fruticans* (Jacq.); l. oval serrate at about the middle, lower l. smaller, *raceme pubescent with crisped hairs not glandular* few-flowered, caps. ovate-attenuate, valves bifid.—*E. B.* 1027. *V. saxatilis* (L.) ed. viii.—Raceme persistently subcorymbose. Ped. long, erect. Fl. large, bright blue. St. much branched, decumbent. woody.—*V. suffruticosa* Sm. is probably. a mistake.—Exposed alpine rocks. P. VII. S.

9. *V. alpína* (L.); l. elliptic or ovate dentate or entire, lower l. smaller, *raceme hairy with patent hairs* not glandular few-flowered, *caps. oval-obovate emarginate* crowned with the very short persistent style.—*Sy. E. B.* 980.—St. decumbent, scarcely rooting, erect, simple except quite at the base, 4—6 in. high. Fl.-raceme corymbose; fr.-raceme dense.—Summits of Highland mountains. P. VII. VIII. S.

10. *V. serpyllifólia* (L.); l. ovate or oval slightly crenate, lower l. smaller and rounder, *raceme long many-flowered, caps. obcordate broader than long* crowned with the long persistent style.—*E. B.* 1075.—Fl. whitish with blue veins. St. rooting below, afterwards erect, 2—6 in. high.—*β. V. humifusa* (Dicks.); st. quite prostrate, racemes shorter. *Sy. E. B.* 979.—Roadsides and damp places. *β.* Mountains. P. V.—VII. E. S. I.

[* *V. peregrína* (L.); l. all obtuse and narrowed below, lower l. obovate-oblong, uppermost l. bractlike exceeding the minute fl., raceme slightly spiked many-flowered lax, ped. very short, caps. obcordate, style very short.—*Sy. E. B.* 977.—St. erect, 2—5 in. high. Fl. white faintly tinged with pink. Caps. smooth, broader than long; lobes rounded. Sep. linear-lanceolate.—Fields. A. V.] S. I.

[* *V. repens* (DC.); *l. all roundly oval,* pan. short few-flowered, ped. exceeding l., cor. 2—3 times as long as the sep., caps. obovate shorter than its style.—*V. tenella* R. xx. 1718.—St. prostrate, rooting; no erect fl.-shoot and raceme. Ped., cal., and caps. glandular-hairy.—A Corsican pl. naturalized at Manchester, York, and Glasgow. P. IV. V.]

11. *V. arven'sis* (L.); l. cordate-ovate crenate, lower l. stalked, uppermost l. lanceolate entire bractlike exceeding the flowers,

raceme slightly spiked many-flowered lax, ped. very short, *caps. obcordate* broader than long compressed ciliate on the keel.—*E. B.* 734.—St. ascending, 2—6 in. or rarely a foot long. Fl. pale blue. Caps. smooth, with rounded lobes which exceed the style. Seeds 12—14. Sep. lanceolate, unequal, sometimes very glandular. [β. *eximia* (Towns.); branches simple from base of st., upper l. bracts and sep. not exceeding caps.]—Gravelly and sandy places. A. IV.—VII. E. S. I.

12. *V. ver'na* (L.); *l. pinnatifid*, lower l. stalked ovate serrate, upper l. lanceolate entire bractlike, *raceme slightly spiked many-flowered lax*, ped. very short, caps. obcordate compressed ciliate on the keel.—*E. B.* 25.—St. erect, 1—3 in. high, simple or branched in the lower part. Caps. smooth or downy, with rounded lobes. Style very short. Seeds 12—14. Sep. linear-lanceolate, unequal.—Sandy heaths. Bury, Thetford, and Mildenhall, Suff. A. V. E.

†† Seeds concave.

13. *V. triphyl'los* (L.); *l. fingered*, lower l. ovate entire or dentate stalked, *raceme slightly spiked lax few-flowered, ped. exceeding the calyx*, caps. obcordate compressed smooth ciliate on the keel.—*E. B.* 26.—St. erect with spreading branches, 4—5 in. high. Fl. deep blue. Ped. usually longer than the leaves. Sep. oval. Known by its spreading st., deeply fingered l., and dark-blue flowers.—Sandy fields. Bury, Mildenhall, Brandon, &c., Suff. York. A. IV. E.

**** *Flowers axillary, solitary. Seeds concave. St. prostrate.*

14. *V. agres'tis* (L.); l. all stalked cordate-ovate incise-serrate, *sep. oval*, stam. inserted at the very bottom of the cor., caps. of 2 turgid keeled lobes, seeds about 6 in a cell.—*E. B. S.* 2603. *R. I.* f. 440.—Caps. hairy all over, or only ciliate on the keel; hairs all straight and glandular. L. usually exceeding the peduncles. Lower cor.-lip white.—[*V. opaca* (Fries), with fewer seeds, hairs on caps. incurved short with a few longer and glandular, stam. inserted in the throat of cor., is probably British.]—Fields and waste places. A. IV.—IX. *Green Field Speedwell.* E. S. I.

15. *V. did'yma* (Ten.); l. all stalked cordate-ovate incise-serrate, *sep. broadly ovate acute*, stam. inserted at the very bottom of the cor., caps. of 2 turgid lobes, seeds 8—12 in a cell.—*E. B.* 783. *R. I.* f. 404, 405. *V. polita* (Fries) ed. viii.—Caps. with short dense glandless hairs and other shorter glandular ones, rarely

glabrous. L. usually falling short of the peduncles. Fl. wholly blue.—Cor. sometimes as large as that of *V. Tournefortii*—Fields and waste places. A. IV.—IX. *Gray Field Speedwell.* E. S. I.

*16. *V. Tournefor'tii* (C. Gmel.); l. all stalked cordate-ovate incise-serrate, *sep. lanceolate-acute*, stam. from the bottom of the cor,. *caps. of 2 divaricate lobes compressed upwards and sharply keeled*, seeds about 8 in a cell.—*E. B. S.* 2769. *V. Buxbaumii* (Ten.) ed. viii.—St. long, hairy. L. falling short of the peduncles. Fl. twice the size of those of the preceding, as large as those of *V. Chamædrys*, blue.—Fields. A. IV.—IX. E. S. I.

17. *V. hederæfólia* (L.); *l. cordate with 5—7 large toothlike lobes* all stalked, sep. cordate-attenuate ciliate, caps. of 2 turgid lobes, seeds 2 in each cell.—*E. B.* 784.—Fl. pale blue.—Fields and banks. A. IV.—VI. *Ivy-leaved Speedwell.* E. S. I.

[*Acan'thus mol'lis* (L.), belonging to the Order *Acanthaceæ*, has been introduced by unknown agency to St. Agnes Isle, Scilly, and Traeth Manaccan, Cornw.]

Order LVIII. LABIATÆ.

Cal. tubular, regular or 2-lipped, persistent. Cor. 2-lipped, upper lip entire or bifid, lower 3-fid, or nearly regular. Stam. 4, didynamous, rarely 2. Ovary free, 4-lobed. Style 1, from the base of and between the lobes; stigma bifid. Fr. in 1—4 small nutlets (not true nuts although so called).—Fl. unsymmetrical, often forming 2 cymes so placed as to resemble a whorl (a vert cillaster). L. opposite.—No stipules.

Tribe I. *MENTHOIDEÆ.* *Cor.* bellshaped, *nearly regular.* Stam. distant, straight, diverging upwards.

1. Mentha. Cor. 4-fid, tube very short. Cal. regular, 5-toothed. *Stam.* 4; anth.-cells parallel.

2. Lycopus. Cor. 4-fid, scarcely longer than the regular 5-toothed calyx. *Stam.* 2; anth.-cells parallel or ultimately diverging; 2 upper stam. imperfect.

Tr. II. *MONARDEÆ.* *Cor.* 2-lipped. Stam. 2, perfect, parallel under the upper lip of the corolla.

3. Salvia. *Filaments with 2 diverging branches*; only one bearing a perfect anth.-cell. Cor. ringent. Cal. tubular, 2-lipped.

Tr. III. *SATUREINEÆ.* Cor. 2-lipped. Stam. 4, distant; cells of anth. separate, diverging; connective dilated.

4. Origanum. Stam. diverging; connective subtriangular. Upper lip of cor. straight, nearly flat; lower patent, 3-fid. *Cal. with* 5 *equal teeth* and 10—13 veins; throat hairy. *Spikes 4-sided,* resembling catkins with imbricate bracts.

5. Thymus. *Tips of stam. patent.* Anth.-cells at first nearly parallel, afterwards diverging; connective subtriangular. Upper lip of cor. straight, nearly flat; lower patent, 3-fid. *Cal. 2-lipped* and 10—13-veined; throat hairy. Fl. whorled, axillary or spiked.

6. Clinopodium. *Tips of stam. converging* under the upper lip of the corolla. Anth.-cells at length diverging; connective subtriangular. Upper lip of cor. straight, nearly flat; lower patent, 3-fid. Cal. 2-lipped.

[*MELISSINEÆ.* Cor. 2-lipped. Stam. distant; anth.-cells connected above.]

[Melissa. Tips of stam. converging under the upper lip of the cor. Anth.-cells diverging. *Upper lip of cor. concave*; lower patent, 3-fid. Cal. 2-lipped; upper lip flat, with 3 teeth, the lateral teeth folded at their midrib.]

Tr. IV. *SCUTELLARIEÆ.* Stam. approaching, parallel under the upper lip of the cor. Cal. 2-lipped, closed in fruit.

7. Scutellaria. Tips of stam. incurved. *Filaments simple.* Anth. of the two longer and *inferior stam.* 1-*celled,* of the shorter and superior 2-celled. Cor. 2-lipped, upper lip concave. Cal. ultimately closed and compressed; lips entire, upper one with a concave scale on its back. Nutlets with a long carpophore.

8. Prunella. Two inferior stam. longest. *Filaments bifid, one branch barren. Anth. all* 2-*celled.* Cor. ringent; upper lip concave, entire. Cal. ultimately closed and compressed; upper lip flat, truncate, 3-toothed; lower bifid.

Tr. V. *NEPETEÆ.* Stam. approaching, parallel under the upper lip of the cor., 2 *inferior shortest.* Cal. tubular.

9. Nepeta. Anth.-cells diverging. Cor. ringent; upper lip flat, straight, emarginate or bifid. Cal. 5-toothed.

Tr. VI. *STACHYDEÆ.* Stam. approaching, parallel under the upper lip of the cor., 2 *inferior longest.* Cal. tubular or bellshaped, spreading in fruit.

* *Stamens exceeding the tube of the corolla.*

10. Melittis. *Anth.* approaching in pairs and *forming a cross,* bursting longitudinally. *Upper lip of the cor. flat,* entire, straight; lower lip with 3 rounded nearly equal lobes. *Cal. membranous,* bellshaped, *ample,* variously lobed.

11. Lamium. Anth. approaching in pairs; cells diverging, bursting longitudinally. *Upper lip of the cor. arched;* lateral lobes of the lower lip minute toothlike or rarely long. Cal. bellshaped, 5-toothed; teeth nearly equal.—*Galeobdolon* (Huds.) has the lobes of the lower lip of the cor. nearly equal and acute.

12. Leonurus. Anth. approaching; *cells nearly parallel,* bursting longitudinally. *Upper lip of cor. nearly flat; lower with 3 blunt lobes.* Cal. tubular, 5-toothed; 2 lower teeth rather the longest. Nutlets flatly truncate.

13. Galeopsis. Anth. approaching in pairs; *cells opposite bursting by 2 valves transversely.* Upper lip of cor. arched; lower lip 3-lobed with 2 teeth on its upper side, lobes unequal. Cal. tubular, 5-toothed; teeth equal or 2 upper ones longest. Nutlets rounded at the end.

14. Stachys. Anth. approaching in pairs; *cells* diverging, *bursting longitudinally.* Upper lip of cor. concave; lower of 3 unequal lobes. Cal. tubular-bellshaped with 5 equal teeth. Nutlets blunt and convex at the end.

15. Ballota. Anth. approaching in pairs; cells diverging, bursting longitudinally. Upper lip of cor. erect, concave, lower 3-lobed, middle lobe cordate. *Cal. funnelshaped* with 5 equal teeth. Nutlets convex and rounded at the end.

** *Stamens falling short of the tube of the corolla.*

16. Marrubium. Anther-cells diverging, bursting longitudinally. Upper lip. of cor. straight, erect, flattish, cloven; lower 3-lobed, middle lobe the largest. Cal. tubular, teeth nearly equal or 2 longer. *Nutlets flatly truncate.*

Tr. VII. *AJUGOIDEÆ.* Cor. with the upper lip very short, or deeply bifid and appearing as if wanting.

17. TEUCRIUM. Stam. parallel, protruded between the lobes of the upper lip of the cor., inferior longest; cells bursting longitudinally. *Cor. with the upper lip deeply bifid, lobes long*; lower lip 3-lobed. Cal. tubular, 5-toothed; the teeth equal or the upper one larger (2-lipped).

18. AJUGA. Stam. parallel, protruded far beyond the upper lip of the cor., inferior longest; cells bursting longitudinally. *Cor. with the upper lip very short*, 2-lobed; lower 3-lobed, much longer than the upper. Cal. ovate-bellshaped, nearly equally 5-cleft.

Tribe I. *Menthoideæ.*

1. MEN'THA *Linn.* Mint.[1]

* *Throat of the calyx naked. General inflorescence determinate. Whorls of fl. in terminal spikes or clusters.*—† Stoles aerial, leafy.

†1. *M. spicáta* (L.); *l. glabrous sessile* lanceolate acute serrate, *spikes lax* cylindrical, bracts subulate, cor. glabrous.—*E. B.* 2424. *Sole Menth.* 5. *M. viridis* (L.) ed. viii. Whorls of the spike rather distant. L. glandular beneath.—β. *M. crispa* (L.); l. deeply cut and crisped.—In marshy places, rare. P. VIII. *Spear-Mint.* E. S.

2. *M. rotundifólia* (Huds.); *l. sessile* roundly ovate *crenate-serrate wrinkled shaggy* beneath, spikes linear cylindrical dense, *bracts lanceolate*, "fr.-cal. not contracted at the mouth," cor. hairy.—*E. B.* 446. *Sole* 3.—Viscid. St. 1—2 ft. high. L. usually with a cordate base. Whorls of spike nearly all close together. Scent acrid.—Waste places, rare. P. VIII. IX. *Round-leaved Mint.* E. S. I.

†† Stoles subterranean.

3. *M. alopecuroïdes* (Hull); *l.* subsessile roundly cordate-oval serrate *wrinkled hairy beneath, spikes conical-cylindrical,* bracts lanceolate, cor. hairy.—*Sy. E. B.* 1021. *M. rotundifolia* Sole 4. *M. velutina* Bab. *M. dulcissima* Dum.—L. not felted beneath. —West of S. and east of E. P. VIII. IX. E. S.

---

[1] See Mr. Baker's valuable paper in *J. of B.* iii. 233.

4. *M. longifólia* (Huds.); *l. subsessile* ovate or lanceolate *serrate silky* beneath, spikes linear-cylindrical dense, *bracts subulate*, "fr.-cal. contracted at the mouth," cor. hairy.—*E. B.* 686. *Sole* 1 & 2. *M. sylvestris* (L.) ed. viii.—L. lanceolate or oblong, more or less hairy, not shaggy. Whorls of spike nearly all close together. Scent sweet. [Extreme forms are *M. mollissima* (Borkh.) with l. softly white-tomentose on both sides, and *M. nemorosa* (Willd.) with shorter oblong l. green pubescent above, slightly tomentose beneath.]—Damp waste ground. P. VIII. IX. *Horse-Mint.* E. S.

*5. *M. piperíta* (L.); *l. stalked* ovate-lanceolate or oblong serrate, upper l. smaller, bracts lanceolate, *spikes lax short blunt interrupted below*, cal. tubular glabrous below with lanceolate subulate teeth.—*E. B.* 687.—St. 1—2 ft. high and l. nearly glabrous or hairy on the veins beneath. Cal. glandular. [β. *vulgaris* (Sole); l. broader, spikes shorter almost capitate.]—Wet places, rare. P. VII. VIII. *Pepper-Mint.* E. S. I.

6. *M. aquat'ica* (L.); l. stalked ovate-acute serrate rounded or subcordate below hairy on both sides, fl.-l. falling short of fl., fl.-whorls in *few axillary and terminal subglobose or ovoid clusters*, cal.-teeth triangular ½ as long as tube.—*M. hirsuta* *E. B.* 447. (*M. citrata* (Ehrh.) *E. B.* 1025 is a glabrous form).—St. 12—18 in. high. Cal. glandular. Moist places. P. VII. VIII. *Capitate Mint.* E. S. I.

7. *M. pubes'cens* (Willd.); l. stalked ovate or ovate-lanceolate serrate hairy above woolly beneath, fl.-l. falling short of the fl., whorls of fl. in subcylindrical *thick dense spikes* interrupted below, cal.-teeth subulate ⅔ length of tube.—*Sy. E. B.* 1026, 1027.—St. 12—18 in. high. I know little of this. Is it distinct from *M. aquatica*? E.

** *Throat of cal. naked. General inflorescence indeterminate. Fl. in axillary distant whorls; none amongst the uppermost l., or shorter than them.*

8. *M. sativa* (L.)[1]; l. nearly sessile ovate or ovate-lanceolate sharply serrate, *upper l.* smaller but similar and *exceeding the fl.*, whorls distant dense, cal. tubular or bell-shaped, cal.-teeth triangular-lanceolate.—St. 1—2 ft. high. Uppermost leaves often above the flowers.—*a. M. sativa* (L.); l. hairy on both

[1] The correct name for this collective species would be *M. gentilis*, L., but *M. verticillata*, L. (=*M. sativa*, L.) and *M. gentilis* are by many regarded as distinct species.—H. & J. G.

sides, ped. cal. and cor. hairy. *E. B.* 448.—*β. M. rubra* (Huds.); l. stalked nearly glabrous, ped. lower part of cal. and cor. glabrous, cal.-teeth hairy, veins of l. purple. *M. gracilis* (Sole) is a slender green form with subsessile leaves. *M. cardiaca* (Baker) differs by having the upper l. sessile and st.-l. nearly glabrous.—*γ. M. pratensis* (Sole); l. rather blunt much veined hairy above, glabrous (except on the veins) beneath, ped. and cal. glabrous, cal.-teeth ciliate.—*δ. M. gentilis* (L.); l. acute with few veins slightly hairy on both sides, upper l. similar, ped. lower part of cal. and cor. glabrous. *E. B.* 2118. —These forms seem to vary into each other.—Wet places. P. VII. VIII. E. S. I.

9. *M. arven'sis* (L.); *l. stalked* ovate bluntly serrate, *upper l. similar and equally large*, whorls distant, *cal.* bellshaped, *teeth triangular as broad as long.*—*a. vulgaris*; l. narrowed below. *M. arvensis* Sm. *E. B.* 2119.—*β. M. agrestis* (Sole); l. roundish subcordate below, upper ones nearly sessile. *E. B.* 2120.—L. very variable in form, from nearly round and blunt to ovate-acute.—Corn-fields. P. VII.—IX. *Corn Mint.* E. S. I.

*** *Throat of the calyx closed with hairs.* PULEGIUM *Opitz.*

10. *M. Pulégium* (L.); l. stalked ovate slightly crenate all similar, whorls all distant globose many-flowered, cal. tubular hispid closed with hairs in the throat.—*E. B.* 1026. *Sole* 23. —St. prostrate. L. often recurved, uppermost axils without flowers. The smallest of our species and remarkably different in habit.—Wet places. P. VIII. IX. *Penny-royal.* E. S. I.

2. LYC'OPUS *Linn.* Gipsywort.

1. *L. europæ'us* (L.); l. stalked ovate-oblong sinuate-dentate or pinnatifid, sterile stam. wanting, nutlets about equalling the tube of the calyx.—*E. B.* 1105.—Subglabrous or pubescent. L. opposite. Fl. small, in dense whorls.—Banks of streams and ditches. P. VII. VIII. E. S. I.

Tribe II. *Monardeæ.*

3. SAL'VIA *Linn.* Sage.

1. *S. Verbenáca* (L.); l. oblong blunt cordate below sinuate and crenate or dentate stalked, *upper l. short broad cordate* sessile clasping, bracts cordate acuminate, *tube of the cor. equalling the*

calyx.—*E. B.* 154.—Varies with the l. incise-dentate. Remarkable for its enlarged very broad sessile upper leaves. St. 1—2 ft. high. Cor. purple, small; upper lip concave, laterally compressed, straight except at the tip. Upper cal.-lip broad; teeth small, converging.—Dry gravelly banks. P. V. VI. *English Clary.* E. S. I.

2. ?*S. clandestína* (L.); l. oblong cordate below sinuate-dentate or incise-dentate stalked, *upper l. oblong acute* sessile scarcely cordate or clasping, bracts cordate acuminate, *tube of the cor. exceeding the calyx.*—St. a foot high. Cor. purple, small. Upper cal.-lip very broad, teeth very small.—Probably distinct from the preceding, but very difficult to characterise. Syme's plant (*Sy. E. B.* 1057) seems different.—Dry gravelly banks, rare. Lizard, Cornwall. Channel Isles. P. VII. E.

3. *S. praten'sis* (L.); l. oblong-ovate cordate below, crenate-dentate stalked, upper l. small, sessile lanceolate acute, bracts cordate acuminate, cor. thrice as long as the calyx.—*E. B.* 153.—St. 1—2 ft. high. Known by its large flowers.—Cobham, Kent. Middleton Stoney, Oxfordshire. P. VII. E.

Tribe III. *Satureineæ.*

4. ORIG'ANUM *Linn.* Marjoram.

1. *O. vulgáre* (L.); l. stalked broadly ovate, bracts ovate exceeding the cal., heads of fl. ovoid or oblong panicled crowded.—*E. B.* 1143.—St. a foot high, corymbose. Bracts usually purple. L. often slightly toothed. Fl. purple.—Spikes sometimes prismatic and oblong. *O. megastachyum* Link, *Sy. E. B.* 1046.—Dry uncultivated places. P. VIII. E. S. I.

5. THY'MUS *Linn.* Thyme.

1. *T. Serpyl'lum* (L.); *st. prostrate creeping,* l. all oblong or lanceolate narrowed into the flat fringed stalks, *flowering shoots ascending,* fl. capitate, upper cal.-lip with 3 short triangular teeth, lower of 2 subulate teeth, *upper cor.-lip oblong.*—*E. B.* 1514.—Forming a cushion with a fringe of prostrate barren shoots which in the next year produce erect fl.-shoots from their lower joinings and are prolonged at the end. L. narrowed below, their lower half and the stalk often fringed, rather conspicuously veined beneath, often narrow. Nutlets globose, mealy. Cor. purple; upper lip conspicuously notched.—[A northern form with broader l. and capitate inflor. has been referred to var. *prostrata* (Hornem).]—Dry heaths. P. VI.—VIII. E. S. I.

2. *T. Chamæ'drys* (Fr.); *stems alike* diffuse ascending, l. all broadly oblong with flat fringed stalks, fl. whorled or capitate, upper cal.-lip with 3 triangular teeth, lower of 2 subulate teeth, *upper cor.-lip semicircular.—E. B. S.* 2992.—Forming a tuft of flowering and barren shoots; the st. of preceding year ending in a fl.-shoot. L. less narrowed into the stalks than in Sp. 1, usually only the stalk is fringed, less prominently veined beneath, broad. Nutlets roundish, subcompressed, with a basal apiculus. Cor. purple; upper lip slightly and obscurely notched. In the large forms the st. is stronger; it is not so in Sp. 1.—Heaths. P. VI.—VIII. E. S. I.

6. Clinopódium *L.* (*Calamintha* (Moench) ed. viii.) Calamint.

* *Fl. in whorls of 2 forked cymes.*

1. *C. Nep'eta* (O. Kuntze); l. ovate serrate pale beneath shortly stalked, cal. rather bell-shaped obscurely 2-lipped its *teeth shortly ciliate* all nearly of the same shape the upper ones slightly shorter and broader, cyme many-flowered its *common stalk about as long as the primary partial stalk.—Thymus* L. *E. B.* 1414.—St. usually many from the crown of the root, 12—18 in. high; branches short, erect. Fl. purple. Hairs in the throat of the cal. protruded.—Dry banks, rare. P. VII. VIII. *Lesser Calamint.* E.

2. *C. Calamin'tha* (O. Kuntze); l. broadly ovate slightly serrate green on both sides on longish stalks, cal. tubular distinctly 2-lipped its *teeth with long cilia* those of the upper lip triangular and ascending, of the lower twice as long and subulate, lower lip of the cor. with distant segments the middle one longest, cyme scarcely forked few-flowered, fl. at an angle with the pedicel.—*Calamintha officinalis* (Moench) ed. viii. *E. B.* 1676.—St. usually solitary, or few from the root, 1—2 ft. high: branches long, ascending. Fl. purplish. Hairs in the throat of the cal. included. Larger in all its parts than the preceding. —*a. Cal. ascendens* (Jord.); ped. of cyme ½ as long as the primary partial stalk.—*β. Cal. off.* var. *Briggsii* (Syme); ped. of lower cymes as long as or longer than the primary partial stalk. *Sy. E. B.* 1051.—Dry banks, rare. *β.* Devon. P. VII.—IX. *Common Calamint.* E. I.

3. *C. grandiflor'um* (O. Kuntze); l. broadly ovate sharply serrate, cal. tubular distinctly 2-lipped, teeth with long cilia those of the upper lip patent or reflexed, *lower lip of the cor. with overlapping segments* all nearly equally long, *cyme many-*

flowered its common stalk about as long as the primary partial stalk.—*Cal. sylvatica* (Bromf.) ed. viii. *E. B. S.* 2897.—Creeping slightly.—St. about 2 ft. high, nearly simple. L. large. Cal.-teeth tinged with purple; hairs in the throat included. Cor. purplish, very large, tube much protruded, middle lobe of lower lip short and broad.—Isle of Wight. Near Torquay. (*J. of B.* xi. 208.) P. VIII.—X. E.

** *Whorls of* 6 *simple separate peduncles.* ACINOS *Moench.*

4. *C. A'cinos* (O. Kuntze); l. ovate subserrate acute with revolute margins, *cal.* tubular *gibbous below* distinctly 2-lipped, upper lip with short triangular teeth lower with subulate teeth all converging in fruit.—*Thymus* L. *E. B.* 411.—St. 6—8 in. long. Fl. blue.—Dry gravelly places and limestone rocks. A. VII. VIII. *Basil.* E. S. I.

*** *Fl. in dense branched axillary clusters. The many setaceous bracts forming a kind of involucre.*

5. *C. vulgáre* (L.); l. ovate rounded below slightly crenate-serrate, clusters equal many-flowered. — *Cal. Clinopodium* (Benth.) ed. viii. *E. B.* 1401.—St. 1—1½ foot high. Fl. purple in 2 or 3 dense whorls, the uppermost terminal.—Dry bushy places. P. VII. VIII. *Wild Basil.* E. S. I.

[*Melissineæ.*]

[MELIS'SA, *Linn.* Balm.]

[*M. officinális* (L.); l. ovate crenate-serrate acute paler beneath, cal. rather bell-shaped slightly ventricose in front distinctly 2-lipped, upper lip flat truncate with three short broad teeth, lower with 2 lanceolate teeth.—*Sy. E. B.* 1053.—St. 2 feet high. Fl. in axillary secund whorls, white.—Escaped in the South. P. VII. VIII.] E.

Tribe IV. *Scutellarieæ.*

7. SCUTELLA'RIA *Linn.* Skull-cap.

1. *S. galericuláta* (L.); l. shortly stalked all oblong-lanceolate cordate below crenate-serrate, fl. axillary opposite secund, calyx without glands.—*E. B.* 523.—Cor. large (¾ in.) blue. St. 6—12 in. high, stout.—A hybrid between this and Sp. 2 has been found.—Banks of rivers and ditches. P. VII. VIII. E. S. I.

2. *S. mínor* (Huds.); l. shortly stalked, lower broadly ovate, intermediate ovate-lanceolate with the base cordate, upper l. lanceolate with a rounded base, fl. axillary opposite secund, cal. pubescent.—*E. B.* 524.—Cor. small, pale pink. St. 4—8 in. high, slender.—[A form with more glandular fl. is var. *glandulosa* (Ar. Benn.).]—Moist heaths and boggy places. P. VII.—IX. E. S. I.

8. PRUNEL'LA *Linn.* Self-heal.

1. *P. vulgáris* (L.); l. stalked oblong-ovate blunt, upper lip of the cal. with short truncate mucronate teeth, lower lip with ovate-lanceolate mucronate teeth.—*E. B.* 961.—About 8 in. high. L. nearly entire or slightly toothed; on Continental specimens sometimes pinnatifid. Fl. blue, rarely white, whorled, crowded into a dense spike, with 2 broad kidney-shaped acuminate bracts under each whorl. Cal. reddish purple.—In damp pastures. P. VII. VIII. E. S. I.

Tribe V. *Nepeteæ.*

9. NEP'ETA *Linn.*

1. *N. Catária* (L.); l. stalked cordate acute incise-serrate whitish-pubescent beneath, cymes dense many-flowered spiked, nutlets smooth and glabrous.—*E. B.* 137.—Fl. white. St. erect, 2—3 feet high, downy or mealy. Stam. at length curved outwards.—Waste places. P. VII. VIII. *Cat-Mint.* E. S. I.

2. *N. Glechóma* (Benth.); l. cordate-reniform crenate, whorls axillary stalked secund 3—4-flowered, cal.-teeth ovate-acuminate awned, nutlets oblong with impressed dots.—*Glechoma hederacea* L. *E. B.* 853.—St. procumbent, creeping. Fl. blue-purple; length of cor.-tube and its hairiness are variable. Anth. in pairs forming a cross.—[A small subglabrous form with short cor. is var. *parviflora* (Benth.).]—Hedges and thickets. P. IV.—VI. *Ground-Ivy.* E. S. I.

Tribe VI. *Stachydeæ.*

10. MELIT'TIS *Linn.* Bastard Balm.

1. *M. Melissophyl'lum* (L.).—*E. B.* 577.—L. oblong-ovate or slightly cordate. Upper lip of the cal. with 2 or 3 teeth. Fl.

purple with a white margin or variegated in different ways, large. St. 1—2 ft. high.—*M. grandiflora* (Sm. *E. B.* 636) is only a slight variety.—Woods in the South. P. V. VI. E.

11. LAMIUM *Linn.* Dead-Nettle.

* *Lower lip of cor. with one large obcordate lobe with* 1—2 *teeth on each side of its base.*

1. *L. amplexicaúle* (L.) ; l. roundish-cordate bluntly incise-crenate, lower l. stalked, upper sessile clasping, *cal.-teeth* longer than the tube (green) at length *connivent,* cor.-tube straight naked within.—*E. B.* 770.—Lower fl.-whorls usually distant. Fl. purple-red. Nutlets small, smooth, three times as long as broad, with a small triangular oblique terminal space. Cor.-tube much exceeding the calyx, slender. The cor. does not always expand, but the anth. are fertile and the fr. is produced.—Sandy and chalky fields. A. V.—VIII. *Henbit.* E. S. I.

2. *L. intermédium* (Fries); l. incise-crenate, lower l. stalked, *upper reniform-cordate* sessile, *cal.-teeth longer than the tube hispid always spreading,* straight cor.-tube with a faint hairy ring within.—*E. B. S.* 2914.—Lower fl.-whorls usually distant. Nutlets large thrice as long as broad, with a large triangular terminal rather oblique space. Tube of the cor. equal, cylindrical. Cal.-teeth usually purple, rigid.—Common in S. rare in E. and I. A. VI.—IX. E. S. I.

3. *L. hy'bridum* (Vill.) ; *l.* cordate incise-dentate *all stalked,* upper broadly ovate crowded, cal.-teeth as long as or longer than the tube always spreading, tube of the cor. straight naked within.—*E. B.* 1933. *L. incisum* (Willd.) ed. viii.—Uppermost l. wedgeshaped below. Fl.-whorls usually all contiguous. Fl. reddish. Cor.-tube equal, cylindrical, sometimes with a faint ring of hairs within. Nutlets smooth.—Cultivated and waste ground. A. IV.—VI. E. S. I.

4. *L. purpúreum* (L.) ; l. cordate crenate serrate all stalked, upper cordate or cordate-ovate crowded, cal.-teeth as long as the tube always spreading, cor.-tube a little curved below with a ring of hairs within.—*E. B.* 769.—Fl.-whorls contiguous. Cor. pale purple, lip spotted with red; tube narrowed below. Nutlets about twice as long as broad, smooth. A form with more deeply-cut leaves, *β. decipiens* (Sond.), is often taken for Sp. 3. —Waste and cultivated ground. A. IV.—VIII. *Red Dead-Nettle.* E. S. I.

5. *L. al'bum* (L.) ; l. cordate-ovate acuminate deeply serrate stalked, cal.-teeth as long as the tube, *cor.-tube* exceeding the calyx *with an oblique ring of hairs within* and narrowed below the ring.—*E. B.* 768.—St. 12—18 in. high. Fl. large, white. —Waste ground. P. V. VI. *White Dead-Nettle.* E. S. I.

[*L. maculátum* (L.); l. cordate-ovate acuminate deeply serrate stalked, cal.-teeth longer than the tube, *cor.-tube* exceeding the calyx *with a transverse ring of hairs within* and narrowed below the ring.—St. 12—18 in. high. Fl. purple. L. marked with white, cordate-ovate (*L. maculatum*) ; or green, triangular-cordate (*L. lævigatum. E. B.* 2550).—An escape. Fifeshire. Clova. P. VI.—VIII.] E.? S.

** *Lower lip of cor. in 3 entire nearly equal acute lobes.* GALEOBDOLON *Huds.* Archangel.

6. *L. Galeob'dolon* (Crantz) ; l. ovate acuminate truncate below coarsely serrate stalked, upper l. lanceolate attenuate below, helmet of the cor. long entire, *lower lip in 3 entire nearly equal lobes.*—*G. luteum* Huds. *E. B.* 787. *G. montanum* Reich. !— Fl. yellow. St. 12—18 in. high. Lower l. coarsely and even doubly serrate.—[*G. luteum* Reich. ! has the l. all ovate-acuminate and the lower ones simply crenate with a minute apiculus.]—Woods and thickets. P. V. VI. E. S. I.

12. LEONU'RUS *Linn.* Motherwort.

†1. *L. Cardíaca* (L.) ; lower l. palmately 5-fid incise-dentate, upper ones 3-lobed entire wedgeshaped below, cor.-tube with an oblique ring of hairs within, helmet nearly flat, lip spreading its middle lobe entire.—*E. B.* 286.—St. 3 feet high. Cor. hairy externally, purple. Fl. in crowded whorls. Cal.-teeth sharp. —Hedges and waste places, rare. P. VIII. E. S.

13. GALEOP'SIS *Linn.* Hemp-Nettle.

1. *G. dúbia* (Leers) ; st. not thickened at joinings with deflexed hairs, l. ovate-lanceolate serrate soft and downy on both sides, upper l. ovate, cal. glandular shaggy, *upper cor.-lip deeply notched.*—*G. villosa* Huds. *E. B.* 2353. *G. ochroleuca* Lam.— Cor. large, pale yellow. St. 10—12 in. high, with gland-tipped hairs on its upper part.—Sandy corn-fields, rare. A. VII. VIII. E.

2. *G. Lad'anum* (L.) ; st. not thickened at joinings soft with deflexed hairs, l. ovate-lanceolate lanceolate or lanceolate-

attenuate at both ends serrate or nearly entire downy on both sides, cal. shaggy with adpressed hairs and a few gland-tipped hairs intermixed, *upper cor.-lip slightly notched.*—*E. B.* 884. *G. angustifolia* (Ehrh.).—Cor. purple variegated with crimson and white, shaggy externally. St. about a foot high. St. and l. varying much in hairiness, pale green or purplish.—β. *G. intermedia* (Vill.); l. broad not narrowed below, regularly toothed, whorls of fl. all separate.—*Sy. E. B.* 1075.—γ. *G. canescens* (Schultz); l. narrow nearly entire, bracts often reflexed at the end, cal. and upper part of st. clothed with patent hairs.—In gravelly and sandy districts. β. Moray. A. VIII. IX. E. S. I.

3. *G. Tet'rahit* (L.); *st. thickened at joinings hispid*, l. oblong-ovate acuminate serrate, cal. tubular, cal.-teeth and tube nearly equal, *cor.-tube equalling the cal.*, upper lip ovate.—*E. B.* 207. —St. 1—2 ft. high. Cor. purplish variegated with white, large; tube slender slightly inflated; middle lobe of lower lip subquadrate, flat, crenulate, blunt or slightly emarginate. Cal.-teeth as long as their tube, which is shorter but quite as broad and more strongly ribbed than in the next species, rather inflated below the mouth. Nutlets wholly green, the oblique top longitudinally marked with veins all springing from the wholly acute inner angle of the nut. L. slightly pubescent above.—β. *G. bifida* (Boenn.); middle lobe of lower cor.-lip oblong emarginate purple with pale ultimately revolute edges, base with a yellow 2-lobed spot and several dots. A more slender plant with paler leaves and smaller flowers. *Sy. E. B.* 1079. —Woods and cultivated ground. A. VII.—IX. E. S. I.

4. *G. specio'sa* (Mill.); st. thickened at the joinings hispid, l. oblong-ovate acuminate serrate, cal. bellshaped, cal.-teeth shorter than the tube, *cor.-tube much exceeding the cal.*, upper lip roundish-oval.—*E. B.* 667. *G. versicolor* (Curt.) ed. viii.—St. 2—3 feet high. Cor.-tube inflated above. Fl. very large, yellow, usually with a broad purple spot upon the lower lip. Cal.-teeth shorter than their tube. Nutlets with the oblique top dark brown, inner angle rounded off almost to the base. Difficult to distinguish upon paper from *G. Tetrahit.*—Cultivated ground. A. VII. VIII. E. S. I.

14. Sta'chys *Linn.* Woundwort.

1. *S. Beton'ica* (Benth.); whorls in oblong interrupted term. spike many-flowered, st. erect, lower l. ovate-oblong with a cordate base crenate blunt with long stalks, upper l. oblong-lanceolate serrate acute subsessile, bracts linear-lanceolate equalling the *nearly glabrous cal.*, stam. falling short of the lip.—

Betonica officinalis L. *E. B.* 1142.—St. 1—2 ft. high. Whorls sometimes considerably separated. Cor. purplish red; tube exserted. The English plant has the round crenate not emarginate lower lip of *B. hirta* (R.).—Woods and thickets. P. VII. VIII. *Betony.* E. S. I.

2. *S. german'ica* (L.); whorls many-flowered, *st.* erect *woolly*, l. oblong-ovate or ovate-lanceolate with a cordate base crenate-serrate stalked *densely silky*, *upper l. lanceolate* acute sessile, *cal. silky*, teeth acute mucronate spinous, bracts equalling the calyx. —*E. B.* 829.—St. 2—3 ft. high. Fl. purple.—Chalky soil, Oxfordshire. B. VII. *Downy Woundwort.* E.

3. *S. alpina* (L.); whorls 6—10-flowered, st. erect *hairy*, upper part glandular, l. ovate-oblong cordate-based obtusely serrate, floral l. *large* sessile *nearly all serrate*, bracts linear-lanceolate *about equalling cal.*, cal.-teeth *broadly* acuminate or mucronate, cor. much exceeding cal. dull purple spotted with white hairy externally.—*J. of B.* xxxvi. (1898) t. 384.—Whorls distant below, lower l. long-stalked crenate.—Woods, Wotton-under-Edge, Glos., *Mr. C. Bucknall.* P. VI.—VII. E.

4. *S. sylvat'ica* (L.); whorls 6—8-flowered, st. erect, *l. cordate-ovate* serrate long-stalked, *floral l. linear entire*, cal.-teeth lanceolate very acute, *bracts minute.*—*E. B.* 416.—Cal.-teeth rather spinous. Petioles and l. nearly equal. Fl. reddish-purple. Nutlets opaque, punctured and irregularly tubercled. L. clothed with scattered adpressed hairs or densely silky on both sides.—Woods and thickets. P. VII. VIII. *Hedge-Woundwort.* E. S. I.

5. *S. palus'tris* (L.); whorls 6—10-flowered, st. erect, *l. linear- or ovate-lanceolate subcordate below acute crenate-serrate nearly sessile*, cal.-teeth lanceolate very acute, bracts minute.—*E. B.* 1675.—St. 1½—2 ft. high. Cal.-teeth rather spinous. Lower l. with very short stalks, uppermost sessile. Fl. dull purple. Nutlets shining, very minutely dotted. [Var. *canescens* (Lange), having l. grey felted on both sides, has been reported.]—*β. S. ambigua* (Sm.); l. stalked ovate-lanceolate cordate below serrate. Petioles sometimes half as long as the leaves. [Considered a hybrid between Sp. 4 and 5.]—River-sides and damp places. P. VII. VIII. *Marsh-Woundwort.* E. S. I.

6. *S. arven'sis* (L.); whorls 4—6-flowered, *st. decumbent* or ascending, *l. ovate-cordate blunt* crenate stalked, *floral l. ovate-oblong* sessile acute, cal.-teeth lanceolate awned, cor. scarcely exceeding the cal., bracts minute.—*E. B.* 1154.—Fl. pale purple. Nutlets covered with minute dots and scattered tubercles.—Corn-fields. A. VIII. IX. E. S. I.

[*S. an'nua* (L.); whorls 4—6-flowered, st. erect, lower l. ovate-oblong blunt crenate-serrate stalked, *floral l. lanceolate* acute, cal.-teeth lanceolate very acute, tube of the cor. exceeding cal., bracts minute.—*E. B. S.* 2669.—Fl. yellowish. Nutlets minutely rough.—Gadshill and Sevenoaks, Kent. A. VIII. IX.] E.

15. BALLO'TA *Linn.* Horehound.

1. *B. nigra* (L.); l. crenate-serrate, bracts linear-subulate, cal. funnelshaped, *cal.-teeth broadly ovate short patent or reflexed.*—*E. B.* 46. *R. I.* f. 1041! *B. fœtida* (Lam.) ed. viii.—St. 2—3 ft. high. Lower l. cordate; upper ovate. Fl. purple or white. Scent pungent. A hard coarse plant. Whole plant including the cor. covered with hairs.—Waste places. P. VII. VIII. E. S. I.

2. *B. ruderális* (Sw.); l. crenate-serrate, bracts linear-subulate, cal. funnelshaped, *cal.-teeth ovate gradually acuminate erect.*—*R. I.* f. 1039!—St. 2—3 ft. high. Lower l. cordate; upper ovate. Fl. purple or white. Scent agreeable. *Very hairy and soft.*[1]—I still with Grenier consider this distinct from Sp. 1.—Waste places, rare. Abundant at Llanwarne, Herefordshire. P. VII. VIII. E.

16. MARRU'BIUM *Linn.* White Horehound.

1. *M. vulgáre* (L.); st. erect hoary, l. ovate narrowed into a petiole or roundish cordate crenate hoary rough, whorls many-flowered, cal.-teeth 10 subulate patent hooked woolly below, their upper half glabrous.—*E. B.* 410.—St. 1—2 ft. high, with many whorls of small whitish flowers. L. sometimes dentate, rarely roundly cordate.—Waste places, rare. P. VIII. IX. E. S. I.

Tribe VII. *Ajugoideæ.*

17. TEU'CRIUM *Linn.*

1. *T. Scorodónia* (L.); st. erect, l. oblong-ovate their base cordate crenate-serrate green on both sides, racemes lateral and terminal one-sided, floral l. ovate acute rather longer than the pedicels, *upper cal.-lip ovate*, lower 4-toothed, cor.-tube exserted.

---

[1] We do not find any correlation between the shape of the cal.-teeth and the hairiness of the plant.—H. & J. G.

—*E. B.* 1543.—St. 1—2 ft. high. L. stalked, with glandular resinous mealiness beneath, wrinkled; sometimes oblong, truncate below or subcordate, coarsely and unequally dentate. Fl. yellowish.—Woods and dry stony places. P. VII. VIII. *Wood-Sage.* E. S. I.

2. *T. Scor'dium* (L.); st. procumbent below, *l. sessile* oblong dentate green on both sides, *floral l. similar*, whorls 2—6-flowered axillary distant, *cal.-teeth short equal.*—*E. B.* 828.—More or less hairy or woolly. St. 1—2 ft. long. L. attenuate or broad or even cordate below. Fl. purple.—Wet places, rare. P. VII. VIII. E. I.

*3. *T. Chamæ'drys* (L.); st. ascending, l. ovate incise-crenate wedgeshaped and entire below green on both sides, *floral l. similar smaller nearly entire, whorls racemose* 5-flowered, cal.-teeth lanceolate nearly equal.—*E. B.* 680.—St. much branched, 6—8 in. long, lower part woody. Fl. purplish. Lower floral l. exactly like the stem-l., upper l. gradually smaller and broader below.—Ruined walls, rare. P. VII. E. [S. I. ?]

‡4. *T. Bótrys* (L.); *l. trifid or pinnatifid* green on both sides, *segm. oblong entire or cut*, floral l. similar, whorls axillary 4—6-flowered, cal. gibbous at base inflated tubular, cal.-teeth lanceolate equal.—*E. B. S.* 2964.—St. erect, about 8 in. high; branches ascending. Fl. many, pale purple.—Box Hill and Selsdon, Surrey. Upper Halling, Kent. A. VIII. IX. E.

18. AJ'UGA *Linn.* Bugle.

1. *A. rep'tans* (L.); fl. whorled, st. solitary with long stoles, l. ovate or obovate entire or crenulate stalked, stem.-l. sessile.—*E. B.* 489.—St. 6—8 in. high. Lower whorls distant, upper ones spiked. Cor.-tube with a ring of hairs within. Fl. blue, rarely white.—Wet places. P. V. VI. *Common Bugle.* E. S. I.

[*A. alpina*, E. B. 477, seems to be *A. genevensis.* It is not stoloniferous, and its upper st.-l. fall short of the flowers.]

2. *A. pyramidális* (L.); fl. whorled most or all of the whorls spiked, st. solitary, l. ovate-oblong entire or crenulate, radical l. attenuate below, stem-l. sessile *upper l. exceeding the fl.*—*E. B.* 1270.—St. about 6. in. high. L. gradually decreasing upwards. Cor.-tube with a ring of hairs within. Fl. bluish purple. Plant hairy. It has subterranean offsets and short autumnal stoles.—Highland pastures, very rare, S. South Isles of Arran, I. P. V. VI. S. I.

3. *A. Chamæ'pitys* (Schreb.); *fl. solitary axillary*, st. much branched spreading, *l. deeply trifid with linear entire segments*, floral l. similar exceeding the flowers.—*E. B.* 77.—Hairy. Lowest l. much broader, toothed rather than 3-lobed. Fl. yellow with dark spots. St. reddish purple, branched, 3—6 in. high.—Sandy and chalky fields. A. V.—VII. *Ground-Pine.* E.

Order LIX. VERBENACEÆ.

Cal. tubular, persistent. Cor. irregular, tubular. Stam. didynamous, or 2. Ovary 2—4-celled; style 1, terminal; stigma bifid. Fr. a caps. or berry, with 2—4 nutlets more or less cohering.—No stipules.

1. VERBENA. Cal. 5-fid. Cor. irregular, 5-lobed, slightly 2-lipped. Stam. included, 4, didynamous, or 2. Capsule dividing into 4 nutlets.

1. VERBENA *Linn.* Vervain.

1. *V. officinális* (L.); st. erect solitary 4-angular, l. ovate oblong trifid or laciniate-multifid rough, spikes filiform somewhat panicled, stam. 4.—*E. B.* 767.—St. rather hispid, 1—2 ft. high. L. lobed and serrate, opposite. Spikes long, slender. Fl. small, distant, pale purple.—Waste ground. P. VII. VIII. E. I.

Order LX. LENTIBULARIACEÆ.

Cal. permanent, inferior, divided. Cor. irregular, 2-lipped spurred. Stam. 2. Ovary free, 1-celled of 2 carpels. Stigma of 2 plates, one smaller or inconspicuous. Caps. 1-celled; placenta large, free, central. Seeds many.—No stipules.

1. PINGUICULA. Cal. 2-lipped, lower bifid, upper of 3 segments. Cor. ringent, spurred. Stam. at base of corolla.

2. UTRICULARIA. Cal. 2-leaved, lower often notched, upper entire. Cor. personate, spurred. Stam. at base of upper cor.-lip.

1. PINGUIC'ULA *Linn.* Butterwort.

1. *P. vulgáris* (L.); spur subulate shorter than the *cor.-segments* which are very unequal *oblong rounded separated* entire.—*E. B.* 70.—L. all radical, fleshy, covered with minute crystalline points, pale green; when the plant is gathered they curve backwards so as to hide the root. Fl. violet. Caps. ovoid, acute.—Bogs. P. V. VI. E. S. I.

2. *P. grandiflóra* (Lam.); spur subulate cylindro-conical often notched as long as *cor.-segments* which are very unequally *broadly obovate roundedcontiguous.*—*E. B.* 2184.—Much larger than *P. vulgaris.* Fl very large, violet. Caps. ovoid, rounded at the end. Length of spur variable.—Bogs. Kerry and Cork. P. V. VI. I.

3. *P. alpína* (L.); *spur conical shorter than unequal border of cor.* and curved towards lower lip, caps. acute, scape glabrous. —*E. B. S.* 2747.—Fl. small, yellowish; spur remarkably short and conical.—Bogs. Skye. Ross. Sutherland. P. VI. S.

4. *P. lusitan'ica* (L.); *spur cylindrical* blunt decurved shorter than the nearly equal limb of the cor., *capsglobose, scape downy.* —*E. B.* 145.—Fl. small, pale yellowish, spur shortand cylindrical.—[*P. villosa* (L.), distinguished from this by its acute spur and obconical capsule, may be expected in the North of Scotland.] —Bogs in the Western parts of the country. P. VI.—IX. E. S. I.

2. Utricula'ria *Linn.* Bladderwort.

* L. spinose-ciliate.

1. *U. vulgáris* (L.); spur conical, roundish 3-lobed, *upper cor.-lip about equalling the bilobed palate, margins of lower lip deflexed at right angles all round,* ped. scarcely thrice as long as ovate bracts thick reflexed with fruit, l. pinnate-multifid, bladders upon the leaves.—*E. B.* 253.—Fl. bright yellow, rather large. Scape 4—6 in. high; fl. lemon-coloured. St. floating in the water. Bladders ⅛ in. long. P. VI.—VIII. *Greater Bladderwort.* E. S. I.

2. *U. májor* (Schmid.); spur directed upwards conical projecting, ovate-oblong blunt or emarginate, *upper cor.-lip* 2 *or* 3 *times as long as the small palate, lower lip with a broad flat spreading margin,* ped. 4—5 times as long as lanceolate bract erect with fr., l. pinnate-multifid, small bladders on leaves.—*J. of B.* xiv. 142. *U. neglecta* (Lehm.) ed. viii.—Closely resembling Sp. 1. Ped. slender. Bladders $\frac{1}{10}$ in. long. Fl. yellow.—Pools. P. VI.—VIII. E. [S. ?] I.

3. *U. intermédia* (Hayne); spur conical adpressed, *upper lip twice as long as the inflated palate, lower lip with a broad flat spreading margin,* ped. 4—5 times as long as ovate bract erect with fruit, l. 3-parted, segments linear forked, *bladders separate from the leaves.*—*E. B.* 2489.—Fl. paler with a much longer

upper lip than in Sp. 1. Bladders on leafless shoots. Increasing by buds at the end of the shoots and seldom flowering.—Ditches and pits, rare. P. VIII. E. S. I.

** L. not ciliate.

4. *U. minor* (L.); *spur very short blunt*, upper lip equalling the palate, lower lip ovate with a nearly flat spreading margin, l. repeatedly forked, bladders upon the leaves.—*E. B.* 254.—Scarcely any spur. Fl. small. Sep. roundish, acuminate. Plant much smaller than either of the others.—Ditches and pits. P. VI.—VIII. *Smaller Bladderwort.* E. S. I.

[5. *U. Bremii* (Heer) probably grows in Moss of Inshoch, Nairn, Loch of Spynie, Elgin and Gordon Moss, Berwicksh. It has larger fl. than Sp. 4, a short conic spur, and orbicular flat lower lip. See *J. of B.* xiv. 142.]

Order LXI. PRIMULACEÆ.

Cal. 4—7-cleft, permanent inferior (except in *Samolus*). Cor. regular, 4—7-fid (or none in *Glaux*). Stam. upon the cor. opposite to its segments. Ovary free, 1-celled, with a free central placenta. Style 1. Stigma capitate. Fr. a capsule. Seeds peltate.—No stipules.

Tr. I. *HOTTONIEÆ.* Ovary superior. Caps.-valves 5. Hyle basal.

1. Hottonia. *Cal.* 5-parted, *divided almost to its base. Valves connected at the top.* Cor. salvershaped, stam. inserted and included in cor.-tube.

Tr. II. *PRIMULEÆ.* Ovary superior. Caps.-valves 5. Hyle ventral.

2. Primula. Cal. tubular, 5-fid. *Cor. salvershaped, tube cylindrical* up to the insertion of the stamens. Stam. 5, inserted and included in the cor.-tube. Caps. many-seeded, 5-valved with 10 teeth.

3. Cyclamen. Cal. bellshaped, half 5-cleft. *Cor. with a short bellshaped tube and 5-parted reflexed limb.* Stam. 5, inserted at the bottom of the cor.-tube, included. Caps. many-seeded.

4. Lysimachia. Cal. 5-parted. *Cor. rotate*, scarcely any tube, limb 5-parted. Stam. 5, at the base of the cor. *Caps. with 5 valves* (in *L. nemorum* sometimes 2-valved or indehiscent, in *L. thyrsiflora* few-seeded).

5. Trientalis. *Cal. 7-parted.* Cor. rotate, *7-parted, tube none.* Stam. 7, inserted at the base of the cor. Caps. many-seeded, opening with 5 revolute fugacious valves. *Seeds with a netlike coat.*

6. Glaux. *Cal. bellshaped*, 5-parted, *coloured. Cor. none.* Stam. 5, inserted at the base of the calyx. Caps. few-seeded (about 10).

Tr. III. *ANAGALLIDEÆ.* Ovary superior. Caps. opening transversely all round. Hyle ventral.

7. Anagallis. Cal. 5-parted. Cor. rotate or funnelshaped, tube none, limb 5-parted. Stam. 5, at the base of the cor. Caps. many-seeded.

8. Centunculus. Cal. 4—5-parted. *Cor. with a subglobose inflated tube* and spreading 4—5-parted limb. Stam. 4 or 5, inserted in the throat of the cor. Caps. many-seeded.—Fl. usually 4-parted.

Tr. IV. *SAMOLEÆ.* Ovary inferior. Caps. opening by valves. Hyle basal.

9. Samolus. *Cal.* 5-parted, its *tube adhering to the lower half of the germen*, persistent. *Cor. salvershaped*, tube short, limb 5-parted with interposed converging scales. Stam. 5, inserted near to the base of the cor.-tube. *Caps.* ½*-inferior*, many-seeded, opening with reflexed teeth.

Tribe I. *Hottonieæ.*

1. Hotto'nia *Linn.* Water-Violet.

1. *H. palus'tris* (L.); fl. whorled stalked upon a long solitary cylindrical stalk, cor. exceeding the calyx, sep. subacute equalling cal.-tube, l. pectinate.—*E. B.* 364.—L. submerged, crowded. Fl. rising above the water, pale pink. Style exceeding the cal., stam. inserted in tube, anth. and filaments about equal in length; or style falling short of cal., stam. inserted at top of tube, filaments 3 or 4 times as long as the anthers.—Ponds and ditches. P. V. VI. E. I.

Tribe II. *Primuleæ.*

2. PRI'MULA *Linn.* Primrose.

* *Cal.-tube angular. L. not mealy.*

1. *P. acau'lis* (L.); *l. oblong-ovate* tapering downwards wrinkled crenate, ped. villose radical 1-flowered, *cal. tubular, teeth lanceolate-subulate* very acute, cor.-limb flat with a circle of scalelike folds at the slightly contracted mouth, *caps. ovate ½ the length of cal.*, long straightish teeth of fr.-cal. meeting at top.—*E. B.* 4. *P. vulgaris* (Huds.) ed. viii.—Young l. reticulate-rugose. Scape rudimentary. Fl. erect. *Cal. villose.*—*β. P. variabilis* (Goup.); l. slightly contracted below, ped. raised on a scape, fl. erect. *Sy. E. B.* 1132, 1133. Often taken for *P. elatior* and called *Oxlip.* A hybrid between Sp. 1 & 2 according to Darwin.—Woods and thickets. P. III.-V. *Primrose.* E. S. I.

2. *P. véris* (L.); *l. ovate abruptly contracted below* then attenuate wrinkled crenate, scape tomentose umbellate many-flowered, *cal. bellshaped, teeth short ovate*, cor.-limb concave with a circle of scalelike folds at the slightly contracted mouth, *caps. oval ½ the length of inflated cal.*, short teeth of fr.-cal. converging.—*E. B.* 5. *P. officinalis* Jacq.—Cal. *tomentose.* Cal.-teeth ⅓ of the length of the tube, blunt or slightly acute. Cor.-segm. cordate.—Probably hybrids between this and Sp. 1 are mistaken for *P. elatior.*—Meadows and pastures. P. IV. V. *Cowslip. Paigle.* E. S. I.

3. *P. elátior* (Jacq.); *l. ovate abruptly contracted below* then attenuate wrinkled denticulate, scape umbellate many-flowered, *cal. tubular, teeth lanceolate* acute, *cor.-limb concave*, segments obcordate-oblong, its *tube not crowned nor contracted at the mouth, caps. linear-oblong exceeding cal.*, teeth of fr.-cal. patent.—*E. B.* 513.—Young l. transversely plicate. Cor.-segm. almost square; limb rarely flat. Outer fl. nodding. Fr. erect. Hybridizes rarely with Sp. 1 & 2.—Clayey woods and meadows in the Eastern Counties. P. IV. V. *Oxlip (true).* E.

** *Cal.-tube not angular. L. mealy beneath.*

4. *P. farinósa* (L.); l. obovate-lanceolate, *cal. oblong-ovate, teeth linear*, cor.-limb flat, *segments obcordate rounded below* distant as long as the tube, caps. twice as long as calyx.—*E. B.* 3.—*Umbellate.* Fl. pale lilac with a yellow centre; but the colour and breadth of segments very variable. Germen obovate.

Stigma capitate. Rarely the scape is wanting and the fl. are amongst the leaves.—North of England and South of Scotland. P. VI. VII. *Bird's-eye Primrose.* E. S.

5. *P. scot'ica* (Hook.); l. obovate-lanceolate, *cal. swollen, teeth short ovate blunt*, cor.-limb flat its *segments broadly obcordate approximate* half the length of the tube, caps. scarcely exceeding calyx.—*E. B. S.* 2608.—Half as large as the preceding. *Umbellate.* Fl. bluish purple with a yellow centre. Germen globose. Stigma with 5 points. Rarely scape wanting.—Sandy heaths of the extreme North of Scotland. P. V.—VIII. S.

3. CYC'LAMEN *Linn.* Sow-bread.

†1. *C. hederæfólium* (Ait.); l. cordate-ovate angular denticulate, cor.-throat 5-angled, sep. ovate acuminate denticulate. —*Sy. E. B.* 1136, 1137.—St. a large depressed tuber. L. after the fl., with wavy white blotches above. Fl. pink with a red base (rarely white in *β. ficariifolium Sy. E. B.* 548). Fr.-ped. rolled up spirally.—Woods near the borders of East Sussex and Kent! *Mr. W. W. Saunders.* P. VIII. E.

4. LYSIMA'CHIA *Linn.* Loose-strife.

1. *L. thyrsiflóra* (L.); *racemes axillary stalked dense*, l. opposite lanceolate.—*E. B.* 176.—St. 1—2 ft. high. Fl. small, very many. Cor. divided almost to its base into narrow segments separated by minute teeth, yellow and as well as the cal. spotted with orange. Stam. combined below into a short ring.—Marshes in the North. P. VI. VII. E. S.

2. *L. vulgáris* (L.); st. erect, *panicles* compound *terminal and axillary, l. ovate or ovate-lanceolate* nearly sessile opposite or 3 or 4 in a whorl, cor.-segments entire with glabrous edges, *stam.* 5 *combined* through $\frac{1}{3}$ of their length.—*E. B.* 761.—St. 2—3 feet high. L. variable in size, shape and pubescence. Panicle much branched or nearly simple. Fl. yellow. Starved forms pass for *L. punctata.*—Sides of rivers and pools. P. VII. E. S. I.

[*L. punctáta* (L.); st. erect, ped. axillary opposite or whorled 1-fl., l. ovate-lanceolate slightly stalked opposite or whorled, upper l. narrower, *cor.-segm.* ovate *glandular-ciliate*, stam. 5 *combined* below.—*Sy. E. B.* 1142. *L. verticillata* Bieb. St. 1—1½ ft. high. St. and l. downy. Sep. narrowly lanceolate. Cor. yellow. Ped. rarely branched, downy, falling short of leaves.—Dulverton, Devon! Hingham, Norf.! P.VI. VII.] E.

[*L. ciliáta* (L.); st. erect, *ped. axillary* opposite or whorled, *l.* opposite *ovate-lanceolate subcordate with ciliate stalks,* cor.-segments roundish crenate, *filaments* 10 *free* 5 *sterile.*—*E. B. S.* 2922.—Fl. yellow. St. 3 feet high?—American. Serbergham, Cumb.! and was there in 1815. P. VI. VII.] E.

3. *L. Nummulária* (L.); *st. prostrate* creeping, *fl. solitary* axillary, sep. cordate-ovate prolonged, *filaments* 5 *glandular connected at the base,* l. opposite roundish or ovate shortly stalked. —*E. B.* 528.—Fl. occasionally in pairs, yellow.—Damp places. P. VI. VII. *Moneywort.* E. S. I.

4. *L. nem'orum* (L.); st. prostrate, fl. axillary solitary, *sep. linear-lanceolate, filaments* 5 *smooth distinct,* l. opposite ovate acute shortly stalked.—*E. B.* 527.—Ped. longer than the l. Caps. 5-valved but usually dividing longitudinally into two parts, sometimes indehiscent. Fl. yellow. Stam. distinct.—Woods and damp shady places. P. VI.—VIII. E. S. I.

5. TRIENTA'LIS *Linn.*

1. *T. europæ'a* (L.); l. oblong-obovate blunt.—*E. B.* 15.—St. 4—6 in. high, with the l. mostly collected at the top. Fl. on slender peduncles, white with a yellow ring. Parts of the fl. and fr. varying from 7 to 9 in each whorl. Valves of the caps. soon falling off.—North of E.; Highlands of S. P. VI. E S.

6. GLAUX *Linn.* Black Saltwort.

1. *G. mari'tima* (L.).—*E. B.* 13.—St. mostly procumbent. L. opposite, ovate, glabrous. Fl. axillary, sessile, pink, with blunt segments. Remarkable in this Order by its want of petals.—Sea-shores and salt marshes. P. VI.—VIII. E. S. I.

Tribe III. *Anagallideæ.*

7. ANAGAL'LIS *Linn.* Pimpernel.

1. *A. arven'sis* (L.); st. procumbent or erect, fl. axillary solitary, cor. rotate, l. opposite sessile ovate or ovate-oblong.—Pet. slightly exceeding the cal., crenate. Filaments distinct.—*a. A. arvensis* (Sm.); st. mostly procumbent, pet. fringed with minute glandular hairs usually scarlet, l. ovate. Fl. sometimes flesh-coloured (*A. carnea* Schrank), wholly white or white with a pink eye. *E. B.* 529.—*β. A. cærulea* (Schreb.); st. mostly erect.

pet. without glandular hairs (usually blue), l. ovate-oblong. *E. B.* 1823. Probably distinct. Mr. Borrer suspected that each varies with red or blue flowers.—Corn-fields and sand-hills by the sea. A.? VI. VII. *Scarlet Pimpernel.* E. S. I.

2. *A. tenel'la* (L.); *st. procumbent* rooting, fl. axillary solitary, l. opposite stalked roundish, *cor. funnelshaped,* pet. much exceeding the calyx entire, *filaments connected below.*—*E. B.* 530. —Ped. long. Fl. rather large, rose-coloured. L. nearly sessile. Spongy bogs. P. VII. VIII. *Bog-Pimpernel.* E. S. I.

8. Centun'culus *Linn.* Bastard Pimpernel.

1. *C. min'imus* (L.); l. alternate ovate acute, fl. nearly sessile, cor. without glands at the base.—*E. B.* 531.—Usually very minute. St. usually prostrate. Cor. very small, pale rose-colour.—Damp sandy and gravelly places.—A. VI. VII. E. S. I.

Tribe IV. *Samoleæ.*

9. Sam'olus *Linn.* Brook-weed.

1. *S. Valeran'di* (L.); l. obovate or roundish blunt, upper l. blunt with a point, racemes many-flowered ultimately elongated, caps. subglobose.—*E. B.* 703.—Remarkable in this Order by its cal. adhering to the germen and by having a crown to the small white corolla.—Damp watery places. P. VII. VIII. E. S. I.

Order LXII. PLUMBAGINACEÆ.

Cal. 5-cleft, persistent, inferior, plicate. Cor. regular, 5-fid or nearly 5-petalous. Stam. 5, hypogynous, or adnate to the base of and opposite to the pet. Ovary free, of 5 carpels, 1-celled, 1-seeded; ovule 1, pendulous by a stalk arising from the bottom of the cell. Styles 5. Fr. a utricle. Seed inverted. Embryo in the axis of farinaceous albumen. Radicle superior.—Stip. 0.

1. Limonium. Fl. spiked. Cal. scarious above. Cor. 5-parted. Styles glabrous. Caps. not bursting.

2. Statice. Fl. in a head with an inverted cylindrical sheath. Styles hairy below. Caps. not bursting.

1. LIMO'NIUM *Mill.* (*Statice* Linn., ed. viii.) Sea-Lavender.

* *L. pinnately veined. Cal.-segm. with intermediate teeth.*

1. *L. vulgáre* (Mill.); l. elliptic-oblong stalked mucronate 1-ribbed strongly veined, st. subterete branched above corymbose, spikelets 1—3-flowered ascending forming *dense* 2-ranked *patent or recurved spikes*, cal.-segments entire acute, *outer bract* pointed *rounded on the back*—*E. B.* 102. *S. Limonium* (L.) ed. viii.—St. 6—18 in. high, usually not branched in its lower half, often not until near the corymbose top. Spikes short; spikelets densely imbricate. Fl. purplish.—*β. pyramidale* (Druce) (= *S. serotina* G. & G., ed. viii.); pan. lax much spreading. *Sy. E. B.* 1157. [A hybrid between Sp. 1 & 2 is reported.]—Muddy salt marshes. *β.* Southern coast. P. VII.—IX. E. S.

2. *L. húmile* (Mill.); l. oblong-lanceolate stalked mucronate 1-ribbed faintly veined, *st.* slightly *angular* usually branched from below the middle panicled, spikelets 1—3-flowered unilateral rather distant forming *lax erect or incurved spikes*, cal.-segments acute denticulate, *outer bract* pointed *keeled on the back*.—*S. Bahusiensis* (Fr.) ed. viii. *S. rariflora* Drej. *E. B. S.* 2917.—St. 6—18 in. high, *not corymbose*, much branched below. Spikes long; spikelets often 1-flowered, not imbricate. Fl. purplish.—Muddy salt marshes. P. VII. VIII. E. S. I.

** *L. not pinnately veined. Cal.-segm. without interm. teeth.*

3. *L. recur'vum* (C. E. Salmon); l. obovate-spathulate narrowed into a broadly winged stalk 3-veined below, st. rigid with thick but *not sterile branches*, spikelets 2—4-flowered densely imbricate 2-ranked forming *linear* thick suberect *spikes*, cal.-segments blunt entire.—*Sy. E. B.* 1160. *S. Dodartii* (Gir.) ed. viii.—St. usually not branched in its lower half; branches often simple, short. Inner bracts obovate, very blunt, with white margins or slightly pink. Fl. purple. Anth. linear.—Rocky shores. Portland. *Henslow*! P. VII. VIII. E.

[*L. lychnidifólium* (O. Kuntze); l. obovate-spathulate usually apiculate with broadly-winged 5—9-*veined* stalks, st. stout tapering upwards very rarely with sterile branches, scales at base of branches *large ovoid-triangular*, spikelets usually 2-flowered densely imbricate 2-ranked forming thick patent or nearly horizontal spikes, cal.-segm. short very obtuse ribs hairy.—*J. of B.* xxxix. (1901) p. 193, t. 422.—Rootstock stout woody. St. branching ⅓ to ⅔ of its length. L. thick leathery. Inner bracts roundish about twice as long as the outer, with bright red band and membranous margin. Anth. oblong.—Rocks, Alderney, *Mr. C. R. P. Andrews.* P. VII.—VIII.]

4. *L. occidentále* (O. Kuntze); l. lanceolate-spathulate rather acute narrowed into a long winged stalk obscurely 3-veined below, scapes slender wavy forked branched from near the base, *few lowest branches sterile,* spikelets 2—4-flowered imbricate 2-ranked forming linear *slender* suberect *spikes,* cal.-segments blunt entire without intermediate teeth.—*S. binervosa* G. E. Sm. *E. B. S.* 2663.—St. usually branching quite from the base; branches repeatedly forked, long, often rough. Inner bracts oval with broad membranous edge, deeply tinged with pink. Anth. oval.—*β. intermedium* (Druce); all the branches flowering, spikelets stouter.—Rocky shores. P. VII. VIII. E. S. I.

5. *L. reticulátum* (Mill.); l. obovate or lanceolate-spathulate narrowed into a petiole, st. branched from near the base granular-rough, *branches many slender repeatedly and acutely forked* uppermost alone bearing dense terminal spikes of 2—3-flowered 2-ranked spikelets, cal.-segments ovate cuspidate denticulate.—*E. B.* 328. *S. caspia* (Willd.) ed. viii.—Remarkable by its much divided sterile branches which fork at an acute angle. L. small.—Muddy sea-shores of Norf., and at Frieston, Linc. P. VII. VIII. E.

2. Stat'ice *Linn.* (*Armeria* Willd., ed. viii.) Thrift.

1. *S. marit'ima* (Mill.); *l. linear,* 1-*veined,* inv.-bracts very blunt 1—3 outer ones mucronate, cal.-segments acute, cal.-tube hairy or pilose-striate.—L. all radical, many, narrow. Fl. rose-coloured or white.—*α.* L. flattish above very slender 1-veined. *A. maritima* (Boiss.), *A. pubescens* Link.—*β. planifolia* (Syme); l. nearly flat above broadly linear blunt, early l. 3-veined. *Sy. E. B.* 1153.—*γ. A. duriuscula* (Bab.); l. subtriquetrous channelled above very slender 1-veined. *A. pubigera* and *A. duriuscula* Bab. A small form with linear flat 1-veined l. but a pubescent scape from Shetland may be *A. sibirica* (Turcz.).—Muddy and rocky sea-shores, also on mountains. β. Scottish Highlands, Ben Lawers! γ. Southern sea-shore. P. IV.—IX. E. S. I.

[*S. plantagin'ea* (All.); *l.* 3—5-*veined* linear-lanceolate with a narrow membranous margin, scapes rough, outer inv.-bracts triangular or lanceolate cuspidate, others ovate or obovate with a broad membranous margin blunt, cal.-segm. with long points. —*E. B. S.* 2928.—Distinguished by its leaves.—Jersey. P. VI. VII.]

Order LXIII. PLANTAGINACEÆ.

Cal. 4-parted, persistent, imbricate, inferior. Cor. 4-parted, regular, scarious. Stam. 4, hypogynous, or at the base of the tube, alternate with the segments of the cor.; filaments at first doubled inwards. Ovary free, of 1 carpel, 1-celled; or 2—4-celled. Seeds peltate or erect. Style 1. Caps. opening transversely (or indehiscent in *Littorella*). Radicle inferior.

1. PLANTAGO. Fl. perfect. Cal. 4-cleft. Cor. with an ovate tube and 4-parted reflexed limb. Stam. on the corolla. Caps. bursting transversely, 2—4-celled, 2—4-seeded.

2. LITTORELLA. Monœcious. Male fl. stalked; sep. 4; tube of the cor. cylindrical, limb 4-parted; stam. hypogynous; filaments very long. Fem. fl. sessile; sep. 3; cor. oblong, narrowed at both ends; style long; caps. 1-seeded.

1. PLANTA'GO *Linn.* Plantain.

* *Placenta* 3—4-*winged, thus forming* 3—4 *cells. Seeds* 1 *in each cell.*

1. *P. Coron'opus* (L.); l. linear pinnatifid or dentate, scape terete, spike slender, bracts subulate from an ovate base erect, *midrib of lateral sep. with a ciliate membranous wing.—E. B.* 892.—Tube of the cor. glabrous. Extremely variable in size and amount of pubescence, woolly or nearly glabrous. L. varying in width, nearly entire, or even doubly pinnatifid. Spikes slender, 1½ in. long, many-flowered; or spherical with 2—6 flowers.—[Extreme forms:—Var. *ceratophyllon* (Rap.), luxuriant with broad 3—5-veined l.; var. *maritima* (G. & G.) with nearly erect fleshy l.; and var. *pygmæa* (Lange), dwarf with short few-flowered spikes and almost entire l.]—Gravelly barren spots near the sea and inland. A. ? VI. VII. *Buck's horn Plantain.* E. S. I.

** *Placenta* 2-*winged. Seeds* 1 *in each cell. Cor.-tube pubescent.*

2. *P. marit'ima* (L.); l. linear channelled fleshy convex on the back, scape terete, spike cylindrical, bracts ovate mucronate, *sep. not winged.—E. B.* 175.—L. usually woolly at their base, nearly flat and broad or linear, toothed or quite entire, glabrous or hairy; 3-veined, veins equidistant. Scape glabrous or hairy. Caps. oblong-conical. Very variable in size. In a Cornish variety the l. are only 1—2 lines in length and semicylindrical,

scapes very short, spikes sometimes only 3- or 4-flowered.—The mountain plant may be (1) *P. serpentina* (Vill.) or (2) *P. alpina* (L.); they are both said to have lateral veins of l. nearer the margin than midrib, (1) to have coriaceous, (2) herbaceous leaves. The var. *lanata* (Edm.), *hirsuta* (Syme), is a small very hairy form from Shetland. *Sy. E. B.* 1167. [f. *pumila* (Kjellm.) is a small form with subglobose few-fld. spikes and scapes exceeding the l.]—Sea-coast and on mountains. P. VI.—IX. E. S. I.

*** *Placenta 2-winged. Seeds* 1 *in each cell. Cor.-tube glabrous.*

3. *P. lanceoláta* (L.); *l. lanceolate attenuate at both ends 5-veined, scape furrowed,* spike ovate or oblong-cylindrical, bracts ovate acute or cuspidate, 2 lateral sep. keeled. *E. B.* 507.—L. nearly glabrous, length 3—12 times the breadth. Anth. and filaments yellow. Bracts and sep. black at the tip. A very variable plant. Spikes globose, scape and l. silky, neck woolly, in sandy places; spikes very long, l. very long and broad, in rich damp soil. In rare cases stoloniferous, and often woolly near the sea (*eriophylla* H. & L.).—Meadows, pastures and sandy places. P. V.—VII. *Ribwort.* E. S. I.

[*P. Timbali* (Jord.); rootstock many-headed, bracts with scarious margins; is sometimes found in cultivated fields.]

4. *P. média* (L.); *l. elliptic-ovate sessile or with short broad stalks* pubescent, scape terete, spike cylindrical, bracts ovate-acuminate, *sep. not keeled.*—*E. B.* 1559.—L. usually lying flat on the ground, sometimes shortly lanceolate and ascending. Anth. yellow, filaments purple.—Meadows and pastures. P. VI.—IX. *Lamb's-tongue.* E. I.

**** *Placenta 2-winged. Seeds* 2—4 *in each cell. Cor.-tube glabrous.*

5. *P. májor* (L.); l. broadly ovate on longish channelled stalks, scape terete, spike long, bracts ovate keeled about as long as the cal., sep. with a prominent dorsal rib.—*E. B.* 1558. *R.* xvii. 1127.—L. ascending. Anth. purple. Seeds about 8.—*P. intermedia* (Gilib.); l. downy coarsely dentate, scapes downy arcuate-ascending, is probably distinct. It is not very rare in England. *R.* xvii. 86.—Fields and waste places. P. VI.—VIII. *Way-bread.* E. S. I.

[*P. arenária* (W. & K.), a branching leafy plant, was found abundantly on Burnham Sand-hills, Som.] E.

2. LITTOREL'LA *Berg.* Shore-weed.

1. *L. jun'cea* (Berg.), *L. lacustris* (L.) ed. viii.—*E. B.* 468.—With runners.—Fl. white. Fertile fl. sessile. Stalks of the male fl. 1—2 in. long. L. all radical, linear, fleshy, somewhat channelled, sometimes hairy above.—Margins of lakes, or under water when it is larger and does not flower. P. VI. VII.
E. S. I.

Division IV. MONOCHLAMYDEÆ.

(Ord. LXIV.—LXXVIII.)

With only a single perianth; that is, the cal. and cor. not distinguishable, or wanting.

[Order LXIV. AMARANTHACEÆ.]

[Perianth 3—5-parted, scarious, persistent. Stam. hypogynous. Ovary free, 1-celled; ovule 1 or several, suspended from a free central seed-stalk. Style 1 or 0. Stigma simple or compound. Embryo curved round central farinaceous albumen. —L. without stipules or sheaths.

[1. AMARANTHUS. Fl. monœcious. Perianth 3—5-parted. Stam. 3—5. Stigmas 3. Caps. 1-celled, 1-seeded.]

1. AMARAN'THUS *Linn.*

[*A. Blitum* (L.); fl. 3-fid, 3-androus, clusters small lateral, the upper ones in a small naked spike, st. diffuse glabrous.—*E. B.* 2212.—Waste places near towns. A. VIII.] E.

[*A. retroflex'us* (L.); fl. 5-fid, 5-androus, spikes large dense compound terminal, bracts lanceolate aristate much exceeding fr., st. stout erect pubescent.—*R. I.* 5. 475.—Waste ground. A. VIII.] E.

Order LXV. CHENOPODIACEÆ.

Perianth 3—5-parted, herbaceous, persistent. Stam. perigynous or hypogynous. Ovary free or adhering to the tube of the perianth; ovule 1, attached to the base of the cell. Styles divided, or rarely 1. Fr. not bursting, dry, membranous, included in the perianth which often becomes enlarged or fleshy. Embryo curved round farinaceous albumen, or spiral, or doubled together without albumen; radicle next the hile.—L. without stipules or sheaths.

Tribe I. *SALSOLEÆ.* Fl. uniform, perfect. Seeds usually without albumen. Embryo spiral. *St. continuous*, leafy. *L. semicylindrical or terete.*

1. Lerchia. *Perianth* 5-parted, *without appendages.* Stam. 5, from the receptacle. Stigmas 2—3, sessile. Pericarp membranous. Seed horizontal or vertical; testa crustaceous.—With bracts.

2. Salsola. Perianth 5-parted, *segments ultimately with a transverse dorsal appendage.* Stam. 5, from an hypogynous ring. Styles 2. Pericarp membranous. Seed horizontal; *testa membranous.*—With bracts.

Tr. II. *BETEÆ.* Fl. uniform, perfect. Seeds with albumen. Embryo curved round the circumference of the seed. *St. continuous*, leafy. *L. flat.*

3. Chenopodium. Perianth 3—5-parted, persistent, unaltered. Stam. 5, from the receptacle. Stigmas 2—3. *Pericarp* thin, *free. Testa crustaceous.* Seed vertical or horizontal. Without bracts.

4. Beta. Perianth 5-parted, persistent. Stam. 5, from a fleshy ring. Styles 2—3. *Pericarp imbedded in and adhering to the fleshy tube of the perianth.* Seed horizontal, attached laterally. *Testa membranous.*

Tr. III. *SALICORNIEÆ.* Fl. uniform, perfect. Seeds and embryo as in Tr. II. *St. jointed*, leafless. *Stam. less than* 5.

5. Salicornia. Perianth fleshy, tumid, undivided, imbedded in an excavation of the rachis. Stam. 1—2. Style very short, stigma bifid. Pericarp membranous. Seed vertical, covered by the persistent perianth.

Tr. IV. *ATRIPLICEÆ.* *Fl. monœcious*, of two forms, rarely perfect. Seeds, embryo, and stem as in Tr. II.

6. Atriplex. Perianth 3—5-parted. Stam. 5. Style 0.—Perianth compressed, of 2 parts not connected above their middle.—Stam. 0. Stigmas 2. *Pericarp membranous free.*

Testa crustaceous. Seed vertical, attached by a lateral hile either near the base or by means of a long seed-stalk in the middle of the side; radicle basal.

7. OBIONE. Perianth 3—5-parted. Stam. 5. Style 0.—*Perianth* of 2 parts *free only at the top,* 3-toothed, wedge-shaped below. Stam. 0. Stigmas 2. *Pericarp very thin, ultimately adhering to the tube of the* perianth. *Testa membranous.* Seed vertical, pendulous from a long seed-stalk; radicle terminal.

Tribe I. *Salsoleæ.*

1. LER'CHIA *Hall.* (*Suæda* Forsk. ed. viii.) Sea-Blite.

1. *L. obtusifolia* ("Steud."); st. erect shrubby, *l. blunt* semicylindrical, styles 3, seeds vertical smooth and shining.—*Suæda fruticosa* (Forsk.) ed. viii. *Salsola* Sm. *E. B.* 635. *Schoberia* Mey.—St. 2—3 feet high, with many erect leafy branches and axillary flowers.—Sandy and shingly shores of south and east of England, rare. P. VII. VIII. E.

2. *L. maritima* (O. Kuntze); st. herbaceous, *l. acute* semicylindrical, styles 2, seeds horizontal netted shining.—*E. B.* 633.—St. erect or procumbent, with many spreading branches. —Muddy sea-shores. A. VII.—IX. E. S. I.

2. SAL'SOLA *Linn.* Saltwort.

1. *S. Kali* (L.); minutely hairy, st. diffuse, l. subulate spinous rough, fl. axillary solitary, segments of the enlarged perianth hard and tough as long as their patent rather coloured roundish wings.—*E. B.* 634.—St. angular, rigid, much branched.—Sandy sea-shores. A. VIII. E. S. I.

Tribe II. *Beteæ.*

3. CHENOPO'DIUM *Linn.* Goose-foot.

* *Flowers all pentamerous.* † Leaves undivided.

1. *C. Vulvaria* (L.); *l. ovate-rhomboidal mealy, fl.* in leafless dense spikes, perianth covering the fr., seed shining slightly rough (very small), *st. diffuse.—E. B.* 1034. *C. olidum* Curt.—L. stalked. Covered with a greasy pulverulent *fetid* substance. —Dry waste places near houses. A. VIII. IX. E. S. I.

2. *C. polysper'mum* (L.); *l. ovate-elliptic*, fl. in axillary leafless cymose racemes, perianth not covering the fr., seeds shining minutely dotted blunt at the edge.—*E. B.* 1480, 1481.—Racemes more or less cymose or spicate. St. erect or procumbent. L. acute or blunt. *C. acutifolium* and *C. polyspermum* are undistinguishable.—Damp waste places. A. VIII. IX. E.

†† Leaves toothed, angled, or lobed.

3. *C. ur'bicum* (L.); *l. triangular* sinuate-dentate or nearly entire their base contracted into the petiole, *spikes erect* nearly leafless compound, perianth not covering the fr., seeds very minutely rough blunt at the edge.—*E. B.* 717.—L. with short triangular teeth. Spike approaching the stem. Seed almost as large as Rape-seed.—*β. C. intermedium* (M. & K.); l. with large acute teeth.—Waste places. A. VIII. E. S. I.?

4. *C. al'bum* (L.); l. *rhomboidal-ovate sinuate-dentate* entire below, upper ones lanceolate nearly entire, fl. in compound branched nearly leafless racemes, perianth covering the fr., *seeds smooth* and shining blunt but keeled at the edge.—*α. C. candicans* (Lam.): fl. in dense spikes shorter than the sinuate-dentate leaves. *E. B.* 1723.—*β. C. viride* (L.); fl. in lax cymose racemes exceeding the nearly entire leaves. *Sy. E. B.* 1189.—*γ. C. paganum* (R.); fl. in lax compound spikes exceeding the sinuate-dentate leaves. *Sy. E. B.* 1190.—Cultivated and waste places. A. VII. VIII. *Fat Hen.* E. S. I.

[*C. opulifólium* (Schrad.), closely allied to Sp. 4, with broadly-rhomboidal-ovate (often broader than long) more or less 3-lobate very obtuse coarsely and unevenly crenate-serrate l. with usually slightly mucronate teeth, rather long slender petioles and very glaucous inflorescence, is occasionally found in waste places. A. VIII. IX.]

5. *C. ficifólium* (Sm.); *l. unequally 3-lobed from a wedgeshaped base*, lobes ascending, *middle lobe long oblong-lanceolate dentate blunt*, upper l. linear-lanceolate entire, fl. in erect nearly leafless cymose racemes, perianth nearly covering the fr., *seeds minutely pitted* shining blunt and not keeled at the edge.—*Sy. E. B.* 1191.—Mealy. L. blunt; middle lobe nearly equally broad throughout. Seeds smaller than in *C. album.*—Cultivated and waste ground, rare. A. VIII. IX. E. I.?

6. *C. murále* (L.); *l. rhomboidal-ovate* unequally and sharply toothed entire below, fl. in divaricately-branched leafless cymes, perianth nearly covering the fr., *seeds minutely granular opaque acute and keeled at the edge.*—*E. B.* 1722.—Waste ground near towns and villages. A. VIII. *Sowbane.* E. I.

7. *C. hyb'ridum* (L.); *l. subcordate angulate-dentate* acuminate, *teeth large distant*, fl. in panicled leafless cymes, perianth not covering the fr., *seeds minutely pitted* opaque blunt and *not keeled* at the edge.—*E. B.* 1919.—Seeds very large. L. with 2—4 large teeth on each side.—Waste places, rare. A. VIII. E. I.

** *Lateral fl. 3—4-merous with vertical seeds, term. fl. 5-merous with horizontal seeds.*

† Stigmas short.

8. *C. rúbrum* (L.); *l. rhomboidal* irregularly toothed and sinuate entire below, fl. in erect compound dense leafy spikes, *seeds* very minute smooth shining *blunt or slightly keeled at the edge.*—*E. B.* 1721.—St. erect and often 1—2 ft. high, leafy to the top. Pericarp very loose. Fl. generally incomplete. Cal. 4- rarely 5-cleft. Stam. 1 or 2.—A variety with much more triangular l., shorter spikes and larger seeds, was found near London.—The var. *pseudo-botryodes* (Wats.) has a prostrate spreading st. 4—5 in. long, and a small panicle. *Sy. E. B.* 1197.—Waste places, particularly salt marshes. A. VIII. IX. E. S. I.

9. *C. botryódes* (Sm.); *l. nearly triangular* slightly toothed, thick and fleshy, fl. in compound dense spikes nearly leafless at the top, *seeds* very minute smooth shining *keeled at the edge.*—*E. B.* 2247.—Succulently brittle when fresh, limp and flaccid when dry. St. erect. Pericarp loose.—Moist sandy places near the sea, on the south-east coast. A. IX. E.

10. *C. glaúcum* (L.); *l. oblong sinuate-dentate mealy beneath*, fl. in erect nearly simple leafless spikes, *seeds* very minute *netted granular* acutely keeled at the edge.—*E. B.* 1454.—St. spreading, often prostrate. Seeds reddish.—Rich waste ground. A. IX. E.

†† Stigmas long.

‡11. *C. Bonus-Henrícus* (L.); *l. triangular-hastate entire*, fl. in compound leafless spikes, seeds smooth and shining.—*E. B.* 1033.—St. a foot high. L. large, dark green, used instead of Spinach. Stig. long. Fr. exceeding perianth.—Waste places near villages. P. V.—VIII. *Allgood. Mercury.* E. S. I.

4. Be'ta *Linn.* Beet.

1. *B. marit'ima* (L.); *st. many from the crown of the root prostrate*, l. triangular-ovate narrowed into a petiole, spikes long

simple leafy, bracts lanceolate exceeding the 2—3-flowered clusters, segments of the perianth with entire keels, *stig. lanceolate.—E. B.* 285.—Root thick, fleshy. St. 6—12 in. long prostrate below, spreading in a circle, the ends ascending.—Sea-shores. P. VII.—IX. *Sea-Beet.* E. S. I.

[*B. vulgaris* (L.), st. solitary erect, stig. ovate, is the cultivated plant.]

Tribe III. *Salicorniece.*

5. SALICOR'NIA *Linn.*[1] Glasswort.

1. *S. herbácea* (L.); herbaceous, joints of st. compressed rather thickened upwards emarginate, branches all flowering, spikes cylindrical tapering stalked, fl. 3 on each side middle fl. much exceeding the others.—*E. B.* 415.—St. usually erect, branched, 3—12 in. high. Stam. 1 or 2.—*β. S. prostrata* (Pall.), *S. procumbens* (Sm.) ed. viii.; st. procumbent, branches divaricate. *E. B.* 2475. [An extreme state is *S. appressa* (Dum.)] Muddy sea-shores. A. VIII.—IX. E. S. I.

2. *S. radícans* (Sm.); st. woody creeping, st.-joints subterete deeply notched scarcely thickened with ascending herbaceous branches some barren, spikes oblong blunt nearly sessile, middle fl. scarcely exceeding the others.—*Sy. E. B.* 1183.—St. procumbent, rooting. [*β. lignosa* (Woods); st. firmer not or but little rooting.]—Muddy sea-shores. P. VIII. IX.

Tribe IV. *Atripliceæ.*

6. A'TRIPLEX *Linn.*[2] Orache.

A. *Fertile fl. of 2 kinds; sep. of fem. fl. distinct, seed vertical; perfect fl. 3—5-parted, seed horizontal.*

[*A. nitens* (Schkuhr), l. shining above silvery-glaucous beneath; also *A. tatarica* (L.) and *A. hortensis* (L.), l. opaque on both sides; are occasionally found.]

B. *Monœcious; sep. of fem. fl. united below.*

* Stem with resinous reddish stripes.—† *Leaves not lobed.*

1. *A. littorális* (L.); st. erect, *l.* linear or oblong-lanceolate

---

[1] See a paper on this genus by Mr. Woods in *Proc. Linn. Soc.* ii. 109.
[2] See *Woods* in *Phytol.* iii. 585, and in *Tourist's Flora*, 315.

entire or toothed, fr.-perianth rhomboidal or triangular toothed tubercled on the back.—*α. vera*; l. nearly or quite entire, fr.-perianth with patent points.—*Sy. E. B.* 1200.—*β. A. marina* (L.); l. broader usually toothed, fr.-perianth shorter with adpressed points. *E. B.* 708.—Salt marshes. A. VII.—IX. E. S. I.

†† *Lower leaves with lateral spreading or ascending lobes.*

2. *A. angustifólia* (Sm. !)[1]; st. erect or prostrate, l. lanceolate entire from an acutely wedgeshaped base, lower l. with 2 ascending lobes, *fr.-perianth rhomboidal* acute *entire with prolonged lateral angles* longer than the fr. and *collected into nearly simple interrupted spikes* the larger leaflike and not tubercled, seeds black and polished.—*E. B.* 1774. *A. patula* Wahl., *St.* 79. 5.—Spikes wandlike with distant clusters of fl., valves of the fr.-perianth netted.—Common. A. VII.—X. E. S. I.

3. *A. erec'ta* (Huds.)[1]; primary st. mostly erect, lower l. ovate-oblong with 2 ascending lobes from a bluntly wedgeshaped base irregularly sinuate-dentate, upper l. lanceolate, *fr.-perianth rhomboidal denticulate* acute more or less muricate on the back scarcely exceeding the fr. and *collected in branched dense many-flowered spikes*, seeds black and polished.—*E. B.* 2223! *Koch in St.* 79. 6.—Lower branches procumbent or ascending. Upper l. mostly entire.—Common upon cultivated land. A. VII.—X. E. S. I.

4. *A. deltoïdea* (Bab.); st. mostly erect, *l.* mostly opposite nearly all *hastate-triangular with spreading lobes*, fr.-perianth ovate-triangular muricate on the back scarcely exceeding the fr. united only at the base, *fr. panicled in dense spikes*, seeds thick black and polished or a few dark brown and larger and with large perianths.—*α*; l. all hastate-triangular toothed, fr.-perianth toothed muricate on back, spikes dense. *E. B. S.* 2860.—*β. salina*; st. and branches prostrate, l. often alternate, uppermost l. lanceolate entire, fr.-perianth slightly toothed or entire.—Cult. and waste ground. *β*. Sea-coast. A. VI.—X. E. I.

5. *A. hastáta* (L.); st. mostly erect, lower *l. hastate-triangular with spreading lobes*, uppermost l. lanceolate entire, *fr.-perianth triangular-rhomboidal* slightly muricate on the back exceeding the fr., *fr. in nearly simple interrupted spikes*, larger seeds dark brown rough compressed, smaller seeds black and shining.—*A. patula* Sm. *E. B.* 936. *A. latifolia* St. 79. 7. *A. Smithii* Sy. —Cultivated and waste ground. A. VI.–X. E. S. I.

---

[1] Sp. 2 & 3 are now usually included under *A. patula* (L.).—H. & J. G.

6. *A. Babingtónii* (Woods); st. spreading procumbent or ascending with spreading branches, *l.* mealy *ovate-triangular* somewhat 3-lobed unequally sinuate-dentate, upper l. lanceolate dentate and often 3-lobed at the base, *fr.-perianth rhomboidal acute toothed* tubercled on the back, *clusters axillary* and terminal few-flowered, seeds minutely tubercular.—*A. rosea* Bab., *E. B. S.* 2880, not *L.*—A very variable plant. Sometimes bright red, at others [var. *virescens* (Lange)] quite green and much more fleshy. Fr.-perianth large, typically forming a diagonal square a little rounded at the lateral angles.—Sea-shore, common. A. VII.—IX. E. S. I.

** Stem buff-coloured, nearly without stripes. Perianth of fr. hard and thick.

7. *A. laciniáta* (L.); st. spreading procumbent with spreading branches, *l. triangular-rhomboidal* sinuate mealy beneath, *spike of male fl. dense naked*, fertile fl. axillary, *fr.-perianths rhomboidal 3-lobed with the lateral lobes truncate* the back 3-ribbed the two lateral ribs often terminating in tubercles, seeds rough opaque.—*A. arenaria* Woods (not Nutt.). *A. farinosa* (Dum.) ed. viii. *E. B.* 165.—Plant hoary throughout. Fr.-perianths large, very broad.—Sandy shores. A. VII.—IX. E. S. I.

7. Obi'one *Gaert.* Sea-Purslane.

1. *O. pedunculáta* (Moq.); st. herbaceous wavy branched, l. obovate entire attenuate below, upper l. narrower, *fr.-perianth long-stalked* wedgeshaped 2-lobed with a small intermediate tooth.—*E. B.* 232. *Atriplex* Sm.—Muddy salt marshes near the East and South coasts; very rare. A. VIII. IX. E.

2. *O. portulacoïdes* (Moq.); st. woody, l. obovate-lanceolate entire attenuate below, *fr.-perianth* inversely triangular rounded below *subsessile* with 3 equal lobes above and muricate on the back.—*E. B.* 261. *Atriplex* Sm.—Common on the sea-shore. P. VIII.—X. E. S. I.

Order LXVI. POLYGONACEÆ.

Perianth 3—6-parted, imbricate. Stam. definite, from the base of the perianth. Ovary 1, free, with 1 erect ovule. Styles and stigmas several. Fr. not bursting, a nut, naked or covered by the enlarged perianth. Embryo inverted, usually on one side of farinaceous albumen; radicle remote from the hile.—Stipules usually cohering in the shape of ocreæ.

1. Rumex. *Perianth 6-parted*; 3 inner segments (pet.) large with fr., connivent. Stam. 6, disposed in pairs. Styles 3. Stigmas multifid. Nut triquetrous, covered by the enlarged inner segments of perianth; embryo lateral.

2. Oxyria. *Perianth 4-parted*; 2 inner segments (pet.) larger. Stam. 6. Stigmas 2, multifid. Nut compressed, with a membranous wing, larger than the persistent segments of perianth; embryo central.

3. Polygonum. *Perianth 5-parted.* Stam. 5—8. Styles 2—3. Stig. capitate. Nut trigonous or compressed; *embryo lateral,* incurved; *cotyledons not contorted.*

[4. Fagopyrum. Perianth 5-parted. Stam. 8. Styles 3. Stig. capitate. Nut trigonous; *embryo central; cotyledons large, leaflike, plicate-twisted.*]

1. Ru'mex *Linn.*[1] Dock.

* *Fl. perfect. Herbage not acid. L. not hastate.*

1. *R. marit'imus* (L.); *enlarged pet. rhomboidal* narrow, each with a lanceolate entire point a prominent narrow oblong tubercle and on each side 2 setaceous *teeth as long as the pet., whorls crowded* many-flowered leafy, l. all linear-lanceolate narrowed at both ends.—*E. B.* 725.—Nut very small: faces elliptic.—Marshes principally near the sea. P. or B. VII. VIII. *Golden Dock.* E. S. I.

2. *R. limósus* (Thuill.); *enlarged pet. ovate-oblong,* with a lanceolate entire point a prominent narrow oblong tubercle and on each side 2 or 3 setaceous *teeth shorter than the pet., whorls distant* many-flowered leafy, root-l. narrowly lanceolate rounded or slightly decurrent below.—*Curt.* i. 63. *Sy. E. B.* 1213. *R. palustris* (Sm.) ed. viii.—Nut 3 times as large as that of *R. maritimus*; faces ovate. Upper l. linear-lanceolate, narrowed below.—Marshy places, rare.—P. VII.—IX. E. S. I.?

3. *R. conglomerátus* (Murr.); *enlarged pet. linear-oblong subacute,* each bearing a large tubercle and entire or *obscurely toothed at the base, whorls distant leafy,* l. oblong pointed, lower l. cordate or rounded at the base.—*R. acutus* Sm. *E. B.* 724.

[1] Hybrids are frequent in this genus.—H. & J. G.

—Nut ovate, acute. Enlarged pet. broadest near the base; the sides nearly parallel. Unopened anth. white. Branches mostly spreading. Uppermost whorls often leafless.—Wet places. P. VI.—VIII. E. S. I.

4. *R. rupes'tris* (Le G.); *enlarged pet. narrow ovate-oblong blunt* all tubercled *entire*, whorls not very distant leafless (except 2 or 3 lowest), l. strapshaped rounded and narrowed at both ends.—*J. of B.* xiv. 2. t. 173.—Nut ovate, acute. Enlarged pet. with nearly parallel sides and very large tubercles.—Shore near Plymouth. P. VI.—VIII. E.

5. *R. sanguin'eus* (L.); *enlarged pet. narrowly oblong blunt entire* only one bearing a tubercle, *whorls distant leafless*, l. ovate-lanceolate, lower l. cordate or rounded at the base.—*E. B.* 1533. —Ped. jointed at their base. Nut ovate-elliptic, acute. Enlarged pet. broadest above their middle. Lowermost whorls often each accompanied by a leaf. Veins of the l. bright red. Branches ascending.—*β. R. viridis* (Sibth.): veins of the l. green. Unopened anth. pale yellow. *R. nemorosus* Schrad.—Woody places, rare. *β.* Woods and roadsides, frequent. P. VI.—VIII. E. S. I.

6. *R. pul'cher* (L.); *enlarged pet. triangular-ovate netted with ribs* toothed below one principally tubercled, branches spreading, whorls mostly leafy, *lower l. fiddleshaped* or cordate-oblong blunt, upper l. lanceolate-acute, st. procumbent.—*E. B.* 1576.—St. straggling. Whorls distant. Nuts ovate, acute.—Dry waste places, local. B.? VII.—IX. *Fiddle-Dock.* E. I.

7. *R. obtusifólius* (L.?); *enlarged pet.* ovate- or oblong-triangular blunt entire or toothed veined one or more tubercled, *lower l. cordate-ovate-oblong* blunt, upper l. oblong or lanceolate, branches ascending.—*a. R. Friesii* (G. & G.); enlarged pet. sinuate-dentate below with subulate spreading teeth and an oblong entire point one principally tubercled. *E. B.* 1999.—*β. R. sylvestris* (Wallr.): enlarged pet. nearly entire all tubercled. *J. of B.* xi. 131.—Whorls usually distant and nearly leafless. Height 2—3 ft. Nut nearly twice as long as broad. A hybrid with Sp. 10, *R. conspersus* (Hartm.), has enlarged pet. broadly cordate toothed membranous netted one tubercled, l. oblong blunt cordate or oblique at the base.—*Sy. E. B.* 1217.—Pastures and waste ground. *β.* By Thames near Putney. P. VII.—IX. E. S. I.

8. *R. acútus* (L.); *enlarged pet. unequal cordate dilated and toothed at the base with a small entire triangular point* one

principally tubercled, l. oblong-lanceolate acute, lower l. slightly cordate below.—*R. pratensis* E. B. S. 2757.—St. and whorls often tinged with dull red. Whorls near together but not crowded, mostly leafless. Unopened anth. white. Nuts elliptic, abundant. One enlarged pet. larger than the others, sometimes all 3 equally tubercled. [Said to be the hybrid *R. obtusifolius*×*crispus*.]—Marshy places, rare. P. VI.—IX. E. S. I.

9. *R. cris'pus* (L.); *enlarged pet. cordate entire or crenulate* one principally tubercled, l. lanceolate acute wavy crisped.—*E. B.* 1998.—Height 2—3 feet. St. and whorls sometimes tinged with bright red. L. narrowed or truncate below. Nut elliptic. Enlarged pet. equal, sometimes, var. *trigranulatus* (Sy.), all equally tubercled. [Var. *subcordatus* (Warr.) has enlarged pet. broader, more or less deltoid. The hybrid *R. crispus*×*aquaticus* (*R. propinquus* Aresch.) occurs in Scotland].—Roadsides, fields, &c. P. VI.—VIII. *Curled Dock.* E. S. I.

[*R. elongatus* (Guss.); enlarged pet. ovate entire one tubercled, l. linear-lanceolate narrowed at both ends not wavy nor crisped. —*J. of B.* xi. 237.—St. 1—2½ ft. high. Rt.-l. 8—12 in. long, 1 in. broad. Of lax habit with distant whorls. A probable form of *R. crispus*.—Surrey side of Thames between Putney and Hammersmith, and Wye at Tintern.]

10. *R. aquat'icus* (L.); *enlarged pet. broadly cordate membranous entire* or wavy *without tubercles*, l. lanceolate, lower l. somewhat cordate, petioles semicylindrical flat and finely margined above.—*E. B. S.* 2698. *R. domesticus* Hartm. *R. longifolius* DC.—Height 3—4 feet. L. very large. Whorls crowded, mostly leafless, forming a large dense lobed panicle. Nut elliptic. —In the North of E. and in S., preferring spots liable to be flooded. P. VII. VIII. E. S.

11. *R. Hydrolap'atheum* (Huds.); *enlarged pet. ovate-triangular* nearly entire *all tubercled, l.* lanceolate acute *tapering below*, petioles flat but not with raised edges.—*E. B.* 2104.—St. 3—5 feet high. L. often more than a foot long, sometimes cordate at the base. Whorls crowded, mostly leafless. Nut elliptic.—Ditches and river-sides. P. VII. VIII. *Great Water-Dock.* E. S. I.

12. *R. maximus* (Schreb.); *enlarged pet. triangular-cordate* denticulate below *all tubercled, radical l. oblong-acute obliquely cordate below,* petioles flat or broadly channelled above with raised edges.—*St.* 73. 16. *J. of B.* xii. t. 140.—St. 3—5 feet

high. L. very long. whorls crowded, mostly leafless.—Near Lewes, Winchester, Kelvedon in Essex, and in Scilly Isles.—Perhaps not distinct from Sp. 11. P. VII. VIII. E.

*13. *R. alpínus* (L.); enlarged pet. cordate-ovate membranous entire or denticulate without tubercles, *l. roundish cordate-blunt with channelled petioles, upper l. ovate.*—*E. B. S.* 2694.—Whorls crowded, mostly leafless. Nut elliptic.—Formerly cultivated. P. VII. *Monk's Rhubarb.* E. S.

** *Fl. diœcious or polygamous. Herbage acid.* Sorrel.

14. *R. Acetósa* (L.); *enlarged pet.* roundish-cordate entire *membranous with a very minute tubercle* at the base, *sep. reflexed,* l. oblong sagittate, stipules laciniate-dentate.—*E. B.* 127.—Height 1—2 feet. Whorls leafless. Nut elliptic with acute angles.—Woods. P. V. VI. E. S. I.

15. *R. Acetosel'la* (L.); *diœcious, petals scarcely enlarged ovate not tubercled closely adpressed to nut, sep. ascending,* l. lanceolate-hastate or linear with entire lobes, stipules torn.—*E. B.* 1674.—Height 6—10 in. Whorls leafless. L. very variable in breadth. [α. *R. angiocarpus* (Murb.) Pet. adhering to nut. β. *R. Acetoselloides* (Bal.). Pet. not adhering to nut.]—Dry gravelly places. P. V.—VII. *Sheep's Sorrel.* E. S. I.

[*R. scutátus* (L.); polygamous, l. hastate-ovate, slightly fiddle-shaped, is naturalized near Edinburgh, &c. *Sy. E. B.* 1222.]

2. OXYR'IA *Hill.* Mountain-Sorrel.

1. *O. dígyna* (Hill)—*E. B.* 910. *O. reniformis* (Hook.) ed. viii.—St. 8—10 in. high, usually leafless. L. radical, reniform, slightly notched at the end; veins radiating from the insertion of the long footstalk. Pedicels thickening upwards, in a spikelike raceme. Permanent pet. not enlarged.—Lofty mountains. P. VII. VIII. E. S. I.

3. POLYG'ONUM *Linn.*

* *Rhizomatous. Stem simple. Ocreæ truncate. Nut triquetrous. Stam.* 8. *Styles* 3. Bistorta.

1. *P. Bistor'ta* (L.); *raceme cylindric dense, l. ovate subcordate,* root-l. with winged footstalks, faces of nut ovate smooth.—*E. B.* 509.—St. 1—1½ foot high. Rhizome large. Fl. flesh-coloured.—Moist meadows, rare. P. VI. and X. *Snakeweed.* E. S. I.

2. *P. vivip'arum* (L.); *raceme slender lax bulbiferous* bearing fl. on its upper part, *l. linear-lanceolate* with revolute margins, *lower l. elliptic* with wingless footstalks, face of nut ovate-lanceolate smooth and shining.—*E. B.* 669.—Height 4—8 in. Fl. on the lower part of the raceme replaced by small red bulbs. Fl. white, sometimes very few in number.—*β. alpinum* (Wahl.) lower l. oval, root much thicker.—Mountain pastures. *β.* Shetland and O. Hebrides. P. VI. VII. E. S. I.

** *Ocreæ truncate. Root fibrous. Nut compressed or subtriquetrous. Stam. 4—8. Styles 2—3. Racemes spikelike.—Fl. purple or white. L. without or with a central dark spot.* Persicaria.

† Creeping. Perennial. Anthers protruded.

3. *P. amphib'ium* (L.); raceme dense ovate-cylindrical, l. stalked ovate-oblong (floating) or oblong-lanceolate or narrow-lanceolate rough at the margins, ocreæ membranous close, nut compressed smooth shining, stam. 5.—*E. B.* 436.—St. long when floating. *Racemes generally solitary and terminal.* Fl. rose-coloured. Stam. protruded or not. Very variable in the form of its leaves according to its habitation.—Floating or growing upon mud or on boggy ground. P. VII.—IX. E. S. I.

†† Root fibrous. Annual. Anthers included.

4. *P. lapathifolium* (L.); raceme oblong cylindrical dense, l. oblong-lanceolate or ovate attenuate at both ends glandular beneath, *ocreæ close not fringed the upper one shortly fringed, peduncles* and perianth *rough with glands, nut compressed* its faces roundish acuminate concave smooth shining scarcely covered by the prominently veined perianth, styles distinct at length divergent and reflexed.—*E. B.* 1382.—Height 1—2 feet. St. sometimes spotted, glandular or glabrous; joinings more or less thickened. Racemes axillary or terminal, 1—2 together. Fl. pale. Bracts auricled.—Waste and damp places. A. VII.—IX. E. S. I.

5. *P. maculátum* (Trim. & Dyer); raceme long slender, l. lanceolate much attenuate at both ends wavy glandular beneath, *ocreæ loose shortly fringed* the floral ones horned, *ped.* and perianths *rough with glands, nut compressed* its faces roundish acuminate concave shining covered by the strongly veined perianth, styles connected below at length divergent and reflexed. —*P. laxum* E. B. S. 2822. *P. nodosum* R I. 496 (not Pers. ?).

Persicaria maculata Gray (1821). *J. of B.* ix. 34.—Smaller than the preceding. St. often prostrate; joinings very thick. Racemes slender and interrupted (the typical form): or thicker and continuous (as in *E. B. S.*), and l. white and woolly beneath. —Damp gravelly places. A. VII.—IX. E. S. I.

6. *P. Persicária* (L.); raceme compact ovate-oblong cylindrical, l. lanceolate flat minutely tubercled, *ocreæ loose strongly fringed, ped.* and perianths *smooth*, nut compressed and gibbous on one side or trigonous its faces roundish acuminate smooth scarcely covered by the obscurely veined perianth, styles connected halfway up at length patent.—*Sy. E. B.* 1237.—St. 1—2 feet high. L. more or less hairy on both sides; sometimes woolly beneath, when it is *P. incanum* of authors. Ped. sometimes slightly hairy.—β. *P. biforme* (Wahl.) (=var. *elatum* G. & G.); racemes slender, much resembling *P. maculatum.*—Waste and damp ground. A. VI.—X. E. S. I.

7. *P. míte* (Schrank); *raceme erect filiform interrupted,* l. lanceolate slightly wavy, *ocreæ* loose funnelshaped pilose, *strongly fringed* without glands, *perianths without glands, nut* (large) *compressed its faces roughish* ovate acute rather shining *convex,* stam. 5, styles connected halfway up.—*E. B. S.* 2867.—St. 1—3 feet high, often much branched. Ocreæ all fringed. Racemes thickening upwards.—Wet places. A. VI.—IX. E. I.

8. *P. Hydrop'iper* (L.); *raceme drooping filiform interrupted,* l. lanceolate wavy, *ocreæ ventricose* glabrous fringed *glandular, perianths glandular,* nut (large) compressed its faces ovate acute rugose-punctate opaque convex, style 2 nearly distinct.—*E. B.* 989.—St. 1—3 feet high. Upper ocreæ funnelshaped, scarcely fringed. Nut rounder than in *P. mite.* Racemes sometimes erect.—Wet places. A. VIII. IX. *Water-Pepper.* E. S. I.

9. *P. mínus* (Huds.); *raceme erect filiform slender lax,* l. linear-lanceolate flat, *ocreæ close* pilose all fringed *without glands, perianths without glands,* nut (small) compressed its faces ovate acute smooth shining convex, styles connected for at least half their length.—*E. B.* 1043.—St. usually procumbent, diffuse. Racemes ascending. Much smaller than *P. Hydropiper*, fl. and fr. only half the size.—Wet gravelly places, rare. A. VIII. IX. E. S. I.

*** *Ocreæ 2-lobed. Root fibrous. Nut triquetrous. Stam.* 8. *Styles* 3. *Fl. axillary,* 1—3 *together.* Avicularia.

10. *P. aviculáre* (L.); l. lanceolate or elliptic stalked, *ocreæ lanceolate* acute *with few distant simple veins* at length torn,

nut striate with raised points opaque about equalling the perianth.—*E. B.* 1252. *Sy. E. B.* 1229—1231.—L. usually blunt, sometimes acute, broad or narrow. Fl. either very distant and scattered, or so much collected as almost to form a leafy spike. St. erect or procumbent. Ocreæ sometimes long and much torn. Nut usually falling short of the perianth, but sometimes slightly exceeding it. Very variable. Some botanists think that it includes several species; but I am quite unable to divide what I have seen similarly.—β. *P. littorale* (Link); st. long, diffuse, prostrate, l. thick broad, nut minutely punctured finely striate. —Waste places. β. Sands by the sea. A. V.—IX. *Knot-grass.* E. S. I.

11. *P. Ráii* (Bab.); *l.* elliptic-lanceolate *flat, ocreæ* lanceolate acute *with few distant simple veins* at length torn, *nut smooth shining exceeding the perianth.*—*E. B. S.* 2805.—St. long, straggling, prostrate. L. bending towards the stem. Filaments broader at the base. Resembling *P. aviculare* in habit, but *P. maritimum* in fruit. It varies with smaller l. and fl.—Sandy sea-shores. A. or P. VIII. IX. E. S. I.

12. *P. marit'imum* (L.); *l.* elliptic-lanceolate coriaceous *with revolute edges, ocreæ lanceolate with many branched veins* at length torn, *nut smooth shining* exceeding the perianth.—*E. B. S.* 2804.—St. procumbent, quite woody below, often much buried. L. convex above, diverging from the stem. Filaments broader at the base.—Sands of the southern sea-shores. Christchurch, Hants. Braunton Burrows, Dev. Falmouth. Channel Islands. P. VIII. IX. E.

[*P. sagittátum* (L.), with reflexed prickles and ovate sagittate l., a native of N. America, has been found in Kerry.]

**** *Ocreæ semicylindrical. Root fibrous. Nut triquetrous. Stam.* 8. *Styles* 3. *Fl. racemose.*

13. *P. Convol'vulus* (L.); st. twining angular, l. cordate-sagittate, *segments of perianth bluntly keeled, nut opaque* striate with minute points.—*E. B.* 941.—St. climbing or prostrate, much shorter than in the next species.—β. *subalatum* (v. Hall); perianth-segm. winged; often taken for Sp. 14.—Cultivated and waste land. A. VII.—IX. *Black Bindweed.* E. S. I.

14. *P. dumetórum* (L.); st. twining striate, l. cordate sagittate, *segments of perianth winged, nut very smooth and shining.* —*E. B. S.* 2811.—St. wiry, climbing to the height of 4 or 5 feet.—Thickets in the South, rare. A. VII.—IX. E.

4. FAGOPY'RUM *Mill.* Buck-wheat.

[*F. sagittátum* (Gilib.); st. erect without prickles, fl. in cymose panicles, stam. 8, l. cordate-sagittate acute, nut triquetrous acute with entire angles.—*Polygonum Fagopyrum* L. *F. esculentum* (Moench) ed. viii.—*E. B.* 1044.—An escape. Sown as food for game. A. VII. VIII.]

Order LXVII. ELÆAGNACEÆ.

Mostly diœcious. Perianth tubular; limb 2—4-toothed, in male fl. 4-parted. Stam. 3 or more, inserted in the throat. Anth. 2-celled, nearly sessile, bursting on the inner side longitudinally. Ovary free, 1-celled, with 1 erect ovule. Fr. crustaceous, enclosed within the fleshy persistent perianth. Albumen thin and fleshy. Radicle inferior.—No stipules.

1. HIPPOPHAE. Diœcious. Male fl. in a sort of catkin; scales ovate, 1-flowered. Perianth of 2 leaves adhering by their points. Stam. 4, with very short filaments.—Female fl. solitary. Perianth tubular, cloven at the summit. Style short. Stigma long. Nut 1-seeded, in the large coloured berrylike perianth.

1. HIPPO'PHAË *Linn.* Sea Buckthorn.

1. *H. Rhamnoïdes* (L.).—*E. B.* 425.—A thorny shrub with linear-lanceolate silvery leaves. Fl. appearing with the young leaves. Fr. orange. Height 4—6 feet.—Sandy spots and cliffs of the South-east and East coasts chiefly. Sh. V. E. *I.

Order LXVIII. THYMELACEÆ.

Perianth tubular, inferior, often coloured, 4—5-cleft. Stam. definite in number, in the tube. Anth. 2-celled, bursting longitudinally. Ovary free, 1-celled; ovule 1, pendulous. Fr. a nut or drupe. Albumen 0 or thin and fleshy. Radicle superior.—No stipules.

1. DAPHNE. Perianth 4-fid, deciduous. Berry fleshy, 1-seeded. Stam. 8, falling short of the perianth, inserted in the tube in 2 rows.

1. DAPH'NE *Linn.*

1. *D. Mezéreum* (L.); fl. about 3 together lateral sessile, tube

hairy, segments ovate acute.—*E. B.* 1381.—Fl. purple or pale red, appearing before the lanceolate l., which are narrowed below. Berries red. A small deciduous shrub.—Woods, rare. Sh. III. *Mezereon.* E.

2. *D. Lauréola* (L.); racemes axillary of about 5 glabrous drooping bracteate fl., l. lanceolate attenuate below glabrous evergreen.—*E. B.* 119.—A small evergreen shrub, 1—3 feet high, slightly branched, naked below. Fl. yellowish green, funnel-shaped. Berries bluish black.—Woods and thickets. Sh. II.—IV. E.

Order LXIX. SANTALACEÆ.

Perianth adnate to the ovary; limb 3—5-fid; æstivation valvate. Stam. 3—5, opposite to and inserted at the base of the segments. Ovary 1-celled; ovules 1—4, pendulous from near the apex of a central placenta. Style 1. Fr. drupaceous, 1-seeded. Embryo in the axis of fleshy albumen.—No stipules.

1. THESIUM. Perianth 4—5-cleft, top- or funnel-shaped. Stam. 5, with a fascicle of hairs at their base. Style 1. Stigma simple. Nut drupaceous, crowned with the persistent perianth.

1. THE'SIUM *Linn.*

1. *T. humifúsum* (DC.); st. procumbent or ascending racemose, racemes branched or simple, fl.-stalks as long as or longer than the fl., fr.-stalks patent, angles of fr.-stalks and edges of bracts and upper l. denticulate-asperous, fr. oval-oblong, l. 1-veined linear.—*R.* xi. 542. *T. linophyllum* Sm. *E. B.* 247.—Parasitic on various pasture plants. Root woody. St. nearly always prostrate, spreading in a circle. L. very slender; veins usually obscure. Bracts 3, lower middle ones exceeding flowers. Perianth-tube very short, open, funnelshaped; segm. triangular, spreading, a tooth on each side. Fr. ovoid, longer than the persistent perianth which is usually inflexed only at the end, strongly ribbed, slightly netted.—Chalky and limestone (oolite) hills. P. VI. VII. E.

[*T. húmile* (Vahl); racemes spiked, *fl. nearly sessile,* fr. strongly ribbed and netted sessile crowned with the very short inflexed perianth, *l. fleshy linear* 1-veined.—*R.* xi. 542.—I gathered two specimens of this plant near Dawlish, Devonshire, in 1829. P. VII. VIII.]

Order LXX. ARISTOLOCHIACEÆ.

Perianth adnate to the ovary below, tubular above, with a lobed dilated usually irregular limb. Stam. 6—12, epigynous. Ovary 3—6-celled. Style simple; stigma radiant. Fr. many-seeded. Seed with a minute embryo at the base of fleshy albumen.—No stipules.

[1. Aristolochia. Perianth tubular, swelling at the base; mouth dilated on one side. Anth. 6, adnate to the short columnar style under the 6-lobed stigma. Caps. 6-celled.]

2. Asarum. Perianth bellshaped, 3-fid. Stam. 12, inserted at the base of the style. Anth. attached to the middle of the filaments. Stigma 6-lobed. Caps. 6-celled.

1. Aristolo'chia *Linn.* Birthwort.

[1. *A. Clematitis* (L.); creeping, st. erect simple, l. cordate stalked glabrous, fl. clustered.—*E. B.* 398. *St.* 6. 16.—Fl. pale yellow.—Established near old ruins. P. VII. VIII.] E.

2. As'arum *Linn.* Asarabacca.

1. *A. europæ'um* (L.); l. reniform blunt.—*E. B.* 1083.—St. short, with 2 evergreen leaves, between which there is a solitary drooping dull-green flower. Perianth-segm. ovate and incurved. Filaments extending beyond the anthers.—Woods and banks, very rare. P. V. E.

Order LXXI. EMPETRACEÆ.

Diœcious. Perianth of hypogynous scales imbricate in several rows. Stam. equal in number to the inner row of scales and alternate with them, free. Ovary free, on a fleshy disk, 2—9-celled. Ovules solitary, ascending. Style 1. Stigma radiant. Fr. fleshy with bony cells. Embryo in the axis of fleshy albumen. Radicle inferior.—No stipules.

1. Empetrum. Cal. 3-parted. Pet. 3. Stam. 3 in the male flowers. Style short; stigma dilated, peltate, with 6—9 rays in the fem. flowers. Berry globose, 1-celled. Seeds 6—9.

1. EMPE'TRUM *Linn.* Crowberry.

1. *E. nigrum* (L.); procumbent, l. linear-oblong their margins meeting in a white line beneath.—*E. B.* 526.—A small leafy heathlike shrub. Fl. axillary, small, purple. Stigma with 9 rays. Berries black.—Mountain heaths. Sh. V. E. S. I.

Order LXXII. EUPHORBIACEÆ.

Fl. monœcious or diœcious. Perianth lobed or 0. Male flowers of 1 or more stamens. Anth. 2-celled. Fem. fl. of 1 superior 2—3-lobed and 2—3-celled ovary. Styles 2—3. Stigma compound or simple. Caps. opening with elasticity; cells 2—3, with 1 or 2 suspended seeds in each. Embryo in fleshy albumen. Radicle superior.—Stip. often present, but very deciduous.

1. BUXUS. Fl. monœcious.—Male. Cal. 3-parted. *Pet.* 2. *Stam.* 4.—Fem. Cal. 4-parted. Pet. 3. Caps. with 3 horns, 3-celled, 6-seeded.

2. EUPHORBIA. Fl. incomplete, collected into monœcious *clusters consisting of* 1 *female and many male flowers.* Involucre bellshaped with 4 or 5 divisions and 4 or 5 alternate glands.—Males consisting of a single stamen upon a pedicel, intermixed with scales and surrounding the female. Fem. of a single pistil. Styles 3. Stigmas bifid. Caps. 3-celled, bursting at the back. Seeds solitary, pendulous.

3. MERCURIALIS. Fl. diœcious or monœcious. Perianth 3-parted.—Male. *Stam.* 9—16.—Fem. Two barren filaments. Style short, forked. *Caps. 2-celled.* Cells 1-seeded, bursting at the back.

1. BUX'US *Linn.* Box.

1. *B. sempervirens* (L.); l. ovate-oblong coriaceous shining above, petioles ciliate, anth. ovate-sagittate.—*E. B.* 1341.—A small bushy tree 10—12 feet in height.—Dry chalky hills especially in Surrey and Kent, rare. T. IV.—VI. E.

2. EUPHOR'BIA *Linn.* Spurge.

A. Leaves with stipules, opposite. Clusters axillary.

1. *E. Pep'lis* (L.); st. procumbent forked, heads solitary, caps.

keeled, seeds smooth, l. opposite stalked half-oblong-heartshaped nearly entire glabrous.—*E. B.* 2002.—St. usually much tinged with purple, glaucous. Inv.-glands rounded.—Loose sand. Southern sea-coast, rare. A. VII.—IX. E. I.

B. Stipules wanting. L. scattered (except in *E. Lathyris*).

* *Involucral lobes roundish or transversely oval.*

2. *E. Helioscópia* (L); umbel of 5 trifid and forked rays, bracts and l. membranous ovate-wedgeshaped serrate upwards, caps. smooth glabrous, seeds netted rugose.—*E. B.* 883.—St. 6—18 in. high.—Waste and cultivated ground. A. VI.—IX. *Sun Spurge.* E. S. I.

3. *E. stric'ta* (L.); umbel of 3—5 each 3—5-fid and forked rays, general and first partial bracts and l. oblong clasping, other bracts broadly cordate blunt with a minute apiculus, inv.-lobes oval, *stam. rarely more than* 2 *in each inv., caps. with prominent cylindrical tubercles*, seeds smooth oval brown and shining, *seed-stalk cordate*.—*E. B S.* 2974.—St. erect, 2—3 ft. high, much branched. Umbel moderate. Inv., caps., and seeds small.—Limestone woods. Gloucestershire and Monmouthshire. A. VII. VIII. E.

4. *E. platyphyl'los* (L.); umbel of 3—5 repeatedly forked rays, l. and general bracts long-obovate with a cordate base, partial bracts all cordate, *stam.* 7—8 *in each inv.*, inv.-lobes oval, *caps. with shortly conical tubercles*, seed obovate brown and shining, *seed-stalk nearly reniform*.—*E. stricta* Sm., *E. B.* 333 (starved).—Plant more diffuse than, and quite distinct from, the preceding, erect, slightly branched. Inv., caps., and seeds twice as large.—Fields. A. VI.—VIII. E.

[*E. dulcis* (L.); umbel of 5 bifid rays, *l. and general bracts* long-obovate blunt *narrowed to the base, partial bracts* triangular-ovate *with a truncate base*, inv.-lobes rounded, caps. with few prominent tubercles.—*R.* v. 135.—Erect. A foot high. Umbel-rays long.—West Bergholt, Essex. Glascoed Dingle, Llansilin. Jedburgh. P. VI.]

5. *E. hiber'na* (L.); umbel of about 5 twice-forked rays, bracts and l. ovate or oblong entire blunt, inv.-lobes reniform, caps. warted glabrous, seeds obovate smooth somewhat shining brownish.—*E. B.* 1337.—Height 1—2 feet. L. broad, usually pilose beneath.—S. and W. of Ireland. Cornw., N. Devon. P. VI. E. I.

6. *E. pilósa* (L.); umbel irregular of about 5 trifid and forked rays, *bracts all elliptic* glabrous, l. broadly lanceolate minutely serrate slightly hairy, inv.-lobes transversely oval, *caps. warted*, seeds obovate minutely punctate smooth.—*E. B. S.* 2787. *E. palustris* Forst. (not Linn.).—St. 2—4 feet high, leafy throughout, annual. Caps. usually hairy.—Shady places near Bath. P. V. VI. E.

[*7. *E. coralloïdes* (L.); umbel of 5 trifid and forked rays, *bracts ovate-oblong* the tertiary ones ovate all hairy, l. lanceolate minutely serrate woolly, inv.-lobes transversely oval, *caps. nearly smooth* woolly, seeds obovate minutely punctate and with faint netted bands.—*E. B. S.* 2837.—Height 2—3 feet. Usually naked below.—Slinfold, Sussex. Introduced. B. ? V. VI.] E.

** *Involucral lobes triangular-lunate or with 2 horns.*

† Seeds smooth. Bracts connate.

8. *E. amygdaloïdes* (L.); umbel of 5 or more forked bifid rays, *bracts rounded connate*, l. ovate-lanceolate hairy beneath entire, inv.-lobes lunate (yellow) with 2 horns, caps with very minute tubercles glabrous, seeds roundish-ovate.—*E. B.* 256. —St. 2—3 feet high, leafy, purple below, biennial.—Woods and thickets. P. III. IV. *Wood-Spurge.* E. I.

†† Seeds smooth. Bracts separate.

[*E. Cyparis'sias* (L.); umbel of many forked rays, partial bracts reniform or cordate blunt entire, l. linear entire or on the barren shoots setaceous many close together, inv.-lobes with 2 horns, caps. tubercular.—*R.* v. 147.—It may be a native at Whitbarrow in Westmoreland. *Rev. W. H. Hawker.* The other stations belong to *E. Esula.*] E.

†9. *E. E'sula* (L.); umbel of many forked rays, partial bracts cordate mucronate, *l. lanceolate or sublinear* narrowed below glabrous *denticulate* scattered, inv.-lobes with 2 horns, "caps. asperous, seeds obovate."—*E. B.* 1399. *R.* v. 146.—St. 12—18 in. high, leafy, with a few axillary leafy branches without flowers. Creeping.—Woods rare. P. VI. VII. E. S.

10. *E. Parálias* (L.); umbel usually of 5 forked rays, bracts rather reniform, *l. coriaceous oblong*, inv.-lobes 5 with 3 or 4 short points, caps. wrinkled.—*E. B.* 195.—Root woody, tough. Flowering st. about a foot high; barren stems shorter, many. L. closely imbricate.—Sandy sea-coasts. P. VIII. IX. *Sea-Spurge.* E. S. I.

††† Seeds rough, tubercled or pitted. L. alternate, scattered.

11. *E. portland'ica* (L.); umbels of 5 forked rays, bracts broadly rhomboidal acuminate, *l. obovate or obovate-lanceolate blunt subapiculate*, inv.-lobes 4 lunate with long horns, caps. rough at the angles, seeds pitted and netted almost white.—*E. B.* 441.—Glaucous, smooth. Scarcely a foot high. L. spreading. Seed-stalk conical, hooded.—Sea-coasts. P. IV.—IX. E. S. I.

12. *E. Pep'lus* (L.); umbel of 3 forked rays, bracts ovate blunt mucronate, *l. broadly ovate stalked*, lower l. suborbicular, inv.-lobes 4 lunate with long horns, caps. smooth with thickened rough keels, seeds ovoid pitted.—*E. B.* 959.—Light green, smooth, erect, 6—10 in. high.—A common weed. A. VII. VIII. *Petty Spurge.* E. S. I.

13. *E. exig'ua* (L.); umbel of 3 forked rays, *bracts lanceolate* acute unequal below, l. linear blunt with a mucro [var. *retusa* (L.)] or acute, inv.-lobes rounded with 2 horns, caps. smooth with slightly thickened and tubercled angles, seeds angular wrinkled. —*E. B.* 1336.—Height 3—6 in., usually branched at the base. Seeds small white.—Cornfields. A. VI.—VIII. E. S. I.

†††† Seeds rough. Leaves opposite.

14. *E. Lath'yrus* (L.); umbel of 3 or 4 forked rays, bracts oblong-ovate attenuate acute, l. linear-oblong sessile, upper l. cordate at the base, inv.-lobes lunate with blunt horns, caps. smooth with a dorsal line.—*E. B.* 2255.—St. solitary, 2—3 feet high, purplish. L. many, in 4 rows.—Truly wild in a few stony and rocky woods, where it appears for 2 or 3 years after the bushes have been cut. Also naturalized on cultivated ground. B. VI. VII. *Caper-Spurge.* E.

3. Mercuria'lis *Linn.* Mercury.

1. *M. peren'nis* (L.); st. simple, l. stalked oblong-lanceolate rough, female fl. on long common stalks, creeping.—*E. B.* 1872.—St. a foot high, usually naked below. Fl. in lax spikes. —β. *M. ovata* (Steud.); l. nearly sessile ovate.—Woods and thickets. β. Hurstpierpoint, Suss. *Mr. Mitten.* P. IV. V. E. S. I.

2. *M. an'nua* (L.); st. branched, l. stalked ovate or ovate-oblong smooth, female fl. nearly sessile, root fibrous.—*E. B.* 559.—Height 6—12 in. Bright green. Much branched.—β. *M. ambigua* (L.); l. lanceolate, fl. whorled male and female intermixed. *E. B. S.* 2816.—Waste and cultivated land. A. VIII. IX. E. I.

Order LXXIII. CERATOPHYLLACEÆ.

Fl. monœcious. Perianth free, in many divisions. Anth. 12—20, sessile, 2-celled, 2-pointed; cells again partially divided.—Perianth usually wanting. Ovary 1-celled. Seed 1, pendulous. Embryo straight; cotyledons 4, alternately smaller.—No stipules.—Only one genus.

1. CERATOPHYL'LUM *Linn.* Hornwort.

1. *C. demer'sum* (L.); *segments of l. linear-filiform*, fr. wingless having a spine on each side near the base and tipped with the curved subulate style.—*E. B.* 947.—L. 2—4 times forked, *dark green*, segments of uppermost l. broader and more rigid. Sometimes the spines are wanting, the style is short, and the surface rough with minute tubercles.—Ponds and ditches, rare. P. VI. VII. E. S. I.

2. *C. submer'sum* (L.); *segments of l. setaceous*, fr. destitute of spines covered when ripe with *cylindrical tubercles* and tipped with the short curved style.—*E. B.* 679.—*L. pale green.* Young fr. smooth.—Ponds and ditches. P. VI. VII. E.

Order LXXIV. CALLITRICHACEÆ.

Fl. mostly monœcious, solitary, minute, usually with 2 white bracts, no perianth. Stam. 1. Filament long. Anth. reniform, 1-celled, opening transversely at the top.—Perianth none. Ovary 1, 4-angled, 4-celled. Styles 2, subulate. Fr. dry, 4-celled, 4-lobed; separating into 4 1-seeded indehiscent carpels. Embryo in axis of fleshy albumen. Radicle superior, long. Cotyledons short, terete.—No stipules.—Only one genus of aquatic plants.

1. CALLIT'RICHE *Linn.* Water-Starwort.

A *L. narrowed at the base*; lower linear, submerged; upper floating, broad, often spathulate and forming a rosette. Fr. of 4 lobes cohering in pairs.—*C. verna* L.

1. *C. vernális* (Koch); fr. nearly sessile longer than broad, *lobes* parallel *bluntly keeled on the back convex on the sides*, styles erect deciduous, bracts straightish deciduous.—*E. B.* 722. *R.* v. 129. *C. verna* (L.) ed. viii.—Floating l. ovate-spathulate, in a rosette, rarely wanting. Fr. small; keels of each pair of lobes converging.—Ponds and slow streams. A. or P. IV.—IX. E. S. I.

2. *C. obtusan'gula* (Le G.); fr. nearly sessile, *lobes* parallel *with rounded edges*, styles erect or spreading persistent, bracts persistent.—L. all obovate, in a rosette. Fr. large, quite blunt and rounded at the edges.—Ditches. A. or P. V.—IX. E. I.

3. *C. polymórpha* (Lönnr.); fr. nearly sessile, its lobes *keeled not winged*; styles erect in fl. persistent and reflexed in fr. *very long*, bracts falcate persistent. Upper l. rosulate, roundly spathulate, lower l. linear emarginate. Fr. about as long as in Sp. 1. Style 2—3 times as long as fr. Much resembles Sp. 4 but fr. smaller and scarcely winged.—Recorded from Surrey and Shetland. A. or P. VI.—IX. E. S.

4. *C. stag'nalis* (Scop.); fr. nearly sessile, *lobes winged on the back*, styles erect in flower persistent and reflexed closely with fruit, bracts falcate persistent.—*E. B. S.* 2864. All l. round-obovate-spathulate, floating-l. in a close rosette, often wanting. Fr. twice as large as in Sp. 1, pale when dry; lobes slightly spreading.—Common, especially on mud. A. or P. V.—IX. E. S. I.

5. *C. intermédia* (Hoffm.); fr. broader than long nearly sessile or stalked, *lobes* parallel *with a blunt dorsal ridge flat on the sides*, styles divaricate and reflexed over the sides of the fruit, bracts falcate very deciduous.—*R.* v. 130. *Sy. E. B.* 1273. *C. hamuláta* (Kütz.) ed. viii.—L. linear; upper ones broader in the middle; uppermost oblong spathulate. Bracts hooked, overlapping, soon falling. Fr. small, the size of that of *C. vernalis*. —*β. C. pedunculata* (DC.); l. linear, upper l. rather broader, fr. stalked. Plant usually very small.—Lakes and streams. *β.* Marshes. A. or P. IV.—IX. E. S. I.

B. *L.* uniform, *enlarged at the base*, linear. Fr. of 4 nearly free lobes.—*C. autumnalis* L.

6. *C. truncáta* (Guss.); fr. shortly stalked, *lobes with a blunt dorsal ridge*, styles reflexed-patent, bracts 0.—*C. autumnalis* Hook. (not L.). *E. B. S.* 2606.—L. all submersed, uniform, very abrupt at the end, translucent green.—Suss., Kent, Wexford, Guernsey. A. or P. IV.—IX. E. I.

7. *C. autumnális* (L.); fr. nearly sessile, *lobes broadly and acutely winged at the back*, styles spreading, bracts 0.—*E. B. S.* 2732.—Fr. 4 times as large as in *C. vernalis*, dark brown, 1 or 2 of the lobes often abortive. L. dark green, all submersed, very abrupt at the end.—Rare. Anglesea. N. of England. Scotland. N. & W. of Ireland. A. or P. VI.—IX. E. S. I.

Order LXXV. URTICACEÆ.

Fl. monœcious, diœcious, or rarely perfect. Perianth inferior, 4—3—6-parted, imbricate; or in the female flowers tubular notched or scalelike and open. Stam. definite, free, inserted at the base of the perianth and opposite to its lobes. Ovary free, 1-celled; ovules solitary. Stigmas 1—2. Fr. not bursting.

1. PARIETARIA. Fl. polygamous, surrounded by an involucre. Perianth bellshaped, 4-parted. Stam. 4. Style filiform.—L. alternate. Hairs simple.

2. URTICA. Fl. monœcious or diœcious. Males in loose racemes; perianth 4-parted; stam. 4. Females in capitate racemes; perianth 2-parted; stigma sessile.—L. opposite. Hairs stinging.

1. PARIETARIA *Linn.* Wall-Pellitory.

1. *P. ramiflóra* (Moench); l. elliptic or elliptic-lanceolate 3-veined above the base, bracts combined into an involucre of two 3-lobed segm., perianth bellshaped equalling the stam. or in the perfect fl. (red) lengthening to twice as long as the stamens.—*E. B.* 879. *Curt.* ii. 203. *P. diffusa* (Koch) ed. viii.—St. prostrate or ascending, simple or branched below. Cymes axillary, dense; primary fem. fl. between the inv.-segm. and with 1 or 2 free bracts; each segm. bears on its face 1 or 3 fl., of which the lateral have bracts.—Old walls. P. VI.—IX E. S. I.

2. URTI'CA *Linn.* Nettle.

‡1. *U. pilulif'era* (L); *l.* opposite ovate ovate-lanceolate or cordate-acuminate *coarsely toothed,* clusters of fr. globose stalked. —*E. B.* 148.—About 2 ft. high. Very venomous.—β. *U. Dodartii* (L.); l. ovate or ovate-lanceolate nearly entire. *Sy. E. B.* 1281.—About towns and villages in the East of England, rare. A. VI.—VIII. *Roman Nettle.* E.

2. *U. úrens* (L.); l. opposite elliptic-ovate serrate, *spikes* axillary *nearly simple two together falling short of the petiole,* seeds oblong.—*E. B.* 1236.—Scarcely a foot high, glabrous, with stinging bristles.—Common weed. A. VI.—IX. E. S. I.

3. *U. dioíca* (L.); l. opposite cordate serrate, *spikes* axillary *panicled exceeding the petiole,* seeds ovate.—*E. B.* 1750.—Creeping. St. 2—3 ft. high. Hairy. Stinging. Dark-green.—β. *angustifolia* (Wimm. & Grab.); l. ovate-lanceolate rounded but hardly cordate at the base.—Common. P. VI.—IX. E. S. I.

Order LXXVI. CANNABINACEÆ.

Fl. diœcious; fem. in catkins. Perianth inferior, of 5 sepals in male, open and scale-like in female flower. Stam. 5, on base of perianth. Ovary free. Ovule solitary. Styles 2, long.—With stipules.

1. Humulus. Fem. fl. open nearly flat, 2 together.

1. Hu'mulus *Linn.* Hop.

1. *H. Lúpulus* (L.).—*E. B.* 427.—Known by its long twining stems, opposite rough 3—5-lobed serrate leaves; ovoid or globular ripe fem. catkins with ovate scales. Male fl. in loose panicles.—A true native in the south of England. P. VII. E.

Order LXXVII. ULMACEÆ.

Fl. bisexual, not in catkins. Perianth inferior, bell-shaped, 3—8-cleft, imbricate. Stam. definite in number, inserted in tube of perianth and opposite to its lobes. Ovary free, 1—2-celled; ovules solitary. Styles 2. Fr. not bursting. With deciduous stipules.

1. Ulmus. Perianth bellshaped, 4—5-cleft, persistent. Stam. 5. Styles 2. Caps. compressed, wings all round.—Fl. before the leaves, clustered; ped. short.

1. Ul'mus *Linn.* Elm.

1. *U. campes'tris* (L.); l. shortly acuminate more or less doubly serrate, ped. short, fl. 4—5-fid, lobes ciliate, fr. obovate or oblong notched, *seed-cavity chiefly above the middle of fr. and near the notch.*—*a. suberosa* (Koch); l. asperous above pubescent beneath. *E. B.* 1886, 2161.—*β. carpinifolia* (Meyer). *U. glabra* (Sm.); l. rather coriaceous shining nearly or quite smooth above glabrous except in the axils of the veins beneath, young l. stip. and fr. with subsessile glands. *E. B.* 2248. *U. stricta* (Lindl.) which has rigid erect close branches is a form of this variety.—*a.* Throughout England. *β.* South of England and Ireland. T. III.—V. *Common Elm.* E. I.

2. *U. glábra* (Huds.); l. much acuminate doubly serrate asperous above rather downy beneath, ped. short, fl. 5—7-fid, lobes ciliate, fr. oblong or roundish notched, *seed-cavity chiefly below the middle of fr. and distant from the notch.*—*E. B.* 1887.

U. montana (With.) ed. viii.—[*U. major* (Sm.) is a large and rough-leaved form with corky growths on the branches, and *U. montana* var. *nitida* (Sy.) has l. shining and glabrous].—Branches large spreading. L. broad. Fr. Hoplike.—Woods and hedges. T. III. IV. *Wych Elm.* E. S. I.

Order LXXVIII. AMENTIFERÆ.

Fl. monœcious or diœcious, rarely perfect. Barren fl. of all our plants in catkins; sometimes with a membranous perianth. Fertile fl. clustered, solitary or in catkins. Ovary with one or several cells. Stigmas 1 or more. Fruit as many as the ovaries, bony or membranous (or drupaceous in *Myrica*). Albumen usually wanting. Embryo straight or curved, plain. Radicle mostly superior.—Young leaves with stipules.

Subord. I. *SALICINEÆ.* Diœcious. Fl. all in catkins. Fr. naked, 2-valved, 1-celled, many-seeded. Seeds erect, hairy.

1. Salix. Catkins consisting of imbricate entire scales. Stam. 1—5. Fr. a 1-celled pod with 1—2 glands at its base. Perianth 0.

2. Populus. Catkins with slashed scales. Stam. 4—30 from a little oblique cupshaped perianth. Fr. almost 2-celled, with a cupshaped perianth.

Subord. II. *MYRICEÆ.* Diœcious. Fl. in catkins. Fr. drupaceous surrounded by scales become fleshy and adherent to the ovary.

3. Myrica. Catkins with concave scales. Stam. 4—8. Stigmas subulate. Drupe 1-celled, 1-seeded. Perianth 0.

Subord. III. *BETULINEÆ.* Monœcious. Fl. all in catkins. Ovary 2-celled, each with 1 ovule. Fr. naked, indehiscent, 1-celled, 1-seeded. Seeds pendulous, not hairy.

4. Betula. Scales of the barren catkins ternate, the middle one bearing the stamens. Perianth 0. Scales of the fertile catkin 3-lobed, 3-flowered, membranous, deciduous. Styles 2, filiform. Fr. with a membranous margin.

5. Alnus. Scales of the barren catkins 3-lobed, 3-flowered. Perianth 4-parted. Scales of the fertile catkin ovate, 2-flowered, coriaceous, persistent. Styles 2. Ovary compressed, 4 minute scales at its base. Fr. not winged.

Subord. IV. *CUPULIFERÆ.* Male fl. in a catkin. Fem. solitary or clustered or spiked. Perianth adnate to the ovary (glans), with a very minute sometimes evanescent limb, surrounded by a coriaceous involucre.

6. FAGUS. Barren catkin globose. Perianth 5- or 6-fid. Stam. 8—15. Fertile fl. 2 together within a 4-lobed prickly involucre. Stigmas 3. Ovaries 3-cornered and 3-celled. Nut by abortion 1—2-seeded.

7. CASTANEA. Barren catkin long, cylindrical. Perianth 6-parted. Stam. 8—20. Fertile fl. 3 within a 4-lobed muricate involucre. Stigmas 6. Ovary 5—8-celled. Nut 1-celled with 1—3 seeds.

8. QUERCUS. Barren catkin long, pendulous, lax. Stam. 5—10. Perianth 5—7-cleft. Fertile fl. solitary, with a cupshaped scaly involucre. Stigmas 3. Ovary 3-celled. Nut 1-celled, 1-seeded, surrounded at the base by the enlarged cupshaped involucre.

9. CORYLUS. Barren catkin long, pendulous, cylindrical. Scales 3-lobed, middle lobe covering the 2 lateral lobes. Stam. 8. *Anth. 1-celled.* Perianth 0. Fertile fl. several, surrounded by a scaly involucre. Styles 2. Nut 1-seeded, enclosed in the enlarged coriaceous cut involucre.

10. CARPINUS. Barren catkin long, cylindrical. Scales roundish. Stam. 5—14. *Anth. 1-celled.* Fertile fl. in a lax catkin. Scales large, leaflike, 3-lobed, 2-flowered. Styles 2. Nut ovate, 1-seeded.

Tribe I. *Salicineæ.*

1. SA'LIX *Linn.*[1] [2] Willow.

Sec. 1. VITISALIX (Dumort.). Catkin and its leafy stalk usually (but not always) deciduous together, lateral, appearing

---

[1] See Dr. Buchanan White, *Scott. Nat.* x. 359, and *Journ. Linn. Soc.* xxvii. 333.

[2] Since the eighth ed. appeared a great advance has been made in the knowledge of this genus. In Dr. White's masterly 'Revision of the British Willows,' *Journ. Linn. Soc.* cited above, most of the former varieties are treated as merely trivial forms, a number of previously-accepted species are referred to hybrids and many fresh hybrids are described. Messrs. E. F. & W. R. Linton have issued an excellent 'Set of British Willows' in which many of the hybrids are represented.—H. & J. G.

with the leaves. Scales of uniform colour. Nectary of 2 pieces or urceolate; germen or stamens from the middle. Vernation convolute.

Subsec. 1. *Lycus* (Dumort.). Stam. 4—8. "Nectary urceolate, undivided." L. glossy, glabrous.—Trees or large shrubs. Stipules soon falling. *Pentandræ* Borr.

1. *S. pentan'dra* (L.); l. ovate-elliptic or ovate-lanceolate acuminate glandular-serrate, "*stip. ovate-oblong straight equal,*" stam. 5 or more, caps ovate-attenuate glabrous, stalk twice as long as the gland, style short, stig. bifid.—*E. B.* 1805.—Height 6—20 feet. Top of petioles glandular. L. fragrant.—Riversides in the North. Sh. or T. V. VI. *Bay-leaved Willow.* E. S. I.

†2. *S. cuspidáta* (Schultz ?)[1]; l. oblong lanceolate acuminate glandular serrate, *stip. half-cordate oblique,* "stam. 3 or 4," caps. ovate attenuate glabrous, stalk 3 or 4 times as long as the gland, style short, stig. emarginate.—*E. B. S.* 2961. *S. Meyeriana* Willd.—Height 20—30 feet. Top of the petioles glandular.—Near Shrewsbury. Sh. or T. VI. E.

Subsec. 2. *Amerina* (Dumort.) Stam. 2 or 3. Nectary of 2 pieces, one between the cal.-scale and germen, the other opposite to it. Naturally trees.

i. *Diandræ.* Stam. 2. Catkin-scales soon falling.—*Fragiles* and *Albæ* Borr.

3. *S. frag'ilis* (L.); l. gradually attenuate very glaucous beneath, stip. ½-cordate, germ. subulate stalked much exceeding scale, style short equalling bifid stigma.—*S. Russelliana* Sm., *E. B.* 1808. *S. viridis* Sy., *E. B.* 1308 (excl. desc.) not Fr. See White in *J. of B.* xxvi. 196. Baker, *J. of B.* xxvi. 249. *S. decipiens* (Sm.)[2] *E. B.* 1937 seems only a slight variety.—Branches polished round smooth. *Bedford Willow.*—Damp meadows and osier-ground. T. IV. V. E. S. (I ?)

4. *S. vir'idis* (Fr.)[3]; l. lanceolate dark shining, stip. ½-cordate, germ. ovate-oblong subsessile equalling scale, style very short. —*S. fragilis* Sm., *E. B.* 1807. *S. viridis* Sy., *E. B.* descript.—Branches very smooth round brown brittle.—Damp meadows. T. IV. V. E. S. (I ?)

---

[1] Considered a hybrid (*S. pentandra* × *fragilis*) by Dr. White.—H. & J. G.

[2] Dr. White refers this doubtfully to *S. triandra* × *fragilis.*—H. & J. G.

[3] *S. fragilis* × *alba*, Dr. White.—H. & J. G.

5. *S. al'ba* (L.); l. elliptic lanceolate glandular-serrate acute silky on both sides when young, stip. minute, *caps. nearly sessile* ovate acuminate glabrous, style short, stigmas thick recurved bifid.—*E. B.* 2430.—Height 50—80 feet. Scales shorter than stam., as long as caps. in *α* and *β*, exceeding them both in *γ*. Branches silky.—*β*. *S. cærulea* (Sm.); l. less silky beneath. *E. B.* 2431.—*γ*. *S. vitellina* (L.); branches bright yellow, l. shorter and broader. *E. B.* 1389.—Wet places. T. V. *White Willow.* E. S. I.

ii. *Triandræ.* Stam. 3. Catkin-scales persistent. L. lanceolate approaching to ovate, glabrous. Catkins lax.—Osiers, naturally trees.

*6. *S. unduláta* (Ehrh.)[1]; l. lanceolate much acuminate serrate glabrous except when young, stip. ½-cordate acute, caps. stalked ovate acuminate, pedicel twice as long as the gland, *style long*, stigma bifid, *scales very shaggy.*—*S. lanceolata* Sm., *E. B.* 1436. —Height 12—15 feet. L. sometimes wavy, often quite silky when young. Germ. glabrous in *S. lanceolata* (or downy in the foreign *S. undulata*).—Lewes, Suss. [and elsewhere] Sh. or T. IV. V. E. I.

7. *S. trian'dra* (L.); l. oblong-lanceolate acute serrate glabrous, stip. ½-cordate blunt, caps. stalked oblong-ovate glabrous, *stigma nearly sessile, scales glabrous.*—*E. B.* 1435.—Height 20 —30 feet. Germen not furrowed. L. narrowing down to the stalk, somewhat paler beneath, or (*S. triandra* Curt.) lanceolate wavy paler and glaucous beneath.—*β*. *S. Hoffmanniana* (Sm.); l. somewhat rounded below ovate-lanceolate, stip. larger, scales shaggy towards their base.—*E. B. S.* 2620.—*γ*. *S. amygdalina* (L.); l. oblong-ovate acute rounded below, caps. ovate tumid furrowed, young shoots furrowed. *E. B.* 1936.—Wet woods and osier-grounds. Sh. or T. IV. V. E. S. I.

Sec. 2. CAPRISALIX (Dumort.). Catkins lateral, sessile, without leaves or with two or three small leaves or leaflike bracts at the base; stalk sometimes elongated in fruit so as to resemble a leafy shoot, but deciduous with the catkin. Catkin-scales often discoloured at the end. Nectary simple (of one piece), on the opposite side of the stam. or germen from the catkin-scale.

---

[1] *S. triandra* × *viminalis*, Dr. White. *S. triandra* × *alba*, E. F. Linton.—H. & J. G.

Subsec. 1. *Helice* (Dumort.). Filament 1 with a 4-celled anther, or forked with 2 anthers each of 2 cells. *Anth. purple, ultimately black.* Nectary cuneate. Catkins bracteate at the base. Vernation equitant.—*Purpureæ* Borr.

8. *S. purpúrea* (L.); l. lanceolate broader upwards acuminate attenuate below finely serrate glabrous, caps. ovate very downy sessile, style very short, *stigm. ovate, anth.* 1, *stip.* 0.—α. *S. purpurea* (Sm.); decumbent, twigs purple, fertile catkins very compact. *E. B.* 1388.—β. *S. Woolgariana* (Borr.); erect, twigs yellowish grey, l. cuneate-lanceolate glaucous beneath, stigmas blunt. *E. B. S.* 2651.—γ. *S. ramulosa* (Borr.); erect, twigs pale yellowish, l. oblong-lanceolate paler beneath, stigmas sessile bifid.—δ. *S. Lambertiana* (Sm.); erect, twigs purplish glaucous, l. oblong-linear-lanceolate slightly narrowed and somewhat rounded below, stigmas ovate emarginate. *E. B.* 1359.—ε. *S. Helix* (L.)[1]; caps. oblong-ovate, *stigmas almost linear* emarginate, twigs pale yellowish or tinged with purple polished. *E. B.* 1343. *Rose Willow.*—Marshes and river-banks. T. III. IV. E. S. I.

9. *S. rúbra* (Huds.)[2]; linear-lanceolate acuminate glabrous green on both sides, caps. oblong-ovate very pubescent, *style long, stigmas ovate undivided, anth.* 2 (first reddish then yellow), *filaments combined below, stip. linear.*—*E. B.* 1145.—Height 10—20 feet. Twigs usually tawny. L. like those of *S. viminalis* but without the white pubescence.—β. *S. Forbiana* (Sm.); l. lanceolate-oblong, style nearly as long as the *linear* "*divided stigmas, anth.* 1." *E. B.* 1344. Twigs greyish yellow. L. rather paler and somewhat glaucous beneath. Stip. broader, variable. Stigma entire in all that I have seen.—Low meadows. T. IV. V. E. S. I.

Subsec. 2. *Vimen* (Dumort.). Stam. 2, mon- or diadelphous. *Anth. becoming yellow.* Nect. linear. Catkin-scales discoloured at the end. Catkins bracteate at the base. Stigmas not sessile. Pubescence of the l. silky.

* *Stipules narrow.*

10. *S. viminális* (L.); *l. linear or linear-lanceolate* obscurely crenate white *silky* and shining beneath, *stip. small* sublanceolate, capsule very shortly stalked lanceolate-subulate, style long, stigmas undivided.—*E. B.* 1898.—Height 10—20 feet.

---

[1] Dr. White refers this doubtfully to *S. purpurea* × *viminalis.*—H. & J. G.

[2] *S. purpurea* × *viminalis*, Dr. White.—H. & J. G.

Branches wandlike, long, slender. Gland longer than the stalk of the caps.—β. *intricata* (Leefe); l. broader, caps. shorter and broader, style very short, stigmas from the first cloven reflexed and entangled.—γ. *stipularis* (Leefe); l. lanceolate, stip. linear-lanceolate denticulate or ½-cordate acuminate, stigmas long.—Wet places. Sh. IV. V. *Common Osier.* E. S. I.

** *Stipules broad.*

11. *S. stipuláris* (Sm.)[1]; l. lanceolate very obscurely crenate white and downy beneath, *stip. ½-cordate acute*, caps. ovate nearly sessile, style very short (*Sm.*) or elongate (*Hook.*), stigmas linear undivided.—*E. B.* 1214.—Height 10—20 feet with upright brittle reddish-brown twigs.—Wet places. Sh. III. E.

12. *S. Smithiána* (Willd.)[1]; *l. long-lanceolate* obscurely crenate white satiny beneath, stip. reniform ½-cordate, *caps. stalked* lanceolate-subulate, pedicel about as long as the gland, style long, stigmas long linear mostly entire.—Twigs erect, somewhat downy brittle.—α. *S. Smithiana* (E. B. 1509); l. rounded at base white with satiny down beneath, stip. small narrow ½-lunate.—β. *S. rugosa* (Leefe); l. greenish white and rather silky beneath, *stip. ½-cordate* acute, style moderate, stigmas linear broad undivided. *S. holosericea* H. & A.—γ. *S. ferruginea* (And.); l. greenish white and rather silky beneath, *stip. ½-ovate or reniform*, style elongate, stigmas linear-oblong undivided. *E. B. S.* 2665. Bushy.—Wet places. Sh. IV. V. E. S. I.

Subsec. 3. *Vetrix* (Dumort.). Stam. 2, free. Anth. becoming fuscous-yellow. Nectary cuneate. Catkin-scales discoloured. Vernation equitant.

i. *Capreæ.* Style short. Stipules reniform, without basal glands. L. rugose, not turning black; pubescence crisped, not silky.—*Cinereæ* Borr.

13. *S. acumináta* (Sm.)[1]; *l.* lanceolate-oblong pointed finely toothed *glaucous ashy and downy* scarcely silky *beneath*, stip. ½-cordate, caps. ovate tapering, style conspicuous, stigmas ovate undivided, *buds downy.*—*E. B.* 1434. *Loud.* 1464.—Height 25—30 feet.—Damp woods and hedges. Sh. or T. IV. E. I.

---

[1] Dr. White refers sp. 11, 12 & 13 to hybrids between *S. viminalis* and the various members of the *Capreæ* section.—H. & J. G.

14. *S. cinérea* (L.); l. elliptic- or lanceolate-obovate pointed subserrate downy and ashy beneath, stip. ½-cordate, caps. lanceolate-subulate, stigmas simple or bifid, *buds downy.*—Height 20—30 feet.—*a. S. cinerea* (Sm.); l. obovate-lanceolate rather thick reddish beneath, stip. ½-cordate acute, style very short thick. *E. B.* 1897.—*β. S. aquatica* (Sm.); l. ovate-elliptic thinner downy and rather glaucous beneath, stip. reniform, style inconspicuous. *E. B.* 1437.—*γ. S. oleifolia*; l. obovate-lanceolate rather rigid downy and rather glaucous beneath, stip. small rounded. *E. B.* 1402.—Varies greatly. These varieties are scarcely distinguishable.—Wet places. T. or Sh. III. IV. *Sallow.* E. S. I.

15. *S. aurita* (L.); *l. obovate* repand-dentate recurved-apiculate much *wrinkled* more or less downy above pubescent beneath, *stip.* roundish or reniform *large stalked,* caps. lanceolate-subulate, stig. generally entire, *buds glabrous* or slightly downy.—*E. B.* 1487.—Height 3—4 feet. Edges of l. deflexed, point hooked. Stigmas and scales often reddish.—Damp woods. Sh. IV. V. E. S. I.

16. *S. Cáprea* (L.); *l. ovate* or elliptic *flat* acute crenate-serrate wavy at the margins deep green with a downy midrib whitish above and cottony beneath, stip. subreniform, caps. lanceolate-subulate, style very short, *buds glabrous.*—*E. B.* 1488.—A small tree, 15—30 feet high. Catkins very thick, blunt. L. mostly large and broad, rarely sublanceolate; spring l. nearly entire.—*β. S. sphacelata* (Sm.); stip. often 0, l. nearly entire. *E. B.* 2333. A bush.—Woods and hedges in dryish and also subalpine places. *β.* Highland valleys. T. IV. V. *Great Sallow.* E. S. I.

ii. *Phylicifoliæ.* Style long. Capsule stalked.

* *Nigricantes.* L. punctate beneath, turning black in drying. Stip. with basal glands.

17. *S. nig'ricans* (Sm.)[1]; young shoots and l. downy, l. ovate-elliptic or lanceolate rather glaucous more or less glaucous beneath thin, point of stip. straight.—Shoots dull-coloured.—*α.* germen and stalk silky, st. erect.—a. *S. cotinifolia* (Sm.); l. roundish elliptic or subcordate. *E. B.* 1403.—b. *S. nigricans* (Sm. proper); l. elliptic-lanceolate. *E. B.* 1213.—c. *S. Forsteriana* (Sm.); l. elliptic-obovate acute, stip. vaulted, catkins long. *E. B.* 2344.—*β. S. rupestris* (Donn); awlshaped germ. and stalk silky or glabrous below, st. trailing, l. elliptic-obovate.

[1] Combined as a subsp. with sp. 18 by Dr. White.—H. & J. G.

E. B. 2342.—γ. germen glabrous its stalk downy or glabrous, st. erect.—d. *S. Andersoniana* (Sm.); l. elliptic-oblong acute, caps. awl-shaped not wrinkled. *E. B.* 2343.—e. *S. damascena* (Forbes); l. broadly elliptic or subrhomboidal acute, caps. not wrinkled. *E. B. S.* 2709.—f. *S. petræa* (And.); l. oblong or sublanceolate, caps. wrinkled near the top. *E. B. S.* 2725.—? *S. hirta* (Sm.); l. elliptic-cordate pointed, "caps. very pubescent" (*Leefe*) or "glabrous" (*Arnott*) not wrinkled. *E. B.* 1404.—*S. floribunda* (Forbes); l. long elliptical glaucous beneath thinly hairy on both sides. Fem. catkins unknown. *E. B.* 2816.—The varieties are scarcely distinguishable.—Sides of streams, osier-grounds, and mountains. Sh. IV.—VI. E. S. I.

** *Virentes.* L. smooth, scarcely any crisped pubescence beneath, not turning black unless gathered when very young.

18. *S. phylicifólia* (L.); shoots and leaves soon quite glabrous dark green above and more or less shining glaucous beneath. Shoots bright chestnut-colour. Stip. none or very small.—α. *germen and stalk silky.*—a. *S. radicans* (Sm.); l. oblong or elliptic-lanceolate, more or less decumbent. *E. B.* 1958.—b. *S. Davalliana* (Sm.); l. oblong-lanceolate pointed, style as long as stigma. *E. B. S.* 2701.—c. *S. Weigeliana* (Willd.); l. roundish or elliptic, style longer than cloven stigma. *E. B. S.* 2656.—d. *S. nitens* (And.); l. elliptic-lanceolate, style longer than undivided stigma. *E. B. S.* 2655.—e. *S. tenuior* (Borr.); l. obovate-lanceolate acuminate, style as long as stigma. *E. B. S.* 2650.—f. *S. tetrapla* (Sm.); lower part of germ. glabrous, style hairy, stigmas divided, l. lanceolate acuminate. *E. B. S.* 2702.—β. *germ. and stalk densely woolly or silky.*—g. *S. laurina* (Sm.)[1]; l. elliptic-oblong suberect, stigm. undivided. *E. B.* 1806.—γ; *lower part of germen and stalk glabrous.*—h. *S. laxiflora* (Borr.); l. broadly obovate-acuminate, style about as long as cloven stigma. *E. B. S.* 2749.—i. *S. propinqua* (Borr.); l. elliptic, style longer than the cloven stigma. *E. B. S.* 2729.—δ; *germen glabrous, stalk hairy.*—k. *S. Borreriana* (Sm.); l. lanceolate attenuate at both ends. *E. B. S.* 2619.—ε; *germen and stalk glabrous.*—l. *S. phillyreifolia* (Borr.); l. elliptic acute at both ends. *E. B. S.* 2660.—m. *S. tenuifolia* (Borr.); l. elliptic with a recurved point. *E. B. S.* 2795.—These plants are scarcely distinguishable by description alone. —By streams especially in mountain valleys. Sh. IV. V.
E. S. I.

[1] *S. phylicifolia* × *Capræa*, Dr. White. *S. phylicifolia* × *cinerea*, E. F. Linton.—H. & J. G.

iii. *Incubaceæ.* Style short. Stipules linear.—*Fuscæ* Bab. *Rosmarinifoliæ* and *Repentes* Borr.

19. *S. rosmarinifólia* (L.)[1]; l. linear-lanceolate silky beneath quite entire or remotely glandular-toothed, *stip. minute lanceolate, germens silky lanceolate*-acuminate *scales short* hairy, "style about as long as the linear divided or entire stigmas."—*E. B.* 1365. *S. Arbuscula* Sm., *E. B.* 1366, exclusive of large leaf., not of *Koch* nor *Fr.*—A slender upright shrub, 1—3 ft. high. Catkins very short, at first drooping. Whole plant becomes nearly black in drying. Style short.—"Several parts of the North." *Sm.* Sh. V. S.?

20. *S. Doniána* (Sm.)[2]; *l. lanceolate or obovate-lanceolate* acute slightly serrate livid with scattered silky hairs beneath, *stip. linear*, germens very silky ovate-oblong longer than the bearded oblong-ovate scales, *style very short, stigmas short emarginate.*—*E. B. S.* 2599.—Shrub about 6 feet high. "Stam. monadelphous, anth. ultimately luteo-fuscous not black." (*Koch.*)—Forfarshire. Sh. V. S.

21. *S. rèpens* (L.); l. elliptic or elliptic-lanceolate acute entire or minutely glandular-serrate glaucous and silky beneath, germ. subconical silky, style rather short, stigmas ovate bifid.—*α. S. repens* (Sm.); l. elliptic-lanceolate with a straight point, st. depressed with short upright branches. *E. B.* 183.—*β. S. fusca* (Sm.); l. oblong-oval straight, st. decumbent below then erect much branched. *E. B.* 1960.—*γ. S. prostrata* (Sm.); l. elliptic-oblong with a twisted point, st. prostrate with long straight branches. *E. B.* 1959.—*δ. S. ascendens* (Sm.); l. elliptic with a recurved point, st. decumbent with long somewhat ascending branches. *E. B.* 1962. *S. parvifolia* Sm., *E. B.* 1961.—*ε. S. incubacea* (L.); l. elliptic-oblong with a twisted point, stip. stalked ovate acute. *E. B. S.* 2600.—*ζ. S. argentea* (Sm.); l. broadly elliptic with a twisted point, stip. stalked oval. *E. B.* 1364.—Fries thinks that *S. fusca* (L.) is different.—Heaths at various elevations. Sh. III. IV. E. S. I.

22. *S. ambig'ua* (Ehrh.)[3]; l. oval obovate or lanceolate slightly toothed with a recurved point somewhat rugose above soft and silky beneath, *stip.* stalked *half-ovate acute*, germens lanceolate-subulate silky, style very short, stigmas short at length cloven. —*E. B. S.* 2733.—*a*; l. oval or obovate moderately hairy.—

---

[1] Considered a form of sp. 21 by Dr. White.—H. & J. G.
[2] *S. purpurea*×*repens*, Dr. White.—H. & J. G.
[3] *S. repens*×*aurita*, Dr. White.—H. & J. G.

β. major; l. obovate very silky on both sides.—*γ. spathulata*; l. obovate or ovate-lanceolate moderately hairy or silky, style somewhat long.—*δ. undulata*; l. ovate-lanceolate.—Gravelly heaths. Sh. V. E. S. I.

iv. *Daphnoideæ*. Style long. Capsule subsessile. Catkins subsessile, bracteate at the base.—*Arbusculæ* Bab.

23. *S. Arbus'cula* (L., Fries); *l.* lanceolate-ovate or ovate *glabrous* smooth glaucous and opaque beneath finely *serrate*, germens oblong-ovate silky, stigmas bifid.—*α. S. carinata* (Sm.); l. ovate finely toothed minutely veined folded into a keel, catkins cylindrical with rounded hairy scales. *E. B.* 1363.—*β. S. prunifolia* (Sm.); l. broadly ovate toothed smooth on both sides, st. erect much branched. *E. B.* 1361.—*γ. S. venulosa* (Sm.); l. ovate toothed naked netted with prominent veins above, st. erect much branched. *E. B.* 1362.—*δ. vacciniifolia* (Sm.); l. lanceolate-ovate serrate smooth and even above silky beneath, st. decumbent. *E. B.* 2341.—Highlands. Sh. IV.—VI. S.

24. *S. Lappónum* (L.); l. lanceolate or elliptic subacuminate *entire cottony* or silky beneath with crisped hairs wrinkled above and when young downy, germens ovate-lanceolate silky, stigmas linear.—*α. S. arenaria* (L.); l. ovate-lanceolate netted and somewhat downy above veined and woolly beneath, style as long as the sessile woolly germen, stigmas linear. *E. B.* 1809. Germen with a long slender reddish style.—*β. S. Stuartiana* (Sm.); l. ovate-lanceolate shaggy above densely silky almost cottony beneath, style as long as the almost sessile woolly germen, stigmas capillary deeply divided. *E. B.* 2586.—[*γ. S. glauca* (Sm.); l. ovate-lanceolate even and nearly smooth above woolly and snow-white beneath, germens sessile woolly, style at first very short with thick ovate stigmas. *E. B.* 1810. Germen blunter. Style elongating and the stigmas becoming linear and deeply cloven as the fruit ripens. This is probably not a native. Smith had it from Mr. Crowe's garden. It is hardly the same as *α* and *β*.—*S. glauca* L., Wahl., Koch, has subterminal catkins with very long leafy stalks and belongs to the next section.]—Helvellyn. Breadalbane, Clova, and other Scottish Mountains. Sh. VI. VII. E. S.

[*S. acutifólia* (Willd.); l. linear-lanceolate acuminate crenate-serrate glabrous glaucous beneath, stip. lanceolate acute sub-½-cordate below, "caps. ovate-conic glabrous sessile, stig. linear-oblong."—*Sy. E. B.* 1366.—Shoots purple with a glaucous bloom. Male catkins short, thick, very hairy. Fertile plant not yet found in England. *S. daphnoides β.* Anders.—North Yorkshire, not native. T. IV.] E.

v. *Chrysanthæ.* Style long. Stigma entire. Capsule sessile. Anth. yellow, scarcely changing colour. Catkins appearing before the leaves, sessile, terminal and lateral, with very shaggy and silky scales. Leaves broad, roundish.

25. *S. lanáta* (L.); *l. broadly oval* pointed entire *shaggy beneath*, stip. oval, catkins with yellow silky hairs, germ. conical glabrous.—*E. B. S.* 2624.—A low (2 ft.) and very beautiful shrub.—Clova and Glen Lochay Mountains. Sh. V. VI. S.

Sec. 3. CHAMELYX (Fries). Catkins on long leafy persistent shoots from the terminal or subterminal buds. Stam. 2. Nectary "of 2 pieces, one between the catkin-scale and germen, the other opposite to it."

i. *Myrsinites.* Catkins at the extremity of the terminal shoot, or of those from the last but one or two of the buds, but in such a manner as to appear to be an elongation of the branch. Small bushy plants.

26. *S. Myrsinites* (L.); l. elliptic or lanceolate serrate shining often hairy with prominent veins, germens subsessile ovate-subulate downy, style long.—St. much branching.—*α. S. arbutifolia* (Sm.); l. ovate or lanceolate rather acute. *S. Myrsinites β.* Sm.—*β. S. Myrsinites* (Sm.); l. elliptic serrate nearly smooth, catkins short, styles short, stigmas cloven. *E. B.* 1360.—Highlands. Sh. VI. S.

27. *S. procum'bens* (Forbes)[1]; l. oval minutely serrate bright green and shining on both sides, *catkins long cylindrical,* germens subsessile ovate lanceolate downy, style short deeply cloven, stigmas bifid.—*E. B. S.* 2753. Scales of the catkin nearly black, longer and more hairy than in *S. Myrsinites.* A low procumbent much-branched shrub.—Highlands. Sh. VI. S.

[Fries states that *S. retusa* (L.) was found by Mr. Winch in Scotland; but there is no such plant in his Herb.]

28. *S. Gráhami* (Baker)[2]; *l.* oval obscurely crenate shining glabrous *silky beneath* netted with veins, *germ. stalked* ovate prolonged *glabrous,* style long, stigma bifid, young branches silky.—*J. of B.* v. t. 66. *Sy. E. B.* 1377.—St. long, trailing.

---

[1] Considered a form of *S. Myrsinites* by Dr. White.—H. & J. G.

[2] Dr. White refers the Scottish plant to *S herbacea*×*phylicifolia* and the Irish plant to *S. herbacea*×*phylicifolia*, subsp. *nigricans.* Mr. E. F. Linton refers the former to *S. herbacea*×*Myrsinites.*—H. & J. G.

Allied to *S. Myrsinites* and *S. retusa* (can it be the plant referred to by Fries?).—Frouvyn, Sutherl. *Graham.* Muckish, Doneg. *D. Moore, J. of B.* ix. 300. Sh. VI.? S. I.

ii. *Reticulatæ.* Catkins opposite to the terminal leaves, with a bud between them.

29. *S. reticuláta* (L.); *l.* nearly roundly oval very blunt *entire* netted with veins and glaucous beneath, *germens* sessile oblong-ovate *downy*, style short, stigmas bifid.—*E. B.* 1908.—A procumbent much branched shrub. St. usually buried. Catkins on long stalks.—Lofty mountains. Sh. VI. S.

30. *S. Sad'leri* (Syme)[1]; l. roundly ovate or subcordate entire smooth and cottony netted and glabrous beneath, *germens glabrous with long woolly stalks,* style long, stigmas linear bifid.—*Edin. Bot. Trans.* xii. 208. t. 1. *J. of B.* xiii. t. 158.—A small almost subterranean prostrate shrub.—I have not seen this plant, and it may be misplaced here.—East of Loch Ceann-Mor, very rare. Sh. VIII.? S.

iii. *Herbaceæ.* Catkins exactly terminal.

31. *S. herbácea* (L.); *l.* round blunt or retuse *serrate* shining glabrous netted with veins, *germens* subsessile ovate-conical *glabrous,* style short, stigmas bifid.—*E. B.* 1907.—A very minute herblike shrub; the st. extend far amongst loose stones on the tops of mountains. Edges and veins of l. hairy.—Alpine situations. Sh. VI. E. S. I.

Index to the Salices.

[1] Dr. White considers this a var. of *S. lanata.* Mr. Linton refers it to *S. herbacea*×*lanata.*—H. & J. G.

2. POP'ULUS *Linn.* Poplar.

1. *P. al'ba* (L.); l.-buds downy, l. roundish-cordate angularly toothed *cottony and snowy white beneath*, l. of the young shoots cordate palmately 5-lobed, *stig.* 2 *linear bifid crosslike* (yellow). —*E. B.* 1618.—With suckers. L. generally lobed. Male catkin-scales hairy.—Damp woods. T. III. IV. *White Poplar. Abele.* E. S.

2. *P. canes'cens* (Sm.); l.-buds downy not viscous, l. roundish obscurely lobed sparingly *cottony and grey beneath*, l. of young shoots cordate-ovate undivided, *stig.* 2 *wedgeshaped* 3—4-*lobed* (purple).—*E. B.* 1619.—With suckers. L. not lobed, except occasionally the youngest. Male catkin-scales hairy.—Damp woods. T. III. IV. *Gray Polar.* E.

3. *P. trem'ula* (L.); l.-buds glabrous slightly viscous, l. nearly round acute serrate *glabrous on both sides*, young l. slightly downy, *stig.* 2 bifid *erect.*—*E. B.* 1909.—A small tree, with suckers. Petioles laterally compressed. Male catkin-scales hairy, palmately cut.—Woods. T. IV. *Aspen.* E. S. I.

4. *P. nigra* (L.); l.-buds glabrous viscous, l. triangular-rhomboidal acuminate serrate *glabrous on both sides*, catkins lax cylindrical, stig. 2 roundish 2-lobed.—*E. B.* 1910.—A large tree, without suckers. Petioles laterally compressed. Young shoots glabrous. Male catkin-scales glabrous.—Damp places and riverbanks. T. III. *Black Poplar.* E.

Tribe II. *Myriceæ.*

3. Myri'ca *Linn.* Sweet Gale.

1. *M. Gále* (L.); l. lanceolate broader upwards serrate, st. shrubby.—*E. B.* 562.—Height 3—4 feet. Bushy. Catkins sessile erect. Fr. with resinous glands. L. fragrant when bruised.—Bogs. Sh. V. *Bog Myrtle.* E. S. I.

Tribe III. *Betulineæ.*

4. Bet'ula *Linn.* Birch.

1. *B. verrucósa* (Ehrh.); l. rhomboid-triangular doubly serrate abruptly acuminate, scales of the fem. catkins 3-lobed, *lateral lobes falcate-reflexed*, fr. obovate.—*Fl. Dan.* 2549. *Sy. E. B.* 1295. *B. alba* Koch. *B. odorata* R. xii. 626.—L. usually glabrous often covered with resinous spots above, always having a manifest tendency to a rhomboidal form, flat beneath with raised veins above. Young shoots mostly with resinous tubercles, often very long and pendulous. Stip. ovate-lanceolate, acute, thrice as long as broad, circinate; sides not deflexed. Buds conical. The catkin-scales and the shape of the l. distinguish this from the next.—Rather common. T. IV. V. *White Birch.* E. S. I.

2. *B. al'ba* (L.); l.-rhomboid-ovate or cordate unequally serrate acute, scales of the fem. catkins 3-lobed, *lateral lobes ascending*, fr. broadly obovate.—*Sy. E. B.* 1296. *R.* xii. 623. *B. glutinosa* (Fr.) ed. viii. *B. pubescens* Koch.—L. usually glabrous, always more or less ovate, flat above with raised veins beneath. Stip. ovate, blunt, twice as long as broad; sides deflexed. Buds ovoid. Not so elegant a tree as the preceding and often little more than a bush. Twigs sometimes pendulous.—*β. B. pubescens* (Erhr.); l. peduncles and young twigs downy.—[Var. *parvifolia* (Regel) is a northern form with smaller often cuneate-based l. and erect catkins.]—Common. T. IV. V. *Common Birch.* E. S. I.

3. *B. intermedia* (Thom.); l. broadly-rhomboidal acute dentate, catkins erect with 3-lobed scales, the lateral lobes rounded porrect broader than interm. lobe, fr. with margin equalling its breadth.—*R.* xii. 624.—A small tree, 6—15 ft. high. L. small. Catkins stalked.—Glen Callater. S.

4. *B. nána* (L.); *l. roundish crenate* glabrous blunt, scales of the fem. catkin digitate-trifid, lobes equal, fr. orbicular with a

very narrow membranous margin.—*E. B.* 2326.—A small procumbent shrub. L. minute. Catkins subsessile, small.—Turfy places in the Highlands. Sh. V. *Dwarf Birch.* S.

5. AL'NUS *Mill.* Alder.

1. *A. glutinósa* (Gaert.); l. roundish blunt wavy serrate glutinous rather abrupt with a wedgeshaped base, axils of the veins beneath downy.—*E. B.* 1508. *R.* xii. 641. *St.* 29. 15. —A moderately large tree. Trunk and branches crooked. Male catkins long and pendent; fem ones short, ovoid or oblong, very persistent.—*β. incisa*; leaves deeply cut.—Wet places and river-banks. *β.* Wigtonshire. *Dr. Balfour.* T. III. E. S. I.

Tribe IV. *Cupuliferæ.*

6. FA'GUS *Linn.* Beech.

1. *F. sylvat'ica* (L.); l. ovate glabrous obscurely dentate ciliate on the edges.—*E. B.* 1846. *R.* xii. 639.—A large tree with triquetrous fruit.—Woods, particularly on calcareous soils. T. III. IV. E.

7. CASTA'NEA *Mill.* Chestnut.

[*C. sativa* (Mill.); l. oblong-lanceolate acuminate mucronate-serrate glabrous on each side.—*Fagus* Sm. *E. B.* 886. *R.* xii. 640. *C. vulgaris* (Lam.) ed. viii.—Height 50—80 feet. A magnificent tree.—A very doubtful native. T. V. *Sweet Chestnut.*] E.

8. QUER'CUS *Linn.* Oak.

1. *Q. Róbur* (L.); l. deciduous stalked obovate-oblong sinuate, lobes blunt, inv. much shorter than the ripe acorn its scales adpressed.—*α. Q. pedunculata* (Ehrh.); young branches glabrous, petioles short, fr.-catkins long-stalked, fr. scattered. *E. B.* 1342. —*β. Q. intermedia* (D. Don); young branches glabrous, petioles short, l. stellate-downy beneath, fr.-catkins shortly stalked, fr. near together. *Mart. Rust.* 11.—*γ. Q. sessiliflora* (Salisb.); young branches downy, petioles long, l. glabrous beneath, fr.-catkins subsessile, fr. near together. *E. B.* 1845.—It is generally supposed by foresters that there are two if not three species of Oak in Britain. I have failed in learning how to distinguish them.—Woods. T. IV. V. E. S. I.

9. Cor'ylus *Linn.* Hazel.

1. *C. Avellána* (L.); stip. oblong blunt, l. roundish-cordate acuminate, involucre of the ovoid fr. bellshaped spreading torn at the margin.—*E. B.* 723.—A shrubby tree. Young twigs hairy and glandular. L. downy beneath. Male catkins long, pendulous. Fem. fl. in ovoid buds. Stigmas bright crimson. — Hedges and copses. Sh. III. IV. *Hazel Nut.* E. S. I.

10. Carpi'nus *Linn.* Hornbeam.

1. *C. Bet'ulus* (L.); scales of the fruit 3-parted, segments lanceolate, the middle one longest.—*E. B.* 2032.—A small tree. L. ovate, acute, plaited when young, deeply and sharply doubly serrate.—Damp clayey woods and hedges. T. V. E.

Division V. GYMNOSPERMÆ.

Ovules and seeds apparently naked. Carpel spread out flat. No calyx, no corolla.

Order LXXIX. CONIFERÆ.

Fl. monœcious or diœcious.—Barren fl. of one or more monadelphous stamens, in a deciduous catkin, about a common axis. Anth. of 2 or more lobes bursting outwards, often ending in a scalelike crest.—Fertile fl. usually in cones, sometimes solitary. Ovary spread open in the shape of a scale and placed in the axil of a membranous bract; in the solitary fl. apparently wanting. Ovules naked in pairs on the face of the ovary and inverted; or (in the solitary fl.) erect. [Or, as some think: ovaries in pairs (or several), inverted, on a scale (which becomes the cone-scale) situated in the axil of a bract; each of 2 connate carpels having together the form of a horseshoe or ring and ultimately producing a wing; ovule between the carpels, its tips exposed; or erect, the 2 carpels uniting in the form of a cup, without a carpellary scale.] Fr. a cone, or solitary seed. Testa hard, crustaceous. Embryo in the axis of fleshy albumen. Radicle next the apex.—Woody tissue marked with circular disks; without medullary rays or proper vascular tissue.

Tribe I. *TAXINEÆ.* Male fl. in catkins. Fem. fl. solitary, naked or bracteate, erect; no carpellary scale.

1. Taxus. Catkins of male fl. oval, scaly below, flowering at the top. Stam. many. Anth. peltate, 3—8-celled. Fem.

fl. scaly below. Style 0. Ovule surrounded at the base by a ring which becomes a fleshy cupshaped disk surrounding the seed.

Tr. II. *CUPRESSINEÆ.* Male fl. in catkins. Anth. 4—7, inserted on the edge of the subpeltate scales. Fem. fl. few, in a small catkin, erect; no carpellary scale.

2. JUNIPERUS. Anth. 4—7, 1-celled, inserted on the lower edge of the scales. Scales of the fem. catkin imbricate, lower ones barren. Ovules 3, surrounded by a 3-fid fleshy involucre formed of the 3 uppermost connate scales of the catkin.

Tr. III. *ABIETINEÆ.* Fl. in catkins. Anth. 2, 1-celled, adnate to the underside of the scales. Fem. fl. a scalelike open ovary in the axil of a membranous scale, bearing two naked ovules pointing towards the axis [or, perhaps, fem. fl. in pairs on an axillary scale.] Fr. winged.

3. PINUS. Male catkins crowded, racemose. Scales of the cone thickened and angular at the end. Fr. with a crustaceous coat.

Tribe I. *Taxineæ.*

1. TAX'US *Linn.* Yew.

1. *T. baccáta* (L.); l. 2-ranked crowded linear acute, fl. axillary sessile.—*E. B.* 746.—A low tree, trunk often attaining a very considerable bulk. Fr. roundish.—*T. fastigiata* (Lindl.) is not even a permanent variety.—Mountainous woods and limestone cliffs. T. III. IV. E. S. I.

Tribe II. *Cupressineæ.*

2. JUNIP'ERUS *Linn.* Juniper.

1. *J. commúnis* (L.); l. 3 in each whorl spreading linear subulate mucronate keeled exceeding the ripe fruit.—*E. B.* 1110.—Fruticose, erect. L. with a broad flat shallow channel above, the keel beneath with a slender furrow. Berries black, tinged with blue, about half the length of the leaves[1]—Dry hills, especially on a calcareous soil. Sh. V. E. S. I.

---

[1] A form intermediate between Sp. 1 and 2 is recorded by Mr. Bennett from Scarp, Outer Hebrides, under the name var. *J. intermedia*, Schur.—H. & J. G.

2. *J. nána* (Willd.); l. 3 in each whorl incurved linear lanceolate mucronate keeled equalling the ripe fruit.—*E. B. S.* 2743.—A prostrate shrub with longer berries and shorter leaves than the last.—Mountains. Sh. V. E. S. I.

Tribe III. *Abietineæ.*

3. PI'NUS *Linn.* Scotch Fir.

1. *P. sylves'tris* (L.); l. in pairs, young cones stalked recurved ovoid-conical, wing thrice as long as the seed.—*E. B.* 2460.—A lofty tree. Cones of this species have been found at considerable depths in the Irish bogs and English fens.—Highlands. T. V. VI. E. S. I.

[*P. Pinaster* (Ait.) *Sy. E. B.* 1381, with stouter branches, longer l., much longer clustered cones, is naturalized about Bournemouth and Poole.]

Class II. MONOCOTYLEDONES.

Stems destitute of medullary rays, consisting of cellular tissue amongst which the vascular tissue is mixed in bundles, increasing by the addition of new matter within. Leaves mostly alternate and sheathing, with parallel simple veins connected by smaller transverse ones, rarely net-veined. Cotyledon one, or if more they are alternate.

Division I. DICTYOGENÆ.

Leaves net-veined. Floral envelopes whorled.

Order LXXX. TRILLIACEÆ.

Perianth inferior, 6- or 8-parted; in 2 whorls; outer whorl or calyx herbaceous; inner or corolla coloured, or in our plant herbaceous. Stam. 6—10. Anth. linear; filaments subulate. Ovary superior, 3—5-celled. Ovules indefinite in number, in two rows in each cell. Styles as many as the cells, distinct; stigmas inconspicuous. Fr. succulent, not bursting. Seeds with a leathery skin. Embryo minute, in fleshy albumen.—Raphidiferous. Often combined with *Liliaceæ*.

1. PARIS. Perianth subherbaceous, 8-parted; 4 inner divisions narrower than the others.—Stam. 8—10. Anth on the middle of the subulate filament.

1. PAR'IS *Linn.* Herb Paris.

1. *P. quadrifólia* (L.); l. usually 4 in a whorl.—*E. B.* 7.—St. 1 ft. high, from the end of a long rhizome, usually with 4, rarely from 3—6, ovate l. at its summit. Fl. solitary, terminal. Sep. lanceolate. Pet. subulate. Styles 4. Berry 4-celled; cells with 4—8 seeds. No root-leaves.—Damp woods. P. V. E. S.

Order LXXXI. DIOSCOREACEÆ.

Perianth superior, petal-like, 6-parted. Stam. 6, inserted on the base of the segments of the perianth. Anth. bursting inwards. Ovary inferior, 3-celled. Ovules 2 in each cell, erect. Style 1. Stigmas 3, reflexed. Fr. baccate, or dry and flat. Embryo minute, quite enclosed in the albumen.

1. TAMUS. Perianth bellshaped; limb 5-parted. Male with 6 stamens. Fem. with the perianth adhering to the ovary and persistent, and very short abortive stamens.

1. TA MUS *Linn.* Black Bryony.

1. *T. commúnis* (L.); l. undivided cordate acute shining.—*E. B.* 91.—Diœcious. Root large, thick, fleshy. St. very long, twining. Racemes axillary, on long stalks. Fl. yellowish-green, regular, small. Berry red.—Hedges and thickets. P. V. VI. E.

Division II. FLORIDÆ.

(Orders LXXXII.—XCVI.)

Leaves parallel-veined, persistent. Floral envelopes whorled, or none.—There is no proper perianth distinguishable in Ord. XCII.—XCVI. in our plants.

Order LXXXII. HYDROCHARIDACEÆ.

Sep. 3, herbaceous. Pet. 3, regular, coloured. Stam. epigynous. Ovary solitary, inferior; placentes parietal sometimes projecting into the centre of the ovary. Stig. 3—6. Fr. dry, or succulent, not bursting, 1- or spuriously many-celled. Seeds many. Albumen 0. Embryo straight, cylindrical.—No raphides.

1. HYDROCHARIS. Diœcious. Cal. 3-parted. Pet. 3. Male with 9—12 stamens in 3 rows surrounding 3 abortive styles. —Fem. with 3 abortive filaments and 3 fleshy scales surrounding the 6 deeply bifid styles. Caps. 6-celled, many-seeded.—L. floating, orbicular.

2. STRATIOTES. Diœcious. Cal. 3-parted. Pet. 3. Male with 12 or more stamens surrounded by many abortive ones. Fem. with 6 deeply bifid styles. Berry 6-celled, many-seeded.—L. swordshaped.

3. ANACHARIS. Diœcious. Fl. from tubular bifid spath. Cal. 3-parted. Pet. 3. "Male. Sep. ovate-oblong. Pet. linear or none. Stam. 9; filaments combined into a column below."—Fem. with a long filiform tube, 3 abortive filaments and ligulate stigmas. Caps. 1-celled, few-seeded.—L. oval-oblong, in whorls.

1. HYDROCH'ARIS *Linn.* Frog-bit.

1. *H. Morsus-ránæ* (L.).—*E. B.* 808. *R.* vii. 62.—Floating. With runners bearing tufts of stalked roundish-reniform entire leaves. Fl. white, delicate, springing from a pellucid membranous sheath. Seeds covered with beautiful prominent spirally twisted cells.—Ponds and ditches. P. VII. VIII. E. I.

2. STRATIO'TES *Linn.* Water-Soldier.

1. *S. Aloïdes* (L.); l. swordshaped-triangular ciliate-spinous. —*E. B.* 379. *R.* vii. 61.—Creeping in the mud. L. many, rigid. Stalk compressed, 5 or 6 in. high, with two l. near its summit. Fl. white, delicate. The plant rises to the surface of the water to flower, and sinks again afterwards.—Fen-ditches in the East of England; naturalized in Scotland and Ireland. P. VII. E. [S. I.]

3. ANACH'ARIS *Rich.* Water-Thyme.

*1. *A. Alsinas'trum* (Bab.); l. 3 in a whorl oval-oblong blunt serrulate, fem. fl. from a tubular bifid spath many times exceeding the sessile germen, sep. and pet. broad nearly equal, stigmas reflexed.—*A. N. H.* ser. 2. i. t. 8. *E. B. S.* 2993. *Elodea canadensis* Rich. ?—St. long, branching; whorls of leaves many, close together. Fl. very small but with a very long slender tube often 2 or 3 in. long. The spath in the axil of a leaflike bract placed within the whorl of leaves. Sep. tinged with green and pink externally, incurved, hooded. Pet. flat, transparent, recurved, oblong. Filaments at first curved outwards, their points placed under the hood of the sepals, afterwards erect, linear, blunt. Anth. 0. Stigmas recurved, linear, or deeply bifid. Sep. pet. and stigmas of about equal length. Style adnate on 3 sides of the tube.—In water. Probably introduced from America. Canals and rivers. P. VII.—IX. E. S. I.

Order LXXXIII. ORCHIDACEÆ.

Sep. 3, usually coloured. Pet. 3, 2 above, 1 below (lip) frequently lobed and spurred and unlike the others. Stam. 3, united in a central column, 2 lateral abortive, or (in *Cypripedium*) the middle one abortive. Pollen powdery or cohering in masses. Ovary 1-celled, inferior, with 3 parietal placentas. Style forming part of the column with the stamens; stigma a viscid space in front of the column. Caps. 3-valved. Seeds very many, minute. Testa loose, netted.—Raphidiferous.

Tribe I. *OPHRYDINEÆ.* Anther wholly adnate to the column. Pollen-masses in divisible lobes which are indefinite in number and waxy, stalked.—Root with 2 fleshy entire or palmate *knobs below the fibres.*

* *Anther-cells with a prolongation of the stigma between their bases.*

1. ORCHIS. Perianth ringent, hooded. Lip 3-lobed, spurred. Glands of the stalks of the pollen-masses in a common pouch.

2. GYMNADENIA. Glands of the pollen-masses without a pouch. Otherwise like *Orchis.*

3. ACERAS. Lip without a spur. Otherwise like *Orchis.*

** *A plate in front of the glands of the pollen-masses.*

4. NEOTINEA. Perianth ringent, hooded. Lip 3-lobed, spurred. Glands of the pollen-masses in separate hollows behind the broad recurved rostellum, which is bounded laterally by 2 semicylindrical ascending processes diverging from its base.

*** *No prolongation of the stigma between the anther-cells, nor plate in front of them.*

5. HABENARIA. Perianth ringent, hooded. Lip 3-lobed or entire, spurred. Glands of pollen-masses naked.

6. OPHRYS. Perianth patent. Lip variously lobed, without a spur. Glands of pollen-masses each in a distinct pouch.

7. HERMINIUM. Perianth bellshaped, segments all erect. Lip 3-lobed, tumid beneath at the base, without a spur. Glands of the stalks of the pollen-masses exserted, naked.

Tr. II. *NEOTTIDEÆ.* Anther attached by its base, persistent. Pollen-masses granular; granules only in a slight state of cohesion; no stalks.—Root of many fleshy fibres, rarely of 2—4 much-thickened ones.

* *Stigma with a prolongation (rostellum) at its top.*

8. PERAMIUM. Perianth ringent. Lip gibbous at the base, entire, included. Stigma subcordate. Rostellum erect, bipartite, with a large squarish appendage between its slender segments.

9. SPIRANTHES. Perianth ringent. Lip channelled, clawed, fringed. Stigma roundish. Rostellum straight, at length bifid, with a long linear appendage between its points.

10. LISTERA. Perianth ringent. Lip deflexed, 2-lobed. Stigma transverse. Rostellum foliaceous, acute, arching over stigma. Column very short.

11. NEOTTIA. Perianth hooded. Lip deflexed, 2-lobed, gibbous at the base. Stigma transverse. Rostellum flat, broad, prominent, entire, without an appendage. Column long.

12. EPIPACTIS. Perianth patent. Lip interrupted; the basal division concave; terminal one (label) larger with 2 projecting plates at its base above. Stigma nearly square. Rostellum short, terminated by a globose appendage. Anth. terminal, erect, sessile, 2-celled. Column short. Germen straight on a twisted stalk.

** *Stigma without a rostellum.*

13. CEPHALANTHERA. Perianth converging (in *C. rubra* spreading). Lip interrupted, the basal division gibbous,

jointed to the recurved label. Stigma transverse. Anth. terminal, erect, movable, shortly and thickly stalked, 2-celled. Column long. Germen sessile, twisted.

Tr. III. *ARETHUSEÆ.* Anther ultimately free, like a lid, deciduous. Pollen in many granules, pulpy or powdery, cohering in 2 stalked masses.—Root (in our plant) of fleshy much-branched fibres.

14. EPIPOGUM. Perianth patent. Lip posterior, erect, large, entire, with a small patent lobe on each side of its base and an erect inflated spur. Anth. tumid, seated in the lobed top of the column. Stigma transverse Germen and stalk not twisted.

Tr. IV. *MALAXIDEÆ.* Anther terminal, free, usually like a lid. Pollen cohering in a definite number of granules and at length waxy and confluent; no stalks.—Root fibrous, with or without a superior bulb.

15. CORALLORRHIZA. Perianth converging. Lip with 2 prominent longitudinal ridges at the base, 3-lobed; lateral lobes small; middle lobe large, slightly emarginate. Spur short or wanting. Stigma triangular. Rostellum wanting; but a large globose appendage. Anth. terminal, 2-celled, opening transversely. Column long. Germen slightly stalked, straight.

16. MALAXIS. Perianth patent. Lip posterior, erect, entire, similar to the pet., smaller than the sepals. Spur 0. Stigma rhomboidal. Rostellum short, entire, acute. Anth. terminal, continuous with the short column, out of the apex of which it appears as if it were excavated, with 2 imperfect cells. Pollen-masses connected at their apex. Germen upon a twisted stalk.

17. LIPARIS. Perianth patent. Lip anterior, erect or oblique, entire, dilated, much larger than the sepals. Spur 0. Stigma roundish. Rostellum wanting; but an appendage consisting of 2 tubercles. Anth. terminal, deciduous, movable like a lid, with 2 distinct cells. Column long. Germen on a twisted stalk.

Tr. V. *CYPRIPEDIEÆ.* Intermediate anther barren petal-like, 2 lateral anthers perfect.

18. CYPRIPEDIUM. Perianth patent. Lip inflated. Column trifid above; the lateral lobes bearing stamens; middle lobe sterile, dilated. Two lower (lateral) sepals combined Germen straight.

Tribe I. *Ophrydineæ.*

1. ORCHIS *Linn.* Orchis.

* *Glands of the pollen-masses separate, lip erect in the bud.*

† Bracts mostly 1-veined, root-knobs undivided.

‡ *Lip 3-lobed, lobes broad and short.*

1. *O. mório* (L.); lip crenulate, middle lobe truncate-emarginate, spur ascending subclavate rather shorter than the germen, *sep. and pet. blunt converging* to form a hood, anth. obovate rather acute.—*E. B.* 2059. *R.* xiii. 363.—St. 6—12 in. high. L. lanceolate, lower spreading, upper adpressed. Fl. few, in a lax spike, purple; sep. (rarely patent) and pet. marked with green veins, converging so as to form a sort of hood; lip pale in the middle, spotted with purple. Fl. sometimes white.—Meadows and pastures. P. V. VI. *Green-winged Orchis.* E. I.

2. *O. mas'cula* (L.); lip crenate, middle lobe emarginate, spur ascending rather longer than the germen, *sep. acute 2 outer reflexed upwards,* pet. converging, anth. obcordate apiculate.—*E. B.* 631. *E. B. S.* 2995. *R.* xiii. 390, 391.—St. a foot high. L. mostly radical, elliptic-lanceolate, usually spotted with purple. Lowest bract often 3-veined. Fl. in a lax spike, purple; centre of the lip whitish at the base, spotted and downy. Sep. and pet. without green veins. Rarely a form may be found without any spur.—Woods and pastures. P. V. *Early purple Orchis.* E. S. I.

‡‡ *Lip pinnately 4-lobed (that is, middle lobe bifid and often with an intermediate tooth).*

3. *O. purpúrea* (Huds.); *basal lobes of cor.-lip linear-oblong, interm. lobe widening gradually upwards* and 2-lobed with an interm. tooth, its lobes denticulate at the tip and broad, spur about ½ as long as the germen, sep. converging into an *ovate hood* including the pet., bracts minute.—*R.* xiii. 378. *O. fusca* Jacq. *O. militaris* Sm., *E. B.* 16.—St. 1—2 ft. high. L. ovate-oblong, blunt. Fl. in a rather dense spike; *helmet dark purple,* variegated; lip paler with raised rough red points, its term. lobe very broad.—Chalky bushy hills in Kent. P. V. E.

4. *O. militáris* (L.); *basal lobes of cor.-lip linear, interm. lobe* broader linear *suddenly widened* and 2-lobed with an interm. tooth at its lobes mostly entire at the tip and broad, spur about ½

as long as the germen, sep. converging into an *ovate-lanceolate hood* including the pet., bracts minute.—*E. B. S.* 2675. *R.* xiii. 376.—Smaller than the preceding. *Helmet ash-coloured or pale purple.* Lip purple, white in the middle with raised rough red spots, with a linear space between the pairs of lobes, and its term. lobes broad.—Chalky hills. Berks., Oxf., Bucks, Herts. P. V. E.

5. *O. Sim'ia* (Lam.); *lobes of lip all long narrow 1-veined* linear with an interm. setaceous tooth, spur ½ as long as the germen, sep. converging into an ovate-lanceolate hood including the pet., bracts minute.—*O. tephrosanthos* E. B. 1873. *R.* xiii. 373.—More slender than the preceding. *Helmet dark purplish.* Lip with remarkably slender segments, dark purple, with or without small rough raised points; a linear space between the pairs of lobes which are about equal in size.—Chalky hills in Berks., Oxf., and Kent. P. V. E.

6. *O. ustuláta* (L.); lobes of lip linear-oblong, *spur ½ the length of the germen,* sep. converging into a roundish hood including the pet., *bracts long.*—*E. B.* 18. *R.* xiii. 368.—St. 4—6 in. high. L. lanceolate, acute. Spike oblong, dense, dark at top, nearly white below. Fl. small, many. Helmet dark purple. Pet. linear-lanceolate, blunt. Lip white with purple spots.—Calcareous hills. P. VI. E.

†† Bracts with 3 or more veins, root-knobs undivided.

[*O. laxiflóra* (Lam.); lip 3-lobed, lateral lobes rounded and crenulate in front longer than the truncate slightly emarginate interm. lobe, spur shorter than the germen cylindrical, 2 lateral sep. reflexed upwards, pet. converging, anth. obcordate apiculate. —*E. B. S.* 2828. *R.* xiii. 393.—St. 1—2 feet high, round, angular and rough upwards. L. lanceolate or linear-lanceolate. Bracts 3—5-veined. Spike long, lax. Fl. bright purple. Some specimens approach *O. palustris.*—Wet meadows and bogs in Jersey and Guernsey. P. V. VI.]

††† Bracts with 3 or more veins, root-knobs palmate.

7. *O. maculáta* (L.); lip 3-lobed flat crenate, spur subulate shorter than the germen, *three sep. patent,* pet. converging, *st. solid,* l. lanceolate-obovate.—*E. B.* 632. *R.* xiii. 407.—St. about a foot high. L. usually spotted with purple; lower blunt or rarely acute, broadest towards their top; upper linear-lanceolate, resembling the bracts. Spike ovate, afterwards elongate; lower bracts exceeding the germen, upper equalling it. Fl. pale purple,

more or less streaked with purple. Lateral veins of bracts inconspicuous. Lip usually flat, deeply 3-lobed, lateral lobes rounded, middle lobe longer and narrower.—[β. *O. ericetorum* (Linton); slender, l. all acuminate narrower carinate, spike shorter, bracts purplish, middle of lower lip short. See *J. of B.* xxxviii. (1900) p. 362. A hybrid between Sp. 7 and 8 is recorded. Towns. Fl. Hants. p. 341.]—Damp woods and pastures. P. V. VI. *Spotted Hand-Orchis.* E. S. I.

8. *O. latifólia* (L.); *lip obscurely 3-lobed its sides ultimately reflexed,* crenate, spur subulate shorter than the germen, *two lateral sep. patent,* middle sep. and pet. converging, *st. hollow, l. lanceolate acute.*—*E. B. S.* 2973. *R.* xiii. 402.—About a foot high. L. often only faintly spotted, spreading, lowest oblong and blunt, from a narrow base but broadest about their middle. Spots on l. ring-shaped. Lower bracts exceeding the flowers. Lip usually only slightly lobed.—Marshes and damp meadows. P. V.—VII. *Marsh-Orchis.* E. S. I.

9. *O. incarnáta* (L.); lip obscurely 3-lobed its sides reflexed crenate, spur subulate shorter than the germen, two lateral sep. patent, middle sep. and pet. converging, st. hollow, *l. narrowed from a broad base* hooded at the end.—*R.* xiii. 397. O. latifolia *E. B.* 2308. *Curt.* ii. 184.—L. erect and approaching the stem, not spotted. Bracts all usually exceeding the flowers.—β. *O. angustifolia* (R.); l. erect-patent linear-lanceolate, upper l. erect, lower bracts equalling the flowers upper ones shorter. *R.* xiii. 394. *O. Traunsteineri* Koch.—Marshes. P. VI. VII. E. S. I.

** *Glands of the pollen-masses united, root-knobs undivided.*

† Spur filiform. ANACAMPTIS Rich.

10. *O. pyramidális* (L.); *lip* with 3 equal lobes and 2 *tubercles at the base above,* lobes oblong truncate, middle lobe sometimes emarginate, spur filiform longer than the germen, lateral sep. ovate-lanceolate acute spreading, bracts 3-veined.—*E. B.* 110.—St. 12—18 in. high. L. linear-lanceolate, acute. Spike pyramidal, afterwards cylindrical. Fl. rose-purple, sometimes white.—Calcareous pastures, rarely on sandhills. P. VII. *Pyramidal Orchis.* E. S. I.

†† Spur conical. HIMANTOGLOSSUM Spr.

11. *O. hircína* (Crantz); lip 3-parted spiral in the bud, lobes linear, middle one very long twisted, lateral much shorter wavy,

spur very short.—*E. B.* 34. *R.* xiii. 359, 360.—St. 2—3 feet high. "Cal. green, spotted with dull purple internally." Lip purplish white and spotted at the base, an inch or more long.—Bushy chalk hills, very rare. Kent. Surrey. Gt. Glenham, Suff.! P. V. *Lizard Orchis.* E.

2. Gymnade'nia *R. Br.*

1. *G. conops'ea* (R. Br.); lip 3-lobed, *lobes equal* entire, blunt, lateral sep. spreading, spur filiform twice as long as the germen, root-knobs palmate.—*E. B.* 10.—St. a foot high. L. linear-lanceolate. Spike cylindrical, elongated. Bracts 3-veined. Fl. rose-purple, fragrant. Pollen-cells open in front and below, stopped below by oblong glutinous valves quite distinct from the stigma, and to the broader ends of which the glands of the pollen-masses are attached. [A hybrid with Sp. 2 is recorded. J. of B. 1898, p. 352.]—Hilly pastures. P. VI. VII. E. S. I.

2. *G. al'bida* (Rich.); lip 3-lobed, *lobes unequal* entire, middle lobe longest and broadest, sep. and lateral pet. converging, spur much shorter than the germen, root-knobs clustered.—*Habenaria* R. Br. *E. B.* 505. *Peristylus* Lindl.—St. 6—12 in. high. L. oblong, blunt; upper lanceolate, acute. Spike elongated, cylindrical, dense. Bracts 3-veined. Fl. small, yellowish white, fragrant.—Mountain-pastures. P. VI. VII. E. S. I.

3. A'ceras *R. Br.*

1. *A. anthropoph'ora* (R. Br.); lip 3-parted, segments linear-filiform, middle one bifid and often with an intermediate tooth.—*E. B.* 29. *R.* xiii. 357.—Root-knobs ovate. Height 8—12 in. Spike long, lax. Fl. greenish yellow. Sep. ovate, acute, converging, margined with purple, including the linear-lanceolate blunt petals.—Dry chalky places. P. VI. *Green-man Orchis.* E.

4. Neotin'ea *Reichenb.*

1. *N. intac'ta* (Reich.); lip 3-lobed, lobes unequal, lateral lobes linear acute falling much short of the broad oblong interm. lobe.—*Sy. E. B.* 1465. *J. of B.* iii. t. 25. *Habenaria* Benth.—Height 2—12 inches. Spike dense, cylindrical. Fl. very delicate, pink. Semilancet-shaped acute sepals converging over the column and narrow petals and exceeding them. Spur very short. The two ascending fleshy lobes of stig. and the flat broad plate between them are characteristic of the plant.—Open limestone pastures. Clare, Galway, Mayo. P. IV. V. I

5. HABENA'RIA *Willd.* Frog Orchis.

* *Spur very short, usually inflated.* CŒLOGLOSSUM Hartm.

1. *H. vir'idis* (R. Br.); spur 2-lobed, *lip. linear flat* 3-pointed middle point the shortest.—*Orchis* Sm., *E. B.* 94. *R.* xiii. 434. *Peristylus* Lindl.—Lip with 3 tubercles at its base, 1 central, 2 lateral. Stigma oblong, slightly emarginate above. Glands of the pollen-masses connected by an elevated transverse line. Sep. and pet. connivent. Root-knobs palmate. Fl. green, lip browner. St. 6—8 in. high.—Pastures. P. VI. VII. *Frog Orchis.* E. S. I.

** *Spur slender.* PLATANTHERA Rich. Butterfly Orchis.

2. *H. bifólia* (R. Br.); spur twice as long as the germen, lip linear entire, pet. connivent blunt, *anth.* oblong truncate, *its cells parallel.*—*E. B. S.* 2806. *R.* xiii. 429.—About a foot high. Root-l. usually 2, elliptic. Stem-l. small, lanceolate, resembling the bracts. Spike slender. Fl. white. Central line between the anth.-cells a furrow in front and a keel behind, stalks of pollen-masses short, gland oval. Stigma truncate, emarginate with pointed lobes.—Heathy places. P. VI. VII. E. S. I.

3. *H. chloroleúca* (Ridl.); spur twice as long as the germen, lip linear entire, pet. connivent blunt, *anth.* truncate *its cells twice as distant at the base as at the top.*—*Orchis bifolia* Sm., *E. B.* 22. *H. chlorantha* (Bab.) ed. viii. *H. montana* R. xiii. 430.—Usually taller and stouter than the preceding. Spike usually lax, but sometimes dense. Fl. larger. Central line between anth.-cells a prominent ridge in front and a groove behind. Space between the bases of the anth.-cells usually, not always, spread open. Stalks of pollen-masses long, gland circular. Stigma very broad, slightly pointed in the middle.—Moist woods and thickets. P. V. VI. E. S. I.

6. O'PHRYS *Linn.*

1. *O. apif'era* (Huds.); lip tumid 5-lobed, 2 lower lobes prominent and with a hairy base, 2 intermediate reflexed truncate, *terminal acute long usually reflexed,* anth. with a hooked point, pet. oblong bluntish downy.—*E. B.* 383.—About a foot high. Fl. few, large, rather distant. Sep. whitish tinged with purple. Lip velvety, brown variegated with yellow. "All the lobes of the lip sometimes reflexed, interm. overlapping term. one."—*β. O. Trollii* (Heg.); term. lobe of lip acute triangular not reflexed. —On calcareous soils. *β.* Reigate. P. VI. VII. *Bee Orchis.* E. I.

2. *O. arachnites* (Reichard); lip somewhat tumid entire or with 4 shallow marginal lobes and a *terminal inflexed flat rather heart-shaped appendage*, anth. with a straight or hooked point, *pet. deltoid downy.*—*E. B. S.* 2596. *R.* xiii. 461.—Sep. pinkish. Lip velvety, dark purple, variegated with yellow; appendage green, never reflexed.—Chalk downs. Folkestone and Sittingbourne, Kent. P. IV.—VI. *Late Spider Orchis.* E

3. *O. aranif'era* (Huds.); lip tumid obscurely 3-lobed, *middle lobe large emarginate without an appendage*, anth. acute, *pet. linear glabrous.*—*E. B.* 65. *R.* xiii. 449.—Smaller than the two preceding and with fewer flowers. Sep. green. Pet. green, quite glabrous. Lip deep brown, hairy with paler or yellowish glabrous lines often resembling the Greek letter Π, entire at the end or notched with or without a central point.—*β. O. fucifera* (Sm.); lip usually undivided often with a gland in the notch, pet. rough. *E. B. S.* 2649.—Chalky places. rare. *β.* Kent, Sussex, and Isle of Wight. P. IV. V. *Spider Orchis.* E.

4. *O. muscif'era* (Huds.); *lip oblong trifid* with a broad pale spot in the centre, *middle lobe long bifid*, anth. short blunt, *pet. filiform.*—*E. B.* 64. *R.* xiii. 447. *St.* 40. 15.—Slender, about a foot high. Sep. green. Lip brownish purple; central spot subquadrate, bluish. Pet. very narrow, purple.—Damp calcareous thickets and pastures. P. V. VI. *Fly Orchis.* E. I.

7. Hermin'ium *R. Br.* Musk Orchis.

1. *H. Monor'chis* (R. Br.); lip 3-lobed, central lobe longest, pet. with a lobe on each side.—*E. B.* 71.—Root-knobs very unequal and distant. L. usually 2, lanceolate. St. about 6 in. high. Sep. ovate, greenish. Spike dense, slender.—Calcareous soil in the South, rare. P. VI. VII. E.

Tribe II. *Neottideæ.*

8. Peramium *Salisb.* (*Goodyera* R. Br. ed. viii.)

1. *P. répens* (Salisb.); l. ovate stalked netted, sep. pet. and lip ovate-lanceolate.—*E. B.* 289. *G. repens* (R. Br.) ed. viii. —St. 6—8 in. high, bearing linear adpressed bracts. Creeping. Whole upper part of the plant covered with minute stalked glands. L. netted with brown. Fl. white, small.—Fir forests of the North. P. VIII. E. S.

9. SPIRAN'THES *Rich.* Lady's Tresses.

1. *S. autumnális* (Rich.); root-fibres few ovate-oblong thick, root-l. ovate-oblong in a lateral cluster, stem-l. like bracts, spike dense.—*E. B.* 541.—St. 4—6 in. high. Spike spiral. Fl. greenish white. Column and lid acute; a blunt ovate membranous process between them on each side.—Dry calcareous and gravelly places. P. VIII. IX. E. I.

2. *S. æstivális* (Rich.); root-fibres few long cylindrical, root-l. oblong-lanceolate round the base of the st., stem-l. narrowly lanceolate, spike lax.—*E. B. S.* 2817.—St. 3—12 in. high. Spike spiral. Fl. with a larger lip. Column and lid acute; the interm. processes lanceolate acute.—Bogs. Between Lyndhurst and Christchurch, Hants. Wire Forest, Worcest. St. Ouen's Pond, Jersey. P. VII. VIII. E.

3. *S. Romanzoffiána* (Cham.); root-fibres few long cylindrical, root-l. linear-lanceolate, stem-l. triangular-lanceolate surrounding the base of the st., bracts shorter than the fl., spike dense 3-ranked, sep. and pet. equal blunt adhering together, lip blunt spathulate.—*Sy. E. B.* 1474. *S. cernua* Bot. Mag. 5277. *S. gemmipara* (Lindl.) ed. viii.—Spike about 1½ in. long. Fl. fragrant. Lateral sep. united at the base. See *Rep. Bot. Congr.* Lond. 176.—Cork, Armagh, Antrim, Londonderry. P. VIII. IX. I.

10. LIS'TERA *R. Br.* Tway-blade.

1. *L. ováta* (R. Br.); l. 2 opposite ovate, lip bifid, column with a crest which includes the anther.—*E. B.* 1548. *St.* 29. 14. —St. 1 ft. high. Spike long, very lax. Fl. small, greenish. L. large.—Woods and pastures. P. V. VI. *Tway-blade.* E. S. I.

2. *L. cordáta* (R. Br.); l. 2 opposite cordate, lip 4-lobed, column without a crest.—*E. B.* 358.—Height 3—5 in. St. slender. Fl. very small, in a lax spike, greenish. Lip with 2 basal and 2 terminal linear lobes.—Turfy mountain moors. P. VI.—VIII. E. S. I.

11. NEOT'TIA *Adans.* Bird's-nest.

1. *N. Nidus-ávis* (Rich.).—*E. B.* 48. *Listera* Sm.—Whole plant pale reddish brown. Root formed of many short thick fleshy fibres from the extremities of which the young plants are produced. (See *Leight. Fl. Shrop.* 434.) St. a foot high, with sheathing brown scales. L. none. Spikes dense, cylindrical, many-flowered. Lip linear-oblong with 2 spreading lobes.—Shady woods. P. ? VI. E. S. I.

12. Epipac'tis *Adans.* Helleborine.

1. *E. latifólia* (All.); st. solitary, l. broadly ovate exceeding the joints, upper l. ovate-oblong, lower bracts exceeding the fl., *label roundish-cordate with a small recurved point* falling short of the broadly ovate sep and pet., basal hunches smooth.—*E. B.* 269.—L. ovate, very broad, the very uppermost sometimes lanceolate-attenuate; lowermost leafless sheaths close. Lower bracts leaflike, lanceolate, attenuate. Fl. green with the lip purple, sometimes all purple. Peduncle shorter than the downy germen. Label of the lip broader than long, crenate. In a slender form of this plant the upper l. are lanceolate, label cordate blunt with a minute apiculus, and sep. ovate-lanceolate. —Mountain woods. P. VII. VIII. E. S. I.

2. *E. média* (Fries!); st. solitary, l. ovate-oblong the upper ones lanceolate acute, lower bracts exceeding the fl. and fr., *label entire triangular-cordate acute* equalling the lanceolate sep. and pet., basal hunches plicate rugose.—*Fr. Herb. Norm.* viii. 65!—Narrower and longer in all its parts than *E. latifolia.* Only the very lowest l. ovate, intermediate lanceolate, upper l. lanceolate-attenuate and merging gradually into the linear-lanceolate bracts; sheaths funnelshaped. Fl. green tinged with purple. Pet. shorter than the downy germen. Label longer than broad, crenate. Fr. abruptly obovate.—Woods. P. VIII. E. S. I.

3. *E. violácea* (Bor.); *st. many clustered,* l. ovate-lanceolate the upper ones narrower passing gradually into slender bracts, *label triangular-ovate acuminate falling short of the ovate-lanceolate sep. and pet.,* hunches plicate-crenate.—*E. B. S.* 2773. *R.* xiii. 486. *E. purpurata* (Sm.)?—Fl. "yellow-green tinged with pink." St. and l. much tinged with purple. Ped. shorter than the downy germen. Label longer than broad, entire, with an attenuate point.—Woods. P. VIII. E.

4. *E. ovális* (Bab.); st. solitary, l. ovate-oblong acute the upper ones lanceolate 1 or 2 lowest bracts exceeding the fl. but falling short of the fr., *label transversely oval mucronate* equalling the ovate acute sep. and pet., basal hunches plicate-rugose. —*E. B. S.* 2884. Helleborine &c. No. 2. *Ray* 383.—L. small; sheaths funnelshaped, rather close. Bracts all much smaller than even the most uppermost leaf. Fl. varying from a dark yellow to blackish red, peduncle shorter than the downy germen. Label transversely oval, crenate, with a small acute point, and elevated folded and tubercularly crenate hunches above. St.

6—18 in. high.—Settle, Yorkshire. Little Doward Hill, Herefordshire. Ormes Head. Durness, Sutherl. Burren, Clare. P. VII. E. S. I.

5. *E. palus'tris* (Crantz); l. lanceolate, bracts falling short of the somewhat drooping fl., *label roundish blunt crenate* equalling the perianth.—*E. B.* 270.—St. 12—18 in. high. Cal. purplish green, pet. and lip white tinged with purple.—Moist places, not rare. P. VII. VIII. E. S. I.

13. CEPHALAN'THERA *Rich.*

1. *C. pal'lens* (Rich.); l. ovate or ovate-lanceolate, *bracts exceeding the glabrous germen,* lip blunt included.—*E. B.* 271. *C. grandiflora* (Bab.) ed. viii.—St. 12—18 in. high. Fl. white; lip marked with several elevated longitudinal lines. Sep. erect, blunt.—Woods, usually on a calcareous soil. P. VI. E. S.

2. *C. ensifólia* (Rich.); l. lanceolate, *bracts much falling short of the glabrous germen,* lip blunt included.—*E. B.* 494.—Height 12—18 in. Fl. white; lip with several elevated white lines and a yellow spot in front. Outer sep. acute.—Woods, rare. P. V. VI. E. S. I.

3. *C. rúbra* (Rich.); l. lanceolate acute, bracts exceeding the *downy germen,* lip acute equalling the pet.—*Sy. E. B.* 1483.—Fl. purple; lip white with a purple margin, marked with many wavy longitudinal lines.—Woods, very rare. In two or three places in Gloucestershire. P. VI. VII. E.

Tribe III. *Arethuseæ.*

14. EPIPO'GUM *S. G. Gmel.*

1. *E. aphyl'lum* (Sw.); *Sy. E. B.* 1486. *Bot. Mag.* 4821. *R.* xiii. 468. *St.* 18. 16.—St. 3—7 in. high, sheathed. L. none. Sep. and pet. narrowly lanceolate, acute, pale yellowish. Middle lobe of lip ovate, furrowed, white, with 4 rows of purple tubercles. Spur very thick. Column short, dilated above the stigma to receive the base of the anther. Appendage triangular.—Damp wood near Tedstone Delamere, Herefordshire. Only once found. Ringwood Chase near Ludlow. *Misses Peele and Lloyd.* P. VIII. E.

Tribe IV. *Malaxideæ.*

15. Corallorrhi'za *Hall.*

1. *C innáta* (R. Br.); spur very short or wanting.—*E. B.* 1547.—Root of thick fleshy much-branched fibres. Height 6—12 in. Spike of few yellow flowers. Sep. and pet. lanceolate acute. Lip oblong, white, with a few purple spots, sometimes with 3 equal lobes.—Boggy woods, rare. P. VII. S.

16. Malax'is *Sw.*

1. *M. paludósa* (Sw.); st. with 3—5 oval concave leaves, lip concave acute.—*E. B.* 72.—St. 1—4 in. high, 5-edged. Sep. ovate, spreading, 2 turning upwards. Lip superior, erect, 3-veined, its base surrounding the column. L. fringed at the end with little bulbs. Forming a small bulb at its base.—This plant and the next rather grow upon the moss as epiphytes than amongst it.—Spongy bogs. P. VIII. IX. E. S. I.

17. Lip'aris *Rich.* (*Sturmia* (Reich.) ed. viii.)

1. *L. Loesel'ii* (Rich.); l. oblong-lanceolate, st. triangular, lip obovate exceeding the petals.—*Malaxis* Sm., *E. B.* 47.—St. 6—10 in. high. Fl. 6—12, in a lax spike, yellowish. Sep. lanceolate. Pet. linear. Forming a large ovate bulb at its base, enclosed in the whitish sheaths of the decayed leaves. An epiphyte?—Spongy bogs in Norf., Suff., and Camb., now very rare. P. VI. E.

Tribe V. *Cypripedieæ.*

18. Cypripe'dium *Linn.* Lady's Slipper.

1. *C. Calcéolus* (L.); st. leafy, middle lobe of the column nearly ovate blunt deflexed, lip slightly depressed falling short of the calyx.—*E. B.* 1.—St. 12—18 in. high, downy, bearing 3 or 4 large ovate pointed leaves. Fl. usually solitary or 2, large; sep. 1—1½ in. long, dark brown; pet. dark brown, rather narrower than sep.; lip 1 in. long, inflated, yellow, netted with darker veins.—Dense Northern woods, very rare. P. V. VI. E.

Order LXXXIV. IRIDACEÆ.

Perianth tubular, 6-parted, petal-like, in 2 often unequal rows Stam. 3, epigynous, opposite the outer segments of the perianth.

Anth. bursting outwards. Ovary inferior, 3-celled. Style 1. Stigmas 3, dilated, often like petals. Caps. 3-celled, 3-valved; valves bearing the dissepiments in the middle. Seeds many. Embryo cylindrical, enclosed in horny or fleshy albumen. Radicle pointing towards the hile.

1. SISYRINCHIUM. Perianth 6-cleft; segments nearly equal, patent. Style short. Stigmas 3, rolled inwards, filiform. Filaments connate below.

2. IRIS. Perianth 6-cleft; alternate segments reflexed. *Stigmas 3, like petals,* covering the stamens.

3. GLADIOLUS. Perianth irregular, 6-cleft, 2-lipped; segments unequal. Style filiform. Stigmas 3, widening upwards. Seeds more or less winged.

4. ROMULEA. Perianth regular, 6-cleft; segments spreading. Style filiform. Stigmas 3, bifid; lobes slender.

5. CROCUS. Perianth regular, funnelshaped with a long tube; limb bellshaped. Style filiform. Stigma 3-fid or 3-parted; lobes widening upwards.

1. SISYRIN'CHIUM *Linn.*

1. *S. angustifólium* (Mill.); scape 2-edged nearly simple nearly leafless, spath 1—6-flowered falling short of the fl., segments of perianth emarginate mucronate [Caps. globular-trigonous] —*Sy. E. B.* 1491. *S. Bermudianum* (L.) ed. viii. *Redoute Lil.* v. 232. St. about 1 foot high. L. grasslike. Spath with about equal lanceolate valves, falling short of the fl. in my Irish specimens. Perianth blue; segments narrowed below. Kerry, Cork, Clare, and Galway. P. VII. I.

†2. *S. californ'icum* (Aiton); scape *simple compressed* broadly winged *leafless,* spath erect outer valve about equalling the fl. sheathing connate at the base 3—5-flowered, per.-segm. oval narrowing below, *caps. ellipsoidal-trigonous.*—*J. of B.* xxxiv. (1896) p. 494, t. 364.—Scapes usually 2. Per. yellow (becoming orange) with dark veins.—Marsh, Rosslare, Wexford. *Rev. E. S. Marshall.* Status uncertain. P. VI. I.

2. I'RIS *Linn.* Flag.

1. *I. Pseud-ac'orus* (L.); l. swordshaped, st. roundish, perianth beardless its inner segments narrower and falling short of the stigmas.—*E. B.* 578.—St. terete, 2—5 feet high. Fl. yellow, or in *I. Bastardi* (Bor.) fl. pale lemon-coloured. Caps. oblong, trigonous, apiculate.—*α*; outer perianth-segm. uniform clear yellow with broadly obovate blade and rather short claw, stigm.

narrow and long.—β. *I. acoriformis* (Bor.); outer perianth-segm. with deeper blotch at the base, the blade suborbicular and claw long narrowish yellow with prominent purplish veins. Yellow of fl. paler than in α.—Wet places. P. VI. VII. *Yellow Flag.* E. S. I.

2. *I. fœtidis'sima* (L.); l. swordshaped, st. compressed, perianth beardless its inner segments about equalling the stigmas. —*E. B.* 596. *R.* ix. 347.—Herb green, not glaucous, yielding an unpleasant smell when bruised. St. 1-angled, 2 ft. high. Fl. lead-coloured or bluish, rarely [var. *citrina* Bromf.] yellow, seeds red.—Woods and thickets. P. V.—VII. *Gladdon.* E. I.

[**I. tuberósa* (L.); l. tetragonal, segments of the perianth acute, root tuberous.—Penzance. Cork.—*E. B. S.* 2818.] E. I.

[*I. spúria* (L.); with linear l., limb of outer per.-segm. roundish, inner segm. and stig. violet, and conspicuously apiculate carp., is apparently established in marshes at Huttoft, Linc.]

3. Gla'diolus *Linn.*

1. *G. illyr'icus* (Koch); corm clothed with nearly parallel fibres netted above with long narrow openings, fl. secund, filaments longer than the anthers, cor.-tube nearly thrice as long as germen, caps. oval emarginate with 3 rounded angles.—*Sy. E. B.* 1493. *J. of B.* i. t. 4. *St.* 83. 3.—Height about 20 in. Corms ovate-acuminate. L. slender, swordshaped. Sheaths 2-edged. Fl. reddish changing to bluish; basal pet. rather acute, exceeding the blunt lower lateral pet.; "edges of upper pet. not covered by the 2 adjoining ones" when the fl. are in their prime. Stigm. narrow below widening from the middle upwards and there fringed.—Amongst *Pteris* in the New Forest. Isle of Wight. *Mr. A. G. More.* P. VII. E.

4. Rom'ulea *Mar.*

1. *R. Colum'næ* (Seb. & Maur); scape 1-flowered usually solitary slightly nodding, l. filiform compressed furrowed recurved, spath exceeding the tube of the cor., style falling short of the stam., stigmas bifid.—*E. B.* 2549. *R.* ix. 354. *Trichonema Bulbocodium* Sm.—A small plant not more than 4 in. high, with a corm. Fl. greenish without, pale with purple stripes and yellow at the base within.—Sandy places. Dawlish Warren, Devon. Jersey and Guernsey. P. III. IV. E.

5. CRO'CUS *Linn.*

* *Scapes enveloped in a tubular sheath.*

1. *C. ver'nus* (All.); l. and fl. at the same time, spath simple, throat of the cor. fringed with hairs, stigma shortly 3-fid, lobes erect wedgeshaped jagged at the end, corm clothed with slender netted fibres.—*E. B.* 344.—Fl. 1—2, violet-purple.—Near Nottingham; and Mendham, Suff. P. III. E.

2. *C. nudiflorus* (Sm.); l. succeeding the fl., spath simple, *stigma in 3 deeply laciniate divisions erect*, corm with a membranous coat.—*E. B.* 491. *C. speciosus* Hook. *E. B. S.* 2752.—L. linear, appearing in March. Fl. solitary, purple. Stigmas only a little higher than the anthers, or rising considerably above them.—Meadows. P. IX. E.

** *Scapes naked.*

[*C. biflórus* (Mill.); l. and fl. at the same time, spath double, *stigma* longer than the stam. erect deeply trifid, *divisions* truncate and *slightly notched* at the end, corm with a membranous coat.—*E. B. S.* 2645. *C. argenteus* (Sab.) ed. viii.—Fl. pale lilac with yellow and purple stripes.—And *C. aúreus* (Sibth.); l. and fl. together, *spath simple, stigma shorter than the stam. shortly 3-fid,* segments truncate or slightly notched at the end, corm coated with compact fibres.—*E. B. S.* 2646.—Fl. yellow.—On site of old garden, Barton, Suff. P. III.]

Order LXXXV. AMARYLLIDACEÆ.

Stam. 6. Anth. bursting inwards. Raphidiferous. Otherwise like *IRIDACEÆ.*

1. NARCISSUS. Perianth tubular below; limb 6-parted, spreading, with equal segments, and bellshaped crown within. Stam. alternately shorter, within the crown.

2. LEUCOJUM. Perianth bellshaped, 6-parted, segments all equal and thickened at their points. Stam. equal.

3. GALANTHUS. Perianth 6-parted, 3 outer segments spreading, 3 inner shorter erect emarginate. Stam. equal, subulate.

1. NARCIS'SUS *Linn.*

1. †*N. biflórus* (Curt.); l. linear acutely keeled with reflexed edges, scape compresed 2-edged striate 1—2-flowered, crown

very short concave crenate at the pale (ultimately white) margin.—*E. B.* 276. *R.* ix. 365.—Pet. of a pale sulphur-colour, sides slightly inflexed.—Sandy fields. A rather doubtful native. P. IV. V. E. I.

[2. *N. poet'icus* (L.); l. linear bluntly keeled, scape compressed 2-edged mostly 1-flowered, crown very short concave crenate at the red margin. *E. B.* 275. *R.* ix. 364.—Pet. white, broadly ovate, crown yellow.—Heathy open fields or a sandy soil. Norf., Kent. P. V.] E.

3. *N. Pseudo-narcis'sus* (L.): l. linear bluntly keeled, scape 2-edged 1-flowered, perianth-segm. scarcely exceeding tube, crown bellshaped crisped at the margin and crenate equalling the perianth, pedicel within the spath short.—*E. B.* 17. *R.* ix. 369.—L. 2 or 3, scarcely a foot long.—Fl. large, yellow.—Open woods and pastures. P. III. IV. *Daffodil.* E. S.

[*N. obvallάris* (Salisb.); "perianth all yellow, its segm. exactly twice as long as tube, crown 6-lobed." Tenby.—*N. incomparab'ilis* (Mill.) *Sy. E. B.* 1502; crown erect ½ as long as perianth, otherwise like Sp. 3. In several places in the South.—These are not native plants.]

2. Leuco'jum *Linn.* Snowflake.

†1. *L. æstivum* (L.); spath many-flowered, style thickened upwards.—*E. B.* 621. *R.* ix. 362.—Bulbous. Height 2—2½ feet. Fl. æstival, white, drooping; tips greenish. L. broadly linear, keeled, hibernal. Scape 2-edged.—Wet meadows in the South and East. P. V. E.

†2. *L. ver'num* (L.); spath 1-flowered, style thickened upwards.—*Sy. E. B.* 1506. *J. of B.* iv. t. 49.—Bulbous. Height 8—10 inches. L. and white drooping flowers vernal.—Near Bridport and Bicester. P. II.—IV. E.

3. Galan'thus *Linn.* Snowdrop.

1. *G. nivάlis* (L.).—*E. B.* 19. *R.* ix. 363.—Bulbous. Fl. solitary, white, drooping; inner segments greenish. L. 2, keeled, broadly linear, glaucous.—Thickets in the West. P. II. III. E.

Order LXXXVI. ALISMACEÆ.

Perianth free, of 6 leaves; 3 inner or all coloured. Stam. 6. Ovaries 3—6 or many, always distinct or ultimately separable; carp. opening at the suture or not at all; seeds 1 or many. Embryo straight or curved; albumen 0.—No raphides.

Suborder I. ALISMOIDEÆ.

Three inner perianth-segm. petal-like. *Seeds* 1—2 *in each cell*, erect or ascending, on the suture. Embryo cylindrical, doubled upon itself; radicle next the hile.

1. ALISMA. Fl. perfect. Stam. 6. *Carp. many, forming a ring or head*, small, 1-seeded, not bursting.—L. ovate or narrow.

2. DAMASONIUM. Fl. perfect. Stam. 6. *Carp.* 6—8, rather large, 2-seeded, combined at the base and *spreading in a radiant manner*. L. cordate-oblong.

3. SAGITTARIA. Fl. monœcious. Male fl. with many stamens. Female fl. with *many* 1-seeded compressed *carpels forming a head*, upon a globose receptacle.—L. sagittate.

Suborder II. BUTOMEÆ.

Three inner perianth-segm. petal-like; three outer herbaceous or slightly coloured. *Seeds many*, minute. Placenta ramified over the inner surface of each carpel. Embryo straight or curved; radicle next the hile.

4. BUTOMUS. Perianth-segm. 6, all coloured, resembling a corolla. Stam. 9. Carpels 6, connected below.

Suborder III. JUNCAGINEÆ.

Perianth uniform, herbaceous, inconspicuous, or 0. Seeds 1—2, erect, close together and at the base of the carpel. Embryo straight; radicle next the hile; plumule coming through a lateral cleft in the embryo.

5. SCHEUCHZERIA. Perianth of 6 reflexed leaves. Stam. 6, with slender filaments. Ovaries 3. Stigma sessile, downy. Carpels compressed, inflated, diverging, 1—2-seeded, free.

6. TRIGLOCHIN. Perianth of 6 erect deciduous leaves. Stam. 6; anth. almost sessile. Ovaries 3—6. Stigmas sessile, feathery. Carp. attached to an angular axis, from which they at length separate at the base.

Suborder I. *Alismoideæ.*

1. Alis'ma *Linn.* Water-Plantain.

1. *A. Plantágo* (L.); fl.-stalk panicled with whorled compound branches, *carp.* ranged in a circle *compressed blunt* obovate, *style below the top of inner edge of carp.*, l. cordate-ovate or lanceolate. —*E. B.* 837. *R.* vii. t. 57.—L. all radical, on long stalks. Submersed l. linear. Fl.-stalks 2—3 ft. high. Fl. pale rose-colour. Sep. ovate-oblong, styles twice as long as ovary.—β. *A. lanceolata* (With.); l. lanceolate narrowed below. *Sy. E. B.* 1438. Sep. ovate, styles equalling the ovary. Perhaps distinct.—By water. P. VII. VIII. E. S. I.

2. *A. ranunculoïdes* (L.); fl.-stalks umbellate, *carp. angular acute* forming a globose squamose head, *style terminal*, l. linear-lancolate acute.—*E. B.* 326. *R.* vii. t. 55.—L. all radical, on long stalks. Fl.-stalks from 3 to 24 in. long, ending in 1 or 2 umbellate whorls of simple peduncles. Fl. pale purple.—β. *A. repens* (Dav.); stoloniferous, producing l. roots and fl. at the nodes, fl. larger. *E. B. S.* 2722.—Turfy bogs. β. By lakes in Wales and Ireland. P. VI. VII. E. S. I.

3. *A. nátans* (L.); st. floating and rooting leafy, *peduncles simple* from the joinings of the stem, *carp.* striate beaked, floating l. stalked oblong blunt, radical leafless petioles broadly subulate.—*E. B.* 775. *R.* vii. t. 54. *Elisma* Buchan.—St. slender, often very long. Root-petioles in small tufts. Fl. rather large, white with a yellow spot.—Lakes, rare. P. VIII. E.

2. Damaso'nium *Hill.* (*Actinocarpus* R. Br. ed. viii.)

1. *D. Alis'ma* (Mill.); stalks with 1—3 whorls of fl., carp. subulate compressed opening longitudinally, l. cordate-oblong. —*E. B.* 1615. *Alisma* L.—L. all radical, floating, on long stalks, 5-veined. Pet. white. Carp. large; with 2 stalked seeds, one from the lower angle erect, one from the upper horizontal.—Pond and ditches, rare. P. VI. VII. E.

3. Sagitta'ria *Linn.* Arrowhead.

1. *S. sagittifólia* (L.); aërial l. arrowshaped with lanceolate straight lobes, fl.-stalk simple, fl. whorled.—*E. B.* 84. *R.* vii. t. 53.—Stoloniferous; each runner ending in a tuber. The l. that rise above the water are remarkably arrowshaped, with the 3 parts nearly equal. The submersed leafless petioles are linear. Fl. white.—Ditches and rivers. P. VIII. E. I.

Suborder II. *Butomeæ.*

4. BU'TOMUS *Linn.* Flowering Rush.

1. *B. umbellátus* (L.).—*E. B.* 651. *R.* vii. t. 58.—Flowerstalk radical, 2—3 feet high, overtopping the leaves, bearing an irregular many-flowered simple umbel with scarious bracts and a membranous 3-leaved involucre. Fl. rose-coloured. L. radical slender, triangular.—Rivers and ponds. P. VI. VII. E. S. I.

Suborder III. *Juncagineæ.*

5. SCHEUCHZE'RIA *Linn.*

1. *S. palus'tris* (L.).—*E. B.* 1801. *St.* 78. 4. *R.* x. 419.—St. 6—8 in. high, erect. L. distichous, few, alternate, semicylindrical, blunt, with a minute pore on the upper side of the apex. Raceme terminal, of about 5 greenish flowers. Caps. about 3, much inflated.—Sphagnous bogs, rare. P. VII. E. S.

6. TRIGLO'CHIN *Linn.* Arrow-grass.

1. *T. marit'imum* (L.); fr. ovoid of 6 combined carpels.—*E. B.* 255. *R.* vii. t. 52.—L. radical, linear. Fl. in a lax simple spike or raceme, greenish.—Muddy salt marshes. P. VII. VIII. E. S. I.

2. *T. palus'tre* (L.); fr. linear angular of 3 combined carpels.—*E. B.* 366. *R.* vii. t. 51.—Slenderer than the preceding but closely resembling it, 8—10 in. high. Stoloniferous.—Marshy places. P. VI. VII. E. S. I.

Order LXXXVII. ASPARAGACEÆ.

Perianth inferior, petal-like, 6-parted or 4—8-parted. Stam. 6 or 4—8, inserted into the receptacle or on the perianth. Anth. bursting inwards. Ovary superior, 3 (rarely 1-) celled. Ovules 1 or many in each cell. Styles 1—3. *Fr. succulent,* not bursting. *Root not bulbous.*—Raphidiferous.

1. ASPARAGUS. Perianth 6-parted, bellshaped, tubular below Stam. 6. Ovary 4-celled; cells 2-ovuled. *Style* 1. *Stigmas* 3, *reflexed.*—Fl. by abortion diœcious.

2. CONVALLARIA. *Perianth* bellshaped, 6-*parted,* deciduous. Ovary 3-celled; cells 2-ovuled. *Stigma blunt,* trigonous. Berry with 1-seeded cells. Fl. jointed to pedicel.

3. POLYGONATUM. *Perianth tubular,* 6-toothed, tardily *deciduous.* Ovary 3-celled; cells 2-ovuled. Stigma blunt, trigonous. Berry with 1—2-seeded cells. Fl. not jointed to pedicel.

4. MAIANTHEMUM. *Perianth* 4-*parted*; segments horizontally patent or reflexed, deciduous. Stam. 4. Style 1, bifid. Stigma blunt. Berry 2-celled; cells 1-seeded.

5. RUSCUS. Diœcious. Perianth 6-parted to the base, persistent. Filaments forming an ovoid tube, on the top of which the 3 anth. are placed.—Fem. the same, but the anthers barren. Style 1. Stigmas capitate. Berry 1-celled, rarely 2-seeded.—Fl. on disk of persistent leaflike shoots.

1. ASPAR'AGUS *Linn.*

1. *A. officinális* (L.); st. herbaceous mostly erect without spines branched, l. clustered terete flexible setaceous.—*E. B.* 339.—Creeping. Stems many.—*a. A. maritimus* (L.); st. prostrate at the base, branches short about 1 ft. long. *A. prostratus* (Dum.); *Bull. S. Bot. Belg.* i. t. 2.—*β. *altilis* L.; st. erect about 3 feet high, branches long.—Sea-coast, rare. Kynance Cove, Cornwall. South coast of Anglesea. Glamorgan. Giltar Point, Pemb. Tramore, Waterf. β. Escape from cultivation. P. VIII. E. I.

2. CONVALLA'RIA *Linn.* Lily of the Valley.

1. *C. majális* (L.).—*E. B.* 1035. *St.* 14. 10.—About a foot high. L. 2, ovate-lanceolate, radical. Scape semicylindrical. Fl. racemose, nodding, pure white, globose-bellshaped, fragrant. —Woods and thickets. P. V. E. S.

3. POLYGONA'TUM *Mill.* Solomon's Seal.

1. *P. verticillátum* (All.); *l. linear-lanceolate whorled,* st. erect angular.—*E. B.* 128. *R.* x. 435.—St. 2 feet high. L. 3—5 in a whorl. Berries red.—Woods. Perthshire. Smalesmouth, Northumberland. P. VI. E. S.

2. *P. officinále* (All.); l. ovate-oblong half-clasping glabrous alternate, *st. angular,* peduncles 1—2-flowered, cor. not narrowed

in the middle, *filaments glabrous.*—*Convallaria Polygonatum* (L.), *E. B.* 280. *R.* x. 434.—Height 1—1½ foot. Berry bluish.—Woods, rare. P. V. E.

3. *P. multiflórum* (All.); l. ovate-oblong half-clasping glabrous alternate, *st. round*, peduncles 1- or many-flowered, cor. narrowed in the middle, *filaments downy.*—*E. B.* 279. *R.* x. 433.—Height 2 feet. Berry bluish.—Woods. P. V. E. S.

4. Maian'themum *Weber*.

1. *M. Convallária* (Web.); st. with 2 alternate stalked triangular-cordate leaves.—*Ger. Herb.* p. 409. *Smilacina, Sy. E. B.* 1510. *R.* x. 436. *M. bifolium* (DC.) ed. viii.—St. 6—8 in. high. Root filiform. L. very deeply cordate. Raceme terminal, resembling a spike. Fl. small; segments reflexed.—Woods. Near Scarborough, in plenty. Howick, Northumb. (now eradicated). Caen Wood, Middlesex. Dingley Wood, Preston; Harwood, Blackburn. *Gerard.* Hunstanworth, Durham. P. V. E.

5. Rus'cus *Linn.* Butcher's Broom.

1. *R. aculeátus* (L.); leaflike flattened shoots ovate-attenuate very acute rigid bearing the fl. upon the middle of their upper surface, fl. 1 rarely 2 with a flat subulate scarious 1-veined bract. *E. B.* 560. *R.* x. 437.—Evergreen. Fl. small. L. very minute.—Thickets. Sh. III. IV. E. S.?

Order LXXXVIII. LILIACEÆ.

Perianth inferior, petal-like, 6-leaved, 6-parted or with 6 teeth. Stam. 6, inserted on the receptacle or on the perianth. Anth. bursting inwards. Ovary superior, 3-celled. Ovules many in each cell. Style 1. Stigmas 3 or 1. *Fr. dry*, capsular, bursting with 3 valves bearing the dissepiment on their middle.

Tribe I. *TULIPEÆ.* Perianth-l. nearly or quite distinct. Cells of caps. many-seeded. Seeds flat (in *Lloydia* angular), placed closely one above another; testa pale or fuscous, not crustaceous.—St. usually more or less leafy. Bulbous.

1. Tulipa. Perianth without nectaries, deciduous. *Anth. erect.* Style 0. Stigma 3-lobed. Seeds flat.

2. Fritillaria. Perianth deciduous; a nectariferous depression at the base of each leaf. *Anth. attached above their base.* Style 3-fid at the apex. Seeds flat.

[3. LILIUM. Perianth deciduous, spreading or reflexed; a longitudinal nectariferous furrow at the base of each leaf. *Anth. attached above their base.* Stigma capitate. Seeds flat.]

4. LLOYDIA. Perianth persistent, patent. Stam. inserted at the base of perianth. *Anth. erect.* Style filiform. Stigma trigonous. Seeds angular above, flat beneath.

Tr. II. *ASPHODELEÆ.* Fl. not jointed to their stalks. Leaves of perianth distinct. Cells of caps. few-seeded. Seeds various in form, usually with a black crustaceous testa.—St. usually leafless. Bulbous.

5. ORNITHOGALUM. Perianth-l. 6, spreading, persistent. Stam. on the receptacle and adhering only slightly to the perianth. *Anth. attached by their backs.*—Fl. white or yellow, never blue.

6. GAGEA. Perianth-l. 6, spreading, persistent. Stam. adhering to the base of the perianth. *Anth. erect.*—Fl. corymbose or umbellate, yellow.

7. SCILLA. Perianth-l. 6, spreading, deciduous. Stam. on the base of the perianth. *Anth. attached by their backs* —Fl. racemose, never white or yellow.

8. ALLIUM. Perianth-l. 6, rather spreading. Stam. at the base of the perianth. *Anth. attached by their backs.*—Fl. umbellate. Spath of 1 or 2 leaves.

Tr. III. *ANTHERICEÆ.* Fl. jointed to their stalks. Leaves of perianth slightly connected below. Cells of caps. few-seeded. Seeds various in form.—Not bulbous.

9. PUBILARIA. Perianth-l. 6, spreading, deciduous. Stam. on the base of the perianth. Filaments bearded. *Anth. attached by their backs.* Caps. 3-celled, 6-seeded.

Tr. IV. *HEMEROCALLIDEÆ.* Leaves of the perianth combined below. Cells of the caps. few-seeded. Seeds various in form; testa (in our plants) black.—Bulbous.

10. ENDYMION. Perianth tubular-bellshaped, of 6 connivent leaves with reflexed points combined below. Stam. inserted below the middle of the perianth; filaments decurrent.

11. MUSCARI. Perianth globose or subcylindrical, narrowed at the mouth, 6-toothed. Stam. inserted at about the middle of the tube; filaments not decurrent.

Tribe I. *Tulipeæ.*

1. Tuli'pa *Linn.* Tulip.

1. *T. sylves'tris* (L.); st. 1-flowered glabrous, fl. at first drooping, tip of segments of perianth and base of stamens hairy. —*E.B.* 63. *St.* 29. 11. *R.* x. 446.—Fl. yellow, rarely produced in a wild state. Chalk-pits in the Eastern Counties. "Meadows near Nottingham and in Yorkshire." P. IV. V. E.

2. Fritilla'ria *Linn.* Fritillary.

1. *F. Meleágris* (L.); st. 1-flowered leafy, l. all alternate linear-lanceolate.—*E. B.* 622. *St.* 18. 4. *R.* x. 442.—About a foot high. Fl. dull red with many dark spots, rarely white.—Meadows and pastures in the East and South. P. V. E.

3. Lil'ium *Linn.* Lily.

[1. *L. Martágon* (L); l. whorled elliptic-lanceolate, st. downy roughish, fl. nodding, perianth reflexed.—*E. B. S.* 2799. *R.* x. 451.—Height 1—1½ foot. Fl. violet-flesh-coloured with dark-purple spots.—Copses. P. VI. VII. *Turk's-cap Lily.*] E.

[*L. pyrenáïcum* (Gouan); l. scattered linear-lanceolate, fl. nodding, perianth reflexed.—*Sy. E. B.* 1517.—About 1 ft. high. Fl. yellow with black dots below.—Between South Molton and Mollond, Devon.] E.

4. Lloyd'ia *Salisb.*

1. *L. alpína* (Salisb.); root-l. semicylindrical, st.-l. dilated below and sheathing, fl. mostly solitary, nectary a transverse plait.—*E. B.* 793. *St.* 28. 2. *R.* x. 440. *L. serotina* (R.) ed. viii. *Anthericum* Sm.—Height 5 or 6 in. St. and l. springing separately from the root. St.-l. several, short. Fl. white with reddish lines internally.—Welsh monntains, very rare. Snowdon. Glyder Fawr. P. VI. E.

Tribe II. *Asphodeleæ.*

5. Ornithog'alum *Linn.* Star of Bethlehem.

‡1. *O. umbellátum* (L.); fl. corymbose, ped. exceeding the linear-lanceolate bracts, filaments lanceolate simple, l. linear

glabrous.—*E. B.* 130. *R.* x. 467.—L. exceeding the stem or filiform and shorter. Height 8—12 in. Fl. white with a broad green longitudinal band externally. Meadows and pastures. P. V. E. S.

2. *O. pyrenaïcum* (L.); fl. in an elongated narrow raceme, ped. at first spreading afterwards erect, bracts lanceolate-acuminate, filaments dilated below with a long point, l. soon fading linear grooved.—*E. B.* 499.—St. leafless, 2—3 feet high. Raceme becoming very long. Fl. greenish white; segments of the perianth variable in breadth. L. withering before the stalk appears, rarely contemporaneous.—Woods. Extremely common near Bath. Sussex. Bedfordshire. P. VI. E.

[3. *O. nútans* (L.); fl. few in a lax nodding raceme, ped. falling short of the bracts, filaments flat membranous trifid, the lateral points acute, middle one very short bearing the anther, l. linear-lanceolate.—*E. B.* 1997. *Albucea* R. x. 473.—Height 9—12 in. Fl. large, white, greenish externally.—Fields and orchards, rare. P. IV. V.] E.

6. GA'GEA *Salisb.*

1. *G. fasciculáris* (Salisb.); radical l. usually solitary linear-lanceolate flat, bracts 2 opposite, peduncles umbellate simple glabrous, segments of the perianth oblong blunt, bulb ovate solitary.—*E. B.* 21. *R.* x. 477. *G. lutea* (Ker) ed. viii. *Ornithogalum* L.—St. about 6 in. high, shorter than the leaves. Bracts lanceolate, one often exceeding the yellow flowers. Bulb often enclosing many small round offsets.—Woods and thickets, rare. P. III. IV. E. S.

7. SCIL'LA *Linn.* Squill.

1. *S. autumnális* (L.); l. linear many, raceme lax, peduncles ascending, *bracts* 0.—*E. B.* 78. *R.* x. 463.—Height 4—6 in. Fl. purplish blue with a green line down the back, in perfection before the l. appear.—Dry pastures in the South and West. P. VIII. *Autumnal Squill.* E.

2. *S. ver'na* (Huds.); l. linear channelled hooded at the end many, raceme few-flowered corymbose, *bracts lanceolate*, as long as or longer than the peduncles.—*E. B.* 23. *R.* x. 463.—Height 4—5 in. Fl. blue. L. as long as or longer than the stalk.—Western and Northern coasts. P. IV. V. *Vernal Squill.* E. S. I.

8. AL'LIUM *Linn.* Garlic.

* *Alternate stamens broader and 3-pointed, the middle point alone bearing an anther.* PORRUM Tourn.

† Stem-leaves flat or keeled, not hollow.

*A. *Ampelop'rasum* (L.); st. leafy below, l. linear, spath long, *umbel globose compact,* stam. exserted, anther-bearing point of 3-pointed filaments as long as the undivided part, bulb compound of 2—4 divisions.—*E. B.* 1657.—Bulb with large offsets within its coats. St. 2—6 feet high. L. long, linear. Spath parting at the base, and falling off in one piece before the fl. open; horn 1—2 in. long. Fl. pale purple, the keel of the outer subemarginate segments greenish and roughish. Germen rather globose; transverse projections at about the middle; lower spaces slightly excavated. *Head-bulbs very rare,* when present small, *the size of peas.*—Steep Holmes Islands in the Severn (remains of former cultivation). Cliffs in Guernsey. P. VIII. E.

†1. *A. Babingtónii* (Borr.); st. leafy below, l. acutely keeled, spath long-pointed, *umbel loose irregular* with hemispherical bulbs, stam. exserted, *anther-bearing point* of 3-pointed filaments rather shorter than the undivided part and *with an incurved tip when young,* bulb compound of few (2) divisions.—*E. B. S.* 2906.—St. 4—6 feet high. L. long, linear, broad. *Heads* large, *with many bulbs about as large as hazelnuts* (a character quite constant in cultivation). Spath usually deciduous in two pieces or persistent, horn 1—2 in long. Fl. pale reddish purple, rather few, more conical than in the preceding, only slightly opening; outer segments with a green keel, ovate-oblong, with callous points, edges and back rough with minute pellucid points; inner segments slightly emarginate and without points. Lengthened stalks 1—2 in. long, bearing *secondary heads,* are usually present. Germen rather conical; transverse projection below the middle; lower spaces rather deeply excavated.—Roundstone sparingly, and South Isles of Aran plentifully, Co. Galway. Cornwall. P. VIII. E. 1.

2. *A. Scorodop'rasum* (L.!); st. leafy below, l. flat, sheaths 2-edged, *spath* short and broad *with a very short point,* umbel globose with many spherical small bulbs, *stam. included* or equalling the perianth, anther-bearing point of 3-pointed filaments shorter than the undivided part or the lateral points, bulb with many purple offsets.—*E. B. S.* 2905. *A. arenarium* Sm.—St. 2—3 feet high. Heads small. Fl. few, purple; segments all with a minute apiculus, outer with the edges and keel

rough. Head-bulbs deep purple.—Sandy woods and fields in the North. P. VI. VII. E. S. I.

†† Stem-leaves hollow.

3. *A. vineále* (L.); st. leafy below, *l. terete* slightly channelled above, spath 1-valved short with a slender long point, umbel globose with many bulbs, stam. exserted, *anther-bearing point of 3-pointed filaments equalling the undivided part and half as long as the lateral points.—E. B.* 1974. *R. I.* t. 404. A. arenarium *Fries.*—St. 2 feet high. L. faded at the time of flowering. Heads of few pale rose-coloured fl. with green keels and long stalks. Head-bulbs small, oval, acute, greenish.—β. *A. compactum* (Thuil.); umbel without fl., head-bulbs with a leaflike point.—Waste ground and dry fields. β is the more common state. P. VII. *Crow-Garlic.* E. S. I.

4. *A. sphærocephʹalum* (L.); st. leafy below, l. subcylindrical channelled above, spath 2-valved short, *umbel* globose *without bulbs,* stam. twice as long as the perianth, *anther-bearing* point of 3-pointed filaments as long as the undivided part *longer than the lateral points,* bulb accompanied by stalked offsets.—*E. B. S.* 2813.—St. 1—2 feet high. L. usually faded before the time of flowering. Heads of many rose-coloured or purple fl.; keels darker and rough.—St. Vincent's Rocks, Bristol. Sands in Jersey. P. VII. E.

** *Stam. all simple, not 3-pointed, connected at the base. Spath 2-valved, one valve with a long point. St.-l. narrow.*

5. *A. oleráceum* (L.); st. leafy below, l. channelled above ribbed beneath, spath with one of the points very long, umbel with bulbs, stam. equalling or shorter than the perianth.—*E. B.* 488.—Height 1—2 feet. L. (of the Bristol plant) thick, fleshy, solid, nearly flat but slightly and broadly channelled above, with 4 ribs beneath. Segments of perianth blunt.—β. *A. complanatum* (Bor.); stam. shorter than the perianth, l. of equal thickness throughout curved upwards at the sides so as to appear channelled with many ribs on each side. *A. carinatum* Sm., *E. B.* 1658.—Borders of fields, rare. β. Mountains in the North. P. VII. VIII. E. S.

6. *A. carinátum* (L.); st. leafy below, l. nearly flat, umbel nearly without bulbs, *stam. much exceeding perianth.*—Height 1—2 feet. L. erect, slightly channelled below, flat in the upper part, slightly furrowed (not keeled) beneath. Perianth-segm. blunt, rose-coloured.—Newark. Seguden, Carse of Gowrie. By Esk above Musselburgh, abundant. P. VIII. E. S.

*** *Stam. all simple and distinct. Spath 2-valved, short. Leaves hollow.*

7. *A. Schœnop'rasum* (L.); st. leafless or with one leaf, l. terete or slightly flattened above subulate, spath ovate pointed about equalling the flowers, umbel many-flowered globose without bulbs, stam. about half the length of the lanceolate segments of the perianth.—*E. B.* 2441.—St. about 6 in. high. L. straight, mostly with smooth ribs. Pet. lanceolate. Barren bulbs with leaves. Fl. pink. Forming dense tufts.—*β. A. sibiricum* (L.); l. curved and bent downwards with crenulate ribs, pet. lanceolate-attenuate, barren bulbs single-leaved, style longer than the young germen. *E. B. S.* 2934. Height 6 in. to 2 feet. Heads large. Possibly a distinct species.—Meadows and pastures in mountainous situations. *β.* Rocks and cliffs near the sea. Tintagel and Rill Head, also between Kynance Cove and Mullion, Cornwall. P. VI. VII. *Chives.* E. S. I.

**** *Stam. all simple. Leaves flat, all radical.*

†8. *A. triquetrum* (L.); st. triquetrous, *l. linear* acutely folded and keeled, spath 2-valved about equalling the erect bulbless lax umbel, stam. half as long as the oblong segments of the perianth.—*E. B. S.* 2963. *R.* x. 503.—Bulb ovate. L. angularly folded, acute. Segments of perianth white with a slender green midrib.—Hedges in Guernsey. Helston, Corn. P. V. VI. E.

9. *A. ursinum* (L.); st. naked triangular, *l. stalked ovate-lanceolate*, spath 2-valved ovate, umbel level-topped lax bulbless.—*E. B.* 122.—Bulb slender, oblong. L. few, broad, smooth, bright green. Stalk one, as tall as or taller than the leaves. Fl. white. Smelling strongly of garlic when bruised.—Damp woods and hedges. P. V. VI. *Ramsons.* E. S. I.

[*A. ambig'uum* (Sm.); *E. B. S.* 2803.—Rochester. Eye Castle Hill, Suff. Not a native; nor is *A. paradoxum* (Don) at Binnig Craig, Linlithgow.]

Tribe III. *Anthericeæ.*

9. PUBILÁRIA *Raf.* (*Simethis* Kunth, ed. viii.)

1. *P. bicolor* (Raf.).—*E. B. S.* 2952.—Root of fleshy fibres. L. linear, flat, or a little keeled upwards. St. and l. enclosed in sheathing scales and surrounded by brown fibres. Fl. panicled. Pet. purple without, white within. Seed-stalks thick, white. Seeds black.—Sandy heaths. Near Bournemouth, Dorset. Near Derrynane, Kerry. P. V. E. I.

Tribe IV. *Hemerocallideæ.*

10. Endym'ion *Dumort.* Blue-bell.

1. *E. nútans* (Dum.); l. linear, raceme nodding, fl. bellshaped cylindrical, tip of the sep. revolute, bracts 2.—*E. B.* 377. *Scilla festalis* (Salisb.). *Agraphis* Link.—Scape about a foot high. Fl. blue, rarely white. Stam. united to the perianth halfway up. L. shorter than the scape.—Woods and thickets. P. V. *English Blue-bell.* E. S. I.

11. Mus'cari *Mill.* Grape-Hyacinth.

1. *M. racemósum* (Mill.); fl. ovoid nodding crowded upper ones nearly sessile abortive, l. linear channelled flaccid recurved at the end.—*E. B.* 1931.—Scape 1 ft. high. Fl. dark blue, scented. Caps. emarginate. Not *M. neglectum* (Guss.), as supposed in ed. 6. The cultivated plant, *M. botryoides,* has globose flowers.—Sandy fields. Plentiful near Cavenham and Pakenham, Suffolk. About Gogmagog Hills, Cambridge. P. IV. V. E.

Order LXXXIX. MELANTHACEÆ.

Perianth inferior, 6—7-parted. Stam. 6, on the receptacle or perianth. Anth. attached below their middle. Ovaries superior, 1 of 3 cells, or 3 of 1 cell more or less connected. Ovules many. Styles 1—3. Fr. bursting inwards, of 3 separate 1-celled follicles, or more or less combined into a 3-celled capsule.

1. Colchicum. Perianth funnelshaped, tube long; limb 6-parted, petal-like. Styles long. Caps. 3, connected throughout, opening at the inner edge, many-seeded — Tuberous.
2. Tofieldia. Perianth 6-leaved. Styles short. Caps. 3, connected to above the middle, 1-celled, opening at the inner edge, many-seeded.—A creeping rhizome.
3. Narthecium. Perianth 6-parted. Style undivided. Caps. 3-celled, loculicidal, many-seeded; seed with a long filiform appendage at each end; placenta basal.

1. Colchicum *Linn.* Meadow-Saffron.

1. *C. autumnále* (L.); l. flat lanceolate erect.—*E. B.* 133.—Root-stock large, tuberous. L. a foot long and often an inch

broad, dark green, smooth, vernal. Fl. bright purple, radical, with very long tubes, autumnal; germen remaining under ground in winter, rises in spring with the leaves. Rarely greenish spring fl. are found.—Meadows. P. IX. X. E. S. I.

2. TOFIEL'DIA *Huds.* Scottish Asphodel.

1. *T. palus'tris*; (Huds.); pedicels with a 3-lobed bract at the base but none at the top.—*E. B.* 536. *T. borealis* Wahl. *St.* 78. 8.—St. 4—8 in. high. L. swordshaped, about 2 in. long, in 2-ranked radical tufts. Fl. in a short dense spike, at first sessile, afterwards slightly stalked. Mountain bogs. P. VII. E. S.

3. NARTHE'CIUM *Huds.* Bog-Asphodel.

1. *N. ossif'ragum* (Huds.); l. swordshaped, pedicels with 1 bract at their base and another above their middle, perianth-segm. linear-oblong exceeding stam. and much falling short of caps.—*E. B.* 535. *St.* 78. 3. *R.* x. 421.—St. 6—8 in. high, slightly leafy, decumbent and rooting below. L. mostly in radical 2-ranked tufts, half the height of the stem. Fl. bright yellow.—Turfy bogs. P. VI. VII. E. S. I.

Order XC. JUNCACEÆ.

Perianth of 6 glume-like scarious segments. Stam. usually 6, on the base of the segments; or 3, opposite to the outer series. Anth. 2-celled, attached by base. Ovary 1—3-celled, superior. Ovules 1, 3, or many in each cell. Style 1, stigmas 3. Fr. capsular, 3-valved, loculicidal. Embryo subcylindrical, very minute, within firm albumen, near the hile.—No raphides.

1. JUNCUS. *Perianth glume-like,* 6-leaved. *Filaments glabrous.* Style undivided. Stigmas 3, filiform. *Caps.* 3-*celled,* 3-valved. Seeds attached to the inner edge of the dissepiments.—L. mostly not flat.
2. JUNCOIDES. Caps. 1-celled, 3-valved, with dissepiments. Seeds 3, at the base of the capsule. Otherwise like *Juncus.* —L. flat, grass-like.

1. JUN'CUS *Linn.* Rush.

* *Barren and fertile stems subulate, with sheathing radical long leaves like the stem, or mucronate sheaths. Seeds with a loose testa forming a sack at each end (appendaged).*

1. *J. marit'imus* (Lam.); sheaths short pale, l. terete sharp-pointed, panicle compound erect, *per.-segm.* equal lanceolate

acute *equalling the elliptic mucronate capsule*, seeds fusiform.—*E. B.* 1725. *R.* ix. 402.—St. erect, 1—2 feet high. Panicle long, lax.—Salt marshes near the sea. P. VII. VIII. E. S. I.

2. *J. acútus* (L.); sheaths long shining, l. terete sharp-pointed, panicle very compound mostly compact, *per.-segm.* equal *half the length of the roundish ovoid caps.*, 3 inner ones blunt with a membranous border, seeds broad-ovate.—*E. B.* 1614. *St.* 71 6. *R.* ix. 401.—St. erect, rigid, with a very sharp rigid point, 3—6 feet high. Panicle dense, corymbose. Fr. twice as large as that of Sp. 1.—Sands on the sea-coast, rare. P. VI. E. I.

** *Barren and fertile stems subulate with sheaths at their bases which are either leafless or bear rudimentary leaves. Seeds with a close testa (not appendaged).*

3. *J. effúsus* (L.); *st. faintly striate soft*, pith continuous, panicle close or diffuse, sheaths dull brown not inflated, *caps.* obovate retuse *not mucronate*, stam. 3.—*E. B.* 836. *R.* ix. 413.—Height 1—2 feet. L. none or minute and slender at the top of sheathing scales. Panicle diffuse, branched; or more or less dense, globose. Anth. oval, short.—Marshy ground. P. VII. E. S. I.

4. *J. conglomerátus* (L.); *st. faintly striate soft*, pith continuous, panicle close or diffuse, from inflated sheaths, *caps.* obovate retuse *mucronate*, stam. 3.—*E. B.* 835. *St.* 71. 3. *R.* ix. 408.—Height 1—2 feet. L. none or minute and slender at the top of the sheathing scales. Panicle globose, dense; or more or less diffuse. Anth. linear. The mucro in the hollowed top of the caps. resembles a little hill bearing the style.—Marshy ground. P. VII. E. S. I.

5. *J. inflex'us* (L.); *st. deeply striate* rigid, *pith interrupted*, panicle loose much branched erect, per.-segm. lanceolate subulate rather exceeding the *elliptic-oblong mucronate capsule*, stam. 6.—*E. B.* 665. *St* 71. 5. *R.* ix. 415. *J. glaucus* (Sibth.) ed. viii.—Panicle ascending, diffuse. Fr. black. St. rigid, slender, glaucous, 12—18 in. high. Sheaths dark. L. none or minute and slender at the top of the scales.—Wet places. P. VII. E. S. I.

6. *J. diffúsus* (Hoppe)[1]; *st. finely striate* rigid, *pith continuous*, panicle loose much branched erect, per.-segm. lanceolate-subulate exceeding the *obovate-blunt mucronate capsule*, stam. 6.—*Sy. E. B.* 1562. *St.* 77. 10. *R.* ix. 414.—Very like the preceding, but the caps much smaller. St. green.—In wet places, rare. P. VII. VIII. E. S. I.

---

[1] Now regarded as a hybrid between Sp. 3 and 5.—H. & J. G.

7. *J. bal'ticus* (Willd.); *st. very faintly striate* rigid, pith continuous, panicle erect slightly branched, per.-segm. ovate-lanceolate acute, *caps. elliptic* scarcely trigonous blunt mucronate.—*E. B. S.* 2621. *St.* 71. 2. *R.* ix. 411.—Creeping widely. L. none or very minute points at the top of the sheathing scales.—Distinguished from *J. arcticus* by its rounded not trigonous capsules; from *J. inflexus* by its far-creeping rhizome, scarcely striate st. and continuous pith.—Sandy and wet sea-coasts of Scotland. P. VII. S.

8. *J. filifor'mis* (L.); st. filiform faintly striate, *panicle simple of few* (about 7) *fl. placed near the middle of the st.*, per.-segm. lanceolate acute, caps. roundish obovate blunt mucronate.—*E. B.* 1175. *St.* 36. 10. *R.* ix. 412.—L. none or as in the preceding plants. St. remarkably slender, about 1 ft. high; small panicles placed very low. Rhizome creeping.—Stony margins of lakes in the North. P. VII. E. S.

*** *Stems mostly leafy, none barren.*

† Fl. capitate or solitary and terminal. Seeds appendaged.

9. *J. castáneus* (Sm.); *st.* with 2—3 *channelled l.*, cymes terminal solitary or 2 or 3, per.-segm. elliptic-lanceolate acute half as long as the ovate-oblong pointed trigonous capsules, *creeping.*—*E. B.* 900. *St.* 71. 14. *R.* ix. 393.—St. 8—12 in. high. Root with lax runners. Leaflike bract exceeding the flowers. Caps. chocolate-coloured, about ¼ inch long. Filaments about twice as long as the anthers.—Micaceous mountain bogs at a great elevation, rare. P. VII. VIII. S.

10. *J. triglúmis* (L.); *st. leafless round, l.* radical subulate *channelled bitubular,* 2 or 3 fl. terminal erect usually equalling the membranous bract, per.-segm. elliptic-oblong blunt falling rather short of the *ovate-oblong blunt mucronate* caps. *cæspitose.*—*E. B.* 899. *St.* 28. 2. *R.* ix. 392.—St. several from one root, 3—6 in. high, perfectly round. Leaflike bract equalling or falling short of the flowers. Caps. chestnut-coloured.—Boggy places on mountains. P. VII. VIII. E. S.

11. *J. biglúmis* (L.); *st. leafless channelled* on one side, *l.* radical subulate *compressed* (*not channelled, nor bitubular*), fl. 2 unilateral upper which is stalked usually falls short of the leaflike bract, per.-segm. oblong blunt falling rather short of the *turbinate retuse trigonous caps.*, root fibrous.—*E. B.* 898.—St. 2—4 in. high, seldom more than one from each root. Caps. light brown with purple margins.—Boggy spots on mountains, rare. P. VIII. S.

12. *J. trif'idus* (L.); *st. with one leaf* on its upper part, basal sheaths awned, upper sheath with a short l., fl. 1—3 *with two setaceous leaflike bracts,* per.-segm. acute, falling short of the rounded elliptic beaked caps., creeping.—*E. B.* 1482. *St.* 71. 12. *R.* ix. 394.—St. crowded, erect, slender, 2—6 in. high. Occasionally the stem-l. is wanting, and sometimes it has a second head in its axil. Perianth and caps. dark brown.—Damp rocky places on mountains. P. VII. VIII. S.

†† Flowers in 1 terminal cluster or 2 one above the other, or in panicled clusters. Seeds not appendaged.

13. *J. capitátus* (Weigel); *st. naked* erect simple, l. radical filiform, head terminal mostly solitary falling short of the setaceous bract, *per.-segm.* unequal, outer ovate-lanceolate, *acuminate-aristate twice as long as the truncate apiculate caps.*, stam 3. —*E. B. S.* 2644.—Plant 1—4 in. high. L. half as long as the stems. Heads large, of 3—10 sessile flowers.—Land's End and Lizard district, Corn.; Guernsey and Jersey. A. VI. VII. E.

14. *J. obtusiflórus* (Ehrh.); 1-leaved st. and internally jointed *l. terete,* panicle repeatedly compound spreading divaricate, *per.-segm. equal blunt* equalling the ovate acute trigonous (pale brown) capsule.—*E. B.* 2144. *R.* ix. 404.—Erect, 2—3 feet high. St. and l. not compressed. Segm. of perianth pale, often purplish, quite blunt or with a small inflexed point.—Marshes, rare. P. VII.—IX. E. S. I.

15. *J. acutiflórus* (Ehrh.); 3—4-leaved st. and internally jointed *l. subcompressed,* panicle compound pyramidal, *per.-segm. acute inner ones longest* all falling rather short of the narrow ovate-acuminate rostrate triquetrous (pale brown) capsule.—*E. B.* 238. *R.* ix. 406.—St. erect, 1½—2 feet high. L slightly compressed. Clusters 5—6-flowered. [β. *multiflorus* (Weihe); clusters fewer larger, per.-segm. equalling fr.]—Boggy places. P. VI.—VIII. E. S. I.

16. *J. articulátus* (L.); 3—6-leaved *st. and* internally jointed *l. compressed,* panicle repeatedly compound erect forked, *per.-segm. equal acute the inner ones blunt* all falling short of the ovate-attenuate mucronate triquetrous (dark brown) capsule.—*E. B.* 2143. *R.* ix. 405. *J. lamprocarpus* (Ehrh.) ed, viii.—St. erect, 12—18 in. high. L. compressed; many internal transverse divisions. Clusters 4—8-flowered.—Boggy places. P. VII. VIII. E. S. I.

17. *J. alpinus* (Vill.); st. 2—3-leaved erect *almost terete,* l.-sheaths with *acute* dorsal angle, pan. compound with nearly erect branches, *per.-segm.* equal *obtuse the outer* 3 *mucronate, fr. oblong-ovoid mucronate.*—*R.* ix. 403.—Resembling Sp. 16 but smaller and more slender, internal transverse divisions of l. less apparent, pan.-branches more erect and slender, fl. fewer (2—5), per.-segm. broader with narrow scarious margin, caps. shorter and blunter.—Mountain bogs. P. VII. VIII. S.

18. *J. nigritel'lus* (D. Don)[1]; 3—4-leaved *st. and* internally jointed *l. nearly cylindrical,* panicle slightly compound erect, sheaths dorsally acute, *per.-segm.* nearly equal (3 inner rather longer and broader) *all acute* falling short of the *linear-oblong* trigonous *beaked* (black) capsule.—*E. B. S.* 2643 (1830), not of *Koch.*—St. erect, 6—12 in. high. L. scarcely compressed. Clusters of more fl. than in Sp. 16. Caps. brown, at length black and glossy, more abruptly pointed than in *J. lamprocarpus.*—Boggy places in the north, and Wells, Norf. P. VII. VIII. E. S. I.

19. *J. bulbósus* (L.); st. filiform, l. setaceous slightly channelled faintly jointed internally, panicle nearly simple irregular long with few distant clusters, per.-segm. equal acute (3 inner rather blunt) nearly equalling the *oblong very blunt mucronate* (pale brown) capsule, anth. as long as their filaments. —*E. B.* 801. *R.* ix. 397. *J. supinus* (Moench) ed. viii. *J. uliginosus* and *J. subverticillatus* Sm.—Extremely variable in size and the direction of its stems, sometimes erect, at others prostrate and rooting at every joint, or [*J. fluitans* (Lam.)] floating. Fl. often viviparous. Stam. 3 or 6.—*β. Kochii*; caps. shorter, stam. 6, filaments nearly twice as long as the elliptic anthers. *J. nigritellus* Koch (1837). *St.* 78. 2.—Boggy and wet places. *β.* Ivy Bridge, Devon (in a bog). Connemara, Galway. P. VI.—VIII. E. S. I.

††† Flowers solitary, remote, or corymbose and forming a terminal panicle. Seeds not appendaged.

20. *J. squarrósus* (L.); *st. leafless* simple, l. linear channelled radical, panicle terminal compound with cymose branches, *per.-segm. ovate-lanceolate* acute or bluntish *equalling the* obovate blunt mucronate *capsule,* anth. 4 times as long as their filaments. —*E. B.* 933. *St.* 36. 11. *R.* ix. 400.—St. erect, 6—12 in. high. L. many, somewhat spreading, rigid, half as long as the stem. Caps. pale brown, shining.—Wet heaths and moors. P. VI. VII. E. S. I.

---

[1] Buchenau considers this an alpine form of Sp. 16.—H. & J. G.

21. *J. compres'sus* (Jacq.); st. with 1 leaf in the middle, l. linear channelled, panicle terminal compound subcymose usually falling short of the bract, *per.-segm. oval-oblong blunt falling short of the roundly obovoid shortly mucronate capsule.—E. B.* 934. *St.* 36. 13. *R.* ix. 399.—St. slender, erect, round and leafy below, naked and compressed above. Floral bracts usually pale. Style half the length of the ovary. Anth. oblong. [A form with condensed panicle is var. *coarctatus* (E. Mey.)].—Damp places. P. VI.—VIII. E. S. I.

22. *J. Gerar'di* (Lois.); st. with one or more leaves, l. linear channelled, panicle terminal compound subcymose usually exceeding the bract, *per.-segm.* oval-oblong blunt *about equalling the oval-oblong strongly mucronate capsule.—J. cœnosus* Bich. Sm. *E. B. S.* 2680. *St.* 71. 8. *R.* ix. 398.—St. trigonous in its upper part. Floral bracts usually shining, brown. Style as long as the ovary. Anth. long.—Salt marshes. P. VI.—VIII. *Mud-Rush.* E. S. I.

23. *J. ten'uis* (Willd.); st. with few l. at the base, l. strongly ribbed with broad sheaths and long slender laminæ, pan. terminal compound *much exceeded by the long slender bracts,* per.-segm. lanceolate *very acute longer than the almost spherical shining caps.,* seeds very small ellipsoid pointed at each end minutely reticulated.—*E. B.* 2174. *J. of B.* xxiii. (1885) p. 1, t. 253.—Cæspitose. Fl. and fr. pale. Style very short. —Sandy ground and roadsides, very local, possibly introduced. P. VII.—IX. E. S. I.

24. *J. bufónius* (L.); st. leafy forked, l. setaceous, fl. solitary unilateral scattered mostly sessile, per.-segm. unequal lanceolate-acuminate exceeding the *oblong blunt capsule, seeds roundly oval.—E. B.* 802. *St.* 36. 12. *R.* ix. 395.—St. 4—8 in. high. Usually with only 1 leaf on the slender stems.—*β. J. fasciculatus* (Bert.); st. shorter (2—3 in. high) and thicker, fl. 2 or 3 together.—Marshy and wet places. A. VII. VIII. *Toad-Rush.* E. S. I.

25. *J. pygmæ'us* (Rich.); st. nearly or quite leafless, l. sub setaceous, fl. in few small clusters sessile, per.-segm. nearly equal narrowly lanceolate exceeding the oblong-acute caps., seeds fusiform-pear-shaped.—*J. of B.* xi. 128.—St. very short. Upper cluster of fl. stalked, naked; lower with a long bract. Seeds much longer than in Sp. 24, apiculate.—Damp spots above Kynance Cove, Cornwall. A. V.? E.

2. Juncoi'des *Adans.* (*Luzula* Cand., ed. viii.) Wood-Rush.

1. *J. sylvat'icum* (O. Kuntze); l. linear-lanceolate hairy, *panicle subcymose doubly compound,* clusters about 3-flowered on

long stalks, per.-segm. bristle-pointed equalling the ovoid-mucronate capsule, filaments very short, seeds minutely tubercled at the end.—*E. B.* 737. *St.* 36. 14. *R.* ix. 390. *L. maxima* DC.—Rhizome woody. St. 12—18 in. high. L. broad, shining, striate, with hairy edges. Panicle much exceeding the leaflike bracts.—β. *gracilis* (Rost., under *Luzula*); rt.-l. 1—3 in. long, pan. simple with a single large term. head, overtopped by drooping 1-headed peduncles.—Sandy places. β. Shetland. *Mr. Beeby.* P. IV.—VI. E. S. I.

2. *J. Fors teri* (O. Kuntze); l. linear hairy, panicle subcymose only slightly branched, *ped. 1-flowered erect* with both fl. and fr., style equalling stam., filaments about as long as the anth., *caps. acute* scarcely falling short of perianth, *seeds with a straight blunt crest.*—*E. B.* 1293. *St.* 77. 2.—St. slender about a foot high. Caps. with 3 acute angles, not suddenly contracted above.—Thickets, rather rare. P. V. E.

3. *J. pilósum* (O. Kuntze); l. lanceolate hairy, panicle subcymose only slightly branched, *peduncles* 1—3-*fl.*, upper ones reflexed after flowering, fl. solitary, style (excluding stigmas) exceeding stam., filaments about half as long as the anth., *caps. blunt* exceeding the perianth, *seeds with a falcate crest.*—*E. B.* 736. *St.* 77. 3. *L. vernalis* (DC.).—St. slender, 6—12 in. high. Caps. ovoid, trigonous, suddenly contracted above.—*L. Borreri* (Bromf.) is a sterile form with shorter and acuter capsules.—[A hybrid with Sp. 2 ?]—Thickets. P. V. E. S. I.

4. *J. campes'tre* (O. Kuntze); l. linear hairy, *panicle of* 3 *or* 4 ovate dense *sessile or stalked clusters*, per.-segm. lanceolate-acuminate exceeding the blunt apiculate caps., *filaments much shorter than the anthers, seeds nearly globular with a basal appendage.*—*E. B.* 672. *St.* 77. 5.—St. 4—6 in. high. Anth. linear, about 6 times as long as the filaments. Large forms often mistaken for Sp. 5.—Pastures and dry places. P. IV. V. E. S. I.

5. *J. multiflórum* (Druce); l. linear hairy, panicle of many ovate dense sessile or stalked clusters, per.-segm. narrowly lanceolate strongly acuminate exceeding the blunt apiculate caps., *filaments more than ½ as long as the anthers, seeds nearly twice as long as broad with a basal appendage.*—*E. B. S.* 2718. *St.* 77. 7. *L. congesta* Sm.—St. 8—20 in. high. Anth. small, rather short. Clusters on elongate drooping stalks, or (*L. congesta* Lej.) subsessile in a rounded lobed head, or a few stalked. [A nearly glabrous mountain form with few heads on short stiff stalks, and with very dark fr., is referred to *L. sudetica* (DC.).]—Wet and turfy places. P. VI. E. S. I.

6. *J. spicátum* (O. Kuntze); l. narrow slightly channelled hairy, *panicle oblong lobed nodding spikelike,* clusters falling short of their bracts, per.-segm. narrow acuminate bristle-pointed, filaments half as long as the anthers, caps. blunt apiculate, seeds oblong with a very slight basal appendage.—*E. B.* 1176.—St. 3—12 in. high. L. short slender. Spike ½—1 in. long, nodding. Partial bracts tapering, bristle-pointed.—Mountains. P. VII. E. S.

7. *J.'arcuátum* (O. Kuntze); l. channelled slightly hairy, *panicle subumbellate* of few 3—5-flowered *clusters on long drooping peduncles,* per.-segm. broadly-lanceolate bristle-pointed, filaments as long as the anthers, caps. roundish-ovate, seeds oblong blunt or apiculate scarcely appendaged below.—*E. B. S.* 2688.—St. slender, 2—5 in. long. L. short, curved, narrowly linear Panicle of 3—5 small clusters, one nearly sessile, the others on long deflexed stalks.—Highest summits of the Cairngorm and Sutherland Mountains. P. VII. ? S.

[*J. nemorósum* (O. Kuntze), *L. albida* (DC.) with doubly compound pan., clusters of 2–4 whitish or pinkish fl., bracts about equalling pan., anth. subsessile; and *J. niv'eum* (O. K.) with less compound pan., clusters of numerous much larger pure white fl., bracts exceeding pan. and filaments about equalling anth., have been found in several places.]

Order XCI. ERIOCAULACEÆ.

Fl. capitate, unisexual. Perianth very delicate, 2—6-parted. Stam. 2—6, if in 2 rows the inner row more developed. Anth. 2-celled. Ovary superior, 2—3-celled. Ovules solitary, pendulous. Dehiscence of caps. loculicidal. Seeds coated with wings or rows of hairs. Embryo lenticular, on the outside of farinaceous albumen, at the end remote from the hile.

1. ERIOCAULON. Fl. in a compact scaly head. Barren fl. in the centre. Perianth 4—6-fid, the inner segments united nearly to their top. Stam. 4—6.—Fertile fl. in the circumference. Perianth deeply 4-parted. Stigmas 2—3. Caps. 2—3-lobed, 2—3-celled; cells 1-seeded.

1. ERIOCAU'LON *Linn.* Pipewort.

1. *E. septanguláre* (With.); scapes 6—8-striate exceeding the cellular compressed subulate glabrous l., fl. 4-cleft hairy at the end as well as the scales, stam. 4, caps. 2-celled.—*E. B.* 733.—Roots of many white jointed fibres. St. varying in height according to the depth of the water. Fertile fl. 4-parted nearly

to the base; 2 lateral divisons keeled, compressed, blunt, fringed, black. Each fl. with a broad blunt black scale in front which is shorter and broader than it.—Hebrides, especially Skye. West coast of Ireland, especially Connemara. P. VIII. S. I.

Order XCII. TYPHACEÆ.

Fl. monœcious, many, closely placed in cylindrical spikes or in dense globose clusters; barren and fertile on different parts of spike, the males uppermost. Perianth of 3 or more scales or hairs. Stam. 1—6, distinct or monadelphous. Anth. erect.—Fertile fl. Ovary free, solitary, 1-celled; ovule 1, pendulous. Style simple. Stigma unilateral. Fr. dry or spongy. Embryo with a cleft on one side.—Raphides abundant.

1. Typha. Spikes long dense cylindrical, upper part male, lower female. Stam. surrounded with hairs. Anth. 3 together on one filament. Ovary surrounded by hairs ultimately stalked.

2. Sparganium. Fl. with a single 3—4-leaved perianth, in distant dense globose heads, the lower bracteate. Stam. free. Fruit dry, sessile.

1. Ty'pha *Linn.* Reed-mace.

1. *T. latifólia* (L.); l. linear nearly flat, *sterile and fertile parts of spikes not separated,* style exceeding the bristles, stig. oblique ovate-lanceolate.—*E. B.* 1455. *R.* ix. 323.—St. 6—7 feet high. L. overtopping the inflorescence, very broad. Spikes very long, sometimes [var. *media* (Sy.)] there is a very short space between them; fertile blackish brown.—[In *T. Shuttleworthii* (Koch), *R.* ix. 322, the style equals the bristles.]—Ponds and lakes. P. VI. VII. E. S. I.

2. *T. angustifólia* (L.); l. linear channelled below, *sterile and fertile parts of spikes a little separated,* style exceeding the bristles, stig. long filiform.—*E. B.* 1456. *R.* ix. 321.—St. 5—6 feet high, much slenderer than in the preceding. L. very narrow overtopping the inflorescence. Spike very long, slender, separated by a flowerless interval of about an inch; fertile reddish brown.—[*T. gracilis* (Schur), *R.* ix. 320, has a rounded spathulate stigma.]—Lakes and ponds. P. VI. VII. E. S. I.

2. Sparga'nium *Linn.* Bur-reed.

1. *S. erect'um* (L.); l. triquetrous at the base, *st. branched*

above, fl. sessile [per.-segm. of female heads broad several-nerved with but slightly enlarged tip], stigma linear, ripe fr. obpyramidal-cuspidate [with short stout beak], seeds few-ribbed.—*E. B.* 744. *R.* ix. 326. *S. ramosum* (Huds.) ed. viii.—St. erect, about 2 ft. high; lower branches with several heads, 1—3 fertile, the rest barren. Heads spherical. L. long linear, erect.—[β. *micro-carpum* (Neum. under *S. ramosum*); smaller, st. less branched, fr. smaller less angular and less abruptly narrowed into a longer beak]. Ditches. P. VI. VII. E. S. I.

2. *S. neglec'tum* (Beeby); root-l. triquetrous at the base, *st. branched above*, per.-segm. of female fl. narrow *usually* 1-*nerved* with much enlarged tip, stigma linear-lanceolate, *ripe fr. oblong-obovoid obscurely angled narrowed gradually into a rather long tapering beak.*—*J. of B.* xxiii. (1885) p. 193, t. 258.—Resembling Sp. 1 in habit, inflorescence less branched and less spreading.—Wet places. P. VI.—VIII. E. S. I.

3. *S. sim'plex* (Huds.); l. trigonous at the base, *st. simple*, stigma linear-subulate, fr. slightly stalked subfusiform, seeds smooth.—*E. B.* 745. *R.* ix. 325.—St. 1—2 ft. high, long, erect or ascending. L. long, often [var. *longissimum* Fries] floating (and then often mistaken for *S. natans*); sheath slightly furrowed, not inflated. Heads many; barren several, sessile; fertile shortly stalked, especially the lowest. Fr. narrowed into a long beak, elliptic-fusiform.--Ditches. P. VII. E. S. I.

4. *S. nátans* (L.); st. simple flaccid, floating l. very long linear flat at the base from a dilated sheathing base, heads many distant, lower fertile heads stalked, *male heads several sessile*, stig. linear-lanceolate, *fr. stalked oblong not longer than its subulate beak.*—*S. affine* Schn. *Sy. E. B.* 1389. *R.* ix. 417.—St. much thicker than in the next plant. L. grass-green, very long. Male heads fewer in our plant than in the Swedish.—Lakes, rare. P. VIII. E. S. I.

5. *S. min'imum* (Fr.); st. simple flaccid, l. linear floating blunt not dilated at the base, heads few racemose or spiked distant, *usually only* 1 *male head*, stig. short oblong-lanceolate, *fr. sessile ovoid shortly beaked.*—*S. natans* Sm. *E. B.* 273. *R.* ix. 324.—St. slender. L. pale, pellucid, long. Lowest fertile heads sometimes very shortly stalked. [Often 2 male heads in Ireland (A. G. More). *S. fluitans* Fr.?]—Lakes and ditches. P. VII. VIII. E. S. I.

Order XCIII. ARACEÆ.

Fl. monœcious or perfect, placed on a spadix and often in a spath. Barren and fertile fl. usually on different parts of the spadix. Perianth none, or of 4—8 scales. Stam. many or definite. Anth. turned outwards. Ovary free, with 1 or more cells. Stigma sessile. Fr. succulent. Embryo slit on one side. *Arum* has abundant raphides, *Acorus* none.

Tribe I. *ORONTEÆ.* Spath like a continuation of stem. Fl. perfect.

1. Acorus. Fl. on a sessile spadix, appearing lateral. Perianth 6-leaved, inferior, persistent. Stam. 6, filiform.

Tr. II. *AREÆ.* Spath convolute at the base. Fl. monœcious. Perianth 0.

2. Arum. Perianth 0. Male fl. of 1 sessile 2-celled anther. Fem. fl. placed lower, of 1 pistil. Top of spadix naked.

1. Ac'orus *Linn.* Sweet Flag.

1. *A. Cal'amus* (L.); st. with a long leaflike prolongation (or spath) beyond the spadix.—*E. B.* 356. *R.* x. 429.—St. 5—6 feet high, resembling the l., swordshaped, flattened. Spadix completely covered by the flowers, 2—3 in. long, lateral. St. and l. sweet-scented when crushed.—In water, rare; except in Norf. and Suff. *Between Lisburn and Moira, Co. Down. P. VI. VII. E. I.

2. A'rum *Linn.* Cuckoo-pint.

1. *A. macul'átum* (L.); *l. vernal* all radical hastate-sagittate with *deflexed lobes*, petiole as long as leaf-limb, spadix club-shaped straight falling short of the spath.—*E. B.* 1298. *R.* vii. 8.—Root tuberous. L. with branching veins, green or spotted with purple. Spath ventricose below and above, constricted in the middle, with inflexed edges when open. *Spadix usually purple* blunt; with ovaries at the base; above them whorls of stamens; then a few filaments, probably abortive pistils; club naked. Berries scarlet, remaining after the rest of the plant has disappeared; seeds mostly 4 or 3, or rarely 2.—Hedge-banks and thickets. P. IV. V. E. S. I.

2. *A. ital'icum* (Mill.); *l. appearing before the winter* all radical triangular-hastate with *divaricate lobes*, petiole longer than leaf-limb, spadix clubshaped straight falling short of the widely spread spath.—*R.* vii. 11.—L. dark blue-green, sometimes with yellowish veins, rarely spotted, blunt. *Spath* ventricose below, opening nearly flat and very broad above, *folding down in front* when fl. are in perfection so as to close the opening like a flap, ultimately

bent or folded double over the *yellow spadix*. Abortive pistils very long, both above and below the stamens.—W. Cornw. S. Devon. I. of Wight. Channel I. E.

Order XCIV. LEMNACEÆ.

Fl. monœcious, 2, in a spath, but without a spadix (rarely found). Perianth 0. Stam. 1—2, distinct. Ovary 1-celled. Style short. Stigma simple. Fr. bladdery, not bursting. Seeds with a coriaceous ribbed testa.—Floating, leaflike, small, proliferous; no distinction of st. or leaf. Fl. very minute.

1. Lemna. Spath membranous, urnshaped. Fl. from lateral cleft of fronds. Stam. 1—2. Anth. 2-celled, didymous (cells bilocular?). Fronds with capillary roots beneath — Increasing chiefly by offsets.

2. Wolffia. Spath 0. Fl. from upper surface of frond. Stam. 1; anth. 1-celled, sessile. Frond very minute, rootless.—Increasing by offsets.

1. Lemna *Linn.* Duckweed.

1. *L. trisul'ca* (L.); *fronds* thin pellucid *elliptic-lanceolate* tailed at one end serrate at the other, roots solitary.—*E. B.* 926. *R.* vii. 15.—Fronds half an inch long, proliferous at right angles. Plants submerged, truly annual, producing autumnal bulblets which survive the winter as in the other species.—In stagnant water. A. VI. E. S. I.

2. *L. minor* (L.); *fronds obovate compressed* opaque, roots solitary blunt.—*E. B.* 1095. *R.* viii. 14.—Fronds 1—2 lines long, nearly flat beneath, of a compact texture.—On stagnant water. A. VI. VII. E. S. I.

3. *L. gib'ba* (L.); *fronds obovate* nearly flat above *hemispherical* and spongy *beneath*, roots solitary blunt. —*E. B.* 1233. *R.* vii. 14. *Telmatophace* Schleid., Endl.—Fronds 1—2 lines long, at first flattish, afterwards remarkably gibbous and cellular beneath.—On stagnant water. A. VI.—VIII. E. S. I.

4. *L. polyrrhiza* (L.); *fronds roundish-obovate, compressed, roots many* clustered acute.—*E. B.* 2458. *R.* vii. 15. *Spirodela* Schleid., Endl.—Fronds half an in. long, green above, purple beneath. Fl. not seen in Britain.—On water. A. E. S. I.

2. WOLFFIA *Horkel.*

1. *W. Michel'ii* (Schleid.); *fronds very small subglobular* flattish above cellular beneath, solitary, young frond separating immediately from the old one.—*Sy. E. B.* 1398. R. vii. 14. *W. arrhiza* (Wimm.) ed. viii.—Frond like a grain of sand, subglobular at all ages, green. Offset from within the base of the old frond.—Fl. not seen in Europe. Ponds near London. A. E.

Order XCV. POTAMOGETONACEÆ.

Fl. perfect or imperfect. Perianth inferior, 4-parted, or 0. Stam. free, 1, 2, or 4. Ovaries 4, distinct, each with 1 ovule and 1 sessile stigma. Fr. a drupe enclosing a hard nut, or a dry nut, not bursting, 1-seeded. Albumen 0. Embryo with a thin skin having a lateral cleft.

1. POTAMOGETON. Fl. perfect. Perianth 4-parted. Anth. 4, sessile, opposite to the divisions of the perianth. Ovaries 4. Styles 0. Drupes 4, sessile.—Fl. sessile, spiked.

2. RUPPIA. Fl. perfect. Perianth 0. Stam. 2, the cells considerably separated; filaments very short, scalelike. Ovaries 4. Styles 0. Nuts 4, with long stalks.—Fl. about 2 together.

3. ZANNICHELLIA. Fl. monœcious, axillary. Barren with 1 stam., and no perianth. Fertile with a bellshaped perianth, persistent style, and peltate stigma. Nuts 2—5 or more, more or less stalked.

1. POTAMOGE'TON[1] *Linn.* Pondweed.

* *L. alternate, floral l. floating and sometimes opposite; stipules free.*

1. *P. nátans* (L.); upper l. stalked coriaceous floating ovate or elliptic folded at the base, *petiole jointed a little below the limb*, lower linear-lanceolate or setaceous, *fr.* (*large*) *rounded on the back when fresh keeled when dry*, peduncle equal.—*E. B.* 1822. *R.* vii. 50.—St. creeping below, simple. *Petioles plane-concave.*

---

[1] See Mr. Fryer's magnificent *Potamogetons of the British Isles*, now in course of publication, and his and Mr. Arthur Bennett's papers in the *Journal of Botany*. We are much indebted to Mr. Fryer for assistance with this genus and we have followed his views in adding several species.
H. & J. G.

L. subcordate below, when pressed flat a ridge is formed on each side of the base, *jointed to their stalks a little below the limb.* Sep. stalked, roundly rhomboidal. Anth.-cells not parallel. *Fr. greenish,* slightly compressed, 1½—2 lin. long.—[Var. *prolixus* (Koch) is an early state having longer and narrower thin and semi-pellucid l.—*P. fluitans* (Roth), *J. of B.* xxvi. (1888) p. 273, is probably *P. natans* × *lucens*, and *P. crassifolius* (Fryer), *J. of B.* xxviii. (1890) p. 321, t. 299, is *P. natans* × *Zizii.*]—Ponds, ditches, and slow streams. P. VI. VII. E. S. I.

2. *P. polygonifólius* (Pourr.); l. all stalked, upper subcoriaceous floating oblong-elliptic or lanceolate subcuspidate, *no leafless petioles,* lower l. linear-lanceolate, *fr. minute blunt and rounded on the back,* peduncle equal.—*P. oblongus* Viv., *E. B. S.* 2849.—St. creeping below. *Petioles* longer than leaves, convex on both sides, *not jointed below the limb.* Lower l. often very narrow. Spikes rather short and irregular. Sep. transversely elliptic, stalked. *Fr. reddish,* scarcely compressed, 1 *lin. long*; a faint keel and lateral ridges when dry. *β. pseudo-fluitans* (Syme); submerged l. membranous linear-lanceolate narrowed at both ends. [*γ. cancellata* (Fryer); floating l. few thinner, submersed l. strongly net-veined.]—Ditches, small streams, and ponds. *β.* Buttermere. Gap of Dunloe, Kerry. P. VII. E. S. I.

3. *P. Griffith'ii* (Ar. Benn.); submersed l. strap-shaped usually sessile or half-clasping wavy somewhat hooded pellucid 9—16-veined, floating l. obovate-lanceolate blunt long-stalked more opaque, stip. long usually exceeding the internodes, ped. slender nearly equal.—*J. of B.* xxi. (1883) t. 235.—St. branched. Floating l. green, submersed l. slightly brownish. Sep. "roundish-oval broader than long." Spikes short. Mr. Fryer considers this a hybrid, *P. polygonifolius* × —— ?—Llyn-an-afon, Carn. *Mr. J. E. Griffith.* P. VI.—VIII. E.

4. *P. Drúcei* (Fryer); submersed l. elliptic-lanceolate pellucid *strongly reticulate-veined throughout very long stalked,* floating l. obovate elliptic or oval narrowed at both ends, stip. very long, fr. broadly obovoid *acutely keeled, keel tubercled with conspicuous angles at the base,* ped. thicker than the st. narrowed at both ends. A very distinct plant, producing but little fruit and possibly a hybrid.—R. Loddon, Berks. *Mr. G. C. Druce.* P. VII. VIII. E.

5. *P. colorátus* (Hornem.); *l. all shortly stalked membranous and pellucid* blunt not cuspidate nor plicate, upper elliptic, lower l. oblong, *fr. minute* rounded on the back when fresh *keeled when dry,* peduncle equal.—*E. B. S.* 2848. *R.* vii. 45. *P. plantagineus* (Ducr.) ed. viii.—St. creeping below, branched, sometimes throwing out long runners from its upper axils. L. all beautifully transparent and netted with veins, the upper often

almost sessile and nearly orbicular, sometimes slightly cuspidate. Petioles plane-concave above. Spikes long, cylindrical. Anth.-cells nearly parallel. Sep. ovate. Fr. greenish, ¾ lin. long. [*P. Billupsii* (Fryer) *J. of B.* xxxi. (1893) p. 353, t. 337–8 is *P. coloratus* × *Zizii*.]—Stagnant peaty water. P. VI. VII. E. S. I.

6. *P. alpínus* (Balb.); *submersed l.* lanceolate narrowed at both ends *subsessile* membranous pellucid *entire not apiculate* with chainlike network near the midrib, floating l. subcoriaceous, obovate blunt narrowed into a short petiole, stip. without wings, fr. acutely keeled, peduncle equal.—*E. B.* 1286. *R.* vii. 32. *P. rufescens* (Schrad.) ed. viii. *P. fluitans* Sm.—St. simple. Upper l. alone slightly coriaceous, often tinged with purple, longer than their stalks; submersed l. all nearly, if not quite, sessile. Sep. "transversely oval."—Ditches and slow streams. P. VII. E. S. I.

7. *P. lanceolátus* (Sm.); *submersed* l. linear-lanceolate sessile *entire acute not apiculate with chainlike network* near the midrib, floating l. subcoriaceous elliptic-lanceolate shortly stalked, stip. almost subulate uppermost broader with 2 dorsal elevated ribs, *peduncles equal* from uppermost floating opposite leaves.—*E. B.* 1985. *J. of B.* xix. t. 217 (not *R.*).—St. very slender, slightly branched. Floating l. not always present with the flowers. No fruit even on Sept. 28. [Mr. Fryer considers this *P. heterophyllus* × *pusillus*.]—By the bridge at Penrbos Lligwy, Anglesea. Burwell Fen, Cambr. Clare. Galway. P. VII.—IX. E. I.

8. *P. sparganifólius* (Laest. !); *submersed l. linear* narrowed at both ends *very long* sessile entire not apiculate *with many parallel veins next the midrib*, floating l. subcoriaceous lanceolate long stalked, stip. very long blunt not winged, fr. "acutely keeled" (Fr.), ped. rather slender not clavate.—*Sy. E. B.* 1403. *P. Kirkii* (Syme).—St. very slender, slightly branched. Floating l. often wanting. Submersed l. sometimes 2 ft. long, very narrow, few-veined with additional parallel veins closely placed next the midrib. L. very much longer than those of *P. lanceolatus* and without the chainlike network, beautifully green when well dried. Exactly the plant of *Fries, H.N.* xii. 75. [Mr. Fryer does not consider this the plant of Laest. but probably *P. natans* × *polygonifolius*.]—In the river at Ma'am, Co. Galway. I.

9. *P. graminifólius* (Fr.); *submersed l. narrowly strapshaped* sessile narrowed at both ends entire not apiculate, floating l. subcoriaceous elliptic long-stalked, stip. long bluntish not winged, fr. obliquely obovate 3-keeled.—*Sy. E. B.* 1404. *P. sali-*

cifolius, ed. viii. (not Wolfg.)[1]—St. slightly branched. Submersed l. with a few additional parallel veins next midvein, subsessile. Floating l. often wanting. Ped. long, slightly thickened upwards. In river Boyne near Navan. [R. Lawne, Killorglin. Pidley Fen, Hunts.] P. E. I.

10. *P. heterophyllus* (Schreb.); *submersed l. lanceolate narrowed at both ends sessile minutely denticulate and apiculate*, floating l. subcoriaceous elliptic stalked, *stip.* broadly lanceolate blunt *with* 2 *stout prominent ribs*, lower ones linear-lanceolate, *fr. blunt on the back, peduncle thicker than the stem swelling upwards.*—*E. B.* 1285. *R.* vii. 41—43. *P. gramineus* Fries, not *Linn.*—St. much branched below. Submersed l. wavy. Lower stip. without the two strong ribs and equally veined, upper ones widely spreading. Dry fr. slightly marked with 3 ridges on the back.—Ponds and ditches. P. VI. VII. E. S. I.

11. *P. nitens* (Web.); *submersed l.* lanceolate *rounded below and half clasping* wavy at the edge, floating l. "coriaceous elliptic stalked," stipules equally veined, (dry) fr. keeled, ped. swelling upwards.—*R.* vii. 34. *Sy. E. B.* 1407. *J. of B.* ii. t. 23. —St. rather wavy, branched below. Subm. l. pellucid, often recurved; upper l. often not coriaceous. Upper stip. large, persistent, nearly ½ as long as the peduncle. Dry fr. with 3 ridges on the back. [Mr. Fryer considers this a series of hybrids *P. graminifolius* × *perfoliatus*, *P. heterophyllus* × *perfoliatus*, and *P. Zizii* × *perfoliatus*, and writes "rarely fruiting, with 1 or 2 drupelets on a spike, probably resulting from a chance fertilization by allied species."]—Lakes near Brandon Mountain, Kerry (*P. curvifolius* Hartm.). Loch Ascog, Bute. River Tay, Perth. Coltfield, Elgin, &c. P. VII. VIII. E. S. I.

12. *P. falcátus* (Fryer); submersed l. sessile *elliptical strap-shaped often not quite symmetrical* with somewhat rounded base, the upper amplexicaul, the margin entire slightly undulated, floating l. stalked elliptic mucronate coriaceous, stip. herbaceous persistent upper broad, fr. small rhomboidal-ovoid acutely keeled with distinct lateral ridges, ped. thicker than the st. scarcely swollen upwards usually shorter than the l.—*J. of B.* xxvii. (1889) p. 65, t. 286.—St. branched from near the base.—Fen ditches, Hunts. P. VI. VII. E.

13. *P. varíans* (Fryer); submersed l. sessile or shortly stalked narrowly lanceolate usually flat, floating l. stalked oval or more or less spathulate apiculate coriaceous, stip. herbaceous persistent narrow bluntish, fr. small nearly circular flattened very acutely keeled with prominent lateral ridges, beak short, ped. slightly thicker than the st. *not thickened upwards from*

---

[1] Mr. Ley's plant from Sellack which according to the Author's MS. is probably *P. salicifolius*, is considered by Mr. Fryer to be a hybrid of unknown origin.—H. & J. G.

the base of the uppermost submersed l.—*J. of B.* xxvii. (1889) p. 33. t. 287. *P. spathulæformis* (Morong.).—St. usually but little branched. Fen ditches. Cambs. P. VIII. E.

** *L. usually all submersed, membranous, lanceolate or broader; stipules free.*

14. *P. Zizii* (Roth); l. pellucid *narrow-lanceolate-oblong* cuspidate *narrowed downwards* minutely denticulate near the end *wavy, stip. long broad* broadly and doubly keeled not winged, fr. with 3 keels, ped. swelling upwards, very long.—*J. of B.* xvii. p. 291, t. 204.—St. long, much branched, lower l. usually recurved. Sometimes with aerial l. short stalked rather coriaceous oblong acute at both ends. Ped. straight, 4—6 in. long, thicker than the stem exceeding the l. terminal. Spike about 1 in. long. [*P. involutus* (Fryer) is *P. Zizii* × *perfoliatus.*]—Lakes, rivers, &c. P. VII. VIII. E. S. I.

15. *P. lucens* (L.); *l.* pellucid oval or lanceolate *stalked minutely denticulate* and cuspidate, *stip. winged*, fr. rounded on the back and slightly keeled when fresh, *peduncle swelling upwards*, spikes cylindrical densely flowered.—*E. B.* 376. *R.* vii. 36.—All the stipules have two prominent wings on their back. Spikes long, about equalling the peduncles.—A form (?) has been found with ovate-lanceolate subacute subcoriaceous floating leaves (var. *coriaceus* Nolte? *R.* 39) which looks different. [Var. *coriaceus* (Nolte) is thought by Mr. Fryer to be probably *P. heterophyllus* × *Zizii*. He writes "upper l. coriaceous like *heterophyllus* but petiole shorter than lamina."] Sometimes (*P. acuminatus* Schum.) the midrib forms a long prominent point. Common in deepish water. P. VI. E. S. I.

16. ?*P. decip'iens* (Nolte); *l.* pellucid *oblong rounded below not hooded* but blunt mucronate and wavy *at the end entire* sessile or uppermost l. sometimes subsessile, stip. not winged[1], fr. "like that of *P. lucens*," ped. equal.—*J. of B.* v. t. 61. *R.* vii. 35.—St. long, much branched. Ped. 1—2 in. long, thicker than the stem. Spikes 1—2 in. long, dense. [*P. lucens* × *perfoliatus* Fryer.]—Canal near Bath. Forfar. Navan Canal, [&c.]. P. VII. E. S. I.

17. *P. longifólius* (Bab. not Gay); *l.* pellucid *long-lanceolate* narrowed below *entire* cuspidate, stip. winged, fr. . . . . ., *ped. long swelling upwards*, spikes with few rather whorled distant flowers.—*E. B. S.* 2847.—St. long, slender. L. distant, *not*

---

[1] I cannot find the wings described by Syme; nor did Mr. Baker. *J. of B.* v. 71.

denticulate, nearly sessile, with very short stalks. Stip. green, lanceolate, with 2 narrow wings on the back. Ped. 6—8 in. long, much thicker than the stem. Spike 1 in. long; fl. rather more whorled than as represented in *E. B. S.* Fr. unknown.—Deep water. Found floating loose in Lough Corrib, Galway. [Mr. Fryer refers this to *P. lucens* and Mr. Bennett to *P. lucens* × *prælongus*.] P. VIII. I.

18. *P. prælon'gus* (Wulf.); *l.* pellucid *oblong-lanceolate half-clasping blunt and hooded at the end entire, stip. not winged, fr.* keeled on the back when fresh *keeled* or winged *when dry*, peduncle very long equal, spikes many-flowered.—*E. B. S.* 2858. *R.* vii. 33.—St. long, growing in deep water. L. with parallel veins adjoining the midrib. Peduncles 6—12 in. long. Spikes 1—2 in. long.—Rivers and ditches, rare. P. V. VI. E. S. I.

19. *P. perfoliátus* (L.); *st. round, l.* pellucid *cordate-ovate* or ovate-lanceolate clasping not hooded, fr. rounded on the back when fresh keeled when dry, peduncle equal.—*E. B.* 168. *R.* vii. 29.—St. long, slightly branched, rather dichotomous at the top. Peduncles rather thick and short. Spikes short.—[A state with l. elongate almost lanceolate has been referred to var. *lanceolatus* (Blytt).]—Lakes and streams. P. VII. E. S. I.

20. *P. cris'pus* (L.); *st. compressed, l.* pellucid *linear-oblong* blunt sessile serrate wavy, *fr. with a long beak* keeled on the back when dry [base of keel with a long projecting tooth, *A. Fryer*], peduncle equal.—*E. B.* 1012. *R.* vii. 29—30.—St. much branched. Ped. long. Spikes few-flowered. Beak as long as the nut. L. usually crisped at the edges, occasionally flat (*P. serratus* L.). [*P. Cooperi* (Fryer) is *P. crispus* × *perfoliatus* and *P. Bennettii* (Fryer) is *P. crispus* × *obtusifolius*.]—Ditches and streams. P. VI. E. S. I.

*** *L. all submersed, alternate, linear-ligulate; stipules free.*

21. *P. zosterifólius* (Schum.); st. flattened, *l.* linear-acuminate *with* 3 *principal and many* close parallel *intermediate veins* occupying the whole surface, *spikes cylindrical upon long ped.*, sep. transversely oval, fr. obovate keeled, style terminal.—*E. B. S.* 2685. *R.* vii. 27. *P. cuspidatus* Sm., *P. compressus* Fries, Hook.—Ped. 2—4 in. long. Spikes 10—15-flowered. Inner edge of fr. rounded; faces a little convex. L. suddenly acuminate or apiculate.—Rivers and lakes. P. VI. E. S.

22. *P. acutifólius* (Link); st. flattened, *l.* linear-acuminate *with* 3 *principal and many* close parallel *intermediate veins* occupying the whole surface, *spikes ovate about as long as the short ped.*, sep. rhomboidal, fr. broadly ½-obovate compressed *inner*

edge straight with a tooth near its base keeled, style facial.—*E. B. S.* 2609. *R.* vii. t. 26.—Ped. very short. Spikes 4—6-flowered. L. gradually acuminate. Styles hooked, continuing the inner edge of the fruit.—Marsh ditches, rare. P. VI. E.

23. *P. obtusifólius* (M. & K.); st. slightly compressed with rounded edges, *l. linear 3-veined, spikes ovate dense continuous* about as long as the short ped., sep. rhomboidal, fr. obovate keeled.—*R.* vii. t. 25. *P. gramineus* Sm., *E. B.* 2253.—St. slender, much branched. Peduncles very short. L. rounded off to a slight point at the end, with oblong network near the midrib, wanting the fine parallel veins of the two preceding species.—Ponds and ditches, rare. P. VI. VII. E. S. I.

24. *P. Fries'ii* (Rupr.); st. slightly compressed, *l. linear 5-veined* suddenly apiculate, *spikes short lax* ½ or ⅔ shorter than the *compressed clavate ped., sep.* transversely *oval,* fr. obliquely ovate bluntly keeled.—*P. compressus* Sm. *E. B.* 418. *R.* vii. 24, not *Fries,* nor *Koch. P. mucronatus* (Schrad.) ed. viii.—Lateral veins nearer together and nearer to the margin of the l. than to the midrib (rarely some l. are 3-veined), no intermediate veins. —Like Sp. 21 rather than Sp. 23.—Ditches. P. VI. VII. E. S. I.

25. *P. pusil'lus* (L.); st. subterete, *l. linear* [usually] *3-veined,* spikes short rather lax ½ or ⅔ shorter than the slender ped., sep. roundish-reniform, *fr. obliquely ovate* bluntly keeled.—*E. B.* 215. *R.* vii. 22.—St. slender. L. narrow, rather acute, without intermediate parallel veins, *the lateral veins equidistant between the midrib and margin.* Ped. scarcely compressed, not thickened. [*β. tenuissimus* (M. & K.); l. very narrow 1-veined, fr. more compressed. *R.* vii. 22, f. 39.—*γ. P. Berchtoldi* (Fieber); l. shorter and broader, fr. thicker more distinctly warty. *R.* vii. 22. f. 37. *δ. Sturrockii* (Ar. Benn.); st. very slender somewhat compressed, l. extremely thin obtuse mucronate 3-5-veined; ped. very slender, beak of fr. short.] Ponds and ditches. P. VI. E. S. I.

26. *P. rútilus* (Wolfg.); st. slender *compressed,* l. linear the lowest (soon decaying) quite obtuse the upper *with long acuminate points,* with 1 pair of well-marked lateral veins, extending almost to the apex, *stip. long often with a long tapering point,* ped. 2—3 times as long as the rather dense spike, fr. oblong elliptic *with obscure keel, inner edge nearly straight.* —*R.* vii. 23. *J. of B.* xxxviii (1900) t. 407.—L. firmer and much more acute and the fr. less oblique than in Sp. 25. Recorded from Sussex, Warw. or Staff., Anglesea. P. VII. E.

27. *P. trichoïdes* (Cham.); st. subterete, *l. subsetaceous 1-veined* finely pointed, spikes short lax long-stalked, fr. transversely reniform obscurely keeled, *straight inner edge with a tooth near its base.*—*Sy. E. B.* 1420. *R.* vii. 21.—St. very

slender, a little thickened below the joinings; branches fasciculate. L. not transversely veined. Floral stip. large. Ped. not thickened. Sep. roundish, stalked. Fr. often warted on the back and with a tubercle on each side at its base when fresh. —Norf. Suff. Camb. Devon. Surrey. P. VII. VIII. E.

**** *L. all submersed, alternate, linear, sheathing.*

28. *P. flabellátus* (Bab.); *lower l.* broadly linear abruptly apiculate or acuminate 3—5-*veined* with transverse veins, *upper l.* narrow acute 3-*veined*, fr. (2 lin. long) broadly ½-obovate *inner edge nearly straight* but gibbous near the top rounded on the back, nut with a prominent keel.—*Sy. E. B.* 1421. *Phytol.* iv. 1158.—Rhizome spreading, from a tuber that has outlived the winter. Floating st. branched, wavy, spreading like a fan. Broad l. usually decayed at the time of flowering. Lateral veins of upper l. at the margin, of lower l. distant from it. Spikes slightly interrupted. Back of fr. without ridges, rounded when fresh; enclosed nut with faint lateral ridges. [A slender maritime form with setaceous l. is var. *scoparius* (Fryer).]—Ponds and ditches chiefly near the sea. B. VI. VII. E. I.

29. *P. pectinátus* (L.); *l. formed of* 2 *interrupted tubes*, lower narrowly linear flattened slightly grooved above, upper setaceous, fr. (2 lin. long) broadly ½-obovate inner edge rather convex rounded on the back, *nut with 2 lateral ridges but* (usually) *no keel.*—*E. B.* 323.—St. branched, forming linear masses. L. very gradually acute, all 1-veined, no marginal veins and scarcely thickened there; upper l. with an oval section. Spikes slightly interrupted. Back of fr. without ridges and rounded when fresh; nut with strong lateral ridges. Varying greatly in length and size of st. and leaves.—*P. marinus* (Huds.) has the stems naked below. *Sy. E. B.* 1423.—Ponds and streams. P. VI. VII. E. S. I.

30. *P. filifor'mis* (Pers.); *l. linear-setaceous* 1-veined with transverse veins, *spikes greatly interrupted, fr.* (1½ lin.) obovate rugose *rounded on the back without keel or ridges* when dry, nut round-backed.—*Sy. E. B.* 1424.—L. like those of the preceding but longer. Fr. smaller. Whorls very distant on the spikes. *Peduncles very long.*—Lakes, rare. P. VI. VII. E. S. I.

***** *L. all opposite, submersed; stipules none.*

31. *P. den'sus* (L.); *l. all opposite* pellucid clasping elliptic-lanceolate or lanceolate, spikes shortly stalked ultimately reflexed.—*E. B.* 397. *R.* vii. t. 28.—L. crowded, rather recurved. Spike 4-flowered. Sep. triangular.—Ditches. P. VI. VII. E. S. I.

2. Ruppia *Linn.* Tassel-Pondweed.

1. *R. marit'ima* (L.); *ped. long spirally twisted*, anth.-cells oblong $1\frac{1}{2}$ times as long as broad, nut ovoid rather obliquely erect.—*E. B.* 136. *R.* vii. 17. *R. spiralis* Hartm.—Whole plant stronger than the next. L. very narrowly linear, formed as in *Potamogeton pectinatus*. *Sheaths large, inflated*.—Salt marshes. P. VII. VIII. E. S. I.

2. *R. rostelláta* (Koch); *ped. short not spiral*, anth.-cells squarish as long as broad, nut gibbous at base obliquely ascending when young.—*Sy. E. B.* 1428. *R.* vii. 17.—Whole plant very slender. L. rather filiform than linear. *Sheaths small, close*.—A very small form (*nana* Bosw.), with very short decurved ped., is found in Orkney. Salt marshes. P. VI.—VIII. E. S. I.

3. Zannichel'lia *Linn.* Horned Pondweed.

1. *Z. palus'tris* (L.); style at least half as long as the divergent nuts.—*E. B.* 1844.—Floating. L. slender, opposite, filiform. Fl. axillary, sessile.—*a. Z. brachystemon* (Gay); *fr.* 2—4 *subsessile* with crenate back, style $\frac{1}{2}$ length of fr., stigma large crenulate, anth. 2-celled.—*β. Z. macrostemon* (Gay); fr. 2—4 sessile rarely crenate, style $\frac{1}{2}$ length of fr., stigma small crenulate, *anth. 4-celled*.—*γ. Z. pedunculata* (Reich.); fr. 2—5 *stalked* muricate, style as long as fr., stigma large crenulate, anth. 2-celled. *Sy. E. B.* 1426. *Z. pedicellata* (Fr.).—I doubt the constancy of these characters.—Stagnant water. *γ*. Brackish water. A. or P. V.—VIII. E. S. I.

2. *Z. polycar'pa* (Nolte); style scarcely $\frac{1}{6}$ of length of converging sessile nuts, *fr.* 5—6 subsessile crenulate on back, *converging, style scarcely $\frac{1}{6}$ length of fr.*, stigma large repand, anth. 2-celled.—Much like Sp. 1.—In water. Orkney[1]. Belfast. A. V.—VIII. S. I.

Order XCVI. NAIADACEÆ.

Fl. monœcious or diœcious, in a spath. Perianth 0. Stam. free, 1—3. Ovary 1, superior, with 1 ovule. Style 1; stigmas 2—4, filiform. Fr. a nut enclosed in the persistent spath, 1-celled, 1-seeded, not bursting. Albumen 0. Embryo with a thin skin having a lateral cleft.

---

[1] Dr. Boswell referred the Orkney plant, with longer styles and less connivent fr., to var. *tenuissima* (Fr.).—H. & J. G.

1. ZOSTERA. Fl. imperfect. Stam. and pistils inserted in two rows upon one side of a spadix. Spath linear, leaflike. Fl. naked. Anth. 1. Ovary 1; style 1; stigmas 2.

2. NAIAS. Fl. imperfect, solitary, sheathed; no perianth. Barren fl. of 1 stamen. Fertile: style short; stigmas 2—4 filiform.

1. ZOSTE'RA *Linn.* Grass-wrack.

1. *Z. nána* (Roth); l. linear 1-veined, ped. filiform, spadix short with 2—5 clasping bands, nuts smooth.—*E. B. S.* 2931. *R.* vii. 2.—Plant small, scarcely 3 in. long. L. slender. Ped. pale-coloured throughout, ½ as broad and quite as long as the inflated oblong-lanceolate spaths. Ripe seed shining black.—Muddy estuaries. P.? VII. VIII. E. S. I.

2. *Z. marína* (L.); l. linear 1—7-veined, spadix without bands, nuts striate.—*E. B.* 467. *R.* vii. 4.—Plant long. L. broad. Ped. rather strong. Ripe seed milky white.—*Z. angustifolia, R.* vii. 3, *Sy. E. B.* 1430, is a slender form growing upon mud, with fewer veins in its l. and rather longer peduncles which are much compressed and green upwards and narrow gradually into the spadix.—In the sea. P.? VII. VIII. E. S. I.

2. NA'IAS *Linn.*

1. *N. marína* (L.); *diœcious*, st. dichotomously branched from near the base, l. linear with *many large spinous teeth*, sheaths rounded nearly or quite entire, fl. solitary male enclosed in a spath, anth. 4-locular, fr. succulent ellipsoid or ovoid narrowing into the stout persistent style.—*J. of B.* xxi. (1883) p. 353, t. 241.—Upper internodes and backs of l. often spinous. Stigmas usually 3.—Hickling Broad and adjacent channels, Norf. A. VII. VIII. E.

2. *N. flex'ilis* (Rostk.); [monœcious] l. very narrow and very minutely denticulate, sheaths ciliate-denticulate.—*Sy. E. B.* 1432.—L. ternate or opposite, 1-veined, pellucid, remotely denticulate. [Anth. 1-locular.] Ovary solitary, axillary, sessile, oblong. Style short.—Deep lakes. A. VIII. S. I.

[*N. minor* (All.), monœcious, with narrow linear spinous-toothed *recurved falcate* l., truncate sheaths, and 1-locular anth., occurs as a Pleistocene fossil in Sussex and should be looked for. See *J. of B.* xxxviii (1900) p. 105, t. 408.]

3. *N. gramin'ea* (Delile); monœcious, st. branched throughout with many simple lateral br., l. tufted narrow linear with numerous minute spinous teeth, sheaths toothed *distinctly auricled, male fl. naked*, anth. 4-locular, stigmas 2, fr. solitary or 2—4 together, narrowly ellipsoid.—*J. of B.* xxii. (1884) p. 305, t. 250.—Canal, Reddish, Lanc., probably introduced with Egyptian cotton. A. VII.—IX. E.

Division III. GLUMIFERÆ.

(Orders XCVII. and XCVIII.)

Leaves parallel-veined, persistent. Floral envelopes imbricate, bractlike.

Order XCVII. CYPERACEÆ.

Fl. perfect or unisexual, each from the axil of a scale (or glume) imbricate on a common axis. Perianth 0, or rarely membranous. Stam. hypogynous (3 or rarely 2 in our plants), with sometimes a row of bristles or abortive filaments. Anth. erect, entire at the apex. Ovary 1-celled, 1 ovule at its base, often surrounded by bristles or enclosed in a bottle-shaped perianth. Style simple, trifid or bifid. Embryo enclosed within the base of the albumen.—L. with entire sheaths.

Tribe I. *CYPEREÆ.* Fl. perfect. Gl. 2-ranked. Perianth 0.

1. Cyperus. Spikelets 2-ranked, many-flowered. Gl. of 1 valve, many, keeled, nearly all with flowers.

2. Schœnus. Spikelets 2-ranked, 2—4-flowered. Gl. 6—9; several lower ones smaller, empty. Bristles few or 0.

Tr. II. *SCIRPEÆ.* Fl. perfect. Gl. imbricate on all sides. Perianth 0.

* *Lowest glumes empty or smaller than the others.*

3. Cladium. Spikelets 1—3-flowered Gl. 5 or 6. Bristles 0. *Nut with a thick fleshy coat,* tipped with the conical base of the style.

4. Rynchospora. Spikelets few-flowered. Gl. 6 or 7. *Bristles about* 6. *Nut* compressed, convex on both sides, *crowned with the dilated base of the style.*

** *Lowest glume empty or larger than the others, sometimes all fertile.*

† Bristles scarely equalling the glume or wanting.

5. Eleocharis. One or 2 lowest gl. broader, empty. Bristles 3—6. Nut compressed, crowned with the persistent dilated base of the style.

6. Scirpus. Glumes nearly equal; or 1 or 2 lowest broader, empty. Bristles about 6 or 0. *Nut* plane-convex or trigonous; *base of the style filiform, not dilated.*

7. Blysmus. Two lowest gl. broader, and empty. *Bristles* 3—6. Nut plane-convex, tipped with the not dilated base of the style. *Spikelets with bracts, alternate, forming a close distichous compound terminal spike.*

†† Bristles ultimately much exceeding the glumes.

8. Eriophorum. Glumes nearly equal, lowest sometimes empty. Bristles ultimately silky. Nut trigonous.

Tr. III. *ELYNEÆ.* Fl. unisexual. Perianth 0 or formed of 1 or 2 scales.

9. Kobresia. Spikes close together. Lower fl. fem., perianth of 1 scale enclosing the germen and covered by the glume. Upper fl. male, without any perianth.

Tr. IV. *CARICEÆ.* Fl. unisexual. Nut completely enclosed in the bottle-shaped perianth.

10. Carex. Fl. in imbricate spikes, each covered by a glume. Female fl. with a single bottle-shaped persistent perianth, 1 style and 2—3 stigmas. Male fl. of 3 stam., without a perianth.

Tribe I. *Cypereæ.*

1. Cype'rus *Linn.* Galingale.

1. *C. lon'gus* (L.); spikelets linear-lanceolate in twice-compound umbels, peduncles of partial umbels erect unequal, stigmas 3, creeping.—*E. B.* 1309. *St.* 52. 10.—St. triangular, 2—3 feet high. Umbel very large, lax, unequal, its base with 2 or 3 long leaves. Glumes brownish red, with green keels and pale margins.—South of England, rare. P. VIII. IX. E.

2. *C. fus'cus* (L.); spikelets linear-lanceolate in small roundish heads at the extremities of the branches, gl. spreading, stigmas 3, root fibrous.—*E. B. S.* 2626. *St.* 52. 5.—A small nearly prostrate plant. Stems many, 2—5 in. long. Heads with 3 unequal l. at the base. Glumes fuscous with green keels.—Formerly at Little Chelsea, Middlesex; Shalford Common, Surrey. Somerset. Dorset. S. Hants. Jersey. A. VIII. IX. E.

2. Schœ'nus *Linn.* Bog-rush.

1. *S. nig'ricans* (L.); st. terete naked, spikelets 5—10 collected into a terminal roundish head overtopped by the lower bract, gl. rough at the keel.—*E. B.* 1121. *St.* 40. 9.—Root of strong black fibres. St. 4—12 in. high, clothed at the base with blackish-brown smooth shining scales some of which terminate in setaceous erect leaves which are shorter than the stem. Bristles variable in number, short, rough, with upward spines. Stigmas 3. Anth. ending in a point. Gl. dark brown or black.—Turfy bogs. P. VI. E. S. I.

2. *S. ferrugin'eus* (L.); st. grooved naked, spikelets *lateral* 1—3 *slender about equalling the erect sheathing bract*, gl. 5—6 to each spikelet reddish-brown smooth on the keel.—*J. of B.* xxiii. (1885) p. 289, t. 261.—Smaller and more slender than Sp. 1. L.-sheaths shorter reddish-brown, lamina usually very short. Bristles about 4. Stigmas 3. Anth. with a short point.—L. Tummel, Perthsh. *Mr. J. Brebner.* P. VI. VII. S.

Tribe II. *Scirpeæ.*

3. Cla'dium *Pat. Br.* Sedge.

1. *C. jamaicens'e* (Crantz); panicles lateral and terminal repeatedly compound, spikelets capitate, st. subterete leafy smooth, l. finely serrate on the margins and keel.—*E. B.* 950. *C. Mariscus* (R. Br.) ed. viii.—Creeping. St. 3—4 feet high. L. very long, rigid, narrowed and triquetrous towards the end. Fl. in each spikelet 1—3; but usually only one nut is produced.—Bogs and fens, rare. P. VII. *Common Sedge.* E. S. I.

4. Rynchos'pora *Vahl.*

1. *R. al'ba* (Vahl); *spikelets* in a compact corymb *about as long as the outer bracts*, stam. 2, bristles with declining teeth, base of the style without teeth.—*E. B.* 985. *St.* 40. 7.—Slightly creeping. St. 6—12 in. high. L. narrowly linear. *Spikelets whitish.* Bristles 9—12. Filaments slender.—β. *sordida*; spikelets brownish, in small oval clusters, often overtopped by the outer bract.—Turfy bogs. P. VII. E. S. I.

2. *R. fus'ca* (Ait.); *spikelets* in an oval head *considerably shorter than the outer bracts*, stam. 3, bristles with ascending teeth, base of the style with erect teeth.—*E. B.* 1575. *St.* 40. 6.—Creeping extensively. St. 6—8 in. high. L. nearly filiform. *Spikelets brown.* Bristles 6. Filaments dilated.—Bogs, rare. South-west of England. Ireland. P. VII. VIII. E. S. I.

5. Eleoc'haris *R. Br.*

Spikes terminal and solitary in all our species.

1. *E. palus'tris* (R. & S.); gl. rather acute, *lowest gl. ½-surrounding the spike*, nut roundish plane-convex with rounded edges smooth crowned with the ovate base of the style and falling short of the 4 bristles, base of *st. clothed with* membranous *almost transversely truncate sheaths.*—*E. B.* 131. *St.* 9. *R.* viii. 297.—Often only slightly creeping, 6—20 in. high. Sheaths with a very blunt point on one side.—Wet and marshy places. P. VI. E. S. I.

2. *E. uniglúmis* (Link); gl. rather acute, *lowest gl. almost surrounding the spike*, nut pearshaped blunt rather compressed with rounded edges very finely punctate-striate crowned with the conical base of the style and falling short of the bristles, base of *st. clothed with transversely truncate sheaths.*—*Sy. E. B.* 1587. *R.* viii. 296.—*Far-creeping*, 6—8 in. high.—Wet sandy places especially near the sea. P. VI. VII. E. S. I.

[*E. Watsóni* (Bab.); nut oblong very blunt a little narrowed below exceeding the bristles.—*A. N. H.* ser. 2. x. 19.—It is probably a form of Sp. 2.—Taynloan, Argyleshire. *D. Balfour.* Murrough of Wicklow. *Dr. D. Moore.*] S. I.

3. *E. multicaúlis* (Sm.); glumes blunt, *nut topshaped triquetrous* smooth crowned with the broad *triquetrous base of the style* equalling the six bristles, base of *st. clothed with obliquely truncate rather acute sheaths.*—*E. B.* 1187. *St.* 78. 11. *R.* viii. 296.—Slightly creeping, about 6 in. high.—Marshy places. P. VII. E. S. I.

4. *E. aciculáris* (R. & S.); glumes blunt, *nut* obovate-oblong compressed longitudinally *ribbed and* transversely striate crowned with the topshaped base of the style, bristles short deciduous.—*E. B.* 749. *St.* 10. *R.* viii. 294.—Root fibrous with slender runners. St. many, slender, erect, 3—4 in. high, when growing in water sometimes a foot long. Spikes very small.—In damp places upon heaths. P. ? A. (Koch) VII. VIII. E. S. I.

6. Scir'pus *Linn.* Club-rush.

* *Bristles* 6. *Spikes many.* Scirpus.

1. *S. marit'imus* (L.); spikes stalked or sessile in a dense terminal cluster, bracts several leaflike, gl. bifid with acute lobes

and a point between them, nut obovate trigonous smooth.—*E. B.* 542. *St.* 13. 3. *R.* viii. 310, 311.—Creeping, sometimes tuberous. St. 1—3 feet high, leafy. Spikes large, sometimes solitary. Stigmas 3, or rarely 2. [β. *conglobatus* (S. F. Gray); spikes all sessile]. A form from Oxlode, Cambs., with spikes 1½ in. long is probably var. *macrostachys* (Reichenb.).—Salt marshes. P. VII. E. S. I.

2. *S. sylvat'icus* (L.); spikes clustered in a large cymose very compound terminal panicle, clusters stalked and sessile, general bracts several leaflike, gl. blunt apiculate, nut obovate bluntly trigonous.—*E. B.* 919. *St.* 36. 8. *R.* viii. 313.—St. 2—3 feet high. Spikes very many, small, greenish, ovate. L. broad, flat. Stigmas 3.—β. *dissitiflorus* (Sond.); spikes mostly solitary and usually stalked.—[*S. radicans* (Schk.), spikes all stalked, glumes not apiculate, stoles long and rooting, may possibly be found.]—Damp woods and banks. P. VII. E. S. I.

3. *S. triquéter* (L.); *st. acutely triquetrous throughout,* spikelets in a small cymose panicle, *gl.* notched mucronate *glabrous* fringed with rounded blunt lobes, nut "roundish-obovate plane-convex smooth."—*E. B.* 1694. *St.* 36. 3. *R.* viii. 305.—St. 3—4 feet high, with concave faces; 1 or 2 long sheaths at the base, the uppermost ending in a *short broad triquetrous leaf.* Lower bract long and rigid, resembling a prolongation of the stem. Spikelets small, stalked and sessile. Anth. with a short beardless point. Stigmas 2.—Muddy banks of the Thames near London; the Arun, Sussex; and the Tamar near Calstock, Cornwall. Shannon estuary below Limerick. P. VIII. E. I.

[*S. americánus* (Pers.); st. acutely triquetrous throughout, *spikelets* few *sessile,* gl. notched mucronate smooth with *acute lobes,* nut roundish-obovate plane-convex smooth.—*E. B. S.* 2819. *R.* viii. 304. *S. Rothii* Hoppe? *St.* 36. 4. *S. pungens* (Vahl) ed. viii.—St. 6—18 in. high, slender; with several sheaths at the base ending in *long narrow keeled leaves.* Lower bract very long and rigid, resembling a prolongation of the stem. Spikelets large, ovate, blunt, all sessile. Anth. with a subulate fringed point. Stigmas 2.—St. Ouen's Pond, Jersey. P. VI. VII.]

4. *S. carinátus* (Sm.); *st.* terete below bluntly *trigonous upwards,* spikes in a small cymose panicle, *gl.* notched mucronate *slightly asperous* and pilose fringed, nut "convex on the back smooth."—*E. B.* 1983. *S. Duvallii* Hoppe, *St.* 36. 2. *R.* viii. 308.—St. 2—4 feet high; with 1 or 2 long sheaths at the base, the uppermost ending in a leaf 3 or 4 in. long. Lower bract

much overtopping the panicle. Stigmas 2.—By rivers, near London and in Sussex, Kent and Cornw. P. VI. VII. E.

5. *S. lacus'tris* (L.); st. terete, spikes in a terminal twice-compound panicle, *gl.* notched mucronate *glabrous* fringed, *nut bluntly trigonous obovate*, stigmas 3.—*E. B.* 666. *St.* 36. 1. *R.* viii. 306.—Far-creeping. St. 4—6 feet high, naked ;' with 1 or 2 long sheaths at the base. *Anth. bearded at the end.* Panicle terminal. The bract sometimes resembles a continuation of the stem. *Sometimes it has long nearly flat floating leaves.*—Rivers and ponds. P. VI. VII. *Bulrush.* E. S. I.

6. ? *S. Tabernæmontáni* (Gm.); st. terete, spikes in a terminal compound panicle, *gl.* notched mucronate *asperous* fringed, *nut* compressed *roundish-oblong* smooth, stigmas 2.—*R.* viii. 307. *S. glaucus* Sm., *E. B.* 2321.—Creeping. St. 2 feet high, with 1 or 2 long sheaths at the base. *Anth. not bearded.* Panicle smaller than in the preceding. Lower bract short. Fr. convex on one side.—Rivers and ponds. P. VI. VII. E. S. I.

** *Bristles* 4—6. *Spike solitary, terminal.* BÆOTHRYON Dietr.

7. *S. cæspitósus* (L.); st. terete striate with *many imbricate leafless acute scales* and sheaths with *short subulate l.* below, spike ovate few-flowered, gl. ovate membranous pointed, 2 *outer gl. as long as the spike enclosing it and ending in long rigid leaflike points*, nut obovate oblong mucronate smooth, bristles longer than the nut with a few erect teeth near the tip.—*E. B.* 1029. *R.* viii. 300.—St. 3—6 in. long, many, erect, many of them barren. Bristles 6.—Barren turfy heaths. P. VI.—VIII. E. S. I.

8. *S. pauciflórus* (Lightf.); st. terete striate with a *few thin narrow leafless scales* and *one tight abrupt leafless sheath* below, spike ovate few-flowered, *gl.* ovate keeled membranous at their edges, 2 *outer gl. blunt shorter than the spike* and enclosing it, *nut* obovate mucronate *finely netted, bristles shorter than the nut with declining teeth.*—*E. B.* 1122. *St.* 10. *R.* viii. 299.—St. 3—10 in. long, erect, many of them barren, soboliferous. Bristles 6.—Boggy moors and heaths. P. VI.—VIII. E. S. I.

9. *S. par'vulus* (R. & S.); st. round many each with one close-pressed leafless sheath, no true leaves, spike oval few-flowered, gl. ovate blunt keeled membranous, 2 outer ones rather longer, *nut* ovate-oblong mucronate *smooth, bristles twice as long as the* nut with declining teeth throughout.—*J. of B.* vi. t. 85. xx. p. 52. *Sy. E. B.* 1591. *R.* viii. 299. *S. nanus* Spreng. not of Poir.—Root fibrous with capillary

stoles ending in ovate-subulate tubers. St. about an inch high, several, with 2 or 3 longitudinal fibres and more or less perfect transverse lines. Bristles 4—6.—Lymington, Hants (extinct). Poole, Dorset. Aveton Gifford, Dev. Wicklow. Kerry. A. VII. E. I.

*** *Bristles* 0.—† Spike solitary. ELEOGITON Link.

10. *S. flúitans* (L.): st. floating branched leafy, l. clustered, fl.-stalks alternate with a sheathing l. at the base, spike terminal ovate few-flowered, gl. blunt keeled membranous at their edges, 2 outer gl. larger shorter than the spike and enclosing it, *nut obovate, stigmas* 2.—*E. B.* 216. *R.* viii. 298.—St. rooting from the lower joinings and spreading to a great extent in a zigzag manner. On mud it is cæspitose with truly sheathing leaves.—Ditches and ponds. P. VI. VII. E. S. I.

†† St. round, leafy at the base. Spikes 1—3. ISOLEPIS R. Br.

11. *S. setáceus* (L.); spikes terminal, lower bract long so as to resemble a short continuation of the st., gl. blunt mucronate, *nut* trigonous obovate *longitudinally ribbed* and transversely striate, stigmas 3.—*E. B.* 1693. *R.* viii. 301.—St. tufted, slender, 3—6 in. high. Spikes small, sessile, considerably shorter than the lower bract. Gl. brown with whitish margins and a green keel.—Wet sandy and gravelly places. P.? VII. E. S. I.

12. *S. cer'nuus* (Vahl); lower bract shorter or slightly longer than the terminal spike, gl. blunt submucronate, *nut* subglobose *rough with minute points*, stigmas 3.—*E. B. S.* 2782. *R.* viii. 301. *S. Savii* (S. & M.) ed. viii.—Closely resembling *S. setaceus*. Spikes varying considerably in length, sometimes 1 in. long. Gl. scarcely mucronate, greenish, usually with a brown spot on the upper part of each side.—In many places near the coast. A. or P. VII. E. S. I.

††† Spikes many, clustered. HOLOSCHŒNUS Link.

13. *S. Holoschœ'nus* (L.); st. round, spikes in dense globular sessile or stalked clusters, lower bract erect long, gl. obovate emarginate mucronate.—*E. B.* 1612.—St. 3—4 feet high, round quite up to the cluster. Upper bract patent or ascending; lower very long, with a flat open white channel. Gl. variegated with fuscous and white, pilose. Anth. with a long entire or toothed point.—Braunton Burrows, Devon. Somerset. P. IX. E.

7. BLYS'MUS *Panz.*

1. *B. compres'sus* (Panz.); spikelets 6—8-flowered, outer gl.

ribbed shorter than the spikelet, bristles 3—6 strong persistent with declining teeth, l. flat rough on the edges and keel.—*E. B.* 791. *R.* viii. 293. *Scirpus caricis* (Retz.).—St. 6—8 in. high. Outer gl. of lowest spikelet with a subulate leaflike point which often overtops the spike. Glumes reddish brown. Nut lenticular, crowned with the long persistent style, shining.—Boggy pastures. P. VI. VII. E. S.

2. *B. rúfus* (Link); spikelets 2—4-flowered, outer gl. smooth as long as the spikelet, bristles 1—6 slender deciduous with patent or ascending teeth, l. channelled not keeled smooth.—*E. B.* 1010. *St.* 85. 7. *R.* viii. 293. *Scirpus* (Schrad.).—St. slender. Gl. dark brown, polished. Nut ovate, with a long beak, opaque. [Length of bract very variable even on the same plant.]—Marshes near the sea on the Northern and Western coasts. P. VII. E. S. I.

8. ERIOPH'ORUM *Linn.* Cotton-grass.

* *Bristles* 4—6, *at length crisped. Spike solitary.*

[1. *E. alpínum* (L.); st. triquetrous rough, l. very short, spike oblong.—*E. B.* 311 (excl. the leafy shoot). *R.* viii. 288. *St.* 10.—A slender elegant plant.—Moss of Restenet, Forfar! (but long lost through drainage). P. VI.] S.

** *Bristles very many, not crisped.*—† Spike solitary.

2. *E. vaginátum* (L.); cæspitose, *st. trigonous above* round below, spike oblong, *nut obovoid,* l. long setaceous triquetrous channelled, upper sheath inflated leafless.—*E. B.* 873. *R.* viii. 289.—Bogs and moors. P. V. *Hare's-tail Cotton-grass.* E. S. I.

†† Spikes more than one.

3. *E. polystáchion* (L.); st. nearly terete, *peduncles smooth,* l. linear channelled their upper half triangular, nut elliptic-acuminate or obovoid triquetrous.—About a foot high and rather slender. L. triangular through more than half their length. Bristles 3 or 4 times as long as the spikes. *E. B.* 564. *R.* viii. 291.—*β. minus*; st. and l. very slender. *E. gracile* Sm., *E. B.* 2402 (not *Koch*). An alpine form has but one nearly sessile spike. *E. capitatum* Don. *E. B.* 2387.—*γ. elatius* (Koch); st. strong tall, l. 2—3 lines broad the triangular part commencing above the middle.—Bogs. *β* on mountains. P. V. VI. *Common Cotton-grass.* E. S. I.

4. *E. latifólium* (Hoppe); st. triquetrous in its upper half, *peduncles asperous,* l. linear, nearly flat contracted above the

middle into a triangular point, *nut pyriform* triquetrous.—*R.* viii. 292. *E. pubescens* Sm., *E. B. S.* 2633. *E. polystachion E. B.* 563.—A slender plant, 12—18 in. high. L. about 2 lines broad; triquetrous point short. Several of the elegant spikes upon longish stalks which are asperous. Bristles 2 or 3 times as long as the spikes.—Bogs, rather rare. P. V. VI. E. S. I.

5. *E. grac'ile* (Koch); st. subtriquetrous, *peduncles downy, l. narrowly linear triquetrous, nut* oblong-*linear* triquetrous.—*E. B. S.* 2886. *R.* vii. 290. *E. triquetrum* Hoppe, *St.* 10. 2.—A tall slender plant. Spikes about 4, most of them on downy not asperous stalks. Gl. with many ribs. Bristles about twice as long as the spike.—Bogs. Dorset. Hants. Surrey. Yorks. P. VI. VII. E.

Tribe III. *Elyneæ.*

9. Kobre'sia *Willd.*

1. *K. caricína* (Willd.).—*E. B.* 1410. *Schk.* Rrr. 161. *R.* viii. 193.—St. erect, 6—12 in. high. L. slender, falling short of the stem. Spikes 4—5. collected at the top of the stem, 6—8-flowered. There is an abortive stam. (?) at the base of the nut; but some authors, considering each fl. a separate spike, think that this represents a second flower.—Moors. Yorkshire. Durham. Perthshire. P. VII. E. S.

Tribe IV. *Cariceæ.*

10. Ca'rex *Linn.*[1]

i. *Monostachyæ.* Spike simple, solitary, terminal. (Sp. 1—5.)

* *Diœcious* or *monœcious with male fl. at the top. Stigmas* 2.

1. *C. dioïca* (L.); usually diœcious, fertile spike ovate dense, *fr. ascending* ovate many-veined angles rough near the top, nut

[1] In the following descriptions, *fruit* means the persistent bottle-shaped perianth, including the ripe nut or true capsule. The *glume* is always taken from the fertile spike unless it is otherwise stated.—*Schk.* refers to the plates of *Schkuhr's Riedgräser*; *H.* to *Hoppe's Caricol. Germ.* in *Sturm Deutschl. Flora*; *R.* to *Reichenbach Ic. Fl. Germ. Cent.* vii.; *B.* to *Boott's Illust. of Carex.* See also *Andersson's Pl. Scand.*, and *Lang* in *Linnæa*, xxiv. 481; and for form, &c., of the nuts *Des Moulins Cat. Dordogne, Suppl. Final.* [See Mr. A. Bennett's papers in *J. of B.*]

roundish oval, terete st. and subsetaceous l. roundish, soboliferous.—*E. B.* 543. *Schk.* A. 1. *H.* a. 1. *R.* 194.—About 6 in. high.—Spongy bogs. P. V. VI. E. S. I.

[2. *C. Davalliána* (Sm.); diœcious, *fr. deflexed* ovate-lanceolate angles rough near the top, nut "linear-oblong," st. and margins of the subsetaceous l. rough, cæspitose.—*E. B.* 2123. *R.* 194.—About 6 in. high.—The true plant did grow on Lansdown near Bath, but is now lost by drainage. P. VI.] E.

3. *C. pulicáris* (L.); half of spike male, *fr.* remote at length *deflexed* oblong-lanceolate compressed, nut linear-oblong plane-convex, l. involute.—*E. B.* 1051. *Schk. A.* 3. *H.* a. 3. *R.* 195. —St. slender, 6—12 in. or more in height, erect, smooth. L. slender, erect. Glumes deciduous. Fr. dark brown.—Bogs. P. VI. E. S. I.

** *Spike male at top. Stigmas* 3.

4. *C. rupes'tris* (All.); half of spike male, *fr. obovate* trigonous with a very short beak adpressed scarcely longer than the persistent gl., nut obovoid acutely triquetrous, l. flat.—*E. B. S.* 2814. *H.* b. 4. *R.* 198.—Soboliferous. St. 3—6 in. high, triquetrous, rough upwards. *L. ending in a wavy* rough slender triangular *point.* Gl. fuscous. Fr. paler.—Lofty mountains. P. VII. S.

5. *C. pauciflóra* (Lightf.); 1—3 *terminal fl. male, fr.* 2—4 *lanceolate-subulate terete* patent or reflexed longer than the deciduous gl., nut linear-oblong trigonous, l. involute.—*E. B.* 2041. *Schk.* A. 4. *H.* b. 1. *R.* 196.—Soboliferous. St. usually about 5 inches high, slender. L. 2 or 3, much shorter than the stem. Fr. pale yellow.—Bogs in the North. P. VI. VII. E. S. I.

ii. *Homostachyæ.* Spikelets in a compound continuous or interrupted spike, male at one or both ends or nearly unisexual. Nut plane-convex or compressed. Stigmas 2. (Sp. 6—24.)

† Spikelets nearly unisexual. Sobole far-creeping.

6. *C. dis'ticha* (Huds.); *spikelets* in an oblong interrupted spike *upper and lower* ones *fertile intermediate mostly male, fr. ovate-lanceolate* veined *narrowly margined* bifid with serrate edges above, nut oval, gl. shorter than the fr. acute its midrib not reaching the *top,* st. with rough angles.—*C. intermedia* Good., *E. B.* 2042. *Schk.* B. 7. *H.* a. 14. *R.* 210.—Height 1—2 feet. Fr. about as long as gl.; usually abortive and twice

as long. Lower bract with a slender leaflike point [sometimes exceeding the spike].—Marshy places. P. V. VI. E. S. I.

7. *C. arenária* (L.) ; *spikelets* in an oblong interrupted spike *upper ones male* lower fertile intermediate male at the end, *fr. ovate* veined *winged* and serrulate *from the middle* to the bifid top of the beak, nut bluntly ovoid, glumes longer than the fr. acuminate, st. rough above.—*E. B.* 921. *Schk.* B. & Dd. 6. *H.* a. 13. *R.* 209.—Height 1 foot. Sobole superficial, very long. Lowermost bracts with slender leaflike points [1]. —Sandy places. P. VI. E. S. I.

†† Spikelets male at the top.—‡ *Sobole far-creeping.*

8. *C. incur'va* (Lightf.); spikelets in a roundish head, fr. inflated broadly ovoid acuminate-rostrate, beak split externally, nut obovoid, st. smooth about equalling the leaves.—*E. B.* 927. *Schk.* Hh. 95. *H.* a. 5. *R.* 199.—St. 2—3 in. high, usually *recurved* so as to bring the large head down to the ground. Beak of fr. usually rather rough. [A tall form with almost erect heads is var. *erecta* (O. F. Lang).]—Sandy shores of the North. P. VI. E. S.

9. *C. chordorrhiza* (Ehrh.); spikelets crowded in a small shortly-ovate head, fr. inflated ovoid brown shining with darker brown veins *narrowing somewhat suddenly* into an inconspicuously bifid beak, nut roundish, *gl. blunt* with a membranous margin, st. smooth *exceeding the l.*—*J. of B.* xxxvi. (1898) p. 73, t. 383.—Stems upright branched 6—12 in. high, the base with very short sheathing l. Bracts acuminate scarious the lowermost about equalling its spike.—Very wet bogs, Altnaharra, Sutherl. *Messrs. E. S. Marshall & W. A. Shoolbred.* P. VII.—VIII. S.

10. *C. divísa* (Huds.); spikelets in a somewhat ovate head, *fr. plane-convex* ovoid many-veined, *beak acutely bifid with finely serrate edges,* nut broadly oblong, gl. with an excurrent rib, *st. roughish at the top.*—*E. B.* 1096.—St. slender, a foot high. Lowermost bract (brown) scarious, often prolonged into a slender green point sometimes overtopping the spike. Fr. veined on both sides. Spike often interrupted below.—Near the sea on the Southern and Eastern coasts. P. V. VI. E. I.

‡‡ *Cæspitose. Spike simply compound. Fr. squarrose, not gibbous.*

11. *C. vulpína* (L.); *spikelets compound* in a cylindrical oblong crowded spike, *fr. ovoid-acuminate* plane-convex veined, beak

---

[1] Specimens with some female fl. in the upper spikes have been named var. *ligerica* (J. Gay) but we have seen none like the French plant.—H. & J. G.

bifid finely serrate, *nut ovoid or oval with a beak constricted at the base,* gl. mucronate shorter than the fruit, st. triquetrous with rough angles, *bracts setaceous.*—*E. B.* 307. *Schk.* C. 10. *R.* 217.—Height 2 feet. St. firm. L. broad. Fr. palish green. Spikelets greenish, bracts long, in shady places. *R.* 216. [*C. nemorosa* (Lumn.) is a form with longer interrupted spikes, long bracts and paler more cuspidate glumes.]—Wet places. P. VI. E. S. I.

12. *C. muricáta* (L.); spikelets contiguous, spike oblong dense, *fr.* ovoid-acuminate plane-convex *obscurely veined* bifid *finely serrate* upper spreading, *nut ovoid* its beak extremely short, gl. mucronate shorter than the fruit, st. triquetrous with rough angles.—*E. B.* 1097. *R.* 215.—Height 1—2 feet. St. slender but strong. L. narrow. Lowermost spikelets not more than their own length distant from each other. Fr. much larger than that of Sp. 13, with a broad flat beak with very sharp edges. *β. virens* (Koch).—Spikes interrupted below [gl. shorter and broader, paler and greener]. *C. muricata* (Hoppe).—Gravelly pastures. P. VI. E. S. I.

13. *C. divul'sa* (Stokes); *spikelets distant* the upper ones nearer together, fr. ovoid acute plane-convex obscurely veined bifid smooth ascending, *beak roughish at the edges, nut ovoid-oblong* compressed its beak extremely short, gl. mucronate shorter than the fruit, st. triquetrous with rough angles above, bracts setaceous.—*E. B.* 629. *Schk.* Dd. & Ww. 89. *H.* a. 16. *R.* 220.—Height 1—2 feet. *St. lax, slender, flaccid.* Spikelets greyish, usually distant, 1 or 2 lowest often lengthened into a short branch. Fr. with a thick green margin slightly rough near the top.—Moist shady places. P. VI. E. I.

‡‡‡ *Cæspitose. Spike compound, often panicled. Fr. ascending, gibbous on its back.*

14. *C. teretius'cula* (Good.); *spikelets forming a dense compound oblong spike, fr. ovate* with 2—5 central ribs on the convex side (back), *beak* bidentate serrulate *split to its base and overlapping on the back,* nut turbinate with a very short beak convex, style not thickened at the base, st. trigonous and rough above.—*E. B.* 1065. *Schk.* D. 19. T. 69. *H.* a. 9. *R.* 222.—Root forming scattered simple tufts. St. 1—2 feet high, slender.—*β. C. Ehrhartiana* (Hoppe); root more cæspitose, spike long rather loose, st. triquetrous above. *Sy. E. B.* 1620.—Boggy meadows, rare. *β* near Manchester [etc.]. P. VI. E. S. I.

[*C. vulpinoïdea* (Mich.) [spikes numerous very dense small-flowered. A N. American species]. Kew. An escape.]

15. *C. paradox'a* (Willd.); *spikes narrowly panicled* lower branches rather distinct, *fr. ovoid with many short elevated ribs* near its base, *beak* obliquely bidentate serrulate *not split nor winged on its back*, nut rhomboidal constricted below doubly convex with a short beak, style slightly enlarged at the base, st. trigonous and rough above.—*E. B. S.* 2896. *Schk.* E. 21. *H.* a. 12. *R.* 222.—Root densely tufted, crowned with the fibrous remains of decayed leaves. St. 1—2 feet high, slender.—Bogs. Near Mullingar, Ireland. Near York. Hoveton, Norfolk. P. VI. VII. E. I.

16. *C. paniculáta* (L.); *spikes panicled* with long diverging branches, *fr.* ovoid faintly *many-veined* with a bifid fringed *beak split to its base and overlapping on the back, nut ovoid blunt* narrowed below *plane-convex, beak slightly thickened upwards, st. triangular.*—*E. B.* 1064. *Schk.* D. 20, Ttt. 163. *H.* a. 19. *R.* 223.—Root forming dense elevated tufts. St. stout, 2—3 feet high. Panicle usually large and spreading or reduced to a slender compound (or even simple) spike. Bracts all much shorter than the spike.—[Var. *rigida* (Blytt) has spikes more rigid darker-coloured and spikelets more crowded.]—Bogs. P. VI. E. S. I.

††† Spikelets male at their base (or at both ends).

‡ *Sobole far-creeping.*

[*C. brizoïdes* (L.) was probably an escape.] E.

‡‡ *Cæspitose. Bracts leaflike. Lower spikelets distant, simple or compound.*

17. *C. Boenninghauseniána* (Weihe); *spikelets* several upper ones simple crowded *lower* distant *alternately branched*, fr. lanceolate plane-convex tapering serrulate *from below the middle*, beak deeply split on one side, nut ovoid elliptic, *gl. equalling the fruit*, root tufted, lower bract at least equalling the spike.—*E. B. S.* 2910. *H.* a. 34. *Kunze Riedg.* 22. *R.* 219.—St. 1—2 feet high, triangular with slightly convex faces and rough edges. Inflorescence often a foot long. Rachis straight, with 3 rough edges. *Gl.* ovate, membranous, *silvery* brown, *smooth.* Lower spikelets wholly male, male at both ends or at either end, Bracts, except the lowest, short. L. channelled.—Marshes and pond-sides, rare. [Probably a hybrid between Sp. 16 & 19.] P. VI. E. S. I.

18. *C. axilláris* (Good.); *spikelets* several upper ones simple crowded *lower* distant *densely compound, fr.* ovoid-lanceolate

plane-convex tapering deeply bifid *serrulate above the middle*, nut obovoid with a beak, *gl. shorter than the fruit*, root tufted, lower bract as long as or longer than the spike.—*E. B.* 993. *H.* a. 33.—St. 1—2 feet high, acutely triangular. Rachis straight, with 3 rough angles. Gl. ovate membranous, brownish; midrib often rough, extending to the point. Divisions of the lower spikelets crowded into the axils of the bracts. Bracts, except the lowest, short. L. flat. Spikelets male at the base or at both ends. [A hybrid between Sp. 11 or 12 & 19 ?]—Marshes, rare. P. VI. E. S. I.

19. *C. remóta* (L.); *spikelets* several *all simple* upper ones crowded lower distant, fr. ovoid-acuminate plane-convex notched at the end serrulate above, nut ovoid with a beak, gl. shorter than the fr., root tufted, bracts long.—*E. B.* 832. *Schk.* E. 23. *H.* a. 33. *R.* 212.—*St.* 1—2 feet high, *trigonous*. *Rachis with 2 rough angles* in its upper part. Gl. oblong, membranous, greenish white; *midrib* smooth, usually *not reaching the point*. Several of the bracts long. L. channelled. Sometimes with a thick rooting densely proliferous rhizome. *C. tenella* (Sm. not Schkr.) is a starved form.—Damp places. P. VI. E. S. I.

‡‡‡ *Cæspitose. Bracts not leaflike. Spikes contiguous or slightly distant.*

† Spikes and leaves glaucous or dusky.

20. *C. echináta* (Murr.); *spikelets about* 4 *globose rather distant*, fr. divergent broadly ovoid acuminate plane-convex striate, *beak bifid* with serrate edges, *nut ovoid* abruptly narrowed below, gl. shorter than the fruit.—*E. B.* 806. *Schk.* C. 14. *H.* a. 28. *R.* 214. *C. stellulata* (Good.) ed. viii.—St. 6—12 in. high, triquetrous, nearly smooth. Lowermost spikelet often with a short bract. Gl. ovate, membranous, reddish with a green keel and white edges. Ripe fr. greenish, stellate. [*C. grypos* (Schk.) is an alpine form with darker brown gl. and conspicuous narrow green keel.]—Boggy places. P. V. VI. E. S. I.

21. *C. elongáta* (L.); *spikelets* many *oblong contiguous*, fr. patent oblong-acuminate plane-convex with many ribs on both sides, beak almost entire with rough edges, *nut linear-oblong* tapering below beak very short style persistent, gl. shorter than the fruit, *bracts none* or one very short.—*E. B.* 1920. *Schk.* E. 25. *H.* a. 32. *R.* 218.—St. 1—2 feet high, triquetrous. Gl. ovate, dark brown with a green keel and whitish edges, blunt, sometimes apiculate. Upper spikelets crowded, lower lax but not distant.—Marshes, rare. P. VI. E. S. I.

22. *C. canes'cens* (L.); spikelets 4—6 elliptic contiguous, *fr. erect acute* plane-convex *faintly striate, beak short notched not split* rough at the edges, *nut elliptic* beak very short, style persistent, glumes ovate shorter than the fruit.—*E. B.* 1386. *Schk.* C. 13. *R.* 206. *C. curta*, Good.—St. a foot high, triquetrous, smooth except at the top. Lowermost spikelet often with a setaceous bract. *Glumes* membranous, *whitish* with a green keel, blunt, apiculate. Fr. whitish. [*C. helvola* (Blytt) (*C. canescens*×*lagopina*?) is recorded from Loch-na-Gar.]—Bogs. P. VI. E. S. I.

23. *C. vit'ilis* (Fr.)[1]; spikelets 4—8 ovate or oblong contiguous, *fr. erect ovate* plane-convex faintly striate, *beak short split to its base externally* rough at the edges, " nut elliptic," glumes ovate shorter than the fruit.—*C. curta β. alpicola* Wahl. *C. Persoonii* Sieb., *R.* 206.—Creeping. St. 6—12 in. high, triquetrous, smooth except at the top. Lowermost spikelet often with a setaceous bract. *Gl.* membranous, *brown* with a white margin. —Mountains. P. VII. VIII. E. S.

⫫ Spikes fuscous.

24. *C. lagopína* (Wahl.); *spikelets* 3—4 roundish-elliptic *contiguous* terminal one longer, fr. erect elliptic acuminate plane-convex narrowed below nearly entire at the point with smooth edges, nut elliptic tipped with the persistent style, gl. ovate nearly as long as the fruit.—*H.* a. 24. *R.* 205. *C. leporina* Linn *Fl. Lap.* not *Fl. Suec.*—*E. B. S.* 2815.—St. 4—8 in. high, smooth, triangular. Glumes reddish with the edges paler. Fr. yellow.— Cairngorms, Aberdeenshire, Invernesshire. P. VIII. S.

25. *C. leporína* (L.); *spikelets about* 6 oval contiguous, *fr.* erect *ovate-attenuate* plane-convex narrowed below *bifid* at the point *with membranous edges serrulate* above, nut elliptic with a short cylindrical beak ending in a persistent style, gl. lanceolate as long as the fruit.—*E. B.* 306. *Schk.* B. 8. *R.* 211. *C. ovalis* (Good.) ed. viii.—St. 1—2 feet high, triangular, smooth, or roughish above. Gl. acute, brown with a paler membranous margin. Fr. yellowish. [Sometimes with crowded spikes (var. *capitata*, Sond.) or, in woods, paler with smaller spikes (var. *argyroglochin* Horn.) or with bracts elongated much exceeding the spike (var. *bracteata* Sy.).]—Meadows. P. VII. E. S. I.

---

[1] Herr Kükenthal apparently refers the Scotch plant to *C. canescens* var. *robustior* (Blytt). See *J. of B.* xxxvi. (1898) p. 75.

iii. *Heterostachyæ.* One or more terminal spikes wholly or rarely only partially male, others axillary fertile. (Sp. 26—74.)

* *Beak of fr. short, entire or emarginate or shortly 2-toothed.*

† Nut plane-convex. Stigmas 2—3. Male spikes 1 or more.

‡ *Stigmas 2. Sheaths of leaves webbed. Fertile spikes erect. L. ultimately revolute-edged; lowest sheaths usually leafless.*

26. *C. eláta* (All.); fertile spikes erect or rarely drooping long cylindrical, bracts auricled lowermost leaflike short, *fr. oblong-elliptic acute compressed veined* closely imbricate, nut roundly obovoid shortly beaked, *sheaths* of l. with *filamentous* network.—*E. B.* 914. *R.* 230. *C. stricta* (Good.) ed. viii. *C. cæspitosa* Gay.—*Densely cæspitose.* St. 2—3 ft. high, triquetrous, rigid. L. short, narrow. Fertile spikes often male at the top. Gl. blunt or acute, equalling or shorter than the fr., narrow, dark purple; keel green. Fr. in 6—9 regular rows.—[*C. turfosa* (Fr.). A stoloniferous plant with filamentous network, intermediate between Sp. 26 and 31, is recorded from Surrey and Cambs.]—Marshes. P. VI. E. S. I.

‡‡ *Stigmas 2. Sheaths not webbed, usually all bearing leaves. Stoloniferous.*

27. *C. acúta* (L.); *male spikes* 1—3, fertile 2—4 slender cylindrical-acuminate erect with fruit, lowermost bract leaflike often overtopping the stem *with long auricles, fr.* oblong lenticular *veined,* nut roundish-obovoid with a short slender beak, gl. acute.—*E. B.* 580. *Schk.* Ee. & Ff. 92. *H.* a. 44. *R.* 231, 232.—St. 2—3 feet high, triquetrous, rough at top. L. broad. Gl. narrow-lanceolate, acute, on the male spikes spathulate-lanceolate, purple with a green keel. Fr. pale, blunt or acute, round or oval, spherical or flattened. Fertile spikes usually with a few male fl. at the end, *nodding with flowers.*—[*C. prolixa* (Fr.), cæspitose, with fr. compressed, more strongly nerved, shorter than the long tapering gl., is recorded from Cambs. and Norf.; and *C. tricostata* (Fr.), stoloniferous, with young fr. ventricose becoming compressed 3-veined, longer than the small obtuse gl., is recorded from N. of Ireland.]—Wet places. P. VI. E. S. I.

28. *C. rig'ida* (Good.); spikes erect, male 1, fertile oval or shortly cylindrical dense subsessile, lower bract leaflike, *fr.* elliptic lenticular *without veins,* nut roundish blunt with a slender beak, st. triquetrous rough towards the top, *l.* curving outwards *with deflexed edges.*—*E. B.* 2047. *H.* a. 40. *R.* 225. *C. saxatilis* L.?—St. 6 in. to a foot or more high. L. broad, flat, keeled,

rigid. Bracts without sheaths. Spikes near together, short, rarely narrowed and laxly flowered below. Gl. about exceeding the fr., purple with a green keel. Nut rather longer than broad.—[β. *inferalpina* (Læst.); taller, l. broader and more erect, br. longer and broader, lower female sp. stalked, elongated tapering below.] —In wet and stony places on mountains. P. VI. VII. E. S. I.

‡‡‡ *Stigmas 2. Sheaths not webbed. Fertile spikes erect. Glumes narrower than the fruit.*

29. *C. aquat'ilis* (Wahl.); *spikes* erect, male 1 or more, fertile 3 or 4 long *narrowed below* lower ones stalked, bracts leaf-like erect overtopping the stem, *fr.* elliptic lenticular *without veins* broader than the gl., nut oblong narrowed below with a short slender beak, *st. trigonous smooth.*—*E. B. S.* 2758.—St. 1—4 ft. high, with convex faces. Spikes rather distant. Gl. usually short, narrower than the yellowish-green fr., reddish purple with a pale midrib [1]. Alpine bogs and riversides. P. VII. E. S. I.

30. *C. salina* (Wahl.); spikes erect or lower ones somewhat drooping, male 2—3, fertile 3—4 on short ped. sometimes male at the top, bracts leaflike equalling the spikes, stigmas 2—3, fr. ovoid compressed with many veins and a short beak, nut obovate narrowed above and below, st. trigonous.—*J. of B.* xxiii. p. 290, t. 262.—St. 1—2 feet high. L. narrow with keel and margin scabrid. Gl. brownish ovate obtuse or mucronate, *lower with an excurrent scabrid midrib.*—A curious plant from Harris is referred by Mr. Bennett to *C. spiculosa* (Fr.) as var. *hebridensis* (Ar. Benn.), but it seems to lack the asperous prolongation of the midrib of the gl. characteristic of that sp.—Extreme N. of Scotland. VII. VIII. S.

31. *C. Goodenóvii* (Gay); spikes erect, male 1 or 2, fertile 3—4 subsessile cylindrical, *bracts with short auricles* leaflike, *fr. elliptic plane-convex* with many veins below and a short entire beak, nut roundish very blunt, st. triquetrous rough towards the

---

[1] Mr. Ar. Bennett (J of B. xxxv. 1897, p. 249) characterizes the following forms:—

1. *cuspidata* (Læst.)...... glumes elongated and cuspidate.
2. *virescens* (And.)...... just the opposite to the last; the gl. ½ the length of the fr. which is very symmetrically arranged.
3. *minor* (Boott). The montane plant with the spikes attenuated at the base.
4. *elatior* (Bab.)=var. *Watsoni* (Sy.). The lowland plant usually with spikes stout, equal and continuous leafy bracts, and tall (3—5 ft.) st.

Mr. Bennett records the hybrids *C. aquatilis*×*salina* (×*C. Grantii*, Ar. Benn.) and *C. aquatilis* × *elata* (× *C. hibernica*, Ar. Benn.).

H. & J. G.

top.—*C. cæspitosa* Sm., *E.B.*, 1507. *C. vulgaris* (Fries) ed. viii. —*St.* about a foot high. *L. slender*, not keeled. Bracts without sheaths. Spikes near together, short. Gl. blunt, shorter than fr., purple; keel slender, pale green. Fr. greenish or with a purple tinge. Nut rather broader than long.—[A slender cæspitose form with narrow convolute l. is *C. vulgaris* var. *juncella* (Fr.).]—Marshes. P. V. VI. E. S. I.

[*C. Gibsóni* (Bab.); *fr. lanceolate acute, nut broadly obovoid.*—*A. N. H.* xi. t. 5.—St. 6—8 in. high. *Perianth nearly twice as long as the nut*, gradually narrowed from below the middle to the top. Perhaps a monstrosity of *C. vulgaris*.—Hebden Bridge, Yorkshire! Not recently found. P. VI.] E.

32. *C. triner'vis* (Degl.); spikes erect, *male* 2—4 (rarely 1), fertile 2—5 *short stout* often male at the top sessile, bracts sheathing with brown auricles, exceeding the spikes, fr. ellipsoid-oblong compressed usually strongly 3—5-veined with very short beak, nut narrowing to the beak, st. *trigonous* strongly ribbed *smooth*.—*Kunze Riedgr.* t. 1.—Stoloniferous. Rootstock stout. St. 5—15 in. cæspitose. L. faintly keeled often exceeding the spikes. Bracts broadening into strongly-ribbed auricled sheaths. Spikes near together very stout. Gl. bluntish narrower than in sp. 31 shorter than or equalling the fr., keel broad greenish-white. Fr. brownish-green, nut longer than broad.—A plant from Ormesby, Norf., 1869, *Mr. H. G. Glasspoole*, has been referred to this sp. P. VI. E.

†† Nut with 3 angles. Stigmas 3, rarely 2.

‡ *Fruit glabrous. Terminal spike male at its base.*

33. *C. alpína* (Sw.); spikes 1—4 roundish or oblong contiguous nearly sessile, *fr. obovoid* veinless rough above with a short notched beak, gl. acutish falling short of fr., nut obovoid triquetrous blunt with a short cylindrical beak, bract scarcely overtopping the spikes, st. triangular rough towards the top.—*B.* t. 356. *C. Vahlii E. B. S.* 2666. *Schk.* Gg. 94 & Ppp. 154. *R.* 235.—St. 6—12 in. high, erect. Gl. brown or black.—Glen Callater and Glen Fiagh, Clova. Glen Lyon. P. VII. E.

34. *C. fus'ca* (All.); spikes 3—4 oblong sessile contiguous, the lowest shortly stalked rather distant, *fr. oval* blunt *compressed* (ultimately *trigonous*) *bidentate* roughish above veined, *gl. cuspidate* lower exceeding fr., *nut* obovoid *trigonous* blunt *apiculate*, lower bract leaflike.—*E. B. S.* 2885. *H.* b. 11. *Schk.* X. & Gg. 76. *R.* 235. *B.* 438. *C. Buxbaumii* (Wahl.) ed. viii. —St. 1—2 feet high, triquetrous, rough. *Sheaths of the l. connected by netlike filaments.* Gl. nearly black, with a green keel prolonged into a cuspidate point. Fr. glaucous green.—Arisaig, W. Inverness. *Mr. W. F. Miller.* Harbour Island near Toom Bridge, Lough Neagh. P. VI. S. I.

35. *C. atráta* (L.); spikes 3—4 ovate-oblong shortly stalked contiguous ultimately drooping, the lowest rather distant and with a longer stalk, *fr. elliptic* veinless *with a short terete slightly notched beak*, glumes acute, *nut elliptic* triquetrous blunt apiculate, lower bract leaflike.—*E. B.* 2044. *Schk.* X. 77. *H.* b. 8. *R.* 237.—St. 1—1½ foot high, triangular, smooth. L. flat, broad. Gl. dark purple; midrib slender, pale. Fr. yellowish. Mr. H. C. Watson mentions a form with the term. spike wholly male and the fem. spikes cylindrical and 2—3 in. below it.—Alpine rocks. P. VI. VII. E. S.

‡‡ *Fruit glabrous. Terminal spike wholly male, solitary.*

36. *C. palles'cens* (L.); fertile *spikes* subpendulous ovate or oblong with exserted stalks *contiguous*, bracts leaflike, gl. mucronate, *fr. ovoid-oblong* convex on both sides veined *blunt*, *beak* 0, *nut linear-elliptic* trigonous.—*E. B.* 2185. *Sy. E. B.* 1657. *Schk.* Kk. 99. *H.* b. 44.—St. slender, triquetrous, rough above 1—1½ foot high. *Spikes* blunt, *pale green*; the barren one sessile, darker.—Marshy places. P. VI. E. S. I.

37. *C. panícea* (L.); *fertile spikes erect remote* subcylindrical on exserted stalks, bracts leaflike sheathing, gl. rather acute, *fr. ovoid subglobose inflated* veinless with a short *terete truncate beak*, nut obovoid-oblong trigonous with a cylindrical beak.—*E. B.* 1505. *Schk.* Ll. 100. *H.* b. 33.—St. 1—2 ft. high, erect, smooth, obtuse-angled. Fertile spikes about 2. Gl. oblong, more or less acute, dark brown with a green keel and membranous pale margins. Lowermost bract about as long as its spike, the rest shorter, sheaths close. [A form with the fruits swollen and purplish towards the top is var. *tumidula* (Læst.).]—Marshy places. P. VI. E. S. I.

38. *C. vagináta* (Tausch !); *fertile spikes erect* remote on exserted stalks, bracts sheathing scarcely leaflike, gl. bluntish, *fr. ovoid triquetrous glabrous* veinless with a short terete *smooth very obliquely truncate and emarginate beak, nut elliptic* triangular "*with a beak slightly thickened upwards.*"—*H.* b. 17, *Kunze Riedgr.* 15. *B.* 478. *C. phæostachya* Sm., *E. B. S.* 2731. *C. Meilichoferi* Sm., *E. B.* 2293.—St. 5—6 in. high, smooth. Fertile spikes 1—2. Glumes bluntish. *Bracts with funnelshaped sheaths.*—Highland mountains. P. VII. S.

39. *C. limósa* (L.); *fertile spikes* 1 or 2 upon very long stalks drooping *ovate densely-flowered* "with occasionally a few *male fl. at their top*," *bracts* auricled *slender* strongly keeled, gl. ovate mucronate, *fr. roundish-obovoid compressed strongly ribbed* with

a very short entire beak, nut obovoid bluntly trigonous with a beak, *l. narrow* linear *channelled rough at the edges* throughout. —*E. B.* 2043. *R.* 238. *B.* 216.—Creeping. St. 1 ft. high. L. and bracts very slender. Gl. purple, usually with a green keel, about as long as the pale fr. which is broadest above the middle. Nut pale.—Spongy bogs. P. VI. E. S. I.

40. *C. magellan'ica* (Lam.) ; *fertile spikes 2 or 3 drooping* upon long stalks oblong densely-flowered *with* occasionally *a few male fl. at their base, bracts* auricled leaflike rather *broad nearly flat*, gl. ovoid-lanceolate attenuate acute, *fr. roundish-ovoid* compressed *faintly ribbed* with a very short entire beak, nut elliptic triangular with a beak, *l.* linear *flat smooth at the edges* except near the tip.—*E. B. S.* 2895. *R.* 238. *B.* 219. *C. irrigua* (Hoppe) ed. viii.—Creeping. St. a foot or more in height. L. and bracts 2 or 3 times as broad as in the preceding. *Gl. wholly purple* usually longer than the pale fr. which is broadest below the middle. Nut pale.—Spongy bogs in the North, rare. P. VI. E. S. I.

41. *C. rariflóra* (Sm.) ; *fertile spikes 2 or 3 drooping* upon long stalks *oblong few-flowered lax, bracts with very short sheaths*, gl. very broad blunt as long as the fr., fr. oblong with 3 blunt angles, *beak extremely short* entire, nut roundish-oblong, l. flat rough-edged towards the tip.—*E. B.* 2516. *B.* 217.—Creeping. St. 6—8 in. high, smooth. *Gl. folded round the fr.*, dark brown; midrib pale, *terminating in a minute apiculus.* Fr. pale, faintly veined; nut darker.—Elevated bogs in the Highlands. P.VI. S.

42. *C. capilláris* (L.) ; *fertile spikes drooping* upon long half-included stalks *few-flowered lax, one bract sheathing several stalks*, gl. blunt, fr. oblong trigonous turgid narrowed below veinless with a slender membranous beak, nut obovoid triquetrous blunt with a short beak.—*E. B.* 2069. *Schk.* O. 56. *H.* b. 53. *R.* 241. —St. very slender, 2—6 in. high, smooth. Peduncles rough, several usually enclosed in one sheath. Gl. short, broad, blunt, midrib not reaching the tip, falling short of the small smooth brown fruit. Nut pale. Root tufted.—Teesdale. Scottish Highlands and near sea-level in Sutherland. P. VI. E. S.

43. *C. strigósa* (Huds.) ; *fertile spikes* drooping about 4 distant rather long *slender lax* lower with exserted stalks, bracts leaflike sheathing, *fr. oblong-lanceolate* narrowed at both ends trigonous *veined with an* obliquely *truncate mouth*, nut elliptic triangular punctured, l. broad.—*E. B.* 994. *Schk.* N. 53.—St. 2 feet high. Sheaths nearly covering the peduncles. Gl. elliptic-lanceolate, diaphanous, greenish down the back.—Groves and thickets, rare. P. V. VI. E. S. I.

44. *C. pen'dula* (Huds.); *fertile spikes* drooping about 5 distant *cylindrical very long densely flowered*, bracts leaflike lower ones with sheaths nearly equalling the flowerstalks upper scarcely sheathing, *fr.* elliptic subtrigonous *tumid with short trigonous emarginate beak*, nut elliptic triangular.—*E. B.* 2315. *Schk.* Q. 60. *R.* 243.—St. 3—6 feet high, round at the angles above. Fertile spikes often 3 or 4 in. long, arched; upper ones frequently with male fl. at the tip. Gl. ovate mucronate, brown with a green keel. Fr. green, ciliate at the mouth.—Damp woods. P. V. E. S. I.

‡‡‡ *Fruit downy, hairy or scabrous. Bracts sheathing. L. in sterile tufts. Male spike solitary.*

45. *C. húmilis* (Leyss.); *fertile spikes* 2 *or* 3 *remote about* 3-*flowered enclosed in the membranous leafless bracts, fr.* obovoid subtrigonous narrowed below *with an entire oblique mouth* not obovoid, trigonous, with a short beak.—*H.* b. 15. *C. clandestina* Good., *E. B.* 2124. *Schk.* K. 43. *R.* 239.—St. about 2 in. high, erect, concealed amongst the leaves. Bracts large, wholly membranous, nearly hiding the fertile spikes. L. all radical, linear, channelled, rough.—Limestone hills in Hants, Wilts, Somerset, and Dorset. P. IV. E.

46. *C. ornithop'oda* (W.); *fertile spikes* 3 *near together* suberect *exceeding male spike*, bracts membranous leafless, *fr. pyriform* trigonous with a subemarginate mouth *exceeding the glumes*, nut elliptic subtrigonous shortly stalked with a minute beak.—*J. of B.* xiii. t. 164. *R.* 240.—St. 3—5 in. high, scarcely exceeding the flat bluntish radical leaves. Bracts much falling short of fertile spikes.—Millers Dale, Derbyshire. P. V.—VII. E.

47. *C. digitáta* (L.); *fertile spikes* 2 *or* 3 *distant linear* erect *lax, bracts* membranous *obliquely truncate, lowermost with a setaceous leaflike point,* fr. obovoid trigonous narrowed below equalling gl., *beak short nearly entire,* "nut elliptic-oblong triangular shortly-stalked and shortly-beaked."—*E. B.* 615. *Schk.* H. 38. *H.* b. 14.—St. 6—8 in. high, erect, sheathed at the base, leafless, taller than the flat radical leaves.—Woods on limestone, rare. P. IV. V. E.

‡‡‡‡ *Fr. hairy or downy or scabrous. Bracts not sheathing* (*except slightly in* No. 52).—† Male spike 1.

48. *C. ver'na* (Chaix); *fertile spikes* 1—3 *oblong-ovoid* near together sessile, bracts clasping the lowest leaflike and slightly

sheathing, *gl. broadly ovate acuminate* with excurrent midrib, *fr. ovoid-rhomboidal* trigonous with an entire mouth, nut obovoid narrowed below trigonous with prominent angles, base of the style surrounded by a cuplike disk, *soboliferous.—E. B.* 1099. *Schk.* F. 27. *H.* b. 24. *R.* 261. *C. præcox* (Jacq.) ed. viii.—St. 3—12 in. high. Lowermost spike often slightly stalked.—Dry places. P. IV. V. E. S. I.

49. *C. ericetórum* (Poll.); *fertile spikes* 1—3 *ovoid* near together sessile, bracts clasping all membranous, *gl. obovate very blunt finely ciliate* midrib not reaching the top, fr. obovoid trigonous with a truncate beak, nut subglobose trigonous with no terminal disk or prominent angles, soboliferous.—*E. B. S.* 2971. *Schk.* J. 42. *H.* b. 26. *R.* 262. *C. ciliata* Willd.—Gl. brown with a broad pale edge, all very blunt. St. 3—6 in. high.—On chalk. Norf., Suff., Cambridge. P. IV. V. E.

50. *C. montána* (L.); *fertile spikes* 1—3 *ovoid* crowded sessile, bracts small membranous the lowest with an awlshaped point, gl. of fertile spikes blunt notched and mucronate, *fr.* narrowed below *oblong-obovoid* trigonous with a short *notched beak,* nut oblong narrowed below with a pyramidal beak, *root fibrous* from a shaggy thick branched rhizome.—*E. B. S.* 2924. *Schk.* F. 29. *H.* b. 21. *R.* 261. *C. collina* Willd.—St. about a span long, slender. L. narrow; sheaths of lower l. red. Gl. very dark; midrib narrowly yellowish. *Fr. hairy,* its beak purple.—Heaths and banks. P. IV. V. E.

51. *C. pilulifera* (L.); *fertile spikes* about 3 *roundish* near together sessile, *bracts* small *lowest scarcely leaflike awlshaped not sheathing,* gl. broadly ovate mucronate, *fr.* stalked *subglobose* with a *short bifid beak, nut subglobose* subtrigonous narrowed below, persistent base of style recurved, *root fibrous.—E. B.* 885. *Schk.* J. 39. *R.* 260.—St. 6—12 in. long, slender, at length decumbent.—*β. longibracteata* (Lange); lowest bract very long leaflike, upper bract long and slender, midrib of gl. excurrent, nut fusiform. Var. *Leesii* (Ridl.) ed. viii. *J. of B.* xix. t. 218.—Wet heaths. *β.* Plumpton, Yorkshire. Glen Callater, Braemar. On rocks in shade. P. V. E. S. I.

52. *C. tomentósa* (L.); *fertile spikes* 1 or 2 nearly sessile *cylindrical blunt, lowermost bracts leaflike with* a very short *sheath,* gl. broadly ovate acute, *fr. obovoid* subtrigonous scarcely beaked *slightly emarginate, nut blunt* trigonous narrowed below with a *short beak constricted at its base, soboliferous.—E. B.* 2046. *Schk.* F. 28. *H.* b. 28. *R.* 263.—*St.* a foot high, with 3 sharp angles,

rough upwards, *erect*. Beak of nut slightly swelling upwards. Fr. with copious white down, mouth very broad.—Water-meadows at Merston Measey, Wilts. Pasture, Fairford, E. Glos. P. VI. E.

†† Male spikes usually more than 1.

53. *C. flac'ca* (Schreb.); fertile spikes 2 or 3 erect or drooping cylindrical densely flowered long-stalked, bracts leaflike scarcely sheathing, gl. ovate acute, fr. blunt elliptic veinless slightly rough entire at the small point, nut roundish-ovate triangular. —*H.* b. 67. *Schk.* O. P. 57. *R.* 269. *E. B.* 1506. *C. recurva* Huds., *C. glauca* (Scop.) ed. viii.—Soboliferous. St. a foot or more in height. Male spikes variable in number. Fertile spikes often have a male fl. at the top.—β. *C. Micheliana* (Sm.); gl. blunt, fr. smaller. *E. B.* 2236.—γ. *C. stictocarpa* (Sm.); fertile spikes ovate, fr. obovate dotted. *E. B. S.* 2772.—Damp places. P. VI. E. S. I.

** *Beak of fr. long, 2-toothed or bifid. Nut with 3 angles. Stigmas 3, rarely 2.* (Sp. 54—74.)

† Beak of fruit terete. Terminal spike male.

54. *C. atrofúsca* (Schk.); *fertile spikes 2 or 4 upon short stalks* ovoid densely flowered, *bracts scarcely leaflike* or sheathing, gl. ovate acute, *fr.* elliptic compressed *rough-edged with a cloven beak*, nut elliptic triangular on a long stalk, root fibrous.—*E. B.* 2404. *Schk.* Y. 82. *H.* b. 47. *C. ustuláta* (Wahl.) ed. viii.—St. 3—10 in. high. L. very short, broad. Gl. dark purple with a slender pale midrib. Fr. dark purple, paler below. Nut fuscous.—Breadalbane Hills, Perthshire. P. VII. S.

†† Beak of fruit plane-convex. Male spike 1, or rarely 2. Fruit glabrous. Bracts sheathing, leaflike.

55. *C. Sadler'i* (Linton); fertile spikes fusiform upper sessile lower ultimately pendulous long-stalked, bracts sheathing leaflike, gl. acute, fr. lanceolate triquetrous gradually narrowed into a ciliate-serrate bifid beak, nut ovate triquetrous, st. triangular smooth.—*C. frigida* ed. viii. *Edin. Bot. Tr.* xii. t. 2. *J. of B.* xiii. t. 159.—St. 3—12 in. high. L. broad, parallel-sided when young.—Above Loch Ceann-mor, Aberdeenshire, and in Glen Dole. P. VIII. S.

56. *C. fláva* (L.); fertile spikes roundish-oval subsessile, lowest spike with a nearly included stalk, bracts leaflike with short sheaths, gl. blunt, *fr.* ovate inflated ribbed smooth *narrowed*

into a deflexed rough-edged bifid *beak, nut* obovoid trigonous *punctate, st. trigonous smooth.—E. B.* 1294. *Schk.* H. 36. *H.* p. 22. *R.* 273.—St. 6—12 in. high. L. broad. Male *spike cylindrical,* blunt; gl. blunt. Fertile spikes usually near together and near the barren spike, sometimes distant; gl. with a green midrib slightly rough and often excurrent at the end. Beak of the fr. curved downwards parallel-sided when young. Lowest fr. declining.[1]—*β. C. lepidocarpa* (Tausch!); barren spikes usually long-stalked, beak of fr. often nearly straight.—*Sy. E. B.* 1673. *Kunze Riedgr.* 13. *R.* 272.—Wet places. P. V. VI. E. S. I.

57. *C. Œderi* (Retz.); *fertile spikes* roundish-ovate subsessile *near together,* lowest spike with an included stalk, bracts leaflike with short sheaths, gl. mucronate, *fr. subglobose* inflated ribbed smooth *suddenly contracted* into an erect narrow rough-edged bifid beak, nut obovoid-trigonous punctate, st. trigonous smooth. —*Sy. E. B.* 1674. *H.* b. 23. *R.* 272.—Much like, but distinct from, *C. flava.* Fertile spikes smaller. Fr. much smaller than those of *C. flava,* more in number, with a shorter and narrower-based beak which is conical when young. Very variable in height.—[Var. *œdocarpa* (And.) has st. recurved, male sp. stalked, fertile spikes distant, fr. more beaked.]—Bogs. P. VI. VII. E. S. I.

58. *C. exten'sa* (Good.); fertile spikes oblong near together subsessile lower one rather distant with a short included stalk, bracts very long leaflike with short sheaths, *gl. mucronate,* fr. ovoid triquetrous ribbed narrowed into a *straight smooth-edged bifid beak, nut oblong-elliptic* triangular smooth.—*E. B.* 833. *Schk.* V. Xx. 72. *H.* b. 32. *R.* 274.—*St.* usually curved, 8—12 in. high, trigonous, *smooth,* usually exceeding the leaves. Barren spike nearly sessile, blunt; gl. blunt. *L. and bracts very narrow,*

---

[1] Mr. Townsend (J. of B. xix. 1881, p. 161) discriminates the following forms —

" *α. genuina.*—L. shorter than the st., male spike nearly or quite sessile, "female spikes contiguous, lowest bract much exceeding male spike; fr. "considerably narrowed towards the base, and gradually narrowing above "into a much deflexed beak, which is as long as the rest of the fr.

" *β. C. lepidocarpa,* Tausch.—St. scabrous; l. narrow, shorter than the "st.; male spike long-stalked, female spikes distant, ovate; fr. crowded, "suborbicular, beak long, strongly deflexed.

" *γ. minor.*—St. shorter than in var. *α*; l. commonly as long as or even "longer than the st., male spike usually stalked, female spikes usually "distant; fr. smaller, suborbicular, more suddenly contracted into a less "deflexed or straight beak, which is shorter than the rest of the fr.

" *δ. argillacea.*—L. broad, as long as or longer than the st.; male spike "short sessile, female spikes contiguous; fr. suborbicular, beak short, "straight." H. & J. G.

convolute, long. [β. *minor* (Syme); smaller, stems shorter than the leaves, fr. less attenuated into a beak.]—Marshes, chiefly near the sea. P. VI. E. S. I.

59. *C. Hornschuchiána* (Hoppe); fertile spikes ovate-oblong distant with stalks exceeding the long sheaths, *glumes ovate* not mucronate, *fr.* ovoid triquetrous rough-edged also *scarious in the notch*, nut obovoid trigonous nearly smooth.—*H.* b. 40. *R.* 252. *C. speirostachya* Sm., *E. B. S.* 2770.—Rootstock often creeping. St. trigonous, rough-edged, about a foot high. Lowest bract often reaching up to or beyond the barren spike.—Boggy places. P. VI. E. S. I.

[*C. xanthocar'pa* (Degl.)[1] "has tufted stems, ovate-acute glumes, spreading fr. not scarious in the notch." It is said to be not uncommon. I do not know it.—*C. fulva* Good., Sm., was an error.—See *J. of B.* xiv. 366.]

60. *C. punctáta* (Gaud.); fertile spikes erect cylindrical with slightly exserted stalks particularly the lowest, bracts sheathing, gl. ovate shortly awned, *fr.* spreading ovoid *tumid obscurely veined pellucidly punctate* with a *linear* bidentate *smooth beak*, *nut ovoid-rhomboidal* narrowed at both ends triangular *rough*.—*H.* b. 37. *Kunze Riedgr.* 6. *R.* 251.—St. smooth, 1—2 feet high, slender. Spikes distant or the upper ones near; stalks slightly (the lowest often greatly) exserted, rough; lowest spike often very distant. Gl. pale red with a broad green longitudinal dorsal band. Fr. pale. Nut brown. Gl. of barren spike blunt. —Marshy places near the sea. P. VI. E. S. I.

61. *C. dis'tans* (L.); fertile spikes remote erect oblong, upper with included stalks, bracts with sheaths, gl. mucronate, *fr.* ascending ovoid trigonous *equally and faintly ribbed* pellucidly punctate smooth, edges of the bifid narrow beak rough, *nut* triquetrous roughish *obovoid narrowed below*.—*E. B.* 1234. *Schk.* T. 68. *H.* b. 42. *R.* 253.—St. smooth, seldom exceeding a foot high, slender. Spikes distant, short; lower ped. half-exserted. Gl. brownish. Fr. yellowish brown, rather inflated; nut yellowish. Male spike cylindrical-clavate, with blunt glumes.—Marshy places, especially near the sea. P. V. E. S. I.

62. *C. biner'vis* (Sm.); fertile spikes remote, upper ones nearer together cylindrical their stalks mostly included, lower long with exserted stalks, bracts sheathing, gl. mucronate, *fr.* ovoid subtriquetrous *with 2 prominent green submarginal ribs on the back* other

---

[1] Generally considered the hybrid *C. flava* × *Hornschuchiana*. H. & J. G.

ribs faint, beak broad bifid rough at the edges, *nut obovoid* roughish.—*E. B.* 1235. *Schk.* Rrr. 160. *H.* b. 39. *R.* 255.—St. triangular, smooth, a foot high. Spikes often very distant; upper stalks often quite included, never much exserted, lower often greatly exserted. Gl. dark purple; midrib greenish yellow. Fr. brown or deeply tinged with purple, 2 prominent ribs always green; nut brown. Male spike with blunt glumes.—Heaths and moors. P. VI. VII. E. S. I.

63. *C. lævigáta* (Sm.); fertile *spikes* remote *cylindrical*, stalks more or less exserted, bracts sheathing, gl. acute, *fr. ovoid attenuate* striate, beak long deeply bifid with rough edges, *nut subpyriform narrowed below* triangular smooth.—*E. B.* 1387. *Schk.* Bbb. 116 & Sss. 162. *H.* b. 38. *R.* 254.—St. smooth, 2—3 feet high. Spikes distant, erect or drooping. Gl. often acute on the male spike, always so on the others, purple with a paler dorsal longitudinal band. Rarely 2 male spikes. Fr. green; nut yellowish. L. broad.—*β. gracilis* (Ar. Benn.); l. much narrower, fertile spikes short, fr. patent smaller but more swollen.—Marshes and wet thickets, rather rare. P. VI. E. S. I.

64. *C. depauperáta* (With.); *fertile spikes* erect remote 3- or 4-*fl.*, stalks exserted, bracts sheathing leaflike, gl. acute, *fr. large nearly globose*, beak long bifid with rough edges, nut elliptic trigonous with bluntish angles.—*E. B.* 1098. *Schk.* M. 50.—St. 1—2 feet high, trigonous, smooth. Gl. of the male spike blunt. Fr. very large and few with many ribs. Spikes very distant.—Dry woods, very rare. P. VI. E.

††† Beak of fr. glabrous, terete or compressed. Male spike 1 (rarely more). Stigmas 3.

65. *C. sylvat'ica* (Huds.); fertile spikes about 4 distant slightly drooping linear with long half-exserted stalks, bracts leaflike sheathing, *fr. elliptic* trigonous *obscurely veined* narrowed into a *long cloven* smooth *beak*, nut obovoid-elliptic-triangular, l. narrower than in the preceding.—*E. B.* 995. *Schk.* Ll. 101. *H.* b. 55. *R.* 242.—St. about 2 feet high, smooth, its top and the fr.-beak rarely a little rough. Sheaths scarcely half equalling the peduncles. Gl. ovate, acute, diaphanous with a green keel.—Damp woods. P. V. E. S. I.

66. *C. Pseudo-cypérus* (L.); fertile spikes about 5 drooping cylindrical densely flowered stalked near together, *bracts* leaflike *scarcely sheathing*, *gl. setaceous rough* dilated at the base, *fr. ovoid-lanceolate* ribbed narrowed into a *deeply bifid beak*, nut

elliptic, st. triquetrous with rough angles.—*E. B.* 242. *Schk.* Mm. 102. *H.* b. 56. *R.* 275.—St. 2—3 feet high. Male spike often with some fertile flowers. Fertile spike 1½—2 in. long. —Damp places, rare. P. VI. E. I.

†††† Beak of hairy fr. terete or compressed with patent cusps. Male spikes 2 or more.

67. *C. filifor'mis* (L.); male spikes 2, *fertile* 3 or 4 remote *erect sessile* oblong, bracts leaflike lowermost slightly sheathing, gl. oblong-ovate cuspidate and ciliate at the point, *fr.* downy oblong-ovoid narrowed into an *obliquely truncate beak* with 2 *lateral points, nut narrowly elliptic* subtrigonous, l. slender channelled.—*E. B.* 904. *Schk.* K. 45. *B.* 132. *R.* 265.—St. 2 feet high. L. with filamentous sheaths below. Lowest spike rarely stalked.—Peat-bogs, rare. P. V. E. S. I.

68. *C. hir'ta* (L.); male spikes 2 or 3, *fertile* 2 or 3 remote erect oblong-*cylindrical stalked, bracts* leaflike the lower *with long sheaths* nearly equalling the peduncles, gl. elliptic-lanceolate with long slender ciliate points, fr. hairy oblong-ovate narrowed into a *deeply divided beak, nut obovate* narrowed below triangular, *l. flat hairy.*—*E. B.* 685. *Schk.* Uu. 108. *H.* b. 58. *R.* 257.—St. 1½—2 feet high, leafy. L. and sheaths shaggy, rarely glabrous. Fr. tawny. Occasionally the spikes are compound at the base and very long-stalked, and the gl. long.—[Var. *ebractsata* (Syme) has crowded spikes, l. and gl. glabrous and bracts without foliaceous laminæ. A monstrosity?]—Wet places. P. IV. *Hammersedge.* E. S. I.

††††† Beak of fr. glabrous, terete, striate, with patent cusps, or 2-toothed. Male spikes many, rarely 1. Bracts not sheathing.

‡ *Male spike* 1. *Stigmas* 2. *Beak of fr.* 2-*toothed.*

69. *C. pul'la* (Good.); male spike 1 (rarely 2), fertile 1—3 roundish-ovoid lower one stalked bracteate sheathless erect, *fr. ovoid obscurely veined inflated, beak short,* nut roundish mucronate, gl. bluntish,—*E. B.* 2045. *Schk.* Cc. 88. *C. vesicaria* v. *alpigena* Fr.—St. 6—8 in. high. Gl. dark purple tipped with white, *midrib dark purple.* Fr. dark purple paler at the base, longer than the gl., stalked.—*C. saxatilis* (Linn. Herb.); but Andersson says that the true plant is *C. rigida.*—Wet parts of the higher Scottish mountains. P. VII. S.

70. *C. Grahámi* (Boott); male spikes 1 or 2 slender acute, fertile 2 or 3 ovate blunt lower one stalked bracteate sheathless,

fr. oblong-ovate strongly ribbed *inflated narrowed into a short bifid beak,* nut oblong compressed triquetrous below not ½ as long as the perianth, gl. acute.—*E. B. S.* 2923.—St. 1—2 ft. high. Gl. fuscous, with the tip and *midrib pale.* Fr. pale or darkish brown, nearly twice as long as the gl., with several strong ribs on each side.—Glen Fiadh, Clova. Ben Cruichben, [Meall Ghaordie &c.], Killin. P. VII. S.

‡‡ *Male spikes many. Stigmas* 3. *Bracts not sheathing.*

71. *C. rostráta* (Stokes); fertile spikes 2—4 remote cylindrical erect stalked, bracts leaflike, *fr. subglobose inflated suddenly narrowed into a long slender beak,* nut obovoid triangular, *st. smooth with blunt angles.*—*E. B.* 780. *Schk.* Tt. 107. *H.* b. 65. *R.* 277. *C. ampullacea* (Good.) ed. viii.—St. 1—2 feet high, *trigonous.* L. glaucous, channelled.—[A small neat form with very compact spikes and short-beaked fr. is *C. ampull.* var. *brunnescens* (And.), and a very luxuriant Irish form with broad l. and large fruit is var. *latifolia* (Aschers.).]—Very wet bogs. P. VI. E. S. I.

72. *C. vesicária* (L.); fertile spikes 2—4 remote cylindrical, bracts leaflike, *fr. ovoid conical inflated gradually narrowed into a subulate bifid beak,* nut elliptic triangular, *st. with acute angles.*—*E. B.* 779. *Schk.* Ss. 106. *H.* b. 64. *R.* 276.—St. 2 feet high, *triangular,* roughish near the top. L. rather broad, green.—*β. involuta* (Bab.); l. narrow folded into a ½-cylinder, midrib of gl. apiculate, fr. narrower.—*Sy. E. B.* 1681. [Possibly a hybrid with sp. 71.]—Wet bogs. *β.* Hale Moss, Manchester. *Mr. J. Sidebotham.* Congleton, Chesh. *Mr. E. Wilson.* P. V. E. S. I.

73. *C. acutifor'mis* (Ehrh.); *gl. of the barren spikes blunt,* anth. apiculate, *fertile spikes* cylindrical *blunt,* gl. acute or mucronate entire, bracts leaflike, *fr. oblong-obovate compressed with a short bifid beak,* nut roundish-obovate triangular, st. with acute angles.—*E. B.* 807. *Schk.* Oo. 103. *C. paludosa* (Good.) ed. viii.—St. 2—3 feet high, angles rough. L. broad. Fr. sometimes recurved at the tip, stig. sometimes 2.—*β. C. spadicea* (Elw., Roth, not Host.); gl. of fertile spikes with a long rough beak. *C. Kochiana* (DC.).—Wet places. P. V. E. S. I.

74. *C. ripária* (Curt.); *gl. of the barren spikes acute,* anth. with a long point, *fertile spikes acute* cylindrical, bracts leaflike, *fr. oblong-ovoid convex on both sides* narrowed into a short broad cloven beak, nut pyriform triangular, st. with acute angles.—*E. B.* 579. *Schk.* Qq. & Rr. 105. *H.* b. 66.—St. 3—4 feet high, angles rough. L. broader than in the preceding.—Wet places. P. V. E. S. I.

Index to the Carices.

Order XCVIII. GRAMINEÆ.

Fl. perfect or unisexual, 1, 2 or more seated bifariously on a common axis which is contained within an involucre of 2 (or 1) valves (glumes) or rarely none, the whole forming a locusta or spikelet. Each fl. of 1 or 2 scales (pales) of which the outer or lower is simple and usually keeled, the inner with 2 veins or keels. Hypogynous scales 2, 3 or none. Stam. hypogynous, 1—6, usually 3. Anth. versatile, notched at both ends. Ovary 1-celled. Styles usually 2, rarely 1 or 3. Embryo on the outside of the albumen and at its base.—L. with split sheaths [1].

Suborder I. CLISANTHEÆ [2].

Flowers closed. Styles or stigmas long, protruded at or near to the top of the flower.

A. *Rachis of inflorescence without lateral excavations.*

Tribe I. *PANICEÆ. Spikelets dorsally compressed,* 1-*flowered,* or with 1 fl. and an *inferior* glumelike *rudiment* or a neuter flower. Lower gl. much the smaller, often rudimentary.

1. Digitaria. *Spikes fingered.* Spikelets in pairs on one side of the flattened rachis, awnless, 1-flowered, with an inferior rudiment. Gl. 2, lower smaller or 0, upper 3-veined. Sterile fl. of one 5—7-veined pale equalling the flower.

[2. Echinochloa. Spikes compound, secund in the whole and in each part. Spikelets on one side of the flattened partial rachis, 2-flowered, inferior fl. rudimentary. *Gl.* 2,

---

[1] Nearly all the genera of this Order are beautifully figured in the *Gen. Fl. Germ. Monocotyl.* vol. i. See also Andersson's *Pl. Scand.* fasc. ii. and Du Mortier in *Bull. Soc. Bot. Belg.* vii. 65.

[Since the last edition appeared a large number of so-called varieties of Grasses have been published as British; many of these we have ignored, being apparently merely trivial and transitory variations characterised by slight differences in stature, habit, or colour, more condensed or extended inflorescence, broader or narrower leaves, greater or less hairiness, and the like.—H. & J. G.]

[2] The suborders of Fries although convenient for us will not apply generally, see *Bentham, Linn. Soc. Journ.* xix. 26. Bentham says practically useless except for grasses in a living state. They are retained here as we use them in that state of the plants.

lower 3-veined, *upper* equalling fl. 5-*veined* mucronate or awned. Lower pale of sterile fl. like and equalling upper glume.]

3. Setaria. *Spike cylindrical*, compound. *Spikelets surrounded by an involucre of bristles*, 2-flowered; inferior fl. rudimentary. Gl. 2, lower 3-veined, upper equalling fl. many-veined. Sterile fl. of 1 pale, like and equalling upper glume.

Tr. II. *CHLORIDEÆ*. *Spikelets* laterally compressed 1-*flowered* in our plants and sometimes with *a superior rudiment*, placed *in 2 rows on one side of a flattened rachis*, or alternate and unilateral.

4. Capriola. Spikes fingered, spreading. Spikelets 1-flowered, awnless, with a superior rudiment. Gl. nearly equal, patent. Pales equal; lower boatshaped, compressed, embracing the inner. Styles long, distinct. Stigmas feathery.

5. Spartina. *Spikes upright*, in a raceme. Spikelets 1-flowered, awnless. Gl. unequal; upper lanceolate. Pales unequal; lower boatshaped, compressed. *Styles* long, *united* halfway up. Stigmas feathery.

6. Mibora. *Inflorescence a somewhat* 1-*sided raceme*. Gl. not keeled blunt. Fl. 1. Pale 1, scarious, very hairy, blunt, not awned.

Tr. III. *PHALARIDEÆ*. Panicled. *Spikelets laterally compressed*, 1-*flowered*, with 1 or 2 glumelike *inferior rudiments*, or 1 or 2 inferior male flowers. Gl. equal covering the flowers. Styles long. Stigmas filiform.

7. Phalaris. Gl. 2, boatshaped, keeled, menbranous, nearly equal, exceeding the flower. Pales coriaceous, unequal, closely investing the fruit. Rudimentary fl. 1—2, scalelike.

8. Anthoxanthum. Gl. 2, unequal, membranous; lower small, 1-veined; upper exceeding the fl., 3-veined. Pales scarious. *Stam.* 2. Rudimentary fl. 2, scalelike, bifid, awned on the back.

9. Hierochloe. Gl. 2, nearly equal, membranous, about equalling the spikelet. Fl. 3; 2 lower male, 3-androus, upper pale with 2 keels; upper fl. perfect, 2-androus, upper pale with 1 keel.

Tr. IV. *PHLEINEÆ.* Inflorescence dense, spikelike. Spikelets laterally compressed, *1-flowered* or with a *superior rudiment.* Gl. nearly equal, covering the flowers. Styles long. Stigmas filiform.

10. PHLEUM. Gl. compressed, keeled, parallel at the midrib, truncate, with a terminal seta, or acute. Fl. 1. Pales 2, membranous; lower 3-veined, blunt, without awns, or with a minute central point.

11. ALOPECURUS. Gl. compressed, connate below, membranous, awnless. Fl. 1. Pale 1, scarious, 5-veined, awned on the back.

Tr. V. *SESLERIEÆ.* Panicle spikelike. Spikelets laterally compressed, with 2 *or more flowers.* Styles 0 or very short. Stigmas very long, filiform.

12. SESLERIA. Spikelets sessile, imbricate all round. Gl. 2—6-flowered, nearly or quite equalling the spikelet. Lower pale keeled, membranous, with a scarious margin, ending in 3 or 5 points; dorsal rib excurrent.

B. *Spikelets sessile in hollows of the rachis.*

Tr. VI. *NARDEÆ.* Spikelets 1-flowered. Gl. 0. Style short. Stigmas filiform.

13. NARDUS. Spikelets in 2 rows on one side of the rachis. Lower pale keeled, tapering into a subulate point. Stigmas long.

Suborder II. EURYANTHEÆ.

Flowers open. Style short. Stigmas protruded near to the bottom of the flower.

A. *Inflorescence panicled or racemose; rachis without lateral excavations.*

Tr. VII. *ORYZEÆ.* Spikelets laterally compressed, 1-flowered. Glumes 0. Stigmas feathery. Pales enclosing but free from the nut.—Pales opening only slightly.

14. HOMALOCENCHRUS. Pales 2, like parchment, compressed keeled, awnless; lower much broader.

Tr. VIII. *STIPACEÆ.* Spikelets cylindric, 1-flowered, without any rudiment. Gl. unequal membranous, enclosing the flowers. Pales hardening, enclosing but free from the nut.

[15. Stipa. Spikelets stalked. Pales coriaceous; outer cylindrical, convolute, evidently jointed to the kneed twisted feathery awn.]

16. Milium. Spikelets stalked. Pales like parchment, awnless; lower ventricose, convex.

Tr. IX. *AGROSTIDEÆ.* Spikelets laterally compressed, 1-flowered, or with a superior rudiment, or many-flowered. Gl. and pales membranous.

† *Gl. falling short of the outer pale. Style long. Stigmas often protruded near the middle of the flower.*

17. Phragmites. Pan. diffuse. Gl. unequal; lower much smaller. Fl. 1—6, awnless, with silky hairs at their base; lower imperfect. Lower pale acuminate, much exceeding the inner, awnless.

†† *Gl. exceeding the flowers. Style short or none.*

18. Ammophila. Pan. spikelike. Gl. nearly equal; lower rather the shorter. Fl. 1, with silky hairs at its base or without a superior rudiment. Lower pale with a very short awn.

19. Calamagrostis. Pan. diffuse. Gl. nearly equal; lower rather the longer. Fl. 1, with silky hairs at its base, with or without a superior rudiment. Lower pale awned.

20. Apera. Pan. loose. Gl. membranous, acute, unarmed; *lower the smaller.* Fl. 1, with hairs at its base, and a pedicel-like rudiment. Pales unequal, scarious; lower with a long subterminal awn.

21. Agrostis. Pan. loose. Gl. membranous, acute, unarmed; *upper smaller.* Fl. 1, with hairs at its base; no rudiment. *Pales unequal,* scarious; dorsal awn falling short of the glumes, or 0.—*A. canina* wants the inner pale.

[22. Lagurus. Pan. spikelike. *Gl.* scarious, *ending in a long fringed bristle.* Fl. 1, with a pedicel-like rudiment. Lower pale ending in 2 long bristles and with a dorsal kneed twisted awn.]

23. POLYPOGON. Pan. close, spikelike. *Gl.* scarious, each *with a long bristle* from just below the notched tip. Fl. 1. Lower pale usually awned from just below the tip.

24. GASTRIDIUM. Pan. close, spikelike. *Gl.* membranous, acute, awnless, *ventricose* at the base, much exceeding the flower. Fl. 1. Lower pale truncate or toothed at the end, with or without a dorsal awn.

Tr. X. *AVENEÆ.* Spikelets with 2 or more flowers; upper often barren. *Gl. equalling or overtopping the flowers.* Lower pale awned. Style short or 0.

25. HOLCUS. Fl. 2; lower perfect, awnless (or very rarely awned); upper usually male, with a dorsal awn. *Pales hardening on the fruit*; tip of lower entire.

26. WEINGAERTNERIA. Fl. 2, perfect, awned. Awn straight, jointed in the middle; *the upper portion clavate*; a tuft of hairs at the joining. Tip of lower pale entire.

27. AIRA. Pan. lax. Fl 2, with or without the rudiment of a third (sometimes perfected). Lower pale denticulate or bifid at the tip, terete on the back. Awn dorsal, kneed, (in *A. cæspitosa* straight). Ovary glabrous. Fr. not crested.

28. TRISETUM. Spikelets crowded. Fl. 2—6. Lower pale with faint lateral veins, ending in 2 acute teeth, awned. Awn dorsal, kneed and twisted. Ovary glabrous. *Fr. neither crested nor furrowed.*

29. AVENA. Fl. 2 or more. Lower pale with lateral veins, awned, ending in 2 points. Awn dorsal, kneed and twisted. Ovary hairy at the top. *Fr. crested and furrowed.*

30. ARRHENATHERUM. Fl. 2, with a rudiment; *lower fl. male* with a long kneed and twisted awn from below its middle; upper with a short straight awn from near the tip. Pales herbaceous, ending in 2 points. Ovary hairy at the top. Fr. oblong, terete, downy, not furrowed.

Tr. XI. *FESTUCEÆ.* Spikelets with 2 or more flowers, upper often barren. *Gl. falling short of the lowest flower. Style short or* 0.

† *Lower pale with nearly paralled veins which do not join to form an awn. Awn* 0. *Styles terminal.*

‡ Lower pale 2—3-fid.

31. SIEGLINGIA. Fl. 2—4. Lower pale rather coriaceous, rounded on the back, bifid with an intermediate broad point. Nut free.

‡‡ Lower pale nearly or quite entire.

32. KŒLERIA. Gl. unequal; upper 2—3-ribbed. Spikelet compressed, 2—5-flowered. Lower pale keeled, acuminate, or with a straight subterminal bristle. Nut free.

33. MELICA. Gl. nearly equal, with lateral ribs, nearly as long as the ovate spikelet consisting of 1 or 2 *flowers rounded* on the back and a clublike rudiment. Pales hardening on the free nut.

34. MOLINIA. *Gl.* unequal, *without lateral ribs*, falling short of the lanceolate spikelet of 2 or 3 semicylindrical flowers and a subulate rudiment. Pales hardening on the free nut.

35. POA. Gl. rather unequal. *Lower pale* 3—5-*veined*, scarious at the tip, *compressed, keeled.* Nut elliptic trigonous, slightly furrowed within, free.

36. GLYCERIA. Gl. unequal, submembranous. *Lower pale with* 5—7 *strong prominent ribs* and a scarious margin, *subcylindrical.* Nut oblong, convex on back, furrowed within, free.

37. SCLEROCHLOA. Gl. unequal, membranous. *Lower pale with* 5 *faint* veins, *cylindrical* below, often keeled at the tip or with a very minute mucro. Nut oblong, convex on back, not furrowed within, free.

38. BRIZA. Gl. nearly equal, broad, 3-ribbed. Fl. 3—8, densely imbricate in a short distichous spikelet. Lower pale boatshaped, heartshaped, blunt, rounded on the back. Gl. and pales membranous with a scarious margin. Nut free.—Lower pale with 7—9 faint veins.

‡‡‡ Lower pale truncate and denticulate at the tip.

39. CATABROSA. Gl. unequal, very short, 1-veined. Flowers usually 2, rounded on the back, distant. Lower pale membranous, with 3 veins ending in teeth which do not quite extend to the edge of the scarious margin. Upper gl. often with 2 short and faint lateral veins.

†† *Lower pale with converging veins, all or* 1—3 *of which combine in the awn.*

40. CYNOSURUS. Gl. nearly equal, scarious, strongly keeled, 1- or more flowered. Lower pale rounded on the back with

a terminal bristle. *Each spikelet with a comb-like bract at its base.* Panicle spikelike, 1-sided. Nut closely coated with the pales.

41. DACTYLIS. Gl. unequal, herbaceous, many-flowered; lower keeled. *Lower pale compressed,* keeled, 5-veined; dorsal vein fringed and excurrent just below the tip as a short awn. *Spikelets crowded, subsecund.* Nut free.

42. FESTUCA. Gl. unequal, herbaceous, many-flowered. *Lower pale rounded* on the back, very acute, *or with the dorsal vein excurrent* at or just below the tip as an awn; *lateral veins slightly converging* and vanishing below the tip. Upper pales minutely ciliate on the ribs. *Styles terminal.* Nut furrowed, adhering to the pales.—Rachis with acute angles. Sheaths of the leaves divided to the base.

43. BROMUS. Gl. unequal, herbaceous, many-flowered; *lower 1-veined, upper 3—5-veined.* Fl. lanceolate, compressed. Lower pale with a long awn, (usually) founded on 3 veins, from below the tip. *Styles below the top of the ovary.* Nut furrowed, adhering to the pales.—Sheaths of the l. divided halfway down.

44. SERRAFALCUS. Gl. unequal, herbaceous, many-flowered, *lower 3—5-veined, upper 7—9-veined.* Fl. oblong, turgid. Lower pale with a short awn, (usually) founded on 3 veins, from below the tip. *Styles below the top of the ovary.* Nut furrowed, adhering to the pales.—Sheaths of the l. scarcely divided halfway down. Spikelets narrower upwards.

B. *Inflorescence spikelike; bearing the spikelets in hollows of the rachis.*

Tr. XII. *HORDEIEÆ.* Spikelets solitary or 2 or 3 together, subsessile on opposite sides of a channelled and toothed jointed rachis. Uppermost fl. often barren. Style very short. (Lateral spikelets often stalked in *Hordeum.*)

† *Spikelets very shortly stalked or subsessile.*

45. BRACHYPODIUM. Gl. opposite, unequal, many-flowered, their edges towards the rachis. Upper pale coarsely fringed on the ribs. — The unequal gl. distinguish this from *Triticum.*

†† *Spikelets quite sessile.*

⸸ Glumes 2.

46. TRITICUM. Spikelets solitary. Gl. opposite, nearly equal, many-flowered, their edges towards the rachis. Inner pale minutely ciliate on the ribs.

47. ELYMUS. *Spikelets 2 or 3 together. Gl.* 2, both on the same side of the spikelet, *without awns or bristles,* with 2 or more perfect flowers.

48. HORDEUM. Spikelets in threes, often partially barren. *Gl.* 2, ending in long bristles; 1 *perfect flower* and a stalk-like rudiment.

49. LEPTURUS. Spikelets solitary, imbedded alternately on opposite sides of the rachis. Gl. 1—2, both on the same side of the spikelet, cartilaginous, covering the one fl. and superior rudiment. Pales scarious. Stigmas feathery.

⸸⸸ Glume solitary, bractlike, or a very small second glume.

50. LOLIUM. Spikelets solitary, placed edgewise on the rachis. Gl. solitary, or that next the rachis very small, with 3 or more flowers.

Suborder I. *Clisantheæ.* Tribe I. *Paniceæ.*

1. DIGITA'RIA *Adans.*

[*D. sanguinális* (Scop.); l. and sheaths hairy, fl. oblong-lanceolate glabrous with downy margins (?).—*E. B.* 849. *P.* 70.—St. ascending, a foot long.—Not a native. A. VIII.] E.

1. *D. humifúsa* (Pers.); l. and sheaths glabrous, fl. ovate downy with glabrous veins.—*E. B. S.* 2613. *P.* 71.—St. mostly procumbent, 4—8 in. long. Spikes usually 3 or 4, springing from nearly the same point. Spikelets in pairs, one on a longer stalk than the other.—Sandy fields, rare. A. VII. VIII. E.

2. ECHINOCH'LOA *Pal. de Beauv.*

[*E. Crus-gal'li* (Beauv.); spikes alternate or opposite, spikelets near together, upper gl. and sterile floret awned or mucronate hispid, rachis hispid.—*E. B.* 876. *P.* 67. *Panicum* L.—Near London. A. VII.] E.

3. SETA'RIA *Pal. de Beauv.*

1. *S. vir'idis* (Beauv.); pan. spikelike, involucral bristles with forward teeth, lower pale smooth.—*Panicum* L., *E. B.* 875. *P.* 68.—London and Norwich. A. VII. VIII. E.

[*S. verticilláta* (Beauv.); pan. spikelike, involucral bristles with declining teeth, lower pale smooth.—*E. B.* 874. *P.* 39. —London and Norwich. A. VII. VIII.] E.

[*S. glaúca* (Beauv.); pan. spikelike, invol. bristles with ascending teeth, pales transversely rugose.—Weybridge, Surr. A. XI.] E.

Tribe II. *Chlorideæ.*

4. CAP'RIOLA *Adans.* (*Cýnodon* Rich, ed. viii.)

1. *C. Dac'tylon* (O. Kuntze); spikes 3—5 digitate, pales smooth their edges and keel slightly ciliate, l. downy beneath, barren shoots prostrate.—*E. B.* 850. *P.* 72.—Creeping. Flowering st. 4—6 in. high, ending in a cluster of spreading many-fl. slender spikes. Spikelets purplish. L. on the long branched barren shoots flat, spreading; on the others usually folded.—Sandy shores. Dorset. Devon. Cornwall. P. VIII. E.

5. SPARTI'NA *Schreb.*

1. *S. stric'ta* (Roth); l. jointed to their sheaths falling short of the spikes, spikes 2—3, *rachis scarcely extending beyond the last spikelet,* outer gl. hairy.—*E. B.* 380.—St. 1—2 feet high. L. narrowing to the base where they easily separate from their sheaths. Spikes pressed close together. A remarkably rigid plant.—Muddy salt marshes. P. VIII. E.

2. *S. Townsendi* (Groves); l. jointed to their sheaths falling short of spike, spikelets 4—9, *rachis produced beyond the spikelets* and flexuose, outer glumes slightly downy. –*J. of B.* xx. 1. t. 225.—St. 1½ to 4 ft. high. L. broadest at the base. Spikelets rather spreading.—Mud-flats. I. of Wight, Hants, Sussex, Kent. P. VIII. E.

3. *S. alterniflóra* (Loisel.); l. continuous with their sheaths equalling or exceeding the 6—8 spikes, spikelets many, *rachis produced beyond the spikelets* and flexuose, outer gl. glabrous.—*E. B. S.* 2812. *P.* 75.—St. 2—3 feet high. L. broadest at the base. Spikes loosely pressed together.—Mud-flats by Southampton Water. P. VIII. E.

6. Mibo'ra *Adans.*

1. *M. ver'na* (Beauv.).—*M. minima* (Desv.) ed. viii. *Knappia* Sm. *E. B.* 1127. P. 73. *Sturmia* Hoppe in *St.* 7. 1. *Chamagrostis* Borkh.—An elegant but very small grass. Root small, fibrous. St. many. L. short, rough. Spikes slender; spikelets 5—10, sessile. Pale shorter than the glumes, hairy, truncate, ragged.—Sandy south-west coast of Anglesea. A. IV. E.

Tribe III. *Phalarideæ.*

7. Phal'aris *Linn.* Reed-grass.

[*P. canarien'sis* (L.); pan. ovoid spikelike, *gl. winged on the keel, wing entire, rudimentary fl. 2* half as long as the fertile fl., pales pilose.—*E. B.* 1310. *P.* 9.—St. 1—2 feet high, ending in a compact compound panicle. Gl. large, pale yellow variegated with green lines and remarkably winged at the back.—Scarcely naturalized. A. VII. *Canary-grass.*]

[*P. minor* (Retz.); pan. cylindric-oblong spikelike, gl. winged on the upper part of the keel, wing *toothed near the apex, rudimentary fl. 1* narrow pilose ⅓ as long as the fertile fl., lower pale pilose quite enclosing the upper, upper much narrower and shorter ciliate on the keel.—*J. of B.* xxxviii. (1900) p. 33, t. 406.—St. with a single spike. More slender than the last, with longer narrower pan., much smaller and more numerous fl., and narrower, toothed, keel to the gl.—Channel I., perhaps native. A. VII.]

[*P. paradox'a* (L.); pan. spikelike, gl. of fertile fl. with a *blunt toothed wing on the keel* many-veined, *rudimentary fl. several.*—St. decumbent below, then ascending, 1—3 ft. high, branched. Lower part of pan. usually barren; branches with about 6 spikes.—Swanage, Dorset. A. VII.] E.

1. *P. arundinácea* (L.); pan. upright with spreading branches, fl. clustered, *gl. not winged,* rudimentary fl. 1 or 2 small hairy.—*E. B.* 402. *P.* 9.—St. 4—5 feet high. Creeping. Pan. 3—4 in. long. Gl. keeled. L. sometimes variegated with white lines. —By water. P. VI. VII. E. S. I.

8. Anthoxan'thum *Linn.* Vernal-grass.

1. *A. odorátum* (L.); pan. spikelike dense oblong, gl. about equalling awns, st. nearly simple below.—*E. B.* 647. *P.* 8.—About a foot high. Panicle lanceolate, dense, or rather interrupted below.—There are two forms or species: (1) with purple anth. in meadows, (2) with dull yellow anth. in woods.—Very common in pastures. P. V. VI. E. S. I.

2. *A. Puel'ii* (Lec. & Lam.); pan. spikelike lax, gl. much falling short of awns, st. many branching from the base.—*J. of B.* xiii. t. 157.—Pan. narrowed to the top. St. 6—12 in. high. —Sandy pastures. A. VII.—IX. E.

9. Hiero'chloe *Gmel.* Holy-grass.

1. *H. boreális* (R. & S.); pan. divaricate, pedicels glabrous, fl. awnless, l. flat.—*E. B. S.* 2641. *P.* 31.—About a foot high. Spikelets ovate, brown.—Forfarshire (but now lost). Thurso, Caithness. *Mr. R. Dick.* Kirkcudbright. P. V. VI. S.

Tribe IV. *Phleineæ.*

10. Phle'um *Linn.* Cat's-tail-grass.

[*P. as'perum* (Jacq.); l.-sheaths slightly inflated, pan. cylindrical, gl. wedgeshaped truncate swelling upwards, keels rough, rudimentary fl. subulate.—*E. B.* 1077. Probably an escape, now lost. A. VII.] E.

1. *P. Boeh'meri* (Wibel); l.-sheaths slightly inflated, pan. cylindrical, *gl. linear-lanceolate* obliquely truncate mucronate, keel ciliate above, rudimentary fl. subulate.—*E. B.* 459. *P.* 80. —St. leafy below, naked upwards, with sterile leafy shoots. Lower pale entire, not awned. Anth. linear-oblong.—Dry chalky fields, rare. P. VII. E.

[*P. Michélli* (All.) was probably a mistake.]

2. *P. arenárium* (L.); l.-sheaths inflated, *pan. oblong,* somewhat *narrowed below,* gl. lanceolate acuminate, keel ciliate above, rudimentary fl. subulate minute.—*E. B.* 222. *St.* 29. 1. *P.* 7.—St. varying greatly in height. Lower pale notched at the summit, ⅓ the length of the glumes. Anth. very small.—Sandy places chiefly near the sea. A. V.—VII. E. S. I.

3. *P. praten'se* (L.); l.-sheaths not inflated, *pan. cylindrical, gl.* oblong *truncate* with an awn of less than half their length, keel ciliate above, no rudimentary flower.—*E. B.* 1076. *P.* 77 & 78.—Slightly creeping (or slightly tuberous, *P. nodosum* L.). Pan. 1—5 in. long. Awns sometimes longer. Lower pale jagged at the summit.—*α. P. pratense* (L.); st. from a prostrate base, l. broad, pan. usually long, anth. purplish.—*β. stoloniferum* (Bab.); many barren leafy prostrate or erect shoots, l. broad, pan. rather short, anth. purplish.—*γ. P. præcox* (Jord.); st. from an ascending base, l. slender, pan. rather short, anth. pale yellow.—Meadows and pastures. P. VI. *Timothy-grass.* E. S. I.

4. *P. alpinum* (L.); l.-sheaths inflated, pan. oblong, gl. truncate glabrous ciliate on the back with a scabrous awn nearly equalling their length, keel ciliate, upper sheath inflated.—*E. B.* 519. *P.* 6. *P. commutatum* Gaud.—Somewhat creeping. St. 6—12 in. high.—Wet alpine moors. P. VII. S.

11. Alopecu'rus *Linn.* Fox-tail-grass.

1. *A. praten'sis* (L.); st. erect smooth, *pan. cylindrical* blunt, gl. acute connected below ciliate downy, pale equalling the glumes, *awn projecting more than half its own length beyond the pale.*—*E. B.* 759. *St.* 8. 1. *P.* 4.—Scarcely creeping.—St. 1—3 feet high. In salt marshes the base of the st. becomes fleshy. Upper sheath slightly inflated. Ligule short, blunt. Pan. 1—3 in. long, branches 4—6-flowered. Anth. yellow. Styles combined. [A hybrid between this and Sp. 3 (*A. hybridus*, Wimm.) has been found in Warwicksh. See *J. of B.* xxxix. (1901) p. 232.]—Rich pastures. P. IV.—VI. E. S. I.

2. *A. alpinus* (Sm.); st. erect smooth, *pan. oblong*, gl. acute connected below hairy, pale equalling the glumes, *awn projecting* ⅓ *of its length beyond the pale.*—*E. B.* 1126. *P.* 4.—Somewhat creeping. St. decumbent at the base, then erect, 9—12 in. high. Ligule short, blunt. Uppermost l. usually (not always) short and broad, ⅓ of the length of its inflated sheath. Gl. connate through about ¼ of length. Awn from about the middle of pale, sometimes wanting. Styles combined. Pan. not exceeding an inch, sometimes rather lax; silky branches 4—6-flowered.—High on mountains. Loch-na-Gar. Cairngorm mountains. Ben Lawers. Clova. P. VII. S.

3. *A. geniculátus* (L.); st. ascending bent at the knots smooth, pan. cylindrical, *gl. blunt* connected below ciliate rather *exceeding the pale*, awn from near the base of the pale and projecting half its length beyond it, *anth. linear.*—*E. B.* 1250. *P.* 5.—Root fibrous. St. about a foot long, branching below. Knots generally (in dry places) oval and fleshy. Upper sheath inflated. Ligule oblong. Gl. membranous at the top except the midrib, often villose or hairy below. *Pale* when laid open *oblong, blunt,* slightly notched. Anth. ultimately violet-yellow. Styles mostly combined. Pan. 1—2 in. long.—*A. pronus* (Mitten)[1] is a prostrate form of this.—Wet places. P. VI. VII. E. S. I.

---

[1] Now known to have been a monstrosity, see *J. of B.* xxxvii. (1899) p. 358.—H. & J. G.

4. *A. æquális* (Sobol.); st. ascending bent at the knots smooth, pan. cylindrical, gl. connected below ciliate rather *falling short of the pale, awn from* just below *the middle of the pale* and scarcely extending beyond it, *anth. short* and broad.—*E. B.* 1467. *P.* 5. *A. fulvus* (Sm.) ed. viii.—St. 1—2 feet long, procumbent below. Ligule oblong. Pan. 2—3 in. long. Anth. at first white, afterwards orange-coloured.—Wet margins of ponds. P. VI.—IX. E. S.

5. *A. bulbósus* (Gouan); st. smooth, pan. cylindrical acuminate, *gl. distinct abruptly acute* downy exceeding the pale, awn from near the base of the pale and projecting half its length beyond it.—*E. B.* 1249. *P.* 76.—St. 1 foot long, ascending or decumbent, in a circular tuft, kneed, the lowermost knots forming *ovate fleshy knobs.* Upper sheath inflated. Ligule oblong. Pale when laid open truncate, emarginate, with 2 small teeth in the middle. Styles combined. Pan. about 1 in. long, less decidedly racemose than in our other species; pedicels usually 1-flowered. —Salt marshes in the South, rare. P. V. VI. E.

6. *A. myosuroïdes* (Huds.); *st.* erect *roughish* above, *pan. tapering* slender, *gl. acute* connected below *nearly glabrous,* awn from near the base of the pale and projecting half its length beyond it.—*E. B.* 848. *P.* 3. *A. agrestis* (L.) ed. viii.—St. 1—2 feet high, slender. Sheaths roughish. Ligule prominent, blunt. Gl. glabrous, but with a row of *fine short cilia on the back,* connate nearly to the middle. Styles combined.—A very troublesome weed. A. IV.—XI. *Black grass.* E.

Tribe V. *Sesleriеæ.*

12. Ses'leria *Scop.* Moor-grass.

1. *S. cærúlea* (Ard.); pan. ovoid slightly 1-sided, outer pale ending in 4 teeth, midrib rough with a short excurrent point, l. abrupt with a minute rough point.—*E. B.* 1613. *P.* 27.—Roots tufted. St. 6—12 in. high. Pan. about ½ in. long, bluish purple. Anth. purple-tipped. Stig. very long, linear. [A yellowish flowered form is *S. luteo-alba* (Opiz.).]—Mountains. [Limestone rocks and pastures.] Banks of the Shannon [&c.]. P. IV. V. E. S. I.

Tribe VI. *Nardeæ.*

13. NAR'DUS *Linn.* Mat-grass.

1. *N. stric'ta* (L.).—*E. B.* 290. *P.* 2.—Tufted. St. and l. erect, slender, rigid. Height 5—8 in. Spike close, slender. Lower pale with a short rough awn, coriaceous, often purplish; upper membranous.—Moors and heaths. P. VII. E. S. I.

Suborder II. *Euryantheæ.* Tribe VII. *Oryzeæ.*

14. HOMALOCEN'CHRUS *Mieg.* (*Leersia* Soland. ed. viii.) Cut-grass.

1. *H. oryzoïdes* (Poll.); pan. patent with wavy branches, spikelets 3-androus half-oval, keel ciliate.—*E. B. S.* 2908.—Creeping. St. 1—2 ft. high. L. broad, rough-edged; uppermost horizontal at the flowering-season. Pan. mostly enclosed in the sheath of the uppermost leaf. The included fl. alone are usually fertile.—Marsh-districts in Sussex, Surrey, Hampshire, and Dorset. P. VIII. IX. E.

Tribe VIII. *Stipaceæ.*

15. STI'PA *Linn.* Feather-grass.

[*S. pennáta* (L.); awn very long twisted feathery its base glabrous.—Awns remarkably long.—Not a native. Common in gardens. P. VI.] E.

16. MIL'IUM *Linn.* Millet-grass.

1. *M. effúsum* (L.); pan. diffuse, pales acute, st. smooth, l. lanceolate-linear.—*E. B.* 1106. *P.* 17.—Stoloniferous. St. 3—4 feet high. Branches of the panicle long, in distant alternate tufts, in flower horizontal, afterwards deflexed.—Damp shady woods. P. VI. E. S. I.

[*M. scábrum* (Merlet); pan. close, pales obtuse, st. scabrid, l. short linear-lanceolate.—*J. of B.* xxxviii. (1900) p. 34, t. 406 B.—A small plant (in Guernsey 1—4 in. high) slightly scabrid throughout. Pan. small with few short flexuous nearly erect br. Glumes 3-veined.—Cliffs, Petit Bot, Guernsey. *Mr. C. R. P. Andrews.* A. IV.]

Tribe IX. *Agrostideæ.*

17. Phragmi'tes *Adans.* Reed.

1. *P. commúnis* (Trin.); pan. diffuse, spikelets 1—6-flowered, fl. exceeding the glumes.—*Arundo* L., *E. B.* 401. *P.* 29.—St. 5—6 feet high, erect. Pan. large, purplish. Spikelets usually 3—6-fl., or 1—2-fl. in *P. nigricans* Dum. L. flat, broad. Soboliferous; soboles rarely aërial and 20—40 ft. long (*Phytol.* t. 146).—Marshes. P. VIII. E. S. I.

18. Ammoph'ila *Host.* (*Psamma* Beauv. ed. viii.) Marram. Bent.

1. *A. arundinácea* (Host); *pan. cylindrical* rather thicker at the middle, gl. and pales linear-lanceolate acute, hairs ⅓ the length of the pale.—*P. arenaria* (R. & S.) ed. viii. *Arundo* L., *E. B.* 520. *P.* 8.—St. erect, stiff, 2—3 feet high. L. rigid, involute, acute, glaucous. Panicle straw-coloured.—Sandy seashore, binding the shifting sands. P. VII. E. S. I.

2. *A. bal'tica* (Link); *pan. lanceolate, gl.* and pales lanceolate-prolonged *very acute,* hairs ½ the length of the pale.—*J. of B.* x. t. 127.—Known by its more lax sublanceolate pan. and very acute glumes.—Ross Links, Northumb. Norfolk. P. VII. E.

19. Calamagros'tis *Adans.* Small Reed.

* *With no superior rudimentary flower.* Calamagrostis Beauv.

1. *C. lanceoláta* (Roth); pan. loose erect, awn very short from the bottom of the notch at the end of the lower pale and scarcely extending beyond it, hairs longer than the pales. —*P.* 84. *Arundo Calamagrostis* L., *E. B.* 2159.—St. slender, 3—4 feet high.—Wet places, rare. P. VII. E. S.

2. *C. epige'jos* (Roth); pan. rather close lobed, *straight awn from about the middle* of the lower pale, hairs longer than the pales.—*E. B.* 403. *P.* 16.—St. 3—5 feet high.—Damp shady places. P. VII. E. S. I.

** *With a superior rudimentary flower.* Deyeuxia Beauv.

3. *C. neglec'ta* (Gaert. M. & S.); pan. close, gl. lanceolate rough on the keel one (at least) 3-ribbed, lower pale nearly as long as the upper gl. deeply notched at the top longer than the hairs, straight awn from below the middle of the pale and scarcely extending beyond it, l. of the barren shoots slender.—

E. B. 2160. *C. stricta*, ed. viii.—St. erect, 2—3 ft. high. L. broad; on the barren shoots much narrower. Uppermost ligule short, blunt; longer and acute in the Irish plant, var. *Hookeri* Sy., which has shorter pan.-branches.—Bogs. Oakmere, Cheshire. By Loch Tay, [*Mr. G. C. Druce.* Referred to var. *borealis* by Prof. Hackel.] Formerly found near Forfar. Lough Neagh. P. VI. VII. E. S. I.

4. *C. strigósa* (Hartm.); pan. close, gl. lanceolate *folding at the tip into a long acuminate point* asperous on the back with 1—2 lateral ribs *much exceeding the pales*, pales deeply jagged at the tip, lower with awn equalling it attached below the middle sometimes near the base, hairs very unequal falling short of the pales, l. slender.—Cæspitose. St. erect, 1½—2 ft. or more high. Young spikes variegated with purple. Upper ligule long. More slender than Sp. 3, with conspicuously longer gl.—Site of Loch Duran, Caithness. *Mr. R. Dick.* P. VII. S.

20. Ape'ra *Adans.* Wind-grass.

1. *A. Spica-ven'ti* (Beauv.); pan. spreading, anth. linear-oblong.—*Agrostis* Sm., *E B.* 951. *Anemagrostis* (Trin.) *P.* 17.—St. 1—2 feet high. Pan. very light and elegant; branches spreading horizontally with flowers. Awn 3 or 4 times exceeding the pale. Rudimentary fl. like a pedicel. A tuft of hair on each side of the inner pale.—Sandy fields, rare. A. VI. VII. E. S.

2. *A. interrup'ta* (Beauv.); pan. close, anth. oval.—*E. B. S.* 2951.—St. 1—2 feet high. Pan.-branches dividing from their base, never spreading. Awn 3 or 4 times exceeding the pale.—Sandy fields. Pampisford and Chippenham, Cambridgeshire. Thetford, Suffolk. Dirleton, S. A. VI. VII. E. S.

21. Agros'tis *Linn.* Bent-grass.

1. *A. setácea* (Curt.); panicle close oblong, branches and pedicels rough, gl. unequal acute, lower pale jagged at the top 4-ribbed, lateral ribs ending in short setæ, kneed and twisted awn from near the base of the pale and twice its length, *l. setaceous involute, sheaths rough*, ligule oblong acute.—*E. B.* 1188. *P.* 83.—Root tufted. L. short, almost capillary. *Pan.-branches short.* Midrib of lower glume rough in its upper half, slightly excurrent. Upper pale very minute, a tuft of hairs at its base.—Dry heaths in the South-west. P. VII. E.

2. *A. canína* (L.); pan. spreading when in flower otherwise close, branches and pedicels rough, gl. unequal acute, lower pale jagged at the top 4-ribbed, kneed and twisted awn from

below the middle of and exceeding the pale, lower l. setaceous tufted, stem-l. narrow flat, *sheaths smooth,* ligule oblong acute. —*E. B.* 1856. *P.* 15.—Trailing leafy shoots. St. decumbent below, then erect. L. narrow; radical involute. Pan.-branches long, slender. Fl. green or purplish. Lower gl. not jagged at the top; midrib rough from rather below the middle. Upper pale 0, or very minute. Awn sometimes very short or rarely [var. *mútica* (Gaud.)] absent.—[A mountain form with simpler pan. and larger fl. is var. *scotica,* Hack.]—Peaty heaths. P. VII. VIII. E. S. I.

3. *A. nígra* (With.); *pan.-branches constantly erect-patent simple below* rigid strongly hispid, gl. nearly equal strongly toothed in upper half of keel, *ligule long* truncate.—*J. of B.* xx. t. 227.—Soboliferous. St. erect. Sheaths rather rough. Ped. toothed. Fl. awnless (?). Ripe anthers pale, ¼ as broad as long.—Borders of fields. P. VII. E. S.

4. *A. vulgáris*[1] (With.); *pan.-branches* constantly patent *branched below,* gl. nearly equal slightly toothed near top of keel, *ligule short* truncate.—*E. B.* 1671.—St. long, ascending or decumbent below and rooting at the knots, sometimes with very long prostrate stoles. Sheaths smooth. Pedicels softly hispid. Fl. rarely awned. Ripe anth. pale, ½ as broad as long.—*β. A. pumila* (Lightf.); cæspitose, st. 2—3 in. high, fl. often awned usually infested with smut.—Rather dry places. P. VII. E. S. I.

5. *A. al'ba* (L.); *pan. compact* after flowering, glumes nearly equal, lower toothed throughout its keel, *ligule long,* acute.—*E. B.* 1189. *P.* 13 & 14.—St. procumbent and rooting below, then erect, often with long prostrate stoles. Sheaths roughish. Pan. spreading with flowers, afterwards close. Pedicels very much toothed. Florets rarely awned.—*β. A. stolonifera* (L.); st. procumbent and rooting, panicle lobed. *E. B.* 1532.—[A glaucous seaside form with narrow pan. and short incurved l. is var. *maritima* (Meyer). Various other forms with contracted pan. have been named var. *coarctata.*]—Fields &c. *β.* Sea-sands. P. VII. *Fiorin-grass.* E. S. I.

22. Lagu'rus *Linn.* Hare's-tail-grass.

[*L. ovátus* (L.).—*E. B.* 1334. *P.* 88.—St. 4—12 in. high. L. broad, lanceolate. Spikes ovate, soft, with long protruded awns. —Sandy places in Guernsey. A. VI. VII.]

---

[1] We have not altered this name to the earlier *A. tenuis* (Sibth.), it being doubtful whether the sp. should not stand as *A. stolonifera* (L.) or *A. pumila* (L.)—H. & J. G.

23. Polypo'gon *Desf.*

1. *P. monspelien'sis* (Desf.); awns more than twice as long as the bluntly and shortly lobed glumes.—*E. B.* 1704. *P.* 11.—Root fibrous. St. a foot or more high. Pan. dense, lobed, pale, silky, often 2 in. long. Gl. linear, hairy. A most beautiful grass.—Salt marshes, rare.—A. VI. VII. E. S.

2. *P. lit'toralis* (Sm.); awns as long as the acute glumes.—*E. B.* 1251. *P.* 81. *R.* vii. 75.—Somewhat creeping. St. a foot or more high. Pan. close, lobed, purplish. Gl. linear-lanceolate.—[Considered by Duval-Jouve and others a hybrid between Sp. 1 and *Agrostis alba.*]—Muddy salt marshes, rare. P. VI. VII. E.

24. Gastrid'ium *Pal. de Beauv.* Nit-grass.

1. *G. austrále* (Beauv.); gl. lanceolate acuminate, lower pale awned, awn rather exceeding the glumes.—*G. lendigerum* (Gaud.) ed. viii. *E. B.* 1107. *P.* 86.—St. 3—12 in. high.—L. roughish at the edges. Ligule oblong. Pan. close, almost spiked, lobed. *Gl. remarkably ventricose and shining at the base.* Pales very small.—Damp places especially near the sea, rare. A. VI.—IX. E.

Tribe X. *Aveneæ.*

25. Hol'cus *Linn.* Soft-grass.

1. *H. lanátus* (L.); upper gl. blunt apiculate, awn smooth except near the tip ultimately curved like a fish-hook and included within the glumes, *sheaths and knots villose.*—*E. B.* 1169. *P.* 21.—Cæspitose. Root fibrous. Height 1—2 feet. Knots not hairy. Inflorescence panicled, often pinkish. Gl. rough. Lower fl. awnless, quite smooth or slightly rough at the point.—Meadows and pastures. P. VII. *Yorkshire Fog.* E. S. I.

2. *H. mol'lis* (L.); upper gl. acute, awn rough throughout ultimately kneed protruding beyond the glumes, l. rough, sheaths glabrous, *knots bearded.*—*E. B.* 1170. *P.* 21 & 22.—Creeping. Height 1—2 feet. St. and l. subglabrous or slightly hairy. Inflorescence not so compact as in the preceding, whitish. Gl. smooth. Lower fl. awnless; but sometimes it has an awn; rarely the upper is perfect. Occasionally the spikelets are much smaller and the plant only 12—18 in. high.—Thickets or open places on a light soil. P. VII. E. S. I.

26. Weingaertne'ria *Bernh.* (*Corynephorus* Beauv. ed. viii.)

1. *W. canes'cens* (Bernh.); pan. rather dense long, gl. exceeding the fl. acuminate, awn from near the base of the pale, l. setaceous.—*Aira* Sm., *E. B.* 1190. *P.* 110.—St. tufted, slender, 6—8 in. high. L. many. Pan. close, spreading with flowers; branches short. Spikelets variegated with purple and white. Anth. dark purple. Lower portion of the *awn* dark yellow, straight, cylindrical, striated lengthwise and slightly twisted; upper part *clavate,* white tinged with purple.—Sandy coasts of Norf., Suff., and Jersey. [Planted near Arisaig, Invern.] P. VI. VII. E.

27. Ai'ra *Linn.* Hair-grass.

* *Lower pale truncate, jagged. Nut free, not furrowed on the back.* Deschampsia Beauv.

† Awn straight.

1. *A. cæspitósa* (L.); pan. spreading, l. flat, *gl. slightly rough at the midrib, awn from below the middle of the pale* and scarcely extending beyond its tip, ped. of second fl. downy or hairy.—*E. B.* 1453. *P.* 23.—Root tufted. St. 1—4 feet high. L. rigid, roughish; their margins involute when dry. Pan.-branches rough. Lower pale with 4 veins in addition to that which ends in the rough awn. Rudiment of third fl. often scarcely, if at all, distinguishable; or half the length of the upper fl. and somewhat clavate.—*β. brevifolia* (Parn.); radical l. short, sheaths and st. smooth, panicle small. *P.* 106. Viviparous states are often called *A. alpina.*—*γ. longiaristata* (Parn.); awns exceeding the fl., sheaths rough. *P.* 105.—*δ. A. alpina* (L.); pan. close, l. mostly involute, *gl. smooth on midrib, awn from above the middle of the pale.* Height 6—12 in. L. narrower. Fl. often viviparous. *E. B.* 2102. *P.* 23.—Meadows, thickets, &c. *β, γ,* and *δ* on mountains. P. VII. E. S. I.

†† Awn bent, twisted at the base.

2. *A. flexuósa* (L.); pan. spreading triply forked with wavy branches, l. very narrow subsetaceous, *awn from near the base of the pale* and exceeding it, pedicel of the second fl. less than ¼ of its length, *ligule short truncate.*—*E. B.* 1519. *P.* 107.—St. erect, slender, about a foot high. L. solid, nearly terete. *Upper sheaths rough from above downwards.* Lower pale notched at the tip.—*A. montana* (Huds. not L.) is a form with more slender and shorter leaves.—[Var. *voirlichensis* (Melv.) has 3 perfect fl. to each spikelet.]—Heathy places. P. VII. E. S. I.

3. *A. setácea* (Huds.); pan. spreading drooping at the end, l. filiform, awn from near the base of the pale and exceeding it, *pedicel of second fl. quite equalling ½ its length, ligule linear-lanceolate.*—*Sy. E. B.* 1733. *A. uliginosa* Weihe. *Desch. discolor* R. & S.—St. erect, slender. L. folded. Sheaths smooth.—Wet turfy bogs. P. VII. VIII. E. S. I.

** *Lower pale bifid. Nut adnate to the pales, furrowed on the back. No rudiment of a third flower.* AIROPSIS Fries.

4. *A. caryophyl'lea* (L.); *pan. spreading* triply forked, spikelets rounded below, awn from below the middle of the pale and extending considerably beyond its attenuate deeply bifid tip.—*E. B.* 812. *P.* 24. *Avena* Koch.—St. 6—12 in. high. L. short and narrow. *Sheaths roughish from below upwards.* Spikelets small, rounded below, chiefly collected at the ends of the branches which are sometimes divaricate with seed.—[*A. aggregata* (Jord.) has numerous st. and spikelets clustered at the ends of the pan.-br. *A. multiculmis* (Dum.), an allied form with the second fl. stalked, is also reported as British.]—Dry gravelly places. A. V. E. S. I.

5. *A. præ'cox* (L.); *pan. spikelike oblong,* spikelets scarcely rounded below, *awn from* below the middle usually *near the base of the pale* and extending considerably beyond its attenuate deeply bifid tip.—*E. B.* 1296. *P.* 25. *Avena* Koch.—Height 1—6 in. Pan. close, oval or oblong. L. very narrow. Sheaths smooth.—Dry and sandy places. A. IV. V. E. S. I.

28. TRISE'TUM *Pers.*

1. *T. praten'se* (Pers.); pan. much branched diffuse equal, gl. very unequal about 3-flowered.—*P.* 54. *E. B.* 952. *T. flavescens* (Beauv.) ed. viii. *Avena* L.—St. about a foot high. Radical l. and sheaths hairy. Spikelets yellowish. Upper gl. oblong-lanceolate, acuminate. Floral axis hairy, hairs short.—[Var. *variegatum* (Gaud. under *Av. flavesc.*) is a mountain form with dark violet gl.]—Fields. P. VII. E. S. I.

29. AVE'NA *Linn.* Oat.

* *Upper gl. 5—9-veined. Spikelets ultimately drooping. Root annual. No lateral clusters of leaves. L. alike on both sides.*

1. *A. fat'ua* (L.); pan. erect, spikelets of about 3 fl., fl. falling short of the gl. hairy at the base, lower pale bifid at the end.—*E. B.* 2221. *P.* 37.—Height 3 feet. Fl. with long fulvous

hairs at their base, by which it may be distinguished from *A. sativa*, the cultivated Oat.—[Sy. E. B. iii. distinguishes *a. pilosissima* (Gray); lower pales densely hairy becoming dark brown. *β. A. intermedia* (Lindg.); lower pales nearly glabrous becoming pale yellowish olive.]—Corn-fields. A. VII. *Wild Oat.* E. S. I.

†2. *A. strigósa* (Schreb.); pan. secund, spikelets of about 2 fl., fl. equalling the gl., lower *pale ending in 2 long straight bristles.*—*E. B.* 1266. *P.* 26.—Height 3 feet. Very like *A. sativa* but readily distinguished by the bristles at the end of the fl.—Corn-fields. A. VII. E. S.

*** Upper gl. 3-veined. Spikelets erect. Root perennial. Lateral clusters of l. barren. L. with raised ribs.*

3. *A. praten'sis* (L.); pan. erect with simple or slightly divided branches, *fl. erect* 3—6 exceeding the glumes, *l. rough.*—*E. B.* 1204. *P.* 52.—Root fibrous. Height nearly 2 feet. St. usually nearly round. L. usually short, narrow, acute. Pan.-branches usually simple with only one spikelet.—*β. longifolia* (Parn.); l. much longer.—*γ. A. alpina* (Sm.); st. often compressed and sheaths keeled, pan.-branches often with several 5—6-fl. spikelets, upper pale less acute, l. broader. I believe that none of these characters are permanent. *E. B.* 2141. *P.* 53.—*A. planiculmis* of *E. B. S.* 2684 may belong to this species; it differs by its greatly compressed st., strongly keeled sheaths and more branched panicle.—In *α* and *β* the lowest fl. sometimes slightly falls short of the longer gl., in *γ* exceeds it.—Dry pastures and mountainous places. P. VI. E. S.

4. *A. pubes'cens* (Huds.); pan. erect nearly simple, fl. erect 2 or 3 scarcely exceeding the glumes, lower *l. and sheaths hairy.*—*E. B.* 1640. *P.* 53.—Creeping slightly. Height 1—2 feet. L. short, rounded behind the tip.—Chalky and limestone districts. P. VI. E. S. I.

30. Arrhenath'erum *Pal. de Beauv.* Oat-grass.

1. *A. avenáceum* (Beauv.); l. flat.—*A. elatius* (M. & K.) ed. viii. *Holcus* Sm., *E. B.* 813. *P.* 25.—Height 2—3 feet. Root fibrous. Knots of the st. glabrous, sometimes downy. Pan. long, ultimately close. Spikelets greenish.—*β. A. bulbosum* (Presl); base of the st. with swollen knobs, knots downy. *P.* 26.—Hedges and pastures. P. VI. E. S. I.

Tribe XI. *Festuceæ.*

31. SIEGLING'IA *Bernh.* (*Triodia*, R. Br. ed. viii.) Heath-grass.

1. *S. decum'bens* (Bernh.); pan. racemose, spikelets few oval, fl. about 4 scarcely extending beyond the glumes without awns.—*E. B.* 792. *P.* 30.—St. 6—12 in. high. L. flat. Sheaths rather hairy. Ligule a tuft of hairs. Spikelets few, 1—7. Gl. smooth, coriaceous. Lower pale with 3 points, 5-ribbed, hairy at the base.—Dry places and heaths. P. VII. E. S. I.

32. KŒLE'RIA *Pers.* Crested Hair-grass.

1. *K. cristáta* (Pers.); pan. compact spikelike interrupted below, lower pale acute, l. narrow rough at the edges ciliate.—*Aira* L., *E. B.* 648. *P.* 19.—Root crowned with the undivided sheaths of the old leaves. St. 6—18 in. high, downy particularly in the upper part. L., gl. and pales downy or glabrous. Gl. finely toothed on the keel. Lower pale finely toothed on the midrib. Sometimes the l. become convolute (*K. albescens* DC.?). In dry places the l. fall short of the st., in damper they are long and often nearly equal it.—[Forms with shorter involute l., slender interrupted spikes and less hairy gl., have been referred to *K. gracilis* (Pers.).]—Dry pastures. A large form on Ben Bulben, Co. Sligo. P. VI. VII. E. S. I.

33. MEL'ICA *Linn.* Melic.

1. *M. nútans* (L.); pan. branched slightly drooping, spikelets erect with 1 perfect glabrous fl., gl. equalling pales, l. flat, *ligule* short blunt *with a slender acuminate lobe* on one side.—*E. B.* 1058. *P.* 18. *M. uniflora* (Retz.) ed. viii.—Shady and rocky woods. P. V. VI. E. S. I.

2. *M. montána* (Huds.); pan. a nearly simple lax secund raceme, spikelets drooping with 2 perfect glabrous fl., gl. falling short of pales, l. flat, *ligule short blunt.*—*E. B.* 1059. *P.* 18. *M. nutans*, ed. viii..—Damp shady woods in hilly districts. P. V. VI. E. S.

34. MOLIN'IA *Schrank.*

1. *M. vária* (Schrank); pan. erect long narrow, spikelets 1—3-fl., lower pale 3-veined awnless, upper part of the st. naked.—*E. B.* 750. *P.* 20. *M. cærulea* (Moench) ed. viii.—St. 1—2 feet high, with only one knot placed near to its base.

L. long, linear, attenuate, all from near the base of the stem. Panicle purplish, close.—*β. M. depauperata* (Lindl.); spikelets 1-fl. few. *P.* 19.—*γ. major*; pan-branches long, spikelets distant of about 3 fl., st. 3—4 feet high.—[A form with broader and blunter gl. and pales is *M. cærulea* var. *obtusa* (Hackel).]—Wet heaths. *β.* Alpine places. P. VII. VIII. E. S. I.

35. Po'a *Linn.* Meadow-grass.

* *Root fibrous, annual. Base of stem sometimes prostrate and rooting. Pan.-branches solitary or in pairs.*

1. *P. an'nua* (L.); pan. spreading erect with a triangular outline, spikelets ovate-oblong of 5 or 6 free fl., lower pale with 5 veins, upper sheath longer than its leaf, ligule oblong acute.—*E. B.* 1141. *P.* 40, 41.—St. ascending or prostrate. L. flaccid, often wavy, broad. Spikelets subsecund with patent or divaricate branches.—[*β. P. supina* (Schrad.); pan. lax its br. deflexed, spikelets larger and blunter, variegated with purple.]—Very common. [*β.* Mountains.] A. III.—IX. E. S. I.

** *Root fibrous, perennial.*

† Lower pan.-branches solitary or in pairs. Dorsal and marginal veins of the lower pale hairy.

2. *P. bulbósa* (L.); pan. close erect, spikelets ovate of 3 or 4 acute webbed[1] fl., lower pale with 3 silky veins, upper sheath below the middle of the st. much longer than its leaf, ligule prominent acute.—*E. B.* 1071. *P.* 89.—Root fibrous. Base of the st. and offsets swollen, bulblike. L. with a narrow white serrate edge. The st. soon wither, and the tubers lie loose until the autumn. Sandy seashore of the South and East. P. IV. V. E.

3. *P. mínor* (Gaud.)[2]; pan. oblong subovate, spikelets of 3 or 4 *webbed fl., lower pale with* 5 *veins* but only 3 hairy, upper sheaf longer than its *leaf* which is *folded and slightly incurved but tapering at the tip,* uppermost knot covered, upper ligule long acute, lower ones short rather blunt.—*P. flexuosa* Sm.,

[1] That is, connected together by fine cottony fibres growing from the base of each flower: when these are wanting, the fl. is said to be *free.*

[2] There is a note in the Author's MS. "Join *laxa* to *alpina.*" Mr. G. C. Druce in *Journ. Linn. Soc.* xxxvi. p. 421, refers Sp. 3 to *P. laxa* (Haenke) and Sp. 4 to *P. alpina,* var. *acutifolia* (Druce).—H. & J. G.

E. B. 1123, not of others.—Root fibrous. St. 6—8 in. high. Fl. rarely viviparous.—Lofty mountains. Loch-na-Gar. P. VII. VIII. S.

4. *P. lax'a* (Haenke!)[1]; pan. lax slightly drooping, spikelets oblong-ovate of 3 free fl., *lower pale with* 3 *hairy veins*, upper sheath longer than its *leaf* which is *flat and taper-pointed*, uppermost knot covered, ligules all long acute.—*P.* 38. *P. stricta Sy. E. B.* 1763. The synonymy of this and Sp. 3 is very doubtful.—Root fibrous. St. 6—12 in. high. Fl. often viviparous.—Lofty mountains. Loch-na-Gar. P. VII. VIII. S.

5. *P. alpína* (L.); pan. erect spreading when in flower, spikelets ovate of 3 or 4 free fl., *lower pale with* 3 *hairy veins*, upper sheath longer than its *leaf* which is *folded and rounded* behind *the tip*, uppermost knot exposed, ligule long pointed.—*E. B.* 1003. *P.* 37 & 94. *P. stricta* Lindb.—Root fibrous, tufted. St. 6—12 in. high, covered with decayed basal sheaths common to it and the tuft of leaves. Fl. often viviparous.—Lofty mountains. P. VI. VII. E. S. I.

6. *P. glaúca* (Sm.); pan. erect slender, spikelets ovate of 2 or 3 acute free fl., *lower pale with* 5 *veins* but only 3 hairy, upper sheath about as long as its leaf which is folded and slightly incurved but tapering at the tip, *uppermost knot near to the base of the stem*, ligule blunt.—*E. B.* 1720.—Root-stock rather creeping. St. 6—12 in. high. *Lowest fl. longer than the large glume.*—*P. cæsia* Sm. *E. B.* 1719 is a very doubtful plant.—Mountains. Ben Lawers. Ben Nevis. Clova. Snowdon. P. VII. E. S.

†† Lower pan.-branches in fives or 2 or 3 together. Dorsal and marginal veins of the lower pale hairy.

7. *P. nemorális* (L.); pan. rather drooping slender, spikelets ovate-lanceolate of 3 or 4 webbed fl., lower pale with 5 veins but only 3 hairy, *upper sheath not longer than its leaf*, uppermost knot at about the middle of the st. exposed, *ligule extremely short truncate.*—*E. B.* 1265. *P.* 36.—Slightly creeping.—St. slender, 1—2 ft. high. Sheaths smooth.—*α*; st. slender weak, pan. lax.—*β. angustifolia* (Parn.); st. and pan. very slender, l. long and narrow, uppermost knot near the pan., spikelets few 1—2-flowered.—*γ. P. coarctata* (Gaud.); st. rigid, pan. close, spikelets 3—5-flowered.—*δ. glaucantha*? ; st. slender, pan. with many long-stalked spikelets, plant glaucous.—[Var. *divaricata* (Sy.) has pan. erect 'distichously unilateral,' spikelets usually 2-flowered, uppermost sheath as long as or longer than its l.]—Shady places. *γ*. On walls. *δ*. Mountains. P. VI. VII. E. S. I.

---

[1] See note to Sp. 3.

8. *P. Parnell'ii* (Bab.); pan. suberect large rather close, oblong, spikelets ovate of 2 or 3 acute *free fl.*, lower pale with 5 veins but only 3 hairy, *upper sheath usually longer than its leaf*, upper knot at about the middle of the st. exposed, *ligule very short* truncate.—*E. B. S.* 2916. *P.* 93.—St. ascending, 1 ft. or more high, compressed; knots 5 or 6, uppermost not above the middle of the stem. Ligule 6 times as broad as long, but longer than that of *P. nemoralis*, to which this plant is perhaps too nearly allied. Occasionally the fl. are slightly webbed.—High Force and other parts of Upper Teesdale. P. VII. E.

9. *P. Balfour'ii* (Parn.)[1]; pan. erect rather spreading, spikelets ovate of 3 or 4 webbed fl., lower pale with 5 veins but only 3 hairy, upper sheath about as long as its leaf, upper two-thirds of the stem without knots, *ligule prominent blunt.*—*P.* 66. *E. B. S.* 2918.—Creeping. St. 3—15 in high; knots about 3, uppermost within the lower third of the stem. Lower fl. as long as the larger glume. *Fl. Dan.* 964? Combined with *P. glauca* by Syme.—*β. P. montana* (Parn.); spikelets few of 2 or 3 free fl. *P. dissitiflora* (R. & S.)?—Tops of mountains. P. VII. E. S.

‡10. *P. palus'tris* (L.); pan. large its br. rough ascending-patent lower subverticillate quinate, spikelets ovate acute of 2—5-webbed fl., *lower pale obscurely 5-veined hairy on the keel and margins towards the base*, sheaths glabrous, *ligule oblong-acute.*—*Fl. Dan.* 2166. *P. serotina* Ehrh. *P. fertilis* (Host).—St. tufted ascending glabrous, l. rather narrow pointed slightly scabrid.—By R. Thames, Kew and Mortlake. By R. Tay below Perth, and Benniebeg Pond near Crieff. R. Boyne below Navan. Probably introduced. P. VI. VII. E. S. I.

††† Lower pan.-branches subverticillate, quinate. Dorsal vein of the lower pale hairy or glabrous; *marginal glabrous.*

11. *P. triviális* (L.); pan. diffuse, spikelets ovate of 2 or 3 acute webbed fl., lower pale with 5 veins, dorsal vein hairy, upper sheath much longer than its leaf, *ligule acute long.*—*E. B.* 1072. *P.* 35.—Root tufted. St. 1—2 ft. high. Sheaths usually slightly rough [or var. *glabra* (Doell.) smooth].—*β. parviflora* (Parn.); spikelets small 1—2-flowered, plant slender.—Moist and shady places. P. VI. E. S. I.

**P. Chaixii* (Gilib.); pan. diffuse, spikelets oval of 3 or rarely 5 acute *not webbed fl., lower pale with 5 glabrous veins*, upper sheath very much longer than its leaf, *ligule blunt very short.*—

[1] Prof. Hackel writes "probably no clear line can be drawn between *P. Balfourii* [as here constituted] and the alpine forms of *P. nemoralis*." See *J. of B.* xxxv. (1897) p. 71.—H. & J. G.

R. i. 90. *P. sudetica* (Haenke) ed. viii.—Rhizomatous, not creeping. St. 2-edged, 2—3 feet high. Sheaths, edges, and midribs of l. rough. L. hooded and apiculate.—In deep shade in several places near Kelso; Birnham, Perthsh. [Berks. Warw. Ayrsh. etc.] P. E. S.

*** *Creeping by long soboles.*

12. *P. praten'sis* (L.); pan. diffuse, spikelets ovate of 3 or 4 webbed fl., *lower pale with* 5 *prominent veins* but only 3 hairy, upper sheath much longer than its leaf, *ligule prominent* blunt. —*E. B.* 1073. *P.* 31—34.—Very variable in size. St. and sheaths nearly always smooth. Fl. strongly webbed.—β. *P. subcærulea* (Sm.); spikelets broader, l. broad and short, upper l. compressed rounded at the end behind. *E. B.* 1004.—γ. *angustifolia* (Gaud.); spikelets small, l. slender long, lower l. involute. [Var. *strigosa* (Gaud.) has a contracted pan. and narrow involute somewhat glaucous l.]—Common. P. VI. VII. E. S. I.

13. *P. compres'sa* (L.); pan. erect or slightly unilateral spreading when in flower otherwise close, spikelets ovate or oblong-ovate of 5—7 blunt slightly webbed fl., *lower pale* 3-*veined, veins hairy*, upper sheath about as long as its leaf, *uppermost knot at about the middle of the stem, ligule short* truncate. —*E. B.* 365. *P.* 37 & 90.—*St.* decumbent at the base, then erect, very much *compressed*, 1—1½ ft. high.—β. *P. polynoda* (Parn.); fl. free, pales with 2 faint interm. veins, uppermost knot higher, ligule rather more prominent. [*P. subcompressa* (Parn.) is intermediate betw. this and the type, having webbed fl. and 5-veined pales.]—*P.* 91—92.—Dry situations. P. VII. E. S. I.

36. Glyce'ria *R. Br.*

1. *G. aquat'ica* (Wahlb.); pan. erect much branched spreading, branches rough, spikelets oblong of 5—10 fl., lower pale blunt, l. smooth with terete sheaths.—*E. B.* 1315. *P.* 44. *G. spectabilis* M. & K.—Creeping. St. 3—6 ft. high, smooth slightly compressed. Sheaths very long. L. long, rough on the edges and keel, never floating. Ligule short. Pan. large; branches angular, slender, branched.—Watery places. P. VII. E. S. I.

2. *G. flúitans* (R. Br.); *pan. secund slightly branched* very long, branches nearly simple roughish, spikelets linear of 7—12 adpressed lanceolate-oblong acute fl., *lower pale nearly thrice as long as broad*, sheaths compressed.—*E. B.* 2975. *P.* 95.—St. ascending, rooting below, or floating. Sheaths nearly smooth,

striate. L. pale green, acute, often floating. Ligule long. Pan. very long, often nearly simple; branches without callosities, ascending, lowermost usually in pairs. Spikelets adpressed. Lower pales rather the shorter, with a triangular central point. *Anthers about* 5 *times as long as broad*, purple, pale yellow when empty.—β. *G. pedicellata*[1] (Towns.); pan.-branches simple roughish, spikelets of 9—13 blunt flowers. L. more acute. Lowermost pan.-branches about in threes. Anth. about 4 times as long as broad, yellow when young. *A. N. H.* ser. 2. v. 105. *Curt. Fl. Lond.* i. 18. Nearer to *plicata* than *fluitans* by its rough furrowed sheath, lower pale 3-toothed and never exceeding the upper.—*G. declinata* (Bréb., Towns.) is a dwarf plant with smooth sheaths, upper exceeding the obtuse-angled 3-toothed lower pale, anth. twice as long as broad and purple. It also is near *plicata*.—Watery places. P. VI.—IX. *Flote-grass.* E. S. I.

3. *G. plicáta* (Fries); pan. compound, branches compound nearly smooth erect with flowers *divaricate with fruit*, spikelets linear of 7—20 oval-oblong rather acute fl., *lower pale twice as long as broad*, sheaths compressed.—*R.* vii. 79. *G. fluitans* Sm., *E. B.* 1520. *P.* 45.—St. ascending, rooting below. Sheaths rough, furrowed. L. glaucous, bluntish, plicate when young. Ligule shorter. Pan. much branched; branches with callosities at the base, lowermost about in fives. Lower pales with 3 teeth at the end. *Anth. about* 3 *times as long as broad*, cream-coloured, fuscous when empty. Stagnant water and wet places. P. VI.—VIII. E. S. ? I

37. SCLEROCH'LOA *Pal. de Beauv.*[2]

* *Glumes with* 3 *veins.*

† Panicle unilateral.

1. *S. marit'ima* (Lindl.); pan. branched, lowermost branches in pairs or single, branches ultimately erect, spikelets linear adpressed 4—8-flowered, rachis terete, *lower pale blunt apiculate, midrib reaching the tip, stoloniferous.*—*E. B.* 1104. *P.* 42. *Glyceria* M. & K.—Root fibrous, with ascending prostrate or rarely rooting leafy stoles. L. involute; the central ridge on their upper surface strongly marked the others

---

[1] Mr. Townsend now regards this as a hybrid, *G. fluitans*×*plicata*.—H. & J. G.

[2] See *Crepin, Notes Fl. Belg.* v. 155—214.

faint. *Anth.* about 6 times as *long* as broad. Lower pale with involute edges. Ligule bluntish.—β. *hispida* (Parn.); st. compressed, rachis furrowed on one side and as well as the pan.-branches rough. *P.* 99.—[γ. *riparia* (Towns.); more slender, spikelets fewer, lower gl. equalling or exceeding middle of lowest fl. on same side, nerves closer together, lower pale with narrower white border.] —Sea-coast, in damp places. P. VI. VII. E. S. I.

2. *S. Bor'reri* (Bab.); pan. branched, branches ultimately erect-patent lowermost generally in fours, spikelets linear 4—7-fl., rachis terete, *lower pale with a rigid apiculus formed by the tip of the dorsal vein,* cæspitose.—*Glyceria*, Bab., *E. B. S.* 2797 (1837). *P.* 98. *G. conferta* Fries (1839).—St. 6—12 in. high. No stoles. L. short, flat, with very long sheaths. Ligule short, truncate. Edges of the lower pale not involute. Spikelets and fl. half the size of those of *S. maritima* and *S. procumbens.* Pan.-branches short, scarcely elongated after flowering, hispid. —Muddy salt marshes. P. VI.—VIII. E. I.

3. *S. procum'bens* (Beauv.); *pan.* ovate-lanceolate *compact distichous* rigid, spikelets linear-lanceolate of about 4 fl., rachis angular, lower pale blunt with an apiculus formed by the tip of the dorsal vein, root fibrous.—*E. B.* 532. *P.* 42. *Glyceria* Dum. *Festuca* Kunth.—St. procumbent (rarely erect when growing in water), rigid. L. flat, with inflated sheaths. Pan. about 2 in. long, with very short rigid branches spreading in 2 rows. Fl. large.—Muddy sea-shores. A. VI. VII. E. S.

†† Panicle regular.

4. *S. dis'tans* (Bab.); pan. branched, branches long ultimately spreading or deflexed lowermost in fours or fives, spikelets linear 3—5-flowered, rachis semiterete rather flat on one side, *lower pale blunt, midrib not reaching to the tip, root fibrous.* —*E. B.* 986. *P.* 41. *Glyceria* Wahl.—Without rooting stoles. St. decumbent below. L. flat, short, with 8—10 equally prominent ridges upon their upper surface. Ligule short and truncate. Edges of lower pale not involute. Spikelets and fl. half the size of those of the preceding.—β. *obtusa* (Parn.); pan. more compound, spikelets about 7-flowered, lower pale truncate and broader, ligule shorter. *P.* 96 & 97.—[Var. *pseudo-procumbens* (W.-Dod), with subunilateral pan. with fewer shorter stiffer ascending pan.-br., larger greener and more strongly-ribbed pales, is apparently a hybrid with Sp. 3. A quite prostrate form with ascending pan.-br. is var. *prostrata* (Beeby).]—Sea-shores and waste sandy places. β. Leicestershire. P. VI.—VIII. E. S. I.

** *Glumes* 1- (*rarely* 3-) *veined.*

5. *S. rig'ida* (Link); pan. lanceolate rigid distichous, spikelets linear acute of 7—10 fl., lower pale blunt with a mucro, upper gl. reaching to the base of the third fl., root fibrous.—*Festuca* Kunth. *Glyceria* Sm., *E. B.* 1371. *P.* 43.—St. slender, wiry, erect. L. nearly flat, acute. Pan. 1—2 in. long, nearly simple. Lower pale faintly veined. Fl. small.—Dry places. A. VI. E. S. I.

6. *S. loliácea* (Woods); pan. racemose narrow rigid secund, spikelets oblong of 8—12 fl., lower pale blunt with a mucro, upper gl. reaching to the base of the fourth fl., root fibrous.—*Triticum* Sm., *E. B.* 221. *P.* 43. *Festuca rottboellioides* Kunth.—St. stout, slightly curved, ascending, 2—6 in. long. L. flat, convolute when dry. Spikelets usually solitary, alternate, all directed to one side; footstalks very short and stout. Lower pale with well-marked marginal veins.—Sandy sea-coasts. A. VI. VII. E. S I.

38. Bri'za *Linn.* Quaking-grass.

1. *B. minor* (L.); spikelets triangular of about 7 fl., *gl. exceeding the lowest fl.*, pan. diffuse, *ligule long* lanceolate acute. —*E. B.* 1316. *P.* 101.—St. very slender, about 1 foot high. Spikelets pale green. Lower pale roundish cordate, cartilaginous, very gibbous in the middle of the back.—Dry and sandy fields in the South-west. A. VII. E.

2. *B. média* (L.); spikelets broadly ovate of about 5 fl., *gl. falling short of the lowest fl.*, pan. diffuse, *ligule truncate very short.*—*E. B.* 340. *P.* 30.—St. slender erect, 1—1½ foot high. Panicle light and elegant, with slender branches. Spikelets usually purplish. L. linear-acuminate. Lower pale roundish oval, cartilaginous, not gibbous.—Pastures. P. VI. E. S. I.

[*B. max'ima* (L.) with few very large ovate many-flowered spikelets is naturalised in Guernsey.]

39. Catabro'sa *Pal. de Beauv.*

1. *C. aquat'ica* (Beauv.); pan. long-pyramidal with half-whorls of patent branches, lower pale 3-ribbed, l. broad linear blunt.—*E. B.* 1557. *P.* 20.—Creeping. St. long, procumbent or floating below. L. flat. Pan.-branches in alternate threes or fives. Spikelets usually 2- (or 3—5-) flowered. Gl. very thin, often purplish. Fl. distant.—β. *minor* (Bab.); st. 2—3 in. high, spikelets mostly 1-flowered.—Ponds and ditches. β. Wet sea-sands. P. VI. VII. E. S. I.

40. Cynosu'rus *Linn.* Dog's-tail-grass.

1. *C. cristátus* (L.); raceme spikelike linear, fl. with a very short awn.—*E. B.* 316. *P.* 28.—St. 12—18 in. high. Spike unilateral, plane-compressed. Spikelets closely placed. Bract comblike.—Pastures. P. VIII. E. S. I.

[*C. echinátus* (L.); raceme contracted close ovate, awns about as long as the pales.—*E. B.* 1333. *P.* 28 & 129.—St. erect, 1—2 feet high. Bract comblike with long points.—Sandy places in Guernsey and Jersey. A. VII.]

41. Dac'tylis *Linn.* Cock's-foot-grass.

1. *D. glomeráta* (L.); pan.-branches with ovate clusters of spikelets, st. erect linear flat with rough margins, root cæspitose.—*E. B.* 335. *P.* 29.—A coarse grass. Pan.-branches long, spreading or divaricate with fl., afterwards adpressed, distant; each bearing an ovate cluster of spikelets; or panicle reduced to one cluster.—Meadows. P. VI. VII. E. S. I.

42. Festu'ca *Linn.*[1] Fescue-grass.

* *Root-leaves very narrow. Ligule with round auricles. Awn terminal.*

† Awn longer than the pale. Gl. very unequal. Vulpia.

‡ *Usually triandrous.*

1. *F. uniglúmis* (Sol.); uppermost sheath far distant from the erect close 2-ranked *simple panicle*, fl. compressed keeled, *gl. very unequal often only one*, larger gl. very long and very acute.—*E. B.* 1430. *P.* 112.—St. 6—12 in. high, erect, leafy nearly to the top. Panicle close, short. Lower gl. usually scarcely distinguishable. Upper gl. setigerous nearly equalling 1st fl. on same side.—Sandy sea-shores. A. VI. E. I.

‡‡ *Usually monandrous.*

2. *F. sciuroïdes* (Roth); uppermost sheath far distant from the *erect-patent oblong* pan., lowermost pan.-branch about equalling ½ pan., fl. terete rough, *gl. unequal as* 2 *to* 1, larger gl. about equalling lowest flower.—*F. bromoides* Sm., *E. B.* 1411.—

---

[1] See Prof. Hackel's exhaustive *Monographia Festucarum europæarum.*—H. & J. G.

Slender. 6—12 in. high. L. linear, involute. [β. *intermedia* (Hack.); pan. longer, sheath covering its base.]—Walls and sandy places. A. ? VI. VII. E. S I.

3. *F. ambig'ua* (Le Gall); uppermost sheath very nearly reaching to the *long* narrow *erect close* pan., lowest pan.-branch equalling more than ⅓ of pan., fl. terete rough, *gl. unequal as* 3—6 *to* 1, larger gl. about equalling ⅓ of lowest flower.—*E. B. S.* 2970 — Slender, 8—12 in. high. L. involute. Probably a glabrous form of *F. ciliata* (Danth.).—Sandy places. Isle of Wight. Kent. Dorset. Suffolk. A. V. VI. E.

4. *F. Myurós* (L.); uppermost sheath reaching to or partly covering the *long* narrow *nodding interrupted pan.*, lowest pan.-branch equalling ¼ of pan., fl. terete rough, *gl. unequal as* 3 to 1, larger gl. about equalling ½ lowest flower.—*E. B.* 1412. *F. pseudo-myurus* (Soy.-Will.).—About a foot high, slender. Pan. very long and narrow.—Walls and sandy places, rare. A. ? VI. VII. E. S.? I.

†† Triandrous. Awn shorter than the pale.

5. *F. ovína* (L.); pan. narrow subsecund close with fr., spikelets 4—6-fl., fl. mostly awned, *l. all setaceous*, sheaths glabrous, *cæspitose.*— *E. B.* 585. *P.* 56, 57.—Very variable. L. short, slightly curved, densely tufted. Fl. glabrous, or (*hispidula* Koch) hairy. Spikelets sometimes changed into leafy shoots.—β. *F. capillata* (Lam.); l. very long setaceous, fl. awnless.—γ. *F. duriuscula* (L.); pan. pyramidal, branches spreading, l. filiform channelled, st.-l. broader. St. and l. stouter. Fries thought it distinct. [Var. *glauca*, Hack. (*F. glauca*, Lam.) has the l. sheaths and fl. more or less glaucous, and var. *supina*, Hack. is a small form with small dense pan., setaceous laminæ and sheaths entire for about ⅓ of their length.]—Dry hilly pastures. β. On mountains. γ. Damper places. P. VI. *Sheep's Fescue-grass.* E. S. I.

6. *F. rúbra* (L.); pan. broadish below subsecund, spikelets 4—10-fl., l. involute-setaceous, st.-l. flat, lowest sheaths hairy, *soboliferous.*—*F. duriuscula* Sm. (in part), *E. B.* 470, *P.* 58—60. —Very variable. Fl. shortly awned, glabrous, hairy or villose. L. variable in length and breadth and the fl. in size. Creeping but cæspitose.—[*F. fallax* (Thuill.) is densely cæspitose and scarcely creeping.]—Common in dry sandy, rarely in wet places. P. VI. E. S. I.

[*F. heterophylla* (Lam.); not stoloniferous, st. taller, root-l. long setaceous densely tufted, st.-l. flat, pan. long lax, considered by Prof. Hackel a subsp. of 6, is recorded from several places, probably sown.]

7. *F. orária* (Dum.)[1]; pan. secund, spikelets 4—10-fl., l. all involute-setaceous, lowest sheaths hairy, far creeping not cæspitose.—*F. rubra* Sm., *E. B.* 2058.—Near *F. rubra*, but mode of growth very different.—Sandy sea-shores. P. VI. E. S.

** *Root-leaves broad and flat. Ligule not auricled. Awn* 0, *or dorsal.* SCHEDONORUS Beauv.

† Uppermost ligule prominent. Lower pale 3-veined.

8. *F. sylvat'ica* (Vill.); pan. erect diffuse much-branched, branches rough, spikelets of 3—4 *awnless acute fl.*, lower pale rough, *dorsal rib serrulate throughout*, l. lanceolate-linear with rough margins.—Poa *P.* 44 & 100. *F. Calamaria* Sm., *E. B.* 1005.—Scarcely creeping. St. 2—4 feet high, covered at the base with imbricate broad acute leafless sheaths, tufted. L. very long, broad, roughish on both sides; uppermost l. smaller. Lower pale very acute; midrib extending nearly to the tip or slightly beyond it. Ovary pilose at the top.—*β. F. decidua* (Sm.); l. narrower. fl. about 2. *E. B.* 2266.—Woods in mountainous districts. P. VII. E. S. I.

†† Uppermost ligule very short. Lower pale 5-veined. BUCETUM Parn.

9. *F. gigantéa* (Vill.); pan. open drooping branched, spikelets of about 5 *awned fl.*, *dorsal rib of lower pale nearly smooth* not extending to the tip but *ending in a rough awn twice as long as the pale*, l. linear-lanceolate.—*E. B.* 1820. *P.* 47.—*Bromus* L. St. 3—4 feet high. L. very long, broad, roughish on both sides, except near the base on the underside. Ligule unequal, auricled. Lower pale roughish, membranous, often bifid at the tip. Top of the ovary glabrous.—*β. F. triflora* (Sm.); pan. smaller and more erect, spikelets scattered of about 3 flowers. *E. B.* 1918.—Moist woods and thickets. P. VII. E. S. I.

10. *F. arundinácea* (Schreb.); panicle diffuse patent, *branches* mostly in pairs *each bearing 2 or more* ovate-oblong spikelets divaricate with fl. or afterwards, spikelets very many of 5—6 closely placed fl., *dorsal rib of lower pale ending at or just below the tip or forming a short awn*, l. linear-lanceolate.—*F. elatior* Sm., *E. B.* 1593. *P.* 46, 47.—St. 2—6 ft. high, forming large

---

[1] *F. oraria* Dum. Agr. Belg. 105 (1823), *F. sabulicola* Duf. (1825) *F. arenaria* Godr. (1855, not Osb.). See *Bull. Belg.* vii. 367.

tufts. L. broad.—*a. F. arundinacea* (Schreb.); pan.-branches divaricate with fl. and fruit. A very large plant, 3—6 ft. high. —*β. F. elatior* (L. ?); pan.-branches shorter "divaricate with fl. afterwards ascending."—*a.* Banks near the sea. *β.* Damp pastures. P. VI. VII. E. S. I.

11. *F. praten'sis* (Huds.): pan. close subsecund, *branches in pairs one bearing a single spikelet the other several never divaricate*, spikelets linear-oblong of 5—10 rather distant fl., dorsal rib of lower pale ending at or just below the tip or forming a very short awn, l. linear-lanceolate.—*E. B.* 1592. *P.* 46. *F. elatior* Koch.—A smaller plant than the preceding. Rachis triangular. Pan.-branches ascending; one of each pair nearly always reduced to a single spikelet. In this and the preceding the pale is blunt or acute according as the midrib is or is not attached up to the tip.—*β. F. loliacea* (Huds.); spikes solitary alternate long slender truly distichous, lower ones stalked, rarely in pairs, upper nearly sessile, fl. distant, lower gl. 5—8-ribbed, veins of lower pale parallel, no awn, rachis flattish. On water meadows spikelets are often all sessile and upper gl. vanishing. [Now usually regarded as a hybrid between *F. pratensis* and *Lolium perenne.*]—*E. B.* 1821. *P.* 45, 113 & 114.—Wet meadows. P. VI. VII. E. S. I.

43. Bro'mus *Linn.*[1]

* *Spikelets broader upwards when in flower, not afterwards. Ribs of upper pale finely fringed.*

1. *B. erec'tus* (Huds.); pan. erect nearly simple, spikelets linear-lanceolate, fl. remote subcylindrical, *lower pale indistinctly 7-veined, lowest fl.* ⅓ *exceeding the upper gl.* and longer than its awn, root-l. very narrow ciliate.—*E. B.* 471. *P.* 51.—St. 2—3 ft. high, erect. Root-l. convolute; upper l. broadest; sheaths somewhat hairy with upward hairs.—*β. villosus*; lower pale hairy.—On dry sandy and chalky soil. P. VI. VII. E. S. I.

2. *B. ramósus* (Huds.); pan. drooping with long divaricate slightly divided branches, spikelets lanceolate, fl. remote linear-lanceolate, *lower pale* pilose on the veins below, lowest fl. twice exceeding the glabrous upper gl. and longer than its awn, *l. broad* hairy.—*E. B.* 1172. *P.* 51. *B. asper* (Murr.) ed. viii. *B. serotinus* (Benek.).—St. 4—5 feet high. L. flat; lower ones broadest; sheaths all with downward hairs; branches 2 together with a semilunar strongly ciliate much decurrent scale at their

[1] Synonymous with the genus *Schedonorus* of Fries, not of Beauv.

base; upper pale glabrous, lower pilose on the veins below, anth. violet.—β. *Benekenii* (Syme); pan. slightly drooping with rather short suberect branches 3—6 together with a semilunar very blunt not ciliate subdecurrent scale at their base, upper gl. ciliate to the top, lower pale pilose throughout; l. flat, upper sheaths subglabrous, lower hispid, anth. golden.—Damp woods and thickets. P.? VII. E. S. I.

** *Spikelets always broader upwards. Ribs of upper pale strongly fringed. Awn long.*

3. *B. ster'ilis* (L.); pan. drooping, branches long slightly divided, spikelets linear-lanceolate, fl. remote, *lower pale* glabrous shorter than its awn *with 7 distinct equidistant ribs*, l. pubescent. —*E. B.* 1030. *P.* 50.—Height 1—2 feet. L. broad, flat.—Waste places. A. VI. E. S. I.

4. *B. madriten'sis* (L.); *panicle erect*, branches short scarcely divided, spikelets lanceolate, fl. linear remote subcylindrical, *lower pale about as long as its awn 7-ribbed, 2 lateral ribs close together, interm. rib faint.*—*E. B.* 1006. *P.* 50.—*a. B. diandrus* (Curt.); st. glabrous, rachis and pedicels rough. St. 6—12 in. high. Remarkable for its erect panicle. Upper pale but little shorter than the lower.—[β. *B. rigidus* (Roth); pan. compact, pedicels very short, upper part of st., pedicels, rachis and gl. pubescent.]—Dry sandy places, rare. [β. Channel Islands.] A. VI. VII. E. S. I.

[*B. tectorum* (L.); like *B. madritensis*, pan. secund drooping, upper pale much the shorter.—Introduced.]

[*B. max'imus* (Desf.); *pan. erect lax at length nodding*, branches slightly divided lengthened after flowering, spikelets downy, *lower pale 7-ribbed about half as long as its awn.*—*E. B. S.* 2820.—Height 1—2 feet. A most beautiful grass.—Sandy places. Channel Islands. A. VIII.]

44. SERRAFAL'CUS *Parlatore*[1].

* *Fl. at first loosely imbricate, afterwards distinct and cylindrical.*

*1. *S. secalinus* (Bab.); pan. loose drooping in fr. slightly compound, simple peduncles about as long as the oblong glabrous spikelets, fl. about as long as the straight awn, lower

---

[1] Corresponds to the genus *Bromus* of Fries; to part of *Bromus* of Sm., Hook., &c.

pale not overlapping the next fl. uniformly rounded at the sides, l. hairy.—*E. B.* 1171. *P.* 49, 121 & 122.—With seed the fl. spread and the spikelets droop. *Top of upper gl.* ½-*way between its base and the top of fourth fl.* (second on the same side). Lower pale not twice as long as broad, longer than the upper, 7-ribbed. —*β. Bromus velutinus* (Sm.); pan. nearly simple, fl. larger downy. *P.* 123.—Corn-fields. A. VI. VII. E. S. I.

** *Fl. closely imbricate even with fruit.*

2. *S. racemósus* (Parl.); pan. long erect usually simple, spikelets ovate rather compressed glossy, fl. imbricate about as long as the straight awn, *lower pale bluntly angular* above the middle, l. and sheaths slightly hairy.—*E. B.* 1079.—*Top of the upper glume* ⅓-*way to the top of the fourth flower.* Lower pale longer than the upper. Anth. 4 times as long as broad. Pan. close with fruit.—Common. B. VI. E. I.

3. *S. commutátus* (Bab.); pan. loose slightly drooping compound, simple peduncles as long as or longer than the oblong-lanceolate spikelets, fl. loose imbricate about as long as the straight awn, *sides of lower pale uniformly rounded* at the sides, l. and sheaths hairy.—*P.* 124. *E. B.* 920.—Lower pale only slightly overlapping the next fl. at the base when in fruit. *Top of upper gl.* ½-*way to top of fourth fl.* Lower pale twice as long as broad, longer than the upper, glabrous or downy, 7-ribbed. Anth. 6 times as long as broad.—*β. multiflorus* (Parn.); more numerous fl., top of upper gl. rather higher. *P.* 125.—Common. B. VI. VII. E. S. I.

4. *S. mol'lis* (Parl.); pan. close erect compound or rarely simple, spikelets ovate rather compressed pubescent, fl. closely imbricate about as long as the straight awn, *sides of lower pale bluntly angular about the middle,* l. and sheaths hairy or downy. —*E. B.* 1078. *P.* 116.—*Top of the upper gl.* ½-*way to the top of the sixth flower;* or a little higher (*ovalis* Parn. 117, with short oval spikelets); or about ½-way to the top of the eighth (*pratensis* Parn. 118, with longer spikelets). Lower pale longer than the upper. Simple ped. not longer than the spikelets. Anth. about thrice as long as broad. Rarely the spicules are glabrous, when it is *B. racemosus* Parn. 119 [=*S. mollis,* var. *glabrescens,* Gren.].—A maritime plant with nearly or quite prostrate st., nearly simple pan., and nearly glabrous pales, seems to be the *B. hordeaceus* (Fries). Another with very short ped. and densely downy spikelets [and divaricate awns (*S. Lloydianus,* G. & G.)] is also found near the sea.—[*β. pseudo-velutinus*;

pan. oblong blunt interrupted with usually single alternate br. bearing 3-5 almost sessile spikelets, *upper pale split to the base.* Perhaps a distinct species, *B. pseudo-velutinus.* *B. interrup'tus* (Druce).]—Common. A.? V. VI. *Lop-grass.* E. S. I.

*5. *S. arven'sis* (Godr.); pan. spreading compound its branches ultimately horizontal, spikelets linear-lanceolate, *pales equal in length lower 7-ribbed* with two prominent ribs near each margin and its sides bluntly angular above the middle, anth. 4 times as long as broad.—*E. B.* 1984. *P.* 126.—Top of the upper gl. ½-way to the top of the fourth flower.—Naturalized in various places from Fife southwards. A. VII. VIII. E. S.

[*S. pat'ulus* (Parl.); pan. spreading compound its branches ultimately deflexed, spikelets lanceolate, pales unequal, lower 7-ribbed with two lateral prominent ribs and its sides bluntly angular above the middle, anth. twice as long as broad.—*P.* 127. —Not naturalized. A. VI.] E.

[*S. squarrósus* (Bab.); pan. drooping simple, spikelets ovate-lanceolate subcompressed, fl. nearly glabrous imbricate compressed, lower pale 9-ribbed and its sides bluntly angular above the middle, awn twisted divaricate, l. pubescent.—*E. B.* 1885. *P.* 128.—Not naturalized. A. VI. VII.] E.

Tribe XII. *Hordeieæ.*

45. Brachypo'dium *Pal. de Beauv.*

1. *B. sylvat'icum* (Beauv.); spike drooping, spikelets (at first) terete alternate distichous, *awns* of the upper fl. *longer than their pales,* l. flat linear-lanceolate flaccid, *root fibrous.*—*E. B.* 729. *P.* 61.—St. usually solitary or 2 or 3 from the same root, erect, 1—2 feet high. Sheaths hairy. Ligule short, blunt, notched or torn. L. ciliate. Pales usually hairy.—Woods and hedges. P. VII. E. S. I.

2. *B. pinnátum* (Beauv.); spike erect, spikelets (at first) terete alternate distichous, *awns* of the upper fl. *shorter than their pales,* l. flat linear-lanceolate rigid, *creeping.*—*E. B.* 730. *P.* 132—137.—St. several, erect, 1—2 feet high. Pales rough or hairy. Sheaths subglabrons. Ligule short, truncate. L. not ciliate.—Sometimes the l. are very narrow and involute, st. very many, spikelets small smooth.—On dry limestone soil. P. VII. E. I.

46. TRIT'ICUM *Linn.* Wheat-grass.

1. *T. caninum* (L.); spike rather close, spikelets 2—5-fl., 3—5-ribbed gl. and lower pales awned [upper pale emarginate], axis and edges of the rachis hispid, l. flat rough on both sides, *root fibrous.*—*E. B.* 1372. *P.* 62. *Agropyron* (Beauv.).—St. erect. Ribs on the upperside of l. very slender. Gl. round on the back, its ribs reaching the tip and joining to form the short awn. Lower pale shorter than its awn; or in an alpine form longer than it.—Banks, rare. P. VII. E. S. I.

2. *T. alpinum* (Don); spike rather close, spikelets 2—6-fl., gl. usually shortly awned strongly 4—6-ribbed with ribs edges tip and awn asperous, lower pale narrowed abruptly and with scarious margins at apex, 4—6-ribbed with awn ¼—½ its length, *upper pale bluntly pointed* densely ciliate on the lateral keels *densely asperous and with a well-marked midrib towards the apex, the two lateral ribs terminating in teeth which fall short of the apex,* axis hairy, rachis ciliate, l. thin flat with many slender ribs, "soboliferous."—*Agropyron Donianum* (Buch.-White).—Variously referred to Sp. 1 and 3, but differing from both in the character of the upper pale; closely allied to the Scandinavian *T. violaceum* (Hornem.).—Rocks, Ben Lawers, *Mr. G. Don.* Rediscovered 1878 by *Mr. J. C. Melvill.* P. VIII. S.

3. *T. répens* (L.); spike rather close, gl. 5—7-ribbed equalling at least ⅔ of the 4—5-fl. spikelet rough on the keel, lower pale acuminate, *axis asperous,* rachis with rough angles not brittle, *l.* mostly flat the *many slender ribs* each bearing a row of deciduous hairs above, *soboliferous.*—*E. B.* 909. *P.* 62. *Agropyron* (Beauv.).—L. at first involute afterwards flat, ribs on upperside not much raised nor nearly hiding the interm. surface of the leaf. Rachis glabrous or downy with forward prickles on the angles. Gl. scarcely keeled, acuminate-subulate; ribs reaching the tip. Pales rarely awned. [The shape of the gl. and pales varies considerably, extreme forms are:—var. *barbatum* (Duv.-Jouv.) "gl. very attenuate subulate or awned, pales long-awned," and var. *obtusum* (Sy.) "gl. obtuse obliquely truncate, pales obtuse with a minute apiculus."]—Common. P. VI. *Couch-grass.* E. S. I.

4. *T. pun'gens* (Pers.); spike close, gl. with 7—9 thick ribs not exceeding ½ the 5—12-fl. spikelet rough on the keel, lower pale acute, axis asperous, rachis nearly or quite smooth not brittle, *l. with involute edges the many thick closely-placed ribs* slightly rough and each bearing a row of acute points above, upper part of l. wholly involute (subulate and rigid), soboliferous.—*Sy. E. B.* 1811. *Agropyron* (R. & S.)—*St. erect,* like a corn-field. Ribs on upperside of l. so broad and so elevated as nearly to hide the interm. part of the leaf. Gl. keeled; ribs

reaching the tip. Lower pale of our plant often awned [*T. littorale* (Host)]. Producing erect barren leafy clustered shoots. [β. *T. pycnanthum* (Godr.); spike shorter and denser, spikelets much compressed, gl. and lower pale obtuse.]—Sea-shores. P. VII. E. I.

5. *T. acútum* (DC.); spike rather lax, *gl. with* 5—7 *slender elevated ribs* blunt or apiculate not exceeding ⅔ of the 5—8-fl. spikelet, lower pale blunt mucronate, axis downy, rachis smooth or slightly rough at the angles not brittle, l. flat or with involute edges the many thick closely placed *ribs rough with minute sharp scattered* points (asperous) above, soboliferous.—*Sy. E. B.* 1812. *T. laxum* Fr. *Agropyron* (R. & S.).—St. prostrate or ascending. Ribs of the l., on each of which there is usually a deciduous row of hairs, not so completely hiding the intervening hollows as in Sp. 3. Gl. keeled; keel often with forward bristles, reaching the tip or forming a slight mucro. Lower pale rarely shortly awned. Producing decumbent and ascending, barren, leafy, clustered shoots.—Sandy sea-shores. P. VII. VIII. E. S. I.

6. *T. jun'ceum* (L.); spike rather loose, *gl. with* 9—11*slender scarcely elevated* ribs blunt equalling at least ⅔ of the 4—8-fl. spikelet smooth on the keel, lower pale blunt rarely mucronate, axis smooth or slightly downy, *rachis brittle* smooth, l. involute with many thick *ribs with much spreading hair above*, soboliferous.—*E. B.* 814. *P.* 63. *Agropyron* (Beauv.).—St. prostrate. The short hairs on the ribs of the l. spread so as to cover the interm. spaces. Rachis easily separating above each spikelet. Gl. rounded or truncate at the tip; ribs not reaching the tip. Producing decumbent barren leafy shoots.—Sandy sea-shores. P. VII. VIII. E. S. I.

47. EL'YMUS *Linn.*

1. *E. arenárius* (L.); spike upright close, rachis flat not winged, gl. lanceolate downy not longer than the spikelets.—*E. B.* 1672. *P.* 64.—Closely resembling *Psamma arenaria*, but readily distinguished by its broad l. and short ligule. Soboliferous. St. 3—4 feet high.—Sandy sea-shores. P. VII. E. S. I.

[*E. geniculátus* (Curt.); spike lax, rachis winged, gl. awl-shaped glabrous longer than the spikelet.—*E. B.* 1586. *P.* 131. —St. 3—4 feet high. Spike 1—2 feet long, usually remarkably bent downwards at the second or third spikelet.—In a salt marsh near Gravesend. *Mr. Dickson*! P. VII.] E.

48. Hor'deum *Linn.* Barley.

1. *H. sylvat'icum* (Huds.); *gl. all awlshaped not ciliate rough, lateral fl. perfect*, middle fl. often barren, lower pale with an awn of twice its length.—*P.* 130. *Elymus europæus* Linn., Sm., *E. B.* 1317.—About 2 ft. high. Spike subcylindrical. Middle fl., if barren, with shorter gl. having involute edges, thus appearing setaceous. The spikelets have a second fl. occasionally. —Woods and thickets on a calcareous soil, rare. P. VII. VIII. E.

2. *H. nodósum* (L.); *gl. all setaceous not ciliate rough, lateral fl. imperfect*, lower pale of fertile middle fl. with an awn of about its length.—*E. B.* 409. *P.* 11. *H. praten'se* (Huds.) ed. viii.—Often 2 ft. high. Spike compressed, erect. Gl. of lateral fl. shorter.—Damp meadows. P. VII. E. S. I.

3. *H. murínum* (L.); *gl. of the middle spikelet linear-lanceolate* ciliate *of the lateral ones setaceous* rough, lateral fl. imperfect.—*E. B.* 1971. *P.* 10.—Spike often slightly nodding, compressed. Height 12—18 in. Awn longer than the lower pale. Lateral gl. sometimes slightly ciliate.—*β. arenarium* (Bab.); lower part of the st. buried, lengthened, branched and rooting, thus appearing to creep.—Waste places. *β.* Loose sand. B. VI. VII. E. S. I.

4. *H. marínum* (Huds.); gl. rough, *inner gl. of the lateral fl. half-ovate* the rest setaceous, lateral fl. imperfect.—*E. B.* 1205. *P.* 10. *H. maritimum* (With.) ed. viii.—Spike thick, erect, subterete.—Pastures and banks near the sea. A. VI. E. S.

49. Leptu'rus *R. Br.*

1. *L. filifor'mis* (Trin.); spike cylindrical-subulate, gl. 2 equalling or slightly exceeding the flowers.—*Rottböellia* Sm., *E. B.* 760. *P.* 2.—St. 2—6 in. long. Spike long, slender straight or slightly curved. [*L. incurvatus* (Trin.) is apparently only found as a ballast plant.]—Gravelly and waste places near the sea. A. VII. E. S.

50. Lo'lium *Linn.* Rye-grass.

1. *L. peren'ne* (L.); with leafy barren shoots, *edges of young l. simply folded*, spikelets 3—11-flowered, gl. equalling the lowest fl., lower pale usually awnless.—*E. B.* 315. *P.* 65.—

St. 12—18 in. high, usually bent at the lower knots. Whole plant rather dark green.—β. *aristatum*; lower pale with a long awn.—γ. *L. tenue* (L.); spikelets few-flowered, l. slender.—Sometimes the spikelets become converted into branches; or the rachis is so much shortened as to produce a broad ovate close distichous spike.—Common. P. VI. *Rye-grass.* E. S. I.

[*L. ital'icum* (A. Braun); with leafy barren shoots, *edges of young l. involute*, spikelets 9—14-flowered, lower pale with a long awn.—*R.* vii. 77. *P.* 138—141.—*St. many, straight, in close tufts*, 1½—3 ft. high. Whole plant, especially the spikelets, paler than in the preceding. Ligule short, abrupt.—*L. multiflorum* (Lam.), perhaps confounded with this, has no barren shoots and is annual.—Cultivated fields. P. VI. *Italian Rye-grass.*] E.

[*L. linic'ola* (Sond.); no barren shoots, spikelets 7—11-flowered exceeding the gl., lower pale longer than its awn or awnless, fl. tumid with fruit.—*E. B.* 2955.—St. erect. Spike slender. Lower pale cartilaginous below, narrower than the upper, tumid in fruit.—Cultivated fields. A. VI. VII.] E.

‡2. *L. temulen'tum* (L.); no barren shoots, *spikelets* about 6-flowered *equalling or shorter than the gl.*, lower pale awned, fl. tumid with fruit.—*E.B.* 1124. *P.* 142.—St. erect. Ligule short. Upper gl. usually present, often bifid.—*α*; awns as long as or longer than the pale.—β. *L. arvense* (With.); fl. 4—5 without or with short awns. *E. B.* 1125.—Cultivated fields. A. VI.—VIII. *Darnel.* E. S. I.

FLOWERLESS PLANTS.

Class III. CRYPTOGAMEÆ.

Substance of the plant of cellular tissue or with a few ducts. No woody fibre. No true flower with stamens and pistils. No distinct embryo, nor cotyledons.

A. Plants with a few ducts amongst the cellular tissue. Producing spores which develop into a prothallus which bears antheridia and archegonia.[1]

Order XCIX. EQUISETACEÆ.

Leafless branched plants with a striate hollow stem; each joint ending in a sheath which conceals the joining and encloses the base of the next joint. Sporules surrounded by elastic clavate filaments and enclosed in capsules arising from the peltate scales of terminal cones or spikes.—Rhizome creeping. Branches whorled. Cuticle abounding in silex. Only one genus.

1. Equise'tum *Linn.* Horse-tail.

* *Fertile stems mostly unbranched and succulent; barren stems with solid whorled branches, appearing later.*

1. *E. arven'se* (L.); sterile st. with 6—19 furrows slightly rough, branches rough with 3 or 4 simple angles, *teeth of sheaths long acute 1-ribbed at the tip,* fertile st. simple with few lax distant sheaths.—*E. B.* 2020. *S.* 1. *H. F.* 60. *N.* 77.—Sterile st. many, procumbent or ascending; with many whorls of roughish solid usually simple branches with deep furrows and

[1] (*N.*) refers to *Newman's British Ferns,* ed. 2 (1844); (*S.*) to *Sowerby's Ferns and Fern allies*; (*H. F.*) to *Hooker's British Ferns. Moore's Handbook of British Ferns,* ed. 3, may be consulted with much advantage, and *Milde in Nov. Act. Soc. Nat. Cur.* vol. xxxii.

3—4 toothed sheaths, their *lowest joint* (including its terminal sheaths) *exceeding the st.-sheath*; general outline narrowed upwards, usually naked at the end. Fertile st. short, with few (4—5) sheaths, appearing before the sterile ones. Sterile and fertile st. distinct.—*β. alpestre* (Wahl.); sterile st. short (2—3 in.) prostrate with ascending terminal point and secund suberect branches.—Damp meadows [cultivated land, &c.] *β.* Mickle Fell, Teesdale. P. IV. E. S. I.

2. *E. praten'se* (Ehrh.); *sterile st.* with about 20 striæ *very rough with prominent points* particularly above, branches simple with 3 or 4 simple angles, *teeth of sheaths* 1-*ribbed but not to the tip*, fertile st. simple with many crowded deeply toothed sheaths. —*E umbrosum*, *H. F.* 59. *S.* 2. *N.* 63. *E. Drummondii E. B. S.* 2777.—Sterile st. 1—1½ ft. high, nearly naked below; with many whorls of slender solid branches in the upper part, having 3- or 4-toothed sheaths, their lowest joint and sheath falling short of the st.-sheath; *general outline remarkably blunt.* Branched fertile st. with larger sheaths and whorls of about 6 branches; simple fertile st. short (4—6 in.), with many loose and still larger yellowish-white sheaths with black prominent ribs upwards and 12—20 teeth.—Wet places, rare. P. IV. E. S. I.

3. *E. max'imum* (Lam.); sterile st. nearly smooth with about 30 striæ and branches, branches rough doubly angular simple, *teeth of sheaths 2-ribbed*, fertile st. simple with many crowded large deeply-toothed sheaths.—*E. Telmateja*, *H. F.* 58. *S.* 3. *N.* 67. *E. fluviatile* Sm. not Linn., *E. B.* 2022.—Sterile st. 3—6 ft. high; occasionally bearing a small terminal spike; furnished from top to bottom with whorls of slender solid branches with 4 longitudinally furrowed angles and 4-toothed sheaths, their lowest joint and sheath exceeding the st.-sheath. Fertile st. stout, 1 foot or more high, with many very long pale-brown sheaths with 30—40 teeth; spike large. [*E. Braunii* (Milde), a form with deeply furrowed sterile st., was collected in Forfar by Mr. W. Gardiner.]—Wet places. P. IV. E. S. I.

** *Sterile and fertile st. subsimilar, contemporaneous, branched.*

4. *E. sylvat'icum* (L.); sterile and fertile st. with about 12 furrows and many whorls of slender *compound spreading or deflexed solid branches*, sheath lax ending in 3 or 4 blunt lobes. —*E. B.* 1874. *H. F.* 61. *S.* 4. *N.* 59.—St. 12—18 in. high. Sheaths of the branches with 3 long *acute teeth each* 1-*ribbed to its tip.* Fertile st. occasionally simple. Spike blunt. General outline of sterile st. usually pyramidal, of fertile abrupt.—

β. E. capillare (Hoffm.); sterile st. with many long slender branches of about equal length, branchlets very fine threadlike, emerald green, 2—3 ft. high.—Wet shady places. P. IV. V. E. S. I.

*** *Stems of one kind, with or without simple hollow whorled branches.*

5. *E. limósum* (L.); stem nearly smooth with many *slight furrows*, teeth of sheaths short rigid acute, *branches simple* whorled or none.—St. 2—4 ft. high. Sheaths rather short. Spike blunt.—*a. E. limosum* (L.); st. smooth, barren st. narrowing gradually upwards, branches short rigid slightly tapering upright and equalling st.-joints from green sheaths often wanting. *H. F.* 62.—*β. E. fluviatile* (L.); st. subglabrous, barren st. with a lax whiplike end, branches long slender tapering lax exceeding st.-joints from dark brown sheaths rarely wanting.—[*E. litorale* (Kühl.), with barren and fertile st. similar, st. more deeply grooved and rougher, and with a smaller central hollow than sp. 5, spores abortive, elaters 0, found at Bisley, Surrey, by Mr. W. H. Beeby, is usually considered a hybrid betw. sp. 1 & 5. See *J. of B.* xxv. (1887) p. 65, t. 273.]—In stagnant water. P. VI. VII. E. S. I.

6. *E. palus'tre* (L.); *st. with* 4—8 *deep furrows branched throughout*, sheaths loose pale with acute wedgeshaped teeth tipped with brown and membranous at the edges, branches simple.—*E. B.* 2021. *H. F.* 63. *S.* 6. *N.* 43, 47 & 49.—St. slightly rough. Barren st. whipshaped at the end. Spike blunt. Sheaths coloured like the st. or paler; teeth brown with nearly transparent edges, ribs furrowed on the back. *Branches* usually barren or (*β. polystachium*) each ending in a spike, *hollow*, with shallow furrows, *lowest joint of branch* (often reduced to its sheath) *falling short of the st.-sheath.* Occasionally (*γ. nudum* DC.) the angles and teeth are fewer and the st. nearly or quite simple and dwarf.—Spongy bogs. *γ.* Sandy places. P. VI. VII. E. S. I.

7. *E. hyemále* (L.); *st. simple* very rough with 14—20 slender furrows, *sheaths close whitish but the top and bottom black, teeth* with slender black-brown *very deciduous tips.*—*E. B.* 915. *H. F.* 64. *S.* 8. *N.* 17.—St. 1—2 ft. high, simple, biennial. its central hollow equalling at least ⅔ of diameter. Spike apiculate. Sheaths widest at their top, at first green with black crenate rim after the teeth have fallen, then entirely black, and ultimately pale in the middle and black above and below.—*β. E. Moorei* (Newm.); *st. annual* very rough with about 12 furrows, *sheaths loose* white with the base black, teeth black-

based rather persistent. *Phytol.* v. 19 (1853). *S.* 12. St. 1—2 ft. high, its central hollow equalling about ⅗ of diameter. Sheaths loose, pearly white; teeth long, usually light brown, whitish above, more persistent and longer.—Damp banks and woods. β. Wicklow. Wexford. P. VII. VIII. E. S. I.

8. *E. trachy'odon* (A. Br.); st. simple or very slightly branched very rough with 8—12 furrows, *sheaths close ultimately wholly black, teeth* slender *persistent.*—*H. F.* 65. *E. Mackaii N.* 24. *S.* 9.—St. 1—3 feet high, simple or with solitary distant branches, biennial, its central hollow equalling ⅓ of diameter. Spike apiculate. Sheaths quite cylindrical, pale green with a black band beneath the teeth but ultimately wholly black. Teeth much more persistent than in the preceding, usually black. —N. and W. of Ireland. P. VII. VIII. I.

9. *E. variegátum* (Schleich.); st. simple or slightly branched very rough with 4—10 furrows, *sheaths slightly enlarged upwards green below black above,* teeth blunt each tipped with a deciduous bristle.—*H. F.* 66. *N.* 31.—St. about a foot high, erect, usually simple except at the base or irregularly branched, its central hollow equalling ⅓ of diameter. Lower half of the sheaths green like the stem, upper part black; teeth persistent ovate, black in the centre, with a white membranous margin. Spike apiculate.—*E. Wilsoni* (Newm. 39. *S.* 10) is a large form [with smoother st. and less prominent angles to the ridges.]—β. *arenarium*; st. procumbent, usually more slender, teeth of the sheaths wedgeshaped. *E. variegatum* Sm., *E. B.* 1987. *S.* 11. —Wet places, or in water. β. Sandy places near the sea. P. VII. VIII. E. S. I.

Order C. FILICES.

Leafy plants with a rhizome or trunk. L. or fronds usually circinate when young (Tribe VII. excepted), simple or divided. Fructification springing from the veins on the underside or at the edge of the l., of 1-celled capsules (thecæ) which are stalked and have an elastic ring or sessile and without a ring [1].

[1] Dr. Boswell in *E. B.* ed. 3, vol. xii, described many varieties of ferns, and these are enumerated in *Lond. Cat.* ed. 9. We have not, however, thought it desirable to include most of them, as they were no doubt known to Prof. Babington, and rejected by him as unimportant. The vegetative organs of ferns are so liable to trivial variations that there is no limit to the number of forms which might be described.—H. & J. G.

* *Capsules with an elastic marginal ring.*

[Suborder I. POLYPODIACEÆ.

Capsules in dorsal or marginal clusters, opening transversely or irregularly. Young fronds circinate.

† *Clusters dorsal. Ring vertical, usually incomplete. Caps. opening transversely.*

Tribe I. *POLYPODIEÆ.* Clusters nearly circular, without an indusium, seated upon the back of the lateral veins.

1. CRYPTOGRAMME. Clusters circular, at length confluent, concealed by the reflexed margin of the frond.—Barren and fertile fronds dissimilar.

2. POLYPODIUM. Clusters circular, naked ; edge of the frond flat, not reflexed.

3. WOODSIA. Clusters circular, with an inferior involucre divided at the edges into many capillary segments.

Tr. II. *ASPIDIEÆ.* Clusters nearly circular, covered by an indusium, seated upon the back of the lateral veins.

4. LASTREA. Indusium reniform, attached by the notch. Veins distinct after leaving the midrib, not uniting with those of the adjoining lobe.

5. POLYSTICHUM. Indusium circular, attached by the centre. Veins distinct after leaving the midrib.

6. CYSTOPTERIS. Indusium attached by its broad hooded case under the clusters, with a long fringed free extremity at first covering the capsules.

Tr. III. *ASPLENIEÆ.* Clusters oblong or linear, usually covered by an indusium opening longitudinally on one side, placed on the side of the lateral veins.

7. ATHYRIUM. Clusters oblong-reniform. Indusium opening towards the central vein or midrib, margin fringed.

8. ASPLENIUM. Clusters long, straight. Indusium opening towards the central vein or midrib, nearly flat.

9. PHYLLITIS. Clusters long. straight, 2 together. Indusia of each pair opening towards each other.

10. CETERACH. Lateral veins anastomosing; clusters attached to their middle on the side next the midrib, except in the lowest. Indusium (?) a narrow nearly erect membrane on the back of the vein. Whole back of the frond covered with chaffy scales.

[11. GYMNOGRAMME. Clusters oblong or linear on both branches of the forked vein, becoming confluent and covering the back of the frond. Indusium wanting.]

Tr. IV. *ADIANTEÆ.* Capsules covered by a marginal or submarginal elongated part of the frond, or by a separated portion of the cuticle resembling an indusium.

12. BLECHNUM. Capsules in a continuous line parallel to the midrib upon a longitudinal anastomosing part of the transverse veins, covered by a continuous scarious indusium.—Barren and fertile fronds dissimilar in our plant.

13. PTERIS. Capsules in a continuous marginal line covered by a continuous indusium formed of the reflexed margin.

14. ADIANTUM. Capsules marginal, oblong or roundish, covered by distinct reflexed portions of the margin of the frond.

†† *Capsule opening irregularly, seated on a receptacle which ends a vein at the edge of the frond. Ring oblique, transverse, complete.*

Tr. V. *HYMENOPHYLLEÆ.*

15. TRICHOMANES. Capsules on a long filiform receptacle within a cupshaped involucre of the same texture with the frond.

16. HYMENOPHYLLUM. Capsules on a narrow subclavate receptacle within a two-valved involucre of the same texture with the frond.

** *Capsules without an elastic ring.*

Suborder II. OSMUNDACEÆ.

Young frond circinate. Rachis woody. Capsules regularly 2-valved, stalked, in clusters at the extremity of the frond.

Tr. VI. *OSMUNDEÆ.*

17. Osmunda. Capsules clustered, arranged in a branched spike terminating the frond.

Suborder III. OPHIOGLOSSACEÆ.

Young fronds straight. Rachis succulent. Capsules regularly 2-valved, sessile, in clusters on a separate branch of the frond.

Tr. VII. *OPHIOGLOSSEÆ.*

18. Botrychium. Capsules distinct, disposed in a compound spike attached to a pinnate or bipinnate frond.

19. Ophioglossum. Capsule connate, disposed in a simple distichous spike attached to an undivided frond.

Suborder I. *Polypodiaceæ.* Tribe I. *Polypodieæ.*

1. Cryptogram'me *R. Br.* Rock-brake.

1. *C. cris'pa* (R. Br.); barren fronds 2—3-pinnate, leaflets wedgeshaped or linear-oblong often bifid at the end, leaflets of the fertile fronds oblong.—*H. F.* 39. *Allosorus* Bernh., *N.* 103. *E. B.* 1160.—Fertile frond nearly triangular. Veins alternate, often forked and each branch ending in a cluster having no indusium but concealed by the reflexed edge of the leaflet. Height 6—12 in. St. slender, very brittle.—Amongst loose stones on mountains. P. VII. *Parsley Fern.* E. S. I.

2. Polypo'dium *Linn.* Polypody.

* *Clusters at the end of a veinlet, other veinlets knobbed at the end and not reaching the edge. Rhizome without fronds at its end. Stipes jointed to rhizome.*

1. *P. vulgáre* (L.); frond deeply pinnatifid, lobes linear-oblong somewhat serrate all parallel upper ones gradually smaller. —*E. B.* 1149. *H. F.* 2. *N.* 111.—Rhizome brown, densely scaly creeping. Fronds strapshaped. Clusters large, on the upper part of the frond. Lateral veins of the pinnæ with 4 branches. Pinnæ occasionally bifid at the end, sometimes deeply serrate or even (*P. cambricum* L.) doubly pinnatifid.—Shady banks, walls and old trees. P. VIII.—X. *Common Polypody.* E. S. I.

* *Veinlets not knobbed but reaching the edge, all usually fertile, clusters near the end of each. Rhizome with fronds at its end. Stipes not jointed to rhizome.*—PHEGOPTERIS *Fée.*

2. *P. Phegop'teris* (L.) ; *fronds pinnate,* pinnæ linear-lanceolate united at the base pinnatifid with linear-oblong blunt lobes, *lowest pair of pinnæ turned downwards and forwards* the rest upwards, clusters marginal.—*E. B.* 2224. *H. F.* 3. *N.* 115. *Pheg. polypodioides* (Fée).—Rhizome nearly black, wiry, slightly scaly, creeping extensively. Fronds triangular. Pinnæ very acute, pointing upwards, rather hairy, connected by their whole width with the rachis ; lowest pair quite distinct, with a minute stalk, standing forwards and pointing from the others. —Damp places, loving the spray of waterfalls. P. VII.—IX. E. S. I.

3. *P. Dryop'teris* (L.) ; *fronds ternate glabrous,* divisions pinnate, pinnæ pinnatifid blunt uppermost nearly entire, clusters marginal.—*E. B.* 616. *H. F.* 4. *N.* 123.—Rhizome black, wiry, creeping, slightly scaly. Stipe slender, brittle. The three divisions of the frond loosely spreading, the middle one rather the largest. Very young fronds *resemble* 3 *little balls on wires.* Not glandular.—Shady mountainous places. P. VI. VII. E. S. I.

4. *P. Robertiánum* (Hoffm.) ; *fronds triangular subternate glandular-mealy,* lower branches pinnate, pinnæ pinnatifid blunt, uppermost nearly entire, clusters marginal.—*H. F.* 5. *P. ca'-careum* Sm., *E. B.* 1525. *N.* 131.—More erect and rigid than the preceding, always covered with very minute stalked glands giving a mealy character to the surface. Frond scarcely 3-fid, the lower branches being much smaller in proportion to the middle one ; all the 3 erect, rigid.—On broken limestone ground. P. V.—VIII. E.

*** *Veinlets not knobbed, scarcely reaching the edge, simple or branched, each bearing a cluster below its end or that of its anterior branch. Rarely there is an oblique curved false indusium. Stipes not jointed to rootstock.*

5. *P. alpes'tre* (Hoppe) ; fronds lanceolate bipinnate, *pinnæ narrow-lanceolate with a broad base contiguous,* pinnules widest at their base acute pinnatifid with serrate lobes and branched veinlets, clusters on upper half.—*H. F.* 6. *S.* 49. *Athyrium* (Milde).—Fronds suberect, 1—3 ft. high. Much like *Athyr. Filix-fœmina.*—High mountain-valleys. P. VII. VIII. S.

6. ? *P. flex'ile* (Moore); fronds linear-lanceolate, bipinnate, *pinnæ ovate-lanceolate distant, pinnules narrow at their base* obovate bluntish serrate and with unbranched veinlets, clusters chiefly on lower half.—*N.* ed 3. 203.—*P. rheticum* Fl. Dan. 2607? *Athyrium* (Sy.).—Stipe short. Fronds much less divided, narrow, elbowed, spreading horizontally.—Abundant in Glen Prosen, Forfarshire. Glen Lyon, Ben Aulder. P. VII. VIII. S.

3. Woods'ia *R. Br.*

1. *W. ilven'sis* (R. Br.); frond lanceolate hairy and chaffy beneath pilose above, *pinnæ oblong or ovate* pinnatifid, lobes very blunt nearly entire.—*E. B. S.* 2616. *H. F.* 8. *N.* 137.—Rhizome tufted. Stipe jointed. Frond 1—5 inches long; pinnæ 4—6 lines long, mostly opposite.—Exposed alpine rocks. Glyder Fawr, N. Wales. Falcon Clints, Durham. White Coombe, Dumfries. P. VII. E. S.

2. *W. hyperbórea* (R. Br.); frond linear-lanceolate or oblong pinnate glabrous or slightly hairy only beneath, *pinnæ triangular* pinnatifid or lobed, lobes 3—7 very blunt nearly entire.—*E. B.* 2023. *H. F.* 7. *N.* 143. *Acrostichum alpinum* Bolt. t. 42.—Rhizome tufted. Stipe jointed. Frond 1—3 inches long; pinnæ mostly alternate, a little longer than broad; pinnules 2—3 lines long.—Exposed alpine rocks. Breadalbane Mts., Perthshire. Clogwyn y Garnedd, Snowdon. P. VII. E. S.

Tribe II. *Aspidieæ.*

4. Las'trea *Presl.* [1]

* *Lateral veins simple or forked. Clusters on the simple veins or either or both branches.*

1. *L. Thelyp'teris* (Presl); rhizome slender far-creeping, fronds pinnate, pinnæ linear-lanceolate pinnatifid slightly downy but *without glands*, lobes oblong, clusters submarginal.—*H. F.* 13. *N.* 183.—Fronds lanceolate; earlier barren with flat lobes;

[1] Adanson's genus *Dryopteris*, established in 1763, apparently included *Polystichum* and *Lastrea*, and there seems no valid reason for its rejection. Some recent authors have adopted it in place of *Lastrea*, retaining *Polystichum*, but this position does not appear to us to be tenable. We have preferred leaving the genera as in the last edition, until the matter has been satisfactorily dealt with.—H. & J. G.

later fertile with revolute-edged lobes; 2 or 3 lowest pairs of pinnæ decreasing in size. Lateral veins alternate, forked, extending to the edge. Clusters at length confluent, midway between the midrib and edge. Height 6—8 inches.—Marshy and boggy places. P. VII. VIII. *Marsh-Fern.* E. S. I.

2. *L. Oreop'teris* (Presl); rhizome thick short, fronds pinnate, pinnæ linear-lanceolate pinnatifid *glandular* beneath gradually decreasing from about the middle of the frond to near the root, lobes oblong flat, clusters marginal.—*H. F.* 14. *E. B.* 1019. *N.* 187.—Fronds remarkably narrowed downwards, rising in a circle from a tufted rhizome, fragrant when bruised. Lobes blunt, entire; lateral veins simple or forked. Height 2—3 feet. Indusium often scarcely distinguishable.—Mountain heaths. P. VII. *Sweet Mountain-Fern.* E. S. I.

** *Lateral veins branched or forked. Cluster upon the first upper lateral veinlet.*

3. *L. Filix-mas* (Presl); *fronds* lanceolate *subbipinnate*, pinnæ linear-lanceolate lowermost pair rather smaller than the second, pinnules oblong blunt or acutish serrate (not spinulose) attached by their whole width or often connected below, *clusters near the midvein.*—*E. B.* 1458. *N.* 198.—Fronds only slightly narrowed downwards and the lowest pinna of considerable size, rising in a circle from a tufted rhizome. Stipe and rachis nearly glabrous, yellow, or densely clothed with purple scales. Indusium very persistent, convex, with no marginal glands. Height 2—4 ft.—*α*; pinnules crowded linear-oblong blunt slightly confluent broad-based adpressed-serrate.—*β*. *Dryopt. Borreri* (Newm.); stipe and rachis very scaly, pinnules truncate subentire at the sides.—*γ*. *affinis* (*Aspidium affine*, Fisch.); pinnules less crowded longer acutish narrow and often slightly auricled at their base, teeth patent lower ones notched. Var. *incisa* Moore.—*δ*. *abbreviata* (*Polyst. abbreviatum*, DC.); clusters near the base of the confluent crowded very blunt pinnules, about 1 ft. high. Not *abbreviata*, Newm.—Woods and banks. *δ*. Cumberland. Yorkshire. Wyck, Glouc. Glen Isla, Forfar. P. VI. VII. *Male Fern.* E. S. I.

4. *L. remo'ta* (Moore); fronds narrowly lanceolate, pinnæ triangular-lanceolate lowermost pair slightly smaller, pinnules acute with a narrow attachment deeply cut, clusters near the midvein.—*H. F.* 22.—Rhizome tufted. Scales ovate-acuminate and subulate. Much resembles *L. spinulosa.* [Perhaps a hybrid betw. sp. 3 and 7 or 8.]—Marshy places. Windermere. P. VIII. IX. E.

5. *L. rig'ida* (Presl) ; *fronds* triangular-lanceolate bipinnate *glandular*, pinnæ triangular-lanceolate lowermost pair not smaller than the second, pinnules oblong blunt lobed and serrate with a narrow attachment, segments 2—5-toothed *not spinulose*, indusium persistent fringed with stalked glands, *stipe clothed with long-pointed 1-coloured scales.—H. F.* 16. *E. B. S.* 2724. *N.* 191.—Fronds erect, lanceolate with the lower pinnæ rather short and triangular, or triangular with the lower pinnæ long; upper pinnæ narrow; all pinnate. Pinnules truncate below. Covered with minute stalked glands. Height 1—2 feet.—Ingleborough, Arnside Knot, and near Settle. P. VII. VIII. E.

6. *L. cristáta* (Presl) ; fronds linear-lanceolate or narrower subbipinnate glabrous, pinnæ short triangular-oblong pinnatifid or pinnate lowermost pair not smaller than the second, *pinnules* oblong blunt or rarely acute serrate *attached by their whole width and connected below* lowermost lobed and subtripinnatifid and superior and inferior nearly equal, stipe with broad ovate acute 1-coloured (pale) scales.—Fronds erect, 2 ft. high. Indusium without marginal glands.—*a*; fronds nearly linear, pinnæ pinnatifid, pinnules blunt; barren fronds broader. *H. F.* 17. *S.* 10. *N.* 203.—*β*. *L. uliginosa* (Newm.) ; fronds linear-lanceolate, pinnæ pinnate, pinnules acute, barren and late fertile fronds lanceolate, pinnæ subpinnate, pinnules blunt.—Bogs and boggy heaths, rare. P. VIII. E.

7. *L. spinulósa* (Presl) ; fronds oblong-lanceolate bipinnate glabrous, pinnæ triangular-oblong or -lanceolate lowermost pair scarcely smaller than the second, *pinnules* ovate-oblong acute incise-serrate *with a narrow attachment* inferior lowermost often largest, stipe with ovate acute 1-coloured (pale) scales, rhizome stout creeping.—*S.* 12. *N.* 203. *H. F.* 18.—Height 3—4 ft.; fronds nearly erect. Upper pinnules narrowed and decurrent below. Indusium without marginal glands.—[Var. *decipiens* (Sy.) has minute glands on the rachis and underside of the pinnæ and the indusium dentate.]—Marshy places and wet woods. P. VIII. IX. E. S. I.

8. *L. dilatáta* (Presl); frond triangular-lanceolate or -ovate bipinnate, lower pinnæ unequally triangular lowest pair not shorter than the second, pinnules oblong with a narrow attachment pinnatifid or pinnate inferior ones largest, segments spinous-serrate, *stipe clothed with long pointed scales with a dark centre and diaphanous margin*, rhizome tufted.—*H. F.* 19. *L. multiflora* N. 215.—Caudex usually erect. Fronds 2—4 ft. high, arched, often drooping, convex, more or less clothed with stalked glands when young; on young or starved plants often

triangular, never so on older and perfect ones. Indusium with marginal glands. The largest scales of the full-grown plant should be examined. "Spores winged and crested."—[Var. *lepidota* (Moore), having the rachis and its br. with numerous "broad cuspidate and narrow piliferous scales," is said to have been found in Yorks.]—*L. collina* (Newm.) has a triangular-ovate-prolonged frond and ovate blunt bluntly mucronate-serrate pinnules. *N.* 223.—*L. glandulosa* (Newm.) has a broad lanceolate frond covered with stalked glands beneath and the scales on the stipe often nearly without the dark centre. *Deak. Fl. Brit.* f. 1612. [Perhaps a hybrid betw. sp. 7 & 8.]—Woods, banks, &c. P. VIII. IX. E. S. I.

9. *L. æ'mula* (Brack.); frond triangular or triangular-ovate bipinnate, lower pinnæ unequally triangular lowest pair longest, pinnules prolonged-triangular with a narrow attachment pinnatifid or pinnate inferior lower ones largest, segments spinous-serrate, *stipe clothed with long narrow laciniate 1-coloured scales.* —*H. F.* 20. *Nephrodium fœnisecii* Lowe! *L. recurva* N. 225. —*Frond* 1—2 feet long; the lower pinnæ much the largest. *Pinnules and segments concave above.* Stipes, rachis, and fronds with many *globose sessile glands.* Sweet-scented.—Rocky shady places. P. VIII. IX. E. S. I.

5. Polys'tichum *Roth.*

1. *P. Lonchitis* (Roth); *fronds rigid* linear *pinnate,* pinnæ not lobed serrate spinous their base auricled above oblique below. —*N.* 163. *H. F.* 9. *E. B.* 797.—Stipe very short. Fronds narrow, very rigid and leathery. Pinnæ overlapping and twisted (most in the Irish, much less so in the Welsh plant), lower ones usually auricled both above and below.—Young simply pinnate fronds of the next species are often much like this plant.—Alpine rocks. P. VII. *Holly Fern.* E. S. I.

2. *P. aculeátum* (Roth); *fronds rigid* linear or lanceolate bipinnate, *pinnules obliquely decurrent.*—*N.* 169. *H. F.* 11. *Aspidum lobatum* Kunze. *A. aculeatum* and *A. lobatum* Sm.— Stipe usually short. Frond 1—2 feet high. First upper pinnule of each pinna longer than the others, its lower side (next the main rachis) usually nearly straight, its upper acutely auricled and forming an acute angle with the lower and with the partial rachis at the point of attachment. In young plants the pinnæ are serrate or pinnatifid or with one or more pinnules distinct. A few of the lowest pinnules are often slightly

stalked, but very differently from those of *P. angulare.*—*A. lobatum* (Sw.) has the pinnæ less divided than in the type of the species and the fronds linear-lanceolate and more rigid. *H. F.* 10.—Hedge-banks. P. VII. VIII. E. S. I.

3. *P. anguláre* (Presl); *fronds lax* drooping lanceolate bipinnate, *pinnules truncate or obtuse-angled below distinctly stalked. N.* 173. *H. F.* 12.—Stipe usually long. First upper pinnule scarcely longer than the others, its lower side rounded below, its upper with a large bluntish auricle and forming an obtuse angle with the lower, at the top of the short stalk which is nearly at right angles with the partial rachis; all short, broad and bluntish; or first upper pinnule longer and deeply pinnatifid, all more acute; or pinnules all narrower and acute.—Sheltered woods and hedge-banks, chiefly in the West. P. VII. VIII. E. S. I.

6. Cystop'teris *Bernh.* Bladder-Fern.

1. *C. frag'ilis* (Bernh.); frond lanceolate bipinnate, pinnæ ovate or ovate-lanceolate, pinnules oblong-ovate or cordate-ovate pinnatifid or cut.—*a. vera*; usually bipinnate, pinnules rather narrowed below, veins ending at tip of term. teeth or if pinnule emarginate in the 2 teeth *not in the notch.* Sporules prickly.—a. *C. anthriscifolia* (Roth); pinnules ovate acute cut, segments oblong toothed. *Cystea fragilis* Sm., *N.* 155.—b. *C. cynapifolia* (Roth); pinnules obovate cut, segments obovate toothed or retuse at the end. *C. fragilis* E. B. 1587.—c. *C. angustata* (Sm.); pinnules lanceolate acute cut, segments lanceolate-oblong cut, teeth acute. *N.* 156.—*β. Cyathea dentata* (Sm.); frond often only subbipinnate, pinnules broadest below blunt bluntly toothed, veins as in *a. H. F.* 23. *N.* 154. *C fragilis* Roth. *Clusters more marginal,* often ultimately confluent. In well-grown plants the pinnules are suddenly widened to their full extent just above their narrow stalklike base. *Spores warted.*—*γ. C. Dickieana* (Sim), pinnæ usually overlapping, pinnules broad blunt, *veins reaching the emarginate end,* clusters scattered, spores verrucose.—*δ. C. alpina* (Desv.); frond bipinnate, pinnæ ovate, pinnules ovate deeply pinnatifid with broadly and shortly linear segments partly cloven, *veins reaching the blunt end. E. B.* 163. *H. F.* 24. Fronds much divided but compact and close.—I have placed these plants under one species with much doubt.—Rocks and walls. *γ.* In a damp cave by the sea near Aberdeen. *δ.* Teesdale. *Mr. J. Backhouse*! P. VII. VIII. E. S. I.

2. *C. montána* (Link); fronds triangular tripinnate, pinnæ and pinnules spreading, lobes pinnatifid with linear notched segments. —*H. F.* 25. *S.* 24.—Fronds in shape like those of *Polypodium Robertianum,* small, short, very finely divided. Lower pair of pinnæ much the largest; their lower larger than their upper pinnules. Stipe long, slender. Rhizome creeping extensively, black.—Breadalbane and Grampian Mountains. P. VII. VIII.? S.

Tribe III. *Asplenieæ.*

7. ATHYR'IUM *Roth.* Lady Fern.

1. *A. Filix-fœm'ina* (Roth); frond lanceolate pinnate or pinnatifid, pinnæ linear, pinnules linear-oblong deeply serrate or pinnatifid.—*H. F.* 35.—*a. A. rhæticum* (Roth); frond bipinnate, pinnules toothed narrowly triangular-lanceolate convex remote not connected toothed, upper ones minute confluent, clusters ultimately confluent. *N.* 245. *A. convexum* Newm. Pinnules long, narrow, with deflexed edges, enclosing the clusters; their attachment broad; segments gradually decreasing from the base of pinnule.—*β. A. Filix-fœmina* (Roth); frond bipinnate, pinnules pinnatifid oblong-lanceolate bluntish flat remote not connected, segments oblong patent with lateral and terminal sharp teeth. *N.* 237. Frond much divided dark green. Pinnules much narrowed at their base. Clusters distinct. *A. latifolium* (Bab.) is apparently only an extreme state of this.—*γ. A. molle* (Roth); frond pinnate, pinnules toothed oblong blunt or slightly pointed flat remote all connected by wing of midrib or lower ones distinct, segments ovate bidentate lowest with 3 teeth uppermost with 1 tooth. *N.* 245. Frond scarcely more than pinnate, bright green. Pinnules attached by a broad decurrent base. Clusters distinct, in 2 rows. Sometimes (*A. trifidum* Roth?) the pinnules are much less connected, moderately cut, have a narrow attachment, and lobes with more but connivent teeth.—There are innumerable subvarieties. —Wet shady places. P. VI. VII. E. S. I.

8. ASPLE'NIUM *Linn.* Spleenwort.

* *Ultimate subdivisions with a distinct midvein.*
ASPLENIUM Newm.

[1. *A. fontánum* (Presl); frond linear-lanceolate bipinnate, pinnæ oblong-ovate, pinnules obovate-cuneate with few spinous-mucronate teeth.—*E. B.* 2024. *H. F.* 34.—Fronds about 4 in. long.—A very doubtful native. Stations want confirmation, to see if it was planted. Formerly on Amersham Church, Bucks! Wybourn, Westm. *Hudson.* Northumberland. *J. Backhouse.* Tany Bwlch and Tremadoc, Merionethshire. Ashford, Hants! P. VI.—IX.] E.

2. *A. lanceolátum* (Huds.); *fronds lanceolate* bipinnate, pinnules obovate deeply and sharply toothed or lobed, clusters *short* nearly marginal.—*E. B.* 240. *H. F.* 32. *N.* 249.—Fronds sometimes nearly linear and simply pinnate, always narrowed at the base. Clusters oblong, ultimately rather confluent into roundish masses.—Rocks and walls, rather rare. P. VI.—IX. E. I.

3. *A. Adiantum-nigrum* (L.); *fronds ovate triangular* or triangular-prolonged twice or thrice pinnate, pinnæ and pinnules triangular sharply toothed, *clusters long* central.—*E. B.* 1950. *N.* 225.—Clusters 2 or 3 times as long as in the preceding, placed near the midrib and ultimately confluent in oblong masses often covering the whole under surface of the pinnule.—*a*; fronds about as long as the stipe ovate-triangular, pinnæ and pinnules triangular-prolonged, ultimate subdivisions blunt.—*β*. *A. Serpentini* (Tausch); pinnules triangular very broad, lobes ovate blunt.—*γ*. *A. acutum* (Bory); fronds much shorter than the stipe triangular-prolonged, pinnæ and pinnules lanceolate-attenuate, ultimate subdivisions very acute. *N.* 231.—Rocks and walls. *β*. "Serpentine rocks of Cabrach, Aberdeenshire." *T. Moore.* λ. South-west of Ireland. P. VI.—IX. *Black Spleenwort.* E. S I.

4. *A. Trichom'anes* (L.); frond linear pinnate, pinnæ roundish-ovate crenate, *veins forked below the clusters.*—*E. B.* 576. *H. F.* 29. *N.* 285.—*Rachis black,* keeled beneath. Pinnæ scarcely oblique; both edges rounded and crenate except at the base, upper often bluntly auricled below.—*β*. *A. anceps* (Sol.?); pinnæ oblong blunt wedgeshaped below upper edge and end crenate-dentate lower entire. Pinnæ oblique, lower edge nearly straight; lower pinna much the smaller.—A curious variety is occasionally found with its pinnæ deeply but irregularly pinnatifid with linear notched segments. [*A. Clermontæ* (Sy.) is apparently a hybrid between this and Sp. 7.]—Rocks and walls. *β*. Killarney. P. V.—X. *Common Spleenwort.* E. S. I.

5. *A. vir'ide* (Huds.); fronds linear pinnate, pinnæ roundish-ovate or rhomboidal crenate, veins simple or *forked beyond the sori.*—*E. B.* 2257. *H. F.* 30.—*Rachis green,* not keeled. Sori at length confluent.—Rocks and mountains. P. VI.—X. E. S. I.

6. *A. marinum* (L.); fronds linear simply pinnate, pinnæ stalked ovate or oblong serrate unequal and wedgeshaped at the base.—*E. B.* 392. *H. F.* 31. *N.* 275.—Varying greatly in size. Sori not confluent.—Maritime rocks. P. VI.—X. E. S. I.

** *Ultimate subdivisions without a distinct midvein.* AMESIUM Newm.

7. *A. Ruta-muraria* (L.); fronds bipinnate *pinnules rhomboid*-wedgeshaped notched or toothed at the end, *indusium jagged.*—*E. B.* 150. *H. F.* 28. *N.* 261.—Fronds 3—4 in. long. [β. *pseudo-germanicum* (Milde); fronds usually little more than pinnate, pinnæ longer narrowly wedgeshaped.]—Rocks and old walls. P. V.—IX. *Wall-Rue.* E. S. I.

8. *A. german'icum* (Weiss); fronds simply and alternately pinnate, pinnæ narrow wedgeshaped, blunt the lowermost ternate, *indusium entire* at the edge.—*H. F.* 27. *N.* 265. *A. alternifolium* Sm., *E. B.* 2258.—Fronds 3—4 in. long.—Rocks, very rare. P. VI.—IX. E. S.

9. *A. septentrionále* (Hull); fronds 2- or 3-cleft, pinnæ very long-lanceolate bifid, indusium entire.—*E. B.* 1017. *H. F.* 26. *N.* 269.—Pinnæ very narrow, narrowing gradually downwards, with 1 or 2 short bifid lateral teeth, and bifid at the end.—Dry clefts of rocks, rare. P. VI.—X. E. S.

9. PHYLLITIS *Hill.* Hart's-tongue.

1. *P. Scolopen'drium* (Greene); frond oblong strapshaped smooth simple with a cordate base, stipe shaggy.—*E. B.* 1150. *H. F.* 37. *N.* 289. *Scolopendrium vulgare* (Sym.) ed. viii.—Fronds 1—2 feet long, acute, often crisped and multifid.—Damp shady places. P. VII. VIII. E. S. I.

10. CETE'RACH [1] *Willd.* Rustyback.

1. *C. officinárum* (Willd.); fronds pinnatifid covered beneath with dense scales, pinnæ alternate or opposite blunt [more or less crenate] sessile.—*Notolepum N.* 293. *Asplenium* L., *H. F.* 36. *E. B.* 1244.—Fronds 3—6 in. long, green and smooth above,

---

[1] Adanson was apparently the first author, after 1753, to adopt this name, but we think he intended it to include the greater part of *Asplenium*, L.—H. & J. G.

wholly covered by very many scales beneath, amongst which the capsules are almost hidden.—Old walls and rocks. P. IV. —X. E. S. I.

11. Gymnogram'me *Desv.*

[*G. leptophyl'la* (Desv.); fronds pinnate or bipinnate glabrous, pinnules wedgeshaped deeply lobed.—*H. F.* 1. *N.* ed. 3. 11. *S.* 48.—Fronds 1—4 in. high. Stipe purple. Pinnæ and pinnules alternate.—Banks. Jersey. [Guernsey.] A. IV.]

Tribe IV. *Adianteæ.*

12. Blech'num *Linn.* Hard Fern.

1. *B. Spicant* (With.); barren fronds pinnatifid with broadly-linear rather blunt pinnæ, fertile frond pinnate with linear acute pinnæ.—*E. B.* 1159. *H. F.* 40. *B. boreale* (Sw.) ed. viii. *Lomaria Spicant* (Desv.). *N.* 89.—Each lateral vein of the fertile pinnæ extends halfway to the edge, than turns at right angles and proceeds up the pinna until it reaches the next vein. Capsules attached in a continuous row to the longitudinal portions of the combined lateral veins.—Stony and heathy places. P. VII. E. S. I.

13. Pte'ris *Linn.* Brakes or Bracken.

1. *P. aquilina* (L.); fronds tripartite, branches bipinnate, pinnules linear-lanceolate the lower ones usually pinnatifid, segments oblong blunt.—*E. B.* 1679. *H. F.* 38. *N.* 93.—Fronds annual, 1—5 feet high, very much divided, with spreading branches. Capsules attached to the marginal vein, lying upon a fine membrane and covered by the membranous continuation of the epidermis. Inferior pinnules pinnatifid or sinuate or entire.—Woods and heaths. P. VII. E. S. I.

14. Adian'tum *Linn.* Maiden-hair.

1. *A. Capillus-Ven'eris* (L.); frond irregular, branches and roundish-wedgeshaped lobed thin pinnules alternate, lobes of the fertile pinnules terminated by a transversely linear-oblong reflexed lobe covering several roundish clusters, sterile lobes serrate.—*E. B.* 1564. *H. F.* 41. *N.* 83.—Rhizome blackish, shaggy. Fronds 6—12 in. high. Stipe and rachis slender, nearly black. Pinnules not jointed to the partial stalks.—Damp rocks near the sea in the south and west. P. V.—IX. E. I.

Tribe V. *Hymenophylleæ.*

15. Trichom'anes *Linn.*

1. *T. radícans* (Sw.); fronds 3 or 4 times pinnatifid glabrous, segments uniform linear, involucres solitary in the axils of the upper segments, receptacle at first included ultimately very prominent.—*E. B.* 1417. *T. speciosum* Willd., *N.* 305.—Fronds rather triangular, very much divided, 4—8 in. long, formed of hard wiry branched ribs each with a rather membranous wing. Rhizome black, downy, very long. Involucres scarcely winged. —The form called *Andrewsii* has lanceolate fronds and winged involucres. *N.* 315.—Very damp shady places, rare. Formerly at Bellbank, Yorkshire! (exactly Bolton's t. 30). N. Wales! Arran! (S.). Killarney! [and other parts of Ireland]. P. IX. X. E. S. I.

16. Hymenophyl'lum *Sm.* Filmy-Fern.

1. *H. tunbridgen'se* (Sm.); fronds pinnate, pinnæ distichous, segments linear undivided or bifid spinously-serrate, *involucre compressed spinously-serrate*, rachis broadly winged.—*E. B.* 162. *H. F.* 43. *N.* 321.—Slender, delicate and small. Rhizome very long, threadshaped. Pinnæ, rachis, and involucres in the same plane. Inv.-valves adpressed throughout the greater part of their length, slightly gibbous at the base.—Amongst moss in damp and shady places. P. VII. E. S. I.

2. *H. Wilsoni* (Hook.); fronds pinnate, pinnæ recurved, segments linear undivided or bifid spinously-serrate, *involucre inflated entire,* rachis slightly bordered.—*E. B. S.* 2686. *H. F.* 44. *H. unilaterale* Bory?—Resembling the preceding, but the pinnæ curve backwards and the involucres forwards. Inv.-valves convex or gibbous throughout, touching only by their edges which are quite entire.—Amongst moss in damp and shady places. P. VII. E. S. I.

Suborder II. *Osmundaceæ.* Tr. VI. *Osmundeæ.*

17. Osmun'da *Linn.* Flowering-Fern.

1. *O. regális* (L.); fronds bipinnate, pinnules oblong nearly entire dilated and slightly auricled at the base, clusters panicled terminal.—*E. B.* 209. *H. F.* 45. *N.* 331.—Fronds erect or drooping, 1—8 feet high.—Boggy places. P. VII.—IX. E. S. I.

Subord. III. *Ophioglossaceæ.* Tr. VII. *Ophioglosseæ.*

18. Botrych'ium *Sw.* Moon-wort.

1. *B. Lunária* (Sw.)[1]; frond pinnate solitary, pinnæ lunate or fanshaped notched or crenate.—*E. B.* 318. *H. F.* 48. *N.* 137. —Height 3—6 in. Pinnæ with veins radiating from the petiole, sometimes deeply notched. Fronds usually solitary, but sometimes two on the same stalk.—*β. B. rutaceum* (Sw.); frond triangular-rhomboidal pinnatifid, pinnæ 7—9 linear incise-serrate decreasing upwards. *N.* ed. 3. 332. See *Milde Fil. Europ.* 195. —Pastures. *β.* Sands of Barry, Forfar. P. V.—VII. E. S. I.

19. Ophioglos'sum *Linn.* Adder's-tongue.

1. *O. vulgátum* (L.); frond ovate blunt, epidermal cells flexuose, spores tubercled.—*E. B.* 108. *H. F.* 46. *N.* 349. —Height 4—12 in., erect. Spike clubshaped, usually rather longer than the frond, sometimes very long.—*β. polyphyllum* (Braun); rhizome often producing two fronds from the same joining, frond oblong-lanceolate narrowed below 2—4 in. high, var. *ambiguum* (C. & G.) ed. viii.—Pastures. *β.* [Merioneth. Shetland. St. Kilda.] Orkney. Scilly. P. V. VI. E. S. I.

2. *O. lusitan'icum* (L.); frond linear-lanceolate, epidermal cells straight, spores smooth.—*H. F.* 47. *N.* ed. 3. 331. *S.* 47.— Height 1—2 inches, erect.—Horn Head, Co. Donegal. Petit Bo Bay, Guernsey. P. I. I.

---

[1] *B. matricæfolium* (Braun) with ovate bipinnatifid sterile fronds having distinct primary veins to the segments, and broadly spreading fertile fronds, is recorded from Stevenston, Ayrsh. (Dr. O. St. Brody), by Mr. Whitwell in *J. of B.* xxxvi. (1898) p. 291, t. 388 B, but some uncertainty attaching to the specimen, the record needs confirmation. Mr. Whitwell follows Moore and Boswell in referring the Sands of Barry plant to *B. lanceolatum* (Angst.).—H. & J. G.

Order CI. MARSILEACEÆ.

Creeping plants with alternate erect leaves, circinate in bud. Fructification consisting of globular coriaceous axillary bodies with 3 or 4 cells and containing sacs including either other bodies that germinate or loose granules.

1. PILULARIA. Capsules solitary, nearly sessile, globose, coriaceous, 2—4-celled. Cells containing bodies of two kinds—granules, and membranes containing minute grains.

1. PILULA'RIA *Linn.* Pillwort.

1. *P. globulif'era* (L.); ped. erect, caps. 4-celled.—*E. B.* 521. *H. F.* 57. *N.* 393.—Rhizome slender, creeping, producing leaves and roots at regular intervals. L. very slender, erect. Caps. pubescent, slightly stalked, axillary, nearly spherical, hairy.—Margins of ponds and lakes. P. VI. VII. E. S. I.

Order CII. LYCOPODIACEÆ.

Leafy plants with simple imbricate leaves, or stemless with erect subulate leaves. Fructification of axillary sessile capsules with 2 or 3 valves and no ring, including minute powdery matter or spores.

* *Capsules not opening.—Leaves radical.*

1. ISOËTES. Caps. in pouches formed of the swollen bases of the leaves. Spores of two kinds attached to filiform receptacles; those of the outer leaves large, of the inner very small.

** *Capsules bursting.—With leafy stems.*

2. LYCOPODIUM. Caps. of one kind, 1-celled, containing many minute spores.

3. SELAGINELLA. Caps. of two kinds, small containing many minute spores, or larger and containing about 4 large spores.

1. Isoëtes *Linn.* Quillwort.

* No persistent leaf-bases. Caps. not wholly covered by the membranous edge of the pouch. Back of swollen leaf-base smooth.—Aquatic.

1. *I. lacus'tris* (L.); l. subulate roundish-quadrangular with 4 longitudinal jointed tubes upright dark green, *larger spores bluntly tubercled rather mealy*, tubercles overtopped by the valve-edges.—*E. B.* 1084. *H. F.* 55. *R.* vii. 1.—Corm with longitudinal furrows. L. slender, broad and flat at the base, but elsewhere between cylindrical and quadrangular, 2—6 in. long.—*I. Morei* (Moore) is a form with exceedingly long leaves. *J. of B.* xvi. t. 199.—Usually cæspitose on the sandy and stony bottom of lakes and pools in hilly districts. P. VI. E. S. I.

2. *I. echinos'pora* (Dur. !); l. subulate roundish-quadrangular with 4 longitudinal jointed tubes patent pale green, *larger spores very acutely tubercled*, tubercles overtopping the valve-edges.—*J. of B.* i. t. 1.—Much like Sp. 1. Corm not furrowed. L. turning yellow, less rigid, flattened and dilated below. Spores covered with long acute spine-like tubercles, not mealy.—Usually solitary on the muddy bottom of pools and lakes in hilly districts. Llanberis, N. Wales. Loch of Park near Aberdeen. Ben Voirlich, Dumbartonshire. [Near Tongue, Sutherl.] Gap of Dunloe, Kerry. Galway. Mayo.] P. VI. E. S. I.

** Corm more or less covered by the persistent hardened leaf-bases. Caps. wholly covered by the membranous edge of the pouch. Swollen leaf-base with a central longitudinal rugose band on the back.—Terrestrial.

[*I. Hys'trix* (Dur.); l. filiform plane-convex obscurely tubular, persistent l.-bases short blackish each with 2 long horns and an interm. tooth, larger spores white and bluntly tubercled.—*I. Duriæi* H. F. 56.—Corm small, rarely if ever quite naked. L. very slender, ultimately very much enlarged at the base to enclose the capsule, 1—2 in. long, annual.—Dampish sandy and stony places. L'Ancresse, Guernsey. *Mr. G. Wolsey.* [Alderney. *Mr. E. D. Marquand.*] P. V. VI.]

2. Lycopo'dium *Linn.* Club-moss.

* *St. creeping, prostrate. Caps in term. spikes, with bractlike leaves.* Lepidotis Palis.

1. *L. clavátum* (L.); l. scattered imbricate incurved hair-pointed, *spikes stalked* 2 or 3 together cylindrical, scales ovate-

triangular membranous finely incise-serrate.—*E. B.* 224. *H. F.* 49. *N.* 353.—St. long. Branches short, ascending. Spikes on long stalks, pale yellow. Scales on the ped. irregularly disposed in whorls.—Heaths. P. VII. VIII. *Common Club-moss.* E. S. I.

2. *L. annot'inum* (L.); l. loosely scattered lanceolate mucronate serrulate, *spikes sessile* solitary terminal, scales roundish shortly acuminate membranous and jagged.—*E. B.* 1727. *N.* 361.—St. very long. Branches rather long, erect, each year's growth marked by a constriction. Spikes cylindrical, greenish yellow, not persistent.—Stony mountains. Very rare in Caernarvonshire and Cumberland. Common in the Highlands of Scotland. P. VIII. E. S.

3. *L. alpínum* (L.); *l. in four rows,* imbricate acute keeled entire, spikes sessile solitary terminal, scales ovate-lanceolate flat, branches erect clustered forked level-topped.—*E. B.* 234. *H. F.* 53. *N.* 365.—St. long prostrate; also a subterranean rhizome. Fertile branches usually twice dichotomous, each division ending in a short cylindrical yellowish-green spike rather thicker than the branch. [Large forms with flattened br. have been mistaken for *L. complanatum* (L.), which may be distinguished by its conspicuously stalked spikes, which are usually not solitary, and its more linear l.]—Stony moors. P. VIII. *Savin-leaved Club-moss.* E. S. I.

4. *L. inundátum* (L.); l. secund linear subulate, spikes terminal sessile leafy solitary upon short erect branches.—*E. B.* 239. *H. F.* 51. *N.* 369.—St. short, prostrate, rooting. Branches few, simple, fertile. *Bracts subulate from* a dilated base.—Boggy heaths. P. VIII. IX. E. S. I.

** *St. decumbent below, then erect. Caps. in axils of upper leaves solitary. No spikes. L. all alike.* PLANANTHUS Palis.

5. *L. Selágo* (L.); *l. in eight rows crowded* uniform linear-lanceolate acuminate, caps. not spiked but in the axils of the common leaves, st. erect forked level-topped.—*E. B.* 233. *H. F.* 54. *N.* 375.—St. short, erect or slightly decumbent, densely leafy. No separate spikes. At the extremity of the stem a few curious viviparous buds may usually be found; they are well illustrated by Mr. Newman (p. 378). Occasionally the stems in sheltered situations become much lengthened.—Heaths, chiefly on mountains. P. VI.—VIII. *Fir Club-moss.* E. S. I.

3. SELAGINEL'LA *Spring.*

1. *S. spinulo'sa* (A.Br.); l. uniform scattered lanceolate ciliate, spikes terminal solitary sessile leafy upon short erect branches. —*E. B.* 1148. *N.* 371.—St. prostrate, much branched, rooting, slender. Flowering branches simple, short, erect. Small spores muricate in 2-valved reniform caps.; large spores papillose in 3—4-valved capsules.—Boggy spots chiefly in mountainous districts. P. VIII. E. S. I.

B. Stems of one or more parallel tubes, verticillately branched. Nucules and globules on the branchlets.

Order CIII. CHARACEÆ.

Leafless branched plants with stems formed of one or more parallel tubes. Two kinds of fructification; round red globules [antheridia] formed of 8 valves, enclosing cells containing granular matter and spiral filaments; oval nucules [oogonia] formed of 1 cell with 5 filaments folded spirally round it and containing minute granules which appear at last to unite into a single seed.—Plants aquatic. The position of this order is very doubtful[1].

1. NITELLA. Globules and nucules at the forking of the branchlets. Crown of nucule of 10 cells in 2 rows lying upon each other, the upper smaller, usually deciduous.—No stipulodes. Stem of one tube.

2. CHARA. Globule taking the place of central bract, below the nucule [except sp. 2]. Crown of nucule of 5 equal cells in one row, persistent.—Stipulodes in 1 or 2 whorls. Usually with cortical cells.

1. NITEL'LA *Agardh.*

* *Globules terminal in the forking of the branchlets. Nucules below the globules. Segments nearly equal.*—NITELLA Braun.

† Branchlets only once divided into 1-celled segments.

1. *N. flex'ilis* (Ag.); monœcious, st. slender flexible transparent, branchlets pointed not mucronate, fertile whorls usually

[1] See Messrs. Groves' valuable paper in the *J. of B.* xviii. and xix., where are figures of nearly all the species.

lax, nucules 2 or 3 together with 8 or 9 spires.—*Atl. Fl. Par.* 40.—Usually slender, flexible, light green, often [annularly] incrusted. —[β. *crassa* (Braun); st. and branchl. stouter, end-segments shorter. γ. *nidifica* (Wallm.); fertile branchl. very short forming compact heads, sterile branchl. often simple.]—Ponds. VI. VII. E.

2. *N. opáca* (Ag.); diœcious, st. slender flexible transparent, branchlets mucronate, fertile whorls usually dense, *nucules* 1—3 together *naked with* 6 *blunt prominent spires. J. of B.* xviii. t. 210.—Rather slender, turning nearly black. Whorls with nucules dense, with globules usually more lax. [β. *attenuata* (H. & J. G.); branchlets longer and much more slender, fertile whorls lax.] —Ponds and ditches. A. V. VI. E. S. I.

3. *N. capitáta* (Ag.); diœcious, st. slender flexible transparent, branchlets mucronate, fertile whorls usually dense, *nucules with acute prominent spires and a mucilaginous coating.*—Slender, yellowish brown. Globules large. Often annularly incrusted.—Cambs. *J. of B.* xxiv. (1886) p. 1, t. 264. A. E.

[*N. syncar'pa* (Chevall.); more slender than *N. opaca* with simple branchlets on the female plant, and *nucules with a mucilaginous coating* and *faintly marked spires*; is probably a native[1].]

†† Branchlets repeatedly divided; terminal segments of 2 or more cells, last usually like a mucro.—Monœcious.

4. *N. translúcens* (Ag.); st. thick equal flexible transparent, last segments of branchlets of 2 or 3 cells, sterile branchlets only once divided (at the end) into 2—4 *minute rays*, nucules small usually in threes with 7 or 8 spires.—*E. B.* 1851.—A very large plant. Sterile branchlets appear simple; the rays resemble minute points at their end.—Deep water. A. VIII. E. S. I.

5. *N. mucronáta* (Miquel); st. equal flexible transparent, branchlets nearly equally forked or trifid *tipped with a minute acutely conical cell, nucules and globules together* at the forkings of the branchlets.—*Atl. Fl. Par.* 40.—Rather thick. Dark green. Secondary rays once or twice forked, terminal subdivisions rather shorter than the others. Nucules with 7 or 8 spires.—Rivers and ditches, rare. A. VII. E. I.

6. *N. grac'ilis* (Ag.); st. equal flexible transparent, slender *branchlets in lax whorls* twice or thrice nearly equally forked

---

[1] We have at present no evidence in support of this; though from its distribution the sp. is likely to occur.—H. & J. G.

their *terminal segments 2- or 3-celled* strongly mucronate, nucules and globules each solitary but together at the forkings of the branchlets.—*E. B.* 2140. *Atl. Fl. Par.* 41.—Very small and slender. Nucules subglobose with 6 or 7 spires.—St. Leonard's Forest, Suss. Salop. Glen Cullen, Co. Dublin. Wicklow. A. IX. E. I.

7. *N. tenuis'sima* (Kütz.); st. equal flexible transparent, *branchlets in very dense compact subglobose whorls* twice or thrice equally forked in 3—6 segments of 2 or 3 cells their terminal cell very slender and acute, nucules and globules each solitary but together at the forkings of the branchlets.—*Atl. Fl. Par.* 41. —Very small and slender with long internodes, 1—4 [—12] in. high, usually dark green. Branchlets *forming little globose compact masses,* often much incrusted. Nucules globose with 9 spires, minute bnt much exceeding the diameter of the branchlets; globules much larger.—Peaty ditches in Fens of Cambridgeshire, Norfolk, and Anglesea. Westmeath. Galway. A. VIII. E. I.

8. *N. confervácea* (Braun); st. very slender equal flexible transparent, branchlets 8 in loose subglobose whorls, twice or rarely 3 times forked in 3—5 segments, ultimate segm. very acute, globules and nucules each solitary but *together at the first forking of the branchlets* with 7 spires, nucules ovate with prominent spires, *surface minutely and irregularly warty.*—*J. of B.* xxviii. (1890) p. 65, t. 296. *N. batrachosperma* (Braun, not Ag.). *N. Nordstedtiana* (H. & J. G.).—Very small. Branchlets less compound and not so slender as in sp. 7, and the internodes shorter. Obbe, I. of Harris. Killarney and Caragh Lakes. VII. VIII. S. I.

9. *N. hyalina* (Ag.); st. flexible transparent, whorls loose subglobose of 8 *primary and about double as many smaller secondary branchlets,* primary 2—3 times forked with 7—10 rays at the first forking, secondary 1—2ce divided or a few simple, nucules ovoid showing 9—10 spires. —*J. of B.* xxxvi. (1898) p. 409, t. 392.—About 6—18 in. high, dark green. At once distinguished from all the other sp. by the presence of secondary branchlets.—The Loe, West Cornwall. *Rev. G. R. Bullock Webster.* VIII. E.

** *Glolules lateral on the forkings of the branchlets, amongst the nucules. Segments unequal. Monœcious.*—TOLYPELLA Leonh.

10. *N. glomeráta* (Chevall.); st. stoutish, sterile *branchlets* 6—12 *simple blunt* 3—5-jointed, fertile branchlets many densely crowded once unequally 3—4-forked with *blunt* 2—5-jointed divisions, globules usually stalked with one or more nucules of 8 or 9 faint spires.—*Atl. Fl. Par.* 41.—Much incrusted and brittle.—Brackish ponds and ditches chiefly. A. IV. E. I.

11. *N. prolif'era* (Kütz.); st. stout, sterile *branchlets* 6—20 *simple acute* 3—5-jointed, fertile branchlets many densely crowded divided into 3—4 acute 2—4-jointed divisions, globules sessile (?) with one or more nucules of 9 or 10 spires.—A very large plant. Branchlets even 7 inches long and very stout.—Canals and ditches, rare. E. I.

12. *N. intricáta* (A. Br.!); st. stout, sterile *branchlets* 6—10 *once or twice divided acute*, fertile branchlets many densely crowded divided once or twice into 4—5-celled acute divisions, globules stalked with one or more nucules of 8—9 spires.—Known by its divided sterile branchlets.—Ponds and ditches. A. IV. V. E. I.

2. Cha'ra *Linn.*

* Stem composed of a single tube.

A. *Stipulodes rudimentary, in one ring. Diœcious.*

1. *C. obtúsa* (Desv.); st. thick equal transparent its lower subterranean branches bearing *white starlike knobs*, branchlets 4—6 of 2 or 3 long joints bearing 1 or 2 unequal 1-jointed bracts, stipulodes exceedingly small, nucules with 9 spires subglobose.—*J. of B.* xix. t. 216. *C. stelligera* (Bauer) ed. viii- *Lychnothamnus stelliger* (Braun).—Very large, resembling *N. translucens.* Nucules with very thick spires and very small crown.—Filby and other Broads, Norf. Devon. Hants. Surrey. VIII. E.

B. *Stipulodes long, in one ring. Monœcious.*

[† *Globules above the nucules.*—Lamprothamnus.]

2. *C. alopecuroïdes* (Del.); st. [sometimes] opaque, 3—5 jointed branchlets 6—8 in a whorl their terminal joints much the shortest and forming an acute point, stipulodes [opposite the branchl.] needleshaped, bracts 5 or 6 at each node equal exceeding the oval nucule of 10—12 spires.—*J. of B.* i. t. 7.—Dark green, 4—8 in. high. Not incrusted.—In brine-pits at Newtown, Isle of Wight. Dorset. A. VIII. E.

†† [Nucules above the globules.]

[*C. Braun'ii* (Gmel.); st. much branched, 4—5-jointed branchlets 6—8 in a whorl, stipulodes alternating with the branchlets, bracts 5—7 usually shorter than the ovate nucules.—*J. of B.* xxii. (1884) p. 1, t. 242. —Often annularly incrusted, 6—12 in. high.—Canal, Reddish, Lanc., doubtless introduced.] E.

** Stem of one central and many cortical tubes (cells).

† *St. with one row of cortical cells to each branchlet. Diœcious.*

3. *C. canes'cens* (Loisel.); st. slender coarsely striate *densely beset* with setaceous patent *clustered spines*, branchlets 8—10 short, stipulodes long, bracts 7—10 whorled slender equal usually exceeding the *narrowly oblong nucules* of 10—12 spires.—*J. of B.* xviii. t. 208. *C. crinita* (Wallr.) ed. viii.—St. rigid, but little branched, rarely incrusted, pale green. Branchlets stout, 6—8-jointed.—Pools near the sea, rare. E. I.

†† *St. with 2 rows of cortical cells to each branchlet.*

a. Monœcious.

4. *C. vulgáris* (L.); st. finely striate, primary cortical cells less prominent than secondary and bearing the few small spines, branchlets 6—9 their upper part without cortical cells, stipulodes small blunt, bracts on inner side of branchlet exceeding the ovoid nucule of 12 or 13 spires.—*E. B.* 336. *C. fœtida* (Braun) ed. viii.—Primary cortical cells collapse as they dry and so place the spines in furrows. Very variable; (*C. longibracteata* Kütz.) branchlets and bracts much longer; (*C. decipiens* Desv.) [var. *papillata* (Wallr.)] spines many spreading long deciduous, secondary cortical cells very prominent; (*C. subverticillata* Nordst. MS.) branchlets spreading with long upper naked joints, not incrusted and few spines, dark green (hardly *C. atrovirens* Lowe); (*C. crassicaulis* Schl.) stout, branchlets stout connivent often ½ naked, bracts short ovate, spines obsolete. —[Var. *melanopyrena* (H. & J. G.); nucleus black instead of brown as in other forms.]—Universal. V. VI. E. S. I.

5. *C. contrária* (A. Br.); st. finely striate, *primary cortical cells more prominent* than the secondary and bearing the spines or tubercles; branchlets 6—9 coated nearly throughout, stipulodes blunt [or acute], bracts 4 (?) scarcely exceeding the nucules of 12 or 13 spires and a conical contracted [or spreading] crown. —[*J. of B.* xix. (1881), t. 224, 2.]—Much like *C. fœtida*. Spines more acute, ultimately on ridges.—[*β. hispidula* (Braun); spine-cells prominent spreading.—*γ.* C. *denudata* (Braun); st. weak and as well as the branchl. almost entirely ecorticate. *J. of B.* xxxiii. (1895), p. 290, t. 350.—A hybrid with sp. 6 occurs in the Norfolk Broads.]—Lakes, pools, and ditches. [*γ*. Westmeath.] VI.—VIII. E. S. I.

6. *C. his'pida* (L.); st. stout thickened upwards rough spirally striate, *secondary cortical cells* larger and more promi-

nent than the primary, many setaceous spreading spines, branchlets 7—9-jointed, stipulodes prominent, bracts on both sides of the branchlet 8—10 exceeding the ovoid nucules of 10 or 11 spires.—*E. B.* 463.—Usually very large, much incrusted, often 3 or 4 feet long. Sterile branchlets 3—4 inches long. Spines variable in quantity and length.—[β. *macracantha* (Braun); spine- and bract-cells very long.—γ. *gymnoteles* (Braun); spine-cells few, branchl. with several ecorticate joints.—δ. *C. horrida* (Wahlst.); small and compact, spine-cells short patent very numerous, branchl. straight spreading.—Subsp. *rudis* (Braun); st. more slender secondary cortical cells very strongly developed almost hiding the primary (spine-bearing) series.]—Ponds and ditches. [δ. Doubtfully British.] VI. E. S. I.

7. *C. polyacan'tha* (Braun); st. stout, *primary cortical cells more prominent* than secondary, many setaceous fascicled spreading spines, 8—10 branchlets of 6—8 joints, stipulodes long slender, bracts 6—10 exceeding the ovoid nucules of 12 or 13 spires.—*J. of B.* xviii. t. 208. *Atl. Fl. Par.* t. 38. B. f. 3.—Very spinous, much incrusted; smaller, more spinous, with shorter branchlets but longer internodes than *C. hispida.*—Ponds and ditches. VI. E. S. I.

8. *C. bal'tica* (Fr.); st. stout finely striate flexible primary cortical cells more prominent than secondary, spines many slender spreading, branchlets 6—9 patent many-jointed, stipulodes slender, bracts whorled long exceeding oblong nucules of 10—15 spires and spreading crown.—[*J. of B.* xix. (1881), t. 224, 1.]—Pale green, *not incrusted. Moderately spinous.* Upper part of branches not coated. Bracts very long. Lowest nodes with bulbils.—Ponds. Kynance Vale, Cornwall. [Dorset. Orkney. Guernsey.] VIII. E.

b. Diœcious.

9. *C. tomentósa* (L.); st. thickened upwards twisted, primary cortical cells very prominent, spines few short stout scattered, branchlets 5—7 rather incurved of 4—6 long joints the upper joints not coated but inflated, *stipulodes usually short and thick,* bracts usually 5 very thick the lateral exceeding the ovoid nucules of 12—14 spires the others falling short of them.—*Hook. Icon. Pl.* 532.—Opaque, greenish white when fresh, usually much incrusted. Spines rather whorled, very short, acute. Upper joints of branchlets remarkably inflated and lengthened in our plant.—Shannon Lakes. I.

††† *St. with 3 rows of cortical cells to each branchlet.*

10. *C. as'pera* (W.); *diœcious,* st. slender flexible with *slender acute spreading spines,* branchlets 6—9 rather incurved or straight,

stipulodes prominent, bracts whorled nearly equal exceeding the ovoid nucules of 13 spires and *spreading crown.*—*E. B. S.* 2738. —Small, in dense masses. Sometimes with smooth bulbils on lower nodes. Length and number of spines variable. Internodes very long.—[β. *capillata* (Braun); spine-cells many very long, plant usually bright clear green.—γ. *subinermis* (Kuetz.); spine-cells very few and inconspicuous, much shorter than in the type.—δ. *lacustris* (H. & J. G.); much smaller 1—4 in. high, branchl. short stout incurved, spine-cells papillate.—Subsp. *desmacantha* (H. & J. G.); st. usually thicker, cortex imperfectly 3-ranked, spine-cells fascicled instead of solitary.]—Lakes and pools. VII. VIII. E. S. I.

11. *C. connivens* (Braun); *diœcious,* st. slender brittle without spines, branchlets usually 8 strongly incurved, *stipulodes scarcely visible,* bracts much falling short of nucules scarcely apparent except with them, nucules small ovoid with 12—14 spires and *conical crown.*—*J. of B.* xviii. t. 207.—Light green. 12 inches high.—Rare. In south [and east] of England. [Wexford.] VII. VIII. E. I.

12. *C. frag'ilis* (Desv.); *monœcious,* st. slender finely striate *no spines,* branchlets 6—9 of 7—9 joints the upper 1—3 joints not coated, lower stipulodes very short, bracts usually 4 usually about equalling the ovoid nucules of 12—14 spires and rather *conical crown.*—*C. pulchella* E. B. S. 2824.—Slender, green, usually not incrusted. Very variable. [A. Primary cortical cells not more prominent than the secondary.] β. *Hedwigii* (Ag.); stronger and larger with very short bracts. *E. B. S.* 2762.—γ. *fulcrata* (Gant.); bracts shorter than the nucules, stip. obsolete.—[δ. *capillacea* (Thuill.); whole plant very slender, branchlets spreading.—ε. *Sturrockii* (H. & J. G.); cortication irregular, branchlets ecorticate. Possibly a hybrid.—B. Subsp. *delicatula* (Braun); primary cortical cells more prominent than the secondary. ζ. small, branchlets usually connivent.—η. *barbata* (Gant.); bracts and stipulodes much longer.]—θ. *C. verrucosa* (Itz.); small, 2—4 in. high [with conical spine-cells], often with compound bulbils at base. —Slow and stagnant waters. [ε. Monk Myre, E. Perth. *Mr. A. Sturrock.*] VI.—VIII. E. S. I.

13. *C. fragif'era* (Dur.); *diœcious,* st. very finely striate *no spines,* branchlets slender 6—9 of many joints upper joints not coated, stipulodes very small, bracts 3—5 half as long as nucule very short with globule, nucules ovoid of 11—13 spires and short *blunt crown.*—*J. of B.* xv. t. 192.—Often flexible and exceedingly slender, 6—12 inches long, bright green, not incrusted. bearing large *compound bulbils* at its base.—West of Cornwall in pools. VII. VIII. E.

"Quanquam multas observaverim plantas et sedulo quidem, tamen non confido me semper veritatem invenisse."—LINK.

APPENDIX.

CONSPECTUS OF GROUPS AND SPECIES OF

RUBI FRUTICOSI.

REPRINTED FROM THE

HANDBOOK OF BRITISH RUBI

BY

The Rev. WILLIAM MOYLE ROGERS, F.L.S.

By kind permission of the Author.

ABBREVIATIONS (not included on page lii) AND EXPLANATIONS.

acic.	.... acicle.	*prklet.*	.... pricklet.
bas.	.... basal (of leaflet in 5-nate leaf).	*rach.*	.... rachis.
interm.	.... intermediate (of leaflet in 5-nate leaf).	*sp. collect.*	.. species collectiva.
lat.	.... lateral (of lt. in 3-nate leaf).	*spn.*	.... specimen.
prk.	.... prickle.	*st.*	.... stem of first year.
		stkd. gl.	.... stalked gland.
		sty.	.... style.

When the name of a subspecies or variety is followed by that of its author *in brackets*, this implies that it was published by him as that of a species.

When the sign ? follows the name of a plant, it points to the fact that some doubt exists as to the identification.

KEYS TO GROUPS AND SPECIES.

1. GENERAL KEY TO *armature* OF STEM AND PANICLE.

Groups I. to VII. and IX.—St.-prk. equal or subequal, almost or quite confined to the angles. Glandular and acicular development absent from Group I. (except occasionally in the form of sessile and rarely subsessile gl. on st.), very rare in Groups II. and III., in the succeeding groups gradually increasing.

Group VIII.—Very transitional. Armature somewhat irregularly mixed; the st.-prk. being usually rather more scattered and unequal than in Groups I. to VII. and IX., while the glandular and acicular development is mostly weak or uncertain on st. *or* pan., or on both st. and pan.

Groups X. to XII.—St.-prk. distinctly unequal. All arms more mixed and scattered.

Groups XIII. and XIV.—Prk. scattered on the round or bluntly angled st., often weak and less unequal than in Groups X. to XII.

2. Conspectus of Groups and Species.

A. Groups I. to V.—St. tall, glabrous or slightly hairy (sometimes more densely hairy in Group V. Discolores), not always rooting. St.-prk. normally equal and confined to angles.

Group I. Suberecti. Root often soboliferous. St. suberect, very rarely rooting at end (never apparently in *R. fissus* and *R. suberectus*); glabrous or very nearly so. Bas. lts. often subsessile. Pan. racemose or subracemose. Sep. externally olive and subglabrescent, with conspicuous white margin. Normally without bloom stkd. gl. or acic.

Section i. Prk. subulate or conical. L. often 6–7-nate. Fr. dark red. (*R. nessensis* Hall.)

R. fissus Lindl. Prk. many slender subulate scattered. L. plicate hairy beneath. Bas. lts. sessile. Stam. and sty. subequal.

R. suberectus Anders. Prk. few short conical with rather long base, confined to angles, sometimes absent. L. plane pale green subglabrous. Bas. lts. subsessile. Stam. exceeding sty.

Section ii. Prk. compressed long-based. L. very rarely 6–7-nate (except in *R. Rogersii*). Fr. black.

R. Rogersii Linton. L. 5–7-nate, finely evenly serrate, greyish-green. Term. lt. long ovate-acuminate. Sty. and stam. subequal. Fr.-sep. loosely reflexed.

R. sulcatus Vest. L. all 5-nate. Term. lt. oval long pointed; bas. stalked. Pan. elongate. Stam. at first exceeding sty. Fr.-sep. reflexed. Very luxuriant.

R. plicatus Wh. & N. L. plicate. Term. lt. broadly ovate-cordate; bas. subsessile. Pan. subracemose, rather short. Stam. and sty. subequal. Fr.-sep. patent.

Var. *Bertramii* G. Braun. Luxuriant. Term. lt. roundish-ovate long-stalked; bas. shortly stalked. Stam. exceeding sty. Approaches *R. sulcatus*, but smaller with broad lts.

Var. *hemistemon* P. J. Muell. Lts. oval greyish-green, hardly plicate, hairy, with long point. Stam. at first much shorter than sty.

Group II. SUBRHAMNIFOLII. St. suberect or arcuate, often rooting, subglabrous. Bas. lts. shortly stalked. Pan. subracemose or composite. Sep. externally hairy, olive or grey, somewhat white-margined. Without bloom. Normally eglandular.

1. St. suberect, sometimes rooting. Sep. externally olive or greyish.

a. Sep. externally olive with conspicuous white margin (sometimes greyish or wholly grey in *R. affinis*).

(i.) Fr.-sep. patent or subpatent. Pan. subracemose above. Pet. obovate or oval.

R. nitidus Wh. & N. Prk. crowded, mostly slender; straight or falcate on st., usually strongly hooked on pan. Term. lt. small oval acute. L. and cal. shining.

Subsp. *opacus* Focke. Prk. usually few straight long. Term. lt. large cordate-acuminate. L. opaque above, softly hairy or felted beneath.

(ii.) Frt.-sep. reflexed. Pan. with cymose branches. Pet. roundish.

R. affinis Wh. & N. St. very tall. Prk. remarkably long, normally straight. L. thick, wavy at edge, often grey beneath. Term. lt. ovate-cordate-acuminate, gradually attenuate. Sep. reflexed throughout.

Var. *Briggsianus* Rogers. St. lower more arcuate. Term lt. oval acute. Pan. strongly developed, less leafy above Sep. loosely reflexed.

b. Sep. externally greyish, less conspicuously white-margined, loosely reflexed or subpatent in fr.

R. integribasis P. J. Muell.? Prk. declining long based. L. only thinly hairy beneath. Term. lt. obovate or oval cuspidate-acuminate. Pan. lax slender, mostly racemose. Whole plant usually slender.

R. cariensis Genev. Prk. strong patent or nearly so. L. deeply incised ashy-felted or softly hairy beneath. Pan. dense compound cylindrical. Strong. (Sometimes abnormally furnished with a few very short stkd. gl. on bracts and ped.)

2. St. soon arcuate but rarely (if ever) rooting. L. very large. Sep. externally grey (or greyish-olive), mostly patent in fr. Very strong plants, but with pan. normally racemose above. Bracts and stip. sometimes very finely gland-ciliate.

R. holerythros Focke. St. lustrous, arcuate almost from the first. L. greyish, softly hairy beneath. Pedicels long. Pet. stam. and (usually) sty. red. Sep. greyish-olive.

R. latifolius Bab. St. very slightly hairy, soon bending quite low. L. dull green, thinly hairy. Pan. narrow, with densely hairy rach. and conspicuous bracts. Sep. grey.

N.B.—The anomalous *R. orthocladus* A. Ley, though in some respects recalling the plants in Group II., is placed among the *Vestiti* on account of its hairy st. and considerable glandular development.

Group III. RHAMNIFOLII. St. arcuate or arcuate-prostrate, rooting, glabrous or very thinly hairy, often much branched. All mature lts. distinctly stalked, strigose above (except in *R. durescens*). Pan. compound, sometimes subracemose above, often densely prickly. Sep. externally grey or whitish, reflexed in fr. (except in *R. carpinifolius*). Without bloom. In a few instances slightly glandular in pan. and even very rarely in st., but normally eglandular.

1. Term. lt. about thrice the length of its stalk, ovate oval or obovate (roundish-obovate-cordate in *R. imbricatus*).

a. L. chiefly 5-nate-digitate.

R. imbricatus Hort. Prk. usually rather small. Lts. convex, wrinkled, mostly imbricate. Pan. narrow above, usually with long strongly ascending racemose lower branches.

R. carpinifolius Wh. & N. Prk. many, strong. Lts. plicate, soft beneath; term. long oval-acuminate. Pan. pyramidal or subracemose. Fr.-sep. patent. Pet. normally white. Very prickly and usually pale in colour.

R. incurvatus Bab. Prk. triangular based. Lts. concave, with thick soft felt beneath; term. broadly ovate-cordate. Pan. long, usually with short branches. Floral organs all pink.

R. Lindleianus Lees. St. glossy. Lts. wavy-edged, mostly narrowing to their base. Pan. dense cylindrical-truncate with patent branches. Pet. white. Fr. small.

b. L. 5-nate pedate.

R. erythrinus Genev. St. dark purple. Lts. convex, obovate-cuspidate-acuminate, usually ashy-felted beneath. Pan. pyramidal with narrow rounded top. Fr. large.

R. durescens W. R. Linton. Lts. oblong-ovate, with principal teeth patent, glabrous above, thinly hairy beneath. Pan. rather short, with patent branches above and broad but hardly truncate top.

2. Term. lt. about twice the length of its stalk, roundish or broadly oval.

R. rhamnifolius Wh. & N. Lts. finely toothed, white-felted beneath; term. often barely twice the length of its stalk, cuspidate cordate. Pan. rather close. Pet. roundish white.

Subsp. *Bakeri* F. A. Lees. Dwarf. Lts. like those of *R. rhamnifolius* but smaller, green and hardly felted beneath, with longer point. Pan. with many long branches; very floriferous. Pet. obovate, pink.

R. nemoralis P. J. Muell. Lts. green on both sides, slightly paler and thinly hairy beneath. Pan. long lax leafy, at first pyramidal then corymbose. Pet. pink.

Var. *glabratus* Bab. Pan. narrow, cylindrical, with subequal lower branches long-pedicelled small fl. and very white-felted rach. and ped.

Var. *Silurum* A. Ley. Term. lt. broader below than in type (*i. e.* oval-roundish instead of obovate-roundish) and more gradually acuminate. Pan. in great part ultra-axillary, truly pyramidal. Pet. faint lilac.

R. Scheutzii Lindeb. St. glabrous or subglabrous, shining. Lts. concolorous, thinly hairy; term. broadly rotund-cuspidate. Pan. long, narrow, leafy. Pet. large, pale lilac. Very prickly.

R. dumnoniensis Bab. St. with some short crisp hair (at first). Prk. long, subulate, crowded. Lts. large, white-felted beneath. Pan. pyramidal. Pet. large, roundish, white.

3. St. hairy for this group. Lts. finely toothed; term. obovate-acuminate, long-stalked. Pan. very long.

R. pulcherrimus Neum. Prk. patent or declining. L. not unfrequently 6–7-nate. Lts. opaque above; term. from 2½ to more than 3 times the height of its stalk. Pan. normally with pink pet. and some stkd. gl.

R. Lindebergii P. J. Muell. Prk. stout, partly falcate or hooked, especially on pan. L. constantly 5-nate. Lts. pale grey-green; term. about twice as long as its stalk. Pan. eglandular, with large white pet.

Group IV. VILLICAULES. Hardly separable from *Rhamnifolii*, but making some approach towards *Silvatici* in st. mostly lower and usually hairy (though often only thinly so and liable to become bald ultimately). Prk. subequal, rarely extending to the faces of st. Without bloom. Pan. usually lax or diffuse, very rarely showing some glandular development.

1. Pan. diffuse, or irregularly branched.

a. St. bluntly angled, thinly hairy. Fr.-sep. reflexed.

R. mercicus Bagnall. Prk. irregularly scattered, strongly declining. Lts. thinly hairy beneath; term. roundish, more than twice the length of its stalk. Pan. sometimes slightly glandular. Pet. white, fading to pink. St. and rach. dark purple.

N.B.—For subsp. *bracteatus* and *chrysoxylon* see below in Sect. 2.

R. leucandrus Focke. Prk. partly patent. Lts. softly hairy beneath. Pan. eglandular. Fl. showy; floral organs all white. (Between *R. affinis* and *R. gratus.*)

b. St. very stout, furrowed, often glabrate. Fr.-sep. patent or clasping.

R. gratus Focke. Prk. short, broad-based, patent or slightly declining. Lts. at first softly hairy, ultimately nearly bare beneath. Pan. with widespreading few-flowered branches. Pet. pink. Luxuriant.

2. Pan. elongate; cylindrical or pyramidal.

a. Pan. pyramidal. Prk. moderate, declining or falcate.

R. bracteatus Bagnall. Lts. greyish, finely closely toothed, often felted beneath; term. obovate. Pan. remarkably elongate and narrowed above, felted, considerably glandular. Bracts many. (Subsp. of *R. mercicus* Bagnall.)

R. rhombifolius Weihe. Lts. rhomboid or ovate, acuminate, green or white-felted beneath. Lower pan.-branches long, strongly ascending. Floral organs usually all red. Sep. reflexed. (Subsp. of *R. villicaulis* Koehl.)

b. Pan. cylindrical, leafy.

(i) Prk. weak, somewhat scattered and unequal. Fr.-sep. subpatent.

R. chrysoxylon Rogers. St. ochreous or fuscous. L. 3–5-nate, pale, with finely incised compound teeth. Pan. with long-stemmed branches. Stkd. gl. scattered, usually few. (Subsp. of *R. mercicus* Bagnall.)

R. calvatus Blox. St. reddish. Lts. convex, harsh beneath, with patent teeth; term. oblong, with short point and cordate base. Pan. long, lax, often glandular. (Subsp. of *R. villicaulis* Koehl.)

(ii.) Prk. strong, confined to angles. Fr.-sep. reflexed.

R. villicaulis Koehl. St. hairy. Prk. long, straight, many.

Lts. ovate or oval acuminate, softly hairy or felted beneath. Pan. with prk. chiefly long, slender, declining. Stam. far exceeding sty.

Subsp. *Selmeri* Lindeb. St. and l. less hairy. Prk. falcate. Lts. concave, roundish-ovate, with wavy edge. Stam. short, often barely equalling sty.

Group V. Discolores. St. erect-arcuate or arcuate-prostrate, thinly or densely pubescent or hairy or glabrous, sometimes pruinose, rarely rooting. Lts. with whitish felt beneath, which becomes greenish in autumn. Pan. hardly narrowed above (except in *R. argentatus*). Usually quite eglandular.

1. St. erect-arcuate, rarely rooting, glabrous or very nearly so, furrowed.

R. ramosus Briggs. L. 3–5-nate-pedate. Lts. convex, shining above, coarsely serrate; term. oblong or obovate cuspidate. Pan. lax. with long branches. Fr. small, poor.

R. thyrsoideus Wimm. L. 5-nate-digitate. Lts. flat or concave, often incised-serrate; term. oval or ovate, acuminate. Pan. elongate, cylindrical above. Fr. fine.

2. St. arcuate-prostrate or climbing, often rooting, hairy or pubescent.

a. St. somewhat pruinose.

R. argentatus P. J. Muell. Prk. (st. and pan.) long, rather unequal, straight or falcate. L. 5-nate-digitate or pedate. Pan. lax, narrowed above, with showy long-pedicelled bright pink fl. and thick villous hair on rachis.

Var. *robustus* (P. J. Muell.). Very strong and prickly. L. thick, irregularly and deeply cut. Pan. less narrowed above.

Var. *clivicola* A. Ley. Prk. weaker. Term. lt. short, roundish-obovate, with long cuspidate or cuspidate-acuminate point. Pan. with long corymbose-cylindrical ultra-axillary part. Pet. faintly pink.

R. rusticanus Merc. St. furrowed, pruinose. Lts. glabrous above, usually cuspidate. Pan. and ped. with close-pressed felt and very broad-based hooked prk. Sep. reflexed (fl. and fr.).

b. St. epruinose, pubescent.

R. pubescens Weihe. L. 5-nate-digitate. Lts. oval or obovate-acuminate, thinly hairy above, irregularly serrate. Pan. elongate, felted, hairy, with hooked prk. Sep. with upcurved tip.

Var. *subinermis* Rogers. L. 3–5-nate-pedate. Pan. quite unarmed or with few declining prk.

B. Groups VI. and VII.—St. mostly arcuate-prostrate and hairy, rarely furnished with a few acic. or stkd. gl. St.-prk. subequal, occasionally somewhat scattered but mostly confined to angles.

Group VI. SILVATICI. St. arcuate-prostrate or climbing, permanently though thinly hairy, usually quite eglandular. Lts. mostly hairy or greenish-felted beneath. Pan. rather frequently furnished with a few stkd. gl. and acic.

1. Pan.-prk. normally acicular, declining; in some forms of *R. lentiginosus* stronger, falcate.

a. Frt.-sep. reflexed. Stam. considerably exceeding sty.

R. silvaticus Wh. & N. L. 5-nate. Pan. rather dense, elongate, with crowded acicular prk. Eglandular. Carp. hairy.

b. Frt.-sep. patent or erect. Stam. and sty. subequal.

(i.) Pan. almost unarmed; its prk. when present very few, subulate.

R. myricæ Focke. L. mostly 3-nate, evenly simply dentate. Pan. rather lax and little branched. Sep. embracing fr.

(ii.) Pan.-prk. many, mostly acicular, declining; in *R. lentiginosus* occasionally stronger, falcate, with some stalked and subsessile glands.

Var. *hesperius* Rogers. L. broad, mostly 5-nate-pedate, with very compound finely pointed teeth. Pan. elongate, pyramidal, compound. Fr.-sep. patent.

R. lentiginosus Lees. L. 3–5-nate, with very finely pointed compound incised teeth. Pan. racemose-truncate above, with long-pedicelled fl. Fr.-sep. erect.

2. Pan.-prk. at most only moderately strong, chiefly declining.

a. Pan. normally eglandular. Fr.-sep. reflexed.

R. macrophyllus Wh. & N. L. 5-nate-digitate, glabrescent above. Term. lt. ovate-cordate, with long acuminate point. Pan. lax, with grey-felted and hairy rach. medium-sized fl. and strongly reflexed sep.

Subsp. *Schlechtendalii* Weihe. L. hairy above, 5-nate-pedate or digitate. Term. lt. obovate-cuneate-cuspidate, with comparatively short point and crowded fine teeth in upper half. Pan. usually short, with broader upper l. larger fl. and less strongly reflexed sep.

b. Pan. somewhat glandular. Fr.-sep. loosely reflexed or subpatent (in *R. Questierii* usually strongly reflexed).

Var. *macrophylloides* (Genev.). Very near *R. Schlechtendalii*, but with laxer glandular pan. subpatent fr.-sep. and more compound incised leaf-toothing.

Var. *amplificatus* Lees. Also very near *R. Schlechtendalii*, but more prickly, with term. lt. longer pointed and more deeply incised, and pan. usually longer with very long lower branches, narrow simple floral l. above and loosely reflexed fr.-sep.

(*R. amphichloros* P. J. Muell. L. 3–5-nate, finely evenly serrate. Term. lt. roundish-oval. Pan. long, lax. Sep. loosely reflexed. Thought by Dr. Focke to be distinguished from *R. amplificatus* by the shape and fine serration of the lts.)

R. Questierii Lefv. & Muell. L. 5-nate-digitate or pedate, concolorous. Lts. oval or obovate-acuminate. Pan. long, lax, normally narrow, with patent upper branches, and whitish-felted rach. and cal. Fr.-sep. strongly reflexed.

c. Pan. usually having a few sunken glands. Fr.-sep. clasping.

R. Salteri Bab. L. chiefly 5-nate-pedate, with compound incised teeth in upper half. Term. lt. long stalked with long point and entire or somewhat cordate base.

(*R. lentiginosus* Lees. See above in Section 1 of this group.)

3. Pan.-prk. very strong, long-based, often hooked.

R. Colemanni Blox. St. with many strong prk. and occasional acic. and stkd. gl. L. 5-nate-pedate. Lts. convex, concolorous. Pan. long, pyramidal. Fr.-sep. loosely reflexed.

Group VII. Vestiti. St. rather densely or very densely hairy, occasionally with a few stkd. gl. and acic. or prklets on the faces. Prk. mostly slender, less constantly equal and less strictly confined to the angles than in Groups I.–VI. and IX. Pan. usually somewhat glandular and aciculate.

1. Stam. not connivent. Carp. hairy.

R. Sprengelii Weihe. St. roundish, hairy, often slightly glandular. L. mostly 3-nate, concolorous. Pan. with long patent branches, usually glandular.

2. Stam. connivent. Carp. glabrous or nearly so.

a. St. very high arching or suberect. Sep. externally olive, with narrow white margin, embracing fr.

R. orthoclados A. Ley. St. with some subsessile glands. L. 3–5-nate-pedate, concolorous. Pan. weak, glandular. Abnormal.

b. St. arcuate-prostrate, rather densely hairy. Pan. elongate. Sep. ashy-felted.

(i.) Term. lt. usually quite 3 times longer than its stalk.

R. micans Gren. & Godr. St. slightly glandular. L. 3–5-nate-pedate. Lts. acuminate, with incised teeth. Fr.-sep. strongly reflexed. Whole plant greyish.

(ii.) Term. lt. from 2 to 2½ or rarely 3 times the length of its stalk.

R. hirtifolius Muell. & Wirtg. St. rarely glandular or aciculate. L. 5-nate. Lts. with shining close hair or greenish felt beneath; term. variable, usually ovate-acuminate. Pet. narrow, pinkish, fugacious. Fr. oblong, with clasping sep.

Var. *danicus* Focke. Lts. with shallow teeth and shining close hair beneath; term. roundish-obovate. Pan.-prk. long. Fl. very showy, white. Fr.-sep. subpatent.

Var. *mollissimus* (Rogers). Lts. flaccid, with fine incised teeth and very soft greyish felt beneath; term. roundish-oval, rarely obovate. Pan.-prk. mostly acicular. Fl. very showy, pale lilac. Fr.-sep. loosely reflexed or subpatent.

(iii.) Term. lt. often 4 times the length of its stalk. Prk. long-pointed.

R. iricus Rogers. Lts. thick, ashy-felted beneath when young; term. broadly ovate-acuminate, with compound teeth. *Pan. broad, truncate. Fl. bright pink.* Sep. reflexed. *Very stout and hairy.*

R. pyramidalis Kalt. L. 5-nate-digitate. Lts. almost velvety beneath, coarsely toothed. Pan. pyramidal, with short patent branches. Fr.-sep. reflexed.

Var. *macranthelos* Marss. Prk. smaller, passing into strong acic. Term. lt. subrotund, cuspidate. Pan.-branches long. Fr.-sep. patent.

c. St. densely hairy; its prk. long, usually with some sunken prklets acic. or stkd. gl. L. very softly hairy.

(i.) Pan.-prk. long, strong, subulate,

R. leucostachys Schleich. Prk. very strong. Lts. and pet. roundish. Stkd. gl. usually few. Fr.-sep. normally reflexed.

Var. *gymnostachys* (Genev.). Lts. and pet. oval or obovate. Pan. elongate, narrow, with acicular prk. as well as long ones.

(*R. vestitiformis* subsp. nov., and *R. adenanthus* Boul. & Gill.) See Group VIII., *Egregii.*

(ii.) Pan.-prk. mostly weak.

R. leucanthemus P. J. Muell.? St.-prk. many, slender, sometimes acicular. Term. lt. broadly ovate-acuminate, with compound finely pointed teeth. Pan. rather lax, with rigid branches. Pet. white. Fr.-sep. subpatent.

(iii.) Pan.-prk. crowded, partly falcate. Fr.-sep. patent or subpatent.

R. lasioclados Focke. St. and pan.-rach. very densely hairy. L. whitish-felted beneath, with rather irregular incised teeth. Term. lt. oval or roundish. Pan. broad, very prickly. Pet. white.

Var. *angustifolius* Rogers. Lts. remarkably long and narrow, with very shallow teeth; term. oblong-oval, long-stalked. Pet. pink.

(*R. Boræanus* Genev.) See Group VIII., *Egregii*.

C. Groups VIII. to XIII.—St. arcuate-prostrate or prostrate. Armature mixed. Bas. lts. distinctly stalked (though only briefly in some of the *Egregii*).

Group VIII. Egregii. 'A group of transition forms' (as Dr. Focke says of his corresponding *Adenophori*), with armature very variable in quantity and distribution, but usually less conspicuously mixed than in the succeeding groups. St. often nearly prostrate and only rarely very hairy, unequally glandular. Bas. lts. sometimes subsessile. *N.B.*—Many of the species placed in this group could no doubt be attached to other groups without great difficulty; but it seems easier to keep the limits of those groups clearly defined if we retain this as 'a middle and collective group.'

1. Bas. lts. distinctly stalked even in summer. (*Adenophori petiolulati* Focke.)

a. St. densely hairy, glaucous. About $\frac{1}{3}$ of pan. ultra-axillary.

R. criniger Linton. L. chiefly 5-nate-pedate, greyish-green. Term. lt. ovate-acuminate, irregularly lobate-serrate. Pan. narrowed above. (Subsp. of *R. Gelertii* Frider.)

b. St. moderately or rather considerably hairy. From $\frac{1}{3}$ to nearly $\frac{1}{2}$ of pan. ultra-axillary.

(i.) L. with wavy edge and compound teeth. St. with scattered prklets. tubercles and stkd. gl. on faces.

R. adenanthus Boul. & Gill. St.-prk. strong, with a good many prklets and stkd. gl. L. 3–5-nate-pedate, ashy-felted beneath, with compound incised teeth. Pan. considerably glandular. Sep. long, patent or ascending on fall of pet.

R. Boræanus Genev. L. 5-nate-pedate and 3-nate. Pan. cylindrical, very weakly armed. Pet. and sty. dark red. Stam. barely equalling sty.

R. curvidens A. Ley. L. large, 5-nate-digitate. Lts. with nearly parallel sides and patent teeth. Pan. large, lax, with many slender acic. and unequal stkd. gl. and usually several simple l. above. Stam. exceeding sty. (Subsp. of *R. anglosaxonicus.*)

R. vestitiformis subsp. nov. St. glaucous. Lts. roundish, with greyish felt and shining hairs beneath. About ½ pan. ultra-axillary; rach. densely villous. Stam. far exceeding sty. (Subsp. of *R. anglosaxonicus.*)

(ii.) L. with shallow even teeth. St. with few or no stout-based prklets. Acic. and stkd. gl. variable.

(*R. pulcherrimus* Neum.) See Group III., *Rhamnifolii.*

R. cinerosus Rogers. Armature (st. and pan.) variable, always considerably mixed and glandular. Term. lt. roundish-obovate. Pan. rather broadly cylindrical, with patent branches.

R. mucronatus Blox. Prk. very slender. Lts. thin, with simple shallow teeth; term. obovate-truncate-mucronate. Pan. lax, usually racemose above, with long-pedicelled fl. and crowded bristles and stkd. gl.

c. St. glabrous or subglabrous.

(i.) Pan. normally as in Sect. *b.* (*i. e.* from ⅓ to nearly ½ ultra-axillary).

(1) L. pale-green and very softly hairy beneath.

nudicaulis var. nov. Stout. St. subglabrous. L. thick, very softly hairy beneath, with compound teeth. Otherwise hardly different from the typical plant as described by Bloxam.

(2) L. normally grey-felted beneath, especially when young.

a. Sep. reflexed in fl. and fr., sometimes only loosely.

R. Gelertii Frider. Stout. Prk. long, subequal. Lts. large, coarsely and irregularly toothed, long-pointed. Pan. very composite, usually broad with rounded top.

β. Sep. partly subpatent or even erect with young fr.

R. raduloides Rogers. Prk. and acic. many, very unequal. Lts. with incised, compound, finely pointed teeth; term. broadly ovate-acuminate. Ultra-axillary pan.-top long, cylindrical. (Subsp. of *R. anglosaxonicus.*)

R. setulosus Rogers. Usually still more prickly and glandular than the last, the armature in extreme examples being quite Koehlerian. Term. lt. obovate-obtusangular. Pan. more leafy above, laxer below. (Subsp. of *R. anglosaxonicus.*)

(ii.) About ¼ of pan. ultra-axillary.

R. anglosaxonicus Gelert. Prk. strong. Prklets many, scattered. Lts. thick, greyish-felted beneath: term. oval, parallel-sided, shortly pointed. Pan. with strongly ascending few-flowered branches.

For subsp. *curvidens, vestitiformis, raduloides* and *setulosus* see above.

R. melanoxylon Muell. & Wirtg. Armature partly strong, but very variable. St. and pan.-rach. blackish-brown or dark purplish-brown. Term. lt. roundish-acuminate, with even toothing.

2. Bas. lts. almost sessile in summer, with stalk becoming a little longer in autumn. (*Adenophori subcorylifolii* Focke.)

a. St. usually subglabrous. Sep. patent or erect on fall of pet. (Often sub-Koehlerian in armature.)

R. infestus Weihe. St. high at first. Prk. mixed; declining, falcate and hooked. Lts. pale and often felted beneath; term. broadly ovate-acuminate. Pet. almost round.

Var. *virgultorum* A. Ley. St. lower. Armature mixed; but falcate and hooked prk. fewer. Lts. shorter, more nearly round. Pan. broad. Sep. ultimately reflexed. (Between *R. infestus* and *R. Borreri.*)

R. Leyanus Rogers. St. shining, pale. Prk. short, declining. Lts. long-pointed, soft and often grey-felted beneath, sharply evenly toothed. Pan. much narrowed in the ultra-axillary top. (Subsp. of *R. Drejeri.* See below.)

b. St. hairy.

(i.) Sep. reflexed in fl. and fr.

R. uncinatus P. J. Muell. Armature all slender. Lts. thin, usually rather softly hairy in exposure, greyish-felted beneath. Pan. narrow, cylindrical, very hairy, with prk. mostly falcate or hooked.

(ii.) Fr.-sep. at first patent or erect.

R. Borreri Bell Salt. St. almost prostrate, yellowish-brown. Prk. crowded. Acic. and stkd. gl. mostly small. Term. lt. obovate-cuspidate. Pan. with broad rounded top and many unequal (chiefly sunken) stkd. gl.

Var. *dentatifolius* Briggs. St. with fewer stkd. gl., furrowed. Lts. nearly oval, with very long gradually-acuminate point.

R. Drejeri G. Jensen. St. fuscous, dull. L. 3–5-nate, concolorous. Lts. shortly pointed, roundish-obovate, with shallow rather irregular teeth, harsh beneath. Pan. long, lax, narrow. Often sub-Koehlerian in armature.

(Subsp. *Leyanus* Rogers.) See above in Subsection *a.*

Subsp. *hibernicus* Rogers. Leaf-toothing loose, sinuate.

Pan. very lax and glandular, slightly armed, much narrowed above. Somewhat intermediate between *R. Drejeri* and *R. Leyanus*.

(*R. podophyllus* P. J. Muell.) See Group IX., *Radulæ*.

Group IX. Radulæ. St. rough with crowded short subequal acic. and stkd. gl. Large prk. subequal (usually equal in typical *R. radula*) and nearly confined to angles. Intermediate prk. absent. While thus strictly limited the members of this group can hardly be confused with those of any other. As in the three succeeding groups, all the pan.-branches except the lowest are usually cymose or umbellate-racemose and not simply racemose. Stkd. gl. and acic. on pan. unequal, mostly short.

1. Stkd. gl. on pan.-rach. and ped. mostly sunken in the patent hairs; well-marked acic. usually few or none. Stkd. gl. and acicular organs on st. crowded.

a. L. with ashy or greenish-grey felt beneath.

R. radula Weihe. Prk. rather few, very strong, all on the angles, about equal. Faces of st. with very short subequal mixed armature. Term. lt. broadly ovate-acuminate. Pan. pyramidal, with strong prk.

Subsp. *anglicanus* Rogers. Prk. weaker, many, more scattered, only subequal. General armature showing rather more range of variation. Term. lt. narrow, obovate or oval, finely toothed. Pan. cylindrical, with weak prk.

Subsp. *echinatoides* Rogers. St. glabrous. Prk. rather more variable, occasionally hooked. Lts. all obovate, with compound incised teeth, often greener and only hairy beneath. Pan. much like that of *R. radula*.

Subsp. *sertiflorus* Muell. & Lefv.? Prk. chiefly short, often hooked. Lts. uniformly narrow, somewhat cuneate, with shallow teeth, soon greenish beneath. Pan. rather narrow, with hooked prk. and small fl.

R. echinatus Lindl. Prk. strong, nearly equal. Lts. thick, incised-lobate, all greyish-felted beneath. Pan. very narrow, with large fl. strongly reflexed sep. and declining prk.

R. rudis Wh. & N. St. glabrous. Prk. many, short. Glands very abundant, sessile and shortly stalked. L. large. Pan. diffuse, with many slender-pedicelled small fl.

b. L. normally green and rather thinly hairy beneath (in vars. *Newbouldii* and *Bloxamianus* sometimes thinly felted).

R. oigocladus Muell. & Lefv.? St. bluntly angled, dark, glaucous, thinly hairy. Term. lt. obovate-cuspidate, much narrowed towards base.

Var. *Newbouldii* (Bab.)? St. furrowed, paler, not glaucous, densely hairy. Term. lt. shorter, roundish-obovate with long point.

Var. *Bloxamianus* Colem. St. glabrous or nearly so, densely glandular. Term. lt. roundish-obovate with rather short point.

R. regillus A. Ley. St. bluntly angled, pale, glaucous, hairy, very leafy. Term. lt. oblong, with long cuspidate point.

(*R. radula*, var. *echinatoides* Rogers.) See above, in Subsection *a*.

2. Stkd. gl. and numerous acic. on pan.-rach. and ped., though rarely long, usually far exceeding the close felt or very short hairs. Stkd. gl. and acic. on st. more thinly and unequally scattered, sometimes nearly absent.

R. podophyllus P. J. Muell. St. bluntly angled, dark, usually subglabrous. Term. lt. oval-oblong, with rather short cuspidate point. Less distinctly Radulan than the other species of the group.

(*R. adenanthus* Boul. & Gill.) See Group VIII., *Egregii*.
(*R. infecundus* Rogers.) See Group XII., *Koehleriani*.
(*R. fuscus* Wh. & N.) See Group XI., *Sub-Bellardiani*.

Group X. Sub-Koehleriani. Distinguished from the true *Radulæ* by the strong larger prk. being somewhat more unequal and less strictly confined to the angles of the st., and by the general armature of prklets, acic., etc., being also rather less uniform in length and stoutness; while they differ from the true *Koehleriani* by the less graduated series of glandtipped organs and the comparative rarity with which such organs exceed the hairs on pan.-rachis and ped.

1. Glandtipped organs on pan.-rach. almost wholly sunken in the patent hairs.

a. Fr.-sep. subpatent, at least for a time. Lts. thick.

(i.) Lts. with whitish felt beneath. Stam. long.

R. Griffithianus Rogers. St. and very unequal prk. deep red. Term. lt. roundish-obovate-acuminate. Pan. rather narrow, lax, nearly cylindrical.

(ii.) Lts. usually softly hairy beneath. Stam. rather short.

R. Babingtonii Bell Salt. St. and subequal prk. fuscous. Term. lt. oblong-oval-cuspidate. Pan. broad, cylindrical-truncate, with patent branches and acicular prk.

R. Bloxamii Lees. St. and rather weak prk. purplish-red. Term. lt. broadly obovate. Pan. broadly pyramidal, round-topped with long ascending lower branches. Sep. patent or subpatent for a time, soon loosely reflexed.

(*R. mutabilis* Genev.) See below in subsection *b.*

b. Fr.-sep. reflexed. Lts. rather thin. Stam. long.

(i.) Lts. thinly hairy beneath.

R. melanodermis Focke. St. and rather unequal prk. blackish-purple. Lts. very plicate, green; term. obovate-truncate. Pan. narrow cylindrical.

R. Babingtonii Bell Salt., var. *phyllothyrsus* (Frider.). St. fuscous. Term. lt. obovate-acuminate. Pan. broad, with strong falcate prk. and several simple l.

(For typical *R. Babingtonii* see above in Subsection *a.*)

(ii.) Lts. softly hairy or with rather close felt beneath,—at least when young.

R. cavatifolius P. J. Muell. St. and rather unequal prk. pale or reddish-brown. Lts. yellowish-green; term. broadly cordate-ovate-acuminate. Pan. nearly cylindrical, truncate.

R. mutabilis Genev. Lts. very long and narrow, deeply incised, with close pale felt beneath; term. narrowed at both ends. Pan. very large, nearly cylindrical.

(For subsp. *nemorosus* Genev. see below, in Section 2.)

2. Glandtipped organs on pan.-rach. more unequal, rather frequently exceeding the short hairs or felt.

a. Fr.-sep. reflexed, with rather close felt.

R. Lejeunei Wh. & N. Lts. thin, light green; term. obovate-rhomboidal. Pan. loosely pyramidal, with slender unequal armature. Pet. broad, bright pink.

Subsp. *ericetorum* (Lefv.). Stronger. Lts. with longer point, grey-felted beneath. Pan. very long, pyramidal-corymbose, more prickly and glandular. Pet. narrow, white or pinkish.

b. Sep. rising on fall of pet.

R. mutabilis Genev., subsp. *nemorosus* Genev. Lts. thick, with subvelvety felt beneath; term. cordate-ovate-acuminate. Pan. pyramidal, interrupted.

(For typical *R. mutabilis* see above, in Section 1.)

Group XI. Sub-Bellardiani. The larger prk. not so strong as in Groups IX. and X. and rather more scattered and unequal. St. often less angular. Near the true *Bellardiani*, but usually stouter with a less graduated series of prk. and other arms and a greater tendency to compound branches in the lower half of the pan. Several of the plants in Section II. of *Bellardiani* however would hardly be out of place among the *Sub-Bellardiani.*

1. Hairs on st. and pan.-rach. dense and fairly long, usually patent. L. mostly 5-nate.

a. Sep. normally reflexed, though often partially patent (or even erect) in fr.

R. fuscus Wh. & N. Prk. rather short, somewhat scattered. Lts. coarsely toothed, softly hairy beneath ; term. rather broad with long point. Pan. narrow, cylindrical, with sunken stkd. gl. and acic.

Var. *nutans* Rogers. Prk. usually more unequal. Lts. more deeply incised, with very attenuate point. Pan. very lax, subracemose above, drooping. Fr.-sep. usually erect at first. (This var. would hardly be out of place in Section 2.)

Var. *macrostachys* (P. J. Muell.). Prk. longer. Lts. whitish-felted beneath. Pan. conspicuously pyramidal, with divaricate branches and blackish-purple villous rach. Stam. comparatively short.

b. Fr.-sep. mostly erect.

Subsp. *obscurus* Kalt. Chief prk. hooked or nearly so. Prklets acic. and stkd. gl. mostly very small. Lts. even-toothed. Pan. long, with deep red pet. and short stam.

2. Hairs on st. abundant, rather short, loose or partly appressed. L. 5-nate or 3–4-nate. Sep. erect or patent on unripe fr., often reflexed ultimately.

a. L. mostly 5-nate.

R. pallidus Wh. & N. Prk. usually confined to angles, mostly subequal. Lts. thin, very irregularly toothed, long and rather narrow. Pan. straggling-pyramidal, with wavy rach. slender ped. and mostly short blackish stkd. gl.

Var. nov. *leptopetalus.* St. most densely clothed with very short acic. and stkd. gl. Lts. with somewhat glaucous tint and sharply incised teeth. Pan. normally broad, with interlacing branches. Pet small, very narrow.

b. L. 3–4-nate, rarely 5-nate.

R. scaber Wh. & N. St. roundish, glaucous, rough with minute acic. Prk. small. Lts. with fine teeth and rather short point. Pan. narrow, with short hair and mostly short stkd. gl.

R. thyrsiger Bab. Hairs on st. partly appressed. Lts. coarsely irregularly toothed, obtusangular-obovate. Pan. long, subracemose, with long-pedicelled fl.

3. Hairs on st. rather thinly scattered or nearly absent. L. mainly 3-nate.

a. Sep. clasping young fr., mostly reflexed ultimately.

R. Lintoni Focke. St. usually subglabrous. Prk. very

slender. Lts. shining, with close fine teeth. Pan. cylindrical, racemose-corymbose, with very unequal stkd. gl.

R. longithyrsiger Bab. Prk. short. Lts. with shallow even teeth, obovate-cuspidate. Pan. pyramidal, racemose or sub-racemose above; ped. rigid, purplish with short stkd. gl. Carp. pubescent-glabrescent.

botryeros var. nov. St. conspicuously glaucous. Prk. and other arms more unequal. L. rather more frequently 4–5-nate. Lts. truncate-mucronate, more narrowed towards base. Pan. more composite, with larger fl. less rigid ped. and hairy carp.

b. Fr.-sep. normally reflexed throughout.

R. foliosus Wh. & N. Prk. many weak subequal. Lts. nearly uniform, usually broadest near middle. Pan. long and narrow, with flexuous rach. and short few-flowered and often fasciculate branches.

Group XII. Koehleriani (*Hystrices* Focke). St. low arching or nearly prostrate or climbing, usually indistinctly angled, with scattered prk. prklets acic. bristles and stkd. gl. of varying form and length. Some prk. very strong and (like all the arms) occasionally glandtipped. L. mostly 5-nate. Pan. composite, sometimes subracemose above; lowest branches racemose; middle and often upper cymose or subumbellate, 3 to many-flowered; rach. and ped. almost as variously armed as st. Usually large strong plants, with more unequal and varied armature than is found in any other group. Distinguished from the *Sub-Koehleriani* by the greater range of variation in the still more scattered armature, the special abundance of strong bristles passing into acic. and the greater number and greater length of the glandtipped organs.

1. Prk. somewhat scattered, but chiefly on angles. Prk. and acic. rarely glandtipped. Intermediate arms only fairly many.

a. Prk. not very unequal, all of moderate length (except occasionally in *R. hostilis*). Stkd. gl. on rach. rarely very long.

R. rosaceus Wh. & N. St. roundish, pubescent or sub-glabrous. Prk. rather short. L. chiefly 3-nate. Lts. broad, hairy on nerves beneath; lat. very gibbous. Pan. broad. Pet. bright pink.

Var. *hystrix* (Wh. & N.). St. angular, more hairy, with stronger more unequal armature. L. often 5-nate. Lts. narrow, pubescent beneath. Pan. narrow. Sep. more erect. Pet. pink or pinkish.

Var. *infecundus* Rogers. Armature nearly Radulan. L.

often 5-nate. Lts. yellowish, very softly hairy beneath. Pan. lax, pyramidal. Sep. rising early and clasping poor fr.

Subsp. *Purchasianus* Rogers. St. roundish, densely hairy. Prk. crowded, stout-based. L. mostly 5-nate-pedate. Lts. narrow, strigose on both sides. Pan. nearly cylindrical. Sep. shaggy, erect in fr.

Subsp. *adornatus* P. J. Muell. St. bluntly angled, pubescent, glaucous. Prk. very strongly declining. L. chiefly 4–5-nate. Term. lt. narrowed very evenly to long point. Pan. very narrow, wavy below. Sep. soon erect.

Subsp. *Powellii* Rogers. Slender. L. small. Lts. very narrow, with long attenuate point and compound incised teeth. Stkd. gl. very unequal; other glandtipped organs usually absent. Pan. very compound, lax, corymbose. Sep. strongly reflexed in fl. and fr.

R. hostilis Muell. & Wirtg. St. bluntly angled. L. chiefly 5-nate, often large. Lts. narrow, with irregular partially compound teeth and very long point. Pan. mostly short and racemose above. Fr.-sep. attenuate, clasping.

b. Prk. extremely unequal, some very long. Stkd. gl. on rach. passing into glandtipped acic.

R. fusco-ater Weihe. Stout and very hairy. L. moderate. Lts. typically roundish-ovate with sharp point, not deeply incised. Armature very strong. Fr.-sep. erect or patent.

2. Prk. acic. bristles and intermediate organs of all kinds very numerous, very unequal, indiscriminately scattered and not unfrequently glandtipped.

a. Pet. white.

R. Koehleri Wh. & N. Armature remarkably unequal, mostly slender, often glandtipped. Lt.-toothing rather coarse. Pan. open, slightly narrowed above. Fr.-sep. reflexed.

Var. *cognatus* (N. E. Brown). Intermediate arms often less numerous. Leaf-toothing almost lobate-sinuate. Pan. broad and much branched. Fr.-sep. at first subpatent.

b. Pet. pink.

Subsp. *dasyphyllus* Rogers. Intermediate arms present in very varying quantity. Lts. thick; teeth compound, partly recurved. Pan. very narrow, interrupted. Fr.-sep. reflexed.

R. plinthostylus Genev. Armature rather sub-Koehlerian. Lts. long, narrowed at both ends, very pale beneath. Pan. large. Fr. small. Fr.-sep. loosely reflexed or subpatent.

R. Marshalli Focke & Rogers. Hairy. Armature exceedingly dense and unequal, almost wholly patent. L. small. Pan. very elongate, with very long lower branches. Sep. subpatent in young fr.

Var. *semiglaber* Rogers. Subglabrous. Armature less uniformly patent and not quite so unequal. Pan. broader. Fl. more showy. Fr.-sep. often clasping.

Group XIII. BELLARDIANI (*Glandulosi* Focke). St. mostly prostrate roundish often pruinose. Armature as scattered as in *Koehleriani* and very varied but weaker, the prk. being usually slender and in some instances almost reduced to acic. L. 3-nate or 4–5-nate-pedate, green on both sides. Bas. lts. stalked. Stip. filiform, rarely narrowly linear. Pan. mostly less composite, occasionally reduced to a raceme. Stkd. gl. and gland-tipped bristles on. pan. usually red or dark purple, mostly very unequal, often considerably exceeding diameter of ped. and hairs on rach. and ped. Usually rather small low-growing plants, occurring chiefly in the woods and thickets of somewhat hilly districts. Many of the sp. very ill-defined and variable. For detailed account of the differences between this group and the *Koehleriani* see Focke, *Syn. R. G.* 355.

1. Stkd. gl. on pan. conspicuously unequal, often twice as long as diameter of ped. Glandtipped bristles often numerous.

a. Fr.-sep. normally erect or patent (sometimes loosely reflexed in *R. Durotrigum*).

(i.) L. chiefly 5-nate.

R. viridis Kalt. Prk. and prklets fairly many, large-based. Lts. softly hairy and subglabrescent beneath, with irregular shallow teeth. Stam. far exceeding sty. Young carp. pubescent.

R. Durotrigum R. P. Murr. Prk. very crowded, slender, long-based. Lts. very slightly hairy on veins beneath, with deeply incised or lobate teeth. Stam. and sty. subequal. Young carp. thinly hairy.

(ii.) L. chiefly or wholly 3-nate.

R. divexiramus P. J. Muell. L. mostly 3-nate. Lts. soft with short hair beneath, with fine closely placed compound teeth; lat. very gibbous. Stam. exceeding sty. Young carp. somewhat pubescent.

R. Bellardii Wh. & N. Prk. acicular, conical. L. 3-nate. Lts. subequal, evenly toothed. Pan. short, racemose above, with long thin ped. Stam. and sty. subequal. Young carp. glabrous.

(*R. flaccidifolius* P. J. Muell.) See below under *R. hirtus.*

(iii.) L. 3–5-nate.

R. serpens Weihe. Prk. short. Lts. unequal, irregularly toothed. Pan. narrow, racemose; sometimes with long racemose branches below. Stam. hardly exceeding sty. Young carp. glabrous.

R. hirtus W. & K. Prk. usually straight, setaceous. L. rarely 4–5-nate. Lts. broad, coarsely toothed, very hairy on veins beneath. Stam. rather exceeding sty. Carp. hairy. Densely glandular and bristly.

Var. *rotundifolius* Bab. Prk. and prklets mostly short, stout-based. Lts. with shallow teeth, subglabrescent beneath. Ped. long, divaricate. Stam. far exceeding sty. Carp. glabrous. Very luxuriant; with yellowish-brown tint.

b. Fr.-sep. reflexed or partly subpatent; occasionally erect.

Subsp. *Kaltenbachii* (Metsch.). Prk. strongly declining, stout-based. Lts. narrow, irregularly toothed. Pan. elongate, pyramidal, drooping. Stam. exceeding sty. Carp. glabrous.

Subsp. *flaccidifolius* (P. J. Muell.). L. chiefly or wholly 3-nate. Lts. subequal, thin, oval, long-pointed. Pan. narrow, lax, wavy. Sep. strongly reflexed in fl. and fr. Stam. unequal.

Subsp. *rubiginosus* (P. J. Muell.). St. and prk. stouter. Lts. thick, with somewhat patent teeth. Pan. narrow, cylindrical. Fr.-sep. loosely reflexed. Stam. usually exceeding sty. Carp. glabrous.

R. horridicaulis P. J. Muell. St. stout. Prk. many, strong. Lts. large, roundish-obovate-mucronate, with broad coarse teeth. Pan. strongly armed and very glandular. See under *R. saxicolus*.

2. Stkd. gl. on pan. almost without exception shorter than diameter of ped. Glandtipped bristles few or none. (Near Group XI.)

a. Fr.-sep. chiefly patent.

R. acutifrons A. Ley. Prk. strongly declining or hooked. L. 3–5-nate. Lts. thinly hairy beneath, with fine somewhat lobate teeth. Pan. very lax, with aggregated ped. Stam. far exceeding sty. Young carp. hairy.

R. saxicolus P. J. Muell. St. angular, subglabrous. L. mostly 5-nate. Lts. with soft short hair beneath, somewhat obovate with short point. Pan. long, with dense patent hair and short glandtipped organs.

R. tereticaulis P. J. Muell. St. densely hairy and aciculate. Prk. slender. Term. lt. obovate-acuminate, rather narrow. Stam. about equalling sty. Carp. glabrous.

b. Fr.-sep. reflexed in fl. and fr.

R. ochrodermis A. Ley. St. stout, very prickly, subglabrous, ochreous. Term. lt. roundish-obovate-mucronate. Stam. exceeding sty.

c. Sep. clasping and almost hiding fr.

R. velatus Lefv. St. with scattered silky hairs. Prk. stout-based. Term. lt. roundish-obovate-acuminate. Stam. exceeding sty. Carp. pubescent.

D. (Group XIV.)—St. low arching or trailing, pruinose. Prk. scattered mostly acicular and straight. Bas. lts. sessile or very nearly so.

Group XIV. Cæsii (*Corylifolii* Focke). St. roundish or slightly angular, with many rooting branches. Intermediate acic. and stkd. gl. few (except in some forms of *R. dumetorum*), sometimes absent. Lts. mostly broad; bas. hardly stalked. Stip. often rather broad. Pet. large roundish. Drupelets usually rather few and large. Fl. early and late. See Focke, *Syn. R. G.* 387.

1. L. 5-nate or 3–4-nate. Stip. linear-lanceolate or filiform (rather broadly lanceolate, in *R. Balfourianus*). Drupelets fairly many, rather large, mostly epruinose.

R. dumetorum Wh. & N. *Sp. collect.* St. and pan.-rach. with many unequal prk. acic. and glandtipped organs. Lts. usually thick. Pan. mostly long, regular, compound, with rach. usually very hairy. Fr.-sep. mostly erect or subpatent.

(For conspectus of *R. dumetorum* forms see below, at the end of this table.)

R. corylifolius Sm. *Sp. collect.* Prk. subequal, irregularly scattered. Glandtipped organs very few. Lts. moderately thick. Pan. somewhat irregular, with few axillary branches and whitish-felted rach. Fr.-sep. normally reflexed.

a. Var. *sublustris* Lees. St. round. Prk. subulate. L. 5–7-nate. Term. lt. broadly ovate-acuminate incised-serrate and (often) lobate.

b. Var. *cyclophyllus* (Lindeb.). St. bluntly angled. Prk. rather short and stout. L. 5-nate. Term. lt. roundish cuspidate-acuminate, not lobate.

R. Balfourianus Blox. Stkd. gl. mostly short, fairly many though unequally scattered. Lts. usually thin, conspicuously wrinkled above. Stip. rather broad. Pan. diffuse, with straggling few-flowered branches. Fr.-sep. clasping. Fl. and fr. large.

2. L. normally 3-nate. Stip. considerably broadened in middle. Drupelets few, very large, conspicuously glaucous.

R. cæsius Linn. Prk. weak short subulate. Stkd. gl. and acic. seldom many, sometimes absent. Lts. thin. Pan. lax, few-flowered. Fr.-sep. clasping. Whole plant weak.

Conspectus of *R. dumetorum* forms.

I. Fr.-sep. partly erect. Stkd. gl. on pan. considerably unequal, partly exceeding hairs.

(1) L. chiefly 5-nate.

a. Var. *ferox* Weihe. Prk. very crowded, mostly long patent. Lts. somewhat roundish. Pan. rather compact, with rounded top and straight rach.

c. Var. *diversifolius* (Lindl.). Prk. very unequal, ranging from very strong ones to prklets and tubercles. Lts. chiefly oval or obovate. Pan. long, leafy, with very narrow top and very short subracemose branches. Strong.

e. Var. *rubriflorus* Purchas. St. slender, very glaucous. Prk. small, chiefly on angles. Pan. straggling, with one or more divaricate many-flowered branches. Pet. rather narrow, purplish. Usually large weak bushes.

(2) L. chiefly 3–4-nate.

(i.) Acic. and stkd. gl. abundant.

b. Var. *britannicus* (Rogers). Prk. often crowded, usually short weak and partly somewhat falcate. Lts. roundish. Pan. much interrupted, with truncate top wavy rach. and lower branches like small secondary pan.

d. Var. *pilosus* Wh. & N. St. and pan.-rach. considerably hairy. Prk. many, rather weak. Pan. leafy nearly to top, cylindrical, with long peduncled branches.

(ii.) Acic. and stkd. gl. comparatively few.

f. Var. *tuberculatus* (Bab.). St. stout. Prk. and prklets very stout-based. Pan. open, cylindrical, with corymbose top. Pet. rather deep pink.

g. Var. *concinnus* Warren. St. rather slender. Prk. broad-based, mostly patent. L. rather small and neat. Pan. elongate, with narrow subracemose top and long distant racemose lower branches. Pet. rather small, pinkish.

II. Fr.-sep. reflexed. Stkd. gl. on pan. short, sunken.

h. Var. *fasciculatus* (P. J. Muell.). St. dark purple, with thinly scattered prklets. L. mostly 5-nate. Lts. with large incised compound teeth. Pan. narrow, with short stkd. gl. sunken in patent hair.

ALPHABETICAL INDEX

OF

THE ORDERS AND GENERA.

*** The names in *italics* are synonymous.

INDEX

TO THE

POPULAR ENGLISH NAMES.

THE END.

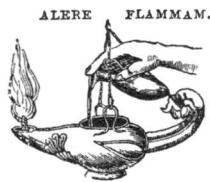

ALERE FLAMMAM.

Printed by TAYLOR and FRANCIS, Red Lion Court, Fleet Street.